Fusarium: Diseases, Biology, and Taxonomy

Edited by
P.E. Nelson, T.A. Toussoun, and R.J. Cook

The Pennsylvania State University Press
University Park and London 1981

Library of Congress Cataloging in Publication Data

Main entry under title:
Fusarium : diseases, biology, and taxonomy.

Includes bibliographical references.
1. Fusarium. 2. Fungous diseases of plants.
I. Nelson, Paul E., 1927– . II. Toussoun,
T. A., 1925– . III. Cook, R. James,
1937– .
SB741.F9F87 632′.34 81-47174
ISBN 0-271-00293-X AACR2

Designed by Dolly Carr

Printed in the United States of America

Contents

PART IV: THE FUNGUS *FUSARIUM:* PHYSIOLOGY AND HISTOPATHOLOGY

PART V: THE FUNGUS *FUSARIUM:* TAXONOMY

Contributors

A. Alabouvette, Station de Recherches sur la Flore pathogène dans le Sol, Institut National de la Recherche Agronomique, Dijon, France

G.M. Armstrong, Department of Plant Pathology, Georgia Agricultural Experiment Station, Experiment, GA

J.K. Armstrong, Department of Plant Pathology, Georgia Agricultural Experiment Station, Experiment, GA

R.R. Baker, Department of Botany and Plant Pathology, Colorado State University, Fort Collins, CO

C.H. Beckman, Department of Plant Pathology and Entomology, University of Rhode Island, Kingston, RI

G.M. Blakeslee, Department of Plant Pathology and School of Forest Resources and Conservation, University of Florida, Gainesville, FL

W.J. Bloomberg, Department of Environment, Canadian Forest Service, Pacific Forest Research Centre, Victoria, B.C., Canada

C. Booth, Commonwealth Mycological Institute, Kew, Surrey, U.K.

D. Bouhot, Station de Recherches sur la Flore pathogène dans le Sol, Institut National de la Recherche Agronomique, Dijon, France

L.W. Burgess, Department of Plant Pathology and Agricultural Entomology, University of Sydney, Australia

D.W. Burke, USDA Vegetable Crops Production, Irrigated Agriculture Research and Extension Center, Prosser, WA

R. Cassini, Station de Pathologie végétale, Institut National de la Recherche Agronomique, Versailles, France

H. Cole, Jr., Department of Plant Pathology, The Pennsylvania State University, University Park, PA

J. Colhoun, Cryptogamic Botany Laboratories, University of Manchester, U.K.

R.J. Cook, USDA Regional Cereal Disease Laboratory, Washington State University, Pullman, WA

M.J. Daboussi-Bareyre, Laboratoire d'Etude et Exploitation du Polymorphisme végétal, associé au CNRS, Université de Paris-Sud, Orsay, France

R.L. Dodman, Queensland Wheat Research Institute, Department of Primary Industries, Toowoomba, Australia

L.D. Dwinell, USDA Forest Service, Southeastern Forest Experiment Station, Athens, GA

D.L. Ebbels, Ministry of Agriculture, Fisheries, and Food, Plant Pathology Laboratory, Harpenden, Hertfordshire, U.K.

R.H. Garber, USDA Cotton Research Station, Shafter, CA

W. Gerlach, Biologische Bundesanstalt für Land- und Forstwirtschaft, Institut für Mikrobiologie, Berlin-Dahlem, West Germany

R.D. Goos, Department of Botany, University of Rhode Island, Kingston, RI
G.J. Griffin, Department of Plant Pathology and Physiology, Virginia Polytechnic Institute and State University, Blacksburg, VA
W.A. Haglund, Northwest Research and Extension Unit, Washington State University, Mount Vernon, WA
C.W. Hesseltine, USDA Northern Regional Research Center, Peoria, IL
R.K. Horst, Department of Plant Pathology, Cornell University, Ithaca, NY
J.P. Jones, Agricultural Research and Education Center, University of Florida, Bradenton, FL
A.J. Kappelman, Jr., USDA Crop Science Research Unit, Auburn University, Auburn, AL
T. Kommedahl, Department of Plant Pathology, University of Minnesota, St. Paul, MN
J.M. Kraft, USDA Vegetable Crops Production, Irrigated Agriculture Research and Extension Center, Prosser, WA
E.G. Kuhlman, USDA Forest Service, Southeastern Forest Experiment Station, Research Triangle Park, NC
R.G. Linderman, USDA Ornamental Plants Research Laboratory, Corvallis, OR
J. Louvet, Station de Recherches sur la Flore pathogène dans le Sol, Institut National de la Recherche Agronomique, Dijon, France
W.E. MacHardy, Department of Botany and Plant Pathology, University of New Hampshire, Durham, NH
A. Maric, Faculty of Agriculture, Institute for Plant Protection, University of Novi Sad, Yugoslavia
P. Mas, Station de Pathologie végétale, Centre de Recherches d'Avignon, INRA, Montfavet, France
P. Mayers, Department of Primary Industries, Toowoomba, Australia
C.M. Messiaen, Laboratoire de Botanique, Ecole Nationale Supérieure Agronomique, INRA, Montpellier, France
P. Molot, Station de Pathologie végétale, Centre de Recherches d'Avignon, INRA, Montfavet, France
P.E. Nelson, Fusarium Research Center, Department of Plant Pathology, The Pennsylvania State University, University Park, PA
J.W. Oswald, President, The Pennsylvania State University, University Park, PA
D. Parisot, Laboratoire d'Etude et Exploitation du Polymorphisme végétal, associé au CNRS, Université de Paris-Sud, Orsay, France
B.W. Pennypacker, Department of Plant Pathology, The Pennsylvania State University, University Park, PA
W. Pont, Kamerunga Research Station, Department of Primary Industries, Redlynch, Australia
J.E. Puhalla, USDA National Cotton Pathology Research Laboratory, College Station, TX
G. Rebell, Department of Ophthalmology, Bascom Palmer Eye Institute, University of Miami, FL
G. Risser, Station d'Amélioration des Plantes maraîchères, Centre de Recherches d'Avignon, INRA, Montfavet, France
F. Rouxel, Station de Recherches sur la Flore pathogène dans le Sol, Institut National de la Recherche Agronomique, Dijon, France
P. Sanders, Department of Plant Pathology, The Pennsylvania State University, University Park, PA
B. Schippers, Phytopathological Laboratory "Willie Commelin Scholten," Baarn, The Netherlands

S.N. Smith, Department of Plant Pathology, University of California, Berkeley, CA
W.C. Snyder, Department of Plant Pathology, University of California, Berkeley, CA
M. Stoner, Department of Biological Sciences, California State Polytechnic University, Pomona, CA
R.H. Stover, United Fruit Company, La Lima, Honduras
S.K. Sun, Department of Plant Pathology, National Chung Hsing University, Taiching, Taiwan, R.O.C.
T.A. Toussoun, Fusarium Research Center, Department of Plant Pathology, The Pennsylvania State University, University Park, PA
G. Toutain, Station de Recherches sur la Flore pathogène dans le Sol, Institut National de la Recherche Agronomique, Dijon, France
W.H. van Eck, Phytopathological Laboratory "Willie Commelin Scholten," Baarn, The Netherlands
R.F. Vesonder, USDA Northern Regional Research Center, Peoria, IL
C.E. Windels, Department of Plant Pathology, University of Minnesota, St. Paul, MN
S.S. Woltz, Agricultural Research and Education Center, University of Florida, Bradenton, FL

Foreword

The spore of the idea for this important and useful book has had an unusual origin and nurture, and I may be the one individual in whom all the strains of connection now come to focus. A brief elaboration seems appropriate as foreword for this publication on a topic relevant to life-sustaining nourishment on this planet.

Over forty years ago I was a student at Berkeley, toiling over my thesis under the masterly direction of my senior professor, W.C. Snyder, who wrote this book's Introduction shortly before his untimely death. I was one of Dr. Snyder's earlier students who was imbued by this leader in the field with a conviction, indeed an excitement, about the importance of the fungus genus *Fusarium*. From Professor Snyder and his inseparable colleague Dr. H.N. Hansen I gained not only knowledge and direction but also a lasting respect and appreciation for the methods and mission of science. Indeed, to this day I think of myself, after thirty years as a university administrator, as first of all a scientist.

Most directly I was infused as a graduate student at Berkeley with a pervasive curiosity about this organism *Fusarium,* its tremendous economic importance, the wide range of its victims, and its appearance throughout the agriculture of the world. Indeed, it became the lens through which I learned to love science and what science endeavors to accomplish. This love affair with *Fusarium* which at that time was more important to me than anything else in my life was abruptly sliced by the incision of World War II and incredible years on a torpedo boat thousands of miles from the fascination of the laboratory in California. The allure resumed after the war and my own responsibilities widened when I became the chairman of plant pathology at Berkeley, welcoming and encouraging new crops of graduate students, some of whom carried on this pervading research interest in California in *Fusarium,* one of the most persistent of the soil-borne pathogens known to plant pathologists. Among these students at Berkeley working at various times with the Hansen-Snyder-Oswald trio were James Tammen, T.A. Toussoun, and Paul E. Nelson. My own particular interests concentrated on the concept of cultivars of *Fusarium,* basic to the categorization of varieties within the species.

The next decades brought a variety of dispersed experiences to all of us away from this focal point in California, but by the 1970s destiny had brought us all to Penn State, still fascinated by *Fusarium,* which was by then the priority concentration of Drs. Toussoun and Nelson. Dr. Tammen, who has since become the Dean of the College of Agriculture at Minnesota, was head of the plant pathology department at The Pennsylvania State University (where I serve as president), and the others are now professors in that department and have organized and operate the Fusarium Research Center. Thus this publication comes to fruition from a common denominator established years ago in California and now reassembled in another academic environ.

Specialists who reach for this book will know already of the importance of the topic of *Fusarium,* recognizing it clearly as one of the most important plant disease fungi throughout the world. A remarkable aspect of *Fusarium* has been its use as a laboratory and field experimental pathogen from which much basic information has been derived concerning how fungi remain in soils year after year and how naturally occurring microflora affect this persistence. The reader will be well rewarded with this volume, thanks to the efforts of Drs. Nelson and Toussoun and their associate, Dr. R. James Cook of the USDA Cereal Disease Laboratory at Washington State University (he, too, studied under Dr. Snyder at Berkeley), for they initiated and directed this project. I am very proud that scientists whom I can claim in part as students at Berkeley and now as faculty at Penn State have become the new leaders in this field of investigation through the establishment of the Fusarium Research Center; I too am proudly a member of the center, using it as the vehicle through which I have kept touch with this scientific interest.

Invited papers from the outstanding researchers of the world have now been brought together in one volume for the stimulation and edification of us all. Many students in the field of Fusarium research will already be familiar with the widely used *Pictorial Guide to the Identification of Fusarium Species* by Toussoun and Nelson (Penn State Press, 1976) which is already in its second edition.

Let me conclude this foreword with a personal word of gratification that the dormant yet vital research interests of a pre-occupied university president were energized by the opportunity of commending this volume to the eager attention of dedicated experts in this field. This project thus becomes the ultimate intersection of a number of lives and careers who had already touched in another place in other years.

Penn State is continually committed to the premise that whatever it does it does well. Certainly the Fusarium Research Center is an example of this, as is our University Press, and it is a delight to me that the important material concerning this engrossing fungus is now being expertly published in a definitive and previously unavailable manner.

John W. Oswald

Preface

The genus *Fusarium* commands a major place in the literature on fungi, yet there are few books that do it justice. Wollenweber & Reinking's *Die Fusarien,* published in 1935, is one such landmark. The present volume marks the first attempt to assemble world knowledge on this genus. The 40 papers by the 60 authors presented herein are a fair representation of what is known today, but this volume is by no means an exhaustive treatment of the subject. The reader who peruses this book, however, will get an insight into the pathology, ecology, physiology, taxonomy, genetics, and histopathology of *Fusarium,* and will understand why it is considered to be ubiquitous and of such importance to humans. The specialist, the student, and others needing material for research projects should find the needed information in these articles and in the 2500 references listed.

If this volume is a milestone in the literature on *Fusarium,* it is because of the authors whose cooperation and efforts we most sincerely appreciate. If this volume has been considered a millstone on occasion by the scientific editors it is simply that it has been 5 years in preparation, that its gestation period has been even longer, and that it has overshot its publication date by about 2 years; thus the literature cited in this volume dates to 1978. Finally, we wish to thank the director of The Pennsylvania State University Press, Chris W. Kentera, and his staff, who did much to alleviate and shorten the birth pangs of this book.

P.E.N., T.A.T., & R.J.C.

Fusarium: Diseases, Biology, and Taxonomy

Introduction

William C. Snyder

Forty-seven years ago I spent a year in Berlin-Dahlem as a post-doctoral student of Dr. H. W. Wollenweber, under the auspices of the U.S. National Research Council. It was one of the most rewarding experiences in my life.

My admiration and respect for Dr. Wollenweber was, and is, tremendous. He was truly a scholar and a gentleman—an amazing scientist who dedicated his life to bringing order to the overwhelming confusion that was developing on the identities of species in his favorite genus, *Fusarium*. No one before nor since has made such an over-all contribution to the scientific treatment of the genus. It is with humble spirit that I pay tribute to this great man.

It was my privilege to live in Dr. Wollenweber's home that year and to sit beside his desk at night with Dr. Otto Reinking while they were writing their book, "Die Fusarien," published in 1935. Both discussed openly and in a calm friendly fashion each issue that arose, and occasionally involved me. What more inspirational and informative experience could a young scientist aspire to at the beginning of a career?

In 1939, five years later, Dr. Wollenweber came to visit us at Berkeley, spending the days on the campus of the University of California and the nights in our home. My esteemed colleague, Dr. H. N. Hansen, and I had prepared many hundreds of single-spored cultures in anticipation of Dr. Wollenweber's coming. Most cultures were in high sporulation, displayed in open white trays covering all the table space in our two adjoining laboratories. These cultures represented a single-spore analysis of a world collection of *Fusarium,* many of which had come from the Wollenweber laboratory. After viewing these cultures for a few days, Dr. Wollenweber said, "I have spent all my life on *Fusarium,* but I have never seen anything like this. You must publish!" Dr. Wollenweber had not used Hansen's single-spore method.

Fusarium The genus *Fusarium* may have been one of the earlier fungi to become established on earth. Once I was shown structures resembling chlamydospores of *Fusarium* in coal which was estimated to date back 200 million years. But certainly if any fungus were to emigrate to another planet, *Fusarium,* as a most adaptable, versatile and pioneering genus, would be a good candidate for establishment out there. Some of its species do well with very little oxygen or water and with unusual mixtures of gases and chemicals, and flourish under extremes of environmental stress.

Fusarium is well adapted to habitation on this earth and is present in our soil, water, air and organic materials. Perhaps more is known about the genus *Fusarium* and its species than is known about other genera of fungi. Yet, the more we learn the more we realize how little we actually know of the genus which Link described 169 years ago. Here I wish to discuss *Fusarium* as I have known it in the present century, since my personal experience does not go back quite to the first 100 years of its recorded existence.

The importance of *Fusarium* to world food production has greatly increased in recent decades. For example, the number of known wilt diseases caused by *formae speciales* of *F. oxysporum* has increased from 24 (13) in 1940 to 76 as listed by Booth (2) in 1971. The worldwide significance of the destructiveness of Fusarium diseases of wheat, barley and corn has become apparent to several Fusarium workers in the past 10 years. Even *Fusarium nivale,* once thought to be economically important only in northern countries, is now recognized as important in some southern regions such as Sicily (8) and Africa. That new *formae speciales* of *F. solani* have been described as having a role in severe infections of human and animal eyes, is causing increasing concern.

Whether the greater awareness of Fusarium pathogens in today's world stems from greater recognition by more Fusarium workers, or from the origination of new pathogenic forms, or both, the fact remains that the genus has emerged as a leading threat to most of the principal food crops of the world, as well as to oil, fiber, ornamental and forest plantings.

Perhaps the nine species concept (13,14,15) has stimulated more pathologists to work with Fusaria. For example, placing all of the vascular wilt Fusaria into the one species, *F. oxysporum,* indicates a relationship between these pathogens; similarly with the *formae speciales* of *F. solani* and the stem and root rots they cause. Pathologists in general understand better now what they are talking about when they read or write or speak of the pathogenic Fusaria. The nomenclature of the components of the species *F. roseum* is precise for pathogens and saprophytes, and for their perfect states. Progress has been made on the pathology of the *Fusarium* spp. in a way not possible before. It is likely that the direction of taxonomy in the future will be towards fewer species than nine.

Communication Meaningful communication is vital to most of man's pursuits and certainly it is so in science. Published records of our findings must be comprehended at least by readers in our own field. In biology a clear understanding of the taxonomy and nomenclature of the pathogen is essential to accuracy, whether we are reporting on its ecology, physiology or pathology.

Satisfactory systems for the classification of organisms are rare. Biologists trained in the science of fungi sometimes recognize genera, but seldom are they able to identify species with certainty. We all depend on other colleagues to identify those fungus species which we do not recognize. Frequently there is disagreement on which taxonomist to follow. This will continue until systems are based on the experimental method instead of upon opinion. Perhaps now is the time to lead the fungus world with a realistic, usable system of designating a *Fusarium* clone for what it is and what it does. An attempt to agree on some precepts on the basic philosophy to taxonomy and nomenclature would be a move in this direction. For example, the following is a list of guidelines or precepts for consideration:

1) Genus and species are based on morphological features. The biological behavior of a fungus is a sub-specific category such as the *forma specialis.* The species epithet tells what a fungus is; the *forma specialis,* what it does.
2) A fungus should be identifiable to species whether it is dead or alive, as long as the required morphological taxonomic characters are present.
3) A species taxon should include recognition of the existence of wild types, mutants, mating types and morphological features influenced by the environment.
4) Only one species epithet should be used for a fungus, no matter how many generic states it may reveal in its life cycle. To give two generic and two species epithets to a fungus is confusing.
5) Shades of pigmentation on artificial media in the laboratory are usually unimportant at

the species level, as are septation frequency, spore measurements as influenced by factors of the environment, sexuality, hermaphroditism, heterothallism vs. homothallism, rate of colony growth, production of metabolites, and pathogenicity.

6) A species description should represent a fungus population as its exists throughout the world. Thus, many isolates from many regions must be studied in order to reveal its variability. All species are variable.
7) Genetic and environmental variability is expressed by all species of fungi, including Fusaria. To ignore such variability is to name as species, various mutants and biotypes.
8) Variability is detectable by experimental analysis, especially of perfect states using, for example, the single-spore method of culture. Species based on an experimental evaluation of variability permits the designation of realistic species while opinionated judgment does not.
9) If certain species cannot be readily identified or distinguished by a biologist then they should be combined until (and if) new findings permit their separation.

Perhaps the biggest problem confronting Fusarium workers today is the identification of the fungi they isolate. There are usually three general steps involved in the development of a classification system. First, analyses of the characteristic features of collections of individuals, looking for differences and similarities in them. Second, the designation of "species," based primarily on the differences which warrant it, in the *opinion* of the investigator—a judgment. And third, the eventual synthesis of these usually narrowly defined "species" into larger species which reflect their genetic variability.

As a rule, the more a group of organisms is studied, the fewer the finally accepted species, whether it be man, or wheat, or dog, or *Fusarium.* A species is made up of different individuals possessing similarities which group them together. Experimental taxonomy clearly demonstrates this. Wollenweber's remarkable life work on *Fusarium* was largely in step three, the synthesis of species already described, when he brought organization out of the chaos of over a thousand species described in the century preceding him. He first studied the basics of the genus *Fusarium* in his "Grundlagen" (1) combining some generic epithets, then brought the number of species down to some 600 in 1931 (20), and finally in the volume "Die Fusarien" (21) reduced the total number of species in *Fusarium* to about 65, with 55 varieties.

There are fads and fancies in the classification of species as there are in other human endeavours. In the 1800's the prevailing philosophy in mycology was that every member of a species must be identical. These species were what we might call clones today. In the first half of the 1900's there developed a consciousness of variability in fungi, and this was shown by Wollenweber in his subsequent treatment of *Fusarium* species in 1935. But not all classification systems for fungi showed this trend. Many instances of clonal speciation are still being published and a current trend seems to be in favor of "splitters" again, rather than the "lumpers." Inevitably the splitters of species will ultimately be re-united.

It is not so important which taxonomic system a pathologist follows, as long as its significance is understood. Also, it is usually unwise to dictate arbitrarily that one system be used exclusively by all writers in a scientific publication. Progress in improved systems of classification and nomenclature depends on the use or rejection of a system by the scientific public. No system can be expected to be acceptable for long periods since new research alters the usefulness of a system, even that great one offered by Wollenweber 45 years ago.

Names We have the anomaly in the ascomycetes, that if an imperfect fungus reveals its perfect state as part of its complete life cycle, then that state may be given a new genus and species name. For example, after the perfect state of *F. solani* was discovered, it was later named *Nectria haematococca* instead of *Hypomyces solani* as Wollenweber chose.

Thus if *F. solani,* with its mycelium and three kinds of vegetative spores, wishes, so to speak, to keep its specific identity, then it must never indulge in its inherent sexual phase. For as soon as the sexual stage appears—contraceptives for fungi were not in vogue at that time—with the formation of beautiful red or white perithecia, the fungus suddenly becomes a different genus and species. What if that concept were applied to beetles or to *Homo sapiens?*

When it is reported in the literature that *"Gibberella zeae"* has been found on some plant, what does it mean? That *Gibberella* perithecia were seen, or that only conidia of *"Fusarium graminearum"* were observed and the author chose to call the fungus by its perfect state name, as some advocate? Is it a pathogen, or a saprophyte, or is its behavior undetermined? This species name does not answer these questions for the pathologist. Does the species *"Gibberella zeae"* refer only to a known pathogen and likewise the species name *"F. graminearum"* always connote pathogenicity? If so, what species names are to be given to a *"F. graminearum"* or to a *"Gibberella zeae"* which is not pathogenic to cereals?

We are greatly in need of a modernization of the International Rules of Nomenclature as they apply to fungi. Rules can be changed, or if not, broken. They are not laws and they are made by man, not God. If the rules are not modified in keeping with new facts and new philosophies then in the name of progress we must move ahead without them.

The variability concept had become a fact in the middle 1900's, yet it is not fully accepted even today. Variability is expressed genetically, basically through mutation, and also by sexuality through combination and re-segregation of chromatin; it is expressed by the differences in the environment to which the clone is subjected. Once the full potential for genetic variability is recognized, many so-called species will melt away. A species description must recognize change, since the expressions of change reveal the fungus itself.

Most speciation in fungi is based on opinion. A fungus looks different, so it must be a different species and so it is named. If mycologists would start their work with the collection of perfect stages, then by simple ascospore clones analyze the breadth of variation, we might be ahead. At present plant pathologists should devise taxonomic systems for groups of fungi which contain pathogens, since they are the ones who must use them.

Fusarium provides an ideal genus for the experimental study of biology as expressed genetically and environmentally. An individual is usually haploid in its vegetative state. The cells are hermaphroditic but not heterocaryotic. It may be either homothallic or heterothallic when it passes briefly through a diploid phase. Its physiologic or biochemical behavior may or may not be evident in its morphologic or cultivar appearance, yet it may be a powerful and economically dangerous organism.

The tremendous variability of *Fusarium* species insures their adaptability and survival in many diverse habitats throughout the world. The pathogens mostly are introduced organisms and once established may remain indefinitely in a locality.

***Fusarium*'s future** The population of *Fusarium* in agricultural fields is often as high as 100,000 propagules per gram of soil, or more, as shown by Smith (11). Most if not all of these clones may be beneficial saprophytes, aggressively decomposing dead organic matter of many kinds. Where a pathogen has been introduced it may reach populations of 4,000 or 5,000 propagules per gram of soil in the case of *F. solani* f. sp. *phaseoli,* or 10,000 or more for *F. roseum* f. sp. *cerealis* and variously for formae speciales of *F. oxysporum,* yet pathogens usually represent only a small percentage of the total *Fusarium* count. To have meaning, a soil count must be combined with proper pathogenicity tests to determine which clones of the *Fusarium* are pathogens.

Air dispersal of *F. moniliforme* by ascospores of the *Gibberella* state on rice may reach very high levels as shown by Sun (17, 22). The genetic variability of this heterothallic pathogen increases its threat as a pathogen in many parts of the world. Likewise, the air-borne spread of *F. nivale* by its *Calonectria* state, and of *F. roseum* f. sp. *cerealis* by its *Gibberella* state, is dramatically destructive in many regions of the world where corn, wheat, barley or rice are principle crops. Water-borne macroconidia disperse pathogens of the species of *F. solani* and of *F. oxysporum,* while insects transport the microcondia of *F. rigidiuscula* and *F. moniliforme.* Man further aids in the spread of Fusarium pathogens in his distribution of infected or infested seeds or other planting materials and by cultural practices.

It seems evident, then, that *Fusarium* is a growing threat to the world crops, more serious now with the shortages of food than at any time before. Losses from the spread of the date palm Bayoud disease have already turned a former date-exporting nation into a date-importing one.

The disastrous destruction of thousands of acres of prime banana plantations of the Cavendish variety in Taiwan by the new race "T" of *F. oxysporum* f. sp. *cubense* poses a threat to all banana culture just as race 1 formerly did. Fusarium wilts of fruits and vegetables are on the increase in many areas.

The disease of the flower pods of cacao caused by *F. rigidiuscula* f. sp. *theobromae* is among the top three pathogens of this crop, world-wide, but is still largely overlooked and thus ignored.

Currently, scab of wheat and barley caused by *Fusarium (Gibberella) roseum* f. sp. *cerealis* is probably the most threatening disease in many areas (9), including mainland China, according to Cook (3), and Korea, as pointed out by Sung (18).

These examples simply serve to highlight the major role Fusarium pathogens are playing in reducing yield and quality of major food crops of the world.

What is the prospect for the future? Research and instruction on *Fusarium* may be kept at a hopelessly low level as they have been in the past. This will insure an increasing spread and a deeper establishment of *Fusarium* pathogens in agricultural regions, more variations and a broader host range. Can we wait until disaster strikes harder? There were 40% losses reported on corn in two or three countries in the 1970's.

An alternative is for us to begin informing governments and granting agencies now of the losses due to *Fusarium* in many crops in most countries, and of the danger of eating or feeding infected products. We are in a better position today than ever before to mount a coordinated attack on *Fusarium* pathogens in spite of our lack of adequate support. We know something about the pathogens, if any, of each species.

We have the rapid, dependable, very simple single-spore method of Hansen (6) which was the most important advance after Wollenweber's book. This, and Hansen's (5) contributions on the genetic nature of *Fusarium* species provide an irrefutable series of facts revealing the extent of variability in nature and in the laboratory.

We have selective media (10) for the rapid isolation of pathogens, methods of culture which produce identifiable spores quickly, and also, mostly good testing methods for pathogenicity.

We now have methods for recognizing soils which are conducive or suppressive to *Fusarium* pathogens and a clue as to the nature of this difference as shown by Smith (12). Today we can measure populations of *Fusarium* in soil, before and after treatments, to detect trends in population numbers in a manner not yet available for most soil-borne organisms. We have other essential technology for our attack.

We now recognize the role of water stress as demonstrated by Cook (4) and something of the different epidemiologies of the diseases in different climates and regions of the world. We have an illustrated guide to *Fusarium* species by Toussoun and Nelson (19) and we have books with much assembled information and guides (7, 16).

Besides all of this, we enjoy a cooperative, concerned spirit among Fusarium workers of the world. Also, an International Newsletter on *Fusarium* has been begun, the pages of which are open to anyone to report briefly any new or interesting findings which may be of interest to Fusariologists. And significantly, although we may have great differences in viewpoints and philosophies among us, I believe there are no real personal animosities and that we are all friends working together for a common cause. And one day it will be as easy as pie for any biologist to identify to the species level any *Fusarium* he or she may encounter!

Literature Cited

1. Appel, O. U., and H. W. Wollenweber. 1910. *In* Arb. Biol. Anst. F. Land U. Forstw. Berlin-Dahlem 8:60.
2. Booth, C. 1971. The Genus *Fusarium.* Commonwealth Mycol. Inst., Kew, Surrey, England. 237 p.
3. Cook, R. J. 1977. Fusarium Diseases of Wheat in the Peoples Republic of China. Plant Pathol. Newsletter, FAO, Fusarium Notes, Rome, 4–6.
4. Cook, R. J., and R. I. Papendick. 1970. Soil water potential as a factor in the ecology of *Fusarium roseum* f. sp. *cerealis* 'Culmorum'. Plant Soil 32:131–145.
5. Hansen, H. N. 1946. Inheritance of sex in fungi. Proc. Nat. Acad. Sci. 32(10):272–273.
6. Hansen, H. N., and R. E. Smith. 1932. The mechanism of variation in imperfect fungi—*Botrytis cinerea.* Phytopathology 22:953–964.
7. Messiaen, C. M. 1959. La systematique du genre *Fusarium* selon Snyder et Hansen. Rev. Pathol. Vegetale Entomol. Agric. France 38:253–266.
8. Piglionica, V., S. Frisullo, and W. C. Snyder. 1974. Observazioni sul "mal del piede" del grano duro. Distribuzione di alciene specie di *Fusarium* nell' Italia meridionale. Ital. Agric. 111:121–123.
9. Piglionica, V., F. Gigante, and S. Frisullo. 1975. Le malattie dei cereali nell'Italia meridionale. Il *Fusarium roseum* f. sp. *cerealis* (1) su Grano duro. Phytopathol. Mediterr. 14:60–68.
10. Smith, S. N. 1965. Qualitative and quantitative comparison of Fusarium populations in cultivated fields and non-cultivated parent soils. Can. J. Bot. 43:939–945.
11. Smith, S. N. 1971. Relationship of inoculum density and soil types to severity of Fusarium wilt of sweet potato. Phytopathology 61:1049–1051.
12. Smith, S. N. 1977. Comparison of germination of pathogenic *Fusarium oxysporum* chlamydospores in host rhizosphere soils conducive and suppressive to wilts. Phytopathology 67:502–509.
13. Snyder, W. C., and H. N. Hansen. 1940. The species concept in *Fusarium.* Amer. J. Bot. 27:64–67.
14. Snyder, W. C., and H. N. Hansen. 1941. The species concept in *Fusarium* with reference to Section Martiella. Amer. J. Bot. 28:738–742.
15. Snyder, W. C., and H. N. Hansen. 1945. The species concept in *Fusarium* with reference to Discolor and other sections. Amer. J. Bot. 32:657–666.
16. Snyder, W. C., and T. A. Toussoun. 1965. Current status of taxonomy in *Fusarium* species and their perfect stages. Symposium: Taxonomy of Pathogenic Fungi. Phytopathology 55:833–837.
17. Sun, S. K. 1975. The disease cycle of rice Bakanae disease in Taiwan. Proc. Nat. Sci. Council, Republic China, 8:245–256.
18. Sung, J., and W. C. Snyder. 1977. *Fusarium (Gibberella) roseum* in cereals in Korea, 1976. Plant Pathol. FAO Newsletter. Fusarium Notes, Rome, 3–4.
19. Toussoun, T. A., and P. E. Nelson. 1976. A Pictorial Guide to the Identification of *Fusarium* Species According to the Taxonomic System of Snyder and Hansen. Second Edition. Pennsylvania State Univ. Press, University Park and London. 44 p.
20. Wollenweber, H. W. 1931. *Fusarium*—Monographie, 516 pp. Julius Springer, Berlin.
21. Wollenweber, H. W., and O. A. Reinking. 1935. Die Fusarien. Paul Parey, Berlin. 355 p.
22. Yu, K.-S. and S.-K. Sun. 1976. Ascospore liberation of *Gibberella fujikuroi* and its contamination of rice grains. Plant Protection Bull. (Taiwan) 18:319–329.

I Diseases Caused by *Fusarium* and Their Control

Prologue

T. A. Toussoun

Fusarium is best known as a plant pathogen. It is natural, therefore, to start this volume with papers on *Fusarium* diseases and their control. Essentially, *Fusarium* causes two types of plant disease: cortical rots and vascular wilts. It is the latter which have caused the most notoriety for they jeopardized, during the first part of the twentieth century, many agricultural industries in the USA and elsewhere. A perusal of the plant pathological literature on cotton wilt, pea wilt, and tomato wilt, to name just three, will show how destructive these diseases were and how close they came to eliminating the industries based on these crops. A reader of this volume can re-live those "old days" in Louvet and Toutain's story of Bayoud. We see how the date palm industry is being threatened; how the wilt, sweeping from West to East across North Africa, is poised on the doorstep of the Middle East, where ⅔ of the world's date production is located. We see the problems to be surmounted, and the difficulties involved in the development and deployment of control methods. The "ultimate weapon"—resistant cultivars—can be most difficult to obtain. This is perhaps most clear in the story of Panama wilt, mentioned in Stover's article on Fusarium diseases in the tropics and dealt with in detail in his reference 37—Banana, Plantain, and Abaca diseases. This disease is presently under control, thanks to the opportune discovery of resistant banana cultivars, except in Taiwan where they have fallen to a new race of the *Fusarium*. Catastrophe could also strike the African oil palm industry and Colhoun's treatment of the subject is well worth reading.

Of course the "old standby's" have not disappeared. All successful tomato cultivars, for example, incorporate resistance to Fusarium wilt. Cotton wilt is cited by Cook as being the most important disease of cotton in the People's Republic of China, occurring wherever the plant is grown; nor is it being ignored in the USA, as is clear from the paper by Smith, Ebbels, Garber, and Kappelman. What prevents these diseases from being the scourges they used to be is, by and large, the use of resistant plant cultivars (see the paper of Kraft, Burke, and Haglund), a job that has no end to it. Jones and Woltz, in their article on *Fusarium*-incited diseases of tomato and potato, and their control, give an insight into the pros and cons of monogenic vs. polygenic resistance and into other methods of control. Some interesting ideas on cross-protection are brought out in the article by Mas, Molot, and Risser on muskmelon wilt.

For the time being, then, Fusarium wilts are not as devastating as they used to be. They have relinquished that position to the other group: the Fusarium cortical rots. Preeminent in this category are the Fusarium diseases of wheat and corn—staple crops for mankind outside of the lowland tropics, in which regions the staple, rice, has its own peculiar Fusarium disease (Sun and Snyder). This book has six articles on Fusarium diseases of cereals and corn, a treatment that attempts to do justice to the subject by covering most of the major areas of the world where these crops are grown. As Maric points out in his paper, wheat grown in eastern Europe and in the USSR represents 31% of the world wheat-producing area. We can add to this Australia (Burgess, Dodman, Pont, and Mayers), China (Cook), the USA (Cook and Kommedahl and Windels). Maric makes the point that more papers have been published in the countries of eastern Europe on the Fusarium diseases of wheat and corn in the last decade than in all the years prior to 1967. Cook reports that scab is the most important disease of wheat and barley in the People's Republic of China. All of these papers deal with that most vexing problem—lack of adequate control. Unlike the case of Fusarium wilt, resistant host cultivars are so far unavailable and the prospects of developing them are rather dim. Stalk rot of corn figures prominently in this scene. It is one of the "modern" Fusarium diseases in that it is caused by a complex of organisms, comprised of more than one *Fusarium* pathogen acting in consort with others and aided and directed by sundry environmental factors. The composition of this complex varies from area to area, and the role of each member in the disease is presently most unclear. An even worse situation exists in the turfgrasses. Here, as Sanders and Cole bring out, Fusarium blight is characterized by an abundance of *Fusarium* on the blighted plants but no one has yet shown it to be the cause. In addition, one of the fungicides that controls the disease proves to be non-toxic to *Fusarium* in vitro.

Linderman deals with Fusarium diseases of flowering bulb crops. These also are cortical rots but interestingly enough belong to *Fusarium oxysporum,* the species whose pathogens heretofore belonged without exception to the Fusarium wilt group. One has recently been found on tomato (Jones and Woltz). One of the striking features of the pathogens in this species is their host specificity in the field. Apparently this trait is kept in the cortical rot *oxysporums,* for Linderman states that the lilly isolates are specific to the tissues such as root, stem, scale, etc.; a most extraordinary situation. While on the subject of ornamentals let us not forget Fusarium wilt of chrysanthemum, a disease so dependent on high temperatures for symptom expression as to be virtually unknown in the USA outside of Florida. It is discussed by Nelson, Horst, and Woltz, who also comment on the variables of plant cultivars and pathogen biotypes that make Fusarium disease of gladiolus such a difficult one to unravel.

Bean root rot, one of the diseases discussed by Kraft, Burke, and Haglund, merits a special place. It is perhaps the most "famous" for it was intensively studied for more than two decades by practically all the plant pathologists interested in soil-borne fungal pathogens in the western USA; it now stands as the model for all soil-borne fungi. Thanks to the studies of bean root rot, fungal behavior in soil, including pathogenesis, is no longer the imponderable phenomenon it once was. How a plant contracts diseases in the soil is now clear.

We have not yet filled out the Fusarium disease picture. There is more to come. As Stover points out, only a small percentage of the tropical *Fusarium* species have been studied. What is most intriguing is his statement that Fusarium wilts of temperate zone crops grown in the tropics are of little consequence. One wonders why, particularly since *Fusarium* species and their perfect states seem to be so abundant in those regions. Bloomberg considers *Fusarium* as an important cause of damping-off in the forest nurseries in all the continents and all the climates of the world. In speaking of these, we could also say that old Fusarium diseases of forest trees do not die and do not fade away. They reappear with new twists, as Dwinell's paper shows. But this is not the limit of the subject, for Rebell discusses the dangers of *Fusarium* to humans as well as to marine turtles, to which we can also add the American lobster and shrimp. One wonders what is left. Well, there are insects and other arthropods. *Fusarium* is known to attack certain scale insects. It is closely associated with others in plant disease yet very little is known. Perhaps we should wait for a second edition to this volume.

1 Bayoud, Fusarium Wilt of Date Palm

J. Louvet and G. Toutain

The date palm (*Phoenix dactylifera*) is an arborescent monocot cultivated for its fruit in irrigated desert regions with a hot, dry climate. It is estimated (6,22,29) that there are presently about 90,000,000 to 95,000,000 trees in production worldwide, mostly in the Middle East (65 to 70 million) and in North Africa (15 to 18 million). The date palm has also been introduced in similar regions of the American and Australian continents. The major disease of this plant is a vascular Fusarium wilt called Bayoud, an Arabic name which describes the bleaching of the fronds of the diseased trees. In Morocco, where the disease has existed for more than a century, most of the date palm growing areas have been attacked and it is estimated that more than 10,000,000 of the most productive trees have been killed (23). The disease also occurs in the western parts of the Algerian Sahara as far east as the meridian of Algiers. On the other hand, to the best of our knowledge, Bayoud does not presently occur in any other area of date palm cultivation in the world. It is therefore localized, but the risk of spread and the seriousness of the disease in these regions where man's livelihood by agriculture is difficult, require thorough surveillance, and research to find good methods of control.

Historical

Beginning in 1919, French researchers working in Morocco and Algeria published the first precise observations on Bayoud. They noted the infectious and epidemic nature of the disease, that the internal symptoms were localized in the vascular system (8), and they isolated the pathogen first called *Neocosmospora vasinfecta* (25), then *Fusarium vasinfectum* (24), and later *Cylindrophora albedinis* (12). Finally Malençon in 1934 (17) definitely identified it as a *Fusarium* of the section Elegans of Wollenweber and called it *F. albedinis,* which in current nomenclature is *F. oxysporum* f. sp. *albedinis* (9).

The mode of penetration of the pathogen in the plants has given rise to controversies during which it had been proposed that it occurred through wounds (12, 18) or through the inflorescences (19). Finally, penetration only by way of the roots, characteristic of Fusarium wilts, was recognized by Malençon (20). Perreau-Leroy (23) later emphasized possible control methods by the use of resistant cultivars. The researches of the last two workers dominated this period of studies on Bayoud.

In the 1960's research was carried on in Morocco, Algeria and France which led to the 1970 meeting in Morocco of the First Congress of Saharan Agronomy; in 1972 the First International Seminar and Workshop on Bayoud occurred in Algiers, during which an International Committee for Bayoud Research presided over by Professor W. C. Snyder was formed (26).

Fig. 1-1. Unilateral symptoms in the crown of a date palm infected with *Fusarium oxysporum* f. sp. *albedinis*. The fronds on the left have the appearance of "wet feathers."

Symptoms

The symptoms of Bayoud are typically those of a Fusarium wilt but present certain peculiarities which are linked to the morphology and the anatomy of the date palm as well as to the ecologic conditions under which it is cultivated. The date palm can live for a century and grow to a height of more than 20 meters. Its columnar trunk is a nonramified stipe whose elongation is governed by a single terminal bud that gives rise to a crown of pinnate leaves. Axillary suckers which are utilized as offshoots develop at the base of the stipe.

Bayoud is a fatal disease and occurs in young trees as well as in trees several decades old. At first disease symptoms usually appear as a whitening and drying of some leaflets from the base up to the apex, and often on one side only of one leaf. These leaflets fold up on the rachis. Later these symptoms appear on the entire affected leaf. At the same time, one or more longitudinal brown discolored bands extend upwards on the leaf rachis. The leaf dies, becomes desiccated, and hangs down along the trunk; at this stage it resembles a wet feather. The symptoms continue to appear on the adjacent leaves on one sector only of the tree (Fig. 1-1). When its terminal bud is finally affected, the tree dies. Very often, the suckers at the base of the trunk do not show symptoms a long time after the death of the tree. They form large clumps which cannot be cultivated. The average time from the appearance of the first symptom to death is generally 6 months to 2 years.

Sometimes the symptoms are different either in their appearance or in their rate of development. When the terminal bud is rapidly infected, trees may die in a few weeks, but others are infected 10 years before they die. These variations mainly depend on climatic and cultural conditions, date palm cultivar reactions, location of affected roots in the tree, and internal progress of the fungus in the trunk.

The below-ground portion of the date palm is made up of several hundred to several thousand cylindrical adventitious roots with diameters ranging between 10 and 15 cm and which can reach lengths of up to 10 m. When a tree is attacked by the Bayoud, only a few of the roots, grouped in a well-delimited zone, take on a red-brown coloration due to the presence of the parasite. After having crossed the parenchyma of the roots, the fungus penetrates the stele. It progresses longitudinally toward the stem, remaining localized in a few vessels of the xylem. The stipe and the rachis of the leaves have a structure which is typical of monocots: numerous fibrovascular bundles, sheathed by a layer of sclerenchyma and dispersed in a background of parenchyma. Only some of these bundles, corresponding to those of the attacked roots, are colonized by the parasite. Cross sections of an infected rachis or stipe show the presence of brown spots (Fig. 1-2) or, in longitudinal sections, streaks. This extreme localization of the parasite to a few roots and then to a few vascular bundles is in contrast to the fatal nature of the disease for the tree.

The Parasite

Fusarium oxysporum f. sp. *albedinis* can be easily isolated, often in pure culture, from vascular bundles that are discolored. On potato dextrose agar, the colonies, composed essentially of mycelium and numerous microconidia, have a salmon pink color. In certain clones a dark violet pigment diffuses into the medium. Macroconidia are rare, which explains the confusion in taxonomy which occurred at the time of the first studies of the fungus. In older cultures chalmydospores and sometime dark blue to black sclerotia are formed.

The pathogen has been found in all the vegetative parts of the infected palms. It has never been found in the inflorescences and as a consequence never been found in fruit, nor in seed. Its presence in the leaflets is very rare.

Microscopic observation of the tissues shows that the colonization of the parenchyma tissues is brought about by intracellular mycelia and then the ascending progression of the parasite in the vessels takes place by mycelia and by microconidia carried by the sap. The latter, just as in the case of Fusarium wilt of banana (1), are stopped by the transverse end walls of the vessels. The spores cross these by germinating and forming new conidia which are likewise stopped by the next end wall. This type of development causes the appearance of brown spots equidistantly spaced the length of the vessels, which are especially noticeable in longitudinal sections of the rachis of the palm fronds (5). Following the death of the tree, the mycelium invades the parenchyma. Macroconidia are rarely formed. On the other hand, numerous chlamydospores are formed characteristically in the lumen of the fibers of the sclerenchyma (Fig. 1-3). There the conditions are very favorable for the survival of these chlamydospores because they are protected in three ways against the outside elements: by their own walls, by the walls of the fibers, and by the grouping of the fibers in dense bundles (16). These chlamydospores in plant residue constitute inoculum which is randomly spread out in the soil, to a depth of 1 m but generally at a low population density (a few dozen chlamydospores per gram of soil).

The identification of isolates of *Fusarium oxysporum* belonging to the forma specialis *albedinis* as well as the determination of their pathogenicity and their virulence are accomplished by inoculating young date palm seedlings (15). These plants present a great genetic heterogeneity because the date palm is dioecious and as a consequence strongly heterozygous. Their individual susceptibility to Bayoud varies therefore from one plant to another. There are no other methods presently available since research on clonal multiplication of the date palm by means of tissue culture in vitro has not yet resulted in the production of plantlets whose susceptibility to the disease would be homogeneous. At present, under practical conditions of date palm culture, vegetative multiplication of clones is done by means of offshoots which weigh more than 10 kg each and which are produced in limited numbers by each mother plant and are not readily utilizable for inoculations.

Even though the young seedlings do not constitute perfect material, their utilization has already yielded some interesting results. Isolates belonging to the forma specialis *albedinis* have been identified in plants and in soil, which has prevented errors in diagnosis. In a date palm tree, isolates obtained from the extremity of the palm fronds are more pathogenic than those obtained from roots or from the stipe (5). This suggests that a selection must take place during the transport of the microconidia in the vessels, at the time when the end walls of the vessels are crossed, allowing for the continued progress of only the most virulent clones, to colonize the upper parts of the trees.

There is nothing to suggest that several races of *F. oxysporum* f. sp. *albedinis* exist in North Africa. In fact, when different infested areas in Morocco and Algeria have been planted with cuttings from known cultivars, no one has observed a modification of the behavior of a cultivar toward Bayoud.

Bayoud has been reproduced by inoculating the root system of date palms of different ages. Even under the most favorable conditions for disease development, the incubation period is always long. Thus, foliar symptoms generally appear 2 to 3 months following transplanting of young seedlings into infested soil, and 5 months to several years after the inoculation of a few sections of roots on 5- to 20-year-old plants. It seems therefore that the trees can harbor the parasite long before the appearance of the first external symptoms. Typical symptoms were also produced by inoculating young date palm plants with isolates obtained from vascular bundles of roots and stems of the henna plant, *Lawsonia inermis,* apparently healthy (15), and from Canary Island palms, *Phoenix canariensis,* showing symptoms of Fusarium wilt (2I). Such plants can therefore play a role in the epidemics of Bayoud by multiplying and protecting inoculum of the pathogen in their tissues.

There are few physiologic characteristics of *F. oxysporum* f. sp. *albedinis* which are known. The optimum temperature for growth of mycelium is between 25 and 28 C. Its carbon and nitrogen requirements do not allow it to be distinguished from other isolates of

Fig. 1-2. Cross section of a trunk of a 20-year-old date palm infected with *Fusarium oxysporum* f. sp. *albedinis.* Only a few vascular bundles are discolored and contain the pathogen (arrows).

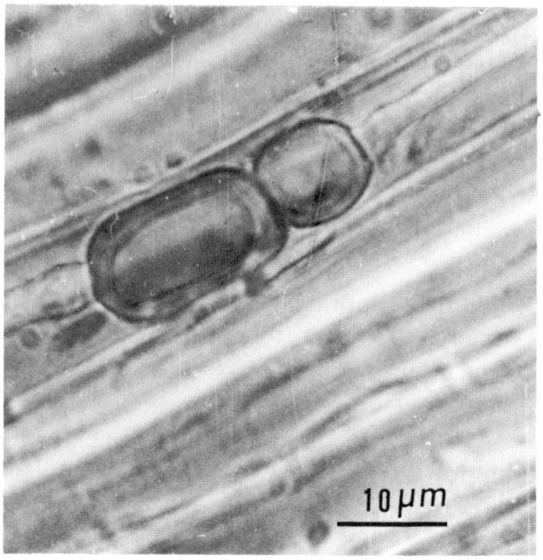

Fig. 1-3. Chlamydospores formed on the mycelium of *Fusarium oxysporum* f. sp. *albedinis* in the lumen of a sclerenchyma fiber in an infected date palm root.

F. oxysporum (2,3). No correlations have been found between pathogenicity of the various isolates and their production of enzymes (7) or toxins as fusaric acid and marasmins (27).

Disease Spread

Beginning with a few attacked trees, the disease spreads in large foci which are particularly visible in plantations uniformly containing susceptible cultivars. Many observations have been made on this spread (4, 5, 10, 13, 30). Since conidia are rarely formed on the surface of the tissues, aerial dissemination is not important. The spread of the parasite from one tree to another seems to take place essentially by root contact. Intensive cultivation, and in particular frequent irrigation, greatly facilitate the development of foci, resulting in disease epidemics. One observes a concurrent increase in the number of trees attacked (Fig. 1–4) and a lessening of the average incubation period for each tree (i.e., the time between symptom expression and death). The conditions which are most favorable for the growth of the date palms are equally most favorable to their rapid destruction by the Bayoud. After a period of drought, irrigation or rain accelerates the movement of sap in the trees and thereby facilitates the activity of the microconidia in the vessels and the development of symptoms. During periods of drought and lack of irrigation, which frequently occurs in the Sahara Desert, the disease develops slowly but the growth and the production of the date palms are reduced also.

Numerous observations made on the chronology of the appearance of Bayoud in the various palm groves of North Africa (28) need to be completed by up to date and precise data. The parasite is spread by man through transporting contaminated vegetative parts or infested soil and the principle mode of dissemination of the pathogen is by infected offshoots. The disease first appeared during the last century in the Draa Valley, a region of the south-western part of Morocco, spread along the valleys, then travelled from oasis

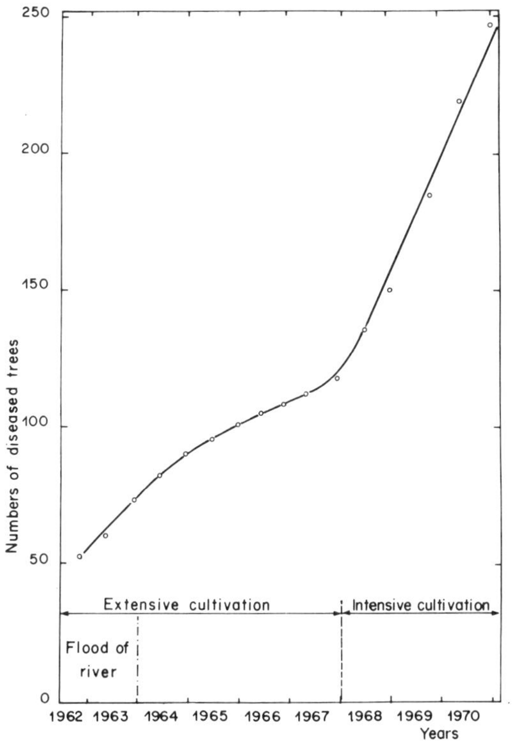

Fig. 1-4. Graph showing the influence of cultural conditions on the spread of Bayoud in a plot planted with 282 palms, spaced 9 m apart, in 1945.

to oasis in the Algerian Sahara. Bayoud presently menaces the eastern part of Algeria and Tunisia which are the most important regions where the best but very susceptible date cultivar, "Deglet Nour," is grown. As a consequence of modern transportation facilities it is not inconceivable that new foci may arise at a distance, for example, in the Middle East.

The progressive spread of Bayoud from a precise point in western Morocco suggests that it is due to a particular clone of *F. oxysporum* which appeared at that particular spot and was transported by man from grove to grove without differentiation into physiologic races. This parasite therefore presents characteristics of a forma specialis having a monotypic origin (3l).

Control Methods

The future of date palm production in North Africa is dependent on the development of effective methods of control against Bayoud. The directions to be taken have been well defined (5, 11, 14, 15, 26).

The first goal is to stop, or at least slow down, the geographic progression of the disease and in particular to try to save the very productive groves of the "Deglet Nour." One must also reconstitute the groves that have been decimated and establish new ones with resistant cultivars, preferably cultivars as good as "Deglet Nour," or "Mehjoul," an important Moroccan cultivar used in export in the past, which has now practically disappeared due to its susceptibility to Bayoud.

Preventive measures require the mapping of disease areas by ground examination or aerial photography, alerting the growers to dangers of dissemination, and the establishment of phytosanitary control measures. These very necessary measures can slow down but not stop the spread of the disease. Eradication is possible only in recently invaded areas, but early diagnosis is difficult due to the long incubation period; many trees may be diseased before the first symptoms appear. In addition, it is difficult to eradicate a pathogen that can persist deep in the soil, and that can cause disease even at low inoculum densities. Control by systemic fungicides also involves technical and economic problems.

The most productive avenue, even though it is difficult and long, is by means of resistant cultivars. Date palms having resistance can be obtained from three sources: 1) by utilizing offshoots of existing clones which are represented by numerous individuals, 2) from individuals obtained from seed during natural propagation, and 3) from individuals selected from seed obtained in breeding trials. Natural selection which occurred in regions which had been infested by Bayoud for long periods of time, has given rise to resistant palms. Some have been vegetatively propagated by Moroccan and Algerian growers. They make up mother blocks for offshoots used in grove regeneration. Others are made up of only a few individuals, often only one. Finally, instead of letting crosses and selection be controlled by chance, for several years a hybridization program has been operated so as to obtain individuals that are resistant, well adapted to the climatic and cultural conditions of the area, and capable of producing an abundance of fruit of good quality. Just as in the case of the improvements of mammals, particular attention is given to the genetic qualities of the males, for each of them can be utilized to fertilize numerous females and thereby have an abundance of offspring. There are certain risks to this program because little is known about the genetic mechanisms of transmission of characters in the date palm, and plant development is of long duration because both sexual reproduction and vegetative multiplication of this plant are very slow.

What is principally needed in order to make these studies more efficient, before new methods of biological control are used, is the improvement of vegetative multiplication by tissue culture in vitro. Not only will this accelerate the development of selection programs in North Africa, but it will also facilitate tests for resistance and the use of cultivars or clones cultivated in countries that are not yet infected but menaced by the Bayoud.

Conclusion

Bayoud is a Fusarium wilt whose characteristics are due to the peculiarities of the date palm, an arborescent monocot, dioecious and strongly heterozygous, whose growth and reproduction are slow. This disease has an economic and social impact of great importance to the Sahara, and has attracted attention to date palm cultivation and to the totality of agricultural problems of the Sahara, an agriculture whose equilibrium is fragile, under climatic conditions which are often hostile, but whose rational intensification has already given promising results. The Bayoud is a scourge, but it is equally a factor in progress.

Acknowledgments

The authors thank T.A. Toussoun and P.E. Nelson for translating and reviewing the manuscript.

Literature Cited

1. Beckman, C.H., M.E. Mace, S. Halmos, and M.W. McGahan. 1961. Physical barriers associated with resistance in Fusarium wilt of bananas. Phytopathology 51 : 507–515.
2. Bounaga, N. 1969. Quelques aspects de la physiologie d'une souche de *Fusarium oxysporum* f. sp. *albedinis* agent de la maladie du Bayoud. Bull. Soc. Hist. Nat. Afr. N. 60 : 137–183.
3. Bounaga, N. 1976. Le palmier dattier et la Fusariose. II. Rapport entre la stéréochimie des oses et quelques uns de leurs dérivés et la croissance du *Fusarium oxysporum* f. sp. *albedinis*. Can. J. Microbiol. 22 : 636–644.
4. Brochard, P., and D. Dubost. 1970. Progression du Bayoud dans la palmerie d'In Salah. Awamia 35 : 143–153.
5. Bulit, J., J. Louvet, D. Bouhot, and G. Toutain. 1967. Recherches sur les Fusarioses. I. Travaux sur le Bayoud, Fusariose du palmier dattier en Afrique du Nord. Ann. Epiphyties 18 : 213–239.
6. Dowson, V.H.W., and F.P. Pansiot. 1965. Improvement of date palm growing. F.A.O. Agric. Study, Dates, Bag. 65, 1. 209p.
7. Dubost, D., L. Kechacha, and B. Rether. 1970. Etude des enzymes pectinolytiques et cellulolytiques d'une souche monospore de *Fusarium oxysporum* f. sp. *albedinis*. Awamia 35 : 195–211.
8. Foex, E., and P. Vayssiere. 1919. Les maladies du dattier au Maroc. J. Agric. Tropic., 19 : 336–339.
9. Gordon, W.L. 1965. Pathogenic strains of *Fusarium oxysporum*. Can. J. Bot. 43 : 1309–1318.
10. Kada, A., and D. Dubost. 1975. Le Bayoud à Ghardaía. Bull. Agric. Sahar., Alger 1, 3 : 29–61.
11. Kellou, R., and D. Dubost. 1974. Organisation de la recherche et de la lutte contre le Bayoud en Algérie. Bull. Agric. Sahar., Alger 1, 1 : 5–13.
12. Killian, C., and R. Maire. 1930. Le Bayoud, maladie du dattier. Bull. Soc. Hist. Nat. Afr. N. 21 : 89–101.
13. Laville, E., and P. Lossois. 1963. Méthode de Van der Plank et mode de propagation du Bayoud. Fruits 18 : 249–253.
14. Louvet, J., and J. Bulit. 1970. Le Bayoud, Fusariose vasculaire du Palmier dattier. Symptômes et nature de la maladie. Moyens de lutte. Awamia 35 : 161–182.
15. Louvet, J., and G. Toutain. 1973. Recherches sur les Fusarioses. VIII. Nouvelles observations sur la Fusariose du palmier dattier et précisions concernant la lutte. Ann. Phytopathol. 5 : 35–52.
16. Louvet, J. 1977. Observations sur la localisation des chlamydospores de *Fusarium oxysporum* dans les tissus des plantes parasitées. Travaux dédiés à G. Viennot-Bourgin, Soc. Franç. Phytopathol., Paris, 193–197.
17. Malençon, G. 1934. Nouvelles observations concernant l'étiologie du Bayoud. Compt. Rend. Acad. Sci., Paris 198 : 1259.
18. Malençon, G. 1934. La question du Bayoud au Maroc. Ann. Crypt. Exot., Paris 7, 2 : 1–41.
19. Malençon, G. 1946. L'infection florale du dattier par le *Fusarium albedinis*. Compt. Rend. Acad. Sci., Paris 223 : 923–925.
20. Malençon, G. 1949. Le Bayoud et la reproduction expérimentale des lésions chez le palmier dattier. Bull. Soc. Hist. Nat. Afr. N. Hors Série 2 : 217–228.
21. Mercier, S., and J. Louvet. 1973. Recherches sur les Fusarioses. X. Une Fusariose vasculaire du palmier des Canaries. Ann. Photopathol. 5 : 203–211.
22. Munier, P. 1973. Le palmier dattier. Maisonneuve et Larose. Paris. 221p.

23. Perreau-Leroy, P. 1958. Le palmier dattier au Maroc. Inst. Recherche sur les Fruits et Agrumes, Paris. 142p.
24. Pinoy, P.E. 1925. Sur la maladie du Bayoud des palmiers de Figuig. Compt. Rend. Soc. Biol. 92 : 137–138.
25. Sergent, E., and M. Beguet. 1921. Sur la nature mycosique d'une nouvelle maladie des dattiers menaçant les oasis marocaines. Compt. Rend. Acad. Sci., Paris 172 : 1624–1627.
26. Snyder, W.C., and A.G. Watson. 1974. Conclusions of Bayoud seminar, Algiers, October 1972, Bull. Agric. Sahar., Alger 1, 1 : 25–30.
27. Surico, G., and A. Graniti. 1977. Produzione di tossine da *Fusarium oxysporum* Schl. f. sp. *albedinis*. Phytopathol. Mediterr., 16: 30–32.
28. Toutain, G. 1965. Note sur l'épidémiologie du Bayoud en Afrique du Nord. Awamia 15 : 37–45.
29. Toutain, G. 1967. Le palmier dattier, culture et production. Awamia 25 : 83–151.
30. Toutain, G. 1970. Observations sur la progression d'un foyer actif de Bayoud dans une plantation régulière de palmier dattier. Awamia. 35 : 155–161.
31. Zadoks, J.C. 1959. On the formation of the physiologic races in plant parasites. Euphytica 8 : 104–116.

2 Vascular Wilt Disease of Oil Palms

John Colhoun

This disease of the oil palm (*Elaeis guineensis* Jacq.) was first reported by Wardlaw (30) as occurring in the Belgian Congo (Zaire) and most of our early knowledge of the disease is based on his work and that of his colleagues in the Cryptogamic Botany Laboratories of the University of Manchester. The disease has later been referred to, particularly by French authors, as "fusariose" or "tracheomycose".

Wardlaw (30) stated that the same species of *Fusarium* had been isolated from diseased palms growing in different localities. Shortly afterwards he established that the species involved was *F. oxysporum* (31). The pathogenicity of isolates of this organism obtained from xylem vessels of typically wilted palms was established by Fraselle (7,8). The name *elaeidis* given to the forma specialis of *F. oxysporum* causing the disease was accredited to Toovey (29).

De Blank and Ferguson (1947, private report) believed that the disease occurred in Nigeria, and this was confirmed by Wardlaw (32), who showed that isolates from palms showing typical symptoms in the Belgian Congo and Nigeria were similar. Wardlaw (32) also stated that in both Nigeria and the Belgian Congo the leaf disease of oil palms known as "patch yellows" was associated with the presence of *F. oxysporum,* but the strain involved was different from that associated with vascular wilt disease. This conclusion was confirmed by Gogoi (9).

An undated report (1) refers to the occurrence in Surinam in 1950 of a disease of oil palms believed to be similar to vascular wilt. The disease has since been reported in the Cameroons (2), Colombia (28), the Ivory Coast and Dahomey (27). No records exist of its occurrence in the important oil palm plantations of Malaysia.

In West Africa the disease is confined to plantations and nurseries, with the vast areas of native groves being free (34). However, according to Prendergast (21), the indigenous West African oil palm as well as *Elaeis melanococca* Gaertner [= *Coroso oleifera* (H.K.B.) Bailey] and *E. madagascariensis* Beccari (? = *E. guineensis*) are susceptible. Recent unpublished work at Manchester indicates that isolates of *F. oxysporum* f. sp. *elaeidis* from wilted oil palms can kill young seedlings of date palms (*Phoenix dactylifera* L.). The pathogenicity to oil palm seedlings of *F. oxysporum* f. sp. *albedinis* isolated from date palms showing typical symptoms of vascular wilt in Algeria is being tested. Disease did not result when oil palm seedlings were inoculated with *F. oxysporum* f. sp. *cubense* from bananas affected with vascular wilt (21).

External Symptoms

The pathogen attacks seedlings in the nursery as well as young and mature palms in plantations. Prendergast (21) has drawn attention to the existence of chronic and acute forms of the disease in adult palms.

When the chronic form occurs the older leaves of adult palms wilt and later turn brown with desiccation occurring from the tip backwards. Leaves tend to break at some distance from the base and hang down forming a sheath around the trunk. In early stages of the attack the affected leaves may be confined to one or two spirals, with the other leaves being normal (Fig. 2-1). The disease usually progresses rather slowly into the crown and finally all the lower leaves become brown or grey, desiccated, broken and hang down around the trunk. The few green leaves which remain show reduced length and small leaflets. A palm may remain in this condition for up to a few years but ultimately the crown of small, green leaves wilts, dies, and falls off to leave a decapitated trunk.

When the acute form of vascular wilt occurs, death of the leaves takes place very rapidly and they may retain their original erect positions on the plant until broken off, usually several feet from the trunk, by wind action. The whole crown rapidly becomes desiccated and the palms usually die within 2 or 3 months. Ultimately, the stipe generally breaks off a little below the crown and falls to the ground. Various intermediate stages between the acute and chronic forms may occur.

When palms 1 to 6 years old are attacked, the first symptom is the appearance of what Prendergast (21) called "lemon frond." This is the development of a bright yellow colour in a leaf in the middle of the crown, but sometimes only a few leaflets in the leaf are chlorotic. The leaflets on the "lemon frond" dry up and the rachis becomes desiccated so that the leaf dies. Other leaves on the plant turn yellow and then grey-brown as they dry up, but breaking of the rachis is rare. Some lower leaves may next show symptoms but usually some young leaves die while many older ones remain green. The youngest leaves are reduced in size and such palms usually die within a year.

According to Prendergast (21), infected plants in the nursery show progressive shortening of the younger leaves so that the crown has a bunched appearance. This is followed by browning and death of the oldest leaves, with the condition progressing inwards until all the leaves have died. In 3-month-old infected seedlings Locke and Colhoun (15) found that there was a slowing down of leaf production, with both leaf number and plant height being reduced.

Internal Symtoms

When providing the first description of the disease Wardlaw (30) suggested that the vascular system of the trunk became infected by way of diseased roots or through wounds in the base of the trunk. After infection the vascular strands in the trunk turn a greyish brown to black instead of retaining their normal pale yellow colour. At first infection is most developed in the peripheral strands in the trunk, but later the central strands also become discoloured. Irregular distribution of the discoloured strands may occur. Vascular symptoms may spread upwards into the crown of the plant and into the bases of the leaves. Discolouration of vascular strands in roots and also in "bulbs" of young plants occurs. According to Kovachich (11), hyphae and chlamydospores occur in the xylem vessels but the fungus was not seen outside the vascular bundles. Discolouration of the vessels is associated with the presence of gum within them and some vessels may be completely blocked by gum in which fungal filaments may be embedded. Tyloses may appear in many vessels. Bachy (4) has suggested that wilting of the leaves results from the mechanical obstruction of the vessels and also from the secretion of toxins.

It has been emphasized by a number of workers (3,4,5,6) that in associating discolouration of xylem vessels with infection it is necessary to distinguish between the vessels and the neighbouring fibres, since the latter can be brown although the vessels are not infected.

In most infected palms many roots show no symptoms, although diseased roots are frequent below the base of the trunk. It can be shown that there is continuity between the infected vascular system of roots and the infected xylem vessels in the trunk (6).

The Infection Process

It is generally agreed that infection takes place through the roots but workers disagree regarding the necessity for root injury before infection can occur. Fraselle (7,8) did not deliberately damage the roots of seedlings prior to or at the time of inoculation but infection readily occurred. Kovachich (12) agreed that roots which had not been artificially damaged could be infected. Prendergast (21) observed that roots seemed to be infected through subsidiary rootlets which had died. On the other hand, Renard (24) stated that the fungus could not penetrate undamaged roots since the hyphae encountered a lignified barrier represented by the hypodermis and rhizodermis. He believed that the hyphae, when in contact with a wound, were incapable of invading the xylem vessels since they were probably stopped by the formation of tyloses following wounding. He found that microconidia were sucked into the sap of the root after wounding and could reach the base of the "bulb" before the root reacted to the wound. If the transverse walls of sound vessels prevented the progression of microconidia upwards, the spores accumulated and germinated, and the germ tubes penetrated the cross wall so that microconidia could then be formed on the other side of the wall.

Locke and Colhoun (16) have recently made a detailed study of the infection process. They showed that infection could occur through roots which were not deliberately damaged even when the plants were grown in nutrient solution throughout the experiment. When seedlings were grown in compost the fungus was found to enter the tips of pneumathodes and the tips of lateral roots or it could enter the loosely packed cortical cells at the base of a pneumathode. It did not enter through the tips of radicles or the tips of lateral roots which were just emerging from the radicle, neither did it enter the cortical cells where a young root or a pneumathode had just erupted through the epidermis.

Pneumathodes were regarded by Locke and Colhoun (16) as providing a frequent infection route. They found that 12 days after infection of young seedlings the fungus was present in the cortical tissue of the pneumathode, while within 30 to 45 days after inoculation it occurred in the cortical cells and xylem of the pneumathode and occasionally deep inside the main radicle. In the cortical cells of the pneumathode the fungus was mainly intercellular, causing damage and separation of the cells. The hyphae were, however, intracellular in the xylem, passing from one cell to another *via* the pits. Although the host reacted to the presence of the pathogen in the pneumathode tissue within the primary root by blocking some cells, presumably with resins, the fungus was not stopped by this barrier in susceptible plants. As the pathogen passed deeper into the radicle along the pneumathode tissue, a second barrier could develop adjacent to the stele of the radicle. In this second barrier there were cells blocked with resins or in which there was tylose formation. The hyphae were able, however, to ramify between the cells of this barrier so that they became separated and the fungus was then able to pass through and enter the xylem of the radicle stele.

Studies of pneumathodes have been made by Yampolsky (35) and Purvis (23), who showed that they are modified secondary roots produced on both aerial and underground roots. Under moist conditions many pneumathodes may form on underground roots. Young pneumathodes possess root caps which are usually lost later and then through expansion of cortical tissue rupture of the epidermis occurs so that the stele and cortex extrude. Ultimately, as the surrounding tissue is sloughed off, the stele becomes more prominent as it extrudes beyond the cortical tissue. The fungus may, therefore, have the opportunity of penetrating the stele directly in its unprotected state or it may penetrate the cortical cells of the pneumathode and so reach the xylem. Having reached the xylem of a pneumathode, the pathogen may spread along it and reach the xylem of the primary root without passing through the cortical cells of the radicle, which it does not seem well adapted to invade.

When Locke and Colhoun (16) examined plants about 9 months after inoculation, hyphae and microconidia of the pathogen were found in the "bulb," with continuous infected vascular tissue from the roots into the "bulb." In roots, the hyphae were confined to the xylem and were not observed in the phloem or cortex of the primary root until a very late stage of the infection process. In the "bulb," hyphae were frequent in the xylem vessels although many strands were not infected (Fig. 2-2). Hyphae were not seen in the cortex of the "bulb." Occasional discoloured vascular strands were seen in the leaf bases in which tyloses were rare although they were frequent in the "bulb." Blocking of the xylem elements, presumably with resins, was seen in roots, "bulbs," and leaves.

Epidemiology

In plantations, palms showing symptoms of vascular wilt may occur singly or in small or large groups. Disease incidence may be substantially determined by the number of infected oil palms previously growing on the land, for Prendergast (21) showed that infection was more frequent in young palms planted close to the sites of diseased palms in a previous stand.

Park (19) showed that *F. oxysporum* was present in soil around palms showing symptoms of vascular wilt and also in wilt-free areas of plantations although the pathogenicity to oil palms of any of the isolates was not established. Renard and Meyer (26) reported

Fig. 2-1 Mature oil palm infected with *Fusarium oxysporum* f. sp. *elaeidis* showing symptoms of vascular wilt.

Fig. 2-2 Transverse section through the "bulb" of an oil palm infected with *Fusarium oxysporum* f. sp. *elaeidis*. Note that many vascular strands are necrotic (X4). (Photograph by permission of "Phytopathogische Zeitschrift" from vol. 88, p. 24).

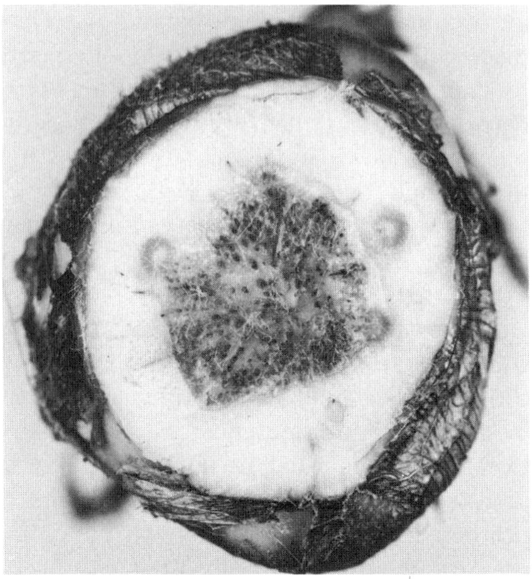

that symptoms of vascular wilt may appear at random in plantations established on savannah soils and Renard et al. (27) have suggested that the pathogen can exist in the forest. These observations raise the question can the pathogen exist in soils which have not previously carried a crop of oil palms? In this connection it must be borne in mind that since *F. oxysporum* f. sp. *elaeideis* may be seed-borne (14) it could be introduced by this means into soil which was not previously contaminated.

Using soil samples collected in Zaire, Locke and Colhoun (15) demonstrated that *F. oxysporum* f. sp. *elaeidis* can be present in soil collected 1.5 m and 4.5 m. from an infected palm to the extent of 1.5×10^3 and 4×10^2 propagules/g of oven dry soil respectively. The pathogen occurred in soil samples from a plantation regarded as being free from vascular wilt disease and here its density was much greater at a depth of 15 cm than at 30 cm. In the limited number of such samples examined the highest number of propagules of the pathogen found to occur was 8×10^2/ g of oven dry soil. When 5×10^2 propagules of the pathogen were introduced into compost, infection readily occurred. It is therefore possible that the pathogen may occur in soils which have not previously borne infected oil palms and infection may develop. This conclusion is supported by data obtained by Locke (13) who isolated the fungus from the roots of oil palm seedlings which had been inoculated with two isolates of *F. oxysporum* obtained from savannah soils in Africa.

Attempts have been made by Renard and Meyer (26) to distinguish between *F. oxysporum* f. sp. *elaeidis* and what were regarded as saprophytic forms of *F. oxysporum* occurring in soil. They concluded that all the strains studied belonged to a single serological group but they found great variability in the antigenic constitution of extracts of strains of the fungus within a phenotype or within a pathogenic forma.

Park (20) showed that the pathogen can survive in the form of chlamydospores in soil without fresh decomposable material. He also demonstrated that under the fluctuating moisture levels found in natural soils spores can be carried in horizontal and vertical directions. Where flooding of plantations occurs it may be assumed that spread of inoculum by flood water is possible. Moreover, since infection can occur readily through pneumathodes (16) wet soils may be expected to promote attacks through inducing development of these organs. There is little evidence available from field observations regarding the effects of soil water on disease incidence, although Prendergast (21) concluded that in one estate in Nigeria there were fewer infected palms in the portion where the soils and drainage were slightly better. Bachy and Fehling (5) reported a relationship between areas where vascular wilt occurred and low ground.

From observations in Nigeria, Prendergast (21) concluded that when vigorous growth of oil palms cannot be maintained epidemics of vascular wilt may occur. Applications of potassium decreased the incidence of the disease and Ollagnier and Renard (18) have confirmed this finding in the Ivory Coast for both susceptible and resistant hybrids.

It is difficult to explain satisfactorily why vascular wilt disease is absent from the large oil palm growing areas of Malaysia. It is possible that the pathogen could have been introduced into that country on seeds but if so the disease has not developed. It has been suggested that differences in climatic conditions between Malaysia and West Africa are important and that perhaps the more even distribution of rainfall over the year in Malaysia may provide less favourable conditions for infection and disease development than those prevailing in West or Central Africa.

Economic Importance

Wardlaw (32) obviously attached considerable economic significance to the disease because of the number of palms which were killed, together with the risk to young palms planted on sites previously occupied by infected specimens. More recently he has drawn attention to the fact that although the disease may appear to progress so slowly over periods of 2 or 3 years that its overall effects may seem tolerable, this is not so when observations are made over longer periods (33). According to Bachy (4), the annual mean loss of palms killed by vascular wilt (presumably in the Ivory Coast) is about 2% but may be as high as 6–8%. In certain plantations in the Belgian Congo, Guldentops (10) found that up to 25% of the oil palms may be dead or dying from attacks of vascular wilt and in some blocks the loss may be as high as 33%. However, he considered that progressive elimination of some individual palms can be compensated for by the increased production of those that remain. Such compensatory effects would not operate readily when large patches are involved and moreover Guldentops recorded that a lowering of yield may occur a few years before external symptoms appear.

Control

In spite of efforts to find a fungicide, either systemic or otherwise, which would provide efficient control, this has not been achieved so far in the long term (8,10,17,25). As already noted, potassium used as a fertilizer serves to reduce attacks (21). Breeding for resistance is, however, likely to be the most successful means of reducing losses. As Wardlaw (33) pointed out, it is necessary to be able to test large numbers of progenies for resistance to vascular wilt using a reliable seedling inoculation procedure. Considerable advances in this direction have been made. Prendergast (22) has devised a method of testing progenies for resistance at the nursery stage. Inoculum was prepared by macerating cultures in liquid medium after 9–14 days growth, diluting the product and pouring a stated quantity over the collar of seedlings (usually with 4 leaves) as they were placed in the planting holes. Seedlings were examined for infection and symptoms of the disease 4–8 months after inoculation. No progeny wholly immune to the disease was found during the study but progenies in a trial could be divided into groups of varying degrees of resistance.

Two methods of testing for resistance, referred to as the pre-nursery and the nursery tests, have been developed by Renard et al. (27) and show similarities to Prendergast's test. In both methods the earth is cleared away from the adventitious roots, which are washed with water; the inoculum suspension (prepared from agar cultures) is poured over the roots and the earth is then replaced. In the pre-nursery test, plants are inoculated when they have 1 to 1½ leaves and symptoms of wilt are recorded 4½ months later when the "bulbs" are cut open to expose discolouration of the vascular strands. In the nursery test, plants are inoculated 7 weeks after being planted in the nursery and are examined 8 months later. In both tests the percentage of wilt-infected plants is obtained for each progeny so that progenies can be classed, in relation to each other, in order of increasing susceptibility. Results obtained by the two tests are similar and it has been shown that certain parents pass to their progeny an ability for notable resistance or greater susceptibility.

A modification of Prendergast's method is also in use at the Nigerian Oil Palm Research Station (Rajagopolan, *personal communication*). The modifications introduced enable inoculation to be effected speedily and with somewhat less root damage.

Locke and Colhoun (15) attempted to devise a testing method which eliminated root damage attributable to the technique and enabled propagules of the pathogen to be introduced at predetermined levels into compost in which germinated seeds were placed. After 4 months the dry weight of the aerial parts of each plant was determined, or if the plants were required for further growth, leaf area was assessed. Comparison of data obtained for plants growing in contaminated and uncontaminated composts enabled those progenies which were significantly reduced in size in the former to be separated from those which showed no statistical reduction. Some progenies showed little reduction in size with high or low levels of contamination of the compost, others showed little reduction at low levels but much reduction at high levels of contamination, and some progenies showed much reduction in size at both levels of contamination. Importance is therefore attached to carrying out tests using at least two known levels of contamination of compost. Such a test should lead to greater comparability being attained between results for the same progenies employed in separate tests.

When resistant or tolerant progenies have been obtained by seedling tests, it is necessary to establish if these progenies remain resistant when grown for many years in plantations where the disease occurs. It is also necessary to establish if progenies show differential reactions to different isolates of the pathogen. Promising results have already been obtained from the selection of resistant seedlings.

Literature Cited

1. Anon. (undated). Jaarverslag over 1950. Versl. Dept. Landb. Suriname, 87 p. (Abstr. Rev. Appl. Mycol. 32: 546–547).
2. Anon. 1960. Ann. Report Cameroons Development Corp. 1959. 29 p. (Abstr. Rev. Appl. Mycol. 40: 148).
3. Bachy, A. 1965. Clef pour la détermination des maladies du palmier à huile. Oléagineux 20: 13–17.
4. Bachy, A. 1970. La fusariose du palmier à huile. Oléagineux 25: 265–267.
5. Bachy, A., and C. Fehling. 1957. La fusariose du palmier à huile en Cote d'Ivoire. J. Agric. Trop. Bot. Appl. 4: 228–240.
6. Bull, R.A. 1954. A preliminary list of oil palm diseases encountered in Nigeria. J. W. Afr. Inst. Oil Palm Res. 1: 53–93.
7. Fraselle, J.W. 1951. Experimental evidence of the pathogenicity of *Fusarium oxysporum* Schl. to the oil palm (*Elaeis guineensis* Jacq.). Nature (Lond.), 167: 447.
8. Fraselle, J.W. 1951. *Fusarium oxysporum* as the cause of vascular wilt disease of the oil palm. Trans. Brit. Mycol. Soc. 34: 492–496.
9. Gogoi, T. 1949. Notes on two strains of *Fusarium oxysporum* from the oil palm in the Belgian Congo. Trans. Brit. Mycol. Soc. 32: 171–178.
10. Guldentops, R.E. 1962. Contribution à l'étude de la trachéomycose du palmier à huile. Parasitica 18: 244–263.
11. Kovachich, W.G. 1948. A preliminary anatomical note on vascular wilt disease of the oil palm (*Elaeis guineensis*). Ann. Bot. (Lond.), N.S. 12: 327–329.
12. Kovachich, W.G. 1953. Private Report of Research Department, Unilever S.A., Yaligimba, Belgian Congo.
13. Locke, T. 1972. A study of vascular wilt disease of oil palm seedlings. Ph.D. Thesis, Univ. Manchester, 120 p.
14. Locke, T., and J. Colhoun. 1973. *Fusarium oxysporum* f. sp. *elaeidis* as a seed-borne pathogen. Trans. Brit. Mycol. Soc. 60: 594–595.
15. Locke, T., and J. Colhoun. 1974. Contributions to a method of testing oil palm seedlings for resistance to *Fusarium oxysporum* Schl. f. sp. *elaeidis* Toovey. Phytopathol. Z. 79: 77–92.
16. Locke, T., and J. Colhoun. 1977. A process of infection of oil palm seedlings by *Fusarium oxysporum* f. sp. *elaeidis*. Phytopathol. Z. 88: 18–28.
17. Moreau, C., and M. Moreau. 1960. Inhibition de la croissance du *Fusarium oxysporum* Schl. par divers fongicides organiques. Rev. Mycol. Paris, 25: 307–310.
18. Ollagnier, M., and J.-L. Renard. 1976. Influence du potassium sur la résistance du palmier à huile à la fusariose. Oléagineux 31: 203–208.
19. Park, D. 1958. The saprophytic status of *Fusarium oxysporum* Schl. causing vascular wilt of oil palm. Ann. Bot. (Lond.), N.S. 22: 19–35.
20. Park, D. 1959. Some aspects of the biology of *Fusarium oxysporum* Schl. in soil. Ann. Bot. (Lond.), 23: 35–49.
21. Prendergast, A.G. 1957. Observations on the epidemiology of vascular wilt disease of the oil palm. J.W. Afr. Inst. Oil Palm Res. 2: 148–175.
22. Prendergast, A.G. 1963. A method of testing oil palm progenies at the nursery stage for resistance to vascular wilt disease caused by *Fusarium oxysporum* Schl. J.W. Afr. Inst. Oil Palm Res. 4: 156–175.
23. Purvis, C. 1956. The root system of the oil palm: its distribution, morphology and anatomy. J.W. Afr. Inst. Oil Palm Res. 1: 60–82.
24. Renard, J.-L. 1970. La fusariose du palmier à huile. Rôle des blessures des racines dans le processus d'infection. Oléagineaux 25: 581–586.
25. Renard, J.-L. 1973. Transport et distribution du bénomyl dans les palmiers à huile au stade de la pépiniere. Oléagineux 28: 557–562.
26. Renard, J.-L., and J.A. Meyer. 1969. Serological study of saprophytic strains of *Fusarium oxysporum* and *F. oxysporum* f. sp. *elaeidis*. Trans. Brit. Mycol. Soc. 53: 455–461.

27. Renard, J.-L., J.-P Gascon, and A. Bachy. 1972. Recherches sur la fusariose du palmier à huile. Oléagineaux 27: 581–591.
28. Sanchez Potes, A. 1966. Enfermedades del Algodomero, del Cototero y de la Palma Africane en Colombia. Acta Agron. Palmira, 16: 1–13. (Abstr. Rev. Appl. Mycol. 46: 630)
29. Toovey, F.W. 1949. Ann. Rep. Agric. Dept., Nigeria, 1948. 75 p.
30. Wardlaw, C.W. 1946. A wilt disease of the oil palm. Nature (Lond.), 158: 56.
31. Wardlaw, C.W. 1946. *Fusarium oxysporum* on the oil palm. Nature (Lond.), 158: 712.
32. Wardlaw, C.W. 1948. Vascular wilt disease of the oil palm in Nigeria. Nature (Lond.), 162: 850–851.
33. Wardlaw, C.W. 1968. Reflections on some diseases and pests of the oil palm. PANS, B. 14: 261–272.
34. Waterson, J.M. 1953. Observations on the influence of some ecological factors on the incidence of oil palm disease in Nigeria. J.W. Afr. Inst. Oil Palm Res. 1: 24–59.
35. Yampolsky, C. 1924. The pneumathodes on the roots of the oil palm (*Elaeis guineensis* Jacq.). Amer. J. Bot. 11: 502–512.

3 Fusarium Wilt of Cotton

S. N. Smith, D. L. Ebbels, R. H. Garber, and A. J. Kappelman, Jr.

Fusarium wilt of cotton (*Gossypium* spp.) is caused by *Fusarium oxysporum* f. sp. *vasinfectum.* It was the first vascular wilt described (5), and in this and other early reports (9) there is an implication that the disease had existed in Alabama and in India for many years before identification of the pathogen. Now, 87 years later, this disease is still causing important yield losses in several places in the world and remains a threat to cotton production in the future. The regions where Fusarium wilt of cotton is most destructive are in southeastern USA (especially Arkansas, Alabama, Georgia, Louisiana and South Carolina), the Nile Valley of Egypt, an area of Tanzania south and east of Lake Victoria, and parts of India. Kelman and Cook (28) have recently reported that it is the most important disease of cotton in the Peoples' Republic of China. The disease is known to occur in most of the major cotton producing areas of the world, notable exceptions being West Africa, Turkey and Australia.

Losses due to Fusarium wilt of cotton vary depending upon (i) host resistance, (ii) general plant health, (iii) inoculum potential, (iv) environmental factors, (v) presence of nematodes, and (vi) use of chemical fertilizers. Heavy losses often occur when susceptible cultivars are grown on infested land. However, more resistant cultivars have been grown since shortly after this disease was first diagnosed, thus reducing disease losses. The recent release and growth in Georgia of 'Atlas' (a high-yielding cultivar with good fiber traits which was highly susceptible to Fusarium wilt) spectacularly demonstrates disease losses which can occur when susceptible cultivars are grown on soils infested with both the wilt *Fusarium* and root knot nematodes.

In some of the earlier reports Fusarium wilt may have been confused with other disorders, such as Verticillium wilt, which was not described on field-grown cotton until 1928 (45). Fusarium wilt has not been found recently at some of the sites mentioned in early reports (14), and at others *F. oxysporum* f. sp. *vasinfectum* was not consistently isolated in all cases (32) or positive pathogenicity tests were not obtained (32, 43). The geographic origin of the pathogen is unknown.

Symptoms

Symptoms of Fusarium wilt can appear at any time from emergence to maturity of the cotton plant. In young *Gossypium hirsutum* L. seedlings, the first symptoms are epinasty, dullness of the cotyledons and other leaves, and virtual cessation of growth. The leaves then become yellow, flaccid, shriveled, turn brown, and drop from the plant. The vascular system is stained brown. Such seedlings die quickly and in some infested fields wilting of young seedlings may be considerable and may be confused with damping off. Very often, however, the pathogen is present only in the root cortex in early infections, and no symptoms are evident. Smith (*unpublished data*) isolated *F. oxysporum* f. sp. *vasinfectum* from roots of over half of the seedlings taken at random from several heavily infested fields 7–14 days after emergence. This indicates that primary infections occur early.

Seedlings of *G. arboreum* L. (10) and *G. barbadense* L. (16) infected with *F. oxysporum* f. sp. *vasinfectum* develop a yellowing along the cotyledon veins (vein-clearing); this symptom is not usually seen in seedlings of *G. hirsutum*.

In older plants first symptoms may appear in the lower leaves; these become yellow and then brown, first at the margins and then often between the veins to form a bold digitate pattern. Ebbels (14) referred to this leaf symptom as "tiger striping." Discrete sections of the leaf commonly become yellowish or wilt and become limp. These areas may be very sharply demarcated, the rest of the lamina remaining turgid, and they may not follow the vein pattern but cut across main veins. With either type of symptom severely diseased plants may be defoliated quickly, leaving leafless stems and sympodia, which may still be green. Such plants are usually severely stunted and the vascular system is stained a dark brown to black.

In older plants, not showing symptoms until well-grown, the vascular staining may be confined to only a small portion of the vascular ring. In California plants of Acala SJ-2 may not show any particular external symptoms even towards the end of the growing season, but when examined, may exhibit stained vascular bundles above the ground level from which the pathogen can be isolated.

The symptoms of Fusarium wilt and Verticillium wilt cannot be reliably distinguished in the field, although there are usually differences. For example, with Fusarium wilt the vascular staining is usually darker and more continuous than with Verticillium wilt. Separating the two diseases according to symptoms is especially difficult when both occur in the same field. It is always desirable to isolate from the diseased plants when diagnosing disease occurrence at a new site, and then test the pathogenicity of the isolates.

Infection

Khadr (29) conducted detailed anatomical investigations on the process of infection. Penetration of roots occurred in the zone 1–4 cm directly behind the root tip. Although this zone contains the root hair region, Khadr (29) did not observe penetration of the root hairs. No penetration took place through meristematic tissue, older regions of the root, or natural wounds caused by eruption of lateral roots. He observed that the path of the infecting hyphae within the cortex is mostly confined to the intercellular spaces. After 4–5 days the hyphae reach the stele and may penetrate the endodermis inter- as well as intracellularly. Within the stele the fungus is confined for the most part to the xylem. Massive aggregations of the fungus hyphae within vessels were found in the root-stem transition zones. Early signs of gum-like depositions were detected on the perforation plates, while heavy depositions occurred only at the late stages of disease.

Etiology of Symptoms

The cause of the wilt symptoms in cotton plants infected with the wilt *Fusarium* is not yet well understood (35). Early theories that the xylem becomes blocked by mycelium are now thought unlikely because water transport would not be decreased sufficiently to cause wilt. The toxin fusaric acid (5-n-butyl-pyridine-2-carboxylic acid) is produced in infected cotton plants (33) and appears to affect the plants' water balance by changing the semipermeable properties of the plasma membrane in leaf cells. However, it does not cause many of the other physiological changes observed in infected plants, such as increased respiration and staining of the xylem, although it decreases photosynthesis (31) and it may also be produced by non-pathogenic strains of the fungus. The senescence-like symptoms which form part of the wilt syndrome may be due to the production of ethylene (52). Vascular browning could occur in various ways; most probably via the oxidation of phenols liberated, possibly, by the β-glycosidase system.

Factors Influencing the Incidence of Infection and Wilt

Although in the field, Fusarium wilt is usually more prevalent in warm weather, Al-Shukri (1) found that 10-week-old plants of *G. barbadense* 'Sakellaridis' were more susceptible when they were inoculated after 5 weeks growth at 15 C (a temperature unfavorable for the growth of cotton) than when held at a higher temperature prior to inoculation. The fact that seedlings are relatively susceptible to infection may be because during the seedling stage in the field air temperatures can be especially cool at night. Other factors which may predispose cotton plants to infection are an imbalance of nutritional requirements (59) and excessive soil moisture levels (1, 18).

Conditions that predispose cotton to infection by *F. oxysporum* f. sp. *vasinfectum* may not be the same as those which most encourage development of wilt after infection. It is reasonable to expect that roots of seedling plants and also stunted or stressed plants (not growing vigorously) are likely to exude more nutrients (amino acids, sugars, etc.) into the rhizosphere. These substances are known to stimulate germination and growth of the fungal resting structures, leading to infection of the roots. Further development of the disease in the plant, however, may depend on warmer temperatures [Young (58) suggested an optimum soil temperature of 30.5 C for development of wilt], as well as other conditions that favor vigorous plant development, and increased transpiration.

The Influence of Nematodes on Incidence of Wilt

Atkinson (5) in the first report of Fusarium wilt of cotton in 1892 noted an association of Fusarium wilt with root-knot nematodes. Although wilt was increased by the occurrence of root-knot, either disorder could occur in the absence of the other. The root-knot nematode commonly associated with Fusarium wilt of *G. hirsutum* (American upland cotton), is *Meloidogyne incognita* (34, 47). However, this species is not mentioned in relation to Fusarium wilt of cotton in Egypt or India. Infections with *Rotylenchulus reniformis* were found to increase Fusarium wilt in upland cotton in the USA (38) and in *G. barbadense* in Egypt (30). Other nematodes implicated in association with wilt of upland cotton are *Belonolaimus gracilis* (19), *B. longicaudatus* (12), and *Pratylenchus brachyurus* (14, 37).

Control of nematodes with soil fumigants often results in considerable decreases in cotton wilt (8, 17, 46) and in large yield increases. However, usually only a part of this benefit of fumigation can be ascribed to control of nematodes or decreases in wilt. Ebbels (14) points out that some of the benefits are derived from greater mineralization of soil nutrients following fumigation. It is also apparent that the application of certain fumigants alters the soil microflora in ways not yet understood. On the other hand, the growing of cotton cultivars with high levels of tolerance to root-knot nematodes has in itself resulted in less disease. Shepherd (44) suggested that damage due to the complex could be largely avoided if cultivars with greater resistance to both root-knot and Fusarium wilt were available.

It is not yet clear in what manner nematodes increase the severity of Fusarium wilt. It has been suggested (36, 43, 47) that penetration by *Fusarium* occurs through those portions of the vessels exposed after the cortex has been damaged by the action of nematodes or other soil organisms. These opinions, however, have been based largely on circumstantial evidence. Perry (42) observed that root-knot larvae do not stimulate infection of roots with *F. oxysporum* f. sp. *vasinfectum* by causing extensive mechanical damage or necrosis in cotton seedling roots. He found no more fungal hyphae near points where larvae had entered roots than elsewhere, nor did the fungus appear preferentially to colonize egg masses or gall tissue. The results of Khadr's histological studies (29) support Perry in that penetration of the epidermis and cortex of the young root was delayed when these tissues were damaged by penetration by root-knot nematodes. Fungus hyphae did not enter at the sites of nematode entry, but instead penetrated undamaged epidermis (intercellularly) a few cells away. Khadr et al. (30) reported in 1972 that physiological changes in host tissues attacked by *Rotylenchulus reniformis* may encourage infection by *F. oxysporum* f. sp. *vasinfectum.* One might expect plants stunted by nematode infection to exude more and different nutrients from their roots than do healthy, vigorously growing plants, and that such plants may be more readily infected, as was described for stunted plants earlier in this chapter. Infections with nematodes, on the other hand, may also interfere with the water uptake and transpiration of a plant, thereby slowing the upward movement of a plant pathogen, as suggested by Baker and Cook (6). The complex requires further investigation for clarification.

Influence of Soil Type on Incidence of Wilt

Typically, Fusarium wilt of upland cotton is most serious on acid, sandy soils, although it has been recorded on alkaline soils of pH 8.1–8.3 (7). In India, however, the problem in *G. arboreum* is most prevalent on black clay soils of pH 7.8–8.3 (32, 51), and Fahmy (16) reported the pathogen in Egypt is most destructive to *G. barbadense* in clay loam soils.

In some soils the disease is more prevalent than in others, and the pathogen itself may be directly influenced by the soil environment. Smith and Snyder (49) found that chlamydospores of *F. oxysporum* pathogens germinated more frequently and grew more vigorously in soils from fields known to be "wilt conducive" compared to soils "suppressive" to Fusarium wilt.

Differentiation and Distribution of Races

To date five races have been described in *F. oxysporum* f. sp. *vasinfectum:* races 1 to 4 by Armstrong (3, 4) and race 5 by Ibrahim (20) as shown in Table 3-1.

Race 2 has a very limited distribution. Furthermore, since Armstrong and Armstrong (2) found that soybean and tobacco, which differentiate races 1 and 2, are not immune to all isolates of race 1, Ebbels (14) suggested that race 2 be regarded as a variant of race 1 with a slightly wider host range, rather than a separate race.

Geographically the races are distributed separately. Race 1 is widespread in the USA (3), East Africa (14), and possibly in Italy (10); race 2 has been reported only in USA (at two sites in S. Carolina); race 3 is in Egypt (3); race 4 is in India (3), and possibly also in the USSR (10); and race 5 is in the Sudan (20, 21).

Table 3-1. Differentiation of races in *Fusarium oxysporum* f. sp. *vasinfectum*

	Differential host species and cultivars					
	Gossypium arboreum	*G. barbadense*		*G. hirsutum*	*Glycine max*	*Nicotiana tabacum*
RACES	'Rozi'	'Ashmouni'	'Sakel'	'Acala'	'Yelredo'	'Gold Dollar'
1 (USA)	R	S	S	S	R	R
2 (USA)	R	S	S	S	S	S
3 (Egypt)	S	R	S	R	n.i.	R
4 (India)	S	R	R	R	n.i.	S
5(Sudan)	S	S	S	R	n.i.	n.i.

S = susceptibility; R = resistance; n.i. = no information.

Isolation from Plants and Soil

The pathogen is easily isolated from diseased host stems and petioles by plating surface-sterilized sections of the tissues on water agar, potato-dextrose agar, or other common media, and incubating them at room temperature. Within 48 hr the fungus grows out from the vascular tissue, usually in pure culture.

Isolations of this pathogen can be made from soil and roots of various plants in the field (50), but this requires more time and effort because of the large populations of other microorganisms. Selective media are useful for such isolations, but one must ascertain that the isolates are indeed the pathogen because nonpathogenic *F. oxysporum* also grow well on these media and are numerous in cultivated field soils and in plant roots. By transferring colonies to more than one medium, comparing colony types and cultures with the known pathogen from the field, f. sp. *vasinfectum* may be fairly well distinguished, but pathogenicity tests are still necessary. Almost all new isolates of the pathogen from California plants or soils are one of two cultural types and are surprisingly uniform.

The Pathogen

Spores and cultures of *F. oxysporum* f. sp. *vasinfectum* are typical for *F. oxysporum*. Furthermore, it seems that the Indian, Egyptian, and American upland races may not necessarily be easily distinguished except by their pathogenicity (16, 32, 56). The cultural variations most frequently occurring among isolates of the forma specialis *vasinfectum* are in numbers of sporodochial masses, amounts of soluble lavender pigment, and the strong odor reminiscent of *Syringa vulgaris* which is produced on various media (32, 50, 56). In laboratory culture, and sometimes even in the plant, "mutations" often quickly alter cultural and spore characteristics of the fungus. Mycelial types, pionnotal types and also more subtle changes occur, which differentiate cultures in the laboratory from the wild types isolated from field soils. Clones capable of long-term survival in soil seem to be relatively few compared to the array of cultural types occurring in the laboratory. Mutations that increase survival in the soil are apparently unusual. Critical research must be conducted with the appropriate clones of *F. oxysporum* f. sp. *vasinfectum*. We recommend that only recent isolates from field soils or infected plants be used.

Host range

It is possible to isolate *F. oxysporum* f. sp. *vasinfectum* from many plant species which have been artificially inoculated (14). Furthermore, this pathogen can be isolated from the roots of many different kinds of crop plants and weeds in infested fields, even though these plants are often symptomless under field conditions. In plants such as grasses, the cortex of fibrous roots is usually invaded to a depth of a few cell layers. As plants senesce, chlamydospores eventually form on the outside of the plant roots, and in the intercellular spaces of the root cortex. There are some hosts other than cotton, most notably okra (*Hibiscus esculentus* L.), in which mild to severe symptoms are induced under field conditions (5). Okra is affected by race 1 of *F. oxysporum* f. sp. *vasinfectum,* although it is not susceptible to races 3 or 4 (3). *Malva parviflora* L. plants, growing as weeds in an infested cotton field, were also infected (50). The pathogen was isolated from the above-ground portions of stems and some plants showed slight external symptoms. Hollyhocks (*Alcea rosea* L.) were very susceptible in inoculation studies in the glasshouse (Smith, *unpublished data*); many of the test plants died, yet there is no record of the disease on this host under natural conditions in gardens. Wood and Ebbels (57) inoculated several native Tanzanian plants of the families Malvaceae, Sterculiaceae, and Tiliaceae and found several of them to be susceptible to *F. oxysporum* f. sp. *vasinfectum*. Susceptibility varied with the age of the plant.

The host range of *F. oxysporum* f. sp. *vasinfectum* includes more than cotton. However, in a discussion of host range we should define whether (i) the disease occurs in the field, (ii) the organism can be isolated from discolored vascular tissue or from internal or surface-sterilized root tissue, or (iii) the disease appears on plants only when usually high inoculum levels are used. Plants act as hosts to the fungus in all of these cases, but the degree of pathogenicity may not be the same.

Mechanisms of resistance

Mechanisms of resistance may operate before or after vascular invasion by the pathogen and include well-known mechanisms such as the formation of lignitubers to contain invading hyphae and occlusion of xylem vessels by gels or tyloses, thus preventing microconidial transport in the xylem. Phytoalexins may also play a part in resisting infection, and the rate at which ethylene is produced in infected plants may be one of the main factors which determine whether susceptible or resistant reactions are displayed (52).

Inheritance of Resistance and Breeding

In spite of studies by several workers over a considerable period of time, the inheritance of resistance to Fusarium wilt in *G. hirsutum* is still undefined. Based on field experiments, Smith and Dick (48) concluded that, where nematodes could be controlled, wilt resistance was regulated by one dominant gene. Jones (23) found that in a cross between Delfos 425 and Half and Half, one major gene without dominance determined resistance in the F_2. Later studies by Jones (24) involving segregating materials from the above cross as well as materials from a cross of Coker 100 Ga. with Half and Half, showed continuous variation from very susceptible to resistant in field resistance to Fusarium wilt in F_2 plants and F_3 line means in each cross. This led to the conclusion that resistance to wilt is a quantitative character. The behavior of these materials led to the additional conclusion that at least two but no more than three gene-pairs were involved, but that additional modifying genes influenced resistance.

More recent glasshouse studies by Kappelman (25) involving segregating generations, backcross materials, and parents from crosses between the susceptible 'Rowden' and six different resistant lines indicated Fusarium wilt resistance was mainly additive. However, these studies were conducted under ambient temperature conditions in the absence of any nematodes. In addition, stems of all plants were needle-inoculated using known concentrations of Fusarium conidia; thus, normal exterior plant resistance factors may have been completely bypassed. However, when Kappelman (26) evaluated 24 lines or cultivars both under similar glasshouse conditions and for several years under field conditions involving high inoculum levels of both Fusarium wilt and nematodes, wilt expression of these materials under glasshouse conditions was highly correlated with their wilt expression under field conditions.

Inheritance of resistance to Fusarium wilt in other *Gossypium* spp. is equally as confusing. Resistance in Sea Island cotton (*G barbadense* L.) was reported controlled by two dominant genes that were additive in effect (48), while resistance in *G. herbaceum* L. to the Indian race of the fungus was reported controlled by one dominant gene (27). However, more extensive work involving crosses and backcrosses up to the F_3 and B_2 generation in both *G. arboreum* and *G. herbaceum* led Kelkar et al. (27) to conclude that resistance was controlled by two complementary dominant genes, but that a third dominant gene, when present, could completely inhibit action of the other two. Earlier work by Jenkins et al. (22) involving crosses in Upland, Hopi, Sea Island, and Egyptian cottons indicated resistance was inherited as a dominant character.

Despite the lack of adequate knowledge of the inheritance and expression of Fusarium wilt resistance, breeding programs and selections for Fusarium wilt-resistant materials have been conducted since the early 1900's (by W. A. Orton and E. L. Rivers) and continue to this day. Although several inoculation techniques and programs are currently available for screening materials for resistance to Fusarium wilt of cotton, most of these are limiting and not entirely suitable for screening large numbers of plants. Development of resistant cultivars is, therefore, dependent on adequate field-screening facilities. Most researchers in the USA and elsewhere do not have such facilities. The only facility in the USA available to all researchers is at Tallassee, Alabama. Since about 1946, cooperators from throughout the USA as well as from several other countries have submitted materials to this unit for evaluation, in what has become known as "The Regional Fusarium Wilt Test." Many resistant lines and cultivars have been developed as a result of this program. However, greater resistance to Fusarium wilt is needed, especially in combination with greater yielding potential and desirable fiber properties.

Means of Survival of the Pathogen in Soil

Infection of host plants, whether or not symptoms are produced, is perhaps the most important means of long-term survival of *F. oxysporum* in field soils. Trujillo (53) observed that both *F. oxysporum* f. sp. *cubense* in infected bananas and *F. oxysporum* f. sp. *lycopersici* in tomato plants grew out of the xylem into surrounding tissue as the plants senesced, and chlamydospores were formed prolifically in the intercellular spaces. The large numbers of chlamydospores of *F.oxysporum* f. sp. *vasinfectum* found in tissues of the symptomless hosts nutsedge and barley (50), probably function as future inoculum, as do chlamydospores in the stems of hosts with symptoms. Colonies of *F. oxysporum* on soil dilution plates almost invariably grow from small pieces of decaying plant debris. In certain of the infested cotton fields in Kern County, California, the numbers of such propagules can reach 5,000 per gram of soil.

Soilborne propagules may also be increased (i) as a result of conidia from sporodochia being washed from the surfaces of infected tissues into the soil by rain or irrigation, and (ii) by further propagation of inoculum that is already present in soil through growth and sporulation. The former has not been observed on cotton plants under the semi-arid conditions of California, but Ebbels (14) reported that in Tanzania, conidia are often found on surfaces of moribund field plants and unburied crop debris. Many of these soilborne conidia may not survive as long as others that are converted to chlamydospores. However, it seems that propagules in plant tissue live longer than do chlamydospores unprotected in soil. Wensley and McKeen (54) reported wide seasonal fluctuations in soil populations of *F oxysporum* f. sp. *melonis;* much of the inoculum in this case was derived from sporulation on the stems of diseased melon. Old and Schippers (40) demonstrated that chlamydospores of *F. solani* f. sp. *cucurbitae* formed in soil from conversion of conidia were far more vulnerable than those produced in pure culture, and were apparently destroyed by soil microflora.

With buried Rossi Cholodny slides, Subramanian (51) observed that hyphae, conidia and even chlamydospores of *F. oxysporum* f. sp. *vasinfectum* were quickly destroyed by the action of soil bacteria, and "that subterranean spread of this fungus is confined to growth internal to the host." The fungus readily colonized host root sections buried in soil at various depths.

Fusarium chlamydospores may increase in number in soil by germination, growth of hyphae, and subsequent formation of new chlamydospores. This may be of especial importance in rhizospheres, furnishing inoculum for multiple infections of plant roots and hypocotyls (11). Nyvall and Haglund (39) reported that with *F. oxysporum* f. sp. *pisi* multiple root infections were necessary for severe disease to develop in susceptible pea cultivars. It seems likely that cotton plants also may require more than one root infection for symptoms to appear because the pathogen can often be isolated from plants showing no external symptoms. Where inoculum may substantially increase in the rhizospheres of susceptible hosts, the chances for multiple infections are enhanced, but the life expectancy of chlamydospores produced in the rhizosphere must be short.

Seed Infection

Internal infection of seeds was first demonstrated by Elliott (15), who found up to 6% infection of cotton seeds from a wilted crop. In India, Kulkarni (32) found 9.9% of seeds from wilted cotton plants were infected, and Perry (41) in Tanzania found 47% of seeds from infected plants carried the pathogen. However, Smith (*unpublished data*) in California found no *F. oxysporum* f. sp. *vasinfectum* in over 3,000 (acid delinted) seeds removed from heavily infected field-grown plants, although peduncles and even a few seed stalks bore the pathogen. Furthermore, few of the California-produced seed contained any Fusaria; in comparison, seed from Mississippi examined by Davis (13) frequently contained *F. roseum, F. moniliforme* or *F. oxysporum.* From this and information supplied by other authors (15, 55) we believe that *F. oxysporum* f. sp. *vasinfectum,* when it is present in seeds, usually enters through penetrations of the seed coats in infested bolls rather than systemically through vascular connections. Since Fusaria are more apt to grow and sporulate in or near bolls under humid rather than dry conditions, one would expect to find more seed infection in moist climates. However, contamination of seed with trash, dust, or soil may also be responsible for transporting the pathogen great distances.

Epilogue

Although Fusarium wilt of cotton was described relatively early in the history of plant pathology, and subsequently formed a basis for recognition of other wilts, many facets of this disease still are not understood. More work is needed to clarify relationships between disease severity and the physiological status of the plant, interactions with nematodes, soil type, etc.

At present in the USA successful control still involves breeding plants for resistance to both Fusarium wilt and root-knot nematodes, together with soil fumigation with nematicides. As costs of petro-chemicals rise, the latter method of control may become less feasible economically. Further, the retention of large populations of the pathogen in the many infested fields that exist increases the probability that new races of the pathogen will also arise to confound plant breeding. We hope that future research will continue to be supported for programs of breeding plants for resistance as well as studies regarding the nature of interrelationships of plants and pathogens in field soils.

Literature Cited

1. Al-Shukri, M. M. 1969. The predisposition of cotton plant to Verticillium and Fusarium wilt diseases by some major environmental factors. J. Bot. U. Arab Repub. 12:13–25.
2. Armstrong, G. M., and J. K. Armstrong. 1953. Fusarium wilt of sesame. Plant Dis. Rep. 37:77–78.
3. Armstrong, G. M., and J. K. Armstrong. 1960. American, Egyptian and Indian cotton-wilt Fusaria: their pathogenicity and relationship to other wilt Fusaria. U.S. Dep. Agric. Tech. Bull. 1219. 19 p.
4. Armstrong, J. K., and G. M. Armstrong. 1958. A race of the cotton-wilt Fusarium causing wilt of Yelredo soybean and flue-cured tobacco. Plant Dis. Rep. 42:147–151.
5. Atkinson, G. F. 1892. Some diseases of cotton. III. Frenching. Ala. Agric. Exp. Stn. Bull. 41:19–29.
6. Baker, K. F., and R. J. Cook. 1974. Biological control of plant pathogens. W. H. Freeman and Co, San Francisco. 433 p.
7. Blank, L. M. 1962. Fusarium wilt of cotton moves west. Plant Dis. Rep. 46:396.

8. Brown, A. G. P. 1969. Plant pathology. Prog. Rep., Exp. Stn. Tanzania Western Cotton Growing Area, 1967–68; 17–22.
9. Butler, E. J. 1910. The wilt disease of pigeon-pea and the parasitism of *Neocosmospora vasinfecta* Smith. India Dep. Agric. Mem. 2(9):1–64.
10. Charudattan, R. 1969. Studies on strains of *Fusarium vasinfectum* Atk. I. On their morphology and pathogenicity on cotton. Proc. Indian Acad. Sci., Sect. B., 70:139–156.
11. Cook, R. J., and W. C. Snyder. 1965. Influence of host exudates on growth and survival of germlings of *Fusarium solani* f. *phaseoli* in soil. Phytopathology 55:1021–1025.
12. Cooper, W. E., and B. B. Brodie. 1963. A comparison of Fusarium wilt indices of cotton varieties with root-knot and sting nematodes as predisposing agents. Phytopathology 53:1077–1080.
13. Davis, R. G. 1976. Relationship of seedborne Fusaria to performance of cotton planting seed, p. 34. *In* Proc. Beltwide Cotton Prod. Res. Conf., Las Vegas, January 1976.
14. Ebbels, D. L. 1975. Fusarium wilt of cotton: A review, with special reference to Tanzania. Cotton Grow. Rev. 52:295–339.
15. Elliott, J. A. 1923. Cotton-wilt: a seed-borne disease. J. Agric. Res. 23:387–393.
16. Fahmy, T. 1927. The Fusarium disease (wilt) of cotton and its control. Phytopathology 17:749–767.
17. Garber, R. H., A. Hyer, E. C. Jorgenson, and S. N. Smith. 1976. Control of Fusarium wilt-root-knot nematode complex in California, p. 30. *In* Proc. Beltwide Cotton Prod. Res. Conf., Las Vegas, January 1976.
18. Gilbert, W. W. 1914. Cotton wilt and root-knot. U.S. Dep. Agric. Farmers' Bull. 625. 21 p.
19. Holdeman, Q. L., and T. W. Graham. 1954. Effect of the sting nematode on expression of Fusarium wilt in cotton. Phytopathology 44:683–685.
20. Ibrahim, F. M. 1966. A new race of cotton wilt Fusarium in the Sudan Gezira. Emp. Cotton Grow. Rev. 43:296–299.
21. Ibrahim, F. M. and O. Khalifa. 1970. Fusarium wilt of cotton in the Gezira, p. 189–194. *In* M. A. Siddig and L. C. Hughes (ed.), Cotton Growth in the Gezira Environment, Wad. Medani: Agric. Res. Corp.
22. Jenkins, W. H., E. E. Hall, and J. O. Ware. 1939. Cooperative breeding, genetic, and varietal studies of cotton. S. C. Agric. Exp. Stn. 52nd Ann. Rept., p. 116–124.
23. Jones, J. E. 1953. The influence of modifying genes on Fusarium wilt resistance in Upland cotton. Amer. Soc. Agron. Abstr., p. 90–91.
24. Jones, J. E. 1961. Inheritance of resistance to Fusarium wilt in Upland cotton. Ph.D. Thesis, Louisiana State Univ., Baton Rouge. (Diss. Abstr. 22:34–35).
25. Kappelman, A. J., Jr. 1971. Inheritance of resistance to Fusarium wilt in cotton. Crop Sci. 11:672–674.
26. Kappelman, A. J., Jr. 1975. Correlation of Fusarium wilt of cotton in the field and greenhouse. Crop Sci. 15:270–272.
27. Kelkar, S. G., S. P. Chowdhari, and N. B. Hiremath. 1947. Inheritance of Fusarium resistance in Indian cottons. Indian Central Cotton Comm. 3rd Conf. Cotton Growing Problems in India, Bombay. p. 125–142.
28. Kelman, A., and R. J. Cook. 1977. Plant Pathology in the People's Republic of China. Annu. Rev. Phytopathol. 15:409–430.
29. Khadr, A. S. 1966. Pathogenesis in Fusarium wilt of cotton. Ph.D. Thesis, Univ. Calif., Berkeley. 87 p.
30. Khadr, A. S., A. A. Salem, and B. A. Oteifa. 1972. Varietal susceptibility and significance of the reniform nematode, *Rotylenchulus reniformis,* in Fusarium wilt of cotton. Plant Dis. Rep. 56:1040–1042.
31. Krishnamani, M. R. S., and M. Lakshmanan. 1976. Photosynthetic changes in Fusarium-infected cotton. Can. J. Bot. 54:1257–1263.
32. Kulkarni, G. S. 1934. Studies in the wilt disease of cotton in the Bombay Presidency. Indian J. Agric. Sci. 4:976–1045.
33. Lakshminaraynan, K., and D. Subramanian. 1955. Is fusaric acid a vivotoxin? Nature, Lond. 176:697–698.
34. Martin, W. J., L. D. Newsom, and J. E. Jones. 1956. Relationship of nematodes to the development of Fusarium wilt in cotton. Phytopathology 46:285–289.
35. Messiaen, C. M., and P. Mas. 1969. Recherches sur les fusarioses. VI. Mise au point sur l'activité parasitaire du *Fusarium oxysporum* et sur divers facteurs rendant les plantes plus ou moins sensibles aux fusarioses vasculaires. Ann. Phytopathol. 1:401–426.
36. Minton, N. A., and E. B. Minton. 1963. Infection relationship between *Meloidogyne incognita acrita* and *Fusarium oxysporum* f. *vasinfectum* in cotton. Phytopathology 53:624 (Abstr.).
37. Mitchell, R. E., and W. M. Powell. 1972. Influence of *Pratylenchus brachyurus* on the incidence of Fusarium wilt in cotton. Phytopathology 62:336–338.
38. Neal, D. C. 1954. The reniform nematode and its relationship to the incidence of Fusarium wilt of cotton at Baton Rouge. Phytopathology 44:447–450.
39. Nyvall, R. F., and W. A. Haglund. 1976. The effect of plant age on severity of pea wilt caused by *Fusarium oxysporum* f. sp. *pisi* Race 5. Phytopathology 66:1093–1096.
40. Old, K. M., and B. Schippers. 1973. Electron microscopical studies of chlamydospores of *Fusarium solani* f. *cucurbitae* formed in natural soil. Soil Biol. Biochem. 5:613–620.
41. Perry, D. A. 1962. Fusarium wilt of cotton in the Lake Province of Tanganyika. Emp. Cotton Grow. Rev. 39:22–26.
42. Perry, D. A. 1963. Interaction of root knot and Fusarium wilt of cotton. Emp. Cotton Grow. Rev. 40:41–47.

43. Rosen, H. R. 1928. A consideration of the pathogenicity of the cotton wilt fungus. *Fusarium vasinfectum.* Phytopathology 18:419–438.
44. Shepherd, R. L. 1975. Control of the root-knot–Fusarium-wilt disease complex with resistant cotton, p. 158. *In* Proc. Beltwide Cotton Prod. Res. Conf., New Orleans 1975.
45. Sherbakoff, C. D. 1928. Wilt caused by *Verticillium alboatrum* Reinke & Berth. Plant Dis. Rep. Suppl. 61:283–284.
46. Smith, A. L. 1948. Control of cotton wilt and nematodes with a soil fumigant. Phytopathology 38:943–947.
47. Smith, A. L. 1953. Fusarium and nematodes on cotton, p. 292–298. *In* A. Steffered (ed.), Plant diseases, the yearbook of Agriculture, U.S. Dept. Agric., Washington, D.C.
48. Smith, A. L., and J. B. Dick. 1960. Inheritance of resistance to Fusarium wilt in Upland and Sea Island cotton as complicated by nematodes under field conditions. Phytopathology 50:44–48.
49. Smith, S. N., and W. C. Snyder. 1972. Germination of *Fusarium oxysporum* chlamydospores in soils favorable and unfavorable to wilt establishment. Phytopathology 62:273–277.
50. Smith, S. N., and W. C. Snyder. 1975. Persistance of *Fusarium oxysporum* f. sp. *vasinfectum* in fields in the absence of cotton. Phytopathology 65:190–196.
51. Subramanian, C. V. 1950. Soil conditions and wilt diseases in plants with special reference to *Fusarium vasinfectum* on cotton. Proc. Indian. Acad. Sci., Sect. B, 31(2):67–102.
52. Talboys, P. W. 1972. Resistance to vascular wilt fungi. Proc. Roy. Soc. Brit. 181:319–332.
53. Trujillo, E. E. 1963. Pathological-anatomical studies of Gros Michel banana wilt. Phytopathology 53:162–166.
54. Wensley, R. N., and C. D. McKeen. 1963. Populations of *Fusarium oxysporum* f. *melonis* and their relation to the wilt potential to two soils. Can. J. Microbiol. 9:237–249.
55. Wickens, G. M. 1964. Fusarium wilt of cotton: seed husk a potential means of dissemination. Emp. Cotton grow. Rev. 41:23–26.
56. Wollenweber, H. W. 1913 Studies on the Fusarium problem. Phytopathology 3:24–49.
57. Wood, C. M., and D. L. Ebbels. 1972. Host range and survival of *Fusarium oxysporum* f. sp. *vasinfectum* in north-western Tanzania. Emp. Cotton Grow. Rev. 49:79–82.
58. Young, V. H. 1928. Cotton wilt studies. I. Relation of soil temperature to the development of cotton wilt. Ark. Agric. Exp. Stn. Bull. 226.
59. Young, V. H., and W. H. Tharp. 1941. Relation of fertilizer balance to potash hunger and the Fusarium wilt of cotton. Ark. Agric. Stn. Bull. 410.

4 Fusarium Diseases of Wheat and Other Small Grains in North America

R. James Cook

Introduction

In 1891, Chester (15) in Delaware, and Arthur (3) in Indiana, reported independently that "scab" was becoming important on wheat in the U.S.A. Until then, scab had been recognized mainly in Europe. Their reports, and those a year later by Detmers (28) in Ohio and Pammel (60) in Iowa, started an era of about 50 years of research in the Corn Belt states of the U.S.A on the nature and control of wheat head blight, scab, seedling blight and foot rot caused by *Fusarium*.

Much of our current knowledge on this disease goes back to the studies reported in this early American literature. For example, Arthur (3) recommended already in his first report that the seeding date in Indiana be moved forward about 2 weeks, to approximately 20 September; this also moved the flowering date forward in the crop and helped the wheat to escape the ascospore inoculum. Arthur was the first to suggest that infection is mainly through anthers at the time of anthesis, an observation subsequently confirmed (2,4,31,64). Selby (70) suggested that the *Gibberella* species on corn stalks was probably the same fungus responsible for scab of wheat, another observaton subsequently confirmed (42,43,50). The classical studies of Selby and Manns (71) demonstrated that head blights of wheat, oats, barley, emmer, spelt, as well as seedling blight and root rots of these and many other crops all were caused by the same *Fusarium*. These and other aspects of the biology and control of the *Fusarium* responsible for wheat scab and corn stalk rot were thoroughly researched by the late 1920s, and the massive literature to that time has been reviewed (4,5,33,53).

Research on the common root and crown rot complex of wheat, barley and oats started in Canada and the U.S.A. about the time research was being discontinued on scab. Much of the credit for this early work goes to Bolley (7), Rose (66), Simmonds (72), Greaney and Machacek (38), Broadfoot (9), Gordon (36), and Gordon and Sprague (37), and has also been reviewed (46,73,74). It is not my intention in this chapter to duplicate earlier reviews, but rather to incorporate all findings, regardless of vintage, into modern understanding of the total picture of Fusarium diseases of wheat and other small grains in North America.

The Pathogens and Their Biology

The *Fusarium* species and subspecies recognized in this chapter as responsible for the diseases of small grains in North America are: *F. roseum* Link emend. Snyd. & Hans.

'Graminearum' [= *F. graminearum* Schawbe, perfect stage, *Gibberella zeae* (Schw.) Petch]; *F. roseum* 'Culmorum' [= *F. culmorum* (W. G. Smith) Sacc. perfect stage unknown]; *F. roseum* 'Avenaceum' [= *F. avenaceum* (Fr.) Sacc., perfect stage, *Gibberella avenacea* Cook]; F. *nivale* (Fr.) Ces. (perfect stage, *Calonectria nivalis* Schaffnit).

The population of *F. roseum* 'Graminearum' in North America can be subdivided into Group I and Group II as reported (34) for Australia. Group I occurs in the Northwest states, on wheat mainly, and has not been observed to produce perithecia. Group II occurs in the Corn Belt and eastern states and forms perithecia readily on host material and sometimes in culture.

A Fusarium population referred to as *F. roseum* 'Crookwell,' and similar to *F. robustum* Gerlach (35), has been found on potato tubers in New South Wales, Australia (13); on wheat in Shenshi Province in the People's Republic of China (chapter 5 below); and on turf in New York State of the U.S.A. (R. W. Smiley, *Personal communication*). Isolates of this population are intermediate between *F. roseum* 'Culmorum' and *F. roseum* 'Graminearum' in conidial morphology and cultural appearance, and are pathogenic to wheat (D. Inglis, *unpublished*). No isolate representing this unique fungus has been recognized thus far on cereals in nature in North America, but probably will be so recognized in time.

Fusarium poae (= *F. tricinctum*) was reported by Gordon (36) to cause leaf blight of durum wheat, but the significance of this *Fusarium* as a pathogen of wheat or other small grains is doubtful. *Fusarium moniliforme* was reported by Henry (41) to cause disease on wheat seedlings grown in an agar culture of the fungus, but there is no evidence that *F. moniliforme* is important on wheat, barley, or oats in nature. *Fusarium* species are highly successful epiphytes and endophytes on and in tissues of the Gramineae, especially on roots (36,37). Most such host-occupant relationships are not particularly pathogenic to the host. Once the tissue dies due to winter injury, drought, age, or other causes, these fungi are virtually guaranteed success as the pioneer colonists of the dead tissues. The ease with which they can then be isolated from such tissue has often been misleading in diagnoses. The most common Fusaria encountered in this way from roots are the numerous saprophytic members of *F. oxysporum* and *F. roseum* sensu Snyder and Hansen. Possibly *Fusarium tricinctum* and *F. moniliforme* are also epiphytic or endophytic on wheat, barley and oats, but not pathogenic.

Fusarium roseum 'Graminearum,' 'Culmorum,' and 'Avenaceum'

The pathogenic populations of *F. roseum* in North America are hereafter referred to as Graminearum (including Groups I and II), Culmorum, and Avenaceum (77). Members of these populations have much in common and are thus treated collectively, but with reference to the peculiarities of each in pathogenicity, means of survival, and ecology.

Pathogenicity Graminearum, Culmorum, and Avenaceum all attack a wide range of plant species in addition to cereals. Selby and Manns (71) showed that Graminearum causes seedling blight of red clover and alfalfa as well as wheat. MacInnes and Fogelman (53) extended the host range of Graminearum to beans, peas, sunflowers, tomatoes, and radishes. Graminearum also causes root rot of soybean (1). Avenaceum is pathogenic to subterranean clover in Victoria, Australia (12,54), alfalfa and sweet clover in Alberta, Canada (26), lentils in the Pacific Northwest U.S.A. (52), and raspberries in Scotland (89). Culmorum has caused various root and basal stem rots of sweet clover and alfalfa (26), of beans (67), asparagus (45), and leek (84). All three Fusaria cause stem rot of carnation (85). Booth (8) gives a more complete host range for these fungi.

Because these Fusarium populations are not physiologically specialized for cereals, the forma specialis *cerealis* designation previously proposed (76) for them within *F. roseum* is of doubtful value. It appears that Graminearum, Culmorum, and Avenaceum are pathogenic to plants quite generally.

Earlier reports claimed pathogenic specialization among the isolates of *Fusarium* pathogenic to wheat (86). Culmorum may be more important than the other Fusaria on oats (19,46,63,72,78), Graminearum Group II on corn (4) and rice (48), Graminearum Group I on wheat (65), and Avenaceum on clover (26,54) and lentils (52), but whether this reflects specialization for the respective hosts, or activity related to ecological requirements is unknown (see next section). More work is needed to verify pathogenic specialization in the cereal Fusaria.

Survival and dissemination The pathogens in *F. roseum* all use infected host tissue for carryover in soil (4,5). Infested refuse may give rise directly to infectious mycelium, or the refuse may serve as a food base for sporulation and dissemination. With Graminearum Group II, the sporulation is probably as common sexually (perithecia) as asexually (sporodochia). Perithecia may even serve as survival structures in soil (88). With Avenaceum, inoculum production is almost entirely asexual, although the perithecial stage has been seen (17). With Culmorum, sporulation is entirely asexual so far as known.

Production of macroconidia occurs on infected host parts or infested crop refuse above ground where light is available. The conidia may serve as water-splash inoculum to produce stem rot, head blight, or scab, or they may enter the soil where either they convert into chlamydospores (18,58) or lyse.

The chlamydospores of Graminearum Group I and of Culmorum formed in macroconidia in soil are double-walled and indistinguishable one from another morphologically (75). On the other hand, the chlamydospores of Graminearum Group I formed in macroconidia are less hardy than the equivalent structures of Culmorum, as evidenced by their greater sensitivity to heat and to rapid drying (75). The more delicate nature of the macroconidial chlamydospores of Graminearum Group I undoubtedly results in more rapid death of this fungus in soil, and may account for why Graminearum Group I is rarely if ever detected by dilution plating of soil in either the Northwest of the U.S.A (75), or in Australia (87), including soil from wheat fields displaying disease caused by this fungus.

Whether macroconidia of Graminearum Group II and of Avenaceum convert into true chlamydospores in soil is unknown. Wearing and Burgess (88) concluded for maize-field soils in eastern Australia that Graminearum Group II does not survive as chlamydospores formed in macroconidia. Conidia of Avenaceum introduced into soil are quickly reduced to one or two center cells that resemble the early stages of chlamydospore formation, but the walls remain thin, and the cells rarely persist longer than 7–10 days in moist soil at 15–25 C (51). Hargreaves and Fox (40) in Scotland described these structures as survival spores but stopped short of calling them chlamydospores. Bennett (6) reported in England that Avenaceum forms true chlamydospores but this is unconfirmed for North American isolates of this fungus.

Culmorum also forms chlamydospores in mycelium in host refuse (18). These chlamydospores have double walls and may have a greater survival value in soil compared with those formed in macroconidia (Inglis, *unpublished*). Graminearum Groups I and II and Avenaceum survive for long periods in crop residue, but whether they also form double-walled chlamydospores in hyphae in the refuse is unknown.

Ecological requirements A basic difference among the populations of Graminearum, Culmorum, and Avenaceum seems to be their respective temperature preferences. Graminearum is the main *Fusarium* on cereals in the warmer regions such as the Corn Belt states of the U.S.A. (4), in California (59), eastern Australia (14), southern Europe, and the lower reaches of the Yangtze River area of the People's Republic of China (chapter 5 below). Graminearum occurs in Washington State, but only in the very warmest areas (18). Culmorum is intermediate in temperature preference, being the main cereal *Fusarium* in the Prairie and Maritime Provinces of Canada, in the Northern and Central Great Plains and Northwest states of the U.S.A., and in England, Holland, and Germany. Avenaceum occurs in the Puget Sound area of the Northwest and adjacent Canada, the panhandle of Idaho and adjacent eastern Washington (17,52), in eastern Canada (26), in Victoria, Australia (12), and other areas characterized by cool climates during the main growing season. In Alberta, Cormack (26) showed that Avenaceum was the dominant *Fusarium* pathogenic to sweet clover and alfalfa, but that Culmorum became important during summer when soils were warmer.

The large geographical areas occupied, respectively, by Graminearum, Culmorum, and Avenaceum, can be characterized not only by temperature but also by cropping practices. Thus, the areas with temperatures favorable to Graminearum are those most suitable for production of corn or rice. It is not surprising, therefore, that Graminearum appears better adapted than Culmorum and Avenaceum to corn and rice. The areas with temperatures suitable for Culmorum are those where wheat, barley and oats are grown. The cool areas favorable to Avenaceum are also the areas typically planted to pasture plants such as grasses and clovers. It appears these fungi are each adapted to crop species typical of their respective temperature niches.

Certain morphological and physiological characteristics of the different Fusarium populations also may reflect an adaptation to circumstance in the respective temperature niches. Areas occupied by Graminearum Group II not only are warm but also suitably humid for Fusarium growth on aerial parts of the host (chapter 22 below); this may account for why Graminearum Group II is the only population characterized by a perithecial stage sufficiently common to have epidemiological significance. Indeed, there is no evidence that perithecia serve this fungus other than for survival (88) and to provide genetically uniform inoculum (10). In addition, isolates from this population form macroconidia maximally at unusally high water potentials (~-1 bar, see chapter 22).

North American isolates of Culmorum must persist and infect roots in relatively dry soil, which may account for the hardy chlamydospore (75) and the ability of this population to grow at very low water potentials (21). Moreover, optimum water potential for production of macroconidia by Culmorum is near -15 bars, considerably lower than for Graminearum Group II (82). The population of Graminearum Group I from dryland wheat-field soils in Washington State never form perithecia under suitably moist conditions, they form chlamydospores like those of Culmorum (75), and have an optimum for asexual sporulation near -15 bars, like Culmorum (82). Possibly this population has undergone natural selection toward the same characters essential to the success of Culmorum. Indeed, except for conidial morphology and a temperature optimum typical of Graminearum, the Group I population in the Northwest is more like Culmorum than Graminearum Group II.

Perhaps an ability to form a perfect stage still exists within the populations of Culmorum and Graminearum Group I and might again be displayed by some isolates given a sufficient number of generations and isolates under conditions suitable for perithecial formation. The perfect stage of Avenaceum was finally found (17) by looking in western Washington where temperatures are ideal for Avenaceum, and also where airborne inoculum is functional due to the rainy conditions in that area.

Graminearum, Culmorum, Avenaceum, and possibly also Crookwell (13) can be considered as ecotypes, that is, members of a species, "each of which differs in genetically based physiologic (and . . . morphologic) features having survival value" (27). If temperature preference is the most basic difference among the three, then they are climatic ecotypes (27) of a species (*F. roseum*) that extends across several climatic zones. Where a single ecotype appears to vary according to habitat, e.g., Culmorum as a cause of scab in Ireland, common root rot in the Great Plains states, and dryland foot rot in the Pacific Northwest states, these are ecophenes (27) "differing in appearance . . . but belonging to essentially homogenous genetic stock." When transplanted to the same habitat, the apparent differences disappear. The cereal Fusaria have a high degree of phenotypic plasticity. They also illustrate that even subtle distinctions in adaptation of a given genetic type to a given environment (in this case, temperature) will be expressed as marked differences in geographic distribution, with the type best adapted becoming almost totally dominant in an area.

Fusarium nivale *Fusarium nivale* is primarily a pathogen of leaves and leaf-sheaths of small-grain crops and grasses. It causes pink snow mold of cereal grains (11) and turf in North America, but not leaf blotch or head blight as reported (57) for other areas.

Fusarium nivale is largely an above-ground pathogen. It uses host refuse for survival in soil and produces no chalmydospores so far as is known. Infections occur at the basal leaf sheath or lower leaves of the plant (16,25), either from mycelium in host refuse, water-splash conidia, or windborne ascospores. Like Graminearum Group II, this fungus is homothallic and the sexual stage, when displayed, is probably for production of inoculum. There is no evidence that isolates responsible for snow mold are physiologically different from those responsible for leaf spots, or Fusarium patch in turf (G. W. Bruehl, *unpublished*).

Whereas the perithecia of *F. roseum* (*Gibberella*) are superficial, blue-black but sometimes buff (17), and occur in clusters around the nodes, those of *F. nivale* (*Calonectria*) are generally submerged in leaf sheath tissues, light to dark brown, with only the apex (and ostiole) protruding through the epidermis (24), and they occur uniformly distributed in the lower leaf-sheath tissues. The submerged habit and pigmentation of the perithecia prompted Müller and Von Arx (55) to transfer the pink snow mold fungus into the *Sphaeriales* and to rename it *Griphosphaeria nivalis*. Booth (8) retains this fungus in the Hyprocreales, but proposed *Micronectriella nivalis* due to the unique habit and color of perithecia. Until further study leaves no doubt of the need for and usefulness of a name change, I am continuing the use of *Calonectria*.

Host Parts Infected by the Cereal Fusaria and the Influence of Environment

The various manifestations of a *Fusarium* attack on cereals (Fig. 4-1) were once all believed to be separate diseases with separate causes. Atanasoff (4) effectively disposed of this theory with his classical studies on root rots, seedling blight, and scab, all of which he showed were caused by the same or separate attacks by one *Fusarium*. The kinds of diseases produced by a given *Fusarium* on small grains can now be described as a function of kind of inoculum, stage of plant growth, and environment.

Head Blight and scab Ascospores (and less commonly, macroconidia) released at the time of anthesis and with high humidity (> 92–94% RH) are needed to produce head blight and scab (2,4,31,64). If the inoculum arrives before or after anthesis, infections may still occur, but the seriousness of the disease is greatly reduced. Anthers (80,83) and pollen (56) are highly stimulatory to growth of the scab *Fusarium* because of compounds such as choline and betaine, which explains the close relationship between flowering and infection (79,81). Whether the dominant scab fungus is Graminearum, Culmorum, or Avenaceum is apparently determined by temperature more than any other factor in the environment.

Seedling blight Seed-borne inoculum (resulting from head blight and scab) in relatively dry soils is usually needed to produce seedling blight of wheat, barley or oats. There is no evidence that seedling blight is a problem due to *Fusarium* in areas with soil-borne inoculum only. For example, in eastern Washington, soil-borne inoculum causes root, crown, and basal stem infections, and plant death near maturity; seedling blight occurs in no more than 2–3% of the cases (18). Seedling blight apparently depends on inoculum on or in the seed, and a weakened seedling due to the shrunken nature of the infected seed.

The classical work of Dickson (30) with Graminearum (probably Group II) showed that temperature was important not only to the pathogen, but also to the host; seedling blight of wheat was greater in warm than cool soils, but for corn infected with the same fungus, seedling blight was greater in cool than in warm soils. Wheat is a cool-temperature crop and apparently has some resistance to Graminearum at the temperatures most ideal for its growth. Corn is a warm-temperature crop and has some resistance to seedling blight in warm soils.

Root, crown, and foot rots There are two routes by which *Fusarium* can cause crown and foot rot. The attack may start above ground from conidia, ascospores or mycelia, in which case the fungus must penetrate the successive layers of leaf sheaths to reach the culm. Generally this happens where the inoculum is trapped in the leaf whorls, where adequate moisture and humidity are available for fungus growth on aerial parts (chapter 22 below), and where pollen collects and is used as a food source by *Fusarium.* Once in the culm, the infection may progress upward one or two internodes, or downward into the crown. The original description of Culmorum by W. G. Smith was in England as a cause of basal culm decay and brown foot rot associated with wet-soil conditions; Culmorum was so convincingly described as a pathogen of above-ground parts of the wheat plant that Detmers (28) incorrectly diagnosed the scab in Delaware as caused by Smith's fungus, when actually Graminearum was the cause in eastern U.S.A. Culmorum and Graminearum have subsequently been shown (18) to cause crown and basal culm rot that starts with root infections from soil-borne inoculum and then extends one to three internodes up the culm. With the latter route leaf sheaths commonly show no evidence of disease, and must be peeled away to reveal the brown foot rot. Both routes may produce plant blight (white heads) just after heading.

A major source of confusion in the literature is that whereas foot rot from above-ground infections requires high humidity, that from below-ground infections requires dry weather and dry soil conditions. Thus, a report from one area that foot rot caused by Culmorum is most severe with wet soil (68) and from another, dry soil (18), may seem contradictory but is not contradictory. The fungus has the same water potential requirement above and below ground, growing best between -15 ($\approx$99%RH) and -90 ($\approx$94%RH) bars (chapter 22 below).

Although any of the four or five populations of *F. roseum* are apparently capable of causing crown and brown stem rot by either of the two routes described above, Culmorum, Graminearum Group I, and possibly Crookwell (13) are the important populations where below-ground infections are concerned. Which of these will dominate is again a function of temperature. Thus, in the Northwest U.S.A., the areas where Culmorum is dominant have a mean high in July of 28.7 C compared with 31.0 C where Graminearum Group I is dominant (75).

Common root rot caused by the various populations of *F. roseum* in the Great Plains states and the Prairie Provinces of Canada is an insidious, chronic destruction of root and crown tissues, but apparently does not often end in premature plant blight and white heads. The severe crown and foot rot as known under dryland conditions in the Northwest has been produced (*unpublished*) experimentally with isolates of Culmorum obtained, respectively, from North Dakota (from a plant with common root rot), Ireland and another from England (from above-ground parts), and from Washington State (from a plant with severe foot rot). Symptoms of crown and basal stem rot beneath bright, healthy leaf sheaths were the same for all four isolates, showing that the different symptoms recognized in different areas are not due to different strains of Culmorum.

Whether root and crown infection remains in the chronic (common root rot) stage, or develop into the more acute (crown and brown foot rot) stage seems to depend on the occurrence of very low plant water potentials (61 and chapter 22 below). For example, wheat in eastern Washington (where Fusarium crown and foot rot is acute) is grown under a Mediterranean-like climate where effective rainfall is virtually negligible from heading onward, where the crop matures on water stored in the soil profile at depths to 180 cm (21), and where plants develop water potentials as low as −50 bars before maturity. In contrast, wheat in the Midwest and the Prairie Provinces is grown under summer rainfall and the exceptionally low plant water potentials probably do not develop. Application of irrigation water to infected wheat plants at about the heading stage was shown (*unpublished*) in field plots to inhibit the onset of severe crown and foot rot.

Production of typical crown and foot rot of wheat is nearly impossible in pots in the greenhouse because the frequent waterings keep the plant water potential too high. Sandford and Broadfoot (69) and Johnston and Greaney (47) concluded that Culmorum was only weakly pathogenic in greenhouse tests; their failure to obtain severe disease with soil-borne inoculum of this fungus probably relates to the lack of plant water stress in their tests and possibly also their expectation of seedling blight rather than the more typical premature plant blight. We now conduct most of our studies with this fungus in field plots where plants are allowed to reach full maturity and where the necessary stress conditions are easier to produce.

Nitrogen fertilization, if done in excess, greatly increases the incidence of crown and foot rot caused by Culmorum, at least under conditions in the summer fallow area in the Northwest (61). A plant with excessive N develops more tillers, wider leaves, and uses the limited soil water supply more quickly. This increases the tendency to stress—and to develop crown and foot rot. The upsurge in importance of this disease in the Northwest traces to the early 1960s and the first use of stiff-strawed and semi-dwarf wheats with ability to respond to nitrogen fertilizer. Earlier fall seeding with these wheats on summer fallow also favored Culmorum—again, by increasing top growth, water use, and water stress, in that order.

Leaf-blotch and snow mold caused by *Fusarium nivale* The many manifestations of an attack by *F. nivale* (Fig. 4-2) are also environment-dependent. Mycelium or conidia from infested refuse on or near the soil surface infect the leaves and leaf sheaths where in contact with the soil. If these plants are covered with snow on unfrozen ground for several weeks or months, the constant 0 to 0.5 C temperature at the soil-snow interface is suitable for continued pathogenesis and snow mold develops. The leaves are destroyed first, followed by the crown; destruction of the latter generally kills the plants.

Plants infected by *F. nivale* in the winter but not exposed to snow mold conditions commonly exhibit a brown decay of the lower leaf sheaths (16) (Fig. 4-2). This decay is similar to the basal sheath and stem disease characteristic of early stages of the "foot rot" caused by populations of *F. roseum* in more humid climates. *Fusarium nivale* rarely penetrates deeper than 2–3 layers of leaf sheath, except possibly under circumstances such as dry soil conditions during the winter months. Usually, the brown leaf sheaths infected by *F. nivale* can be peeled away to reveal healthy culms. Moreover, plants with these symptoms show no apparent ill-effects of this attack, possibly because these leaves are only of temporary use to the plant anyway.

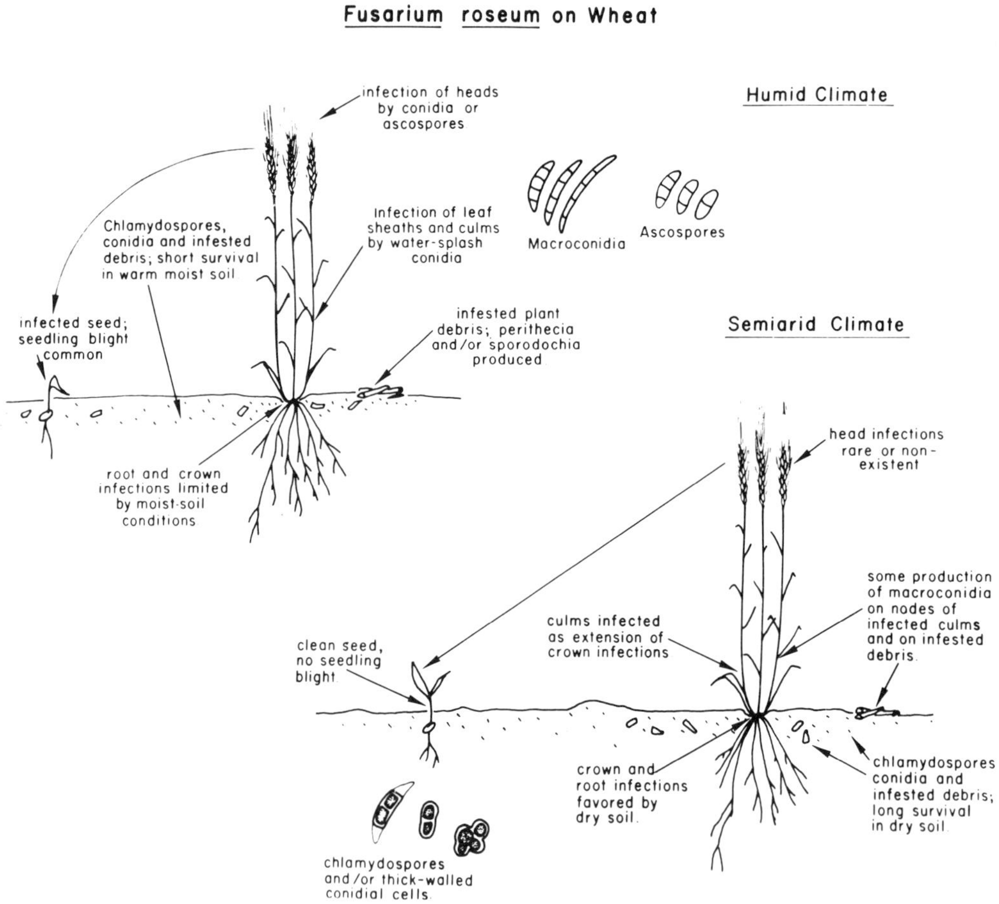

Figure 4-1. The life cycle and associated disease manifestations of *Fusarium roseum* on wheat under different climates and conditions.

On the other hand, these infected lower leaf sheaths are the site for formation of perithecia the following spring (24,57). Cook and Bruehl (24) found no evidence that ascospore inoculum produced snow mold; this is not surprising since the perithecial stage is timed for spring and summer infections at later stages of plant growth, rather than fall and winter infections on plants in the seedling or rosette stage. The full life cycle of this fungus on winter cereals can be summarized as (Fig. 4-2): i) establishment in lower leaf sheaths and blades of young plants in the fall and winter; ii) formation of perithecia in the spring during the warming period but while conditions are still cool; iii) production of leaf spots and possibly even head blight (57), if conditions continue cool and moist (preferably foggy). On this basis, pink snow mold, although the most infamous manifestation of an attack on cereal grain plants by *F. nivale,* is probably the anomaly in the life cycle of this fungus.

Only a few areas are known in agriculture where *F. nivale* exhibits its full life cycle. These include Scotland and northern England (57), the Toluca Valley near Mexico City where CIMMYT wheats are grown, and the lower reaches of the Yangtze River area in the People's Republic of China (chapter 5 below). All of these areas are characterized by little or no snow cover during the winter, but long, cool wet periods during stem elongation, heading, and maturity the following spring and early summer. Pink snow mold occurs in the northern counties and higher elevations of the northwest states, and in the northeastern states of the U.S.A.; these areas all have prolonged snow cover on unfrozen ground (11).

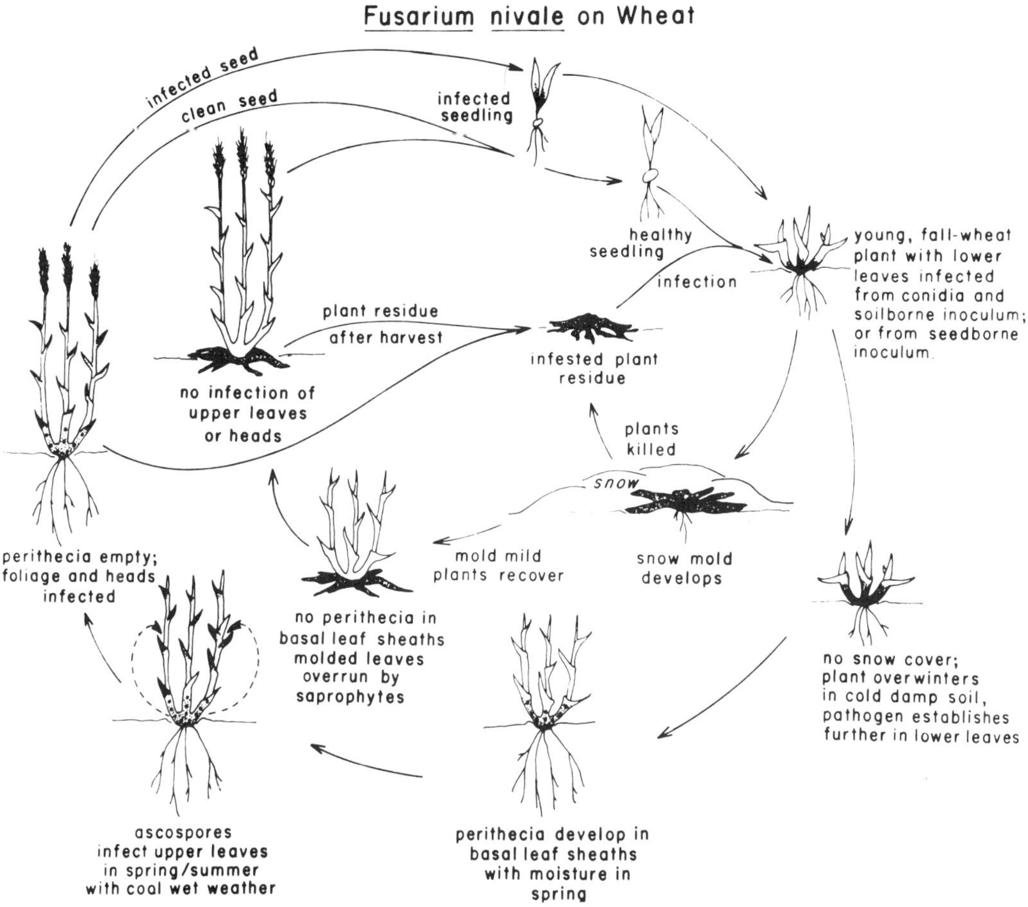

Figure 4-2. The life cycle and associated disease manifestations of *Fusarium nivale* on wheat under different climates and conditions.

Control by Management and Plant Breeding

Arthur (3) suggested the first management practice for scab control when he recommended that the seeding date be moved forward from October to September, so that the wheat would flower ahead of the *Fusarium* inoculum. Crop rotation, particularly the avoidance of wheat after corn, also traces back to the first studies by workers in the Corn Belt states and the demonstration that Graminearum is short-lived in soil and crop refuse in the absence of a host crop. In the midwest, U.S.A., corn is now rotated with soybeans in Iowa, southern Minnesota, and southern Wisconsin, while wheat has become concentrated in the Great Plains states of North Dakota, Montana, Kansas, Nebraska, Oklahoma, Texas, and Colorado.

Crop rotation is particularly effective against pink snow mold, presumably because this fungus carries over in the more fragile leaf tissues and does not form chlamydospores. In the Northwest, a spring wheat or barley crop inserted into the otherwise winter wheat/fallow/winter wheat rotation will significantly reduce snow mold in a following winter wheat crop (Bruehl, *personal communication*). Clover or alfalfa in the rotation also helps to control snow mold. On the other hand, snow mold has been observed to increase in Idaho on winter wheat when barley straw was added to the soil surface prior to snow (44); apparently, *F. nivale* may colonize barley straw saprophytically and thereby acquire a greater food base for production of disease.

Crop rotation is of little benefit in the control of root, crown, and foot rots caused by

Culmorum. The chlamydospores formed by Culmorum are not quickly eliminated from dryland areas (75). On the other hand, oats must be avoided in rotation with wheat in eastern Wahington because they greatly increase populations of this *Fusarium* (19). Parasitized culms of oats support more rapid asexual sporulation by Culmorum which subsequently adds to the population of soil-borne propagules. Population increases of up to ten-fold have been measured following one or two oat crops. Rain or dew needed for sporulation on culms may last for only 2–3 days in eastern Washington; the tendency to sporulate more rapidly on oats than on wheat thus becomes very significant in eastern Washington. Where moisture is not limiting to sporulation, there is no evidence that oats contribute more inoculum than wheat.

The use of soil and crop managment practices that reduce or delay plant water stress will effectively control crown and foot rot caused by Culmorum under Northwest conditons. The most important of these is to use less nitrogen fertilizer, and the second most important is to seed late (October rather than September). The pathogen infects the roots and crown tissues under a wide variety of conditions, but without water stress in the host, does not progress beyond a slow decay. Reduced nitrogen and delayed seeding both allow the plant to develop more in balance with the limited supply of soil water. Both practices also offer the means to manage Culmorum even after the pathogen has become established inside the plant.

Clean tillage (maximal burial of the crop residue) is probably useful in scab control since perithecia and sporodochia form and release inoculum only on the infested residue left on the soil surface. On the other hand, the added inoculum from surface residues might be offset by the cooler, wetter soil conditions associated with surface residues (23). Conceivably, Fusarium seedling blight would be less under such conditions. Doupnik et al. (32) have shown for Fusarium (*F. moniliforme*) stalk rot of sorghum in western Nebraska that reduced tillage (increased surface residues) greatly reduced stalk rot and increased yields compared with tillage systems where residues were buried. Theoretically, the same should hold for crown and foot rot of wheat caused by *Fusarium*. The benefits probably will be the greatest in the midwestern states where rain falls in the summer and where reduced tillage and increased surface residues increase water storage and reduce evaporative loss of water from the soil. In the Northwest, where rain falls mainly in the winter, some tillage is needed to maintain a mulch during the warm, dry fallow season and thus reduce evaporative loss (62).

Culmorum is a highly effective saprophytic colonist of straw provided the straws are not already occupied by other fungi (20). Leaving the straw as a mulch on the soil surface allows the various molds to establish and these subsequently prevent colonization by Culmorum. This is one more reason why the stubble-mulch or a minimum tillage system in wheat production should help reduce the importance of Culmorum. There is even the possibility that above-ground, airborne saprophytes can replace *Fusarium* in straw under conditions of temperature and/or water potential unfavorable to continued metabolism and hence possession of the residue by the *Fusarium*. Conditions needed to displace a pathogen in residue with saprophytes have been reviewed (10,22,23).

Considering that the cereal Fusaria cause disease of most Gramineae, and even of carnations, legumes, and other plant species, there is little prospect for good or even moderate genetic resistance in wheat, barley, or oats to these pathogens. On the other hand, much progress has been made in breeding cultivars with traits that allow escape or tolerance to Fusarium damage in wheat under field conditions.

With scab, wheat cultivars that flower early, flower in the boot, or flower without exposing anthers show remarkably little damage compared to cultivars that flower with anthers exposed at the time of ascospore release. Selection for this trait in wheat has been suggested (29,83) as an approach in breeding for scab control.

Cultivars with ability to emerge quickly and with good seedling vigor are most resistant to seedling blight caused by Culmorum (39).

With dryland foot rot, several cultivars of soft white winter and spring wheats are now in use in the Northwest that avoid (or tolerate) plant water stress when seeded early on summer fallow (21). These cultivars become infected when grown in fields infested with Culmorum but show little damage from the infections. The better cultivars include Luke, Sprague, Paha, and Moro; all are more "resistant" to dryland foot rot than the once popular Gaines and Nugaines. These better wheats have deeper roots, a greater diffusive resistance to vapor loss from the leaf, or they use the limited soil water supplies more slowly (Papendick and Cook, *unpublished*).

The cultivar trait used to reduce damage from snow mold is ability to store carbohydrates in the crown tissue and to produce new shoots from the crowns in the spring, even if all foliage is destroyed (49). The cultivar, Sprague, was developed with this capability by G. W. Bruehl and associates. Previous commercial wheats depleted their carbohydrate reserves under mold conditions and had little or no reserve for recovery.

The various cultivar traits used to reduce Fusarium damage in one area or against one kind of attack may not be useful in another area or against another kind of attack. Thus, a cultivar with "resistance" to scab because of flowering traits in one area may not protect against scab in another area where day-length, temperature, or time of inoculum production all are different. A cultivar "resistant" to crown and foot rot because of water use characteristics may not be resistant to scab, and vice versa; nor will such cultivars necessarily protect against crown and foot rot in another area where planting dates and other factors are different. These observations demonstrate the value of extensive cultivar testing *in situ,* where the problems occur, rather than at remote locations or under controlled conditions where nature is not duplicated. These observations and accomplishments also demonstrate, however, that cultivar modifications, singly or in combination with the proper management, can reduce damage caused by *Fusarium.*

Literature Cited

1. Agarwal, D. K. 1976. Fusarium root-rot of soybean—a new record from India. Indian Phytopathol. 29:471.
2. Andersen, A. C. 1948. The development of *Gibberella zeae* headblight of wheat. Phytopathology. 38:595–611.
3. Arthur, J. C. 1891. Wheat Scab. Ind. Agric. Exp. Stn. Bull. 36:129–132.
4. Atanasoff, D. 1920. Fusarium blight (scab) of wheat and other cereals. J. Agric. Res. 20:1–32.
5. Atanasoff, D. 1923. Fusarium blight of the cereal crops. Mededeelingen van de Landbauwhoogeschool 27:1–132.
6. Bennett, F. T. 1928. On two species of *Fusarium, F. culmorum* (W.G. Sm.) Sacc. and *F. avenaceum* (Fries) Sacc., as parasites of cereals. Ann. Appl. Biol. 15:213–244.
7. Bolley, H. L. 1913. Wheat: soil troubles and seed deterioration; causes of soil sickness in wheat land; possible methods of control; cropping methods with wheat. N.D. Agric. Exp. Stn. Bull. 107. 96 p.
8. Booth, C. 1972. The Genus *Fusarium.* Commonwealth Mycol. Inst., Kew, Surrey. 237 p.
9. Broadfoot, W. C. 1934. Studies on foot and root rot of wheat. IV. Effect of crop rotation and cultural practice on the relative prevalence of *Helminthosporium sativum* and *Fusarium* spp. as indicated by isolations from wheat plants. Can J. Res. 10:115–124.
10. Bruehl, G. W. 1976. Management of food resources by fungal colonists of cultivated soils. Annu. Rev. Phythopathol. 14:247–264.
11. Bruehl, G. W., R. Sprague, W. R. Fischer, M. Nagamitsu, W. L. Nelson, and O. A. Vogel. 1966. Snow molds of winter wheat in Washington. Wash. Agric. Exp. Sta. Bull. 677. 21 p.
12. Burgess, L. W., H. J. Ogle, J. P. Edgerton, L. L. Stubbs, and P. E. Nelson. 1973. Biology of fungi associated with root rot of subterranean clover in Victoria. Proc. Royal Soc. Vic. 86:19–29.
13. Burgess. L. W., P. E. Nelson, and T. A. Toussoun. 1980. Characteristics of a newly recognized population, *Fusarium roseum* 'Crookwell.' (In press)
14. Burgess, L. W., A. H. Wearing, and T. A. Toussoun. 1975. Surveys of Fusaria associated with

crown rot of wheat in eastern Australia. Australian J. Agric. Res. 26:791–799.
15. Chester, F. D. 1891. The scab of wheat. Del. Agric. Exp. Stn. 3rd Ann. Rpt. 1890, p. 89–90.
16. Colhoun, J., G. S. Taylor, and C. S. Millar. 1963. Disease of cereals caused by *Fusarium nivale.* Nature 200:597.
17. Cook, R. J. 1967. *Gibberella avenacea* sp. n., perfect stage of *Fusarium roseum* f. sp. *cerealis* 'Avenaceum.' Phytopathology 57;732–736.
18. Cook, R. J. 1968. Fusarium root and foot rot of cereals in the Pacific Northwest. Phytopathology 58:127–131.
19. Cook, R. J. 1968. Influence of oats on soil-borne populations of *Fusarium roseum* f. sp. *cerealis* 'Culmorum.' Phytopathology 58:957–960.
20. Cook, R. J. 1970. Factors affecting colonization of wheat straw by *Fusarium roseum* f. sp. *cerealis* 'Culmorum.' Phytopathology 60:1672–1676.
21. Cook, R. J. 1973. Influence of low plant and soil water potentials on diseases caused by soilborne fungi. Phytopathology 63:451–458.
22. Cook, R. J. 1977. Management of the Associated Microbiota, p. 145–166. *In* J. G. Horsfall and E. B. Cowling (ed.), Plant Disease, An Advanced Treatise, Vol. I, How Disease Is Managed, Academic Press, New York. 488 p.
23. Cook, R. J., M. G. Boosalis, and B. Doupnik. 1977. Influence of Crop Residues on Plant Disease, p. 147–163. *In* Crop Residue Management Systems, Amer. Soc. Agron, Madison, WI. 248 p.
24. Cook, R. J., and G. W. Bruehl. 1966. *Calonectria nivalis,* perfect stage of *Fusarium nivale,* occurs in the field in North America. Phytopathology 56:1100–1101.
25. Cook, R. J., and G. W. Bruehl. 1968. Ecology and possible significance of perithecia of *Calonectria nivalis* in the Pacific Northwest. Phytopathology 58:702–703.
26. Cormack, M. W. 1937. *Fusarium* spp. as root parasites of alfalfa and sweet clover in Alberta. Can. J. Res. Sec. C. 15:493–510.
27. Daubenmire, R. F. 1974. Plants and Environment. A Textbook of Plant Autecology. 2nd Ed. John Wiley and Sons, New York. 422 p.
28. Detmers, F. 1892. Scab of wheat. Ohio Agric. Exp. Stn. Res. Bull. 44:147–149.
29. Dickson, J. G. 1922. Breeding strains to resist wheat scab. Wisc. Agric. Exp. Stn. Bull. 339:36.
30. Dickson, J. G. 1923. Influence of soil temperature and moisture on the development of the seedling-blight of wheat and corn caused by *Gibberella saubinetii.* J. Agric. Res. 23:837–870.
31. Dickson, J. G., H. Johann, and G. Wineland. 1921. Second progress report on the Fusarium blight (scab) of wheat. Phytopathology 11:35.
32. Doupnik, B., M. G. Boosalis, G. Wicks, and D. Smika. 1975. Ecofallow reduces stalk rot in grain sorghum. Phytopathology 65:1021–1022.
33. Eide, C. J. 1935. The pathogenicity and genetics of *Gibberella saubinetii* (Mont.) Sacc. Minn. Tech. Bull. 106. 67 p.
34. Francis, R. G., and L. W. Burgess. 1977. Characteristics of two populations of *Fusarium roseum* 'Graminearum' in eastern Australia. Trans. Brit. Mycol. Soc. 68:421–427.
35. Gerlach, W. 1977. *Fusarium robustum* spec. nov., der Erreger einer Stamm fäule an *Araucaria angustifolia* (Bertol.) O. Kuntze in Argentinien? Phytopathol. Z. 88:29–37.
36. Gordon, W. L. 1939. *Fusarium* species associated with disease of cereals in Manitoba. Phytopathology 29:7–8.
37. Gordon, W. L., and R. Sprague. 1941. Species of *Fusarium* associated with rootrots of the Gramineae in the Northern Great Plains. Plant Dis. Rep. 25:168–180.
38. Greaney, F. J., and J. E. Machacek. 1934. Studies of the control of root-rot diseases of cereals caused by *Fusarium culmorum* (W. G. Sm.) Sacc. and *Helminthosporium sativum* P., K., and B. I. Field methods with root-rot diseases. Sci. Agric. 15:228–240.
39. Greaney, F. J., J. E. Machacek, and C. L. Johnson. 1938. Varietal resistance of wheat and oats to root rot caused by *Fusarium culmorum* and *Helminthosporium sativum.* Sci. Agric. 18:500–523.
40. Hargreaves, A. J., and R. H. Fox. 1977. Survival of *Fusarium avenaceum* in soil. Trans. Brit. Mycol. Soc. 69:425–428.
41. Henry, A. W. 1923. The pathogenicity of *Fusarium moniliforme* Sheldon on cereals. Phytopathology 13:52.
42. Hoffer, G. N., A. G. Johnson, and D. Atanasoff. 1918. Corn-root rot and wheat scab. J. Agric. Res. 14:611–612.
43. Holbert, J. R., J. F. Trost, and G. N. Hoffer. 1919. Wheat scabs as affected by systems of rotation. Phytopathology 9:45–47.
44. Huber, D. M., and G. R. Anderson. 1976. Effect of organic residues on snowmold of winter wheat. Phytopathology 66:1028–1032.
45. Inglis, D. A. 1978. The association of two *Fusarium* species with asparagus seed. M. S. Thesis. Wash. State Univ., Pullman. 73 p.
46. Johnson, E. C. 1914. A study of some imperfect fungi isolated from wheat, oats, and barley plants. J. Agric. Res. 1:475–490.
47. Johnston, C. L., and F. J. Greaney. 1942. Studies on the pathogenicity of *Fusarium* species associated with root rot of wheat. Phytopathology 32:670–684.
48. Kasai, M. 1923. Cultural studies with *Gibberella saubinetii* (Mont.) Sacc. which is parasitic on rice plants. Ber. Ohara Inst. Landw. Forsch 2:259–272.
49. Kiyomoto, R. K., and G. W. Bruehl. 1977. Carbohydrate accumulation and depletion by winter cereals differing in resistance to *Typhula idahoensis.* Phytopathology 67: 206–211.
50. Koehler, B., J. G. Dickson, and J. R. Holbert. 1924. Wheat scab and corn rootrot caused by *Gibberella saubinetii* in relation to crop successions. J. Agric. Res. 27:861–880.

51. Lin, Y. S. 1975. Fusarium root rot of lentils in the Pacific Northwest. M. S. Thesis, Wash. State U., Pullman. 60 p.
52. Lin, Y.S., and R. J. Cook. 1977. Root rot of lentils caused by *Fusarium roseum* 'Avenaceum.' Plant Dis. Rep. 61:752–755.
53. MacInnes, J., and R. Fogelman. 1923. Wheat scab in Minnesota. Minn. Agric. Exp. Sta. Tech. Bull. 18. 32 p.
54. McGee, J. C., and A. W. Kellock. 1974. *Fusarium avenaceum,* a seedborne pathogen of subterranean clover roots. Australian J. Agric. Res. 25:549–557.
55. Müller, E., and J. A. Von Arx. 1955. Einige Beiträge zur Systematik und Synonymie der Pilze. Phytopathol. Z. 24:353–372.
56. Naik, D. M., and L. V. Busch. 1978. Stimulation of *Fusarium graminearum* by Maize pollen. Can. J. Bot. 56:1113–1117.
57. Noble, M., and I. G. Montgomerie. 1956. *Griphosphaeria nivalis* (Schaffnit) Müller & Von Arx and *Leptosphaeria avenaria* Weber on oats. Trans. Brit. Mycol. Soc. 39:449–459.
58. Nyvall, R. F. 1970. Chlamydospores of *Fusarium roseum* 'Graminearum' as survival structures. Phytopathology 60:1175–1177.
59. Oswald, J. W. 1949. Cultural variation, taxonomy and pathogenicity of *Fusarium* species associated with cereal root rots. Phytopathology 39:359–376.
60. Pammel, L. H. 1892. Some diseases of plants common to Iowa cereals. Iowa Agric. Exp. Stn. Bull. 18:502.
61. Papendick, R. I., and R. J. Cook. 1974. Plant water stress and development of Fusarium foot rot in wheat subjected to different cultural practices. Phytopathology 64:358–363.
62. Papendick, R. I., M. J. Lindstrom, and V. L. Cochran. 1973. Soil mulch effects on seedbed temperature and water during fallow in eastern Washington. Proc. Soil Sci. Soc. Amer. 37:307–314.
63. Parkinson, D., and C. G. C. Chesters. 1958. Occurrence of *Fusarium culmorum* (W. G. Sm.) Sacc. in the rhizosphere of oats. Nature 181:1746–1747.
64. Pugh, G., W. H. Johann, and J. G. Dickson. 1933. Factors affecting infection of wheat heads by *Gibberella saubinetii.* J. Agric. Res. 46:771–797.
65. Purss, G. S. 1971. Pathogenic specialization in *Fusarium graminearum.* Australian J. Agric. Res. 22:553.
66. Rose, J. P. 1924. *Fusarium culmorum* in Oregon, its varieties and strains that cause disease of cereals and grasses. Phytopathology 14:49.
67. Russell, P. E., and P. W. Warburg. 1976. *Fusarium culmorum* on *Phaseolus vulgaris.* Plant Pathol. 25:55–56.
68. Russell, T. A. 1932. Observations on foot-rot diseases of cereals. Trans. Brit. Mycol. Soc. 16:253–269.
69. Sandford, G. B., and W. C. Broadfoot. 1934. On the prevalence of pathogenic forms of *Helminthosporium sativum* and *Fusarium culmorum* in the soil of wheat fields and the relation to the root rot problem. Can. J. Res. 10:264–274.
70. Selby, A. D. 1898. Some diseases of wheat and oats. Ohio Agric. Exp. Sta. Res. Bull. 97:31–43.
71. Selby, A. D., and T. F. Manns. 1909. Studies in diseases of cereals and grasses. Ohio Agric. Exp. Sta. Res. Bull. 203:187–236.
72. Simmonds, P. M. 1928. Studies in cereal diseases. III. Seedling blight and foot-rots of oats caused by *Fusarium culmorum* (W. G. Sm.) Sacc. Can. Dept. Agric. Bull. 105. 43 p.
73. Simmonds, P. M. 1928. A review of the investigations conducted in western Canada on root rots of cereals. Sci. Agric. 19:565–582.
74. Simmonds, P. M. 1941. Rootrots of cereals. Bot. Rev. 7:308–332.
75. Sitton, J. W., and R. J. Cook. 1981. Comparative morphology and survival ability of Chlamydospores of *Fusarium roseum* 'Culmorum' and 'Graminearum.' Phytopathology 71: 85–90.
76. Snyder, W. C., and H. N. Hansen. 1945. The species concept in *Fusarium* with reference to discolor and other sections. Amer. J. Bot. 32:657–666.
77. Snyder, W. C., H. N. Hansen, and J. W. Oswald. 1957. Cultivars of the fungus, *Fusarium.* J. Madras Univ., B. 27:185–195.
78. Sprague, R. 1939. Leaf reddening in oats in Oregon. Ore. Agric. Exp. Sta. Inform. Circ. 208. 3 p.
79. Strange, R. N., A. Deramo, and H. Smith. 1978. Virulence enhancement of *Fusarium graminearum* by choline and betaine and of *Botrytis cinerea* by other constitutents of wheat germ. Trans. Brit. Mycol. Soc. 70:201–207.
80. Strange, R. N., and H. Smith. 1971. A fungal growth stimulant in anthers which predisposes wheat to attack by *Fusarium graminearum.* Physiol. Plant Pathol. 1:141–150.
81. Strange, R. N., and H. Smith. 1978. Specificity of choline and betaine as stimulants of *Fusarium graminearum.* Trans. Brit. Mycol. Soc. 70:187–192.
82. Sung, J. M., and R. J. Cook. 1980. Effect of water potential on reproduction and spore germination by *Fusarium roseum* 'Graminearum,' Culmorum,' 'Avenaceum.' Phytopathology 70: (In press).
83. Takegami, S. 1957. Studies on the resistance of wheat varieties to *Gibberella zeae* (Schw.) Petch (head blight) and its mechanism. 1. Varietal differences of the position of florets in the wheat spikelet attacked by head blight incipiently and the relationship between the existence of anther corpses in florets and the infection by head blight. Sci. Rep. Faculty Agric., Okyama Univ. 10:33–42.
84. Tamietti, G., and A. Garibaldi. 1977. Osservazione: su un marciume pedale del porro da *Fusarium culmorum* (W. G. Smith) Sacc. Rivista di Patologia Vegetale 13:69–75.
85. Tammen, J. 1958. Pathogenicity of *Fusarium roseum* to carnations and to wheat. Phytopathology 48:423–426.

86. Tu,C. 1929. Physiologic specialization in *Fusarium* spp. causing headblight of small grains. Phytopathology 19:143–154.
87. Wearing, A. H., and L. W. Burgess. 1977. Distribution of *Fusarium roseum* 'Graminearum' Group I and its mode of survival in eastern Australian wheat belt soils. Trans. Brit. Mycol. Soc. 69:429–442.
88. Wearing, A. H., and L. W. Burgess. 1978. Distribution and mode of survival of *Fusarium roseum* 'Graminearum' Group II in maize soils of eastern Australia. Trans Brit. Mycol. Soc. 70:480–486.
89. Williamson, B., and H. J. Hargreaves. 1977. Report of the Scottish Horticultural Research Institute, 1976.

5 Fusarium Diseases in the People's Republic of China

R. James Cook

The Fusarium diseases of greatest importance in the People's Republic of China (PRC), according to plant pathologists and mycologists consulted in that country during visits in 1976 (1,5) and again in 1979 (as a member of the U.S. Biological Control Work Panel), are the Fusarium diseases of cereal crops and the Fusarium wilts, particularly Fusarium wilt of cotton.

Fusarium wilt of cotton, caused by *F. oxysporum* f. sp. *vasinfectum*, is the most important disease of cotton in the PRC. The disease occurs in all of the 15–20 provinces where the crop is grown (1). Cotton is very important in the economy of that country; and loss from disease has a bearing on the balance of payments for the PRC as well as domestic availability of fiber for their large textile industry. Cotton is grown in the Yangtze river area and westward to Shaanxi province in central China, and from the Beijing area in the north all the way into the southern-most provinces. Like rice, cotton is a summer crop and is grown between April and December.

At least three races of the *F. oxysporum* f. sp. *vasinfectum* have been identified in the PRC. Resistant cultivars are available, but are grown only when the disease potential is high. Apparently, the peasants prefer the older, susceptible cultivars because of their higher yield potential in the absence of wilt. A common control is the "wet-dry" method, where paddy rice is used every 3rd or 4th year in rotation with cotton, to eliminate inoculum of the pathogen. Chloropicrin is also used to eliminate the pathogen, but only as a spot treatment in some heavily infested fields.

Of the Fusarium diseases of cereals, scab caused by *F. roseum* 'Graminearum' is the most important (4,5,6). In fact, scab may be more important than any other disease of wheat and barley in the PRC at the present time. The disease occurs in the Yangtze river area around Nanjing and Shanghai, as far west as Shaanxi province, north to Hebei province, and in most other provinces where winter wheat is grown. According to specialists in the PRC, scab and corn stalk rot have also now become serious in Jilin province (Northeast China). *Fusarium roseum* 'Avenaceum' also causes root rot of millet and occasionally of soybeans in Jilin province.

Cultures of cereal Fusaria were kindly given to me in 1979 by plant pathologists working in the Beijing and Shanghai municipalities, and in Jilin, Jiangsu, Zheijang, and Guangdong provinces. The cultures represented collections from most major cereal-growing areas in China, including Shaanxi province in central China. Graminearum, Culmorum, and Avenaceum all occur in China, but of the 18 cultures provided, 17 matched the description of Graminearum Group II (3). One from Shaanxi province matched the description of *F. roseum* 'Crookwell' (2).

As with all other areas in the world, the importance of scab in the PRC relates to the climate. There are two contrasting climates, a cold dry period during the winter months and a warm wet period through the summer months. The cold dry climate results from air masses that move southeasterly from Mongolia, and the warm wet climate from air movements northwesterly from the South China sea. Provinces in the southern part of China have warm temperatures most of the year, while provinces in the north have only 3–5 months of summer. Since wheat and barley are grown mostly during the cold dry season, these crops should escape scab, but unfortunately warm wet conditions occur before the crop is ripe. The disease is especially serious in the Yangtze river area because the climate becomes warm and wet while the crop is still in the boot stage. Perithecia of *Gibberella* form by anthesis and thus timing for scab is ideal.

The *Gibberella* stage is formed on stems or stalks of either rice or corn. The development of scab in a corn-wheat or corn-barley rotation is a familiar story, but its development in rice-wheat or rice-barley rotations is less well known. Perithecia of *Gibberella zeae* develop in profusion on rice stems just as they do on corn stalks. Their development on rice straw helps explain why rotation to paddy rice does not control Fusarium scab of wheat or barley in China. The fungus apparently does little damage to the rice stem.

Several approaches to control of scab are used in the PRC. One approach is to breed earlier-maturing wheat and barley cultivars to escape the scab. The cultivars of these crops in the PRC are currently among the earliest maturing in the world, but even earlier maturity is desired and possible. Along with earlier maturity, wheat and barley cultivars have been identified that flower during the boot stage and do not expose their anthers; such cultivars tend to escape infection. Improved soil drainage is used in the Shanghai and Nanjing areas to keep the soil surface dry as long as possible and thus to delay development of perithecia. The most effective method used at present is an application of methyl 2-benzimidizole-carbamate (Bavistin or MCB) to the heads at flowering and again 7 days later. Spore traps are used on some communes and research stations in the Shanghai area as a means to monitor ascospore numbers in the air and help decide when and where to apply fungicide.

Fusarium nivale is also important on wheat in the PRC, causing leaf blotch and head infections. The primary inoculum responsible for these infections is ascospores of *Calonectria nivalis,* formed in the lower leaf sheaths of the host in March-April. This disease is most evident in the Yangtze river area and westward to Shaanxi province where moisture and temperature during spring are ideal for the fungus. The climate also becomes ideal for formation of *Gibberella* perithecia, and consequently perithecia of both fungi sometimes occur on the same plant.

The formation of perithecia of *C. nivalis* in the lower leaf sheaths of wheat in the spring and the subsequent development of leaf blotches and head infections from ascospore inoculum is a display of the full life cycle of *C. nivalis,* as opposed to snow mold which involves only a small part of the life cycle of this fungus (chapter 22 below). Economically the leaf blotch phase from ascospore inoculum in the PRC is not as important as snow mold in the Pacific Northwest of the U.S.A., but mycologically the leaf blotch phase is the more interesting of the two.

The bakanae disease of rice caused by *Gibberella fujikuroi* (*F. moniliforme*) was observed by the Plant Studies Delegation (1) in Guangdong province in the southern part of the PRC. In general, this disease is rare in the PRC, possibly because of the control by organic mercury or formalin seed treatments. A seedling blight of rice in Jilin province (Northeast China), serious when the crop is sown early in the cool dry soil, is attributed, in part, to *F. moniliforme.*

A very complete listing of the *Fusarium* species and their host plants or habitats in the PRC is available (7) and should be consulted for details on *Fusarium* in the PRC.

Acknowledgements

I acknowledge with deep appreciation the many helpful discussions on *Fusarium* with plant pathologists and mycologists in the People's Republic of China, including those at: Mycological Laboratory, Institute of Microbiology, Academia Sinica, Beijing; Jilin Provincial Academy of Agricultural Sciences, Gongzhuling, Jilin; Suzhou Institute of Agricultural Sciences, Swzhou region in Jiangsu Province; the Jiangsu Province Agricultural Sciences Research Institute, Nanjing; Shanghai Academy of Agricultural Sciences, Shanghai; Shaanxi Province Academy of Agricultural and Forestry Sciences, Wugong; Northwest College of Agriculture, Wugong; Zhejiang Academy of Agricultural Sciences, Hangzhou; Zhejiang Agricultural College, Hangzhou; and the Guangdong Provincial Academy of Agricultural Sciences, Guangzhou.

Literature Cited

1. Anonymous. 1975. Plant Studies in the People's Republic of China: A trip report of the American Plant Studies Delegation. Nat. Acad. Sci., Wash. D.C. 205 p.
2. Burgess, L. W., P. E. Nelson, and T. A. Toussoun. 1980. Characteristics of a newly recognized population, *Fusarium roseum* 'Crookwell.' Trans. Brit. Mycol. Soc. In press.
3. Francis, R. G., and L. W. Burgess. 1977. Characteristics of two populations of *Fusarium roseum* 'Graminearum' in eastern Australia. Trans Brit. Mycol. Soc. 68:421–427.
4. Johnson, V. A., and H. L. Beemer, Jr. (Eds.) 1977. Wheat in the People's Republic of China: A trip report of the American Wheat Studies Delegation. Nat. Acad. Sci., Wash. D. C. 190 p.
5. Kelman, A., and R. J. Cook. 1977. Plant Pathology in the People's Republic of China. Annu. Rev. Phytopathol. 17:409–429.
6. Roelfs, A. P. 1977. Foliar fungal diseases of wheat in the People's Republic of China. Plant Dis. Rep. 61:836–841.
7. Tai, F. L. 1979. Sylloge Fungorum Sinicorum. Science Press, Academia Sinica, Peking, China. 1527 pp.

6 Fusarium Diseases of Cereals in Western Europe

R. Cassini

The Fusarium diseases of cereals have been known in Europe for a long time. With the exception of important losses caused by these diseases in corn production, their importance and effect on production of cereals such as wheat, barley, oats, and rye is still unclear. The report presented here is by no means complete; it only is meant to give an overview of European cereal culture with respect to the diseases caused by this important group of fungi.

In addition to our own experience obtained in 20 years of study, we have referred principally to the work of Colhoun, Cook, and their co-workers in Great Britain (3,4,14,15,16,17,20,32); Jamalainen in Finland (28); Zogg and Hani in Switzerland (25,39); de Tempe and Feekes in Holland (21,37); Parmentier in Belgium (34); Piglionica in Italy (35); and Barrière, Cassini, Messiaen, and Molot in France (1,2,5,6,7,8,9,10,11,12,31,33).

We apologize in advance to our colleagues if this overview is given with too much of a French viewpoint, but it is based on personal observations and on the work which the author knows best. We have not dealt with work which is directed exclusively toward fungicidal treatments, whether they be seed or seedling treatments.

Cereals

Wheat, barley, oats, and rye are grown in Europe over about 25 million ha. In comparison with the cultivation of cereals in North America or the U.S.S.R., the culture of cereals in Western Europe is characterized by being very intensive. The yields are very high and are constantly increasing (about 0.1 tons/ha for 20 years) and investments are also high and increasing. In Great Britain and France "clubs" exist whose declared goal is to reach and even to surpass yields of 10 tons/ha. Under these conditions it is important to limit the development of disease, not only to assure a maximum yield, but also to reduce as much as possible all the factors which could cause variation in yield. In this way, keeping in mind the fact that the price of cereals is generally fixed at the start of the growing season, the grower has a good estimate of his return.

The situation, which is relatively favorable from the economic point of view, involves the use of cultural practices which depend strongly on the use of pesticides and a reduction or elimination of rotation, leading to the repeated growing of cereals in the same soil, i.e., monoculture. Continual monoculture of plants that are susceptible to the same pathogens results in a degradation of sanitation, which then becomes a limiting factor in yield. This phenomenon is one of the key factors in the development of the Fusarium diseases of cereals, which were of secondary importance 30 or so years ago, and at present are among the most important diseases. It is probably Guyot (24) who has provided the most detailed and accurate description of both the symptoms of and the *Fusarium* species responsible for the attacks on cereals. With the exception of minor details of classification and synonomy, the descriptions of the species and their variability are perfectly usable.

In wheat, as in the other cereals, there is a disease of roots and crowns which can extend to the stubble, and a disease of the heads which can attack individual grains, part of the head, or even the entire head. Root and crown rot causes a scalding of the head and sometimes even a lack of grain development. On the other hand the disease of the head adds to the scalding by contaminating the seed, which can limit germination or reduce the quality of the grain by the production of mycotoxins, as in the case of *F. roseum*. When the seeds are contaminated, seedling emergence may be unaffected (*F. roseum*) or strongly affected (*F. nivale*).

Fusarium nivale This fungus is the principal causal agent of Fusarium diseases in the colder regions of Northern Europe and the Alpine regions on winter cereals. The classic symptom produced by this species is "Nivale Mildew" which destroys the plants during tillering at the end of winter, after the snow has melted. The fungus survives in soil and residue even when it is buried. Detection of *F. nivale* is difficult because direct isolation is difficult, if not impossible, and one must trap or bait the fungus. Seed treatment of seed planted in contaminated soil is often useless because the treatment does not last long enough.

Conditions favorable to disease development do not occur all at the same time (insufficiently low temperatures, absence of snow) and under these conditions the plant survives with only a few, slight symptoms at the base of the tillers, accompanied by a violet lesion at the level of a node where the leaf sheath arises. Following shooting the first sheath often is covered with a multitude of perithecia which, under the climatic conditions of the Parisian basin, mature and liberate their ascospores. Beginning in mid-May these ascospores infect the heads when their liberation coincides with flowering of the wheat. Trapping of ascospores in the air enables one to establish a positive correlation between spore frequency in the air and intensity of head blight.

Depending on humidity and temperature, head blight can be almost invisible and be noticed solely by a loss in weight per thousand seeds accompanied by disease of the caryopsis. A discoloration of the back of some glumes bordered by a brown margin can also occur, as well as the withering of certain portions of the head at the base of which one can find pink sporodochia. In this case the color of the fructifications is often insufficient to distinguish these attacks from those of *F. roseum*, and only examination of the macroconidia under the microscope can enable one to decide which species of *Fusarium* is involved.

In all cases these attacks cause yield losses and more or less deep infections of the caryopses. If the latter are to be used for seeding, it is absolutely necessary to use a seed fungicide in order to limit pre-emergence damping-off. The detection of seed contamination is very simple. Surface-disinfected seed is placed on an agar medium (malt-agar) in petri dishes and held at 15 C. In 4–5 days the colonies of *F. nivale* appear; they are not overgrown by saprophytes or *F. roseum*, whose growth is more rapid at higher temperatures. This laboratory observation helps to explain the variable results obtained following natural or artificial inoculation of the heads. *Fusarium nivale* is a weak competitor, and if the temperature is too high (greater than 15–18 C during the day) when inoculation occurs, inoculation may very well be aborted by saprophytes or other species of *Fusarium*.

Fungicidal treatments during growth are useful against *F. nivale* providing they are applied at the time the perithecia mature, in which case the inoculum level is reduced, or at flowering, resulting in protection of the stamens which are the infection courts for infection of the seed. We have already seen that seed disinfection is often insufficient in contaminated soil, but nevertheless it is necessary to do it systematically for it limits the supply of inoculum in the soil and is always useful in preventing the growth of the fungus from contaminated seed. Differences in susceptibility of cultivars is not marked in wheat and in any case is not used in selection, which remains completely empirical. The breeder merely eliminates the most susceptible plants.

Fusarium roseum Included under this designation are diseases caused by *F. roseum* var. *culmorum, F. roseum* var. *graminearum* and *F. roseum* var. *avenaceum* (30). We should mention here that the last named variety plays a relatively minor role as a pathogen on cereals. Whether these are varieties of one species or different species is a question for mycologists. For the cereal pathologist these fungi individually, or more rarely in association, are the cause of the disease.

The most important of the diseases caused by these fungi in Europe is head blight, the most spectacular symptom. This is what the grower sees between flowering and maturity where groups of grains wither and take on a more or less pink color. The disease is common on all cereals and on most of the native grasses (*Lolium* spp., *Agropyron* spp. etc.) and occurs in north Holland as well as in southern Italy. Only the frequency and the intensity of the attacks vary. The disease of the foot and the root described by Cook (18) and Guyot (24) also occurs but its incidence is not as great as in Washington State, probably because water stress on plants is not as great. It is an insidious disease, for on good land it is not obvious and in addition is confused with other foot rots. On the other hand in light, shallow, poorly drained soils, it is one of the most important limiting factors in production of cereals in France. This point of view is not often shared, but for our part we consider it as being largely responsible for yield loss due to scald, which occurs when a period of dry hot weather precedes maturation.

The behavior of the fungi responsible for the Fusarium diseases of cereals is on the border between parasitism and saprophytism, i.e., they are not particularly good at either. This peculiarity makes a study of the disease difficult, for one cannot easily master the techniques of artifical inoculation of young plants. Cook (18) has shown clearly that the importance of the damage is proportionate to the inoculum in soil. Starting with his observations, we believe that by determing the level of inoculum in the soil one can predict the level of disease and determine the effect of cereal monoculture, as well as various aspects of working the soil such as irrigation, etc., on disease incidence. From experiments including rotations of corn-wheat, corn-hard wheat, wheat-hard wheat-corn, etc., and monocultures, with or without irrigation, with normal tillage, minimum tillage, or direct drilling, we have obtained results (5,12) which can be sumarized as follows:

i) The number of propagules of *Fusarium* as shown by soil analysis is of value only for that given soil: one can have serious losses with 500 propagules/g in one soil and insignificant loss with 2500 propagules/g in another type of soil.

ii) Cereal rotations bring about an increase in propagules. Hard wheat and oats favor this increase more than soft wheat (*Triticum aestivum*) or corn. *Fusarium roseum* var. *graminearum* is more frequent after corn than after cereals. The increase in inoculum is not infinite and an upper limit is quickly attained (after 4 years in our tests) after which the situation is stabilized. We should note, however, that H. Zogg and J. A. Amiet (*unpublished data*) have shown the phenomenon of decline in the *Fusarium* inoculum that is analogous to that known with *Gaeumannomyces graminis*. This limit seems to correspond to the level of a certain fraction of the non-decomposed organic matter in the soil. The analysis of this fraction, more easily done than that for *Fusarium* inoculum, allows for a good estimate of the approximate level of inoculum of the fungi (R. Cassini, *unpublished data*).

iii) If one compares the three techniques of working the soil (normal tillage, minimum tillage, direct drilling), the anaylsis obtained from the first 5 cm of soil shows that surface working of the soil is the most favorable for conservation of inoculum.

iv) Finally, the effect of irrigation on growth, especially through parasitic lodging in corn, is not detectable by soil anaylsis.

All this indicates that the succession of plants susceptible to the same pathogen increases and sustains inoculum level, particularly if one uses the techniques of minimum tillage. Inoculum level is not the limiting factor on disease development in the main cereal growing regions, but the climatic conditions and variable responses are. In order to have a good forecasting system for foot rot, one must study the conditions that permit the fungus to enter and develop within the plant, especially the dynamics of water in the soil and in the plant. The problem is much simpler in the case of head blight. Since the inoculum is always plentiful, warm, humid weather (24 C), with rain at the time of flowering so that the spores of the fungus are spread, is all that is needed for the disease to become serious rapidly. Yield losses can be high if the disease occurs before the dough stage. On the other hand, even with apparent symptoms, yield losses are insignificant when the plant is attacked after this stage. Early attacks during flowering or at the milk stage bring about the destruction of the embryo. In the case of later attacks the mycelium remains localized in the pericarp. In both cases grain quality and germination are affected.

Method of Control As with *F. nivale,* seed disinfection is useful, since it reduces the inoculum level in the soil and diminishes disease at emergence. There are few methods of direct control against foot and root rot other than seed treatment, which can assure at best good germination, and fungicidal treatments applied at tillering, which can only slow down the progress of the fungus and limit its development. In the case of head blight the results of fungicidal treatments are irregular. When the risks are very great, as they are during warm humid weather with rain during flowering, treatment becomes imperative. The available fungicide treatments are not sufficiently long lasting to protect the head from flowering to maturity; nevertheless, if they are applied at the right time with sufficient pressure and dispersion, yield losses can be reduced. These treatments are preventive and they are useless after symptoms have appeared.

Cultivar Behavior and Resistance There is no direct relationship between susceptibility to foot rot and head rot. The more or less saprophytic behavior of the fungus makes the work of selecting against foot rot very difficult, since the selection must occur under experimental conditions that are difficult to control. On the other hand, flower infection is easily done experimentally and this allows a good evaluation of cultivar differences in resistance and susceptibility. Selection work against head blight is presently underway in France and in Quebec, Canada, where this disease is by far the most damaging to spring wheat because of the weather conditions in July in the wheat-growing area of the Montreal plain.

Corn

The amount of corn grown in western Europe is relatively modest, especially if one compares it to that of North America. With the exception of Italy, Spain, and southwest France, where corn has been grown for at least 4 centuries, corn growing has been practiced for only 20 years. In France between 1960 and 1975 the amount of land devoted to corn has grown from 300,000 ha to 3 million ha and its cultivation has spread to West Germany, Belgium, Great Britain, and even to Holland and Denmark. This northward spread has been made possible by the development of early hybrids between lines of French "flints" and North American "dents."

Corn has developed essentially in the place of row crops (i.e., those crops that need cultivation) and pastures. It is grown either in monoculture or, much more frequently, in rotation with wheat or rye-grain where cattle are raised.

In Great Britain corn is used as a "break up" in intensive cereal rotation, for it is not susceptible to the majority of cereal diseases (20). Unfortunately, in the parasitic complex characterized by stalk rot, seedling blight, root rot, and ear rot, the Fusaria have a large role. Here again *F. roseum* var. *graminearum* (*Gibberella zeae*) is most commonly isolated in Italy and southern France, while *F. roseum* var. *culmorum* is largely dominant in the other regions. One should note, however, that *F. roseum* var. *graminearum* has followed corn in its northward progress and that it is now much more common than 20 years ago. *Fusarium moniliforme* is also associated with ear rot and with wilt after flowering in regions with warm summers.

Ear rot In addition to the varieties of *F. roseum* mentioned earlier and *F. moniliforme, F. tricinctum* is also involved in ear rot. These diseases generally are not important: selection, even if it is empirical, is effective, and it is often those cultivars with ears poorly covered by the spathes which are susceptible. In addition, the risks of toxicity to the seed by the production of mycotoxins have been insignificant. This favorable situation is due to the fact that, in France at least, corn growers are rarely also cattle raisers and therefore the harvested seed is dried rapidly to 14% humidity and sold to the cattlemen. Under these conditions there is no further development of the *Fusarium*. It is possible that the increased cost of drying will require that harvested ears be dried in cribs again; under such conditions a recurrence of the problems caused by the development of *Fusarium* on ears in the cribs and the resultant spoilage may occur.

Stalk rot Stalk rot, whether or not it is associated with lodging, is without doubt the most important corn disease in western Europe, and probably the only one that is really damaging to early corn. In France losses are about 10% per year (1,8). In Great Britain, Cook (20) estimates that between 1973 and 1975 the median loss in seed weight ranged from 18.7% to 11.2% in 39 fields under observation. Losses of 20 to 25% have been regularly recorded in the U.S.A., and these are further aggravated when stalk rot is accompanied by lodging, since many ears fall to the ground.

Numerous studies have been made on this disease throughout the world but no clear conclusions can be drawn. One of the biggest problems is that the disease cannot be reproduced at will. The results are unreliable, except perhaps when they are done in new soil where cereals or other susceptible crops have never been grown; but even under these conditions the results are unpredictable. The inoculation of seedlings is easier, but, as in the case of wheat, the relationship between susceptibility of seedlings and of the adult plant is not clear.

The rapid spread of corn into areas where it had never been cultivated has allowed for interesting observations, but the interpretation of these observations is still incomplete. It has been observed particularly that when corn is introduced in an intensive cereal rotation it can be attacked severely by Fusaria in the first year. This suggests that in such a rotation there are clones of *Fusarium* in the soil capable of attacking corn.

In the course of the continued development of cultivars it has been observed that certain corn hybrids considered resistant to the disease became susceptible. This suggests the development of clones of the fungus more specialized to corn, but unfortunately this cannot be verified experimentally; it also suggests an increase in the inoculum level in the soil, which has been verified, and thus the likelihood of a threshold above which a given hybrid becomes susceptible. This latter point is very difficult to demonstrate in the field.

Resistance and susceptibility Molot (33) has tried to relate the percentage of stalks lodged at harvest with the total level of phenolic compounds in the stalks at fertilization, with or without acid hydrolysis (phenols in the form of heterosides and free phenols) and has obtained significant correlation. He believes that following fertilization there is a hydrolysis of the heterosides by the fungal beta-glucosidases and a more or less abundant liberation of phenols which can affect the development of the fungus. The resistant genotypes correspond to those which at the end of the growth period have an acid content—free phenols susceptible of being hydrolysized—that is greater than the threshold of inhibition to the *Fusarium*. Unfortunately, the doses are too sensitive to be useful in a selection program.

Basing results solely on the mechanical strength of stalks, Chang et al. (13) have obtained very important gains in developing resistance in three recurring cycles of selection in two synthetic populations. Similar results were obtained by Fourie and Robberts (22), but these methods don't take into account the early rotting of the stem which can lead to a yield loss through scald. It follows that these methods are not totally effective in the development of resistant cultivars.

The idea of a selection for polyvalent resistance advanced by Hooker (26) and Messiaen et al. (31) appears to be incorrect today. Selection against *Diplodia zeae* in the U.S.A. between 1952 and 1975 has led to a repression of this fungus and its replacement by a complex that is dominated by *F. roseum* and *Colletotrichum graminicola* (27). Incidentally, this is precisely what also happened in southwest France.

Rouhani et al. (36) believe that through artifical inoculation made of the stalk with *Diplodia zeae, Colletotrichum graminicola,* and various Fusaria, one isolates the role of roots in resistance. Based on the idea that a primary parasitic fungal complex of the roots served as a portal of entry for *Fusarium,* Rouhani et al. (36) studied this complex. Their initial results have shown that none of the fungi isolated from roots was capable on its own of reproducing completely the successive symptoms observed in nature, and they are now attempting to reconstitute those associations which may lead to stalk rot symptoms.

Barrière (1) and Barrière et al. (2) consider that the symptoms of stalk rot result from the invasion of plants that are more or less genetically or physiologically senescent and are receptive to a complex of fungi, which they characterize as opportunists, in which *F. roseum* var. *graminearum* or *F. roseum* var. *culmorum* dominate, depending on the geographic region. After having established a scale to measure the intensity of the attacks at the base of stalks that are split longitudinally, they propose a sorting of the resistant genotypes that has the advantage of not incorporating types susceptible to rot but mechanically resistant. On the other hand, it would appear that the behavior of the hybrids is not predictable based on parental behavior.

Analysis of variance was computed using Griffing's (23) method for diallel crosses or an interaction model analysis for test crosses. General and specific combining ability effects for disease evaluation are highly significant, and some very early inbreds have a very good general combining ability value for resistance. Breeding for general combining ability would be possible by using families from the F_3 generation onto susceptible and resistant early testers.

Stalk rot is practically never seen prior to maturity; the date each year when the first symptom appears is variable, occurring at dry matter levels of the seed which also differ, but always after most of the reserves have migrated to the seed (immature plants are never attacked). Barrière (1) has noticed that in the north of France the first attacks always occur following a period of low temperatures (especially night temperatures below 7 C). He relates this effect of low temperature without freezing to the phenomenon of "slow chilling injury" described by Levitt (29), which leads to a premature senescence of the plant and thereby increases its susceptibility.

Continuing their investigations Barrière et al. (2) have studied the development of dried matter in a series of hybrids. Their results indicate that resistant cultivars differ from susceptible cultivars in having a lower ratio of seed dry matter to whole-plant dry matter; the ratio of stem dry matter to whole-plant dry matter is greater from the 45th day after mid-silking. Such results, and similar ones given by Wall & Mortimore (38), must be proved to be good generalizations before they can be used in selection procedures.

Therefore it appears that stalk rot cannot be approached solely from the phytopathological point of view and an understanding of it cannot be limited to the study of Fusaria or other pathogens. It must be looked at through study of the relationship of the plant to its environment and through physiological and genetic studies of relations between the ear, where the reserves accumulate at the end of the growing season, and the rest of the plant.

"Fusariosis" of corn is a disease of maturity and senescence and therefore all stresses that accelerate this senescence make the plant more susceptible. Among these stresses one can cite drying and low temperatures (chilling injury), which act on the plant but which can also have an action through the intermediary of the parasite. The work of Cook and Pappendick, as summarized by Cook (19), on the effect of low soil water potential on *Fusarium* supports this idea. Recently Messiaen (*personal communication*) observed that *F. roseum* var. *graminearum* becomes very aggressive on corn at 12 C. Other causes of stress are attacks on the roots, the stalk, the ear, or the leaves by pathogens, insects, or nematodes. It is commonplace to note a definite increase in stalk rot following an attack on the leaves by *Helminthosporium turcicum,* but it is less easy to observe attacks of *Pythium* or *Phaeocytosporella* on the roots.

Taking into account these various elements and the impossibilities of using chemical control methods, a selection method which is less empirical than the one practiced now is possible. Besides purely mechanical resistance, progenitors should be sought which have the following characteristics: i) they senesce little or as late as possible, taking into account the growth demands of the various regions; ii) they show resistance to environmental stresses such as drying and low temperatures; iii) they show resistance to the various pathogens and insects of the stalk and the leaves (one should give special attention to root pathogens which at first may seem to be minor); iv) they show an optimum of energy transfer for any given locality and have a stalk capable of accumulating sufficient reserves for the seed in the ear, while retaining enough reserve to prevent early senescence.

The difficulties in the development of these characteristics are great. Numerous studies on the development of dry matter are both necessary and very difficult. Diallel tests and crosses with various tester strains that would permit an understanding of the heredity of various characters are even more difficult. However, the pursuit of these studies would enable us to better master the problems posed by the "Fusariosis" of corn in the future.

Literature Cited

1. Barrière, Y. 1979. Sélection du mais pour la résistance à la pourriture de tiges. Etude de génotypes précoces. Ann. Amelior. Plantes 29:289–304.
2. Barrière, Y., A. Panouille, and R. Cassini. 1980. Relations source-puits, et sélection du mais pour la résistance à la pourriture des tiges. Ann. Amelior. Plantes 30: (In Press).
3. Booth, R.H., and G.S. Taylor. 1976. Fusarium diseases of cereals. X. Straw debris as a source of inoculum for infection of wheat by *Fusarium nivale* in the field. Trans. Brit. Mycol. Soc. 66:71–75.
4. Booth, R.H., and G.S. Taylor. 1976. Fusarium diseases of cereals. XI. Growth and saprophytic activity of *Fusarium nivale* in soil. Trans Brit. Mycol. Soc. 66:77–83.
5. Cassini, R. 1967. A propos des dégâts provogués par *Fusarium roseum* (Link) Sn. et H. dans les cultures des céréales du Bassin parisien. Compt. Rend. Acad. Agric. Fr. 53:858–867.
6. Cassini, R. 1969. Les fusarioses du blé. Bull. Tech. Information Min. Agric. Fr. 244:819–822.
7. Cassini, R. 1970. Sur l'importance de la contamination des semences dans l'apparition et le développement de la maladie du pied des céréales due à *Fusarium roseum* (Link) Sn. et H. Ann. Acad. Sci. Fenn. A, IV Biol. 168:28–30.

8. Cassini, R. 1973. Influences des techniques de culture sur le développement des maladies des céréales. J. d'Etudes sur la lutte contre les maladies des céréales. Paris. 17–34.
9. Cassini, R. 1973. Etat actuel des principales maladies du mais. Phytiatrie-Phytopharmacie 22:7–18.
10. Cassini, R. 1975. Problèmes de pathologie sous les rotations céréalières. Bull. DEPP 5:161–151.
11. Cassini, R., and C.M. Messiaen. 1976. Problems associated with *Fusarium roseum* on wheat in relation to the intensive cultivation of maize. *In* Cereal Section, Conference on Disease Resistance, EUCARPIA, Munkebjerg/Vejle, Denmark. (Abstr.). 2 p.
12. Cassini, R., D. Guerin, and L. Lescar. 1977. Incidence de la simplification du travail du sol sur le développement des champignons parasites de céréales. J. d'Etude sur la lutte contre les maladies des cereales. Paris. 219–229.
13. Chang, H.S., P.J. Loesch, and M.S. Zuber. 1976. Effects of recurrent selection for crushing strength on morphological and anatomical stalk traits in corn. Crop. Sci. 16:621–625.
14. Colhoun, J. 1970. Epidemiology of seed-borne Fusarium diseases of cereals. Ann. Acad. Sci. Fenn. A, IV Biol. 168:31–36.
15. Colhoun, J. 1971. Cereals, p. 181–225. *In* J. Western (ed.), Diseases of crop plants, Macmillan Press, London. 404 p.
16. Colhoun, J., and D. Park. 1964. Fusarium diseases of cereals. I. Infection of wheat plants, with particular reference to the effects of soil moisture and temperature on seedling infection. Trans. Brit. Mycol. Soc. 47:559–572.
17. Colhoun, J., G.S. Taylor, and R. Tomlinson. 1968. Fusarium diseases of cereals. II. Infection of seedlings by *F. culmorum* and *F. avenaceum* in relation to environmental factors. Trans. Brit. Mycol. Soc. 51:397–404.
18. Cook, R.J. 1968. Fusarium root and foot rot of cereals in the Pacific northwest. Phytopathology 58:127–131.
19. Cook, R.J. 1973. Influence of low plant and soil water potentials on diseases caused by soilborne fungi. Phytopathology 63:451–458.
20. Cook, R.J. 1978. The incidence of stalk rot (*Fusarium* spp.) on maize hybrids and its effect on yield of maize in Britain. Ann. Appl. Biol. 88:23–30.
21. Feekes, W., and D.T. Wieten. 1967. De veredeling van tarwe op resistentie tegen *Septoria* en *Fusarium*. Tech. Ber. Nederl. Graancentrum 17:43–68.
22. Fourie, A.P., and P.J. Robberts. 1977. Selection criteria for stalk-rot quality in *Zea mays* L. Agroplantes 9:55–60.
23. Griffing, B. 1956. Concept of general and specific combining ability in relation to diallel crossing systems. Australian J. Biol. Sci. 9:463–493.
24. Guyot, A.L. 1934. Observations sur quelques maladies fusariennes des céréales en France. Rev. Pathol. Veg. Entomol. Agric. 21:143–186.
25. Hani, F. 1980. Uber Getreidefusariosen in der Schweiz: Saatgutbefall, Ahrenbefall und Bodenkontamination. Z. Pflkrankh. Pflschutz. 87:257–280.
26. Hooker, A.L. 1956. Association of resistance to several seedling, root, stalk, and ear diseases in corn. Phytopathology 46:379–384.
27. Hooker, A.L., and D.G. White. 1976. Prevalence of corn stalk rot fungi in Illinois. Plant Dis. Rep. 60:1032–1034.
28. Jamalainen, E.A. 1974. Resistance in winter cereals and grasses to low-temperature parasitic fungi. Annu. Rev. Phytopathol. 12:281–302.
29. Levitt, J. 1972. Responses of plants to environmental stresses. Academic Press, New York. 697 p.
30. Messiaen, C.M., and R. Cassini. 1968. Récherches sur les fusarioses. IV. La systématique des *Fusarium*. Ann. Epiphyties 19:387–454.
31. Messiaen, C.M., R. Lafon, and P. Molot. 1959. Necroses de racines, pourritures de tiges et verse parasitaire du Mais. Ann. Epiphyties 10:441–474.
32. Millar, C.S., and J. Colhoun. 1969. Fusarium disease of cereals. VI. Epidemiology of *Fusarium nivale* on wheat. Trans. Brit. Mycol. Soc. 52:195–204.
33. Molot, P.M. 1969. Recherches sur la résistance du mais à l'helminthosporiose et aux fusarioses, II. Facteurs de résistance. Ann. Phytopathol. 1:353–366.
34. Parmentier, G. 1973. Septorioses et Fusarioses de maladies de l'epides céréales. Note Tech. 3/12 CRA. Gembloux.
35. Piglionica, V. 1978. Rapporti sull' attivita svolta nell'ambito del subprogetto "Fitoatris del Frumento del Mais e del sorgo." Instituto Pathologia Vegetale. Univ. Studi Bari. 429 p.
36. Rouhani, H., P. Davet, B. Poinso, A. Beyries, and C.M. Messiaen. 1979. Inventaire et evaluation du pouvoir pathogène des composents de la microflore fongique sur racines de mais en France. Ann. Phytopathol. 11:69–93.
37. Tempe, J. de. 1970. Testing cereal seeds for Fusarium infection in the Netherlands 1. Proc. Int. Seed Test. Ass. 35:193–206.
38. Wall, R.E., and C.G. Mortimore. 1965. The growth pattern of corn in relation to resistance to root and stalk rot. Can. J. Bot. 43:1277–1283.
39. Zogg, H. 1976. Probleme der Weizenmonokultur (Fruchtfolgeversuche). Schweiz. Landw. Forschung 15:431–439.

7 Fusarium Diseases of Wheat, Maize and Grain Sorghum in Eastern Australia

L.W. Burgess, R.L. Dodman, W. Pont, and P. Mayers

The most common Fusarium pathogens of wheat (*Triticum aestivum* L.), maize (*Zea mays* L.) and grain sorghum [*Sorghum bicolor* (L.) Moench] in eastern Australia are *Fusarium roseum* 'Graminearum' Group 1, *F. roseum* Graminearum Group 2 and *F. moniliforme* respectively. Because each of these crops is susceptible to several Fusaria, we attribute this pattern of host/pathogen combinations to the influence of the environment in addition to the intrinsic pathogenic ability of each fungus. Therefore a brief summary of environmental features is given below. Comments on agronomic practices are included in the introduction to each crop.

Australia is a large, relatively flat, island-continent lying across the mid latitudes. The Great Dividing Range, the only topographic feature of consequence, is situated in eastern Australia and extends from northern Queensland to south-western Victoria. It parallels and is adjacent to the east coast. Although this range is quite rugged in some areas, a significant proportion of the tablelands is suitable for pastoral purposes and, to a lesser extent, for cultivation. The arid interior is a dominant feature of the geography, with approximately 75% of the continent being too dry for crop production.

The tropical north and northeast of Australia is characterised by a hot wet summer (monsoon season) and a warm dry winter. The southern areas have a Mediterranean climate, with a hot dry summer and a cool wet winter. The east coast, south of the Tropic of Capricorn, receives summer rain from the south-east trade winds. In the major areas where cereals are grown the rainfall is distinctly seasonal and relatively unreliable and is not a good guide to available soil moisture because evaporation rates are high. Representative rainfall and temperature data are presented in the accompanying figures.

Wheat is the major crop grown in each of the three states of eastern Australia. The wheat belt extends from the central highlands of Queensland to north-western Victoria. Essentially, it is situated between the more rugged areas of the Dividing Range and the margin of the dry inland areas. Sorghum is confined mainly to the more subtropical areas of the wheat belt. Maize is grown in limited areas in Queensland and New South Wales, basically as a dryland crop in the wetter areas, but also is grown as an irrigated crop in dry inland districts.

For further information on the Australian environment, the reader is referred to Davidson (9) and "The Australian Environment" (33).

Taxonomic Comments

The nomenclature adopted in this review is based on the nine species system of Snyder and Hansen as illustrated by Toussoun and Nelson (34).

The two populations within *F. roseum* 'Graminearum' are referred to as Groups 1 and 2 as defined by Burgess et al. (5). Further characteristics of the two groups have been described by Frances (13) and Frances and Burgess (15). Members of Group 1 are mostly soil-borne, cause crown rot of cereals and grasses and rarely form perithecia in nature. In contrast, members of Group 2 are mostly air-borne and cause a variety of diseases of aerial plant parts (viz., stalk and cob rot of maize; head blight of wheat, barley, and oats; and stub-dieback of carnations) and normally form perithecia abundantly in nature.

On the basis of cross-inoculation studies, Purss (29,31) postulated that pathogenic specialization exists within Graminearum. He demonstrated that isolates of Group 1 from wheat affected by crown rot and isolates of Group 2 from maize affected by stalk rot caused typical stalk rot of maize. However, only Group 1 isolates from wheat affected by crown rot caused typical crown rot of wheat. Some isolates of Group 1 from barley and several grasses, affected by crown rot, also caused the typical crown rot symptoms in wheat. He àlso found that both Group 1 and Group 2 isolates caused head blight.

Wheat

Introduction The eastern wheat belt of Australia extends from the central highlands of Queensland (the Tropic of Capricorn) through New South Wales to northwestern Victoria (latitude S37°), a distance of approximately 1500 km (Fig. 7-1). It is bounded in the east and west by mean winter rainfalls (May-October) of about 50 cm and 15 cm respectively. The terrain varies from the gently undulating country of the slopes of the Great Dividing Range to the flat plains of the western areas.

The eastern wheat belt can be divided into two major agro-climatic regions (Fig. 7-1). The northern region (Region 1) is characterised by summer-dominant rainfall whereas the southern region (Region 2) is characterised by winter-dominant rainfall and lower winter temperatures. The boundary represents a gradual transition zone. Although water deficits are not unusual in either region, they are more common in Region 1 because of the summer dominance of the rainfall.

Wheat is sown from late autumn to mid-winter depending on the locality and seasonal conditions. The crop is harvested in late spring or early summer. About 4.5 million ha is grown annually in eastern Australia. All cultivars are essentially spring wheats and most show a degree of vernalization response. A different spectrum of cultivars is grown in each region. The normal growing season in Region 1 is from May to November and in Region 2 from May to December, but the planting date is dictated by available soil moisture and can be as late as August in most areas. Except in the most northern areas, very early planting (March-April) of cultivars in current use is prevented by the threat of frost damage at flowering. Temperatures do not normally limit production but may retard growth in winter.

Wheat is grown on a variety of soils in both regions (20). In Queensland black earths occupy most of the Darling Downs and several other areas but light soils occupy a considerable portion of the western wheat areas. In northwestern New South Wales, black earths predominate but grey clays and light soils also occur in some areas. In contrast, in Region 2 in New South Wales, red brown earths predominate but solonized brown (mallee) soils are common in the southwest. Mallee soils are light textured and, in Victoria, occupy a major portion of the northwestern wheat area. Grey and brown clays and red brown earths are common in the remaining wheat areas of Victoria.

Current trends in wheat production include the introduction of semi-dwarf cultivars, the adoption of stubble retention for soil erosion control and the use of nitrogen fertilizer. Each of these practices could affect the incidence of soilborne diseases. The reader is referred to the review by Brown (1) for information on wheat diseases in Australia.

Fusarium Diseases Crown rot, caused by Graminearum Group 1, has been regarded as the only important Fusarium disease in the eastern wheat belt for at least 20 years. It was first recorded on the Darling Downs in Queensland in 1951 by McKnight and Hart (21), who suggested, however, that the disease had probably been present much earlier. It was subsequently reported in New South Wales (23) and Victoria (27). Although it has caused significant losses in individual fields in most areas (5), it is more prevalent on the Darling Downs and in northwestern New South Wales. Estimates of total losses from crown rot are not available. However, McKnight (*personal communication*) estimated an overall loss in Queensland of approximately 5% in 1956 and again in 1957. In 1976, although no assessment of losses for the whole wheat area was made, we did record losses of up to 26% in individual fields. These determinations were based on random samples of approximately 100 plants. Visual estimates of losses based on percentages of whiteheads were similar. Such estimates of yield reduction do not take into account losses due to reduced grain quality.

Prior to the late 1950s *Fusarium roseum* 'Culmorum' was regarded as an important member of the complex of fungi associated with foot and root rot problems of wheat in Region 2 (6, 16, 17, 25). Estimates of losses attributable directly to Culmorum prior to 1970 are not available, but foot and root rots generally were considered to cause serious losses in some years (6, 16). In contrast to these earlier reports, recent surveys indicate that foot and root rot caused by this fungus are, at present, of little consequence (5, 7, 27). We cannot explain this apparent decline in the incidence of foot and root rot caused by Culmorum in Region 2.

Culmorum was not reported from Region 1 until 1965, when it was isolated from wheat affected by foot rot in a field on the Darling Downs (G.S. Purss and G.B. Wildermuth, *personal communication*). These workers subsequently isolated the fungus from wheat and barley at two additional sites in 1969.

In 1972, 1973, and 1974, surveys were made (5) to clarify the relative importance of Graminearum Group 1, Culmorum, and *Fusarium roseum* 'Avenaceum' as crown and foot rot pathogens of wheat in eastern Australia. Stem bases of diseased plants were collected from fields selected at random in most areas of the wheat belt. Fungi were isolated from the stem bases and identified using the same procedures for each survey. Graminearum Group 1 was the dominant member of *F. roseum* associated with crown rot in all areas.

Avenaceum was not isolated from sampling sites affected by crown rot on the Darling Downs in Queensland. However, it was isolated from one site (<1%) in 1973 and one site (<1%) in 1974 in the southern areas of Region 1. It was found at four sites (7%) in 1973 and at seven sites (12%) in 1974 from Region 2 in New South Wales. It was isolated from only one site (3%) in Victoria, in 1974. Avenaceum was more common in the wet seasons and was usually associated with plants affected by root rot caused by *Gaeumannomyces graminis*. Plants affected by Avenaceum and *G. graminis* show symptoms similar to mild symptoms of crown rot caused by Graminearum, in addition to root rot. We do not regard Avenaceum as an important cause of crown rot. Chambers (7) commonly isolated Avenaceum from lesions on wheat roots from most areas of the Victorian wheat belt but considered it a weak pathogen. This fungus is a recognised root-rot pathogen of subterranean clover (*Trifolium subterraneum* L.) (4, 18) and barrel medic (*Medicago truncatula* Gaertn. var. *truncatula*) (19) which are common pasture legumes in south-eastern Australia. Avenaceum was isolated only from wheat areas where these legumes are common.

Colonization of wheat roots could contribute to the successful persistence of Avenaceum during the wheat phase in pasture-wheat rotations.

Culmorum was isolated from only three sites (11%) affected by crown rot on the Darling Downs in 1973. It was not found in New South Wales. In Victoria it was isolated from only three (10%) of sites affected by crown rot in 1972. In 1976 Burgess, Dodman and Tio (*unpublished data*) assessed the relative frequency of isolation of Graminearum Group 1 and Culmorum from plants affected by crown rot (foot rot) collected from 35 sites on the Darling Downs and 91 sites in New South Wales. In this survey 25 plants were collected, where possible, from each site, compared with a maximum of only five plants per site in the previous surveys by Burgess et al. (5). Graminearum Group 1 was again the dominant fungus isolated from the Darling Downs and New South Wales and was recovered from all sampling sites affected by crown rot. Culmorum was isolated from 11 sites (32%) on the Darling Downs but not from New South Wales. There is no obvious reason for this difference in the incidence of Culmorum. The apparent increase in the number of sites from which Culmorum was isolated on the Darling Downs, in comparison to the earlier surveys, is attributed to the increase in sampling intensity.

Head blight, caused by Graminearum Group 2, occurs in wet seasons in the higher rainfall areas of Region 1 but generally is not significant except in the Kingaroy district of Queensland, where wheat is a minor crop. Graminearum Group 2 is also commonly associated with stalk rot of maize in the Kingaroy district. Graminearum Group 1 is isolated occasionally from spikelets affected by head blight in Region 1 and it is not unusual to find crown rot caused by members of Group 1 in the same field as head blight caused by members of Group 2. On several occasions we have observed discrete brown lesions on one or two internodes of wheat stems during or after wet weather. Graminearum Group 2 has been isolated from these lesions.

Crown Rot

Infection, colonization, symptoms and effects on yield There have been no histological studies of the mode of infection and subsequent colonization of wheat by Graminearum Group 1. The conclusions which follow, concerning parasitic activity, have been inferred from the results of systematic isolations from various parts of plants sampled at regular intervals during the growing season (28; U. Roongruangsree and L.W. Burgess, *unpublished data;* G.B. Wildermuth and R.L. Dodman, *unpublished data*). Infection normally occurs through the subcrown internode, the coleoptile, or the crown. However, infection through the scutellum and via senescent leaf sheaths in the crown region is also significant. Subsequently, the fungus progressively colonizes the crown region, stem bases, lower-leaf sheaths and the proximal regions of the nodal (crown) and seminal roots. Although the fungus is occasionally isolated from discrete root lesions, these are not regarded as primary infection sites. Plants are susceptible to infection at all stages of growth. Infected plants can be extensively colonized without showing visible symptoms.

The symptoms of crown rot of wheat caused by Graminearum Group 1 have been described by McKnight and Hart (21), Purss (28, 29) and Wildermuth (37) and are similar to those caused by Culmorum (8) except that Culmorum also causes root rot. The symptoms are also similar in other cereal and grass hosts, but the severity varies with host susceptibility (29). Necrosis of the coleoptile, subcrown internode, and basal leaf sheaths is usually the first symptom of infection in wheat. Extensive colonization of the crown and stem bases is usually associated with dark-brown necrosis of the crown and a honey-brown discoloration of the stem bases. This discoloration is usually uniform and distinctive. Although colonization and subsequent discoloration of the stem is usually restricted to the lower three internodes (28), it may extend to the sixth internode. Colonization of the head via the stem, however, is rare. Dense white mycelium usually develops in the lumen of the colonized region of the stem. Under moist conditions extensive mycelium develops externally on the crown tissues, particularly the lower-leaf sheaths, together with the formation of salmon-pink spore masses which are also common on the nodes.

Crown rot causes premature ripening (white-head formation) through disruption of the translocation system. This leads to yield losses, with effects ranging from no grain formed through various degrees of shrivelled grain (28) to slight but significant reductions in the weight of individual grains. If extensive colonization of the crown occurs late in the growing season only one or some of the stems, usually the youngest, will collapse. Although severe symptoms of crown rot are most commonly observed after flowering, the fungus can cause death of seedlings and of plants at intermediate growth stages. Plants severely affected in the seedling or tillering stages are often distinguished by chlorosis of the leaf blade and sheath. Stunting is not a common symptom of crown rot. Occasionally fertile perithecia of Graminearum Group 1 are found on the lower-leaf sheaths and lower nodes of severely affected plants.

Seedling death caused by Graminearum Group 1 occurs in dry soils and results from infection from soil-borne inoculum. In contrast, seedling blight caused by Graminearum Group 2 is due to seed-borne inoculum and is common when seed from crops affected by head blight is planted. Generally, seedling blight caused by Graminearum Group 2 is not common.

Epidemiology The incidence and severity of crown rot are affected by soil moisture, soil type, topography and time of planting, and are favoured by low soil moisture (5). The disease is usually not severe in crops grown under optimum soil moisture, even in fields where the pathogen has previously been prevalent, but isolation studies indicate that a significant proportion of the plants in such crops will be infected, and some extensively colonized, by Graminearum Group 1 (R.L. Dodman and L.W. Burgess, *unpublished data*).

Although crown rot has been recorded on all major soil types, it occurs principally on soils of heavy texture or on light-textured soils with an impervious layer in the profile (5, 21, 37). Severe crown rot has not been observed on well-drained sandy soils such as the mallee soils of northwestern Victoria and southwestern New South Wales (L.W. Burgess, *unpublished data*) or the light-textured soils in Queensland (37; R.L. Dodman, *unpublished data*).

Crown rot is more prevalent on the plains than on undulating terrain. Within a field the disease is more common in low-lying areas (5), and it sometimes occurs in distinct patches in fields where localised depressions, known as gilgais, are common. In one such field, McKnight and Hart (21) demonstrated that the severely diseased patches were invariably lower than adjacent areas where diseased plants occurred at random. We believe that soil type and topography affect the incidence and severity of crown rot indirectly, by their effect on the water relations of the plant.

Purss (30) found that the incidence of infected plants (as determined by isolation) and crown rot severity were greater in early-planted (May) compared with late-planted (August) plots, but that the magnitude of the difference varied from season to season. The effect of planting time on the incidence of infected plants and disease severity did not appear to be linked to soil temperature or moisture. Although the incidence of infected plants increased progressively throughout the growing season, the greatest increase occurred between the rosette and late flowering stages (30). Presumably, early planting, by prolonging the parasitic phase, will favour more extensive colonization of the infected plant. In addition, early planting increases the risk of moisture stress which favours crown rot. In practice the planting time is dictated by available soil moisture.

Graminearum Group 1 also causes crown rot of many gramineous hosts including barley

(*Hordeum vulgare* L.), oats (*Avena sativa* L.), rye (*Secale cereale* L.), barley grass (*Hordeum leporinum* Link.), common wheat grass (*Agropyron scabrum* (Labill.) Beauv.), *Bromus unioloides* H.B.K., *Danthonia linkii* Kunth and *Phalaris paradoxa* L. (2, 21, 29).

In the absence of suitable conditions for parasitism, Wearing and Burgess (36) found that Graminearum Group 1 survives as hyphae in identifiable stubble (crown and stem bases) and in small debris, the products of stubble disintegration. These authors also detected hyphal fragments and macroconidia in wheat soils but concluded that these propagules did not constitute a significant mode of survival. Chlamydospores were not observed, although some macroconidia were modified and had thick cell walls. Burgess and Griffin (3) found that the fungus could survive for at least 2 years in old stubble, but probably only a minor proportion of the inoculum survives for longer than 12–18 months. The absence of wheat or other hosts for 12–18 months causes a significant reduction in the severity of crown rot in the subsequent wheat crop. The pathogen has also been isolated from debris sieved from virgin pasture soils within the wheat belt (L.W. Burgess, *unpublished data*). The disease has been recorded frequently in wheat crops grown on virgin soils (21, L.W. Burgess, *unpublished data*).

The two senior authors have initiated a long-term study to assess the influence of cultural practices on crown rot. Disease severity is being assessed annually at approximately 200 sampling sites in commercial wheat fields in Queensland (Darling Downs) and New South Wales. This study will enable us to assess the impact of new cultural practices such as stubble retention, the use of nitrogen fertilizer and the use of semi-dwarf cultivars (including crown rot-tolerant cultivars) on crown rot severity. In addition the relative importance of Graminearum Group 1 and Culmorum is being monitored by culturing diseased plants from each site.

Control by cultural practices Crop rotation and long-fallow (12–18 months) are the most effective methods of control. These methods are usually economically justified only where losses from crown rot are quite severe. The wide host range of the pathogen and economic considerations make the choice of an alternative crop rather difficult. Significant reductions in crown rot have been observed after rotation to linseed (*Linum usitatissimum* L.) and safflower (*Carthamus tinctorius* L.) (winter crops) and to sorghum (summer crop). Dryland sorghum is restricted to the more reliable summer rainfall areas of Region 1. Rotation to a natural pasture is not recommended because many common grasses are susceptible.

Stubble burning has been a common practice and eliminates the inoculum in the aboveground part of the stubble. This can be significant in seasons that favour extensive colonization of the stems of infected plants. However, burning does not eliminate the fungus in the crown tissue. Stubble retention is becoming a popular practice, especially in Region 1, to minimize soil erosion and to improve water infiltration. This innovation could lead to a gradual increase in inoculum.

Control by host resistance The first report of a differential reaction of wheat cultivars to crown rot in Australia was made by McKnight and Hart (21). They examined 18 cultivars in Queensland and found that although none was completely resistant to crown rot, at least four cultivars showed low levels of disease relative to the other cultivars. Purss (28) continued this work with a detailed examination of 10 cultivars. He concluded that two of these (Gala and Mengavi) possessed a reasonable level of field tolerance. Purss (28) found white-head production and crown symptoms just prior to maturity to be the most suitable criteria for assessing tolerance. Further investigations by Purss (32) showed that no greater levels of resistance could be found in a collection of *Triticum* species and *Triticales,* while Wildermuth and Purss (38) found that several other wheat cultivars, including Gluyas Early and Mexico 234, had resistance equal to that of Gala.

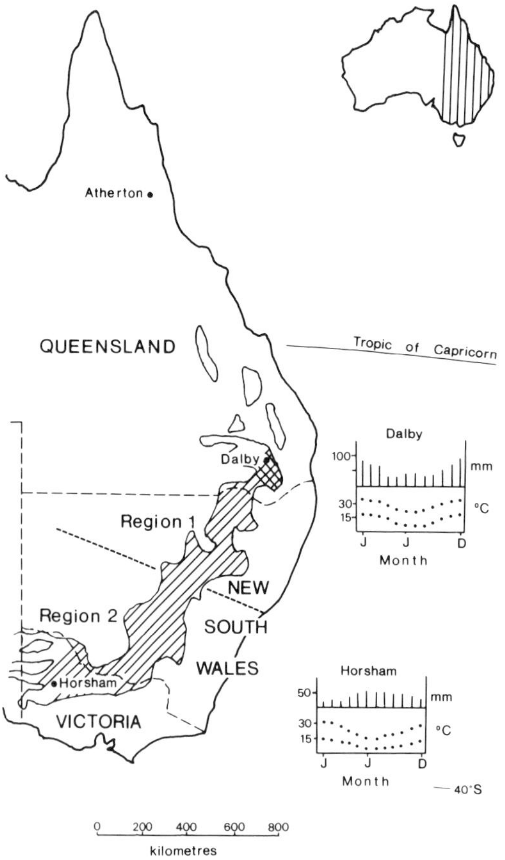

Fig. 7-1. Bounded areas delineate wheat areas of eastern Australia. Hatched zone indicates areas where crown rot caused by 'Graminearum' Group 1 has been recorded regularly. The Darling Downs of Queensland is indicated by double-hatching. The climatic data in the graphs represent mean monthly rainfall (mm) and mean monthly maximum and minimum temperatures (C).

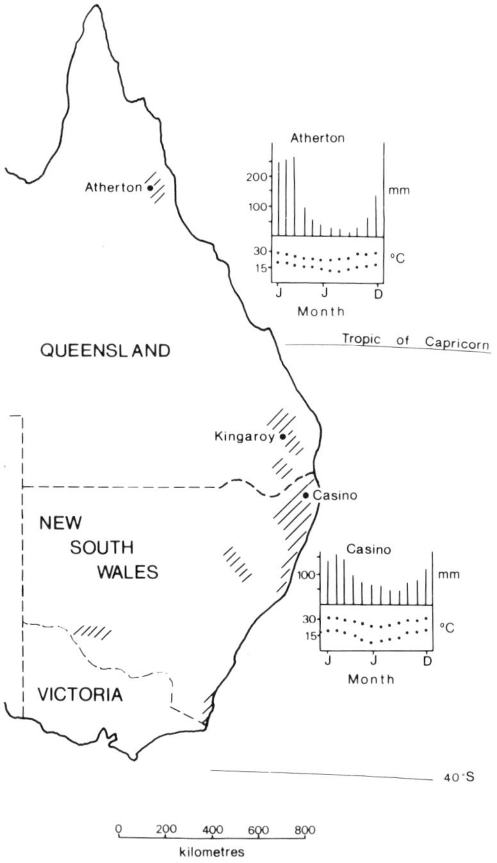

Fig. 7-2. Maize growing areas of eastern Australia are indicated by hatching. The climatic data in the graphs represent mean monthly rainfall (mm) and mean monthly maximum and minimum temperatures (C).

Genetic control of resistance to crown rot was examined in a diallel cross involving four resistant and four susceptible cultivars. Analysis of the data from the diallel cross and an examination of the progeny from some specific crosses indicated that there are at least two genes controlling crown rot development, with resistance being recessive and susceptibility being dominant (R.L. Dodman and G.B. Wildermuth, *unpublished data*). Potential cultivars for Region 1 (Queensland) are now screened prior to release, in an attempt to provide the farmer with reliable information on the reaction of all cultivars to crown rot.

Maize

Introduction Maize, a summer crop, is grown mainly in New South Wales and Queensland where it is used for the production of feed grains and, to a lesser extent, for forage and ensilage. It is a relatively minor crop compared to sorghum but is important on a local basis. Approximately 100,000 ha is grown annually in areas scattered from the tropical areas of northern Queensland to the temperate areas of southern New South Wales (Fig. 7-2). As a dryland crop it is restricted to areas with a high summer rainfall, whereas sorghum is grown in areas with a slightly lower summer rainfall. In Queensland maize is grown mainly as a dryland crop, but in New South Wales it is grown as a dryland crop on the coastal and tableland areas and as an irrigated crop in inland areas where rainfall is lower and less reliable.

The majority of hybrids grown in coastal areas of New South Wales and Queensland are derived from locally adapted cultivars. They show significant resistance to a range of leaf, stalk, and ear diseases and insect pests, but they are late maturing (6 months), rendering them unsuitable for the temperate, inland areas. In contrast, hybrids grown in the inland irrigation areas of New South Wales and on the Darling Downs are derived from inbreds used for the production of hybrids for the corn-belt in the U.S.A.. These hybrids have a high yield potential and are earlier maturing (5 months) but are usually more susceptible to disease and insect problems than the hybrids used in the wet coastal areas. For further information on maize and grain sorghum production see McWhirter (22).

Fusarium pathogens and their importance Three Fusaria, Graminearum Group 2, *F. moniliforme* and *F. moniliforme* 'Subglutinans,' are recognised pathogens of maize in eastern Australia and have been implicated in stalk, ear, and root rot diseases as well as seedling blight. Subglutinans was first reported as a pathogen of maize by Edwards (10) working in New South Wales. There is, however, a paucity of quantitative data on the economic importance of these pathogens and their relative roles in each disease from season to season. The heterogenous and dispersed nature of the maize-growing areas has made assessments difficult. In general the diseases are more common in Queensland and the coastal and tableland areas of New South Wales. There is no evidence that they cause significant losses in the drier inland irrigation areas of New South Wales. Pont (*unpublished data*) has estimated yield losses as high as 40% from lodging caused by stalk rot in individual fields in the Atherton area of north Queensland.

Frances and Burgess (14) surveyed the Fusaria associated with stalk rot in eastern Australia. The inland irrigation areas of New South Wales were not surveyed. Graminearum Group 2 was the most commonly isolated *Fusarium* from each area. Subglutinans was also commonly isolated, especially from the more temperate areas. *Fusarium moniliforme* was not commonly isolated. Further surveys are needed to clarify the relative importance of these fungi particularly with respect to seedling and root rot diseases.

Seedling blight Because of their ability to cause symptomless internal infection, Subglutinans and *F. moniliforme* are the most common contaminants of commercial maize seed (12). Seed-borne inoculum can cause severe seedling blight when climatic conditions are unfavourable for germination and seedling growth. Edwards (11) reported that Subglutinans was the most common pathogen isolated from the mesocotyl of seedlings affected by seedling blight in the wetter areas of New South Wales. The relative contribution of seed-borne versus soil-borne inoculum of these fungi in the initiation of seedling blight has not been assessed.

Root rot Results of the isolation studies by Frances and Burgess (15) suggest that Graminearum Group 2 colonizes only the exposed brace roots and does not cause root rot per se. Pont (*unpublished data*) considers that root rot caused by *F. moniliforme* could be responsible for significant yield reductions in the Atherton Tablelands of northern Queensland.

Stalk rot Stalk rot is the most important disease caused by the Fusaria in maize. The descriptive information concerning this disease and ear rots has been based on work by Pont (26; *unpublished data*). Stalk rot usually becomes obvious after tasselling, originating in the lower part of the stalk. It subsequently extends both up the stem and down into the proximal regions of the brace roots. In addition it is not uncommon to find discrete lesions at any point on the stem, often associated with the leaf junction where insect damage has occurred. Severe rot causes premature senescence of stalks, which turn light brown externally while the internal tissues become shredded and often develop a pink or red colour. In wet weather these stalks are rapidly colonized by saprophytic micro-organisms. The three pathogenic Fusaria sometimes develop spore masses on infected stalks, especially on leaf sheaths or at the nodes. In wet weather Graminearum Group 2 forms abundant perithecia on infected stalks, particularly the nodes, leaf sheaths, and the husks. In northern Queensland perithecia of this fungus and of *F. moniliforme* are occasionally found on the same part of an infected stalk. Edwards (11) reported that Subglutinans also formed perithecia on old maize stalks in the wetter areas of New South Wales. Plants affected by stalk rot are very susceptible to lodging by wind or during harvest. Lodging is particularly serious when harvesting is delayed.

The mode of infection and subsequent colonization of maize stalks by these fungi have not been studied under Australian conditions. Francis and Burgess (15; *unpublished data*) compared the frequency of isolation of Graminearum Group 2 from the various parts of mature maize stalks collected in the field in New South Wales and found that this fungus was most commonly isolated from the stem and crown and rarely from distal parts of the roots.

Ear rot This disease is a common problem and is often associated with ear worm or other insect damage. Although Graminearum Group 2 usually causes a generalised ear rot which starts at the tip, the rot occasionally originates at the butt of the ear. This fungus causes a pronounced reddish discolouration of the rotted grain and husk tissues and produces a growth of pinkish-white to red mycelium on the surface of colonized grain. Perithecia are often formed abundantly on the outside of husks in wet conditions.

Fusarium moniliforme and Subglutinans may cause a generalised ear rot which usually originates at the tip end where insect damage has occurred. More commonly these two species cause rotting of individual grains or localised areas of grain. These fungi form conspicuous, pale-pink to white mycelium between the grains and under the husk. They commonly infect the grain internally at the embryo end, usually without causing visible symptoms (12). Edwards (12) suggested that internal infection occurs through the silks or from adjacent cob tissues. Occasionally, internally infected grain can be recognised by white streaks in the seed coat and the presence of "long-stringy attachments" at the tips (12).

Overseasoning In the absence of living maize plants, Graminearum Group 2 persists as a pathogen of other gramineous hosts (L.W. Burgess, *unpublished data*), as perithecia on old maize stubble remaining on the soil surface, as hyphae in stubble and small debris in soil (35), and in infected seed (11). Ascospores are regarded as the main source of primary inoculum. Using the dilution plate technique Wearing (35) rarely isolated Graminearum Group 2 from soils where it was common as a stalk rot pathogen. He concluded that this fungus did not form chlamydospores in these soils.

Although Wearing (35) regularly isolated *F. moniliforme* and Subglutinans from maize soils using the soil dilution plate technique, he did not assess their mode of survival. Little is known of the host range of these fungi, but they are recognised pathogens of sorghum. However, sorghum and maize are grown in close proximity or in rotation with each other only in the inland irrigation areas and on the Darling Downs. *Fusarium moniliforme* (W. Pont, *unpublished data*) and Subglutinans form perithecia on old corn stalks (11) but their importance in the disease cycle is not known.

Control The Fusarium diseases of maize are difficult to control. There are very few resistant crops which can be grown economically in rotation with maize for disease control. Some of the current hybrids possess moderate field resistance to stalk rot caused by these fungi. Growers are advised to chop and deep-plough stubble to prevent perithecia of Graminearum Group 2 from providing a source of primary inoculum. However, wind-blown ascospores from perithecia on adjacent pasture grass stubble such as senescent kikuyu stolons (*Pennisetum clandestinum* Hochst. ex. Chiov.) could obviate the effectiveness of these measures. Similar measures are recommended for *F. moniliforme* and Subglutinans.

Grain Sorghum

Introduction Production of grain sorghum, and summer cropping in general, are increasing in importance relative to winter cropping in the Queensland and New South Wales portions of the wheat belt. In Queensland, the area of grain sorghum more than trebled between 1966 and 1971, from 134,670 ha to 423,200 ha, but has since stabilized around 340,000 ha. In Queensland sorghum is grown mainly in the central and southern areas. More than 75% of the sorghum from the southern area is grown on the Darling Downs. In New South Wales the area harvested expanded from 36,000 ha in 1966 to 160,000 ha in 1976. In Region 1, sorghum is grown mainly as a dryland crop, whereas in Region 2 in New South Wales it is predominantly an irrigated crop.

Many diseases of sorghum in Queensland have been attributed to *F. moniliforme*. This conclusion is based on a consistent, if not exclusive, association between symptoms and isolations over several seasons and on the results of some pathogenicity tests. These diseases are as follows: embryo death and seed dry rot, pre- and post-emergent seedling blights, root rot, stalk rot, peduncle and rachis blight, leaf sheath blotch, weak neck, and axillary-bud and tiller blight (P. Mayers, *unpublished data*). Although Subglutinans and Graminearum Group 2 have been isolated occasionally from sorghum, there is no evidence that they are important pathogens. Graminearum Group 1 has not been isolated from sorghum.

Seed problems Infection by *F. moniliforme* resulting in seed dry rot and/or embryo death is an important cause of inviable seed and unthrifty seedlings in grain sorghum in Queensland. Levels of more than 80% inviable seed have been recorded for inbred lines imported from overseas and examined prior to quarantine treatment. However, internal infection of commercial hybrid seed batches in Queensland rarely exceeds 5% and infected seed is usually viable. Seed killed by *F. moniliforme* has a pinched and sometimes shrivelled appearance externally, while internally the embryo is discoloured and the endosperm chalky with areas of pale pink to mauve discoloration.

Seedling blights Pre- and post-emergent seedling blights are usually associated with conditions which delay seedling germination and emergence such as low soil temperatures, very wet soil, a receding moisture profile, high evapotranspiration, deep planting (>6 cm) and poor seed-bed tilth.

In cool, wet conditions *F. moniliforme* (as soil-borne or to a lesser extent seed-borne inoculum) is a common cause of seedling blight, alone or in association with species of *Pythium* and *Drechslera* (P. Mayers, *unpublished data*). These fungi infect the seminal root system and the mesocotyl causing severe lesion development. Following infection of the mesocotyl these fungi cause a lesion to develop in the epidermis and cortex which can cincture the stele before nodal root development is able to meet the transpirational demand of the shoot (P. Mayers, *unpublished data*). Mesocotyl rot is frequently associated with post-emergent death. The incidence of seedling death is increased when crusting of the soil and high surface-soil temperatures prevent the penetration of nodal roots (P. Mayers, *unpublished data*).

Seedling death in dry soil is usually caused by *F. moniliforme* in association with *Penicillium* sp. which cause seminal-root and mesocotyl rot (L.W. Burgess, *unpublished data*). This type of seedling blight is more common in light-textured soils in which the moisture profile recedes quickly in dry windy weather, conditions which also cause high transpiration rates.

Control of seedling blights is based largely on cultural practices which increase the rate of seedling germination and secondary root development. Thus improved stand establishment has been achieved by planting when soil temperatures at planting depth exceed a daily minimum of 15 C, planting at a uniform depth of not more than 6 cm, the use of press wheels on the seed drill to improve seed-soil contact, and the use of a broad-spectrum protectant fungicide on the seed.

Root and stalk rots Root rot of seedlings and of advanced and mature plants is common in Queensland and is caused by *F. moniliforme* and *Periconia circinata* (24). These fungi have also been implicated in root rot of sorghum in New South Wales (D.S. Trimboli and L.W. Burgess, *unpublished data*). This root rot complex is believed to be a major factor predisposing mature sorghum plants to accelerated senescence and hence stalk rot development and lodging.

Stalk and root rot almost invariably lead to lodging and have thus resulted in serious losses to individual farmers in some seasons. These losses often exceed 30%, and in highly susceptible hybrid cultivars approach 100% (P. Mayers, *unpublished data*). Estimates of total losses from lodging, however, are not available.

Fusarium moniliforme commonly causes discrete infections of roots and stalks in southern and central Queensland in most seasons. However, the fungus causes severe stalk rot only in those plants whose photosynthate sink is large relative to their capacity for photosynthesis and which are subjected to prolonged moisture stress during the grain filling period (P. Mayers, *unpublished data*).

There are no economically feasible control measures available for root and stalk rot. However, some hybrids have gained a reputation for standability under conditions which favour these diseases.

Peduncle and rachis blight Peduncle infection, and to a lesser extent rachis infection, with *F. moniliforme* has been observed in crops throughout Queensland, usually in association with damage caused by aphids (P. Mayers, *unpublished data*). The pathogen can be isolated from red discoloured epidermal lesions and the underlying, reddish-brown, parenchyma and vascular tissues. The vascular tissue may be colonized above and below the infection court. Grain pinching occurs when rachis blight affects the distal florets during the grain filling stage. Intensive peduncle infection at or below the flag-leaf ligule frequently results in a unilateral weakness and breakage of the peduncle, a condition known as weak neck. This disease has affected up to 5% of heads in individual crops between anthesis and the soft-dough stage and is prevalent during warm, showery weather.

Axillary-bud blight Axillary buds and tillers at pre-anthesis stages of development can be blighted by *F. moniliforme*. Vascular traces which connect vascular bundles in the stem with those in axillary buds and leaf sheaths are an important mode of fungus entry into intermediate and upper nodes of the sorghum stem (P. Mayers, *unpublished data*). Lodging of sorghum at nodes high on the stem facilitates sprouting of grain in heads which touch the ground. This type of lodging occurs commonly in Queensland and makes harvesting difficult. It is distinct from basal node lodging which can also be caused by the charcoal stalk-rot pathogen, *Macrophomina phaseoli*.

Leaf sheath blotch Leaf sheath blotch, a disease of minor importance, develops on the adaxial surface and is caused by *F. moniliforme*. These lesions are characteristically diffuse, purple-red, and up to a few centimeters in diameter. Under field conditions, leaf-sheath blotches coalesce and extensive reddening occurs, particularly in areas previously colonized by aphids. Premature senescence of leaves can result.

General comments In the absence of effective host resistance the diseases caused by *F. moniliforme* will be difficult to control. The use of crop rotation is of doubtful value because the fungus has commonly been isolated from sorghum soils and from wheat, maize and pasture soils in areas where sorghum is grown (35, 36; L.W. Burgess, *unpublished data*). The mode of survival of the fungus in these soils has not been studied.

We have not found perithecia of *F. moniliforme* on infected sorghum or on sorghum stubble although perithecia of this fungus have been found on maize (W. Pont, *unpublished data*) and some isolates from maize have formed perithecia in culture (L.W. Burgess, *unpublished data*). Infection of above-ground parts of the plant is apparently initiated by conidia.

Addendum

Crown rot was common in most wheat areas of eastern Australia in 1977. The disease caused a loss of approximately $5M in Region 1 in New South Wales. This estimate was based on data collected during an intensive survey of the area by the senior author. Yield reductions exceeding 90% were recorded in some fields while in the lower areas of such fields 100% loss was usual. Losses of 10–25% were common.

The severity of crown rot in Region 1 in New South Wales in 1977 is attributed to the coincidence of high inoculum levels and a dry season. The high inoculum levels were a result of a gradual increase over recent growing seasons characterised by periods of moisture stress. Although soil moisture was generally adequate at planting in 1977, most areas received less than 50 mm of rain during the growing season. The dry conditions favoured the development of severe crown rot and in addition prevented the development of nodal roots in most crops. Plants were therefore entirely dependent on water absorbed by the primary roots and translocated through the sub-crown internode. Thus, only limited crown rot caused complete disruption to translocation and subsequent plant death. Significant differences in crown rot severity were observed between cultivars in adjacent blocks within the one field. Crown rot was not severe in fields which had been rotated recently to sorghum or other resistant crops or which had been long-fallowed in the last 3 years. Reasonable yields were obtained in these fields. Severe crown rot was also observed in barley grass (*H. leporinum*) and wild canary grass (*P. paradoxa*) in native pastures.

Crown rot was also common in Region 1 in Queensland where moisture stress was generally more severe than in New South Wales. However losses caused by the disease could not be differentiated from the effects of moisture stress per se. Detailed information on crown rot severity is not available for Region 2.

Literature Cited

1. Brown, J.F. 1975. Diseases of wheat—their incidence and control, p. 304–363. *In* A. Lazenby and E.M. Matheson, (ed.), Australian field crops, Vol. 1: Wheat and other temperate cereals, Angus and Robertson, Sydney.
2. Burgess, L.W. 1967. Ecology of some fungi causing root and crown rots of wheat. Ph.D. Thesis, Univ. Sydney. 122 p.
3. Burgess, L.W., and D.M. Griffin. 1968. The recovery of *Gibberella zeae* from wheat straws. Australian J. Exp. Agric. Anim. Husb. 8: 364–370.
4. Burgess, L.W., H.J. Ogle, J.P. Edgerton, L.L. Stubbs and P.E. Nelson. 1973. The biology of fungi associated with root rot of subterranean clover in Victoria. Proc. Roy. Soc. Vict. 86: 19–28.

5. Burgess, L.W., A.H. Wearing, and T.A. Toussoun. 1975. Surveys of Fusaria associated with crown rot of wheat in eastern Australia. Australian J. Agric. Res. 26: 791–799.
6. Butler, F.C. 1961. Root and foot rot diseases of wheat. N.S.W. Dep. Agric. Sci. Bull. No. 77. 98 p.
7. Chambers, S.C. 1972. *Fusarium* species associated with wheat in Victoria. Australian J. Exp. Agric. Anim. Husb. 12: 433–436.
8. Cook, R.J. 1968. Fusarium root and foot rot of cereals in the Pacific Northwest. Phytopathology 58: 127–131.
9. Davidson, B.R. 1975. Development of Australian Agriculture, 1. Developments up to 1914. Agric. Environ. 2: 251–281.
10. Edwards, E. T. 1933. A new Fusarium disease of maize. N.S.W. Dep. Agric. Gazette 44: 895.
11. Edwards, E.T. 1935. Studies on *Gibberella fujikuroi* (Saw.) Wr var. *subglutinans* N. comb. The hitherto undescribed ascigerous stage of *Fusarium moniliforme* var. *subglutinans* Wr and Rg and on its pathogenicity on maize, *Zea mays* L. In New South Wales. M.Sc. Agr. Thesis. Univ. Sydney. 112 p.
12. Edwards, E.T. 1936. Maize seed selection and disease control. The problem of internal seed-borne infection. N.S.W. Dep. Agric. Gazette 47: 303–305.
13. Francis, R.G. 1976. Characteristics of two populations of *Fusarium roseum* 'Graminearum' in eastern Australia. Ph.D. Thesis, Univ. Sydney. 124 p.
14. Francis, R.G., and L.W. Burgess. 1975. Surveys of Fusaria and other fungi associated with stalk rot of maize in eastern Australia. Australian J. Agric. Res. 26: 801–807.
15. Francis, R.G., and L.W. Burgess. 1977. Characteristics of two populations of *Fusarium roseum* 'Graminearum' in eastern Australia. Trans. Brit. Mycol. Soc. 68: 421–427.
16. Geach, W.L. 1932. Foot and root rots of wheat in Australia. *Fusarium culmorum* (W.G. Sm.) Sacc. as a causal organism. J. Council Sci. Industrial Res. Australia 5: 123–128.
17. Hynes, H.J. 1935. Studies on Helminthosporium root-rot of wheat and other cereals. Part 1. Economic importance, symptoms and causal organisms. Part 2. Physiological specialisation in *Helminthosporium* spp. N.S.W. Dep. Agric. Sci. Bull. 47. 39 p.
18. Kellock, A.W., and D. McGee. 1972. A fungus that rots the roots of subterranean clover. J. Dep. Agric., Vict. 70: 112–113.
19. Kollmorgen, J.F. 1974. The pathogenicity of *Fusarium avenaceum* to wheat and legumes and its association with crop rotations. Australian J. Exp. Agric. Anim. Husb. 14: 572–576.
20. McGarity, J.W. 1975. Soils of the Australian wheat-growing areas, p. 227–255. *In* A. Lazenby and E.M. Matheson, (ed.), Australian field crops, Vol. 1: Wheat and other temperate cereals, Angus and Robertson, Sydney.
21. McKnight, T., and J. Hart. 1966. Some field observations on crown rot disease of wheat caused by *Fusarium graminearum*. Qld. J. Agric. Anim. Sci. 23: 373–378.
22. McWhirter, K.S. 1972. An appraisal of present and future production and breeding problems of maize, sorghum and millet. Paper 5–10, Australian Specialist Conference on Crops of Potential Economic Importance. C.S.I.R.O., Sydney, N.S.W.
23. Magee, C.J. 1957. Foot rot and "scab" of wheat. Commonwealth Phytopathol. News 3: 26.
24. Mayers, P.E. 1976. The first recordings of milo disease and *Periconia circinata* on sorghums in Australia. Australian Plant Pathol. Soc. Newsletter 5: 59–60.
25. Millikan, C.R. 1942. Studies on soil conditions in relation to root-rot of cereals. Proc. Roy. Soc. Vict. (N.S.) 54: 145–195.
26. Pont, W. 1963. Maize diseases are common in north Queensland. Qld. Agric. J. 89:357–365.
27. Price, R.D. 1970. Stunted patches and deadheads in Victorian cereal crops. Dep. Agric. Vict. Tech. Publ. 23. 165 p.
28. Purss, G.S. 1966. Studies of varietal resistance to crown rot of wheat caused by *Fusarium graminearum* Schw. Qld. J. Agric. Anim. Sci. 23: 475–498.
29. Purss. G.S. 1969. The relationship between strains of *Fusarium graminearum* Schwabe causing crown rot of various gramineous hosts and stalk rot of maize in Queensland. Australian J. Agric. Res. 20: 257–264.
30. Purss, G.S. 1971. Effect of planting time on the incidence of crown rot (*Gibberella zeae*) in wheat. Australian J. Exp. Agric. Anim. Husb. 11: 85–89.
31. Purss, G.S. 1971. Pathogenic specialization in *Fusarium graminearum*. Australian J. Agric. Res. 22: 553–561.
32. Purss, G.S. 1971. Reaction of a collection of *Triticum* species and Triticales to crown rot (*Gibberella zeae*). Qld. J. Agric. Anim. Sci. 28: 131–135.
33. Leeper, G.W., ed. 1970. The Australian environment, (4th ed.). C.S.I.R.O., Australia with Melbourne Univ. Press. 163 p.
34. Toussoun, T.A., and P.E. Nelson. 1976. A pictorial guide to the identification of *Fusarium* species. (2nd ed.) Pa. State Univ. Press, University Park, Pennsylvania.
35. Wearing, A.H. 1976. Studies on the saprophytic behaviour of *Fusarium roseum* 'Graminearum'. Ph.D. Thesis. Univ. Sydney. 145 p.
36. Wearing, A.H., and L.W. Burgess. 1977. Distribution of *Fusarium roseum* 'Graminearum' Group 1 and its mode of survival in eastern Australian wheat belt soils. Trans. Brit. Mycol. Soc. 69:429–442.
37. Wildermuth, G.B. 1972. Studies on crown rot (*Gibberella zeae*) of wheat in Queensland. M.Sc. Thesis, Univ. Queensland. 204 p.
38. Wildermuth, G.B., and G.S. Purss. 1971. Further sources of field resistance to crown rot (*Gibberella zeae*) of cereals in Queensland. Australian J. Exp. Agric. Anim. Husb. 11: 455–459.

8 Fusarium Diseases of Wheat and Corn in Eastern Europe and the Soviet Union

A. Maric

Wheat is the principal staple food in Eastern Europe and the Soviet Union. In 1974 wheat was grown on 59,900,000 ha in the U.S.S.R., 2,400,000 ha in Romania, 1,970,000 ha in Poland, 1,850,000 ha in Yugoslavia, 1,270,000 ha in Czechoslovakia and 950,000 ha in Bulgaria. This amounts to a total of 70,000,000 ha or 31% of the total land area devoted to wheat production in the world. This crop is grown under a wide range of soils and climate, from Central Europe to the far borders of Eastern Siberia. The largest areas are in the U.S.S.R., where 70% of the crop is spring wheat. Winter wheat is predominant in the western regions of the U.S.S.R. and other countries of Eastern Europe. The average yield of wheat per ha varies widely in individual countries and years and ranges from 1540 kg/ha in the U.S.S.R. to 3860 kg/ha in Czechoslovakia.

Corn, in relation to wheat, is grown on a smaller land area, amounting to 13,090,000 ha. This crop is concentrated mainly in the southern regions of the Soviet Union with 5,200,000 ha, in Romania with 3,210,000 ha, in Yugoslavia with 2,390,000 ha, in Hungary with 1,480,000 ha and in Bulgaria with 650,000 ha.

Diseases are an important limiting factor in the production of wheat and corn in Eastern Europe and the Soviet Union. Based on frequency and severity, during the last 10 years, Fusarium diseases have become the most important diseases of corn and wheat in the U.S.S.R. and other countries of Eastern Europe. In the last decade more papers were published on these diseases than in the entire period prior to 1967. The increase of the incidence of Fusarium diseases can be attributed to intensive growing practices (better soil tillage; higher rates of nutrients, particularly nitrogen; increased density of the plant population; etc.) and the introduction of high-yielding genotypes of wheat and corn that are more susceptible to these diseases. The principal aim in wheat breeding up to now has been to produce resistance to *Puccinia graminis* and *P. recondita,* and in corn breeding, to *Helminthosporium turcicum.* In this article, I would like to present the most important results of research on Fusarium diseases of wheat and corn in Eastern Europe and the Soviet Union.

Fusarium Diseases of Wheat

Distribution and damage The most common Fusarium disease is root rot. This is a complex disease that ranges from seedling blight to premature blight at the flowering and grain-ripening stages of plants. A number of papers on this disease have been published in the last few years, particularly in the U.S.S.R. where the emphasis has been on common root rot caused by *Helminthosporium sativum* and *Fusarium* sp. In the U.S.S.R. (50, 51) (Fig. 8-1), wheat in the European part of the Soviet Union, as well as in the regions between the Black and Caspian seas, is attacked mainly by *Fusarium* sp., while mixed infections of *Fusarium* sp. and *H. sativum* occur in some parts of the Ukraine, the Krasnodar region and in the major parts of Siberia. The damage caused by common root rot has been very important in recent years, particularly under conditions unfavourable for wheat growing, mainly because of the deficiency or uneven soil moisture, especially in the first part of the growing season. High temperatures at the beginning of vegetative growth cause drying of the superficial soil layer which reduces the formation of secondary roots and the water supply of plants and they become weak and more susceptible to parasites. Korshunova et al. (51) point out that common root rot on newly cultivated wastelands in Kazakhstan often reduces the yields 36–75%. In the Altai region the losses are at least 15%, and in the Krasnodar region a severe occurrence of this disease was recorded in 1961–62 in about 25% of wheat areas, the losses being 20–40%. Some workers (109) report that root rot causes more damage than all other diseases combined together.

Seedling blight and snow mold are generally less important than other types of damage caused by *Fusarium* on wheat, but in some regions these diseases cause considerable damage. In the Irkutsk region 5–7% of the seedlings are blighted every year (109). In Uzbekistan 25–30% of the young plants are blighted in some years (20). Snow mold on winter wheat occurs sporadically in other countries of Eastern Europe also, but causes slight damage, and this only in some localities.

The most widespread diseases causing the highest losses are root and stem rot which cause premature dying of plants. Latent infection also occurs without obvious symptoms on plants, but infection in combination with unfavourable conditions for wheat development causes the reduction of yield and makes the quality of the grain worse. There is no precise definition of the condition of latent infection. This is a particular relationship between the pathogen and the host when the infected plant does not die but the yield is decreased because of smaller heads and the lower number and weight of kernels. Numerous data on the distribution and damage of common root rot in different regions of the Soviet Union exist, but it is difficult to make a distinction between root rot caused by *Fusarium* and that caused by *H. sativum*. In Belorussia Fusarium root rot is widespread both on winter wheat and spring wheat. Years and localities in which 80–90% of the plants are affected by these diseases are common (76), and in the individual years the losses amount to about 15% (79). Severe damage caused by *Fusarium* sp. on winter wheat in the Ukraine does not exceed 1–1.5% but the loss caused by the latent infection is considerably higher (109). In the period between 1966 and 1968 losses caused by these diseases in the Harkov region were estimated to be 3.7–10.9% (94). On the forest-steppe land of the Ukraine, latent infection by *Fusarium* is most widespread (74), while in the southeast part of this republic the average infection from these diseases varies from 50 to 60% (83). In the central steppe of the Ukraine the mixed infections of Fusarium and Helminthosporium root rot are most widespread, causing yield reductions up to 41%. Fusarium root rot of wheat is the most important disease in the Primorska region, where the number of diseased plants varies from 8 to 39% (2). In the most important wheat-growing areas in the U.S.S.R., such as west Siberia, and the Zaural, Altai and Krasnojarsk regions where spring crops prevail, mixed infections of *H. sativum* and *Fusarium* sp. are most common, with the losses ranging from 3.5 to 7%, and in the individual years as high as 14 to 15% (50). In 1961, 1962, 1964, and 1967 these diseases caused yield reductions up to 50% in many collective farms of the Altai region (96). Common root rot (*H. sativum* and *Fusarium*) appears on spring wheat in the Irkutsk region on about 27% of the plants and in the individual fields even as high as 67% (93).

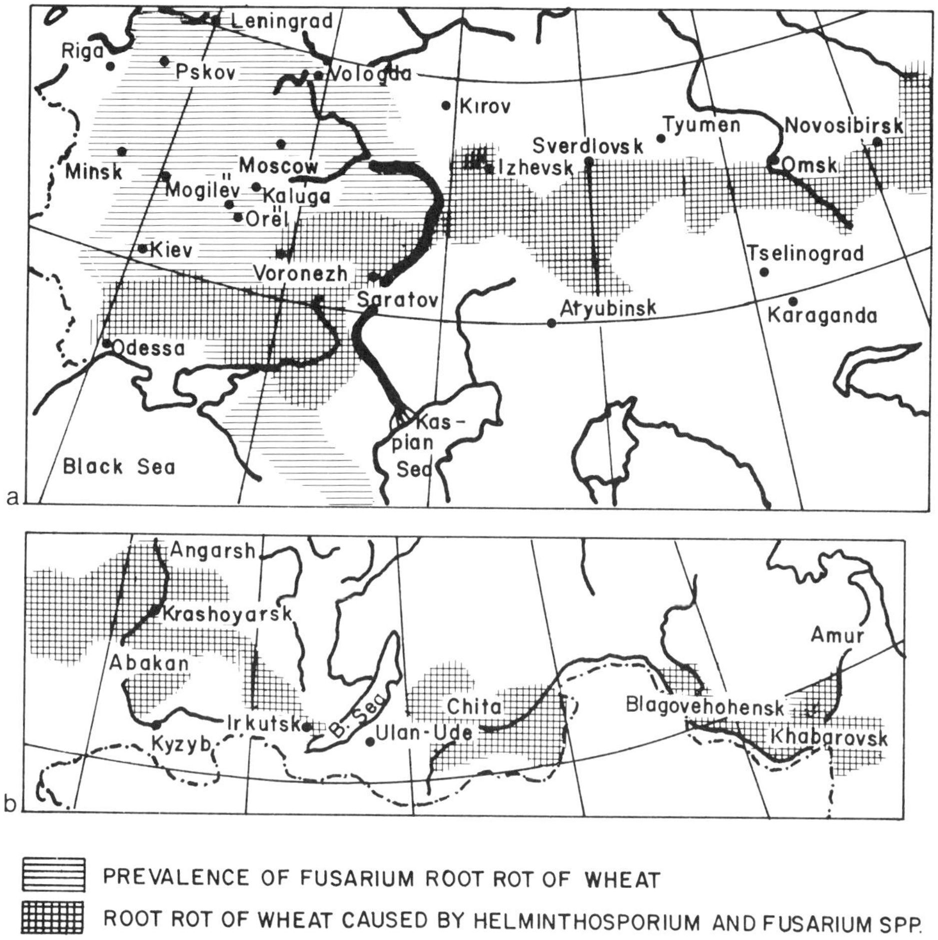

Fig. 8-1 The distribution of root and foot rot of wheat in the Soviet Union. a) European Soviet Union and the Western Siberian Plain. b) Eastern Siberia.

Vetrov et al. (109) report that *Fusarium* sp. attack the primary and secondary roots of spring wheat in Siberia in 80% of the cases. They also report that *H. sativum* invades the base of the stem in 61% of the cases and *Fusarium* in 36% of the cases. As the plant grows, the incidence of infection by *H. sativum* increases, so that in the milk stage, this fungus is represented in 85% and *Fusarium* sp. only 5–6% of the infections. In the central and south parts of the Ural, the below-ground part of the seedling is infected mainly by *H. sativum* (73–98%) and rarely by *Fusarium*. Later, the tissue is invaded by *Fusarium* sp. so that *Fusarium* sp. are represented in 70–80% of the infections at the end of growing season.

Fusarium root rot accompanied by shrivelling of kernels in heads has occurred on winter wheat in Yugoslavia during recent years. A severe outbreak of this disease was observed in 1969 and again in 1970 when up to 70% of the plants were infected in many fields (37). Although this disease has not been studied sufficiently, it is probably caused by *Fusarium,* because in addition to other fungi, *F. graminearum* has been isolated often from the stems of diseased plants.

Fusarium head blight occurs in the Soviet Union and in most countries of east Europe and is frequently severe in west Siberia and the Primorska region where infection varies from 0.5 to 22.5% (52). In the individual regions of the Krasnodar district Fusarium head blight occurs on as many as 50% of the plants (94, 95). A disease often mentioned in the

Soviet literature is "black germ," the etiology of which includes an involvement of *Fusarium* sp. (72). Fusarium head blight caused by *F. graminearum* is common on winter wheat in Yugoslavia and considerable damage was observed on some introduced cultivars (36). In the last 18 years of observations (106) severe outbreaks of Fusarium head blight were registered in Bulgaria during 1957, 1961, 1964, 1970, and 1975. Fusarium head blight also is one of the most important diseases in Romania; in epidemic years this disease causes losses of about 40% in the individual regions of the country and up to 70% in some fields (73). Fusarium diseases of wheat and particularly Fusarium head blight in recent years have caused considerable loss on wheat in Hungary; losses in 1970 were particularly heavy when on many fields the yield was reduced 40 to 50% (53). The fact is, however, that Fusarium diseases of wheat in these countries have been studied less than in the U.S.S.R., and the least attention has been paid to Fusarium root and stem rot of wheat.

Fusarium nivale is described mainly as the cause of snow mold in Eastern Europe and the Soviet Union (76, 107, 108). Seedling blight is more widespread, however, particularly on spring wheat, and is caused by other *Fusarium* sp. This type of disease in non-chernozem regions of the Soviet Union is caused chiefly by *F. graminearum, F. culmorum, F. avenaceum,* and *F. gibbosum* (51). In Uzbekistan *F. culmorum, F. gibbosum,* and *F. moniliforme* var. *subglutinans* are recovered most often from wheat beginning with seedling emergence and continuing to the tillering stage (20). Seedling blight of spring wheat in the central and south Ural is caused chiefly by *H. sativum* and rarely by *Fusarium* sp. Similar results are presented for some regions of the Ukraine (85). In the first stages of wheat development in east Siberia the most prevalent fungi are *Fusarium* sp., among which the most frequently isolated are: *F. avenaceum, F. oxysporum, F. sambucinum,* and *F. solani* (109). Wheat blight before emergence in the Saratovska and Volgogradska regions is caused mainly by *F. culmorum* and *F. sporotrichiella,* and rarely by *F. oxysporum* and *F. gibbosum* (111).

Fusarium sp. are often saprophytes in wheat or they can occur as symbionts forming mycorrhiza on the roots of plants. These fungi become parasites under conditions unfavorable for wheat development (51). In the recent Soviet literature it is reported that a great number of *Fusarium* species present in tissues are associated with common root rot of wheat, but the composition of this fungus flora differs from region to region. For instance, in the central belt of the U.S.S.R. it is considered that Fusarium root rot is caused by the following species: *F. culmorum, F. sambucinum, F. oxysporum, F. gibbosum, F. sporotrichiella* and *F. graminearum* (51); in Moldavia, *F. heterosporum, F. graminearum, F. avenaceum* var. *herbarum, F. javanicum* var. *radicicola, F. gibbosum* var. *bulatum, F. moniliforme* var. *lactis* and *F. solani* (92); and in the southeastern Ukraine, *F. sambucinum, F. gibbosum, F. solani, F. gibbosum* var. *acuminatum, F. javanicum, F. javanicum* var. *radicicola, F. sambucinum* var. *trichothecioides, F. sambucinum* var. *minus, F. oxysporum* var. *orthoceras* and *F. semitectum* (85). However, there are few data on the pathogenicity of these fungi. In some trials the highest virulence on wheat has been expressed by *F. culmorum* (20) and in others by *F. oxysporum* and *F. gibbosum* (111). Zakharova (111) points out that *F. culmorum* and *F. sporotrichiella* cause severe seedling blight before emergence while *F. oxysporum* and *F. gibbosum* cause only slight necrosis of the roots. At tillering and heading the first two fungi cause premature dying of 36–42% of plants and the latter two only 4–8%. In some trials *F. culmorum* showed the highest degree of pathogenicity (38, 39). A 1:10 dilution of culture filtrates of different *Fusarium* sp. including *F. sporotrichiella* var. *tricinctum* (19) has been shown to reduce or retard germination of wheat seed (85). The infiltration of plants by the fungus metabolites changes the water balance in the plant and retards growth so that infected plants die prematurely (52). There are relatively few data on the composition of *Fusarium* sp. on the head and grain of wheat. Fusarium head blight in Belorussia is caused by *F. culmorum* and *F. graminearum* (76).

According to an investigation carried out in Poland, in the wheat rhizosphere the most widespread fungi were *Fusarium* sp. Among nine isolated species pathogenic to wheat, *F. culmorum* occurred most often (12).

The effect of environmental conditions on the occurrence of Fusarium diseases of wheat
Snow mold caused by *F. nivale* is widespread on the weakened crops of winter wheat when the snow is melting under high relative humidity (90–100%) and at air temperatures between 5 and 10 C. This is particularly true in years with heavy snow cover and slow melting of snow in the spring. Seedling blight at high temperatures is caused by *F. culmorum* and *F. avenaceum* (76).

There are a lot of data in the Soviet literature about the influence of ecological factors on the development of common root rot of wheat (*Fusarium* sp. and *H. sativum*). Because of the frequent occurrence of mixed infections, the analyses of the environmental conditions refer mostly to both pathogens. Therefore, it is difficult to draw proper conclusions on the effect of ecological factors on the development of Fusarium root rot of wheat alone.

The greatest influence in the development of Fusarium diseases of wheat has been attributed to the rate and distribution of rainfall and the moisture supply to plants. For example, *H. sativum* (1972) prevailed on spring wheat in the Primorska region, under conditions of deficient soil moisture and high temperatures during the first part of growing season. In a moist year (1971) *Fusarium* was more prevalent (2). It was shown experimentally and with artificial inoculation on spring wheat that a rapid decrease of soil moisture from 62 down to 25% of the maximum soil water capacity at tillering and heading caused extreme shrivelling of kernels in heads. This happened particularly when inoculation was made with *F. oxysporum* and *F. gibbosum,* when 55–70% of the plants died (111). In diseased plants there occurred thickening and darkening of xylem, epidermal and parenchyma cells, and formation of tyloses; sometimes the xylem was plugged by gum-like materials. As the soil moisture decreased the pathogenic ability of these fungi increased considerably. The infiltration of plants with fungus metabolites was given as the cause of retarded growth and premature dying of plants as well as the diseases resulting from latent infection of plants (111).

Zakharova (111) points out that low temperatures at the time of wheat emergence favor the development of the root system and inhibit the development of parasites. High temperatures (15–20 C) increase the number of diseased plants one and one-half times when inoculated with *F. gibbosum* and *F. oxysporum* and two times when inoculated with *F. culmorum* and *F. sporotrichiella.* She (111) used these observations to explain the low occurrence of Fusarium root rot in early sown fields of spring wheat and the severe infection in early sown fields of winter wheat.

Fusarium root rot, according to some authors, is more prevalent in regions with uniform distribution of precipitation or in the moist years, when the disease is particularly damaging on the seedlings of spring wheat (49, 51). In the steppe regions with insufficient and uneven amounts of precipitation (north Kazakhstan, west and east Siberia, the middle Volga) the common root rot is caused mainly by *H. sativum,* with *Fusarium* sp. of only secondary importance (3, 49). *Fusarium* sp. prevail in the relatively moist forest-steppe and mountainous regions.

On soils with an unfavorable water regime, successful growing of wheat is more dependent on timely rains that usually are rare and uneven in many regions of the Soviet Union. Plant resistance to root rot under conditions of deficient soil moisture decreases due to the low content of available phosphorus (52).

Seedling blight of winter wheat occurs rarely under conditions favorable for germination and emergence (22). Severe drought in the spring results in more widespread occurrence (30–80%) of root rot, which is especially the case on soils with insufficient available soil water. An outbreak of Fusarium root rot was observed in 1966, a year characterized by even distribution of precipitation (94). In some regions with mixed infections, common root rot is more severe in the steppe, which has lower total precipitation during the growing season (220 mm), than in the regions with a higher total (311 mm) and a more even distribution of precipitation (101). In some investigations the relationship between the lack of precipitation and the intensity of the occurrence of common root rot of spring wheat was not established. According to these authors, the lowest number of infections (27%) were in a favorable year for wheat development (1970) when there was sufficient rainfall and high temperatures. Very severe infestations occurred in 1971, a year characterized by large amounts of precipitation, particularly in June, and low temperatures. Similar occurrence of Fusarium root rot was observed with identical weather conditions in 1969. In 1972, which was a dry year, severe infections (50%) were registered in the steppe subregions.

A more severe occurrence of common root rot of wheat in the Stavropolska region, where *Fusarium* and *H. sativum* are of equal importance, is considered to be connected with wind velocity. Meleshko (65) points out that fields protected with border rows had only one-half to one-fifth the amount of disease.

It is probable that ecological conditions have the greatest influence on the prevalence of one or the other type of common root rot in the individual regions or in the individual stages of wheat development.

In the Ukraine the percentage of *Fusarium* sp. in plant tissues on winter wheat in the autumn period (tillering) was 50–60%, and in the spring it increased in the wax and ripening stages to 80–100% of the fungus flora (85). Fusarium head blight is severe in the years with frequent rainfall between flowering and maturity of wheat. Severe infection of heads caused by *Fusarium* sp. occurred at flowering of wheat with high relative humidity (over 70%) and temperatures over 18 C during three decades (52).

The effect of irrigation on Fusarium diseases of wheat According to the data of Morshchackii (71), the resistance of winter wheat to root rot increases considerably under irrigation. For example, on the irrigated areas of winter wheat in the central steppe of the Ukraine, the occurrence of common root rot was 20–40% lower compared with dryland crops. On the irrigated lands of winter wheat in the south Ukraine, the occurrence of *Fusarium* sp. was most frequent (47, 48).

Under conditions of the lower Povolozje (Volgograd region) the distribution of root rot on spring wheat and the composition of the fungus flora in the diseased plants are identical with dry and irrigation farming. However, as observed by Malikova (56) in 1969, only 21% of the plants in irrigated fields showed evidence of damage, compared to 69% of plants with severe symptoms in dry-farmed areas. In 1971, the incidence of severe symptoms was 14 and 41%, respectively, in irrigated and dry-farmed fields. In the first year of the investigations, in dry farming a severe infection was recorded on 80% of the plants and on only 4.5% of the plants on irrigated land. In the second year, these values were 42 and 11% respectively. However, this author did not find such regularity in winter wheat. Grisenko (26) in the Ukraine has also shown that proper irrigation decreases the number of infected plants significantly, and the system of soil tillage which provides the best conservation of moisture during the growing season also reduces Fusarium root rot of wheat (21, 52). Likewise in Yugoslavia, with the same composition of microflora and identical numbers of diseased plants, premature dying of plants of winter wheat is more severe and damage is greater in dry-farming areas compared with irrigated lands.

Problems of resistance to Fusarium in wheat Spring and winter wheat cultivars in this area vary in their susceptibility to Fusarium (1, 10, 37, 67, 73, 98). Within *Triticum aestivum,* cultivars of the hard (high-protein) or bread-type are most susceptible, followed by soft (low-protein) cultivars. The more Fusarium resistant species are *T. polbi, T. compactum* and *T. timopheevi* (23). There are certain differences in the degree of susceptibility within cultivars of *T. durum* and *T. vulgare* as well as certain early-maturing local and introduced cultivars. The greater susceptibility of cultivars of the hard-wheat type is explained by the fact that these cultivars form fewer secondary roots in the tillering stage compared with soft-wheat cultivars, and therefore become weak and susceptible to infection at the time of moisture deficiency (51).

In wet years the early-maturing cultivars, primarily those requiring higher temperatures for maturation, are more affected by diseases than the middle- and late-maturing cultivars. However, under drought conditions the infections are less severe on early-maturing cultivars because they make better use of the winter moisture (23). The local wheat cultivars, particularly those adapted to the soil and climate conditions, are less affected than introduced cultivars. This is shown by numerous examples of cultivars from different countries. Neofitova (76) reports that in Belorussia, wheat cultivars introduced from other regions of the Soviet Union are affected more severely by Fusarium diseases than are local cultivars. However, the local and introduced cultivars do not show any differences in the degree of susceptibility under conditions of artificial inoculation. Because of the high susceptibility to Fusarium root rot and head blight of some Soviet cultivars, they were excluded from cultivation very soon after they had been introduced into large areas in Yugoslavia.

There are many difficulties in breeding for resistance to *Fusarium* because of the wide host range of these fungi and because of their variability and ability to adapt to the different ecological conditions. In addition, sources of resistance to other pathogens responsible for root rots also must be found.

It is of special importance to develop cultivars of spring wheat resistant to common root rot (*Fusarium* and *H. sativum*) for use in large areas of the Soviet Union, particularly for those regions with variable weather. Under such conditions the cultivars must be not only resistant to fungi and their toxins, but also to drought, and they must have high vitality, the ability to form a strong root system, resistance to lodging, and high yields (52). By selecting the healthiest plants under conditions of severe infestation, cultivars two or three times more resistant than the initial material can be found according to Korshunova et al. (50).

The effect of preceding crops and application of fertilizers on the occurrence of Fusarium diseases The occurrence of Fusarium diseases of wheat in Belorussia is four or five times more severe in monoculture than in crop rotation with nonsusceptible crops (44, 76). According to some (20), Fusarium root rot of wheat becomes more severe after cereal crops and especially after fallow. On the other hand, in Voivodina (Yugoslavia), wheat yields are often better in monoculture than when wheat is rotated with other crops. This is because wheat and corn occupy about 80% of the area and in this region *F. graminearum* is the main cause of Fusarium diseases of both plants. It was found in the Soviet Union (110) that higher amounts of Fusarium inoculum form on the organic residues of other plants and therefore a severe outbreak of Fusarium diseases can be expected with regular crop rotation. For instance, it was found that *F. avenaceum, F. culmorum,* and *F. herbarum* cause root rot of clover and are highly pathogenic to wheat when it follows as the next crop. The most prevalent species on the corn residues were *F. graminearum* and *F. culmorum;* on peas, *F. avenaceum, F. oxysporum,* and *F. solani;* and on potato, *F. solani* and *F. culmorum.* With regard to fertilizer use, the Soviet literature indicates that higher rates and the proper intake of the NPK nutrients in soil reduce Fusarium root rot of wheat

(2, 14, 45, 93, 112). Special attention has been devoted to applying higher rates of phosphorus to reduce disease (52). The application of nitrogen only increases Fusarium head blight (10, 53, 73).

The effect of seed and date of sowing on the development of the disease In the Kazakhstan, west Siberia, and Primorski regions in the Soviet Union, wheat seed is regularly and severely infested by different fungi, including *Fusarium* sp. (52). In west Siberia the stage of seed formation and maturity of wheat ideal for contamination very often coincides with long rainy periods and high relative humidity. In Eastern Europe wheat seed is very often highly contaminated by *Fusarium* sp. This happens particularly in the years when epidemics of Fusarium head blight occur (37, 46, 53). Infected seed is generally shriveled, germinates more slowly or not at all, and produces seedlings of low vigor. Such seed is used often for sowing and thus special attention has been devoted to methods for seed disinfestation.

In recent years, numerous investigations have been carried out concerning the various protective and systemic fungicides as well as the biopreparates for seed dressings. Testing also has been carried out under conditions of mixed infections. Chulkina (13, 15) reported that by use of seed treatments the rate of germination increased by 10% in the field, the development of primary and secondary roots was improved, root rot decreased to one-half or one-third that of untreated seed, and yields increased about 10%. Roshchina (93) reports that seed treatment with quinolate (Cu-8-hydroxyquinolate) or captan increased the germinability by 10% and reduced root infection 13–22%. In some trials, when the seed was treated by thiram and the biopreparate trihodermin, root rot was decreased 60–70% and the yield increased (97). Seed treatment with a combination of thiram and oxytins or benomyl helped in protecting wheat against Fusarium root rot (90). By applying biopreparate to the soil at sowing time, both the rate of germination and seedling vigor of winter wheat increased and root rot decreased (43). Seed treatment by the antibiotics polymicin and trichotecin reduced root rot of spring wheat to half or even less that of untreated plants and increased yield up to 30% (17). Certain investigations show that some bacteria or their metabolitic products have a stimulatory effect on the growth of wheat and increase plant resistance to root rot (87).

Date of sowing is one of the most important factors affecting the severity of the occurrence of Fusarium diseases of wheat. However, there are contradictory reports on the effect of sowing date on Fusarium diseases of wheat, probably because of the different climate and soil conditions under which wheat is grown, whether it is spring or winter wheat, etc. It is necessary in every region to find the optimum date of sowing so that wheat can germinate, emerge, and develop under the most favorable conditions of soil moisture, air humidity, and temperature, thereby increasing resistance to Fusarium diseases. In the Irkutsk region, late crops of spring wheat are more severely infected by Fusarium head blight because grain formation and maturity coincide with the humid periods (93). In contrast, early crops of winter wheat are damaged the most by this disease. Grisenko (26) recorded head infections of 18 to 22% on wheat sown on the optimal date and only 5 to 7% in the later crops. Kukedi (53) reported 35% infected heads in Hungary on the crops sown in early September, and 10% on crops sown in the middle of October.

Fusarium Diseases of Corn

In most of the regions of East-European countries and the Soviet Union, *Fusarium* is the most important pathogen of corn. Fusarium diseases have been studied less on corn than on wheat, particularly in the U.S.S.R. In this part of the world *Fusarium* sp. contribute to seed rot and seedling blight, but are even more important as the cause of stem rot and ear rot. In some regions of Yugoslavia, Romania and Bulgaria a disease designated as redness of corn occurs; the etiology of this disease involves *Fusarium* sp.

Seedling diseases occur regularly and cause great losses in some years. *Fusarium* sp. are frequently mentioned as the cause of seed rotting after sowing and seedling diseases, but there are no precise data on their importance and damage. Only in the western regions of Yugoslavia are *Fusarium* sp. mentioned as the prevalent cause of corn seedling blight (69).

The composition of the fungus flora in corn seed has been studied extensively. In recent years *Fusarium* sp. have been found more frequently on seed because the infection of ears during the growing season has become more frequent (30, 31, 36, 59, 61, 63, 75, 82). This is due to the high susceptibility of single crosses which, in the last 10 years, have been grown on large areas in these countries. Seed infections are especially early and severe in some years, and germinability and seedling vigor are reduced accordingly. Such contaminated seed is often used for sowing. These fungi are external on the seed and regularly penetrate into the seed. Within *Fusarium*, the most frequent species in corn seed are *F. graminearum, F. moniliforme* and *F. moniliforme* var. *subglutinans*.

There is general agreement that chemical disinfestation of corn seed is a useful measure for protecting seed during germination and emergence of corn. Our investigations show that the use of untreated seed with severe internal infection by *Fusarium* sp. results in poorer germination and emergence of corn in cold climate tests and under cold, wet conditions in the field (63). This internal seed infection is less damaging to the seed or seedling in soil at higher temperatures. After testing more than 40 different chemicals, including fungicides and insectofungicides, there is little doubt that seed disinfestation improves its germinability, seedling vigor and the initial development of the corn seedling. Fungicides (thiram, mercury, captan, maneb, quinolate) have proved particularly effective, while the application of lindane, aldrin, or dieldrin, alone or in combination with fungicides, retards the growth of seedlings. These insecticides reduce root development and weaken the seedlings under conditions unfavorable for corn development. The negative effect of insecticides was more pronounced in single crosses than in double cross hybrids. The minimum protection of seed and seedling was achieved by systemic fungicides (benomyl, carboxin, oxycarboxin).

Fusarium stalk rot Undoubtedly the most harmful disease of corn in this part of the world, Fusarium stalk rot's most common manifestations are: (i) weak infections characterized by root and stem necrosis which hastens the maturity of plants; (ii) early wilting of plants in the milk and wax stages; and (iii) lodging, which is the most destructive form of corn stalk rot and appears every year in varying degrees, especially on susceptible hybrids.

Fusarium stalk rot of corn is widespread in all regions of the Soviet Union (24). The number of diseased plants on the flint cultivars averages about 18%, whereas on sweet corn infections average up to 35% of the plants. Losses caused by this disease are generally estimated at about 10–12%. Fusarium wilt of corn often occurs in Romania (34). In the southern and southeastern parts of this country, in 1958 and 1959, more than 40% of plants died prematurely. During a 2-year investigation of the effect of this disease on different single crosses in Yugoslavia, losses in plants not lodged varied from 12 to 20%, but in lodged plants the loss was between 17 and 35% (64). The yield reduction of a moderately resistant hybrid during a 4-year period varied between 2 and 20 q/ha (2–19%). Fusarium stalk rot of corn also causes considerable damage in Hungary and Bulgaria (77, 91, 103).

The most important causes of corn stalk rot in this part of the world are *F. gramine-*

arum, F. moniliforme, and *F. moniliforme* var. *subglutinans* (24, 33, 35, 58, 64, 84, 99, 100). According to our studies these fungi invade the corn stalk mainly after flowering. During the milk and wax stages *F. moniliforme* is most prevalent, and by the end of the growing season, *F. graminearum* begins to increase (64); thus, depending on the date of sampling, the microflora recovered from a corn stem may vary considerably.

Temperature is probably one of the important factors affecting the time fungi invade the stem. This is shown by investigations conducted in vitro, where *F. moniliforme* developed best at 30 C and *F. graminearum* at 25 C. The same picture emerged from pathogenicity tests with these fungi on corn at the milk stage and under controlled environmental conditions (64). *Fusarium moniliforme* invades stem tissues during the summer, in the warmest part of the growing season, but when the temperature decreases in the autumn, changes occur which favor the presence of *F. graminearum.* In root exudates there is a lot of food for fungi and therefore, according to Grisenko (25), *Fusarium* sp. do not cause rot in the first part of the growing season of corn. The root exudates are produced more intensively when the plants are young and vital. Besides, this author considers that during the growing season changes occur in the pathogenic ability of fungus. We found great variability in the pathogenic ability of different isolates of *F. graminearum.* This fungus belongs to the group of the most pathogenic parasites that cause stalk rot of corn.

Infested corn residues and other susceptible plant remains are probably the most important sources of infection in the epidemiology of corn stalk rot. Perfect states of these fungi are rarely formed. During our many years of investigation we have found no correlation between seed infection and frequency of stalk rot. In this part of the world, wheat is very frequently grown in rotation with corn, which may explain the heavy occurrence of Fusarium diseases on corn, and vice versa.

According to investigations conducted in the Soviet Union (24), severe stalk rot occurred in years with low rainfall in July and August (1963, 47 mm; 1966, 87 mm). There was considerably less disease in the more humid summers (1964, 136 mm; 1965, 137 mm). We obtained similar results in a 4-year experiment in Yugoslavia (64). This disease was most severe in 1972, when soil moisture was deficient in the first part of the growing season, and also in 1973, when drought extended throughout the whole growing season. The disease was worst when drought was accompanied by high temperatures. Very slight infestations were observed in 1974 and 1975, when rains were abundant and favorably distributed over the growing season. It is obvious a soil moisture deficiency at the flowering and milk stages leads to a decrease in plant vigor and consequently a decrease in resistance to the stalk rot fungi. Irrigation in the drought years considerably decreased the severity of this disease (27).

Our trials over many years have shown that different agrotechnical practices have some effect on the development of stalk rot. Crops sown earlier are more susceptible to stalk rot than those sown later, chiefly because the early-sown crop is more exposed to water stress and high temperatures during the susceptible stage (flowering) of growth. The incidence of disease goes up with increases in plant density in drought years, due to the competition of the plants for water. Increasing the amount of nitrogen to 100 kg per ha on the degraded chernozem results in remarkable yield increases as well as increased susceptibility of corn to stem rot when compared to the nonfertilized plot. A further increase in nitrogen (110–270 kg per ha) gives no yield increase, but rather increases the number of diseased plants. Corn in monoculture for 12 years did not increase the infestation in comparison to a corn-wheat rotation. Other trials on the effect of time of sowing, fertilization, and crop rotation on the development of this disease have given similar results (18).

Apparently there are no lines or hybrids with complete resistance to stalk rot of corn (70, 89, 99, 105). There are, however, considerable differences in the degree of tolerance of breeding material to this disease.

The lack of an accurate method of artificial inoculation, in addition to other problems, has hindered the development of corn hybrids with resistance to this disease. Our investigations have shown that by applying the "tooth-pick" method for artificial inoculation of the stem, it is not possible to duplicate the plant-parasite interaction as it occurs under natural conditions. Mesterhazy (66) suggests that breeding for resistance to Fusarium stalk rot should begin with artificial inoculation and plant selection at the seedling stage. The activity of enzymes of *Fusarium* sp. in corn stems differs among lines and hybrids of corn, and some workers see this as a possible method in breeding for resistance (102, 103, 104). The presence of higher amounts of phenol and certain amino acids (alanine and histidine) in root exudates explains, in some cases, the nature of resistance of corn to the penetration of the parasite into host tissues (25). However, until a simple and efficient artificial method of inoculation has been developed, it seems more convenient and accurate to test lines and hybrids under natural conditions of infection, especially in dry regions. Variables tested should include early sowing, higher plant density, and high rates of nitrogen application.

Redness of corn This is a special disease which we discovered for the first time in the eastern regions of Yugoslavia in 1957, and its etiology is connected with *Fusarium* sp. (40, 41, 57, 58, 60, 62). This disease has since been reported in Romania and Bulgaria (32, 34, 35). It occurs temporarily in epidemic proportions in individual smaller regions of these countries, with losses averaging between 10 and 30%.

The symptoms of the disease first become apparent after flowering, usually in the milk and wax stages, and are evident as a reddish-blue color on leaves, leaf sheaths, and husks. Until then the plants appear normal and develop normally. After the green parts of plants have changed color, the plants wither and die, and chaffy ears are formed. Chaffy ears occur in varying degrees depending on disease occurrence and plant wilting.

There are several hypotheses regarding the cause of this disease. On the basis of numerous isolations and studies of histological changes we have concluded that redness of corn is caused by *F. graminearum* and *F. moniliforme* var. *subglutinans* under specific environmental conditions. Plant tissues, especially nodes, are invaded by these fungi at the time of the appearance of disease symptoms. Fungi isolated from such plants and used in inoculation trials are pathogenic on seedlings and on the roots and stems of adult plants. However, in spite of having used different methods of artificial inoculation, we have not reproduced all the symptoms of the disease. Only in cases of heavier infection of the cob after artificial inoculation with *F. graminearum* has an intense red-blue color appeared on the upper parts of plants, accompanied by subsequent drying of leaves. The xylem vessel elements are more or less plugged with fungus mycelia and gum-like material. We found by staining and radioactive phosphorus that the xylem vessels of the stem of diseased plants have lost some of the normal function of transporting water and minerals. In one experiment the infected plants absorbed 62% less water than the healthy ones during 24 hr. As the result of pathologic changes in the xylem vessels of the stem, the transportation of material from the leaf to the internodes became difficult and sugar thus accumulated in these organs and formed anthocyanins. Important changes in the water balance of infected plants sooner or later cause wilting of the corn. The diseased plants die sooner in dry periods with higher temperatures than in wet years. Some symptoms of redness of corn are very similar to certain types of stalk rot (wilting, chaffy ears), which shows the similarity of these two diseases. Lines and hybrids differ in the degree of susceptibility to redness of corn. Crops sown earlier have more of this disease and neither crop rotation, nor the application of different macro and micro elements to soil or as a spray on the plants, nor the density of plant population has an effect on the development of redness of corn.

Fusarium ear rot This disease occurs regularly in the field in this part of the world, though in the past the highest losses were recorded mainly in cribs (11, 29, 59, 61, 68, 75, 88). In the last 10 years this disease has occurred in epidemic proportions more frequently in the field, causing enormous damage, particularly in some east European countries. In 1972, severe Fusarium ear rot was found on 10–30% of plants in many localities of Bulgaria. As many as 70% of the plants were affected by this disease in certain fields. Similar losses occurred in Hungary and Romania in certain years (8, 70, 105). Heavy infestations of ear rot were recorded in Yugoslavia in 3 of the past 10 years (61; A. Maric and F. Balazh, *unpublished data*), with direct losses estimated to be about 5–10% in 1968 and about 20% in 1972. Severe Fusarium ear rot occurred in cribs during the winter of 1974/75, causing a loss of 14%. In the countries mentioned above, particularly in Hungary and Yugoslavia, the occurrence of Fusarium toxicosis in swine has been widespread.

The most important causes of Fusarium ear rot are *F. graminearum, F. moniliforme,* and *F. moniliforme* var. *subglutinans*. Besides these, the ear may be also attacked, although rarely, by *F. sporotrichiella, F. sporotrichiella* var. *poae, F. semitectum* var. *majus,* or *F. oxysporum* var. *orthoceras* (7, 8). The composition of *Fusarium* species on the ear varies with the cultivar and the ecological conditions. Thus, for instance, during a 12-year period (1955 to 1968) of growing American double cross hybrids in Yugoslavia, ear rot occurred chiefly on corn in cribs and was caused mainly by *F. moniliforme* and *F. moniliforme* var. *subglutinans* (59). However, from 1968 to 1976 Fusarium ear rot was considerably more prevalent in the field and was caused mainly by *F. graminearum* (61; A. Maric and F. Balazh, *unpublished data*). This was primarily due to the introduction of high-yielding but susceptible domestic single cross hybrids. The susceptibility of hybrids, in addition to the high moisture content of the cob and mild winters, is probably an important factor affecting the intensity of development of *F. graminearum* on the ears stored in cribs.

There are numerous factors affecting the development of Fusarium ear rot of corn under field conditions. The most important one is undoubtedly the susceptibility of hybrids. Analyzing the severity of infestation on several thousand hectares of corn in the years of epidemics, we noticed that the loss caused by Fusarium ear rot on resistant hybrids averaged 1–2% and on the susceptible ones 25–30%. There were also big differences in the degree of infection within one hybrid among the individual growing regions. The years of widespread disease were characterized by severe drought and high temperatures in the first part of growing season, particularly at the flowering stage (mid-July) of corn, and by frequent rains in the later period. On light and porous soils, crops suffered to a great extent from lack of moisture, losing vigor and thus becoming susceptible to the disease. The infections were considerably less severe on low lands and under irrigation because of less water stress during the critical periods. Fusarium ear rot did not appear in the years with abundant rains and good distribution of precipitation during the growing season (1974, 1975).

Fusarium ear rot also occurred more severely on early-sown crops (45–65%) than on late ones (1–10%). Crops sown earlier suffered more from drought at the flowering stage, and rains that came at silking favored infection. Silking of late-sown corn occurred in the moist period when plants were more resistant. It seems that the prediction of this disease can be made on the basis of plant vitality and weather conditions at the silking stage of corn.

Investigations conducted under controlled environmental conditions showed that temperature (20, 25, 30 C) had no differential influence on the development of Fusarium ear rot (*F. graminearum, F. moniliforme* var. *subglutinans*). It is quite possible, however, that temperature together with other factors, particularly soil moisture, has a direct effect on the host and thus its reaction to the parasites. In our trials we found no clear results on the influence of fertilization on the development of this disease. In most cases the unfertilized plots were most severely attacked. Poposcu (88) achieved similar results by growing corn in monoculture. In more dense populations of plants, Fusarium ear rot is more severe, especially when the corn plants suffer from a shortage of soil moisture in the critical periods.

In cribs, we have noticed that the disease develops more severely in corn with over 25% moisture during a mild winter without heavy frosts and winds. *Fusarium* sp. usually disappear from artificially dried corn grain stored in silos.

Fusarium ear rot has been considered of minor importance until now because it rarely appeared under field conditions on the local cultivars and American double cross hybrids. Today this problem is being more intensively studied, particularly in connection with breeding for resistance. It is possible to find genotypes with satisfactory resistance to this disease (29, 70, 105; A. Maric and F. Balazh, *unpublished data*), but no lines that are entirely resistant have been found. There are certain differences in the degree of susceptibility to Fusarium ear rot among the commercial and experimental hybrids, but in practice the susceptible material prevails.

Artificial inoculation by the "toothpick" method and injection of a fungal spore suspension into the ear are used in the breeding program. *Fusarium graminearum* is used almost exclusively. Corn at the beginning of the silking period is highly susceptible to this fungus, but host resistance later increases considerably. We achieved the best differentiation of genotypes for resistance by inoculating ears 10 to 15 days after the beginning of silking. By introducing inoculum on the top of ear, in most cases we obtained real differences in the reaction of genotypes to this disease. Poor results were obtained only in years when inoculation was followed by a long drought period. In such cases the introduction of inoculum into the middle part of ear was suitable, but this method was very drastic in the years with rainy summers. More work is needed to find a simple and efficient method of artificial inoculation of the ear.

We decided to use *F. graminearum* in our breeding work because this fungus is the most frequent cause of ear rot and shows a higher pathogenicity than other *Fusarium* sp. Within the same species, however, there is a great variability, especially in pathogenicity. The present results indicate that it is better to use isolates of medium pathogenicity in the breeding work, because with a highly pathogenic isolate it is not possible to differentiate among corn genotypes in their susceptibility to ear rot. With the increase of pathogenicity of the isolate, namely the intensity of ear infection, the number of perithecia formed on ear husks also increases.

Heavy appearance of Fusarium ear rot in recent years has led to extensive investigations of corn mycotoxins and mycotoxicosis in domestic animals, particularly in Hungary and Yugoslavia (4, 5, 6, 7, 9, 16, 28, 42, 54, 55, 78, 80, 81, 86, 113). In the years of epidemics it was determined by the chemical method that zearalenone was present in both corn in cribs and on artificially dried grain stored in silos. Numerous experiments were carried out with domestic animals fed moldy corn. The occurrence of mycotoxicosis was most widespread on swine. Besides the estrogenic effect, refusal to eat infested grain resulted in loss of weight of the animal and enormous losses in swine breeding. Fusarium toxicoses have become one of the most important problems in the nutrition of swine in Eastern Europe. Probably this problem can be solved only by creating corn hybrids resistant to this disease.

Fusarium graminearum is the most important producer of mycotoxins in corn. According to the investigations conducted in Hungary, other *Fusarium* sp., found rarely on the ear, also produce mycotoxins with estrogenic effects (7).

Acknowledgment

I am greatly obliged to Professor P. E. Nelson and to Dr. R. J. Cook for their critical reading of the manuscript and for their substantial editorial and linguistic aid.

Literature Cited

1. Bäkesi, P., and K. Hinfner. 1971. The occurrence of *Fusarium* species on ears of winter wheat of different varieties. Novenyvedelem 7:353–356 (In Hungarian).
2. Barbajanova, I. A. 1973. Kornevie gnili pshenici i jachmenjav uslovijah Primorskogo kraja. Bjul. n.-i. Instit. him. sel.hoz. 8:18–21.
3. Bedina, S.F. 1974. Kornevaja gnil jarovoi pshenici v Kazahstane. Kazah. auil.sharuashilik riliminin harabashisi, Vestn.s.-h.nauki Kazahstana 9:40–44.
4. Bilai, V. I. 1973. Toxicity of *Fusarium* (Section Sporotrichiella) to plants and animals. Abstr. No. 0788 in Abstracts of Papers, 2nd Int.Cong. Plant Pathology, 5–12 September, Minneapolis, Minnesota (unpaged).
5. Bilai, V. I., and V. M. Pidoplichko. 1970. Toksikoobrazujushchie mikroskopiecheskie gribi. Naukova Dumka. Kiev.
6. Birone, G. M. 1975. Kukurican karosito *Fusarium* fajok elterjedese es toxikologiai viszgalata. Kandidatusi ertekezes tezisei, Gödöllö, 21 p.
7. Birone, M., F. Gyorgy, and Gy. Feher. 1972. Study on toxic effect for white rats of *Fusarium* species damaging maize. Magyar Allatorvosok Lapja, Budapest, 11:597–604 (In Hungarian).
8. Birone, G. M., G. Feher, and F. Gyuru. 1974. A kukurican eloskodo *Fusarium* gombak toxikus hatasanak vizsgalata. Agr. Egyetem Közl. Gödöllö, 289–308.
9. Brodnik, T., and N. Klemenc. 1972. Mikroorganizmi v krmilih, mikotoksikoze pri domachih zivalih. Ljubljana.
10. Capetti, E. 1974. The behaviour of several winter wheat varieties to Fusarium ear rot. An. Inst. de cer. pen. Protectia Plantelor. 10:57–62 (In Romanian).
11. Cheremisinov, N. A. 1962. Fuzarioz semjan i pochatkov kukuruza. Bot. Zhur. 47:461–472.
12. Chrzanovski, J. 1976. Mikoflora ryzosfery pszenicy. Hod. Rosl. Akl. Nasien. 20:19–45.
13. Chulkina, V. A. 1972. Vlianie protravlianija semjan na razvitie obiknovenoj kornevoi gnili i urozaj zernovih kultur. Bjul. Sib. NII him. s.-h. 1-2:48–53.
14. Chulkina, V. A. 1973. Sostojanie i perspektivi isledovanii fuzariozno-gelmintosporioznih zabolevanii hlebnih zlakov v Sibiri. Bjuleten Sib. n-isl. inst. hem. seb. hoz. 8:1–4.
15. Chulkina, V. A. 1974. Glubina zadelki semjan i razvitie obiknovenoi kornevoi gnili hlebnih zlakov v Sibiri. Sib. vestn. s.-h.nauki. 2:40–45, 128.
16. Danko, G., and O. Toth. 1969. A *Fusarium graminearum* (Schwab.) altal eloidezett toxikozis serteseknel szarvasmarhanal es juhoknal. Debrec. Agrartud. Fiosk. Tud. Kozl. 15:3–18.
17. Fedorinchik, N. S., and S. F. Buga. 1969. Ispolzovanie antibiotikov polimicina i trihotecina dlja biologicheskoj borbi s kornevimi gniljami pshenici v Krasnojarskom krae. Bjul. Vses.n.-i.in-ta zashchita ras. 1 (13):14–19.
18. Florja, M. B. 1974. Roli agropriemov v borbe so steblevimi gniljami i puzirchatoi golovnie kukuruzi. Sb. nauch. tr. Krasnodar. NIIs.h. 7:45–52.
19. Gavrilov, A. A. 1969. Vlijanie produktov ziznedejatelnosti *Fusarium sporotrichiella* var. *tricinctum* na rost prorostkov ozimoi pshenici. Nauch. tr. Stavropolsk. s-h. in-ta. 32:136–138.
20. Goldstein, L. E., and G. K. Baigulova. 1972. Wheat root rot occuring on nonirrigated lands in Uzbekistan. Mikol. fitopatol. 6:524–528 (In Russian).
21. Gorodilova, L. M., and S. I. Shevcov. 1969. Kornevaja gnil pshenici pri razlichitih sposobah obrabotki pochvi. Vestn. s. h. nauki. 11:64–66.
22. Grigorev, M. F. 1972. On the root rot of wheat. Vest. sel. Hoz. nauk. 9:60–65 (In Russian).
23. Grigorev, M. F. 1975. K vaprosu luenki ustoichivosti ozimoi pshenici k vazbuditeljam kornevih gnilei. Nauch. trud. Vashniil. Moskva. Kolos. 238–246.
24. Grisenko, G. V. 1968. Causal organisms of corn stalk rot and peculiaries of its pathogenesis. Mikol. fitopatol. 2:315–322 (In Russian).
25. Grisenko, G. V. 1971. Stablevie i kornevie gnili i ih vzaimosvjaz s mikofloroi rizosferi kornevoi sistemi kukuruzi. Dokl. Vashnil. 11:19–20.
26. Grisenko, G. V. 1972. Bolezni pshenici i kukuruzi pri oroshenii stepnih raionov USSR. Mikol. fitopatol. 6:386–390.
27. Grisenko, G. V., and Z. S. Pavuk. 1972. Bolezni kukuruzi v uslovjah oroshenija. Zern. i kormovie kulturi na oroshaem. zemljah. Kishinev. Shtiinca, 156–159.
28. Gyula, D., and O. Toth. 1969. Toxicity caused by *Fusarium graminearum* in hog, cattle and sheep. Debreceni Agrardrud. Foisk Tud. Kozl. 15:3–18 (In Hungarian).
29. Hatmanu, M., T. Sapunaru, V. Rusanovschi, T. Petrovici, and D. Caea. 1967. Comportarea unui sortiment de hibrizi si linii de porumb fata de atacul Fuzariozei si al sfredelitorului porumbuli. Inst. central de cercet. agricole. Bucuresti 24:263–271.
30. Hulea, A., D. Cebotaru, D. Mihuta, I. Comes, and L. Gheonea. 1961. Establishing the effectiveness of a new series of fungicides to control the rotting of maize seed in soil. An. Inst. centr. cerc. Agric. 29, ser.B.: 385–395 (In Romanian).
31. Hulea, A., and M. Raianu. 1964. Efficacité des produits fongicides contre la pourriture des graines de mais, en fonction de la micoflore du sol. 8 Int. congr. soil Sci., Bucharest, Romania. 3:907–915.
32. Hulea, A. 1965. Fusarioza porumbului. Inst. central cerc. Agricol. Bucharest 17:37–42.
33. Hulea, A., M. Tircomnicu, and M. Hatmanu. 1967. Contributii la studiul putregaiului radacinilor si tulpinilor de porumb in conditiile din Romania.

An. Inst. de cercet. pentru Protectia Plant. Bucuresti. 5:65–81.

34. Hulea, A., S. Bunescu, J. Sandru, M. Tircomnicu, G. H. Piticas, and R. Schmid. 1968. Investigations on stalk root rot of maize under the environmental conditions prevailing in Romania. II. Distribution, symptoms and effect on maize yields. Ann. Res. Inst. Plant Protection 6:52–65 (In Romanian).
35. Hulea, A., C. Tusa, S. Bunescu, M. Tircomnicu, and M. Hatmanu. 1969. Aspecte privind etiologija fuzariozei spicelor de griu si secara si prevenirea ei. An. Inst. de cerc. pentru Protectia Plantelor 7:69–82.
36. Jovichevich, B. 1969. Prilog poznavanju mikoflore semena kukuruza u Vojvodini. Savremena poljoprivreda. N.Sad 5–6:483–487.
37. Jovichevich, B. 1970. The occurence of black ear and poor grain of wheat. Plant Protection, Beograd. 109:247–253 (In Serbocroatian).
38. Kachalova, Z. P., and N. D. Jacenko. 1972. O patogenosti fuzariev (*Fusarium* sp.) vozbuditelei kornevih gnilei pshenici. Dokl. Mosk. s-h. akad. im. A. Timirjazeva 182:127–132.
39. Kachalova, Z. P., and E. A. Makarenko. 1974. Kornevije gnili jarovoi pshenici v Kirovskoi oblasti. Dokl. Mosk. s. -h. akad. im. K. A. Timirjazeva 204:167–172.
40. Kastori, R., I. Vereshbaranji., and A. Maric. 1969. The effect of redness of maize on physical and chemical properties of grain. Proc. Inst. Agric. Res., N.Sad 7:137–144.
41. Kastori, R., and A. Maric. 1970. Ascendant transport of ^{32}P in corn affected by redness. Contemporary Agric., N.Sad 17:523–529.
42. Klemenc, N., and T. Brodnik. 1972. Mikroorganizmi v zhivalnih, mikotoksikoze pri domachih zhivalnih. Kmetski Inst. Slovenije, Ljubljana.
43. Kolodijchuk, V. D. 1974. Ispolzovanije Trichoderma koningi protiv vozbuditelei fuzarioznoj gnili ozimoi pshenici. Nauk. praci Ukrai. s. g. Akadem. 88:147–149.
44. Kolodijchuk, V. D. 1974. Predshesvenniki i kornevaja gnil ozimoi pshenici. Nauk praci Ukrai. s. g. Akadem. 123:205–209.
45. Kolosov, L. I. 1967. Porazhenije ozimoi pshenici kornevimi gnilami v zavisnosti ot predshestvennikov i udobrenija. Dokl. Mosk. s. h. akad. im. K. A. Timirjazeva 131:239–245.
46. Konszky, D. E., and L. Pasti. 1971. Data to the infestation of cereal seeds by *Fusarium*. Novenyvedelem 7:393–397 (In Hungarian).
47. Kornienko, V.Ju. 1970. Mikoflora kornevoi gnili ozimoi pshenici v uslovjah oroshenija na juge USSR. Ukrain. bot. zhur. 27:783–785, 807 (In Ukrainian).
48. Kornienko, V.Ju. 1972. Vlijanije srokov i norm viseva v uslovjah oroshaemogo zemledelija juga USSR na razvitije kornevih gnilei ozimoi pshenici. Zahist. roslin. Resp. mizvid. temat nauk. zb. 15:75–77 (In Ukrainian).
49. Korshunova, A. F. 1972. Pathogenic root-infecting fungi on cereal crops. Proc. All-Union Sci. Res. Inst. Plant Protection 33:145–148 (In Russian).
50. Korshunova, A. F., R. I. Shchekochikina, and L. A. Makarova. 1965. Root rot and grain diseases in wheat. Proc. All-Union Sci. Res. Inst. Plant Protection 25:122–126 (In Russian).
51. Korshunova, A. F., A. E. Chumakova, and R. I. Shchekochikina. 1966. Zashchita pshenici ot kornevih gnilei. Leningrad, Kolos, 94 p.
52. Korshunova, A. F., S. M. Tupnevic, G. A. Kraeva, and L. M. Gorodilova, 1974. Kornevie gnili jarovoi pshenici. Kolos, Leningrad, 63 p.
53. Kükedi, E. 1972. Ujabb adatok az öszibuza 1970–71. evi fuzariumos fertözött segehez. Novenyvedelem 8:289–294.
54. Loncharevich, A., D. Milich., M. Antonijevic, and T. Dujin. 1970. Poremechaji u svinja izazvani smeshama sa kukuruzom zharazhenim nekim gljivama i plesnima. Dok. za teh. i tehniku u poljop. 2:1–3.
55. Loncharevich, A., V. Penchich., H. Smiljakovich, and S. Gotovchevich. 1972. Mikotoksikoza svinja prouzrokovana gljivicama roda Fusarium. Dok. za teh. i tehniku u poljop. 9–10:5–9.
56. Malikova, A. V. 1974. Fungal diseases of wheat crops in irrigated lands. Mikol. Fitopatol. 8:63–64 (In Russian).
57. Maric, A., and Z. Kosovac. 1959. Proucavanje uzroka i shtetnosti crvenila kukuruza u Vojvodini. Savr. polj. 7:1028–1043.
58. Maric, A., Z. Kosovac, and Z. Jovanov. 1962. Further investigation of redness and root and stem rot of corn in Voivodina. Ann.Sci.work Agric. Facul. N.Sad 6:1–18 (In Serbocroatian).
59. Maric, A., and M. Panic. 1963. Fusarium ear rot of maize in Yugoslavia. Contemporary Agric. N.Sad 9:628–636 (In Serbocroatian).
60. Maric, A., and B. Mojsovich. 1964. The redness of corn. Histological changes in the stem of diseased plants. Ann. Sci. work Agric. Faculty N.Sad 8:1–7 (In Serbocroatian).
61. Maric, A., Zh. Markovic, and P. Drezgic. 1969. Epiphytotic appearance of corn cob blight in the course of 1968 and the influence of some agrotechnical measures on the intensity of infestation. Plant Protection, Beograd 20:15–28 (In Serbocroatian).
62. Maric, A., and R. Kastori. 1969. Conductivity of the vascular bundles and water regime in plants diseased with corn redness. Contemporary Agric. N.Sad. 16:259–266 (In Serbocroatian).
63. Maric, A., and Z. Klokochar. 1970. The value of current pesticides in control of corn seed borne infection (*F. graminearum, N. oryzae*) in cold test and in field conditions. Plant Protection, Beograd 21: 269–283 (In Serbocroatian).
64. Maric, A., F. Balazh, and Zh. Markovich. 1976. The effect of temperature and rainfall on the development and harmful effect of stem rot of

maize (*F. moniliforme, F. graminearum*). Agric. Consp. Sci. 39:213–221 (In Serbocroatian).

65. Meleshko, A.P. 1971. Kornevaja gnil ozimoi pshenici na poljah, zashchishchennih lesnimi polosami v krajne zasushlivoi zone Stavropolskogo kraja. Nauch. tr. Stavrop. s. h. in-ta. 34:234–238.
66. Mesterhazy, A. 1974. A kukurica fiatalkori fogekonysaga Fusarium fajokal szemben, kulonos tekintettel a *F. graminearum* Schl. Novenytermeles 23: 273–281.
67. Mesterhazy, A. 1974. Fusarium diseases of wheat and triticale in South-east Hungary. Cereal Res. Comm. 2:167–173.
68. Milatovich, I. 1967. Ear rot of maize in Croatia. Poljoprivredna znanstvena smotra 24:3–9 (In Serbocroatian).
69. Milatovich, I. 1969. Root rot and stalk rot of corn in Croatia. Zbornik radova Savetovanja o novijim dostign. u zashtiti bilja. Zagreb 181–186 (In Serbocroatian).
70. Mong, N., and A. Hulea. 1974. Assessment of the resistance to Fusarium infection of several corn lines and single crosses. Ann. Res. Inst. Plant Protection. 10:97–110 (In Romanian).
71. Morshchackij, A.A. 1971. Ozdorovlenie ozimoi pshenici ot kornevih gnilei pri oroshenii. Zroshchuvane zamnerobstvo. Resp. mizvid. temat. nauk, zb. 11:28–31 (In Ukrainian).
72. Morshchackij, A.A. 1972. O razvitiji chernogo zarodisha ozimoi pshenici v uslovjah Centralnoj Stepi, Zahist. roslin. Rsep. mizvid. temat. nauki. zb. 16:59–62 (In Ukrainian).
73. Munteanu, I., T. Muresan, and V. Tataru. 1972. Fusarium wilt in wheat and integrated disease control in Romania. Acta Agr. Acad. sci. hung. 21:17–29.
74. Musatova, L.P. 1971. Poteri urozhaja ozimoi pshenici ot kornevoi gnili v uslovjah Lepostepi Ukraini. Nauk. praci. Ukrai. s. g. Akadem. 33:116–119 (In Ukrainian).
75. Nemlienko, F.E., K.Ja.Kalashnikov, and G.V. Grisenko, 1964. Bolezni kukuruzi. Trud. Vses. nauch, isled. in-ta. zashch. rast. 22:201–213.
76. Neofitova,V.K. 1973. Fuzarioznie zabolevanija zernovih kultur v Beloruskii i obsnovanie mer borbi s nimi. Bjul. Sib. n. i. in-ta. him. sel. hoz. 8:13–16.
77. Nikolova, V., M. Markov, and P. Popov. 1965. Neprijateli i bolesti na carevicata. Zemizdat, Sofija, pp. 95.
78. Nowar, El.S. 1975. Moldy corn naturally contaminated with mycotoxins in feeding animals and poultry. Thesis, Faculty Agric. Novi Sad, Yugoslavia, 240 p.
79. Obuhovich, E.M. 1971. Rasprostranenie, tipi pojavlenija i vredonostnost kornevih gnilei pshenici v BSSR. Puti povish. urozajnosti polevih kultur. Mezved. temat.sb. 1:167–173.
80. Ozhegovich, L. 1970. Trovanje svinja plesnivim kukuruzom F-2 (zearalenone) Fuzaritoksikoza. Veterinaria 19: 525–531.
81. Ozhegovich, L., and V.Vukovich. 1972. Zearalenon (F-2) Fusarium-toxikoze der Schwine in Yugoslavien. Mykosen. 15.
82. Penchich, V., H.Smiljakovich, and M. Alimpich. 1973. Uticaj veshtachkog sushenja na mikofloru zrna kukuruza. Savremena poljoprivreda 9–10:1–11.
83. Peresipkin, V.F., and V.M. Pidoplichko. 1970. Osobennosti razvitija kornevoi gnili ozimoi pshenici v uslovijah jugo-vostoka USSR. Vesnik s.g. nauki. 5:32–36 (In Ukrainian).
84. Perisic, M. 1953. Prilog prouchavanju *Gibberella zeae* (Schw) Petch. u nas. Zbor. rad. Polj. fak., Beograd, 1:188–197.
85. Pidoplichko, V.M. 1970. Mikoflora kornevoi gnili ozimoi pshenici v uslovjah jugo-vostoka USSR. Mikrobiol. Zh.2:215–220 (In Ukrainian).
86. Pidoplichko, V.M. 1970. O toksichnosti gribov roda Fusarium-vozbuditelei gnili ozimoi pshenici. Mikrob. Zh. 32:700–704 (In Ukrainian).
87. Pidoplichko, V.M., and O.D. Garagulja. 1974. Vlijanie bakterii-antagonistov na razvitie kornevoi gnili pshenici. Mikrobiol. Zh. 36:599–602 (In Ukrainian).
88. Popescu, V.P. 1961. Povedenie *Gibberella zeae* (Schw) Petch. na gibridah i sortah kukuruzi vozdeljaemih na raznih agrofonah. Lucrari stiint Inst. agrar. Cluj.17:209–215 (In Romanian).
89. Popescu, V. 1963. Metodi ispitivanja ustoichivosti kukuruzi protiv krasnoi pleseni stabloi i pochatkov, vizivajemoi, *Gibberella zeae* (Schw) Petch. Stud. si cer. agron. Acad. RPR Fil. Cluj. 14:253–259 (In Romanian).
90. Popov, V.J., and E.M. Kumachova. 1972. Vlijanie fungicidov na vozbuditelei kornevoi gnili pshenici. Hem.v.s.h.10:33–35.
91. Popov, A. 1974. Fuzariozi po carevicata. Rastitel. zashch. 22:40–44.
92. Popushoj, S.I., M.Sh. Grinberg, G. Zh. Prostakova., D.E. Kogan, and E.F. Khripunova. 1976. Species composition of fungi associated with root rots on winter wheat crops at tillering. Mikol. Fitopatol. 10:356–368 (In Russian).
93. Roshchina, F.V. 1973. Agrotechnicheskie meri borbi s kornevimi gnilami pshenici v Irkutskoj oblasti. Bjul.Sib.n.-isl. Inst.hem.sel.hoz.8:21–23.
94. Shcekochikina, R.I., and A.F. Korshunova. 1969. Kornevie gnili i bolezni zerna pshenici. Trud. vses. nauch. isled. Inst. zashch. rast.34:155–160.
95. Shcekochikina, R.I., and A.F. Korshunova. 1969. Kornevie gnili zernovih. Trud. vses. n.-i. Inst. zasht. rasten. 28:143–148.
96. Shevchenko, F.P., and P.G. Alinovskii. 1973. Iz opita vnedrenia kompleksa mer zashchiti pshenici ot kornevih gnilei. Bjul. Sib. n.-isl. Inst. hem. sel. hoz. 8:16–18.
97. Shirokov, A.J. 1973. Bolezni semjan zernovih kultur v Omskoj oblasti. Bjul. Sib. n.-isl. Inst. hem. sel. hozj. 8:41–44.
98. Sidrova, S.F. 1970. Rol sorta v porazhenii ozimoi pshenici i rzi snezhnoj plesnju. Bjul.VNII zashchita. rast. 16:38–40.

99. Smiljakovich, H. 1972. Maize rot in Yugoslavia. Actas III Congr. Un. Phitopathol. Med. Oieras: 530–533.

100. Smiljakovich, H., M. Martinovich,and G. Grujichich. 1975. Uticaj fitopatogenih nematoda na pojavu trulezhi kukuruza. Agron. glasnik 1–4:153–158.

101. Spirikova, M.E. 1973. Vlijanie predshestvennikov na razvitie kornevih gnilei zernovih kultur. Bjul. n.-i. Inst. him. sel. hoz.8:27–28.

102. Szecsi, A. 1970. Role of enzymes in Fusarium causing stalk rot of maize. Novenyvedelem. 6:365–368 (In Hungarian).

103. Szecsi, A. 1972. *Fusarium roseum* var. *graminearum* es *Fusarium roseum* var. *culmorum* izolatumok szaharas aktivitasa talajkulturaban. Novenytermeles 21:309–312.

104. Szecsi, A. 1973. Saccharase activity of *Fusarium roseum* var. *graminearum* and *F. roseum* var. *culmorum* in soil culture. Phytopathol. Z. 78:227–231.

105. Tircomnicu, M., and O. Cosmin. 1974. Studies of the reaction of some corn hybrids to *Fusarium moniliforme* (Sheld) and *Fusarium graminearum* (Schw.) infections. Ann. Res. Inst. Plant Protection 10:119–127 (In Romanian).

106. Todorova, V. 1975. Vazmozhna li e borbata sreshchu fuzariozata po klasovete chrez obezzarazjavane na semenata. Rastitel. zashch. 9:21–24.

107. Tupenevich, S.M. 1964. Viprevanie ozimih hlebov. Trud. Vses. nauch. Inst. zasht. rast. 22:195–201.

108. Tupenevich, S.M. 1974. Kornevie gnili jarovoi pshenici. Leningrad skoe otdelenie, Leningrad, Kolos, 62 p.

109. Vetrov, Yu.F., A.F. Korshunova, A.A. Morshchatsky, M.K. Khokhryakov, and V.A. Chulkina. 1971. Root rots of cereals in the USSR. Mikol. Fitopatol. 5:148–155 (In Russian).

110. Volchkova, E.V. 1973. Importance of preceding crops on developing causal agents of wheat root rot in the north-Ossetian. ASSR. Mikol. Fitopatol. 7:541–542 (In Russian).

111. Zakharova, T.I. 1971. Role of fungi belonging to *Fusarium* genus in the formation of the hollow ears in spring wheat. Proc. All-Union Sci. Inst. Plant Protection 32:92–99 (In Russian).

112. Zhukova, R.V. 1973. Kornevie gnili jarovoi pshenici v Bashkirskoj ASSR. Bjul.VNII zashchita rast. 25:54–59.

113. Woler, L., M. Birone, and M. Koppanyi. 1971. Toxicological investigation on *Fusarium graminearum* (Schw) causing the mouldyness of the maize. Novenyvedelem 7:443–450 (In Hungarian).

9 Root-, Stalk-, and Ear-Infecting *Fusarium* Species on Corn in the USA*

Thor Kommedahl and Carol E. Windels

One or more species of *Fusarium* can be isolated from nearly every vegetative or reproductive part of corn in every state where corn is grown and in every season. They may occur singly or as part of a complex of *Fusarium* sp. or other fungi. They produce seedling blight, root rot, stalk rot, and ear rot. The damage may be negligible or extensive, depending upon environmental conditions and hybrid. Frequently *Fusarium* sp. infect corn and cause damage when the plant is under stress from biotic, climatic or edaphic causes.

The literature on stalk rot of corn is voluminous and has been reviewed by Christensen and Wilcoxson (7). Koehler has published data and summaries for stalk rot (32) and ear rot (31) of corn. Because of the thoroughness of these bulletins and monograph, the emphasis in this chapter will be on work done primarily in the past decade, or on work not covered by those references. Also, papers in which only the genus *Fusarium* is mentioned are not covered. Species named in this chapter are as given by Snyder and Hansen (62,63,64) and Snyder et al. (65).

Fusarium moniliforme

Kernel infection This species is the most frequently isolated *Fusarium* sp. in shelled corn. Seed lots with 100% infection are not uncommon. Literature from the early years illustrates wide prevalence. Hoppe (23) in 1938 reported this species to be the most important fungus on shelled corn in all states west of the Mississippi River as well as in Illinois, Indiana, Kentucky, Maryland, and Tennessee. The average infection of the 1937 crop varied from 9% in Ohio to 91% in Texas. In the next season, *F. moniliforme* predominated in samples from southern, west-central, and western states, with 75% infected kernels in Kansas and Nebraska and only 13% in the east-central states (24). Manns and Adams (44) reported infected kernels in lots from 17 states that ranged from 1 to 80%, averaging 30%. More recently, *F. moniliforme* has been reported as the predominant *Fusarium* sp. on kernels in Indiana (70), Iowa (20), Minnesota (50) and Nebraska (66).

Work in Indiana (73,77) and Iowa (20,41) indicates that high-lysine hybrids are more susceptible to kernel infection by *F. moniliforme,* but this can vary with the inbred or hybrid (50,77). Incidence of infection was reported to be higher in kernels from Texas male-sterile hybrids than in normal ones in Florida (76) but not in Minnesota (50). Differences in results may be accounted for by differences in hybrids, fungus isolates, and environments.

*Paper No. 9978, Scientific Journal Series, Minnesota Agricultural Experiment Station, St. Paul, MN 55108

Ear rots Although *F. moniliforme* occurs almost universally in seed lots of corn produced in the Corn Belt, ear rots are far less common (1–5%). However, of the ear rots, *F. moniliforme* is the most frequent cause (31). Rainfall during August, September, and October, but especially September, is the most important factor influencing ear rot prevalence (31). Rain and wind spread inoculum that is present on corn plants, crop debris, and corn in cribs (51). Also, picnic beetles can carry propagules from ears buried the previous season to ears of the new crop (82).

Ears are most susceptible to *F. monilforme* 2 days after silking and to a lessening extent for an additional 40 days according to Koehler (31). However, Warren (77) found infection was greatest 4–14 days after pollination, and least 21 or more days after pollination. It enters more ears than any other fungus, even when ears are covered by husks and free from bird or worm damage (29). Rot is usually confined to the tip, or in places where ear worms or corn borers feed, or where ears are wet. When tips of ears were covered by husks, percentage of infected kernels increased from 3 to 41 from September to mid-October, whereas exposed tips of ears increased from 19 to 72% during that same period. The fungus entered the base of the cob in only 2 of 1240 ears (29). By the ninth week after pollination, *F. moniliforme* was isolated from cob pith and kernels of the same ear (21).

The fungus appears to enter through the silks, grow over the pericarp, and then grow into the kernel at the hilar end; hyphae then grow into the cavity between the pedicel and the black layer (29,31). To a decreasingly lesser extent, hyphae were found in the embryo, floury endosperm, and horny endosperm (29,66).

Boling and Grogan (5) found that resistance to ear rot could be developed by reciprocal recurrent selection or by a simple backcross program. Some lines of corn are extremely susceptible to *F. moniliforme* ear rot, e.g., the brown midrib single cross (46).

Seedling blight Because kernels are so frequently infected with *F. moniliforme,* one might suspect that planting such kernels would lead to seedling blight. Many early investigators tested this hypothesis and the results have been controversial. When infected kernels were planted in sterile sand under adverse conditions, necrotic roots and stems and some blight occurred. Some concluded that *F. moniliforme* was a weak pathogen but others doubted its pathogenicity.

Foley (17) and Sumner (66) found that hyphae grew from infected kernels into roots, mesocotyls, and crowns of seedlings when infected kernels were sown in sterile sand in growth chambers. On the other hand, Kucharek and Kommedahl (39) planted two seed lots with a high incidence of infected kernels and two with a very low incidence in the field and found no differences in seedling blight, root rot, or infected mesocotyls, crowns, roots, nodes, internodes, or ears. Similarly, Gulya et al. (20) found no significant correlations between kernel infection by *Fusarium* and percentage germination or percentage of abnormal seedlings.

In experiments where inoculum was applied to soil, temperature has been found to affect pathogenicity. Seedlings were blighted when temperatures were below 24 C. Dry weights of seedlings were reduced by 28% for roots and 50% for shoots in field soil in the greenhouse at 22 C but not at 27 C (52). Similarly, Palmer and MacDonald (53) reported significant reductions in root and shoot weights of seedlings at 25 C.

Voorhees (75) observed that hyphae penetrate the root epidermis or ruptures made in the cortex by emerging roots. The fungus can grow into the stele in the mesocotyl region where adventitious roots grow, or grow into the plumule where shoots break through the coleoptile. Hyphae can also enter through natural openings in the cotyledonary plate and into the scutellum and endosperm of the germinating seedlings.

Toxins, perhaps "moniliformin" (8), were found to favor seedling blight by inhibiting root development (19). Pathogenicity of a given isolate depended upon its ability to produce the toxin (19).

The questionable pathogenicity of *F. moniliforme* may also be attributed to differences in resistance. For example, Djakamihardja et al. (13) evaluated 69 inbreds and 14 single crosses for resistance to seedling blight and they obtained differences in emergence, stand, and seedling vigor. Crosses among resistant inbreds were affected less by *F. moniliforme* than were crosses among susceptible ones (13). Highest resistance was obtained if the resistant hybrid served as the female parent. Additive gene action and maternal effects were more important than dominant gene action in the inheritance of resistance to seedling blight (42,43).

Root rot Although *F. moniliforme* is not regarded as a root rot fungus of corn, there are several reports of *F. moniliforme* isolated from roots of seedlings and older plants (14,17,80,81). Roots may become infected as they grow into buried, infected stalk residue (48), or from rootworm injuries (52). Leaf diseases predisposed roots to infection by *F. moniliforme* (16).

Stalk rot Porter (59) had shown early that *F. moniliforme* was present in nodal tissue of stalks from 24 states. Christensen and Wilcoxson (7) further described the distribution of stalk rot in the USA. *Fusarium moniliforme* has been considered to be the predominant stalk rot fungus in at least 10 states: Florida, Idaho, Iowa, Minnesota, Nebraska, New Jersey, North Carolina, Pennsylvania, South Carolina, and Virginia. It was second to *F. roseum* 'Graminearum' in Illinois, New York, and Ohio.

Incidence of stalks infected with *F. moniliforme* appears to increase during the season. In Iowa, Foley (17) isolated it from 75% of the plants in July and 100% by October. Infected tissue showed no disease symptoms in July or August and rotted stalks were first detected from 10 September to 13 October (18). Similarly, in Minnesota, 15% of the stalks were infected by mid-July and 54% by mid-September, 1974 (33). In September, 1976, 60–100% of the stalks were infected and the average of 76% exceeded that in 1973 (34). In 1977, 93% of the stalks were infected (38). Increases in populations of *F. moniliforme* in stalks was also noted by Young and Kucharek (84). Leonard and Thompson (40) found *F. moniliforme* in 54% of stalks in the Coastal Plain and 33% on the Piedmont, in North Carolina, showing the importance of location.

Most investigators list *F. moniliforme* without distinguishing it from 'Subglutinans,' although both have been identified on maturing and mature plants (27). Ayers et al., (3) isolated *F. moniliforme* from 64 and 88%, and Subglutinans from 8 and 32%, of the samples in two seasons. In a 4-year period in Pennsylvania, 85% of all samples yielded *F. moniliforme;* Subglutinans increased from 8 to 37% during that time (68).

Fusarium moniliforme has been isolated from nearly all vegetative parts of the corn plant (17,39). Foley (17) isolated *F. moniliforme* in decreasing frequence from leaf sheaths, nodes and internodes, and axillary buds. Roane (60) reported that this fungus accounted for 90% of the stalk breakage at the nodes. Propagules of *F. moniliforme* deposited on leaf surfaces by wind and rain (51) are washed into sheaths where they infect nodes.

Many factors affect stalk rot, and these have been summarized by Christensen and Wilcoxson (7), and by Koehler (32). They include hail, soil moisture, stalk strength, crop rotations, fertilizer, and the European corn borer. In addition, Sumner (67) reported that in Nebraska, *F. moniliforme* was more prevalent on irrigated than on dryland farms, and where field capacity was greater than 50%. In Iowa stalk rot from *F. moniliforme* was positively correlated with pith condition ratings and negatively with stalk strength (1). Nitrogen applications increased and potassium decreased pith condition ratings 10 days after silking.

Survival on stalks and in soil *Fusarium moniliforme* was more common in cornfield than in contiguous prairie soils but made up less than 2% of the population of *Fusarium* sp. in soil (36,80). The cultivar Subglutinans was rarely isolated from soil (80).

The lack of a chlamydospore as a survival structure probably accounts for its low incidence in soil. That it is present at all is due to its survival in fragments of host material. Short sections of hyphae develop in stalk fragments and the hyphal walls thicken in the parenchyma and sclerenchyma cells of the host—these thickened hyphal cells apparently function as chlamydospores (48,49). Still, *F. moniliforme* constituted less than 2% of the colonies isolated from corn residues (78). Fungus survival in corn residues was greater at 30 cm depth than at the soil surface (49).

Fusarium moniliforme has been reported to survive as perithecia in Florida (75), New Jersey (71), Ohio (71) and Minnesota (11), but not in Illinois (32). Ullstrup (71) found perithecia of *F. moniliforme* 'Subglutinans' also.

Fusarium roseum

Kernel infection *Fusarium roseum* 'Graminearum' is seldom isolated from sound corn kernels (30). Koehler (30), in a 14-year period, reported Graminearum on no more than 1% of shelled corn in Illinois. However, frequency of isolation varies with the year and region. *Fusarium roseum* 'Graminearum' was confined to the humid sections of the Corn Belt east of the Mississippi River in the 1937 crop (23). The only state that year in which this cultivar was the leading fungus in kernels was Ohio, where 20% of the kernels were infected. In contrast, Graminearum was most prevalent in states along the Atlantic seaboard in 1938 (24) and 1940 (25). In a year when *F. moniliforme* was frequently isolated in Indiana (45% infected kernels), Graminearum occurred in only 3% of the kernels from 20 counties surveyed in 1970 (70).

Ear rot *Fusarium roseum* 'Graminearum' causes ear rot mainly in the Atlantic Coast states (24,25). When it occurs elsewhere, conditions are unusually favorable for ear rot and it may appear suddenly in a field (31).

As with *F. moniliforme,* infection appears to start at the tip of the ear and seldom involves the whole ear (29,31). It was isolated from the tip, cob pith, and base of the ear, but never from kernels in the middle of the ear, by 9 weeks after pollination (21).

Although Graminearum has been the principal cultivar pathogenic to rotted ears, Acuminatum (72) and Culmorum (6) have proved pathogenic also. Graminearum isolated from corn or small grains caused ear rot upon inoculation but Culmorum from small grains did not (6).

Shelled corn containing kernels rotted by Graminearum are toxic when fed to swine and other nonruminants (31). Such animals refuse to eat grain when 10% or even fewer kernels are rotted. Similar reports from Indiana (10) indicate that three active metabolites are produced by the fungus: one causes uterine hypertrophy, one induces emetic activity, and one causes swine to refuse infected kernels as feed.

The mycotoxin zearalenone (F-2) is produced by isolates of Graminearum (6,15) and Culmorum (6) that are pathogenic to corn. Nonpathogenic isolates and cultivars (Avenaceum, Equiseti, and Gibbosum) which do not cause ear rot, do not produce zearalenone (6). Zearalenone may also enhance production of perithecia in Graminearum (83).

Seedling blight Of the *F. roseum* cultivars, 'Graminearum' is the one that usually causes seedling blight. However, pathogenicity is dependent on temperature and isolate. A soil temperature below 10 C after planting will almost invariably result in infection and, if prolonged, in blight. Seedling blight from Graminearum was found to occur from 8 to 20 C but not above 24 C (12). However, isolates were found that reduced seedling stands more at 25 C than at 15 or 20 C (9). In addition, Graminearum was reported to cause seedling blight at 18, 24 and 29 C, but damage was greater at the lower temperatures (78).

During infection, hyphae penetrate root epidermis and enter cortex and stele, and occasionally cause thickened cortical walls (28). The endodermis of semi-resistant cultivars of corn acts as a partial or even complete barrier to hyphae (57).

Cultivars other than Graminearum are seldom implicated in seedling blight. However, Culmorum caused 50% incidence of diseased seedlings and the disease severity increased with atrazine in soil (58). However, as plants age, other cultivars begin to colonize root surfaces. In a 2-year field study in Minnesota, Equiseti was found on 30–50% of roots by August, and Acuminatum in 10% or less by September, while Graminearum was isolated only infrequently (37,81). At another Minnesota location, Graminearum was the predominant cultivar on roots (78).

Stalk rot Many organisms can cause stalk rot and *F. roseum* 'Graminearum' is one of the important causes (7). It frequently causes severe stalk rot in the New England, mid-Atlantic, and northern Corn Belt states (32), and occurs from North Carolina (40) to Idaho (61). Several references point to its widespread occurrence in Pennsylvania (7) and recent reports show its prevalence at 22–25% in 1970–71 (3) to 67% of samples in 1973 (68). Ullstrup (74) reported it as unusually severe in 1965, 1967, and especially 1974.

Early reports in Illinois, summarized by Koehler (32), showed considerable variation with season in incidence of infected stalks. Hooker and White (22) reported 38% infected stalks as an average from 1948 to 1952 and 95% for 1975.

In earlier reports from Minnesota (11), stalk rot from Graminearum occurred in 80 to 98% of samples; however, more recently, Graminearum-infected stalks were not more than 20% in 1973 (35), 1976 (34) and 1977 (38). Yet, 40–100% of stalks were infected with Graminearum after harvest (31). However, in 1978, Graminearum was isolated from 36% of standing stalks before stalk rot was apparent in Minnesota (Windels and Kommedahl, *unpublished data*). Associated with Graminearum are many other fungi, some of which are much more effective than *F. roseum* in rotting mature stalks; such fungi include *Trichoderma viride, Penicillium urticae,* and *Helminthosporium pedicellatum* (2).

The spread of *F. roseum* 'Graminearum' in the first and fourth internode following inoculation was highly related to the area of dead parenchyma cells and the pith core density (54). Living cells limited fungus growth. Injury to the plant, especially root damage, changed the cell death rate in stalks, which was followed by a similar change in stalk rot response (55). As cells in the nodal tissues died, plants shifted from resistance to susceptibility to stalk rot and in the growth of hyphae from roots to stalks and from internode to internode (55). Susceptibility to basal stalk rot and nodal rot was correlated with the time of occurrence and extent of cell death in nodal tissues (56).

Many factors affect susceptibility to stalk rot and these have been reviewed by Christensen and Wilcoxson (7) and Koehler (32). Stalk rot is often a response to stress in which planting date, plant density, previous crop, fertilization, plant injury, and other diseases and pests make it possible for fungi such as *F. roseum* to invade the host. This fungus is not a strong pathogen but causes damage as host cells weaken and die. The fact that early-maturing bybrids are more susceptible than late-maturing hybrids (32) may reflect differences in rates of cell death. Resistance to stalk rot is apparently quantitative and conditioned by multiple factors (32). White (79) found that resistance to Graminearum was correlated with resistance to *Diplodia zeae* in both early-maturing (r = 0.87) and late-maturing (r = 0.81) inbred groups.

Cultivars of *F. roseum* have been isolated from infected and rotted roots and stalks and these include: Acuminatum (4,34,35,37,38), Avenaceum (3,34,38), Equiseti (34,35,37,38), Gibbosum (3) and Graminearum (37,38). These cultivars can occur in stalks while they are still green and without symptoms (34,35,37,81). However, the role of cultivars other than Graminearum in the disease complex is not clear.

The mycotoxins zearalenone, deoxynivalenol, and T-2 toxin were found in corn stalks standing in the field, reported by Mirocha et al. (45).

Survival in residues and soil *Fusarium roseum* is frequently isolated from soils (36), and occurs three times more often in cornfield than in nearby prairie soils; Equiseti and Gibbosum (=Acuminatum) were the predominant cultivars (80). Graminearum is infrequently present, probably because of its lack of chlamydospores as survival structures, except occasionally in host material (47,62).

When residues of corn were examined, *F. roseum* 'Avenaceaum' was the most prevalent cultivar, followed in order by 'Graminearum,' 'Culmorum,' and 'Equiseti' (78). As long as the residues retain their integrity, at least in part, these cultivars can survive in buried host tissue and serve as inoculum for the next planting. Graminearum also forms perithecia abundantly on crop debris.

Fusarium tricinctum

This species is not an important pathogen of corn. It was isolated from 0.5% of kernels from five locations in Minnesota in 1975 (69). Occasional fields in Minnesota have sustained severe ear infection, especially among inbreds; one such field was observed in early October, 1968. Two ears from this field were found and fed to two rats: both rats died in 1 week after each had consumed 40 g of this feed as the sole ration (C. M. Christensen, *unpublished data*).

Fusarium tricinctum was isolated occasionally from corn roots (52,81). We isolated it from stalks in 7 of 12 counties in southern Minnesota in 1976 (34), and surveys in 1974 and 1975 show up to 15% stalk infection (35, 81) and 24% in 1977 (38). Ayers et al. (3) also found it in stalks in Pennsylvania. This species produced stalk rot upon inoculation (52).

This species probably overwinters on crop debris in or on soil (about 20 propagules/g soil) in cornfields (80), making up 4% of the *Fusarium* isolates in soil and less than 2% of isolates in corn residues (78).

Fusarium oxysporum

This species is infrequently isolated from kernels and ears of corn. Of 18 seed lots produced in Minnesota, only 3% of the kernels were infected with *F. oxysporum* (69).

In New York, Edmunds et al. (14) reported *F. oxysporum* and *F. moniliforme* to be the dominant *Fusarium* species in roots. In Minnesota, *F. oxysporum* has been found in 65% (78) to 100% (81) of the roots by August. In Ohio also, it was isolated late in the season (4). In roots damaged by rootworms, *F. oxysporum* made up 95% of the *Fusarium* isolates (52). This species made up 60% of the *Fusarium* species in the rhizosphere of corn in Minnesota (78) and was found "numerically dominant" in the rhizosphere of corn in Indiana (26).

In surveys in Minnesota, *F. oxysporum* was absent in stalks in July, 1973, but was isolated with increasing frequency from August through October (35). In September of 1975 (81) and 1977 (38), it was isolated from 50% and in 1976 (34) from 23 to 66% of the stalks. It occurred also in plants with stalk rot in Pennsylvania (3).

It is not certain that *F. oxysporum* is pathogenic to corn. Seedling growth at 27 and 29 C, with *F. oxysporum* present in soil, was less than that at 18 and 24 C (52,78). Also, 12 of 19 isolates caused stalk rot upon inoculation in the field (52). The fungus survives in corn fragments over winter, as shown by its isolation from 68% of corn refuse and 55% from soil samples containing bits of organic matter (78).

Fusarium solani

This species occurs mainly on roots of corn. It appears to be part of a community of organisms in seedling roots that reaches its zenith at silking time (84). It was isolated from 50 to 80% of the roots, and constituted 20% of the *Fusarium* isolates in three different seasons in Minnesota (78,81), and made up 21% of the *Fusarium* isolates in the rhizosphere (78).

Fusarium solani occurs infrequently in stalks. Ayers et al. (3) found it in rotted stalks in Pennsylvania in 1970 and 1971. In Minnesota (1972–1977), it was isolated from 1 to 28% of the stalks (34,35,38,81). This species survives in soil (80) and in corn residues (78). Its pathogenicity to corn is questionable. Warren and Kommedahl (78) reported that *F. solani,* alone or combined with *F. oxysporum,* did not affect seedling growth at 18 or 24 C but significantly reduced growth at 29 C.

Miscellaneous *Fusarium* Species

Of the remaining species of *Fusarium,* only *F. episphaeria* has been reported on corn. Zeller (85) reported that *F. merismoides* (=*F. episphaeria*) on stalks in Oregon, caused decay in the field and in silage. *Fusarium lateritium, F. nivale,* and *F. rigidiusculum* have not been reported to occur on any parts of the corn plant.

Summary

Fusarium moniliforme is the most widely prevalent and economically important *Fusarium* sp. on corn in the USA. *Fusarium roseum* 'Graminearum' is next in importance, but *F. episphaeria, F. oxysporum, F. solani,* and *F. tricinctum* are of little importance except as parts of disease complexes. *Fusarium lateritium, F. nivale,* and *F. rigidiusculum* have not been reported on corn. *Fusarium* sp. do much greater damage to stalks than to ears and roots. *Fusarium roseum* and *F. tricinctum* produce mycotoxins in kernels and stalks, and *F. moniliforme* produces a phytotoxin.

In general, *Fusarium* sp. are probably not strong pathogens but damage corn only when it is under stress; their advance in the host is determined largely by that stress and the rate of cell death in tissues. The wide host range on cereals and grasses and their effective methods of survival in soil and residues insure an abundance of inoculum. Thus wherever corn is grown in the USA, there will always exist the potential for a *Fusarium* problem on seedlings, roots, stalks, or ears, depending upon environmental conditions current at each stage in the development of corn.

Literature Cited

1. Abney, T. S., and D. C. Foley. 1971. Influence of nutrition on stalk rot development of *Zea mays.* Phytopathology 61:1125–1129.
2. Amosu, J. O., and A. L. Hooker. 1970. Weight loss of mature corn stalk tissue induced by twelve fungi. Phytopathology 60:1790–1793.
3. Ayers, J. E., P. E. Nelson, and R. A. Krause. 1972. Fungi associated with stalk rot in Pennsylvania in 1970 and 1971. Plant Dis. Rep. 56:836–839.
4. Balakrishna, R., and A.F. Schmitthenner. 1976. Prevalence and virulence of *Pythium* and other fungi associated with root rot of corn. Proc. Amer. Phytopathol. Soc. 3:292 (Abstr.).
5. Boling, M. B., and C. O. Grogan. 1965. Gene action affecting host resistance to Fusarium ear rot of maize. Crop Sci. 5:305–307.
6. Caldwell, R. W., and J. Tuite. 1970. Zearalenone production in field corn in Indiana. Phytopathology 60:1696–1697.
7. Christensen, J. J., and R. D. Wilcoxson. 1966. Stalk rot of corn. Amer. Phytopathol. Soc. Monogr. No. 3. St. Paul. 59 p.
8. Cole, R. J., J. W. Kirksey, H. G. Cutler, B. L. Doupnik, and J. C. Peckham. 1973. Toxin from *Fusarium moniliforme:* effects on plants and animals. Science 179:1324–1326.

9. Covey, R. P. 1959. The effect of 3 isolates of *Fusarium graminearum* and herbicides on seedling blight of corn. Phytopathology 49:537 (Abstr.).
10. Curtin, T. M., and J. Tuite. 1966. Emesis and refusal of feed in swine associated with *Gibberella zeae*-infected corn. Life Sci. 5:1937–1944.
11. Devay, J. E., R. P. Covey, and P. N. Nair. 1957. Corn diseases and their importance in Minnesota in 1956. Plant Dis. Rep. 41:505–507.
12. Dickson, J. G. 1923. Influence of soil temperature and moisture on the development of the seedling blight of wheat and corn caused by *Gibberella saubinetii*. J. Agric. Res. 23:837–870.
13. Djakamihardja, S., G. E. Scott, and M. C. Futrell. 1970. Seedling reaction of inbreds and single crosses of maize to *Fusarium moniliforme*. Plant Dis. Rep. 54:307–310.
14. Edmunds, J. E., C. W. Boothroyd, and W. F. Mai. 1967. Soil fumigation with D-D for control of *Pratylenchus penetrans* in corn. Plant Dis. Rep. 51:15–19.
15. Eugenio, C. P., C. M. Christensen, and C. J. Mirocha. 1970. Factors affecting production of the mycotoxin F-2 by *Fusarium roseum*. Phytopathology 60:1055–1057.
16. Fajemisin, J. M., and A. L. Hooker. 1974. Top weight, root weight, and root rot of corn seedlings as influenced by three Helminthosporium leaf blights. Plant Dis. Rep. 58:313–317.
17. Foley, D. C. 1962. Systemic infection of corn by *Fusarium moniliforme*. Phytopathology 52:870–872.
18. Foley, D. C. 1969. Stalk deterioration of plants susceptible to corn stalk rot. Phytopathology 59:620–627.
19. Futrell, M. C., and M. Kilgore. 1969. Poor stands of corn and reduction of root growth caused by *Fusarium moniliforme*. Plant Dis. Rep. 53:213–215.
20. Gulya, T. J., Jr., C. A. Martinson, and L. H. Tiffany. 1979. Ear-rotting fungi associated with opaque-2 maize. Plant Dis. Rep. 63:370–373.
21. Hesseltine, C. W., and R. J. Bothast. 1977. Mold development in ears of corn from tasseling to harvest. Mycologia 69:328–340.
22. Hooker, A. L., and D. G. White. 1976. Prevalence of corn stalk rot fungi in Illinois. Plant Dis. Rep. 60:1032–1034.
23. Hoppe, P. E. 1938. Relative prevalence and geographic distribution of various ear-rot fungi in the 1937 corn crop. Plant Dis. Rep. 22:234–241.
24. Hoppe, P. E. 1939. Relative prevalence and geographic distribution of various ear-rot fungi in the 1938 corn crop. Plant Dis. Rep. 23:142–148.
25. Hoppe, P. E. 1941. Relative prevalence and geographic distribution of various ear-rot fungi in the 1940 corn crop. Plant Dis. Rep. 25:148–152.
26. Hornby, D., and A. J. Ullstrup. 1967. Fungal populations associated with maize roots. Composition and comparison of mycofloras from genotypes differing in root rot resistance. Phytopathology 57:869–875.
27. Kingsland, G. C., and C. C. Wernham. 1962. Etiology of stalk rots of corn in Pennsylvania. Phytopathology 52:519–523.
28. Kisiel, M., K. Deubert, and B. M. Zuckerman. 1969. The effect of *Tylenchus agricola* and *Tylenchorhynchus claytoni* on root rot of corn caused by *Fusarium roseum* and *Pythium ultimum*. Phytopathology 59:1387–1390.
29. Koehler, B. 1942. Natural mode of entrance of fungi into corn ears and some symptoms that indicate infection. J. Agric. Res. 64:421–442.
30. Koehler, B. 1954. Some conditions influencing the results from corn seed treatment tests. Phytopathology 44:575–583.
31. Koehler, B. 1959. Corn ear rots in Illinois. Illinois Agric. Exp. Stn. Bull. 639. 87 p.
32. Koehler, B. 1960. Cornstalk rots in Illinois. Illinois Agric. Exp. Stn. Bull. 658. 90 p.
33. Kommedahl, T., and C. E. Windels. 1975. Relative importance of *Fusarium moniliforme* and *F. roseum* 'Graminearum' as stalk rot pathogens of corn in Minnesota. Proc. Amer. Phytopathol. Soc. 2:61 (Abstr.).
34. Kommedahl, T., and C. E. Windels. 1977. Fusarium stalk rot and common smut in cornfields of southern Minnesota in 1976. Plant Dis. Rep. 61:259–261.
35. Kommedahl, T., C. E. Windels, and H. G. Johnson. 1974. Corn stalk rot survey methods and results in Minnesota in 1973. Plant Dis. Rep. 58:363–366.
36. Kommedahl, T., C. E. Windels, and D. S. Lang. 1975. Comparison of Fusarium populations in grasslands of Minnesota and Iceland. Mycologia 67:38–44.
37. Kommedahl, T., C. E. Windels, and R. E. Stucker. 1979. Occurrence of *Fusarium* species in roots and stalks of symptomless corn plants during the growing season. Phytopathology 69:961–966.
38. Kommedahl, T., C. E. Windels, and H. B. Wiley. 1978. *Fusarium*-infected stalks and other diseases of corn in Minnesota in 1977. Plant Dis. Rep. 62:692–694.
39. Kucharek, T. A., and T. Kommedahl. 1966. Kernel infection and corn stalk rot caused by *Fusarium moniliforme*. Phytopathology 56:983–984.
40. Leonard, K. J., and D. L. Thompson. 1969. Corn stalk rot fungi in North Carolina. Plant Dis. Rep. 53:718–720.
41. Loesch, P. J., Jr., D. C. Foley, and D. F. Cox. 1976. Comparative resistance of opaque-2 and normal inbred lines of maize to ear-rotting pathogens. Crop Sci. 16:841–842.
42. Lunsford, J. N., M. C. Futrell, and G. E. Scott. 1975. Maternal influence on response of corn to *Fusarium moniliforme*. Phytopathology 65:223–225.
43. Lunsford, J. N., M. C. Futrell, and G. E. Scott. 1976. Maternal effects and type of gene action conditioning resistance to *Fusarium moniliforme* seedling blight in maize. Crop Sci. 16:105–107.

44. Manns, T. F., and J. F. Adams. 1923. Parasitic fungi internal of seed corn. J. Agric. Res. 23:495–524.
45. Mirocha, C. J., B. Schauerhamer, C. M. Christensen, and T. Kommedahl. 1979. Zearalenone, vomitoxin and T-2 toxin associated with stalk rot in corn. Appl. Environ. Microbiol. 38:557–558.
46. Nicholson, R. L., L. F. Bauman, and H. L. Warren. 1976. Association of *Fusarium moniliforme* with brown midrib maize. Plant Dis. Rep. 60:908–910.
47. Nyvall, R. F. 1970. Chlamydospores of *Fusarium roseum* 'Graminearum' as survival structures. Phytopathology 60:1175–1177.
48. Nyvall, R. F., and T. Kommedahl. 1968. Individual thickened hyphae as survival structures of *Fusarium moniliforme* in corn. Phytopathology 58:1704–1707.
49. Nyvall, R. F., and T. Kommedahl. 1970. Saprophytism and survival of *Fusarium moniliforme* in corn stalks. Phytopathology 60:1233–1235.
50. Ooka, J. J., and T. Kommedahl. 1977. Kernels infected with *Fusarium moniliforme* in corn cultivars with opaque-2 endosperm or male-sterile cytoplasm. Plant Dis. Rep. 61:162–165.
51. Ooka, J. J., and T. Kommedahl. 1977. Wind and rain dispersal of *Fusarium moniliforme* in corn fields. Phytopathology 67:1023–1026.
52. Palmer, L. T., and T. Kommedahl. 1969. Root-infecting *Fusarium* species in relation to rootworm infestations in corn. Phytopathology 59:1613–1617.
53. Palmer, L. T., and D. H. MacDonald. 1974. Interaction of *Fusarium* spp. and certain plant parasitic nematodes on maize. Phytopathology 64:14–17.
54. Pappelis, A. J. 1965. Relationship of seasonal changes in pith condition ratings and density to Gibberella stalk rot of corn. Phytopathology 55:623–626.
55. Pappelis, A. J. 1970. Effect of root and leaf injury on cell death and stalk rot susceptibility in corn. Phytopathology 60:355–357.
56. Pappelis, A. J., and L. V. Boone. 1966. Effect of planting date on stalk rot susceptibility and cell death in corn. Phytopathology 56:829–831.
57. Pearson, N. L. 1931. Parasitism of *Gibberella saubinetii* on corn seedlings. J. Agric. Res. 43:569–596.
58. Percich, J. A., and J. L. Lockwood. 1975. Influence of atrazine on the severity of Fusarium root rot in pea and corn. Phytopathology 65:154–159.
59. Porter, C. L. 1927. A study of the fungous flora of the nodal tissues of the corn plant. Phytopathology 17:563–568.
60. Roane, C. 1950. Observations on corn diseases in Virginia from 1947 to 1950. Plant Dis. Rep. 34:394–396.
61. Simpson, W. R. 1967. Influence of cultural practices on the incidence of stalk rot of corn. Plant Dis. Rep. 51:540–542.
62. Snyder, W. C., and H. N. Hansen. 1940. The species concept in *Fusarium*. Amer. J. Bot. 28:64–67.
63. Snyder, W. C., and H. N. Hansen. 1941. The species concept in *Fusarium* with reference to section Martiella. Amer. J. Bot. 27:738–742.
64. Snyder, W. C., and H. N. Hansen. 1945. The species concept in *Fusarium* with reference to Discolor and other sections. Amer. J. Bot. 32:637–666.
65. Snyder, W. C., H. N. Hansen, and J. W. Oswald. 1957. Cultivars of the fungus, *Fusarium*. J. Madras Univ., B. 27:185–192.
66. Sumner, D. R. 1968. Ecology of corn stalk rot in Nebraska. Phytopathology 58:755–760.
67. Sumner, D. R. 1968. The effect of soil moisture on corn stalk rot. Phytopathology 58:761–765.
68. Switkin, C., J. E. Ayers, and P. E. Nelson. 1974. *Fusarium* species associated with corn stalk rot samples in Pennsylvania from 1970–1974. Proc. Amer. Phytopathol. Soc. 1:145 (Abstr.).
69. Tjokrosudarmo, A. 1976. Relationship of kernel infecting fungi to stalk rot of corn. M. S. Thesis, Univ. Minn. 73 p.
70. Tuite, J., and R. W. Caldwell. 1971. Infection of corn seed with *Helminthosporium maydis* and other fungi in 1970. Plant Dis. Rep. 55:387–389.
71. Ullstrup, A. J. 1936. The occurrence of *Gibberella fujikuroi* var. *subglutinans* in the United States. Phytopathology 26:685–693.
72. Ullstrup, A. J. 1964. A nonpigmented form of *Gibberella roseum* forma *cerealis* on corn in Indiana. Mycologia 56:110–113.
73. Ullstrup, A. J. 1971. Hyper-susceptibility of high-lysine corn to kernel and ear rots. Plant Dis. Rep. 55:1046.
74. Ullstrup, A. J. 1974. Corn diseases in the United States and their control. U. S. Dept. Agric. Handbook No. 199. 56 p.
75. Voorhees, R. K. 1934. Histological studies of a seedling disease of corn caused by *Gibberella moniliformis*. J. Agric. Res. 49:1009–1015.
76. Warmke, H. E., and N. C. Schenck. 1971. Occurrence of *Fusarium moniliforme* and *Helminthosporium maydis* on and in corn seed as related to T cytoplasm. Plant Dis. Rep. 55:486–489.
77. Warren, H. L. 1978. Comparison of normal and high-lysine maize inbreds for resistance to kernel rot caused by *Fusarium moniliforme*. Phytopathology 68:1331–1335.
78. Warren, H. L., and T. Kommedahl. 1973. Prevalence and pathogenicity to corn of *Fusarium* species from roots, rhizosphere, residues, and soil. Phytopathology 63:1288–1290.
79. White, D. G. 1978. Correlation of stalk rot reactions caused by *Diplodia maydis* and *Gibberella zeae*. Plant Dis. Rep. 62:1016–1018.
80. Windels, C. E., and T. Kommedahl. 1974. Population differences in indigenous *Fusarium* species by corn culture of prairie soil. Amer. J. Bot. 61:141–145.

81. Windels, C. E., and T. Kommedahl. 1976. *Fusarium* species in roots and stalks of corn in Minnesota in 1974 and 1975. Proc. Amer. Phytopathol. Soc. 3:292 (Abstr.).
82. Windels, C. E., M. B. Windels, and T. Kommedahl. 1976. Association of *Fusarium* species with picnic beetles on corn ears. Phytopathology 66:328–331.
83. Wolf, J. C., and C. J. Mirocha. 1973. Regulation of sexual reproduction in *Gibberella zeae* (*Fusarium roseum* 'Graminearum') by F-2 (Zearalenone). Can. J. Microbiol. 19:725–734.
84. Young, T. R., and T. A. Kucharek. 1977. Succession of fungal communities in roots and stalks of hybrid field corn grown in Florida. Plant Dis. Rep. 61:76–80.
85. Zeller, S. M. 1929. Contributions to our knowledge of Oregon fungi—III. Mycologia 21:97–111.

10 The Bakanae Disease of the Rice Plant

Shou-kung Sun and William C. Snyder

Introduction

The bakanae disease is one of the oldest rice diseases in the Orient. It was first reported in 1898 by Hori, who mentioned blast, brown spot, and bakanae as the three major diseases of rice in Japan at the time (8). Hori also identified *Fusarium heterosporum* as the causal agent of bakanae disease. Sawada in Taiwan found the ascigerous stage of the rice bakanae fungus in 1919 and named the fungus *Lisea fujikuroi.* The second word of the binomial was adopted in memory of Fujikuro, who first found the disease in 1918 (31). A comprehensive study of the bakanae disease was done in 1931 by Ito and Kimura, who identified the causal fungus as *Gibberella,* and its conidial stage as *Fusarium moniliforme.* We think that the perfect stage should be called *Gibberella moniliformis* with *G. fujikuroi* as a synonym.

Extensive studies of the bakanae disease of rice carried out in Japan included symptomalogy (5, 12, 13, 15, 17, 24, 31), the bakanae phenomonon (5, 13, 14, 15, 16, 30, 32), pathogenicity (12, 13), soil temperature and disease incidence (13, 18, 34), and survival of the pathogen (11, 13, 15). Since Yabuta and Hayashi isolated the growth regulator Gibberellin from the filtrate of a fungus culture in 1939 (47), all studies of the disease were concentrated on the production, purification, and physiology of Gibberellins. The disease itself was not extensively studied; only a few articles were written from 1939 to 1950, and less than 20 papers were published in Japan, Taiwan, and Thailand from 1960 to 1976. A general review of the rice bakanae disease was made by Ou (26). The present paper summarizes recent investigations on spore liberation, the disease cycle, heterothallism, survival, and inoculum potential of the rice bakanae fungus.

The Disease Symptoms

The most common symptom is the elongation of the diseased plant stems. "Bakanae" is a Japanese word meaning bad seedlings (30); in Taiwan, diseased plants are called "Male seedlings," which means sterile big plants, and Chinese farmers call diseased plants "White poles." Diseased plants are slender, pale yellowish in color, and taller than their undiseased counterparts (Fig. 10-1, A). Diseased seedlings usually die during or after transplanting, but healthy seedlings may be infected in the paddy field sporadically after transplanting. Besides elongation, bakanae symptoms include leaves bending over and the production of adventious roots at nodes on the lower portions of the stems. White mycelium grows out of the nodes and spreads over the lower half of the stems. Eventually the plants die, covered with dull white to light pink mycelium and sporodochia (some macroconidia, but mostly microconidia). If it rains or becomes very humid, diseased stems turn

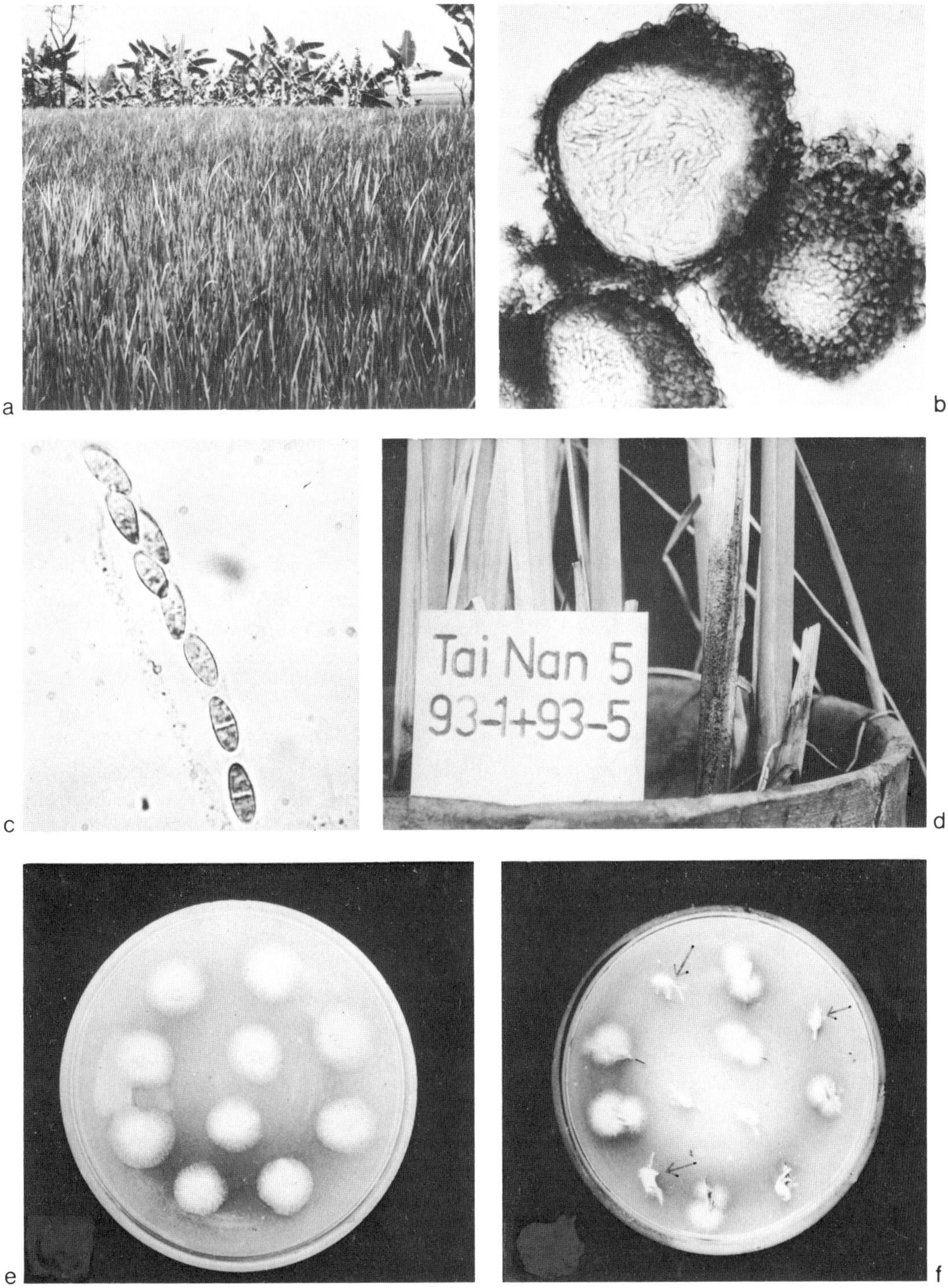

Fig. 10-1. A-F. A) Symptoms of the bakanae disease on rice plants in the field showing taller, slender, and yellowish infected rice plants. B) Perithecia of *Gibberella moniliformis*. C) Ascus containing ascospores of *G. moniliformis*. D) A mating of compatible isolates 93-1 and 93-5 showing the production of perithecia of *G. moniliformis* on rice seedlings. E) Rice seed collected in the field and plated on PCNB medium showing growth of *Fusarium moniliforme* from 100% of the seed. F) Plating the outer husk and inner part of the rice grain (arrows) showing that the fungus grows only from the husks.

Table 10-1. Sizes of macroconidia, microconidia, perithecia, and ascospores of *Gibberella moniliformis* (=*F. moniliforme*)

	Sawada (30) µm	Ito & Kimura (13) µm	Wollenweber & Reinking (26) µm	The authors µm
Microconidia	6–20 × 3–4.5	9.92–9.68 × 3.5–2.88		8.59 × 3.67
Macroconidia	29–56 × 3.2–4.5	28.75–48.75 × 3–3.75	32–50 × 2.7–3.5	28.75–52.5 × 2.5–4.5
Perithecia	240–360 × 180–420	239.88 × 225.18	280–330 × 220–280	283.2 × 247.2
Ascospores	12–24 × 6–9	15.53 × 5.08	14–18 × 4.4–7.0	13.93 × 5.84

bluish-black, and blue-black perithecia (Fig. 10-1, B, C) form on the surface of the stems. About 50% of the infected plants produce perithecia (49). When the weather is dry during harvesting time, only a few or no perithecia can be found.

In Japan, panicles are often infected and referred to as "Pink panicles" (38). We found that some late-infected plants produced pink panicles, but some of the pink panicles are caused by *Fusarium roseum.*

The Causal Agent

Ascigerous stage: *Gibberella moniliformis* (Sheld.) Wineland (1924) [Syn.: *Lisea fujikuroi* Sawada (1919), *G. fujikuroi* (Saw.) Ito et Kimura (1931), *G. fujikuroi* (Saw.) Wollenweber (1931)]; conidial stage: *Fusarium moniliforme* (Sheld.) emend. Snyd. & Hans. (1945) (39).

The morphology of the rice bakanae fungus has been previously described in detail (1, 13, 26, 30). Measurements of spores and perithecia sizes are summarized in Table 10-1.

Microconidiophores are single, lateral, subulate phialides formed from aerial hyphae. Microconidia are formed in chains or agglutinated in false heads. They are fusiform or clavate, mostly 1-celled, occasionally 2-celled and colorless.

Macroconidiophores consist of a basal cell bearing 2–3 apical phialides which produce macroconidia. The macroconidia are delicate and slightly sickle-shaped, with a sharply curved apical cell and a pedi-celled foot cell. They may be 0–7 septate, but 80% are 3 septate.

Chlamydospores are absent. Thick-walled hyphal cells or macroconidia cells are usually formed from old cultures. Dark blue irregular globose sclerotia are present in culture.

Growth on potato dextrose agar (PDA) is rather rapid. Mycelium is white at first, later becoming white to light pink in color. Salmon-orange or carrot-red sporodochia appear among mycelial masses 3 weeks later if the cultures are put under continuous light at temperatures of 20–23 C. If the cultures are kept in the dark or at a temperature above 28 C, no sporodochia are formed, and the mycelium grows poorly as a pale white mass. Under unfavorable conditions, cultures may mutate to mycelial or pionnotal types without sporodochia; on pionnotal cultures a dull purple color develops.

Physiology of the fungus Although Richard's and Knop's solutions were commonly used by early workers, the rice bakanae fungus is more easily isolated on selective media, such as Pentachloronitrobenzene (PCNB) (21), then transfered to PDA (2). The optimum temperature for growth is 27–30 C, maximum 36–40 C, and minimum 7–8 C (6, 18, 22, 36). For better sporulation, cultures must be kept under continuous light at 20–23 C (2, 48) and single cultures must be selected after each transfer in order to maintain wild type characteristics in culture (2, 48). A recent study by Yu and Sun (51) showed that rice straw-soybean meal agar and Sach's agar plus rice straw favor the production of microconidia, while glucose casamino acid agar and rice straw casamino acid agar are good for macroconidia production (51). Xylose is the most suitable carbohydrate for macroconidia production. If the amount of xylose is doubled, e.g., 16g/liter, considerable numbers of microconidia can also be produced. Asparagine and glutamic acid are the amino acids that are favorable for sporulation. Good sporulation was obtained with a C:N ratio from 80 to 320, optimum 160. For most media, light is indispensable for sporulation, but by using Tochinai agar (peptone 10g, maltose 20g, KH_2PO_4 0.5g, $MgSO_4 \cdot 7H_2O$ 0.25g, distilled water 1 liter), a lot of macroconidia were produced in the dark. This indicated that certain substances may substitute for light for sporulation.

Survival Ito and Kimura (13) reported on the viability of the mycelium in infected stems and grains under laboratory conditions. They recovered 75% of the fungus after a storage period of 1,190 days. Macroconidia on rice stems remained viable under room conditions in Hokkaido for 2 years, but under field conditions, both mycelium and spores lost their viability within 5 to 6 months. Iguchi reported that the fungus may remain alive until the next season in Japan if kept indoors (11). Kanjansoon (15) found the fungus was viable in seeds and in infected plants after 4 to 10 months at room temperature and after more than 3 years at 7 C. Chang found that both conidia and ascospores survived for about 4 months on stems under room conditions as well as in soil in the field (2). Lim (20) and Nyvall and Kommedahl (24) demonstrated that *F. moniliforme* survives in infected corn tissue in soil by thick-walled hyphae (the so-called functional chlamydospores) on plant debris, thus it is not surprising that the fungus does not survive long in soil.

Heterothallism and Sexuality The heterothallic nature of sexual reproduction in *G. moniliformis* on rice was suggested by Ito (13) and Kurosawa (19), but they did not succeed in producing perithecia in culture. This was not confirmed until Snyder and Sun (40) and Chang and Sun (3) were able simultaneously in Berkeley and Taiwan to obtain perithecia by mating two compatible ascospores in Sach's medium plus gas-sterilized rice straws. Hsieh et al. (10) provided further evidence that there are three mating groups in *F. moniliforme,* and that the isolates of *F. moniliforme* from rice cannot mate with those from corn and sugarcane and vice versa. Actually, in 1924 Wineland found perithecia in her culture of *F. moniliforme* from corn by mating two compatible strains (46).

In all the crosses of single ascospores of rice isolates of *F. moniliforme,* none has been found to be self-fertile. Two alleles, *A* and *a,* control compatibility reactions. It was found that half of the 8 ascospores in an ascus are *A* and the other half are *a* (Table 10-2). *A* and *a* among a group of single conidial isolates are in the proportion of approximately 1:1. Rice seedlings inoculated with two compatible isolates exhibit fertile perithecia on stems when the rice plants reach maturity (Fig. 10-1, D).

Functional sexuality between compatible isolates is determined by reciprocal crosses on Sach's medium plus rice straws. If perithecia appear earlier when conidia from one culture are poured onto another culture than when the role of the two cultures is reversed, the culture providing the conidia is a male strain, and the receptor culture is a female strain.

Table 10-2. Distribution of mating types (*A* and *a*) among ascospores in a single ascus of *Gibberella moniliformis (G. fujikuroi)*

Isolate	Mating types of ascospores in an ascus: A	a	Remarks
48	1, 3, 4, 7	2, 5, 6	No. 8 ascospore was lost.
91	1, 2, 3	5, 6	No. 4, 7, 8 ascospores were lost.
93	1, 2, 7	4, 5, 6, 8	No. 3 ascospore was lost.
ED	1, 5, 7, 8	2, 3, 4	No. 6 was uncertain, probably *a*.

Table 10-3. Results of crosses between unisexual and bisexual strains of *Gibberella moniliformis (G. fujikuroi)*

Culture being conidiated			Culture providing conidia			Production of
Isolate	Mating type	Sexuality	Isolate	Mating type	Sexuality	perithecia
48-1	A	⚥	48-2	a	♂	Yes
48-6	a	♀	48-1	A	⚥	Yes
48-1	A	⚥	48-5	a	♂	Yes
48-6	a	♀	48-3	A	♀	No
48-6	a	♀	48-4	A	♀	No
93-1	A	⚥	93-5	a	⚥	Yes
93-5	a	⚥	93-1	A	⚥	Yes
93-5	a	⚥	93-2	A	⚥	Yes
93-5	a	⚥	93-7	A	♂	Yes
93-6	a	♂	93-7	A	♂	No
93-8	a	⚥	ED-1	A	♂	Yes
93-6	a	♂	ED-1	A	♂	No
ED-2	a	♀	ED-1	A	♂	Yes
ED-3	a	♀	ED-1	A	♂	Yes
ED-7	A	♀	93-5	a	⚥	Yes
ED-7	A	♀	ED-2	a	♀	No

If perithecia appear at the same time in both cultures, they are hermaphrodites. Chang and Sun (3) found that in 35 single ascospore isolates, 8 were hermaphrodites, 10 were unisexual males, 10 were unisexual females and the sexual status of 7 was unknown (Table 10-3). The unknown isolates may have been neuters or else perithecia were not produced because of unknown factors. There is evidence that the percentage of hermaphrodites is larger than this in fresh isolates of the fungus. Mutation from hermaphrodites to unisexuals takes place in laboratory cultures.

Spore Liberation

Sasaki (29) found that conidia were liberated at night, usually from 5 PM to 9 PM. No conidia were collected during the day, except during rainy days when conidia were discharged (29). Sun (41) and Yu and Sun (49) reported a pattern of nocturnal liberation of ascospores of *G. moniliformis (G. fujikuroi)* occurring primarily after midnight. However, ascospore discharge may occur after rainfall even during the daytime (Fig. 10-2). In a paddy field, perithecia begin to discharge ascospores from the heading stage of rice plants to harvest time of each crop, a total period of about one month. No ascospores were trapped after harvest, but when rice straw was heaped in the paddy field, spores could be collected 2 weeks after harvest. Yu and Sun (49) found that each perithecium contains about 126–135 asci and more than 1,000 ascospores, but an average of only 990 ascospores are discharged into the air. The liberated spores are carried away by air currents, so rice grains are commonly contaminated (Fig. 10-1, E). Sun (41) reported that about 11–14 propagules were found on each grain in moderately infected rice fields. Ascospore discharge requires high humidity after midnight, and when relative humidity was maintained at 33% under room conditions, ascospores oozed out of the perithecia instead of being forcibly ejected (49). Ascospore oozing from perithecia can be found in the field during dry and fine days. The ooze can be washed off into the soil or may fall into the paddy field along with rice straw, thus contaminating the soil.

Inoculum Potential

The pathogen is both seed-borne (2, 5, 13, 15, 27, 28, 31, 35, 37, 41) and soil-borne (2, 15, 45, 50). Hemmi et al. (5) reported that heavily infected reddish-colored seeds gave

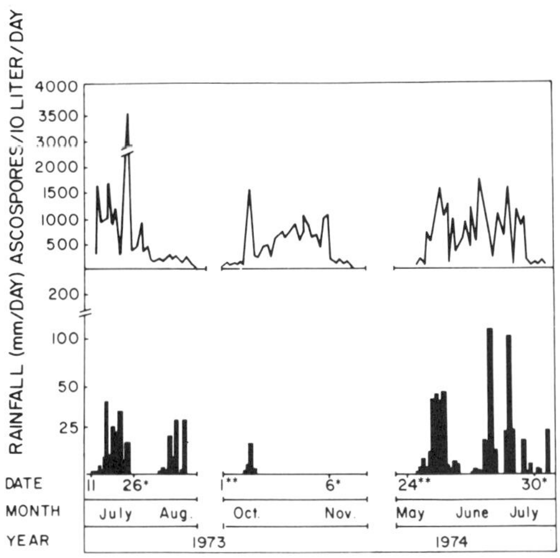

Fig. 10-2. Ascospore liberation of *Gibberella moniliformis* from July 11, 1973 to July 10, 1974 in Taichung, Taiwan. * = Harvesting period; ** = Heading stage.

rise to stunted seedlings, while Seto (33) found some stunted seedlings in the field instead of bakanae symptoms. Viswanath-Reddy (45) reported that when seeds were heavily infected, both stunted and tall-slender symptoms were exhibited in the field due to the interaction of two metabolites, Fusaric acid and Gibberellic acid. In Taiwan, this phenomenon can be found in the nursery beds, but stunting is much less common than the occurrence of tall-slender seedlings.

Yu and Sun (50), using the serial dilution-end point method (25, 42), set up a series of infested soils containing different levels of propagule density, representing different inoculum potential indexes (IPI). When sterilized seeds were planted in the infested soils of different dilutions, several types of symptoms appeared. Seedling blight, dwarf seedlings, and yellow seedlings appeared in soils with a fairly high IPI and bakanae symptoms were produced only in soil with moderate propagule density (Table 10-4). If a soil has an initial IPI of up to 256, the IPI is zero at the end of 9 weeks; the lower the IPI, the faster it declines to zero. At zero IPI, seedlings are not infected but *F. moniliforme* can be isolated from the soil. With the same level of inoculum density, the incubation period for symptom

Table 10-4. Relationship of inoculum density to symptom expression and the incubation period of the rice bakanae pathogen

Dilution of infested soil	Propagules per g of soil	Bakanae (cm) and other symptoms appearing on 20-day-old rice seedlings after two incubation periods: 6 days	12 days
1/1	66×10^4	14.4^{d*} (Seedling blight)	14.9^{g} (SB)
1/2	33×10^4	15.4^{c} (SB & dwarf)	18.4^{fg} (SB)
1/4	16.5×10^4	17.0^{bc} (Yellow seedlings)	17.3^{g} (SB)
1/8	8.25×10^4	16.8^{bc} (YS)	19.4^{efg} (SB)
1/16	4.13×10^4	17.8^{b} (YS)	26.0^{bcde}(YS)
1/32	2.06×10^4	17.2^{bc} (YS)	25.7^{cde} (YS)
1/64	10.3×10^3	20.9^{a} (Bakanae)	30.0^{abd} (YS)
1/128	5.16×10^3	20.7^{a} (Bakanae)	32.6^{ab} (Bakanae)
1/256	2.58×10^3	17.4^{b} (Normal)	35.5^{a} (Bakanae)
1/512	12.9×10^2	17.4^{b} (Normal)	33.3^{a} (Bakanae)
1/1024	5.45×10^2	16.8^{bc} (Normal)	32.5^{abc} (Bakanae)
1/2048	3.22×10^2	17.0^{bc} (Normal)	31.3^{abd} (Normal)
1/4096	1.61×10^2	16.9^{bc} (Normal)	29.3^{abcd} (Normal)
1/8192	8.06×10^1	16.8^{bc} (Normal)	24.5^{def} (Normal)
Check	0	17.2^{bc} (Normal)	24.6^{de} (Normal)

* Different letters indicate significant differences at the 1% level (by Duncan's significant difference).

expresion varies with the age of the seedlings, e.g., 6-day incubation period for 2-day-old seedlings and 12 days for 20-day-old seedlings.

Disease Cycle

Earlier Japanese investigators found that seeds infected at the flowering stage developed pink panicles. The fungus might also be isolated from healthy seeds collected from an infested paddy field. Umehara reported that leafhoppers might be vectors for the dissemination of the rice bakanae fungus (43, 44). Recent studies in Taiwan indicated that grain contamination is due principally to air-borne ascospores, but contamination may also occur from conidia during the harvesting operation (41). Chang and Sun (3) demonstrated by carefully separating the husk from the inside rice grains and placing them on PCNB agar (Fig. 10-1, F), that the fungus spores are on the surface of seeds only. Although most seedlots yield 100% *F. moniliforme* on agar plates, only 33% bakanae disease incidence occurred when infested seeds were planted in the soil. Panicle infection occurs in Taiwan, but not as frequently as in Japan. Most grains of "pink panicles" are empty and are usually eliminated during seed treatment.

Conidia and perithecia on diseased stems are often washed into soil by rain, and diseased plants and stubble are often discarded in the field. Therefore, paddy soil is commonly contaminated by *F. moniliforme.* Healthy seedlings and seeds, even sterilized seeds, can be infected sporadically from soil. Sun (41) found that 4.8 to 4.8×10^3 propagules per g of soil were capable of infecting healthy plants. The same author reported that $1.1–2.4 \times 10^3$ propagules per g of soil were found in paddy soil near diseased plants.

Seto (37) reported that the fungus can also infect the branches of the panicles. Sun (41) and Yu (48) tried several times to infect the above-ground parts of the rice plants, and except for the panicles, found that the other plant parts were resistant to infection. Infection takes place very easily below ground, but roots and crowns of older plants also appear to be resistant to infection.

Although the fungus survived for 1,190 days under room conditions, it may remain alive in soil only for 6 months (11, 13, 15). *Fusarium moniliforme* survives in soil by means of thick-walled hyphae or macroconidia which function as chlamydospores (20, 24) and their longevity in soil is about 4 months in Taiwan (2, 50). The population of fungus propagules decreased within 9 weeks to a level that was not capable of inducing bakanae disease (50). Since the two rice crops are about 2–3 months apart in Taiwan, in early plantings even sterilized seeds may be infected at random in the soil.

After the fungus penetrates the plant through roots or crowns, it becomes systemic and moves upward. Nisikado and Kimura (23) found microconidia and mycelia in the vascular bundles distributed discontinuously. Chang (2) isolated the fungus in diseased plants from the roots up to the lower portion of the panicle stalk. Panicle infection is due to air-borne ascospores. The fungus produces conidia, mostly microconidia, beginning on diseased stems, first around the nodes and then on the entire stem. Perithecia begin to appear on the lower portion of the diseased stem and rain or high humidity is required for formation and maturation of perithecia. In Taiwan, more perithecia can be found in the first rice crop during the rainy season in June and July than in the second crop during the beginning of the dry season (October and November). Whether conidia and ascospores can cause new infections remains unknown. The authors found no secondary cycle of the disease in a single crop period (Fig. 10-3).

Control of Bakanae Disease of Rice

Japanese plant pathologists proved that grains were infected, and the pathogen has been frequently isolated from seeds (5, 13, 15, 35). Thus seed treatment is of prime importance

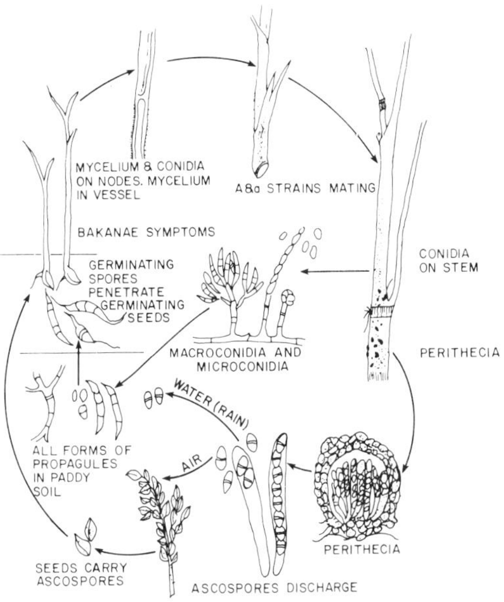

Fig. 10-3. The disease cycle of the rice 'bakanae' disease caused by *Gibberella moniliformis.*

(9). Brine water was recommended for seed treatment in order to eliminate the lightweight diseased seeds, and formalin was also recommended for seed disinfestation (31). After the discovery that mercuric compounds were effective for rice blast control (4, 7), they were also shown to be effective against the bakanae disease. Several mercuric compounds such as Granosan and Ceresan-lime were used widely in Japan and Taiwan in rice cultivation until 1970, when mercuric residues were found to be harmful to humans and further use was discontinued. Other chemicals have been screened as substitutes for mercury, and Yu's tests indicated that Benlate-T, 1:1,000 dilution, was as effective as the mercury compounds for seed disinfestation (48). At present, Benlate or Benlate-T are widely used for rice seed treatment in Taiwan.

Literature Cited

1. Booth, C. 1971. The genus *Fusarium.* Commonwealth Mycol. Inst., Kew, Surrey, England. 237 p.
2. Chang, I. C. 1973. Studies on the perfect stage and ecology of the rice bakanae fungus. M. S. Thesis, Nat. Chung Hsing Univ., Taichung, Taiwan. 64 p.
3. Chang, I. C., and S. K. Sun. 1975. The perfect stage of *Fusarium moniliforme.* J. Agric. Res. China 24: 11–20.
4. Chen, C. C., and C. C. Chien. 1959. Studies on the effect of some fungicides on the three important causal fungi of rice diseases. J. Agric. Assoc. China, N. S. 28: 39–48.
5. Hemmi, T., T. Seto, and J. Ikaya. 1931. Studies on the 'bakanae' disease of the rice plant. II. On the infection of rice by *Lisea fujikuroi* Sawada and *Gibberella saubinetii* (Mont.) Sacc. in the flowering period (In Japanese, English summary). Forschn. Geb. Pflkrankh. Kyoto 1: 99–110.
6. Hemmi, T., and Z. Aoyagi. 1941. Ecological studies on important fungi pathogenic to the crops in the Far East. 1. Germination of macroconidia of *Gibberella fujikuroi* in relation to some environmental factors (In Japanese, English summary). Ann. Phytopathol. Soc. Japan 11: 66–80.
7. Hirada, E., and S. Sato. 1931. The control of rice bakanae disease by seed treatment with mercuric compounds (In Japanese). Korean Bull. 13(3): 1–14.
8. Hori, S. 1898. Researches on 'bakanae' disease of the rice plant (In Japanese). Nojishikenjyo Seiseki 12: 110–119.
9. Hoschino, Y. 1955. A study on the rice-seed disinfection with special reference to the bakanae disease (In Japanese, English summary). Jubilee Publ. Commem. 60th Birthday Prof. Y. Tochinai and T. Fukushi: 290–299, Sapporo, Japan.
10. Hsieh, W. H., S. N. Smith, and W. C. Snyder. 1977. Mating groups in *Fusarium moniliforme.* Phytopathology 67: 1041–1043.
11. Iguchi, S. 1964. Overwintering of bakanae disease fungus (In Japanese). Ann. Rep. Soc. Plant Protection North Japan 15: 39.
12. Imura, J. 1940. On the angles between blade and culms in the accelerated rice seedlings caused by

Gibberella fujikuroi. Ann. Phytopathol. Soc. Japan 10: 45–48.
13. Ito, S., and J. Kimura. 1931. Studies on the bakanae disease of the rice plant (In Japanese, English summary). Rep. Hokkaido Agric. Exp. Stn. 27:1–95.
14. Ito, S., and S. Shimada. 1931. On the nature of the growth promoting substance excreted by the bakanae fungus. Ann. Phytopathol. Soc. Japan 2: 322–338.
15. Kanjansoon, P. 1965. Studies on the bakanae disease of rice in Thailand. Dr. Agric. Thesis, Tokyo Univ. Japan. 120 p.
16. Kurosawa, E. 1926. Experimental studies on the filtrate of the causal fungus of the bakanae disease of the rice plant (In Japanese). Trans. Nat. Hist. Soc. Formosa 16: 213–217.
17. Kurosawa, E. 1928. Studies on the symptoms and the causal organism of rice bakanae disease. Trans. Nat. Hist. Soc. Formosa 18: 230–247.
18. Kurosawa, E. 1929. On the cultural characters of the bakanae disease fungi on various nutrient media and the temperature of their development (In Japanese). Rep. Nat. Hist. Soc. Formosa 19: 150–179.
19. Kurosawa, E. 1934. Studies on bakanae symptoms and perfect stage of *Gibberella fujikuroi* and related fungi (In Japanese, English summary). Ann. Phytopathol. Soc. Japan 2: 276–277. (Abstr.).
20. Lim, G. 1967. Fusarium populations in rice field soils. Phytopathology 57: 1152–1153.
21. Nash, S. M., and W. C. Snyder. 1962. Quantitative estimation by plate counts of propagules of the bean root rot Fusarium in field soils. Phytopathology 52: 567–572.
22. Nisikado, Y., and H. Matsumoto. 1933. Studies on the physiological specialization of *Gibberella fujikuroi*, the causal fungus of the rice bakanae disease (In Japanese, English summary). Trans. Tottori Soc. Agric. Sci. 4: 200–211.
23. Nisikado, Y., and K. Kimura. 1941. A contribution to the pathological anatomy of rice plants affected by *Gibberella fujikuroi* (Saw.) Wr. 1. Ber. Chara. Inst. Landw. Forsch. 8: 421–426.
24. Nyvall, R. F., and T. Kommedahl. 1968. Individual thickened hyphae as survival structures of *Fusarium moniliforme* in corn. Phytopathology 58: 1704–1707.
25. Onesirosan, P. F. 1971. The survival of *Phytophthora palmivora* in a cacao plantation during the dry season. Phytopathology 61: 975–977.
26. Ou, S. H. 1972. Rice diseases. Commonwealth Mycol. Inst., Kew, Surrey, England, 368 p.
27. Rajagopalan, K., and K. Bhuvaneswari. 1964. Effect of germination of seeds and host exudation during germination on foot rot disease of rice. Phytopathol. Z. 50: 221–226.
28. Reyes, G. M. 1939. Rice diseases and methods of control. Philippine J. Agric. 10: 419–436.
29. Sasaki, T. 1971. Conidia discharge of rice bakanae fungus (In Japanese). Ann. Phytopathol. Soc. Japan 37: 163. (Abstr.).
30. Sawada, K. 1919. Descriptive catalogue of Formosan fungi. Part I. Special Rep. No. 19, Agric. Exp. Stn. Formosa. 695 p.
31. Sawada, K., and E. Kurosawa. 1924. Control of rice bakanae disease (In Japanese). Formosan Agric. Dept. Special Bull. 21. 11 p.
32. Seto, F. 1928. Studies on the bakanae disease of the rice plant. 1. A consideration of the occurrence of the bakanae disease and the bakanae phenomonon. Ann. Phytopathol. Soc. Japan 2: 118–139.
33. Seto, F. 1932. Experimentelle untersuchungen die hemmende und die beschleunigende Wirkung des Erregers der sogenannten 'Bakanae'-Krankheit, *Lisea fujikuroi* Sawada, auf dos Wachstum der Reikeimlinge. Mem. Coll. Agric. Kyoto Imp. Univ. 18: 1–23.
34. Seto, F. 1933. Unterschungen Über die 'Bakanae'-Krankheit der Reispflanze. IV. Über die Beziehungen zwischen der Bodentemperaturen und dem Kankheitsbefall bei Bodeninfektion (In Japanese, German summary). Forsch. Geb. Pflkrankh. Kyoto 2: 138–153.
35. Seto, F. 1934. Diseased seeds as the primary inoculum of the bakanae disease of the rice plant. Ann. Phytopathol. Soc. Japan 4: 61–63. (Abstr.).
36. Seto, F. 1935. Beitrage zur Kenntnis der 'Bakanae'-Krankheit der Reispflanze. Mem. Coll. Agric. Kyoto Univ. 36: 1–81.
37. Seto, F. 1937. Studies on the bakanae disease of the rice plant. V. On the mode of infection of rice by *Gibberella fujikuroi* (Saw.) Wr. during and after the flowering period and its relation to the occurrence of the so-called 'Bakanae' seedlings (In Japanese, English summary). Forsch. Geb. Pflkrankh. Kyoto 3: 43–57.
38. Shimada, S. 1932. Further studies on the nature of the growth promoting substance excreted by the bakanae fungus. Ann. Phytopathol. Soc. Japan 2: 442–452.
39. Snyder, W. C., and H. N. Hansen. 1945. The species concept in *Fusarium* with reference to discolor and other sections. Amer. J. Bot. 32: 657–666.
40. Snyder, W. C., and S. K. Sun. 1971. Heterothallism in *Fusarium moniliforme*. Report of US-ROC Cooperative Science Seminar on Plant Root Diseases, Univ. California, Berkeley, Calif. 41 p.
41. Sun, S. K. 1975. The disease cycle of rice bakanae disease in Taiwan. Proc. Nat. Sci. Council 8(2): 245–256.
42. Tsao, P. H. 1960. A serial dilution end-point method for estimating disease potentials of citrus Phytophthora in soil. Phytopathology 50: 717–724.
43. Umehara, Y. 1974. Leafhoppers as vectors for the dissemination of rice bakanae fungus (In Japanese). Ann. Phytopathol. Soc. Japan 40: 189. (Abstr.).

44. Umehara, Y. 1975. Distribution of *Gibberella fujikuroi* in the stem of rice plant (In Japanese). Ann. Phytopathol. Soc. Japan 41: 246. (Abstr.).
45. Viswanath-Reddy, M. 1965. Inoculum potential and foot-rot of rice (*Oryza sativa* L.). Phytopathol. Z. 53: 197–200.
46. Wineland, Grace O. 1924. An ascigerous stage and synonomy for *Fusarium moniliforme*. J. Agric. Res. 28: 909–922.
47. Yabuta, T., and T. Hayashi. 1939. Biochemical studies on the bakanae fungus of rice. II. Isolation of 'Gibberellin,' the active principle which makes the rice seedlings grow slenderly (In Japanese). J. Agric. Chem. Soc. Japan 15: 257–266.
48. Yu, K. S. 1975. Factors influencing the incidence of rice bakanae disease. M. S. Thesis, Nat. Chung Hsing Univ., Taichung, Taiwan. 89 p.
49. Yu, K. S., and S. K. Sun. 1976. Ascospore liberation of *Gibberella fujikuroi* and its contamination of rice grains (In Chinese, English summary). Plant Protection Bull. 18: 319–329.
50. Yu, K. S., and S. K. Sun. 1977. Studies on inoculum potential and incubation period of rice bakanae disease (In Chinese, English summary). Plant Protection Bull. 19: 245–250.
51. Yu, K. S., and S. K. Sun. 1978. Influence of nutrition and light on sporulation of the rice bakanae fungus, *Fusarium moniliforme* (In Chinese, English summary). Plant Protection Bull. 20: 141–150.

11 Fusarium Diseases in the Tropics

R. H. Stover

The tropics include about one-quarter of the earth's land mass, the area between 23° N and S of the equator. There are two broad classes of crops grown in this area: crops common to both temperate and tropical areas and crops that grow well only in a tropical climate where mean minimum monthly temperatures do not fall below 10 C. The latter include major tropical exports such as coffee, tea, cocoa, bananas, pineapple, rubber, palm oil, copra, and cane sugar.

To gather information on Fusarium diseases in the tropics, the Review of Plant Pathology (30, 31, 32), symposia (3, 4, 45), reviews (19, 28, 38, 44), and books (9, 16, 22, 26, 37, 41, 42, 48) on tropical crops were consulted. For tropical *Fusarium* species the only modern treatments are those of Gordon (13, 14) and, more recently, Booth's comprehensive monograph (5). Fusarium diseases of major crops grown in both temperate and tropical climates are discussed below under the type of disease caused: wilts, root and stem rots, cankers, galls, and leaf diseases. Field diseases of tropical crops are listed in Table 11-1. Post-harvest and storage diseases are discussed under Rots of Fruit, Seed, and Storage Tissue.

In reviewing the literature, especially that pertaining to disease lists, a distinction must be made between those *Fusarium* spp. found "associated" with a certain disease or merely identified on a crop, and those species where pathogenicity studies have been undertaken. *Fusarium* was frequently listed as being "associated" with a disease, almost invariably along with one or more other common fungi. Only a small percentage of the *Fusarium* spp. listed in tropical areas have been adequately studied with respect to pathogenicity.

Vascular Wilts

Tropical and temperate zone crops Prior to 1950, the most frequent reference to wilt encountered in the tropics was cotton wilt. With the development of resistant cultivars, reference to cotton wilt is now rarely found in tropical literature. Other important wilt diseases in temperate areas damaging tobacco, tomato, sweet potato, and melons are rarely found in the tropics. In fact, of all the vegetables and fruits common to both temperate and tropical areas I know of no widespread fusarial wilt disease occurring in the tropics. In pulses, *F. oxysporum* is reported to cause localized outbreaks of wilt in cow peas and chick peas in Ethiopia.

Tropical crops A perusal of Table 11-1 shows that 10 of the 25 diseases are attributed to *F. oxysporum*. However, apart from banana wilt (Panama disease) and vanilla wilt and root rot, most of the wilt diseases are of restricted distribution. None of the diseases caused by *F. oxysporum* are major limiting factors in the production of important crops

Table 11-1. Field Diseases of Tropical Crops Caused by *Fusarium* Species

Food and Beverage Crops		
African oil palm	Fusarium wilt (16)*	*F. oxysporum* f. sp. *elaeidis*
	Patch yellows (16)*	*F. oxysporum*
Abaca	Fusarium wilt (37)	*F. oxysporum* f. sp. *cubense*
Bambarra groundnut	Fusarium wilt (10)*	*F. oxysporum* f. sp. *voandzeiae*
Banana and plantain	Fusarium wilt (37)	*F. oxysporum* f. sp. *cubense*
	Heart rot (37)	*F. moniliforme*
Cacao	Canker and die-back (5, 42, 43)	*F. tumidum, F. lateritium* *Calonectria rigidiuscula*
	Cushion galls (7, 42)	*C. rigidiuscula*
Celosia argentea	Leaf spot (1)*	*F. lateritium (Gibberella baccata)*
Coffee	Root rot (35, 48)*	*F. oxysporum* f. sp. *coffeae*
	Canker and die-back (23, 48)	*F. lateritium, C. rigidiuscula*
	Bark diseases (34)	*F. stilboides (G. stilboides)*
	Tracheomycosis (6, 48)*	*F. xylarioides (G. xylarioides)*
Ginger (*Zingiber officinali)*	Yellows (46)*	*F. oxysporum* f. sp. *zingiberi*
Granadilla (*Passiflora* sp.)	Wilt (9)*	*F. oxysporum* f. sp. *passiflorae*
Guava (*Psidium* sp.)	Wilt (9)*	*F. oxysporum* f. sp. *psidii*
Mango	Flower galls (11, 40)	*F. moniliforme*
Sugarcane	Top distortion (22) (pokhah boeng)	*F. moniliforme*
	Stem rot (20, 22)	*F. moniliforme*
Vanilla	Wilt and root rot (2)	*F. oxysporum* f. sp. *vanillae* *F. solani*
Tree Crops		
Gliricidia sepium	Canker (36)*	*C. rigidiuscula*
Pinus caribaea	Pitch canker (5)	*F. lateritium* f. sp. *pini*
Teak	Canker (5)	*F. solani*

*Diseases restricted geographically.

such as oil palm, abaca, bananas, and coffee. Vanilla wilt and root rot is a serious disease wherever vanilla is grown and high-quality resistant clones have not yet been developed (2). Oil palm wilt is confined to West Africa and a major epidemic has not developed (16). The wilt-susceptible Gros Michel banana was replaced with the wilt-resistant Cavendish cultivars (37). These cultivars have not been damaged by wilt except in Southern Taiwan, where a new race of *F. oxysporum* f. sp. *cubense* has developed. Coffee root rot and wilt is common in Puerto Rico, where incidence seldom exceeds 6% of the plants in most farms (35). Coffee tracheomycosis causes serious losses in parts of Africa on *Coffeae liberica* and *C. canephora* (6). Wilts of ginger (46), granadilla (9), and guava (9) have been reported only from Hawaii, Australia, and India, respectively. *Fusarium oxysporum* causes a locally important wilt of bambarra groundnut (*Voandzeia subterranea*) in Tanzania (10).

Root Rots

Tropical and temperate zone crops The major cereal crops grown in both tropical and temperate climates are rice, corn, and sorghum. In the tropics root rots caused by *Fusarium* spp. are of minor importance and references are rarely found in the literature. Padwick (28) rates losses from Bakanae disease (*Gibberella fujikuroi*) on rice as small. Ou (27) stated Bakanae disease was of moderate importance in the Asian tropics and of minor importance in the African and American tropics.

Pulses, of which red beans and cowpeas are probably the most important, are widely grown in all areas of the tropics and are the major source of protein. Again, there were few references to widespread root rots caused by *Fusarium* where proof of pathogenicity was obtained. Localized root rot outbreaks caused by *F. solani* are reported on *Phaseolus vulgaris* from Malawi and *F. lateritium* var. *uncinatum* is reported to cause a wilt of pigeon pea in Tanzania. *Fusarium oxysporum* is one of four fungi responsible for seedling mortality and root rot in chick-pea in Ethiopia.

The absence of important Fusarium root rots among grains and pulses in most areas is likely the result of local selection of cultivars with considerable genetic diversity and disease tolerance. Most areas are planted with landrace cultivars in which natural selection for resistance or tolerance to local pathogens has occurred over the decades. With the more recent development of higher yielding, genetically uniform cultivars requiring heavy fertilization, and in some instances irrigation, root rots could increase in importance. In addition to varietal resistance, Gordon (13) noted that the following important cereal root rot fungi were not found in his survey in the West Indies: *F. poae, F. avenaceum, F. acuminatum, F. culmorum,* and *F. graminearum*. The Commonwealth Mycological Institute, however, has collections of *F. graminearum* from corn, rice, and other hosts from tropical Australia, Africa, and Southeast Asia and it is commonly associated with rots of grain.

Tropical crops Fusarium root rots not involving *F. oxysporum* and vascular invasion of above-ground parts are uncommon in the tropics, although the well-known root rot species of temperate areas, *F. solani,* is common in tropical soils. *Fusarium solani* may be the primary invader followed by *F. oxysporum* in vanilla wilt and root rot (2).

Fusarium oxysporum, F. moniliforme, and *F. solani* are common inhabitants of nematode lesions on banana and plantain caused by *Radopholus similis* and *Pratylenchus coffeae* (37). *Fusarium oxysporum* is also common in nematode lesions on citrus roots. Apparently, *Fusarium* spp. can be important components of nematode–root rot complexes in the tropics, but by themselves are weak root pathogens under normal growing conditions.

Rots of Fruit, Seed, and Storage Tissue

Fruit *Fusarium* spp. alone are seldom found in rots of tropical fruit following harvest. They are primarily wound invaders and compete with other fungi in colonization of damaged and adjacent sound tissue. *Colletotrichum* and *Botryodiplodia* are commonly associated fungi. The crown and peduncles (necks) of boxed bananas are subject to rot in transit and *Fusarium semitectum* and *F. moniliforme* are often present along with *Acremonium* spp. *(Cephalosporium), Colletotrichum,* and other fungi (21, 47). *Fusarium graminearum* was obtained frequently but in a small proportion of total isolates from crown rot of bananas in the Windward Islands (15). *Fusarium solani* causes a post-harvest dry rot of papaya in Hawaii (17); incidence is moderate. It was also observed to cause a rot of young fruit in the field, causing abscission from the peduncle. Occasionally, incidence reached 30–40%. Joffe (18) isolated seven species of *Fusarium* from avocado, banana, and citrus in Israel and showed some isolates were good wound invaders. In other subtropical and tropical areas *Fusarium* spp. are not considered important post-harvest pathogens of citrus and avocados.

Grains and pulses Corn ears in the tropics that have been injured by insects and birds are often invaded by *F. moniliforme (G. fujikuroi)*. Ear rot of corn is probably the most serious grain disease in the tropics, especially in highland areas. Sorghum grain can also be affected in wet weather. However, since sorghum is usually grown as a dry season crop and matures during dry weather, grain is seldom damaged (41). Scab of rice grain caused by *G. saubinetii* is found occasionally during wet weather. It is not a major cause of losses, however (26).

Fusarium spp. are common invaders of peanut hulls. The hulls of over-mature peanuts and peanuts harvested during wet weather have high *Fusarium* populations of which *F. solani, F. oxysporum, F. semitectum,* and *F. moniliforme* are most common (24). However, there is no evidence that *Fusarium* spp. are the primary cause of seed injury. In India, *Fusarium* spp. comprised only 6% of the flora of damaged seed, contrasted to 60% for *Aspergillus* (39).

Root crops and vegetables *Fusarium* spp. are often present in post-harvest rots of tropical root and vegetable crops along with other common fungi such as *Botryodiplodia, Botrytis, Sclerotium, Penicillium, Aspergillus,* and *Rhizopus.* On the Chicago market *F. oxysporum* was considered important in the rotting of apio *(Arracacia xanthorrhiza),* yams (*Dioscorea* spp.), ginger, and taro *(Colocasia esculenta)* (8). *Fusarium solani* was involved in rotting of yautia (*Xanthosoma* sp.) and squash. In Nigeria *F. moniliforme* was associated with soft rot of yams, along with other fungi that entered through wounds (25). In the Solomon Islands *F. solani* was associated with other fungi in causing a spongy black rot of taro corms when humidity was high (12).

Stem or Stalk Rots, Cankers, Die-Back

Stalk rot of corn and sorghum caused by *F. moniliforme (G. fujikuroi)* is reported occasionally from the tropics. Stalk rot is not a cause of important losses, except in lowland areas during the wet "summer" season (29). *Gibberella zeae* stalk rot is seldom found in tropical areas, except at high elevations where it is of minor importance (29). Stalk rot is more important in temperate climates where higher levels of nitrogen, known to favor the disease, are used, and higher populations of genetically uniform cultivars are planted.

Several species of *Fusarium* are associated with canker and die-back of cocoa (Table 11-1). Disease is serious only following capsid attacks and the problem is considered primarily entomological (23, 42, 43). Canker and die-back of coffee occur primarily on old trees and under poor growing conditions and the two associated *Fusarium* spp. (Table 11-1) are not considered strong pathogens. Bark diseases of coffee caused by *F. stilboides* can cause important losses in East and Central Africa (34).

Fusarium moniliforme causes a rot of sugarcane stems following planting (20) and sometimes the standing stalks. It is considered an important cause of losses in some areas (22). A top distortion of disease called "pokhah boeng" is caused by *F. moniliforme.* A similar top distortion sometimes occurs on sorghum (41). Neither disease is widespread and serious losses of cane have occurred only in Java; resistant cultivars are available (22).

Three *Fusarium* spp. are associated with cankers on tree crops in the tropics (Table 11-1). Only pitch canker on *Pinus caribaea* is considered a serious cause of damage.

Galls

Gall diseases caused by *Fusarium* sp. (Table 11-1) have been reported on cacao (7, 42) and mango (11, 40). No information is available as to how extensive the damage caused by cushion galls is in the major cacao-growing areas of the tropics. Some fields in Central America had as high as 50% of the flowers galled (42). In India *F. moniliforme* is associated with flower galls on mango. The disease is widespread throughout the country and all cultivars are affected (40). A similar disease has been reported from Brazil (11).

Leaf Diseases

Fusarium moniliforme can cause a heart rot of the folded, unemerged leaf of Gros Michel bananas and abaca (37). Heart rot is a minor disease and causes very little damage. *Fusarium moniliforme* along with an *Erwinia* sp. is consistently associated with a lethal spear rot disease of African oil palm in Central America and Colombia. Rotting of the young, unfurled heart leaf spreads down to the apical growing point and destroys the plant. Lethal spear rot has destroyed more than 50% of the trees in plantations near the Atlantic coast of Colombia and Central America. Occasionally, under hot, wet conditions, *F. moniliforme* has attacked corn foliage (33). *Fusarium oxysporum* is the cause of a minor leaf spot disease of oil palm in Africa called patch yellows (16). In Nigeria, *F. lateritium* causes a leaf spot of the leafy vegetable *Celosia* (1).

Discussion

Padwick (28) classified losses caused by plant diseases in Great Britain's former colonies as: A, over 10%; B, important losses under 10%; and C, small or very small losses. Among the three major tropical grain crops, the following Fusarium diseases were recorded: Bakanae disease on rice, C; ear rot of corn, A; and a sorghum root rot (only in Bechuanaland), B. Millet, an important tropical dry-land grain, had no Fusarium diseases listed. No Fusarium diseases were reported on beans or the major root crops: sweet potatoes, yams, and cassava. As noted previously, *Fusarium* is associated with storage diseases of these important tropical food crops along with other fungi. Among the major export crops, no Fusarium diseases were reported on peanuts, sugarcane, tea, cocoa, coffee, cotton, rubber, and sisal. On oil palm, wilt rates B and patch yellows C. In addition, a survey of symposia on tropical plant pathology (3, 4, 45) shows only Fusarium wilt of bananas to be among the major tropical diseases discussed. Thurston (44) lists no *Fusarium* in his review of threatening tropical plant diseases. Thus far, the evidence indicates Fusarium diseases are not major causes of losses of important tropical food crops in comparison with the host of other pathogens that attack these crops, especially fungus foliage diseases and viruses. This situation could change as new, more intensive production systems for increased yields are introduced along with new, genetically uniform cultivars.

Of the tropical fusarial plant diseases, those caused by *F. oxysporum* and *F. moniliforme* are the most important. These two species account for 60% of the 25 tropical plant diseases listed in Table 11-1. The important Fusarium pathogens of cereal and pulse crops in temperate climates are not as destructive in the tropics, and in most tropical areas *Fusarium* spp. identified on these crops have not been studied. Also, the common wilt diseases of vegetables, cotton, and tobacco are seldom damaging in the tropics. In part, this is because resistant cultivars developed in temperate areas are now grown. In contrast to Fusarium wilt, bacterial wilt *(Pseudomonas solanacearum)* is widespread and destructive on Solanaceae and *Musa*.

Fusarium spp. with a perithecial stage are much more common inhabitants of aboveground plant parts in tropical areas than in temperate areas. They are frequently associated with senescent or necrotic tissue, and with cankers and die-back in tree crops, but seldom initiate an attack alone on sound tissue or vigorous trees. Plants subjected to stress as a result of disease, wounding, old age, poor growing conditions, and insect attacks are the most susceptible.

Among the most interesting of tropical Fusaria are the two species causing galls and flower proliferation. Little is known about the physiology of gall induction by *F. moniliforme* and *C. rigidiuscula*.

Acknowledgements.

The author is grateful to Dr. C. Booth and Dr. I. W. Buddenhagen for valuable information on tropical Fusaria. Dr. D. J. Allen kindly provided information on diseases of pulses in Africa. Dr. W.C. Snyder and Dr. Shirley Nash Cook made helpful suggestions, including information on the species attacking pepper and cocoa in Brazil.

Literature Cited

1. Afanide, B., S. A. Mabadeje, and S. H. Z. Naqvi. 1976. Fusarium leaf spot of *Celosia argentea* L. in Nigeria. Trans. Brit. Mycol. Soc. 66: 505–507.
2. Alconero, R., and A. G. Santiago. 1969. Fusaria pathogenic to Vanilla. Plant Dis. Rep. 53: 854–857.
3. American Phytopathological Society. 1962. Symposium on Tropical Plant Pathology. Phytopathology 52: 928–953.
4. American Phytopathological Society. 1973. Symposium on Tropical Pathology. Phytopathology 63: 1436–1454.
5. Booth, C. 1971. The genus *Fusarium*. Commonwealth Mycol. Inst., Kew, Surrey, England. 237 p.
6. Booth, C., and J. M. Waterston. 1964. *Gibberella xylarioides*. No. 24. *In* Descriptions of pathogenic fungi and bacteria, Commonwealth Mycol. Inst., Kew, Surrey, England.
7. Brunt, A. A. F., and A. L. Wharton. 1962. Etiology of a gall disease of cocoa in Ghana caused by *Calonectria rigidiuscula* (Berk & Br.) Sacc. Ann. Appl. Biol. 50: 283–289.
8. Burton, C. L. 1970. Diseases of tropical vegetables on the Chicago market. Trop. Agric. (Trinidad) 47: 303–313.
9. Cook, A. A. 1975. Diseases of tropical and subtropical fruits and nuts. Hafner Press, New York. 317 p.
10. Ebbels, D. L., and R. V. Billington. 1972. Fusarium wilt of *Voandzeia subterranea* in Tanzania. Trans. Brit. Mycol. Soc. 58: 336–338.
11. Fletchtmann, C. H. W., H. Kunati, J. C. Medcalf, and J. Ferre. 1970. Preliminary observations on malformation in mango (*Mangifera indica* L.), inflorescences, and the fungus, insects, and mites found on them. Anais da Escola Superior de Agricultura "Luiz de Queiroz" 27: 281–285. (Rev. Plant Pathol. 52: 162. 1973).
12. Gollifer, D. E., and R. H. Booth. 1973. Storage losses of taro corms in the British Solomon Islands Protectorate. Ann. Appl. Biol. 73: 349–356.
13. Gordon, W. L. 1956. The taxonomy and habitats of the *Fusarium* species in Trinidad, B. W. I. Can. J. Bot. 34: 847–864.
14. Gordon, W. L. 1960. The taxonomy and habitats of *Fusarium* species from tropical and temperate regions. Can. J. Bot. 38: 643–658.
15. Grifee, P. J. 1976. Fungi associated with crown rot of boxed bananas in the Windward Islands. Phytopathol. Z. 85: 149–158.
16. Hartley, C. W. S. 1967. The oil palm. Longmans, London. 706 p.
17. Hunter, J. E., and I. W. Buddenhagen. 1972. Incidence, epidemiology and control of fruit diseases of papaya in Hawaii. Trop. Agric. (Trinidad) 49: 61–71.
18. Joffe, A. Z. 1972. Fusaria isolated from avocado, banana, and citrus fruit in Israel and their pathogenicity. Plant Dis. Rep. 56: 963–966.
19. Kearns, H. G. H. 1963. Crop protection in the tropics. Ann. Appl. Biol. 51: 353–360.
20. Liu, L. J., and J. Mignucci. 1971. Effects of *Fusarium* spp. on germination and stem rot of sugarcane in Puerto Rico. J. Agric. Univ. Puerto Rico 55: 426–434.
21. Lukezic, F. L., W. J. Kaiser, and M. M. Martinez. 1967. The incidence of crown rot of boxed bananas in relation to microbial populations of the crown tissue. Can. J. Bot. 45: 413–421.
22. Martin, J. P., E. V. Abbott, and C. G. Hughes (eds.). 1961. Sugar cane diseases of the world. Vol. 1. Elsevier Publishing Co., New York. 542 p.
23. Meiffren, M., and M. Belin. 1960. Cafe-Cocoa-Thé. 4: 150–158.
24. Mercer, P. C. 1977. A pod rot of peanuts in Malawi. Plant Dis. Rep. 61: 51–55.
25. Ogundana, S. K., S. H. Z. Naqvi, and J. A. Ekundayo. 1970. Fungi associated with soft rot of yams (*Dioscorea* spp.) in storage in Nigeria. Trans. Brit. Mycol. Soc. 54: 445–451.
26. Ou, S. H. 1972. Rice diseases. Commonwealth Mycol. Inst., Kew Surrey, England. 368 p.
27. Ou, S. H. 1973. Contrasting pathological problems of rice under tropical and temperate climate. Abstr. No. 1046 *in* Abstracts of papers, 2nd Int. Congr. Plant Pathol., Minneapolis, Minn., U. S. A. Sept. 5–12, 1973.
28. Padwick, G. W. 1956. Losses caused by plant diseases in the colonies. Phytopathol. Pap. 1. Commonwealth Mycol. Inst., Kew, Surrey, England. 60 p.
29. Renfro, B. L., and A. J. Ullstrup. 1976. A comparison of maize diseases in temperate and in tropical environments. Pest Abstr. News Sum. (PANS) 22:491–98.
30. Review Applied Mycology. 1968. A bibliography of lists of plant diseases and fungi. Rev. Plant Pathol. 1. Africa 47: 553–558.
31. Review Plant Pathology. 1970. A bibliography of lists of plant diseases and fungi. Rev. Plant Pathol 49: 103–108.
32. Review Plant Pathology. 1971. A bibliography of lists of plant diseases and fungi. Rev. Plant Pathol.. 50: 1–7.
33. Schieber, E., and A. S. Muller. 1968. A leaf blight of corn (*Zea mays*) incited by *Fusarium moniliforme*. Phytopathology 58: 554. (Abstr.).
34. Siddiqui, M. A., and D. C. M. Corbett. 1963. Coffee bark diseases in Nyasaland. I. Pathogenicity, description, and identity of the causal organism. Trans. Brit. Mycol. Soc. 46: 91–101.
35. Singh-Dhaluval, T., J. H. Lopez Rosa, G. Steiner, L. Igaravidez, and A. Torres Sepulveda. 1963. Studies on coffee root rot and horticultural practices for its amelioration. Univ. Puerto Rico Agric. Expt. Stn. Tech. Paper 36. 30 p.
36. Stevenson, J. A. and F. L. Wellman. 1944. A preliminary account of the plant diseases of El Salvador. J. Washington Acad. Sci. 34: 259–268.
37. Stover, R. H. 1972. Banana, plantain, and abaca diseases. Commonwealth Mycol. Inst., Kew, Surrey, England. 316 p.

38. Stover, R. H. 1977. Fungicidal control of plant diseases in the tropics. *In* M. R. Siegel and H. D. Sisler (ed.), Antifungal compounds I, Marcel Dekker Inc., New York.
39. Subrahmanyam, P., and A. S. Rao. 1976. Fungi associated with concealed damage of groundnut. Trans. Brit. Mycol. Soc. 66: 551–552.
40. Summanwar, A. S., S. P. Raychaudhuri, and S. C. Phatak. 1966. Association of the fungus *Fusarium moniliforme* Sheld. with the malformation in mango (*Mangifera indica* L.). Indian Phytopathol. 19: 227–228.
41. Tarr, S. A. J. 1962. Diseases of sorghum, sudan grass and broom corn. Commonwealth Mycol. Inst., Kew, Surrey, England. 380 p.
42. Thorold, C. A. 1975. Diseases of cocoa. Clarendon Press, Oxford. 423 p.
43. Thresh, J. M. 1960. Capsids as a factor influencing the effect of swollen-shoot disease on cacao in Nigeria. Emp. J. Exp. Agric. 28: 193–200.
44. Thurston, H. D. 1973. Threatening plant diseases. Annu. Rev. Phytopathol. 11: 27–44.
45. Toussoun, T. A., R. V. Bega, and P. E. Nelson (eds.). 1970. Part VII. Root diseases of tropical plantation crops, p. 177–200. *In* Root diseases and soil-borne pathogens, Univ. California Press, Berkeley.
46. Trujillo, E. E. 1963. Fusarium yellows and rhizome rot of common ginger. Phytopathology 53: 1370–1371.
47. Wallbridge, A., and J. A. Pinegar. 1975. Fungi associated with crown rot disease of bananas from St. Lucia in the Windward Islands. Trans. Brit. Mycol. Soc. 64: 247–254.
48. Wellman, F. L. 1961. Coffee, botany, cultivation, and utilization. Leonard Hill Ltd., London, and Interscience Publishers Inc., New York. 488 p.

12 Fusarium Diseases of Ornamental Plants

Paul E. Nelson, R. K. Horst, and S. S. Woltz

The three diseases discussed in this chapter have been selected to illustrate several ways in which *Fusarium* species damage ornamental plants. Fusarium wilt of chrysanthemum, caused by *Fusarium oxysporum* f. sp. *chrysanthemi,* is an unusual vascular wilt disease because symptoms begin on the youngest foliage first and later progress to the older foliage. This symptom syndrome is especially striking on a susceptible cultivar such as Yellow Delaware and is unusual because symptoms of most Fusarium wilt diseases appear first on the lower or oldest foliage and later progress to the younger foliage.

Fusarium stem rot and stub dieback of carnation caused by *F. roseum* 'Graminearum' is a cortical rot of stems and roots. In addition to the obvious damage resulting from stem lesions and stub dieback, the pathogen may also be responsible for a reduction in the number of flowers cut, a delay in cropping time, and a reduction in flower grade.

The Fusarium disease of gladiolus, caused by *F. oxysporum* f. sp. *gladioli,* has been considered as a vascular wilt, a root rot, and a rot of stored corms. It is not clear whether there are several phases of the disease, all caused by the same pathogen, or whether there are several pathogens involved in this disease complex. The disease can be devastating in gladiolus grown for flower production and for corm production.

Fusarium Wilt of Chrysanthemum

The first report of Fusarium wilt of *Chrysanthemum morifolium* was by Brown (11) in 1939. The pathogen is listed as *Fusarium oxysporum* f. sp. *callistephi* in Agricultural Handbook 165 (62). Toop (59), working with the susceptible cultivar Encore, also reported *F. oxysporum* f. sp. *callistephi* as the causal agent. However, Horst (26) tested the isolate Toop used for pathogenicity on aster and found that it was non-pathogenic on three cultivars of aster. In 1965 Armstrong and Armstrong (2) reported that the disease on chrysanthemum was not caused by *F. oxysporum* f. sp. *callistephi* but rather by *F. oxysporum* f. sp. *tracheiphilum* race 1. They also worked with the cultivar Encore. In 1966 Littrell (32) reported on Fusarium wilt of cultivar Yellow Delaware and indicated that the isolate of *Fusarium* used in his study appeared to be a different biotype from the Fusaria previously reported to attack chrysanthemum. In 1970 Armstrong et al. (3) described a new forma specialis, *F. oxysporum* f. sp. *chrysanthemi* from the cultivar Yellow Delaware which is different from *F. oxysporum* f. sp. *tracheiphilum* race 1. Apparently two formae speciales of *F. oxysporum* attack chrystanthemum and they can be separated on the basis that f. sp. *tracheiphilum* race 1 attacks the cultivar Encore but not Yellow Delaware while f. sp. *chrysanthemi* attacks both cultivars.

The symptoms of Fusarium wilt on chrysanthemum are different from symptoms seen on most plants attacked by formae speciales of *F. oxysporum.* A good description of the symptoms on chrysanthemum is given by Engelhard and Woltz (17) and the description given below is taken from their paper. The initial symptoms on some susceptible cultivars exemplified by Yellow Delaware include unilateral chlorosis of one or more leaves at or near the stem apex and a slight to pronounced curvature of the chlorotic leaves and the stem towards the affected side of the plant. This is in direct contrast to the initial symptoms of wilting and chlorosis of the lower older leaves on most plants infected with Fusarium wilt fungi. As the disease on chrysanthemum progresses, chlorosis of affected leaves becomes more general and severely affected leaves wilt. Wilted leaves occur initially on the most severely affected side of the plant, but as the disease progresses the entire plant wilts and dies. More resistant cultivars may show only leaf chlorosis, stunting of leaves, and a reduced rate of growth. These symptoms can be so mild that they resemble a nutritional disorder or so severe that after a period of exhibiting leaf chlorosis and stem necrosis, death of the plant occurs. Black necrosis of the stem may develop and sometimes occurs as only a streak up one side of the stem. In some cases the black necrosis may occur in the upper parts of the stem and have no externally visible connection with the base or apex of the plant.

Vascular discoloration occurs in the stem and the leaves. When the stem is cut, a brown to reddish-brown discoloration of the vascular tissue may be evident when foliar symptoms occur or in symptomless plants. Externally visible vascular discoloration in the veins of affected leaves may occur on infected Yellow Delaware plants.

Initial symptoms may occur as early as 7 days after inoculation on a very susceptible cultivar such as Yellow Delaware but may not occur until 35 days after inoculation on a more resistant cultivar such as Bluechip. In some cases Yellow Delaware plants were dead 28 days after inoculation, while Bluechip had only mild symptoms 8 weeks after inoculation (17).

Fusarium wilt of chrysanthemum is often difficult to diagnose because of the variability of the symptoms expressed by different chrysanthemum cultivars and because the conditions that favor disease development and symptom expression have not been well understood in the past. The symptoms of Fusarium wilt may resemble those exhibited by plants affected by Pythium root rot, nutritional deficiencies, or excess water (17).

Engelhard and Woltz (18) studied the reaction of 47 cultivars of chrysanthemum that were inoculated with *F. oxysporum* f. sp. *chrysanthemi* and f. sp. *tracheiphilum* race 1. Of the cultivars inoculated with f. sp. *chrysanthemi*, 16 developed foliage symptoms and 24 developed vascular discoloration and the pathogen was consistently recovered from both groups of plants. In addition the pathogen was recovered from 11 cultivars that developed neither foliage symptoms or vascular discoloration. Of the cultivars inoculated with f. sp. *tracheiphilum* race 1, nine developed foliage symptoms and 15 developed vascular discoloration and the pathogen was consistently recovered from both groups of plants. In addition the pathogen was recovered from eight cultivars that developed neither foliage symptoms or vascular discoloration. Both pathogens were recovered more frequently from the basal stem portion of symptomless plants. Both pathogens could also be recovered from the bases of leaf petioles on plants with foliage symptoms.

Environmental factors such as temperature, nitrogen reaction, and soil reaction also are major factors influencing disease development (18). The most severe symptom expression occurs when soil temperatures are 28 C or higher. During periods of lower air and soil temperatures, the pathogens may be present in plants that appear to be symptomless but may be causing reduced plant vigor and growth in them. When soil temperatures are below 22 C, even very susceptible cultivars express mild disease symptoms or remain symptomless.

Woltz and Engelhard (68) showed that the use of nitrate-nitrogen as a fertilizer decreased the severity of symptom expression when compared with a fertilizer of half ammonium and half nitrate-nitrogen. The use of nitrate-nitrogen as a fertilizer together with calcium hydroxide to raise the soil pH decreased the severity of symptom expression additively. The severity of symptom expression generally decreased with increasing soil pH.

Fusarium wilt of chrysanthemum causes the most serious losses in warm tropical climates. However, even under these environmental conditions, infected plants of susceptible cultivars may remain symptomless. Cuttings taken from such plants serve as an efficient means of spreading the pathogens. The pathogens can also be spread in pieces of stem and leaf tissues from infected plants (18).

The Fusarium wilt pathogens can also survive for long periods of time in soil or on plant debris in soil (17). Therefore treatment of soil with steam or chemicals is an essential control measure. However, the basic control practices are dependent on a balanced management system utilizing the proper fertilizers, maintaining the proper soil pH, and utilizing systemic fungicides (16, 19). Complete control of disease symptoms can be obtained by using a high-lime, all-nitrate-nitrogen cultural regime along with drenches of a systemic fungicide such as benomyl. These conditions suppress symptom expression and are unfavorable for disease development, in contrast to an all-ammonium-nitrogen, low-pH regime without the use of a systemic fungicide, which results in the death of plants of susceptible cultivars in a very short time.

Fusarium Stem Rot and Stub Dieback of Carnation

Fusarium stem rot on florists' carnation *(Dianthus caryophyllus)* occurs worldwide wherever carnations are grown (23, 45, 49, 50, 58). The disease has been called Wilt, Stem Rot, Dieback, Branch Rot, Crown Rot, Collar Rot, Basal Rot, Basal Stem Rot, and Stem Fusariosis (23, 24, 49, 63, 64, 65). Three species of *Fusarium* have usually been reported to cause stem rot—*F. culmorum, F. avenaceum (Gibberella avenacea),* and *F. graminearum (G. zeae)* (10, 49). These three species have been grouped in the literature on this disease under the name *F. roseum* according to the classification of Snyder and Hansen (55), which has unfortunately resulted in some confusion on the specific identification of the pathogen. There was also some confusion in early reports with carnation wilt diseases associated with Fusarium stem rot. Clarification was given by Wickens (65) in a key to the identification of the separate carnation diseases based on symptoms. Fusarium stem rot was reserved for the non-vascular disease whereas the vascular wilts were termed Fusarium wilt caused by *Fusarium oxysporum* f. sp. *dianthi* and *Verticillium* or *Phialophora* wilt caused by *Phialophora cinerescens.*

Fusarium stem rot on carnations has been known since early 1900; the collar-rot and crown-rot stages were considered most important during production (63, 65), whereas the cutting rot stage was important during propagation (23, 64). Later reports described the importance of the basal or crown-rot stage (1, 8, 30, 31) and the significance of infested or infected cuttings (7, 27, 46, 47, 52, 54). The role of the stub dieback stage has also been described (49).

The symptoms of Fusarium stem rot are quite distinct. Four phases are distinguishable based on infection court and symptomatology. Fusarium basal rot causes serious loss of cuttings during propagation (53). Infection may be severe or mild depending on the environment and the amount of inoculum. Severely infected cuttings exhibit complete rotting at the base and up the stem, whereas mildly infected cuttings exhibit rotting on one side of the stem with root development on the opposite side. The typical rot symptom is reddish brown, and pink crusts of sporodochia may appear at the soil line or in leaf axils. Mild infections may appear on normally rooted cuttings as small brown spots at the base of the stem that serve as a source of inoculum for further disease development (53).

Fusarium basal stem rot has been described by numerous investigators (5, 6, 8, 25).

Symptoms appear after planting as a wilting or stunting resulting from a reddish-brown cortical rot. Plants may develop with no obvious external symptoms if the cortical rot does not completely girdle the stem until plants are placed under stress, i.e., flowering, temperature, or moisture (8, 48, 63).

Fusarium branch rot development is similar to basal stem rot except the site of infection is in branch junctions. Reddish-brown lesions may develop through the branch and into the main stem. First the branch wilts and dies, followed by wilting of other branches as the disease develops (23, 63).

Fusarium stub dieback infections develop at wounds left when flowers are cut or plants pruned (56). Symptoms first appear as a shrinking of the stub accompanied by dark discoloration and death of leaves at the top node. Reddish-brown lesions can be found internally in split stems. The pathogen is capable of growing down the main stem and killing the entire plant under favorable conditions.

The pathogen remains viable in infected tissues for many weeks (13, 63) and in soil either as chlamydospores or in plant debris for long periods of time (10, 51). Thus the source of inoculum is infested soil, plant debris, and contaminated tools (7, 23, 25, 65). Wounds are required for infection; however, wounds are unavoidably provided during planting, cutting removal, and flower cuts. Masses of characteristic pink sporodochia containing macroconidia are produced on infected tissues, most commonly under moist conditions (49, 65). Such conditions are frequently found within the canopy of mature carnation plantings. The macroconidia are spread by splashing water, on tools, or on dust particles. Splash droplets containing conidia may travel as far as 70 cm horizontally and as high as 10 to 20 cm above the soil surface (22). *Fusarium* species have been isolated from dust particles collected in greenhouses (49). Thus splash droplets and dust particles may distribute the pathogen throughout the greenhouse. Perithecia may also be formed on tissues infected with *F. graminearum (G. zeae)* under favorable environmental conditions. Optimum temperature for initiation of perithecia is 28 C and optimum temperature for ascospore release is 16 C (61). Ascospores are released from perithecia into the air under humid conditions or following wetting of perithecia or substrate (28, 60). The pathogen grows both laterally and longitudinally in the stem after infection. The rate of infection is affected by temperature (5), cultivar (5, 6, 65), and culture (15).

There is a significant correlation between increasing temperature from 13 to 26 C and disease severity. Between 13 and 18 C the severity of disease is not significantly different; however, infection occurs at these low temperatures without significant development of visible symptoms (56). The nature of losses to *Fusarium* are not only plant and branch death but also reduction in numbers of flowers cut, delay in cropping time, and reduction in flower grade (56). Thus if infection occurs at low temperatures without significant plant or stem death, subtle losses in flower production may still occur, with severity increasing with changing environmental conditions. Cultural practices which provide increased disease development are: rooted cutting storage, plants in reproductive growth, short stubs left after pruning (long stubs are not as susceptible), and contaminated cutting tools. Older stubs are more susceptible than young stubs, and more infection is found among plants in closely spaced plantings than in widely spaced plantings. There is good evidence, too, for resistance in many carnation cultivars to Fusarium stem rot (57). Unfortunately, some cultivars resistant to Fusarium stem rot are among those which McCain (39) lists as susceptible to Fusarium wilt. It is interesting that commercial cultivars used in the eastern United States, where Fusarium stem rot is a serious problem, are cultivars listed as more resistant to the disease. Therefore the choice of cultivars may well have been influenced by the Fusarium disease complex.

Control of Fusarium stem rot is somewhat complex and difficult with the type of cultural procedures now commonly used for growing carnations. Rooted carnation cuttings are planted directly in greenhouse benches and remain in place for the entire flowering season. In addition, 2- and 3-year crops are often used in this system, i.e., the plants remain in place for 2 and 3 years. There are numerous cultural advantages to this practice, but the limiting factor is often Fusarium stem rot. Also, the mother block system for production of cuttings from culture-indexed (indexed for wilt pathogens) reselected foundation stock has improved the quality of cuttings available to the grower; however, the production of such stock is concentrated in the hands of a few, large, specialist propagators. Unfortunately, the mother block system favors buildup of the Fusarium stem rot pathogen (15), which necessitates elaborate preventative control measures that are not completely successful (8, 27). Systemic fungicides have provided some control, along with simpler procedures (7, 31, 46, 47). Furthermore, biological control offers some interesting possibilities (43, 44).

Fusarium Disease of Gladiolus

Fusarium disease of gladiolus (38, 42) has been reported repeatedly as being caused by *Fusarium oxysporum* f. sp. *gladioli* (4, 9, 12, 14, 20, 33). The disease may have different phases according to the predominant symptom expression, for example, corm rot, root rot, or vascular wilt disease (36). Plants are attacked by the organism in varying ways apparently regulated by the sequence of events in infection, nutrition, cultivar resistance, and environmental conditions (21, 29, 34, 35, 37, 41, 66). Disease expression in a specific situation is usually characterized by a portion of the following list of symptoms attributed to the disease. Root infection shows as individual brown lesions on a root or generally rotted roots. Rotting of the roots may extend into and through vascular bundles in the corm. Corm rot may occur in storage or as the corm is growing. Storage tissue rot is called brown rot because of the color of the diseased tissue. Basal plate infection may occur, resulting in basal rot. Corm infection may result in discoloration or decomposition of the vascular bundles, creating a vascular type disease syndrome. The corm, when examined in the field, may have wiry, brown roots and dark brown root traces where tissue has decomposed and dried out, leaving tunnels in the corm. Leaf infection, usually below the soil, may be independent of or associated with corm infections. Leaf "yellows," a chlorotic condition caused by *Fusarium,* may or may not be present in diseased plants. One-sided infection of the corm often results in curving of leaves away from the rotted side of the corm. Flower spike characters may be modified by slight infections.

Questions have arisen as to potential specialization of biotypes of the fungus involved in the various phases of the disease. The phases, as described above, primarily are those involving root, corm, vascular, or leaf tissue. Also, the question arises as to whether species other than *F. oxysporum* could be involved. Complete definitive answers are not available at present.

The question of specialization by *F. oxysporum* f. sp. *gladioli* has been partially resolved. It does not appear from the literature (12, 20) that any clear-cut differences have been established for *F. oxysporum* pathogens of gladiolus in regard to the pattern of disease symptoms produced or specificity for a particular type of tissue of the plant. The results are not, however, definitive; it most likely will be difficult to establish firm information in this area because of wide variability in the characteristics of the host material and the pathogen. Gladiolus corm stocks available to investigators have generally harbored a variable microflora population, commonly including *F. oxysporum.* Previous corm treatments together with a varied chemical and physical environment during experiments often result in a variable response. Variability in cultures of *F. oxysporum* is well substantiated (9, 12, 40). Gladiolus cultivar variability further confuses the issue in that resistance to the different phases of the disease is not predictable.

Controlled inoculation procedures have been used minimally in research on the gladiolus Fusarium disease (9, 12, 14, 20, 67, 69). This is due to complications from competitive native microflora in plant material and soils, as well as to the fact that the experimental corm stock often is infected with diverse forms of *Fusarium*. Corm stocks free of pathogens and other microflora have not been readily available to researchers. Only stock grown gnotobiotically, from seedling and tissue culture propagules, would be free of complicating microorganisms. These will be available more readily with the passage of time, permitting more closely controlled, definitive research.

Studies with Florida soils and a number of horticultural crops (discussed in Chapter 31, below) have indicated that the imposition of cultural variations in soil management can be employed to greatly reduce the effects of *F. oxysporum* wilt diseases. Results from experiments with gladiolus have not been decisive (70), but adequately demonstrate the possibility of reducing Fusarium disease losses by amending soils with lime sufficiently to maintain pH levels in the approximate range of 6.5 to 7.0. More acid reactions are associated with disease development. Nitrogen furnished as 90% nitrate, 10% ammoniacal is much less supportive of disease than nitrogen furnished in reciprocal proportions. High rates of nitrogen fertilization are more favorable to corm rot than low rates.

As will be discussed in the chapter on Fusarium nutrition (chap. 31), it is possible to severely limit development of the inoculum potential of *F. oxysporum* by adjusting soil fertility regimes, especially in soils of low native fertility. This procedure appears to hold the greatest potential benefit in gladiolus culture. Plantings of disease-free propagation material should be maintained at a very low internal inoculum potential by growing corms in "clean" soil with maintained "clean" sanitation practices augmented by fertility regimes that are inimical to multiplication and virulence-development of *F. oxysporum*. In a secondary manner, these regimes apparently retard the proliferation of *Fusarium* in gladiolus plants, corms, and cormels. Current cultural practices involve avoidance of practices favoring *Fusarium*, such as the excessive use of nitrogenous fertilizers, especially in the organic and ammoniacal forms, and the use of adequate liming levels.

The possibility has been investigated that other *Fusarium* species in addition to *F. oxysporum* might cause diseases of gladiolus (71). Twelve isolates from gladiolus, potato, and caladium were used to inoculate large lots of the cultivar Pink Friendship gladiolus corms obtained from commercial culture. Corms were inoculated prior to placing them in cold storage. Upon removal from storage, corms were examined and appeared free from storage rot. Six of the twelve isolates caused economic crop failures in that from eight replications of 40 corms each, inadequate numbers of plants emerged to continue cultural operations. Four of the remaining isolates significantly reduced corm or flower yield compared to the control series. One *Fusarium* species (*F. moniliforme* 'Subglutinans') apparently favored crop growth, perhaps by antagonism to the indigenous Fusaria present in the corm stock. In addition to *F. oxysporum*, four other species or cultivars appeared to be pathogenic. These were: *F. moniliforme* 'Subglutinans,' *F. roseum* 'Sambucinum,' *F. roseum* 'Culmorum' and *F. solani*.

The *F. oxysporum* disease certainly is more variable than most plant diseases for a number of reasons. The pathogen is extremely variable, occurring in an everchanging array of biotypes. The gladiolus plant and corm are subject to a wide variety of physiological changes during normal culture and normally support a varied internal microflora population. Since gladiolus cultivars are bred to meet exacting horticultural requirements, selection of parent material is severely limited. Plant breeding thereby suffers constraints in use of genetically resistant plant material. Soil, the most heterogeneous environment available, is the common meeting ground for pathogen (*F. oxysporum*) and host (gladiolus). It is apparent from the foregoing points that such complicating factors must be taken into account in gladiolus-*Fusarium* research. As developments in methodology are forthcoming, the disease will be better understood.

Literature Cited

1. Andreucci, E. 1958. Indagini su una grave alterazione del Garofano produtta da *Fusarium roseum* (Lk.) S. & H. Nuovo G. Bot.-Ital. NS 65:163–195.
2. Armstrong, G. M., and Joanne K. Armstrong. 1965. Wilt of chrysanthemum caused by race 1 of the cowpea Fusarium. Plant Dis. Rep. 673–676.
3. Armstrong, G. M., Joanne K. Armstrong, and R. H. Littrell. 1970. Wilt of chrysanthemum caused by *Fusarium oxysporum* f. sp. *chrysanthemi*, forma specialis nov. Phytopathology 60:496–498.
4. Aycock, R., and F. A. Haasis. 1963. Corm treatments for control of Fusarium disease of gladiolus. N. C. Agric. Exp. Stn. Tech. Bull. No. 154. 24 p.
5. Baker, R. 1955. Resistance of some carnation varieties to Fusarium stem rot. Colo. Flower Growers Bull. 64:2–3.
6. Baker, R. 1955. Resistance of some carnations to Fusarium stem rot in the nurse bed. Colo. Flower Growers Bull. 73:1–2.
7. Baker, R., and N. Denoyer. 1973. Fusarium stem rot of carnations: Control using systemic fungicides as sprays on mother blocks. Colo. Flower Growers Bull. 272:2–3.
8. Baker, R., and J. Tammen. 1954. Fusarium stem rot of carnation. Colo. Flower Growers Bull. 58:1–3.
9. Bald, J. G., T. Suzuki, and A. Doye. 1971. Pathogenicity of *Fusarium oxysporum* to Easter lily, narcissus and gladiolus. Ann. Appl. Biol. 67:331–342.
10. Booth, C. 1971. The genus *Fusarium*. Commonw. Mycol. Inst., Kew, Surrey, England. 237 p.
11. Brown, C. C. 1939. Contribution toward a host index to plant diseases in Oklahoma. Oklahoma Agric. Exp. Stn. Mimeo. Circ. 33. 73 p.
12. Bruhn, C. 1955. Untersuchungen uber die Fusarium-krankheit der gladiolen. Phytopathol. Z. 25:1–38.
13. Burgess, L. W., and D. W. Griffin. 1968. The recovery of *Gibberella zeae* from wheat straws. Australian J. Exp. Agric. Anim. Husb. 8:364–370.
14. Buxton, E. W., and N. F. Robertson. 1953. The Fusarium yellows disease of gladiolus. Plant Pathol. 2:61–64.
15. Dimock, A. W. 1958. Reports on carnation diseases. Carnation Craft 43:1–5.
16. Engelhard, A. W. 1974. Effect of pH and nitrogen source on the effectiveness of systemic fungicides in controlling Fusarium wilt of shrysanthemums. Proc. Amer. Phytopathol. Soc. 1:121 (Abstr.).
17. Engelhard, A. W., and S. S. Woltz. 1971. Fusarium wilt of chrysanthemum: symptomatology and cultivar reactions. Proc. Florida State Hort. Soc. 84:351–354.
18. Engelhard, A. W., and S. S. Woltz. 1973. Pathogenesis and dissemination of the Fusarium wilt pathogens of chrysanthemum. Phytopathology 63:441 (Abstr.).
19. Engelhard, A. W., and S. S. Woltz. 1973. Fusarium wilt of chrysanthemum: complete control of symptoms with an integrated fungicide-lime-nitrate regime. Phytopathology 63:1256–1259.
20. Forsberg, J. L. 1955. Fusarium disease of gladiolus: Its causal agent. Bull. Illinois Nat. History Survey 26:477–503.
21. Gould, C. J. 1949. Influence of climate on incidence of Fusarium rot and dry rot in gladiolus corm. Phytopathology 39:8 (Abstr).
22. Gregory, P. E., E. J. Guthrie, and M. E. Bunce. 1959. Experiments on splash dispersal of fungus spores. J. Gen. Microbiol. 20:328–354.
23. Guba, E. F. 1945. Carnation wilt diseases and their control. Mass. Agric. Exp. Stn. Bull. 427. 64 p.
24. Hellmers, E. 1960. Nellikens rodhalofusariose, stab-fusariose og hvidkarfusariose some aarsager til nedvisning of drivhusnelliker. Horticultura 14:90–129.
25. Holley, W. D., and R. Baker. 1963. Carnation production. Wm. C. Brown & Co., Dubuque, Iowa. 142 p.
26. Horst, R. K. 1965. Pathogenic and enzymatic variations in *Fusarium oxysporum* f. *callistephi*. Phytopathology 55:848–851.
27. Horst, R. K., and P. E. Nelson. 1968. Losses from Fusarium stem rot caused by *Fusarium roseum* in commercial production of cuttings of carnation, *Dianthus caryophyllus*. Plant Dis. Rep. 52:840–843.
28. Ishii, H. 1961. Studies in the epidemiology of scab disease of wheat and barley caused by *Gibberella zeae* (Schw.) Petch. Tokushima Agric. Exp. Stn. Bull. 1961:1–121.
29. Jones, R. K., and J. M. Jenkins, Jr. 1975. Evaluation of resistance in *Gladious* sp. to *Fusarium oxysporum* f. sp. *gladioli*. Phytopathology 65:481–484.
30. Kinnaman, H. R., and R. Baker. 1973. Fusarium stem rot of carnations: Uptake of benomyl by mature plants. Colo. Flower Growers Bull. 275:1–4.
31. Kinnaman, H. R., and R. Baker. 1974. Fusarium stem rot of carnations: Inhibition of *Fusarium roseum* in benomyl treated plants. Colo. Flower Growers Bull. 293:1–3.
32. Litttrell, R. H. 1966. Effects of nitrogen nutrition on susceptibility of chrysanthemum to an apparently new biotype of *Fusarium oxysporum*. Plant Dis. Rep. 50:882–884.
33. Magie, R. O. 1953. Some fungi that attack gladiolus, p. 601–607. *In* A. Stefferud (ed.), Plant Diseases, the yearbook of agriculture, 1953, U.S. Dept. Agric., Washington, D.C.
34. Magie, R. O. 1960. Breeding gladiolus for Florida. Proc. Fla. State Hort. Soc. 73:375–378.
35. Magie, R. O. 1971. Carbon dioxide treatment of gladiolus corms reveals latent Fusarium infection. Plant Dis. Rep. 55:340–341.

36. Magie, R. O., and W. G. Cowperthwaite. 1954. Commercial gladiolus production in Florida. Fla. Agric. Exp. Stn. Bull. 535. 67 p.
37. Marshall, B. H. 1953. Relation of wound periderm in gladiolus corms to penetration by *Fusarium oxysporum* f. *gladioli.* Phytopathology 43:425–431.
38. Massey, L. M. 1926. Fusarium rot of gladiolus corms. Phytopathology 16:509–523.
39. McCain, A. H. 1974. Evaluations of resistance of carnation varieties to Fusarium wilt. A progress report. Calif. Plant Pathol. 20:2–3.
40. McClellan, W. D. 1945. Pathogenicity of the vascular Fusarium of gladiolus to some additional irridaceous plants. Phytopathology 35:921–930.
41. McClellan, W. D., and N. W. Stuart. 1947. The influence of nutrition on Fusarium basal rot of narcissus and on Fusarium yellows of gladiolus. Amer. J. Bot. 34:88–93.
42. McCulloch, L. 1944. A vascular disease of gladiolus caused by *Fusarium.* Phytopathology 34:263–287.
43. Meyers, J. A. 1971. Biological control of Fusarium stem rot. Colo. Flower Growers Bull. 252:1–2.
44. Michael, A. H., and P. E. Nelson. 1972. Antagonistic effect of soil bacterial on *Fusarium roseum* 'Culmorum' from carnation. Phytopathology 62:1052–1056.
45. Moreau, M. 1953. La Fusariose de l'oeillet dans la région parisienne. Rev. Hort. (Paris) 125:930–932.
46. Nash, C. T., and R. Baker. 1972. Fusarium stem rot of carnations: Control using systemic fungicides in rooting hormone. Colo. Flower Growers Bull. 272:1–2.
47. Nash, C. T., and R. Baker. 1973. Fusarium stem rot of carnations: Chemotherapeutic control in propagated cuttings from mother blocks treated with systemic fungicides. Colo. Flower Growers Bull. 273:3–6.
48. Nelson, P. E. 1960. Fusarium stem rot of carnations. New York State Flower Growers Bull. 171:1–3.
49. Nelson, P. E., B. W. Pennypacker, T. A. Toussoun, and R. K. Horst. 1975. Fusarium stub dieback of carnation. Phytopathology 65: 575–581.
50. Nilsson, G. L. 1962. A survey of carnation diseases in south Sweden. Plant Dis. Rep. 46:152–155.
51. Nyvall, R. F. 1970. Chlamydospores of *F. roseum* 'Graminearum' as survival structures. Phytopathology 60:1175–1177.
52. Petersen, L. J., and R. Baker. 1959. Control of Fusarium stem rot of carnations: II. The use of dips and drenches. Plant Dis. Rep. 43:1209–1212.
53. Phillips, D. J. 1962. Histochemical and morphological studies of carnation stem rot. Phytopathology 52:323–328.
54. Phillips, D. J. 1965. *Fusarium roseum* and the carnation shoot tip. Colo. Flower Grower Bull. 182:1–3.
55. Snyder, W. C., and H. N. Hansen. 1945. The species concept in *Fusarium* with reference to discolor and other sections. Amer. J. Bot. 32:657–666.
56. Stack, R. W. 1976. Fusarium stub dieback of carnation: Etiology and epidemiology. Ph.D. Thesis, Cornell Univ., Ithaca, N.Y. 189 p.
57. Stack, R. W., R. K. Horst, P. E. Nelson, and R. W. Langhans. 1976. Differential susceptibility to Fusarium stub dieback in carnation cultivars. J. Amer. Soc. Hort. Sci. 101:654–657.
58. Tammen, J. 1954. Relation of various clones of *Fusarium roseum* to the etiology of the foot diseases of *Dianthus caryophyllus* and *Triticum vulgare.* Ph.D. Thesis. Univ. Calif., Berkeley. 118 p.
59. Toop, E. W. 1963. The effect of pre-inoculation treatment of rooted chrysanthemum cuttings on subsequent vascular wilt development. Plant Dis. Rep. 47:284–287.
60. Tschanz, A. T., R. K. Horst, and P. E. Nelson. 1975. Ecological aspects of ascospore discharge in *Gibberella zeae.* Phytophatology 54:597–599.
61. Tschanz, A. T., R. K. Horst, and P. E. Nelson. 1976. The effect of environment on sexual reproduction of *Gibberella zeae.* Mycologia 68:327–340.
62. U. S. Dept. Agric. 1960. Index of plant diseases in the United States. Agric. Handbook No. 165. 531 p.
63. White, H. L. 1929. Wilt diseases of the carnation. 14 Ann. Rep. Cheshunt Exp. Stn. 1928:62–75.
64. White, H. L. 1938. Stem-rot and wilt of the perpetual flowering carnation. Sci. Hort. 6:86–92.
65. Wickens, G. M. 1935. Wilt, stem rot and dieback of the perpetual flowering carnation. Ann Appl. Biol. 22:630–683.
66. Wilfret, G. J., and S. S. Woltz. 1973. Susceptibility of gladiolus cultivars to *Fusarium oxysporum* f. sp. *gladioli* Snyd. & Hans. at different temperatures. Proc. Fla. State Hort. Soc. 86:376–378.
67. Woltz, S. S. 1974. Gladiolus Fusarium disease: Assay of soil-borne inoculum potential and cultivar susceptibility. Plant Dis. Rep. 58:184–187.
68. Woltz, S. S., and A. W. Engelhard. 1973. Fusarium wilt of chrysanthemum: Effect of nitrogen source and lime on disease development. Phytopathology 63:1555–1557.
69. Woltz, S. S., and R. O. Magie. 1973. Gladiolus corm rot: A method of cross indexing pathogen isolates and host cultivars for virulence-susceptibility reactions. Plant Dis. Rep. 57:957–960.
70. Woltz, S. S., and R. O. Magie. 1975. Gladiolus Fusarium disease reduction by soil fertility adjustments. Proc. Fla. State Hort. Soc. 88:559–562.
71. Woltz, S. S., R. O. Magie, Constance Switkin, P. E. Nelson, and T. A. Toussoun. 1977. Gladiolus disease response to pre-storage corm inoculation with *Fusarium* species. Plant. Dis. Rep. 62:134–137.

13 Fusarium Diseases of Flowering Bulb Crops

R. G. Linderman

Introduction

Flowering bulbous plants grow almost everywhere in the world, either as native plants or as introduced cultivated forms. For the few countries that are major commercial producers, cultivated ornamental bulbous crops constitute an agricultural industry of great economic significance (23, 54).

Among the most important of the diseases that occur on flowering bulb crops are those caused by species of *Fusarium*. Fusarium diseases that occur on the major true flowering bulbs, i.e., *Narcissus, Tulipa, Iris,* and *Lilium,* as compared to corms of *Gladiolus, Crocus,* and *Freesia* (54), are the subjects of this chapter. Emphasis will be placed on those aspects of infection and epidemiology that are unusual or unique because of the bulbous habit of these plants, and the special conditions that occur during field growing, harvest, shipping, and storage, and ultimately during forcing into flower.

While a number of *Fusarium* species have been isolated from flowering bulbs, only two have been demonstrated to cause disease. *Fusarium roseum* (60) was shown by Beaumont and Buddin (8) to cause a leaf blight on tulips in glasshouses and in the field in Europe when temperatures and humidity were high. Gould (20) observed this disease occasionally on leaves and stems of field-grown tulips in the northwestern U.S.A., but it does not appear to be significant.

The main pathogenic species of *Fusarium* on flowering bulb crops is *F. oxysporum* (59) *(=F. bulbigenum),* which will be given primary emphasis in this chapter.

The Bulbous Habit

Bulb morphology Rees (54) describes a true bulb as "an organ consisting of a short stem bearing a number of swollen fleshy leaf bases or scale leaves, with or without a tunic, the whole enclosing the next year's bud." The bulbous habit occurs almost exclusively in monocotyledons, which typically lack a cambium. Despite the absence of a cambium, monocotyledonous plants have cell clusters at various locations that remain meristematic and provide opportunity for development of adventitious buds. These meristematic cells are generally localized at nodes so that adventitious structures such as lateral buds and roots occur only at nodes. The result is the formation of a compressed area of underground stem nodes and internodes capable of giving rise to roots and shoots. This area is called the basal plate and anatomically is made up of a composite of foliar, root, and axial tissues. This area is the prime target for pathogenic *F. oxysporum* to infect, resulting in the disease called "basal rot."

All the bulb plants considered here do not form the same type of bulb (54). The tulip bulb type is composed entirely of concentric scales (leaf-like organs, as compared to true, photosynthetic leaves, which may produce aerial green parts) separated by very short internodes. The outer scale or "tunic" starts out white and fleshy like other scales, but shortly after the top senesces, it darkens and becomes tough and papery, probably due to transfer of starch food reserve materials to the outermost daughter bulb. The importance of this outer scale transformation will be discussed later in regard to resistance to *Fusarium* infection. In narcissus the bulb is made up of both scales and leaf bases, the scales surrounding the foliage leaf bases and flower axis. In bulbous iris, the bulb is made up of scales and leaf bases. The bulb is enclosed at maturity by a number of brown fibrous tunics, inside of which are white swollen scales whose edges just meet. The lily bulb is comprised of densely crowded, spirally arranged thick scale leaves which are not concentric as in the other types mentioned. Basal roots are formed from the lower surface of the basal plate; stem roots and stem bulblets are formed on the stem above the bulb but generally below the soil surface (54).

Bulb propagation The bulbous habit provides for easy propagation of daughter bulbs at numerous sites on the plant such as bulb scales, foliage leaf bases, and along the underground portion of the stem. In narcissus, each terminal daughter bulb in the branching system is replaced by two daughter bulbs. Daughter bulb production can be induced artificially by procedures called "scooping" or "cross-cutting." The base of the bulb is wounded to kill the shoot to remove apical dominance, and numerous daughter bulbs will then form on the cut edges of the scales.

In tulips and iris, daughter bulbs are borne in the axils of the scales which enclose an axis bearing leaves and terminating in a flower. Each mother bulb planted in the autumn dies and is replaced by daughter bulbs.

In lilies, bulblets form on the below-ground portion of the stem or at the base of scales that have been removed from the bulb. The principle involved in lily scale propagation is the same as for narcissus. Scales are removed from the bulb to remove apical dominance, and new bulblets usually form near the base of the scale. According to Rees (54), the outer and middle scales are preferred because they produce more bulblets. The use of outer scales that have been exposed to soil containing *Fusarium* may result in early infections of scale bulblets, since these scales are more often infected with *F. oxysporum* capable of causing basal rot than the inner scales (R. G. Linderman, *unpublished data*).

Narcissus Basal Rot

History and Etiology Moore (51) has reviewed the history of this disease. The first reported disease occurred during the hot summer of 1911 when the bases and scales of narcissus bulbs rotted and turned a red- or grey-brown in storage in England and the Netherlands (36, 68). However, before the determination of *F. oxysporum (=F. bulbigenum)* as the causal agent of the disease (36, 68), growers were well acquainted with the very troublesome "basal rot" of narcissus bulbs. Accordingly, Dod (19) described a basal rot of narcissus in 1894 that he attributed to excessive N fertilizer. Affected plants were stunted, the tips of the leaves turned brown, and the flowers developed imperfectly. At digging, the basal plates of affected bulbs were soft and rotten and few or no roots were present (36, 68). In addition, Westerdijk (68) showed that *Fusarium* could rot narcissus bulbs more rapidly at 26 or 30 C than at 18 or 20 C.

In the next few years, confusion occurred about the true cause of basal rot, since nematodes were frequently found to be closely associated with the bulb rot. However, Westerdijk (69) and Beaumont and Hodson (9) distinguished between the symptoms caused by *Fusarium* and the nematodes. *Fusarium*-infected narcissus in the absence of the nematode were observed, suggesting that the fungus was often followed by the nematode (9).

Basal rot of narcissus became an even more important disease during the mid- to late 1920s as narcissus culture expanded from Europe to other countries. The concommitant occurrence of the disease in new areas of narcissus culture provided a clue to early workers in the U.S.A. (49, 50, 64) that the causal agent was carried in or on the bulbs. Control efforts from that point on were primarily attempts to disinfect infected bulbs. Wedgeworth (64) successfully controlled the disease with some of the early mercuric fungicides and concluded that the organism causing narcissus root rot could be carried on the bulbs. After that, control efforts were aimed largely at disinfecting bulbs. It was known that narcissus bulbs could also carry bulb nematodes and bulb flies, and the U. S. Federal Quarantine Act accordingly required that all bulbs imported into the U.S.A. be hot water treated to remove those pests. However, U.S. growers soon began to complain of increased basal rot following the hot water treatment. Miles (49, 50) showed experimentally that their claims were justified, that basal rot was increased by treatment in hot water (43.3 C for 2.5–3 hr.), or in water at ambient temperature. He concluded that water at 43.3 C did not kill the pathogen, but served to effectively disseminate spores to otherwise healthy bulbs. The addition of mercuric compounds to the hot water dip tank did effectively disinfect bulbs of nematodes, bulb flies, and the basal rot pathogen.

Beaumont (7) stated that the enhancement of narcissus bulb rotting after hot water treatment was not due to *Fusarium*, but to secondary organisms. He challenged Miles' (49, 50) conclusion that the bath spread *Fusarium* spores, since direct evidence that spores were present in the water was not presented.

The subject of the etiology of the narcissus basal rot disease was not completely settled in the U.S.A until Weiss (65) reported that the *Fusarium* associated with the disease was primary, beginning its infection of the roots and basal plate in the field, and that subsequent handling and storage only modified that initial infection. Further, he reported that the *Fusarium* consistently isolated from diseased specimens was pathogenic only on narcissus and that cultivars differed in susceptibility to basal rot. His test showed also that microconidia of the pathogen could survive in water at 43.3 C for at least 8 hr (66).

Gregory (24) reported on the Fusarium bulb rot of narcissus as it occurred in the U.S. Hawker (26, 27, 29) confirmed and added to the results of Gregory (24) and McWhorter and Weiss (48); these basic findings have been modified only slightly by subsequent work.

Symptoms and disease development The first sign of basal rot of narcissus is a reddish-brown discoloration of the basal plate (Fig. 13-1). Infected bulbs feel soft at first, but later during storage become brittle, shrunken, and mummified (29). White hyphae are often visible between infected scales (24).

In the field infected plants show basal rot symptoms, and roots are few or lacking. In severe cases, above-ground symptoms consist of dwarfed leaves, flowers, and flower stalks. In less severe cases, above-ground plant parts may appear normal. When infected bulbs are planted, poor stands result, and plants are weak with yellow leaves. Such plants usually do not flower (24). Planting bulbs with patches of visible decay often results in many grass-like leaves emerging from secondary bulbs that result from splitting due to infection of the basal plate (24).

Gregory (24) reported that infection usually occurred through the basal plate, but he found no evidence that basal infections were the result of field infections which began in the roots and then progressed upward into the basal plate. Hawker (26) showed that when inoculations occurred in autumn, when moisture was adequate and temperatures were rather high, the pathogen could penetrate and destroy roots of all cultivars tested. Further, the pathogen subsequently grew up the roots into the basal plate. She also showed (26, 28) that bulbs may become infected via older, dying roots near the end of the growing season when conditions are favorable. McWhorter and Weiss (48) suggested that infections by this pathogen could occur through injuries of the bulb scales. Hawker (27) also confirmed Gregory's work (24) showing that basal rot is enhanced by hot water treatment if treatment is done in the autumn. When she added spores to the treatment tanks, penetration usually occurred directly into the basal plate. She also demonstrated that during the normal treatment time (late August to early September) bulbs are in their deepest dormancy and exhibit minimum susceptibility (26).

Gregory (24) and Hawker (27) both demonstrated that high humidity and temperatures of 27 C or above during storage aggravate the disease, and that rapid drying after lifting reduced losses. Hawker (27) explained the growers' observations that early planting often reduced losses, on the basis of soil temperatures in the field being lower than those in warehouse storage, thus reducing field infections.

Hawker (29) pointed out that the pathogen is able to exist saprophytically in soil for an unknown period of time. Further, bulbs which rot in storage are almost certainly contaminated or infected in the field, but symptoms occur only when temperatures increase during transit or handling. She noted that losses were least when bulbs were lifted in dry weather from a light soil.

From the early 1940s to the present, numerous papers have been published on narcissus basal rot, but these reports relate primarily to fungicidal control trials. In 1947, McClellan and Stuart (46) reported the results of studies on the influence of nutrition on the disease. The results along with those of an earlier report (62) indicated that certain growth-regulating substances, naphthaline acetamide, indolebutyric acid, indoleacetic acid, naphthaline acetic acid, uric acid, guanidine, and allantoin, used because of their potential to increase narcissus flower and bulb production, enhanced growth of *F. oxysporum* f. sp. *narcissi* in culture and increased disease incidence on narcissus when applied to bulbs before planting. Further, when narcissus bulbs were grown in the field and fertilized with combinations of inorganic nitrogen, phosphorus, and potassium, the most basal rot occurred with nitrogen and phosphorus amendment, and the least with potassium. The same relationship held for the pathogen grown in culture. Basal rot enhancement by amendment with organic fertilizers probably was due to the presence of some of the growth-regulating substances mentioned above in such fertilizers (46).

While most of the early workers pointed out that narcissus basal rot was most severe under warm, moist conditions, no one collected data to support those observations in the field until McClellan did so in 1952 (43). He showed that root infections and subsequent basal rot was more severe at soil temperatures of 23.9 and 29.4 C than at 18.3 C, though infection occurred at 12.8 C or below.

Cultivar susceptibility and host specificity Early workers (24, 29, 48) pointed out that basal rot occurs primarily on the large trumpet cultivars, but most of the Incomparabilis, Barri, and Leedsi cultivars are resistant. Weiss (65) and McWhorter and Weiss (48) reported that the narcissus basal rot pathogen was pathogenic only on narcissus. This is one of the earliest reports of host specialization of the Fusaria on flowering bulb crops. In 1940, Snyder and Hansen (59) published their revision of *Fusarium* taxonomy, and established that the narcissus basal rot pathogen was a specialized form of *F. oxysporum* which they designated *F. oxysporum* f. sp. *narcissi*.

Disease control Control of narcissus basal rot, as summarized by Hawker (29) could be achieved by careful control of conditions during storage, transport, and handling; by lifting in dry weather, if possible, along with early cleaning and drying; by incorporating 0.2–0.5% formalin in the hot water treatment tanks; or by cold fungicidal treatments soon after lifting, but independent of hot water treatments (48, 67); and by discarding heavily infested stocks.

Iris Basal Rot

History, etiology, and symptomotology Creager (17) reported a root and bulb rot of Spanish and Dutch iris on Long Island, N.Y., in 1929, observing that the pathogen entered the roots and progressed up to the basal plate and scales. He isolated and demonstrated the pathogenicity of *F. oxysporum.* No further description of the disease occurred until 1950 when Gould (21) described the symptoms and distribution of the disease in the northwestern U.S. Diseased plants had stunted and yellow leaves, with few or no roots. Infection first occurred in the roots, and the pathogen moved up into the base of the bulb, which became brown or reddish-brown and shrunken with a somewhat firm dry rot, with a definite margin between healthy and decayed tissue.

McClellan (42) reported that the vascular *Fusarium* of gladiolus was pathogenic to some other iridaceous plants, including iris, and caused stunting of plants and a dark basal rot of the bulbs and rotted roots. Browned vascular tissues were noted in the bulb scales, occasionally extending upward into the stem, but usually discoloration was confined to the basal plate. He also reported the close association of a *Penicillium* sp. with bulbs with *Fusarium* basal rot. Often the *Penicillium* sp. was fruiting freely between rotting bulb scales; and in such cases the bulbs often developed a soft rot instead of the typical firm rot that occurred with *Fusarium* alone.

Apt (2) induced the typical basal rot disease on iris, but also reported that the stunted leaves on infected plants became chlorotic and badly curled or twisted. When bulbs infected with *Fusarium* are forced, this twisting or "buck-horn" symptom is often very pronounced and distinctive (Fig. 13-2).

Cultivar susceptibility and host specificity Cultivars of iris are known to vary widely in their susceptibility to Fusarium basal rot. For example, De Wit is notably susceptible, whereas Wedgewood is decidedly resistant (22). In general, yellow cultivars are more susceptible than blue and white cultivars.

McClellan (42) also isolated a *Fusarium* from bulbous iris from Oregon and Texas which was similar to the gladiolus pathogen, but neither isolate would cause infection on gladiolus. Thus, iris isolates of *Fusarium* caused basal rot of iris, but not gladiolus; the gladiolus *Fusarium,* however, caused basal rot on iris. In 1958, Apt (2) explored this relationship more fully and concluded that *Fusarium* isolates from genera of the Iridaceae (*Crocus, Gladiolus,* and *Iris)* were not only highly pathogenic on their original host, but were also pathogenic on the other two genera in the family, although to a lesser degree. He proposed that the name *F. oxysporum* f. sp. *gladioli* be used for the pathogen on all three hosts. Perhaps the two isolates McClellan (42) used were less pathogenic or had greater host specificity than the one used by Apt (2).

Control To control basal rot of iris, Gould (22) advocated avoiding bruising during handling and careful selection of planting stock, along with hot water treatment at 43.5 C for 4 hr. Vigodsky (63) recently reported on hot water treatment of iris bulbs to eliminate the basal rot pathogen referred to as *F. oxysporum* f. sp. *iridis.* In that study, hot water treatment at 57 C for 30 min eliminated *Fusarium* in iris bulbs, but the treatment also reduced sprouting by more than 50%. Pretreatment of bulbs at 30 C or 35 C for 2 or 4 weeks prior to hot water treatment increased the tolerance to the heat treatment as evidenced by increased percent sprouting; bulbs pretreated at 25 C were all killed by the heat treatment. Vigodsky (63) indicated that increased temperature hardiness might be related to retarded respiration that occurs at 30 C and above, as well as other endogenous factors in the bulb.

Tulip Basal Rot

History and etiology Fusarium basal rot of tulips was not described in any detail until 1955 when Slootweg (58) described the disease, and designated *F. oxysporum* as the causal agent. He was able to reproduce the disease by planting tulip bulbs in infested soil. He was not able to obtain infection with basal rot isolates from gladiolus and montbretias, and suggested that tulips may have their own strain of the pathogen. Apt (2) confirmed the specificity of the tulip pathogen and proposed that it be called *F. oxysporum* f. sp. *tulipae*.

In 1967 Schenk (55) concluded that the rapid increase in disease incidence was the result of revolutionary changes in cultural methods that provided conditions more favorable for the spread and infection of *Fusarium* in tulips. He cited changes in growing practices such as shorter rotations, mechanical lifting, less careful culling of diseased stocks, a later date of lifting, and changes in handling methods leading to more damage during harvest and storage, as prime factors. Another factor mentioned was the possibility that more virulent strains of *Fusarium* had developed.

Symptoms and disease development Gould (22) described basal rot of tulips from Washington state as primarily a storage disease, both in bulb sheds and in warehouses. The disease in Holland was called "zuur" or sour disease by Slootweg (58) because of the acid smell of the early soft rot. Infected bulbs in storage are a dull white and the rotted area is firm, shrunken, and sometimes zonate. Mycelia and spores of the pathogen are often visible under the outer husk, usually at the base of the bulb. The diseased area eventually turns chalky unless secondary organisms invade and cause a soft rot.

Bulb losses in storage are usually greatest when moisture of the packing medium and humidity are both high. In the field, leaves on plants from diseased bulbs turn red, wilt, and often die (22), and the bulb itself will be rotted at the base. If the root crown has been damaged by the fungus before planting, the bulb will sprout weakly, if at all. Damaged bulbs deteriorate by spring, and the basal rot will appear gray-brown to green in cross section. Under the Netherlands' cool climatic conditions (11, 55), slightly infected bulbs never show symptoms in the field at harvest, whereas in growing areas with a warmer spring climate, such plants would be dead at harvest. Within a few weeks after harvest, small, gray, somewhat sunken spots with a dark edge may be apparent on the outer membrane. Such spots may be isolated or in groups, anywhere on the bulb, and may enlarge rapidly, often developing a concentric pattern. The pathogen also grows into the bulb so that the entire bulb can be involved (Fig. 13-3).

When infected bulbs are forced in the greenhouse at 15 C or higher following cooling at 5 C, the resulting plant may be stunted or may suddenly become yellow and wilt. Examination of the bulb will show a typical gray-brown basal rot that may extend upward into the stem as dark vascular streaks (56).

Bergman showed (12) that the pathogen can survive at least 5 years in artificially infested soil without a host. Further, enough inoculum is carried into uninfested soil on heavily diseased stock to induce significant disease for 2 years afterward.

Numerous strains occur in any natural soil population, strains which may differ in pathogenicity and in sensitivity to a specific temperature and to some chemicals. There also appears to be some slight variation in susceptibility among tulip cultivars (12).

Most researchers and growers agree that soil temperature is one of the most important factors influencing infection. In years when soil temperatures are high in the weeks before lifting, a very heavy infection may be expected. Schenk (55) cited Bergman's unpublished results showing that the critical minimal temperature for infection is near 16 C, and the number of infections increases above that. Bergman (11) concluded, therefore, that in the Netherlands under natural conditions, field infections occur almost exclusively in the last few weeks before lifting. Bergman (13) also showed that during that same period, there is a rapid breakdown of a pre-infection resistance factor in the skin of the tulip bulb which was later isolated, identified, and called "tulipalin" (14). Tulipalin forms a barrier to early infection of the underlying bulb tissue.

Bergman (11) also studied the mode of penetration of *Fusarium* into tulips and concluded that it is able to infect roots and from there grow into the bulb base. However, under natural conditions, the fungus more often infects the fleshy outer scale of the new bulb during the last weeks before the skin turns brown. Often these infections do not appear on freshly harvested bulbs but show later on bulbs that have been in storage several weeks.

Bergman (11) examined infected roots microscopically and found that the pathogen was growing in the root cortex, destroying cortical parenchyma completely, but rarely penetrating the xylem vessels. Examination of scale infections showed that the fungus grew both inter- and intracellularly in the disorganized tissue of the storage parenchyma without exhibiting any preference for vascular bundles. He therefore concluded that *F. oxysporum* f. sp. *tulipae* is not a vascular parasite, in contrast to other non-bulb crop strains of the species which show a definite preference for vascular tissues.

Bergman (10, 12) also demonstrated the importance of the storage period regarding spread and infection of *Fusarium* in tulips. He showed that the fungus sporulates readily on infected bulbs, and these spores, conidia and/or chlamydospores, are quite tolerant of drying and can survive on the bulb surface or in dust in the storage room. The spores germinate quickly and can infect healthy bulb tissue even if held in storage only one day. Thus, sanitation as a control measure cannot be over-emphasized.

Fusarium Diseases of Lilies

History, etiology, and disease development Although Charles and Martin (16) and Slate (57) reported a disease of lily caused by a species of *Fusarium*, an accurate description of the basal rot disease of lily was not published until Imle (33, 34, 35) described and named the pathogen *F. oxysporum* f. sp. *lilii* due to the specificity of the strain for lilies. Imle (34, 35) reported that infection by the pathogen usually occurred at the basal part of the bulb where the scales and basal plate join, although entrance probably occurred through the old outer scales, through tissue breaks resulting from handling, through wounds from root emergence, or through the old stem scars. He did not report root infections followed by growth of the pathogen upward into the basal plate. Brownish or black rot areas on the side of the bulb indicate the early stage of infection, which may then expand into the basal plate and base of the scales, eventually involving the entire basal plate and all the scales. When dug, such bulbs fall apart (Fig. 13-4 A, B). The roots on infected bulbs may be inferior to those of sound bulbs, although not so in early bulb infections. Infected bulbs tend to make many new bulblets, usually on severed scales due to the removal of apical dominance. Such bulblets, however, usually form at the infected end of the scale, and thus also readily become infected. Often this diseased bulb may be nearly destroyed before any top symptoms appear, especially on those lilies that tend to produce abundant stem roots which can support a vigorous top. If top symptoms do appear they consist of a premature yellowing, purpling, or browning of the lower leaves and stem. Often, if the main bulb has been destroyed, stem bulblet production is increased.

Imle (34) reported that hyphae of the pathogen invade tissue both inter- and intracellularly, often killing cells in advance. He reported that the pathogen can invade the vascular system of the scales and basal leaves, although it is primarily a rot-producing *Fusarium* and not a wilt-producing organism. He made no mention of invasion of vascular tissue in the stem.

1

2

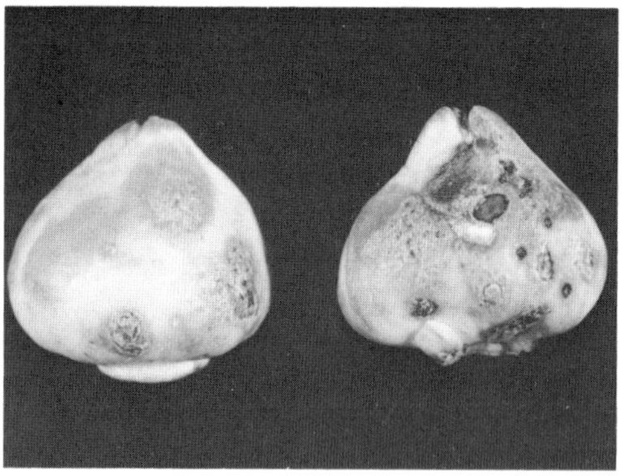

3

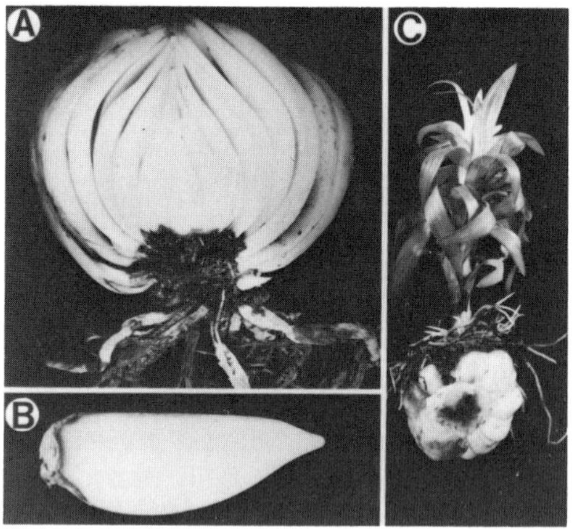

4

Fig. 13-1. Basal rot of narcissus caused by *Fusarium oxysporum* f. sp. *narcissi.*

Fig. 13-2. Twisted foliage symptom of the basal rot disease of bulbous iris caused by *Fusarium oxysporum* f. sp. *gladioli* (left) compared to uninoculated control plants (right).

Fig. 13-3. Tulip bulbs infected with *Fusarium oxysporum* f. sp. *tulipae* showing gray-brown lesions scattered over the bulb scales.

Fig. 13-4. Basal rot of lilies caused by *Fusarium oxysporum* f. sp. *lilii* showing the total destruction of the basal plate (A) with decay extending into the base of the scales (B); (C) decay of the basal plate and root origins of lily bulbs involving a complex of organisms including *Fusarium oxysporum* f. sp. *lilii* and *Cylindrocarpon* sp.

Imle (34) reported that *F. oxysporum* f. sp. *lilii* is disseminated primarily in or on bulbs. The pathogen was repeatedly found in the dry brown tissues on the edge of scales from diseased bulbs. This finding is especially significant if such scales are used for scale bulblet production, since the latter readily become infected soon after planting. The pathogen is also readily disseminated by conidia and/or chlamydospores that can be carried along in soil on the surface of bulbs, tools, equipment, or packing crates,

Bald and Solberg (5) noted an apparent synergistic association of *F. oxysporum* from lily scales with *Pseudomonas,* especially where scale tip rot was concerned. The author *(unpublished data)* has repeatedly isolated *F. oxysporum* and *Cylindrocarpon* sp. from infected basal plates of both Easter and garden lilies, but the isolation pattern usually suggests that *Fusarium* was the primary pathogen and *Cylindrocarpon* was secondary. However, the latter fungus may have contributed substantially to the total disease syndrome (Fig. 13-4C).

Cultivar susceptibility and host specificity Imle (34) catalogued those lily species that were highly, moderately, and only slightly susceptible to the pathogen. Only *Lilium hansonii, L. sargentiae,* and *L. maximowiczii* were highly resistant in the bulb stage, and seedlings of all the species he tested were highly susceptible.

As mentioned earlier, Imle (34) determined that the lily *Fusarium* was quite specific for lilies on the basis of wound inoculations with the lily pathogen into the basal plate of several species each in the Liliaceae and Iridaceae, and *Narcissus* in the Amaryllidaceae. The bulbs were subsequently held either in moist chambers or planted into soil. Likewise, pathogenic *F. oxysporum* isolates from tomato, narcissus, crocus, gladiolus, spinach, freesia, and onion did not infect *L. formosanum* seedlings.

While other bulb Fusaria are as host specific as *F. oxysporum* f. sp. *lilii,* most infect primarily the basal plate and/or scales, sometimes getting there by first invading the roots and progressing upward. The lily isolates of *F. oxysporum* apparently can also infect the stem and roots (4, 30, 34, 47). In fact, Bald and Chandler (4), McWhorter (47), and Bald et al. (6) reported that isolates from roots, basal plate, scales and stem appear identical in culture but represent a range of virulence depending on which tissue is inoculated. For example, root rot or stem lesion isolates caused slight basal rot, but caused severe root or stem lesions; also isolates from the same original tissue, i.e., basal plate, roots, or stem, varied in their virulence even when inoculated back into the type of tissue from which they were isolated (6). Some virulent isolates that could cause basal rot as well as root rot advanced through the scales and basal plate by intercellular hyphae, while mildly pathogenic isolates were able to infect only superficially. Bald et al. (6) therefore considered those pathogens of lilies, gladiolus, and narcissus capable of intercellular penetration of non-vascular tissue or vascular parenchyma a stage in the evolution of the truly vascular habit in the Fusaria.

Soil ecology Relatively few studies have been conducted on populations of *F. oxysporum* f. sp. *lilii* in field soils with regard to numbers and survival of propagules. The author *(unpublished data)* surveyed Easter lily fields and assayed *Fusarium* populations on the Nash-Snyder selective medium (52), and found no correlation between numbers of *F. oxysporum* propagules and disease incidence, presumably because the total population is comprised of both pathogenic and saprophytic isolates that are not readily separated in culture.

Hammerschlag and Linderman (25) examined the effects of organic acids from pine needles on *F. oxysporum* f. sp. *lilii* chlamydospore germination and formation of replacement chlamydospores, with attention toward survival and ultimate population decline. Several organic acids, especially shikimic and quinic acids, greatly stimulated chlamydospore germination without the usual formation of replacement chlamydospores. The response was proportional to the organic acid concentration, and was most pronounced at low pH. They hypothesized that these organic acid amendments could result in the eventual decline of the pathogen. The author has had personal communications with several growers who maintain that amendment of *F. oxysporum*-infested soils with pine needles or oak leaves, extracts of which can stimulate chlamydospore germination, does result in reduced basal rot disease. These reports remain unconfirmed experimentally.

Control Control measures for the Fusaria infecting lilies have largely focused on bulb-dip treatments in chemicals (34, 44, 45). McWhorter, (47) pointed out that hot water dips without formalin aggravated lily basal rot, although hot water was effectively used to eradicate foliar nematodes. He felt that none of the surface chemicals would be adequate to reach the internal infections, and that more attention should be paid to treating the soil, such as with fumigants (41). With the advent of the systemic benzimidazole fungicides in the early 1970s came hope of bulb treatments that could penetrate tissues and reach internal infections. These materials are currently being used by some growers, but they are not completely effective (15, 38, 40, 53, 58). Boontjes (15) did report increased yields with *L. speciosum* following hot water treatment of bulbs for 2 hr at 39 C followed by a 30 min dip in benomyl.

In 1956, Bald and Chandler (4) advocated continued work on improvement of culturing techniques to produce lilies free of known fungal and bacterial pathogens with a combination of fungicides, hot water, and surface sterilization of scales used for bulblet production. Those methods could only be successful if some means were developed to avoid recontamination of planting stock. More effort should also be placed on selecting or breeding lilies for commercial production with resistance to *Fusarium* (39).

Physiological Considerations

The infection of flowering bulbs by *Fusarium* greatly alters the physiology of the plant in the field, as evidenced by stunting, altered flowering, twisted foliage, and premature senescence. These symptoms strongly suggest that ethylene is produced in the disease reaction (1, 31). Hitchcock et al. (31) induced many of these symptoms in lily, narcissus, tulip, and hyacinth with ethylene added to the atmosphere. Williamson (70) also has demonstrated that ethylene was a metabolic product of diseased or injured plants, although none of the diseases he studied involved *Fusarium* or flowering bulb crops. Of special concern in these physiological disease reactions is whether ethylene is produced in the internal atmosphere of the bulbs in sufficient concentration to induce the growth alterations observed as disease symptoms. Staby and DeHertogh (61) analyzed the ethylene in the internal atmosphere in bulbs of *Tulipa, Iris, Narcissus, Lilium* and *Hyacinthus* and concluded that physiologically active levels of ethylene do occur therein. Further, they showed that treatment of the bulbs with the fungicide benomyl greatly reduced the ethylene levels, suggesting that ethylene was a product of the host-parasite interaction, possibly involving *Fusarium.* Ilag and Curtis (32) showed that many fungi can produce ethylene in culture, but they did not test any Fusaria. Kamerbeek (37) reported that *F. oxysporum* f. sp. *tulipae* was capable of producing large amounts of ethylene, and DeMunk (18) reported that *Fusarium*-infected tulips were the source of ethylene causing bud necrosis in tulips. The author *(unpublished data)* has measured significant amounts produced by *F. oxysporum* f. sp. *lilii.* Kamerbeek (37) also reported that ethylene production was oxygen dependent, but not correlated with mycelial growth. Further, ethylene prevented the synthesis of the antifungal compound tulipalin in host tissue, supporting the hypothesis that ethylene plays an important role in the host-parasite relationships (3).

Summary and Conclusions

Fusarium oxysporum causes significant disease on flowering bulb crops, i.e., *Narcissus, Tulipa, Iris,* and *Lilium,* wherever these crops are grown commercially and in many home gardens. The diseases are most pronounced in areas that are warm and humid during the growing season. The diseases do occur in cooler areas, but disease development is slower and symptoms are reduced or do not occur.

Infections by *F. oxysporum* on most flowering bulb crops probably begin in the field, either in the roots, scales, or basal plate. However, most infections eventually involve the basal plate, and may be severe enough to destroy the whole bulb. The diseases usually progress rapidly in geographic areas where temperatures and moisture are high, so that symptoms may occur by the end of the growing season. In many cases, infections which begin in the field do not result in symptoms until after harvest, either in storage or transit, or after replant for forcing into flower. Thus, many opportunities exist along the way for environment and handling to influence both the rate and severity of the disease.

The bulb Fusaria, in general, are considered by most workers to be rot Fusaria that show no special preference for vascular tissues, and they do not cause typical vascular wilt disease.

Pathogenesis may well involve the production of ethylene by the pathogen, the host, or both, since the usual disease symptoms of stunting, altered flowering, yellowing, premature senescence, and foliage twisting are all typical ethylene reactions.

The pathogenic isolates of *F. oxysporum* on flowering bulbs show a high degree of host specificity (except between members of the Iridaceae), although superimposed on that specificity is usually a range in isolate virulence, even from the same field. There also appears to be a tissue specificity within isolates of *F. oxysporum* f. sp. *lilii,* i.e., there are basal rot, scale rot, root rot, and stem lesion isolates that can cause severe disease only on the lily tissue from which they were isolated.

It is generally agreed that high nitrogen fertilizer, high temperatures, and low soil pH will increase these bulb diseases in the field. Organic amendments that contain large amounts of fertilizer enhance the diseases; amendments with materials like pine needles or oak leaves may, on the other hand, reduce disease.

Dissemination of these pathogenic strains of *F. oxysporum* is largely by shipping bulbs carrying the pathogen in a latent state from one growing area to another. Spores of these fungi can also be disseminated in soil, on tools and crates, and probably even in the air. Under normal field practices, these pathogens probably can survive in the soil, as chlamydospores or in organic debris, in sufficient numbers to cause disease for many subsequent years. Once introduced into the soil, the only immediate remedy is soil fumigation. Planting stock free of *Fusarium* in the first placc is the best control practice.

Effort to control these diseases has emphasized chemical eradication of the pathogen from infected bulbs. Treatment with systemic fungicides has given substantial yield increases, but *F. oxysporum* can often develop a tolerance to such materials. Most growers rightfully place considerable emphasis on careful roguing and culling of infected plants or bulbs to maintain a low incidence of diseases in their planting stock. The need is great for developing commercially acceptable *Fusarium*-resistant cultivars for the future.

Literature Cited

1. Abeles, F. B. 1973. Ethylene in plant biology. New York and London. Academic Press, 302 p.
2. Apt, W. J. 1958. Studies on the Fusarium diseases of bulbous ornamental crops. Ph.D. Thesis, Washington State Univ., Pullman. 88 p.
3. Archer, S. A., and E. C. Hislop. 1975. Ethylene in host-pathogen relationships. Ann. Appl. Biol. 81:121–126.
4. Bald, J. G., and P. A. Chandler. 1957. Reduction of the root rot complex on croft lilies by fungicidal

treatment and propagation from bulb scales. Phytopathology 47: 285–291.
5. Bald, J. G., and R. A. Solberg. 1960. Antagonism and synergism among organisms associated with scale tip rot of lilies. Phytopathology 50: 615–620.
6. Bald, J. G., T. Suzuki, and A. Doyle. 1971. Pathogenicity of *Fusarium oxysporum* to Easter lily, narcissus and gladiolus. Ann. Appl. Biol. 67: 331–342.
7. Beaumont, A. 1935. Diseases of narcissi and tulips. Sci. Hort. 3: 184–191.
8. Beaumont, A., and W. Buddin. 1938. Notes on the *Fusarium avenaceum* attacking the leaves of tulips in glasshouses. Trans. Brit. Mycol. Soc. 22: 113–115.
9. Beaumont, A., and W. E. H. Hodson. 1926. Second Annual Report, Dept. Plant Pathol., Seale-Hayne Agric. College, Newton Abbot, Devon, Eng.
10. Bergman, B. H. H. 1964. Het drogen van tulpen en de invloed op het optreded van *Fusarium* ("het zuur"). Praktijkmeded. Lab. BloembollOnderg. Lisse 13.
11. Bergman, B. H. H. 1965. Field infection of tulip bulbs by *Fusarium oxysporum*. Neth. J. Plant Pathol. 71: 129–135.
12. Bergman, B. H. H. 1966. Het zuur in tulpen is een besmettelijke ziekte. Weekbl. BloembollCult. 76: 698–699.
13. Bergman, B. H. H. 1966. Presence of a substance in the white skin of young tulip bulbs which inhibits growth of *Fusarium oxysporum*. Neth. J. Plant Pathol. 72: 222–230.
14. Bergman, B. H. H., J. C. M. Beijersbergen, J. C. Overeem, and A. K. Sijpesteijn. 1967. Isolation and identification of α-methylene-butyrolactone, a fungitoxic substance from tulips. Recl. Trav. Chim. Pays.-Bas. 86: 709–714.
15. Boontjes, J. 1970. A new chemical to control old disease in lilies. North Amer. Lily Soc. Yearbook 23: 74–79.
16. Charles, V. K., and G. H. Martin. 1928. Some disease of lilies. Plant Dis. Rep. 12: 82.
17. Creager, D. B. 1933. Fusarium basal rot of bulbous iris. Phytopathology 23: 7 (Abstr.).
18. DeMunk, W. J. 1975. Ethylene and abnormal developmental patterns in tulips. Ann. Appl. Biol. 81: 107.
19. Dod, C. W. 1894. Basal rot in daffodils. Gard. Chron. III 15: 379.
20. Gould, C. J. 1950. Fusarium leaf blight of tulips. Phytopathology 40: 965 (Abstr.).
21. Gould, C. J. 1950. Disease of bulbous iris. Wash. State Coll. Ext. Bull. 424 32 p.
22. Gould, C. J. 1957. Bulbous iris—Fungus diseases, p. 8. *In* C. J. Gould (ed.), Bulb Growing and Forcing, Northwest Bulb Growers Assoc., Mt. Vernon, WA.
23. Gould, C. J. 1967. World production of bulbs. Florists Rev. 140: 14–16, 70–71.
24. Gregory, P. H. 1932. The Fusarium bulb rot of narcissus. Ann. Appl. Biol. 19: 475–514.
25. Hammerschlag, F., and R. G. Linderman. 1975. Effects of five acids that occur in pine needles on Fusarium chlamydospore germination in nonsterile soil. Phytopathology 65: 1120–1124.
26. Hawker, L. E. 1935. Further experiments on the Fusarium bulb rot of Narcissus. Ann. Appl. Biol. 22: 684–708.
27. Hawker, L. E. 1940. Experiments on the control of basal rot of narcissus bulbs caused by *Fusarium bulbigenum* Cke. & Mass. Ann. Appl. Biol. 27: 205–217.
28. Hawker, L. E. 1943. Notes on the basal rot of narcissus. II. Infection of bulbs through dying roots in summer. Ann. Appl. Biol. 30: 325–326.
29. Hawker, L. E. 1946. Basal rot of Narcissus due to *Fusarium bulbigenum* Cke. & Mass. Daffodil Tulip Yearbook, Roy. Hort. Soc. 12: 78–83.
30. Hawker, L. E., and B. Singh. 1943. A disease of lilies caused by *Fusarium bulbigenum* Cke. & Mass. Trans. Brit. Mycol. Soc. 26: 116–126.
31. Hitchcock, A. E., W. Crocker, and P. W. Zimmerman. 1932. Effect of illuminating gas on the lily, narcissus, tulip, and hyacinth. Contrib. Boyce Thompson Inst. 4: 155–176.
32. Ilag, L., and R. W. Curtis. 1968. Production of ethylene by fungi. Science 159: 1357–1358.
33. Imle, E. P. 1940. A bulb disease of lilies caused by *Fusarium* spp. Phytopathology 30: 11.
34. Imle, E. P. 1942. The basal rot disease of lilies. Ph.D. Thesis, Cornell Univ., Ithaca, NY. 102 p.
35. Imle, E. P. 1942. Bulb rot diseases of lilies. Amer. Lily Yearbook 1942: 30–41.
36. Jacob, J. 1911. Daffodil bulbs and the heat. The Garden 75: 523, 593.
37. Kamerbeek, G. A. 1975. Physiology of ethylene production by *Fusarium* and possible consequences in the host-parasite relation in tulip bulbs. Ann. Appl. Biol. 81: 126.
38. Magie, R. O., and G. J. Wilfret. 1974. Tolerance of *Fusarium oxysporum* f. sp. *gladioli* to benzimidazole fungicides. Plant Dis. Rep. 58: 256–259.
39. Magines, E. A., and J. D. Smith. 1971. Scale test to assess susceptibility of lilies to various bulb rot organisms. North Amer. Lily Soc. Yearbook. 24: 20–28.
40. McDaniels, L. H. 1975. Notes on the use of Benlate for lily disease control. North Amer. Lily Soc. Yearbook 28: 58–61.
41. MacDaniels, L. H., and C. Hogger. 1972. Soil fumigation tests with lilies. North Amer. Lily Soc. Yearbook 25: 44–52.
42. McClellan, W. D. 1945. Pathogenicity of the vascular *Fusarium* of gladiolus to some additional iridaceous plants. Phytopathology 35: 921–930.
43. McClellan, W. D. 1952. Effect of temperature on the severity of Fusarium basal rot of Narcissus. Phytopathology 42: 407–412.
44. McClellan, W. D., and N. W. Stuart. 1944. The

use of fungicides and growth substances in the control of Fusarium scale rot of lilies. Phytopathology 34: 966–975.
45. McClellan, W. D., and N. W. Stuart. 1946. Treatment of lily scales reduces rot and increases bulb yields. Amer. Lily Yearbook 1946: 18–21.
46. McClellan, W. D., and N. W. Stuart. 1947. The influence of nutrition on Fusarium basal rot of narcissus and on Fusarium yellows of gladiolus. Amer. J. Bot. 34: 88–93.
47. McWhorter, F. P. 1963. Non-virus diseases of lilies (*Lilium* spp.). North Amer. Lily Soc. Yearbook 16: 54–88.
48. McWhorter, F. P., and F. Weiss. 1932. Disease of narcissus. Ore. Agric. Exp. Stn. Bull. 304. 41 p.
49. Miles, L. E. 1932. Control of basal rot of narcissus. Phytopathology 22: 19 (Abstr.).
50. Miles, L. E. 1932. Control of basal-rot of narcissus. Miss. Agric. Exp. Stn. Tech. Bull. 19: 1–12.
51. Moore, W. C. 1939. Diseases of bulbs. Ministry Agric., Fisheries, Bull. 117. London, England. 176 p.
52. Nash, S. M., and W. C. Snyder. 1962. Quantitative estimations by plate counts of propagules of the bean root rot Fusarium in field soils. Phytopathology 52: 567–572.
53. Raabe, R. D., and J. H. Hurlimann. 1971. Control of Easter lily root rots with fungicides. North Amer. Liy Soc. Yearbook 24: 11–19.
54. Rees, A. R. 1972. The Growth of Bulbs. Applied aspects of the physiology of ornamental bulbous crop plants. Academic Press, London and New York. 311 p.
55. Schenk, P. K. 1967. Phytopathological consequences of changing agricultural methods. V. Bulb crops. Neth. J. Plant Pathol. 73: 152–163.
56. Schenck, P. K., and B. H. H. Bergman. 1969. Uncommon disease symptoms caused by *Fusarium oxysporum* in tulips forced in the glasshouse after pre-cooling at 5°C. Neth. J. Plant Pathol. 75: 100–104.
57. Slate, G. L. 1936. Disease among the lilies. Horticulture 14: 96–97.
58. Slotweg, A. F. G. 1955. Zure tulpen. De Hobako. (The Netherlands) Sept. 23, 1955, p. 4, 5.
59. Snyder, W. C. and H. N. Hansen. 1940. The species concept in *Fusarium*. Amer. J. Bot. 27: 64–67.
60. Snyder, W. C., and H. N. Hansen. 1945. The species concept in *Fusarium* with reference to discolor and other sections. Amer. J. Bot. 32: 657–666.
61. Staby, G. L., and A. A. DeHertogh, 1970. The detection of ethylene in the internal atmosphere of bulbs. HortScience 5: 399–400.
62. Stuart, N. W., and W. D. McClellan. 1943. Severity of Narcissus basal rot increased by the use of synthetic hormones and nitrogen bases. Science 97: 15.
63. Vigodsky, H. 1970. Hardening of iris bulbs for hot-water treatment. J. Hort. Sci. 45: 87–97.
64. Wedgeworth, H. H. 1928. Experiments on the control of a narcissus root-rot. Miss. Agric. Exp. Stn. Cir. 79: 1–4.
65. Weiss, F. 1929. The basal rot of narcissus bulbs caused by *Fusarium* sp. Phytopathology 19: 99–100 (Abstr.).
66. Weiss, F. 1929. The relation of the hot-water treatment of narcissus bults to basal rot. Phytopathology 19: 100 (Abstr.).
67. Weiss, F., F. A. Haasis, and C. E. Williamson. 1942. Prestorage disinfection of narcissus bulbs. Phytopathology 32: 199–205.
68. Westerdijk, J. 1911. Bloembollen ziekten. Jversl. Phytopathol. Lab. "Scholten" 1911: 16–20.
69. Westerdijk, J. 1917. Ziekten der Narcissen. Jversl. Phytopathol. Lab. "Scholten" 1916: 3–7.
70. Williamson, C. E. 1950. Ethylene, a metabolic product of diseased or injured plants. Phytopathology 40: 205–208.

14 Fusarium Diseases of Beans, Peas, and Lentils

J. M. Kraft, D. W. Burke, and W. A. Haglund

Beans *(Phaseolus vulgaris)*, peas *(Pisum sativum)*, and lentils *(Lens esculenta)* are among the most important and widely grown food crops in the Leguminosae. These crops have been cultivated since ancient times and are important sources of protein. Fusarium diseases are serious factors limiting production and are prevalent wherever each crop is grown throughout the world.

Fusarium Diseases of Beans

Bean Root Rot

Fusarium root rot of bean has been studied widely, for its own sake and as a model for establishing microbiological principles relating to many soil-borne fungus diseases. The disease was first described by Burkholder in western New York State in 1916 (29,30) where as many as 90% of the plants in several counties were infected. This disease, often called "dry root rot of bean," has been considered a serious disease in most bean-producing areas of the world. Initial symptoms are reddish streaks on the hypocotyl and tap root which are evident about 1 week after plant emergence (Fig. 14-1). This reddish discoloration increases and coalesces to eventually cover the entire below-ground stem and root system, giving them a brown, corky appearance.

Roots of bean plants which are widely spaced and in soil conducive to good growth become large and vigorous in spite of the presence of *Fusarium* (15). In conditions unfavorable for optimum root growth, *Fusarium* reduces root volume and efficiency. The primary roots may be killed and adventitious roots arising from the hypocotyl just below the soil surface will keep the plant alive and productive, if soil moisture is adequate.

Crop losses from Fusarium root rot of beans can be severe. The disease in central Washington (27) reduced seed yields among 12 cultivars from 6% (181 kg/ha) in relatively resistant Sutter Pink to 43% (2141 kg/ha) in highly susceptible Red Mexican UI-36, where productivity was measured in adjacent noninfested and *Fusarium*-infested fields. In New York (101) yields were increased 25–50% by chemical controls. In Colorado (68) yield from dryland pinto beans was reduced as much as 84% in 1971. Estimated losses of 11–26% were attributed to root rot in Nebraska (121).

Biology and ecology of the organism Fusarium root rot of bean is caused by *Fusarium solani* f. sp. *phaseoli.* Most isolates of this fungus produce appressed pseudopionnotal colonies which are blue or blue-green on carbohydrate media. A few isolates are near white to buff and a few are fluffy. Macroconidia are abundant and microconidia rare.

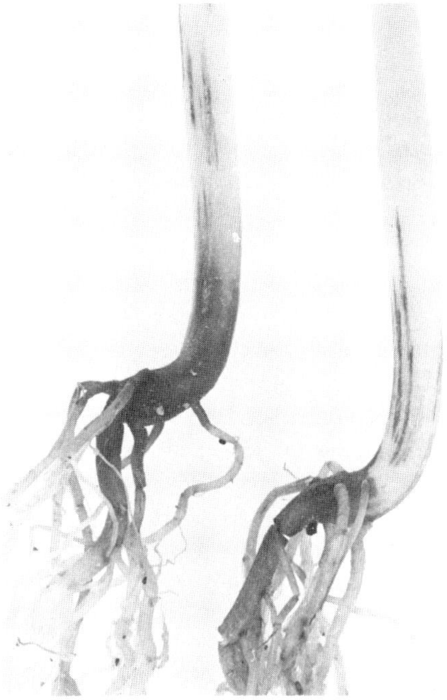

Fig. 14-1. Bean seedlings with typical symptoms of infection by *Fusarium solani* f. sp. *phaseoli* on roots and hypocotyls.

Chlamydospores are terminal, single, cantenulate or intercalary. Variation among isolates is evident in culture, but variation in pathogenicity has not been clearly demonstrated. Some isolates appear to be more virulent than others because of more profuse sporulation.

Conversion of macroconidia and mycelia of *F. solani* f. sp. *phaseoli* to chlamydospores in different soils was demonstrated by Burke (16), and Nash et al. (96,97) proved that chlamydospores were the survival structures of this fungus in nature. Except in certain soils suppressive to *F. solani* f. sp. *phaseoli* (16), this fungus becomes uniformly dispersed in the tilled layer of fields during the growth of only two or three crops of beans (23,97) and is most likely introduced initially through dust on seed (98). Propagules and clonal types are less numerous below the plowed layer. In sandy and silt loam soils of central Washington bean roots as well as the *Fusarium* were found to be largely confined to the plowed layer. In fact this *Fusarium* could rarely be isolated from soil even 1 cm below the plowed layer. It could be isolated, however, from bean roots that had penetrated soil 76 cm or more (23).

Chlamydospores of *F. solani* f. sp. *phaseoli* germinate readily when near germinating bean seed or root tips, sites of abundant exudation of amino acids and sugars (116). Mature bean root tissue was shown to have little or no effect on stimulating chlamydospore germination (116) and germ tubes are quickly lysed by other microorganisms unless susceptible host tissues are readily available (17, 41). Chlamydospores of *F. solani* f. sp. *phaseoli* also germinate and reproduce in soil near seeds and roots of many non-susceptible plants and other organic matter (115). Thus recycled, the survival in soil of this fungus is almost interminable, and it has been shown to survive in soil planted continuously with non-host crops for over 30 years (1).

Probably one of the greatest contributions to the study of this pathogen and *Fusarium* spp. in general was the development of a selective medium for the isolation of *Fusarium* propagules from soil by Nash and Snyder (97), which permits differentiation among species and even among host-specific forms of the same species (Fig. 14-2). Modifications of this medium (104) have made it even more useful for isolation from certain types of soil.

Host-parasite relations Earlier literature on this aspect of bean root rot was thoroughly reviewed by Christou and Snyder (36) who found that *F. solani* f. sp. *phaseoli* penetrates the bean plant directly or through stomata or wounds. Infection hyphae proliferate in the intercellular spaces of the cortex, until they are stopped by the endodermis. Under conditions of high soil moisture, conidia may be produced on sporodochia emerging from stomata near the soil surface. As infected tissues degenerate, the mycelium converts to chlamydospores.

Christou and Snyder (36) concluded that *F. solani* f. sp. *phaseoli* is primarily a hypocotyl invader rather than a root pathogen, because the fungus grew less profusely in roots than in the hypocotyl. This observation was in accord with the popular opinion that Fusarium root rot of bean is mainly a "foot rot" and therefore should be controllable by localized seed or seed-furrow treatments. In contrast, Burke (21), working in bean fields with differing-sized "islands" of *Fusarium*-infested and non-infested soil, demonstrated that hypocotyl infection alone was not important in influencing the effect of the disease on seed yields and that the disease was serious only when a large part of the root system was damaged by *Fusarium*. In fact, a typically rotted hypocotyl would support a productive plant, if the root system was functional. Conversely, plants with severely rotted root systems failed to yield well even if the hypocotyl was protected from infection. Obviously, seed treatments or other localized treatments, unless they provide systemic protection, cannot be expected to control Fusarium root rot of beans as it occurs in the field.

Control-cultural practices The extent of yield reduction by Fusarium root rot in beans is influenced by previous cropping, plant spacing, soil temperatures, soil moisture level, soil compaction, soil aeration, and other factors such as soil fertility and environmental stresses which affect pod and seed set and development (18). Most soil management decisions affect bean plant growth and yield, and the results of these decisions can be much more important if Fusarium root rot is a problem.

Timing of planting dates to take advantage of favorable germination temperatures can greatly influence production in *Fusarium*-infested fields (14). Widely spaced plants usually grow more vigorously and have less obvious root rot than closely spaced plants (15,26). Nevertheless, yields in *Fusarium*-infested land are best where plant populations are sufficient to give complete ground cover, even though close spacing may be required to accomplish this (26).

Soil Compaction results in a serious aggravation of the Fusarium root rot problem (19,25,91,100). In some soils compaction by tractor wheels and by tillage implements creates layers in the soil barely penetrable by *Fusarium*-infected roots. Thus restricted, infected roots become rotted severely, reduced in volume, and less capable of extending into sources of water and nutrients sufficient for optimum yields. Soil compaction can be reduced and soil moisture relations improved by planting beans after small grain or alfalfa crops (25,42). In central Washington, Fusarium root rot is largely counteracted when small grain residues are plowed down at least 30 days before beans are planted, with sufficient N to provide for microbial decomposition of the straw (25). Previous cropping with potatoes, sugar beets (D. W. Burke, *unpublished data*), and cabbage (100) actually aggravates the root rot problem because of soil compaction resulting from the harvesting operations.

Another method of reducing soil compaction is to subsoil or chisel before planting. The subsoiling should extend into the soil deeply enough to break compacted layers created by plowing or by the disking in of herbicides (25). However, the effects of subsoiling are negated when followed by other tillage operations prior to planting. Double-disking for the incorporation of herbicides in moist seed beds has no doubt increased the root rot problem in some areas.

Herbicides themselves may also increase root rot. Wyse et al. (133) reported that Navy bean seedlings were predisposed to Fusarium root rot when grown in soil treated with the herbicide EPTC (S-ethyl dipropylthiocarbamate). Wyse concluded that EPTC altered the hypocotyl surface by restricting cuticle formation, which in turn increased exudation of electrolytes, amino acids, and sugars and allowed for easier direct penetration by the pathogen.

Because the principal effect of root rot is in reducing the volume and efficiency of root systems, optimum soil moisture is another critical factor in reducing the effects of Fusarium root rot in beans (25,91). On the other hand, soil saturation or flooding with water, even though temporary, greatly restricts oxygen diffusion to the roots, resulting in reduced root growth and predisposition to severe root rot (92,93). Saturation of the soil in the entire root zone also negates any prospective benefits from subsoiling.

The level and form of nitrogen in the soil have been found to affect root rot in greenhouse and laboratory experiments (62,130). In the field, however, where nitrogen form and distribution changes rapidly, Burke and Nelson (27,28) found that neither form nor amount of nitrogen applied in nitrogen-deficient soil affected the disease.

Fusarium solani f. sp. *phaseoli* does not readily invade lesions on hosts or non-hosts as a secondary invader following other fungi (80,99) or nematodes (63, D. W. Burke, *unpublished data*).

Control of bean root rot in various soils by incorporation of small grain (86,103,120) and corn (61) residues with high C:N ratios has frequently been attributed to the stimulation of microbial populations which either compete with the *Fusarium* for available nitrogen or otherwise antagonize the pathogen. Decomposition products of some plant residues are injurious to bean roots and predispose them to increased root rot (13,122).

Resistance The literature has been thoroughly reviewed relative to breeding beans for Fusarium root rot resistance (12,118). More recently, Boomstra and Bliss (11) reported a dozen or more additional sources of resistance. Also, a number of wild *Phaseolus* spp. have been found to have a level of resistance at least equal to that of P.I. 203598, the most widely used parent (O. Norvell and D. W. Burke, *unpublished data*). Probably the most highly resistant selection known is NY 2114–12, a *Phaseolus vulgaris* x *P. coccineus* hybrid developed in New York by Wallace et al. (129).

Pierre (106) found that *F. solani* f. sp. *phaseoli* utilized pectolytic but not cellulolytic enzymes in its pathogenesis of bean. Pierre and Wilkinson (107) reported that the initial reaction of susceptible and resistant beans to *F. solani* f. sp. *phaseoli* was similar. As infection progressed, however, a more rapid accumulation of brown material occurred in resistant lines in advance of the infection hyphae. They believe resistance is associated with some chemical product of the pathogen-host interaction which restricts the spread of infection and lesion size. Unlike Huber (60), they considered periderm formation secondary in importance as a factor in resistance to this fungus.

Inheritance of resistance to *F. solani* f. sp. *phaseoli* is based on additive effects of 2–7 genes (5,12). Resistance is usually associated with late maturation, small seeds, and large, indeterminate vines (129). Some notable exceptions are the very early maturing, short-vine California cultivar, Sutter Pink (27), and the new pink cultivars, Viva, Roza, and Gloria, developed by Burke (20), which have both Sutter Pink and one of the principal sources of resistance, P.I. 203958, in their parentage. By the use of the early Pink cultivars in combination with derivatives of the other sources of resistance, effective resistance is being established in early-maturing, short-vine Pinto, Red, and White beans as well (D. W. Burke, *unpublished data*).

Bean Yellows

Bean yellows, caused by *Fusarium oxysporum* f. sp. *phaseoli* (69), was originally observed by Harter (57) on dry beans in California in 1928. Since then the disease had been recognized in Colorado, Idaho, Montana, South Carolina, Costa Rica, and England (2,3,48,135). This disease has not, however, become economically important. Known hosts of this pathogen are beans, cowpeas *(Vigna sinensis)*, soybeans *(Glycine max)*, and lima beans *(P. lunatas)*. Echandi (48) found that all commercial cultivars of *P. vulgaris* in Costa Rica were susceptible.

The pathogen invades and discolors the vascular bundles of the stem, peduncles and petioles. Infected plants do not wilt. Symptoms consist primarily of a progressive yellowing of the leaves beginning with the lower ones. Diseased plants are readily recognized by the distinctive yellow foliage. Spores of the pathogen, deposited on the seed at harvest (57), are killed by standard chemical seed treatments.

Fusarium Diseases of Peas

Pea Root Rot

Fusarium root rot of peas, caused by *Fusarium solani* f. sp. *pisi*, was first reported as a serious pathogen in Minnesota in 1918 (8) and in Wisconsin in 1923 (64). The disease was also observed in Europe at approximately the same time (31,32). Fusarium root rot is serious in the Pacific Northwest in dryland and irrigated areas. It is distinct from Fusarium wilt and may occur in conjunction with other diseases of peas (82,114,127,134).

In pea seedlings the initial center of attack by this pathogen is the cotyledonary attachment area, below-ground epicotyl, and upper taproot (34,82,127). Infection extends upward to the soil line and downward to the roots. The degree of root damage depends on soil conditions (J. M. Kraft, *personal observation*). A red discoloration of the vascular system may occur in the root but usually does not progress above the soil line. Aboveground symptoms are not readily defined but consist primarily of yellowing of the basal foliage and stunted growth.

Data on crop losses due to this pathogen are scarce. However, Kraft and Berry (79) reduced yields 30% in artifically infested field plots compared with non-infested plots in the same field. Basu et al. (6) determined that yield losses in processing peas in five Canadian provinces average 35–57% in experimental plots. It is evident that soil type and moisture relationships greatly influence the severity of Fusarium root rot of peas, as with other cortical diseases. As an example, Burke et al. (22) found in Wisconsin pea fields that root rot was a serious problem only in those fields where *Aphanomyces euteiches* was prevalent, even though *F. solani* f. sp. *pisi* was prevalent in all fields. The soil types and their moisture relationships did not favor severe damage by *F. solani* f. sp. *pisi*.

The organism In culture this pathogen produces sporodochia which are blue-green to buff in color (64,127). Macroconidia are primarily 3 septate, 4.4 to 5μm by 27 to 40μm, curved and hyaline. Microconidia are less abundant, except in liquid culture where they are numerous. Chlamydospores, produced in the mycelium or by conversion of conidia, are abundant, intercalary, terminal, single or cantenulate. The optimum temperature for growth on agar is about 30 C. However, the disease will develop at 18 C and above, with an optimum of 25–30 C (82,127).

Reichle et al. (110) first reported that *F. solani* f. sp. *pisi* produces a *Hypomyces*-perfect stage and is heterothallic. They further reported that differences in virulence exist between ascospore pairs within each ascus. *Hypomyces* perithecia have been found in nature only on diseased mulberry (*Morus* sp.) branches in Japan. Matuo and Snyder (88) demonstrated that *F. solani* f. sp. *pisi* was identical, by mating tests, with the pathogen causing branch blight of mulberry trees and root rot of gensing (*Panax* sp.). Cook et al. (40)

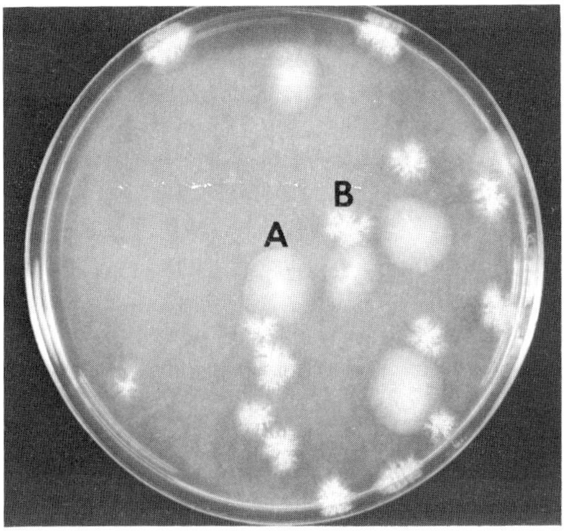

Fig. 14-2. Typical colonies of *Fusarium solani* f. sp. *pisi* (A), and f. sp. *phaseoli* (B) on Nash and Snyder's selective medium (97).

Fig. 14-3. Four-week-old pea plants with symptoms of infection by *Fusarium solani* f. sp. *pisi* on left, control plants on right.

reported that the occurrence of compatible mating types of *F. solani* f. sp. *pisi* in Australia, New Zealand, and the United States was most likely due to the international movement of the fungus on commercial pea seed in dust. Matuo and Snyder (89) concluded that *F. solani* f. sp. *pisi* could be classified into one of four groups (group B) of *F. solani,* based on the morphology of macroconidia.

Host-parasite relations Bywater (34) stated that *F. solani* f. sp. *pisi* normally infects the cotyledons, foot, and root of peas (Fig. 14-3). Initial infection, after an infection thallus formed, was often via the stomates on the epicotyl. All root tissue was reported to be susceptible to hyphal penetration but, as in bean root rot (36), spread of infection was slower in roots than in stem tissue.

Disease severity, due to *F. solani* f. sp. *pisi,* is increased by soil moisture stress (82). Cook and Flentje (39) found maximum chlamydospore germination of *F. solani* f. sp. *pisi* 20 hr after seeds were planted at a soil moisture content of 8.7%. Seed germination, with its corresponding release of nutrients into the surrounding soil, was the important factor that triggered chlamydospore germination and survival near pea seed.

The amount of the toxin, isomarticin, produced by an isolate of *F. solani* f. sp. *pisi* is positively correlated with the virulence of that isolate (70). Dorn (47) found that chemically induced mutants of the fungus that lost their ability to produce isomarticin were less virulent than the wild type.

Interactions with other organisms Combined inocula of *F. solani* f. sp. *pisi* and *Pythium ultimum* produced slightly more damage to pea roots grown in pots than the individual pathogens did in a range of several soil moisture conditions, indicating that the effects of the two fungi were additive (82). *Fusarium solani* f. sp. *pisi* and *F. oxysporum* f. sp. *pisi* race 2, together in soil, were found to be more than additive, causing greater disease severity than either pathogen alone (132). Virus diseases, such as bean yellows and common pea mosaic, increase susceptiblity to root rot caused by *F. solani* f. sp. *pisi* by increasing root exudation (7,49).

Attempts to inhibit pathogenic strains of *F. solani* f. sp. *pisi* with avirulent strains of *Fusarium* have been unsuccessful (80,112). Buxton and Perry, however, (33) found that severity of Fusarium wilt caused by *F. oxysporum* f. sp. *pisi* race 1 was decreased by simultaneous inoculation with *F. solani* f. sp. *pisi* and race 1, and previous inoculation with *F. solani* f. sp. *pisi* was even more suppressive of wilt. Delay of wilt symptoms was believed due to physical damage to the root by the root-rotting pathogen, which reduced the number of infection sites available to the wilt pathogen (105). Similarly, *Pythium ultimum* has been observed to reduce pea wilt caused by *F. oxysporum* f. sp. *pisi* races 1 and 5 (J. M. Kraft, *unpublished data*). In contrast, Kerr (71) found that *P. ultimum* increased the severity of wilt caused by *F. oxysporum* f. sp. *pisi* race 2.

Cultural variants of *F. solani* f. sp. *pisi* and *F. oxysporum* f. sp. *pisi* race 2, which overlap in morphology and symptom expression, have been discussed (10,58). In fact, Bolton and Donaldson (10) suggested that these two organisms constitute a single species. The authors have seen no evidence, from repeated field and greenhouse isolations and observations, to indicate *F. solani* f. sp. *pisi* and *F. oxysporum* f. sp. *pisi* race 2 are one species. In addition, production of a perfect stage through crossing "wild type" isolates of both species has not been reported.

In the field root rots usually involve more than one pathogen. For instance, during 6 and 15 years monoculture of peas and beans, respectively, in two fields of the same soil type, populations of *Rhizoctonia solani, Pythium ultimum,* and *Thielaviopsis basicola,* equally pathogenic to beans and peas, accumulated in each field in addition to the respective host-specific forms of *F. solani* (24). When both peas and beans were then grown in each field for 2 years, it was obvious that the host-specific *F. solani* f. sp. was essential to any significant yield reduciton, i.e., the pea made excellent growth and production in the bean field and the beans were excellent in the pea field, in spite of obvious infection by the three fungi other than the host-specific *F. solani* f. sp. In contrast, both crop species were severely damaged in the fields of their previous monoculture, where the disease complex included their respective host-specific forms of *F. solani* (24). In a third year of planting of the two species in both fields, it was obvious that destructive populations of both the bean and pea forms of *F. solani* were active in both fields. In a later study, Kraft and Burke (80) found further evidence that the two *F. solani* f. sp. acted entirely independently. Macroconidia of *F. solani* f. sp. *pisi* and *F. solani* f. sp. *phaseoli* germinated equally well near germinating bean or pea seeds whether or not the other *Fusarium* was present. Furthermore, no suppression of or increase in symptom severity occurred when beans or peas were planted in soil infested with both pathogens at varying inoculum levels in comparison with plants exposed only to the respective host-specific pathogens. *Fusarium solani* f. sp. *pisi* increased in population in the rhizosphere of either bean or pea and could be isolated from lesions on either host; in contrast, *F. solani* f. sp. *phaseoli* was more host-specific.

Mechanisms of resistance Kraft and Roberts (83) found *F. solani* f. sp. *pisi* did not sporulate as readily in seedling exudates from resistant Plant Introduction (P.I.) lines as in exudates from the susceptible cultivar, Dark Skin Perfection. In addition, *F. solani* f. sp. *pisi* did not increase as rapidly in population in rhizosphere soil from resistant pea roots as in rhizosphere soil from susceptible plant roots. More recently, Kraft (76) found that peas with pigmented seeds, both resistant and susceptible to root rot, produced like amounts of phenols in exudates from germinating seeds and seedlings. However, only exudates from resistant P.I. accessions inhibited sporulation of *F. solani* f. sp. *pisi in vitro* and *in vivo.* Further studies (78) revealed that both resistant and susceptible P.I. accessions, with the "A" gene for anthocyanin production, contained the fungistatic pigment delphinidin which is located primarily in the seed coat. *Fusarium solani* f. sp. *pisi* was able to overcome this fungistatic effect in susceptible accessions bearing the "A" gene because they, unlike the resistant accessions, exuded free sugars along with delphinidin during imbibition and early seedling growth. Unfortunately, seedling resistance is not always correlated with physiologic or cellular resistance in older plants (J. M. Kraft, *unpublished data*). Cellular resistance is present in some resistant P.I. accessions, however, as evidenced by the fact that lesions caused by *F. solani* f. sp. *pisi* on epicotyls are fewer and coalesce less rapidly than on more susceptible plants (76). Whether this restriction in lesion size and fungal infection is due to the formation of pisatin (35,43,44,45,95) or other inhibitory phytoalexins is questionable. In any case, an active role of such compounds in the resistance of peas to fungal pathogens has yet to be demonstrated (109). Because physiologic resistance to *F. solani* f. sp. *pisi* in peas does exist in certain pea selections (73,76,77,83,85), it would seem logical to study the host-pathogen response in those selections to determine if pisatin or other phytoalexins play a role in that resistance. Perhaps cellular resistance to spread of infection is due to the ability of the pea plant to inactivate or metabolize the Fusarium toxin isomarticin rather than to the formation of a phytoalexin.

Genetic resistance to Fusarium root rot was reported by Knavel (74) to be dominant and affected also by cytoplasmic factors. In peas grown in soil artificially infested with *F. solani* f. sp. *pisi* and *P. ultimum,* Muehlbauer and Kraft (94) observed a correlation between disease indices for the two fungi which indicated that resistance to both pathogens is governed by the same gene factors. There are no currently available commercial cultivars resistant to *F. solani* f. sp. *pisi;* however, some obsolete cultivars such as Wando and Horal have been reported to be tolerant (51,111). Because germ plasm with high levels of resistance to this pathogen has not been found, developing effective levels of resistance has been difficult. However, in Prosser, Washington (77), resistant seedlings from segregating pea hybrid populations exposed to high concentrations of chlamydospore inoculum of *F. solani* f. sp. *pisi* (20,000–40,000 propagules per gram of soil) are now being selected and transplanted 6–10 days after emergence. By this rapid technique, resistant segregants have been obtained which demonstrate effective levels of resistance in the field. By continued intercrossing and testing it may be possible to accumulate genes to provide a fairly high level of resistance.

Control Fusarium root rot and other cortical diseases of peas, like bean root rot, are favored by environmental conditions which are adverse to root growth. Consequently, such practices as good tillage procedures, which prevent or reduce soil compaction (J. M. Kraft, *unpublished data*) and promote good soil moisture conditions, and the use of high-quality seed (87,117) will help to greatly reduce the problem. Crop rotation, in which peas are not planted back on the same field more than once in 5 years (111,114), aids in minimizing disease occurrence. Maintenance of good soil fertility is also important. Drilling of fertilizers at time of planting has been reported to reduce losses from root rot on heavy soils (111).

Pea Wilt

Fusarium wilt of peas, caused by *Fusarium oxysporum* f. sp. *pisi* race 1, was first reported in Wisconsin in 1924 (65,84). Reports in 1928 and 1932 established that this disease was prevalent in all growing areas of the United States and was one of the most important diseases affecting processing, dry and seed pea production fields (66,126). Resistance to this disease was quickly found and was attributed to a single dominant gene (124,125,126). The introduction of wilt-resistant cultivars resulted in complete control of this disease in the commercial pea-growing industry. By 1942, a pea disease survey in Wisconsin failed to reveal race 1 of the wilt *Fusarium* in any commercial field (128). Furthermore, race 1 was not found to be an economic problem in the United States again until 1972, when the disease appeared in eastern Washington in a field planted to a susceptible cultivar (81). Race 1 has not been eliminated, but the disease is under control through the use of resistant cultivars (127).

Fusarium oxysporum f. sp. *pisi* race 2 was described in 1933 and has been reported from most pea-growing areas of the world (119). Like race 1, resistance to race 2 was quickly found and resistance was again attributed to a single dominant gene (56). However, economic losses due to this disease have not been sufficient in the eyes of the pea industry to mandate the growing of resistant cultivars (134).

In 1963, a Fusarium wilt was observed in one field in Skagit County, Washington, and in 1967 it was determined that this was a new race of Fusarium wilt (52). This disease, which by 1969 had spread to more than 240 fields encompassing 4047 ha, was designated *F. oxysporum* f. sp. *pisi* race 5 in 1970 (55). Yet another race of pea wilt *Fusarium,* capable of wilting pea cultivars resistant to races 1, 2, or 5, was also detected in 1970. This wilt, caused by a *F. oxysporum* tentatively designated race 6 (new race), has been detected in approximately 100 fields in western Washington, from the Canadian border to the Montesano area in southwestern Washington. The occurrence of the new strain was simultaneously observed in four locations in western Washington, geographically separated to such an extent as to preclude its having been originally disseminated to all these locations by wind, water, man's movement of equipment, vines, or contaminated soil. Seed contamination is considered a possible means of the original spread.

In 1974, *F. oxysporum* f. sp. *pisi* was reclassified into 11 races, including previously described races 1, 2, and 5 (4). Later, Kraft and Haglund obtained isolates of these reclassified races from the American Type Culture Collection, Rockville, Maryland, for study. All 11 isolates were determined to be either race 1 or 2 on the basis of standard inoculation procedures with 8 differential cultivars of known genetic resistances to races 1, 2, and 5. Hubbeling in 1974 (59) found that races previously described as 3 and 4 (9,113) are actually both race 2. In view of this evidence, it appears that isolates previously designated as races 3, 4, and 6 to 11 of *F. oxysporum* f. sp. *pisi,* in reality, are race 1 and 2 types that vary only in virulence and not specific pathogenicity. In the evaluation of pathogenic forms of *F. oxysporum* f. sp. *pisi* it is essential to use differential cultivars of pea with known genes for resistance together with a standard inoculation procedure, if consistent and reproducible results are to be obtained (53).

In northwest Washington and southwest British Columbia (Canada), where peas have been growing essentially in monoculture for many years in the presence of races 1 and 2 of *F. oxysporum* f. sp. *pisi,* new strains or races of Fusarium wilt are evolving that cause severe economic losses in race 1 resistant peas. Cultivars are currently available with resistance to races 1, 5, and/or 6, but combined resistance to all races is not currently available in an acceptable commercial cultivar. By determining the predominant race in a given field, the appropriate resistant cultivar can be selected for that field. However, epidemiology and symptoms of the disease caused by races 1, 2, 5, and 6 on susceptible cultivars are indistinguishable (55,81,84,114,127). Consequently, an accurate classification of the predominant race of Fusarium wilt in each field is dependent upon a soil-sample bioassay to determine the dominant race of the wilt *Fusarium* in that field. The bioassay consists of the following: i) Four liters of soil are collected from the wilted areas of a field and screened. ii) To eliminate seedling and root rot due to *Pythium* sp. and *Aphanomyces euteiches,* 125 mg per liter of Dexon (paradimethylaminobenzenediazo sodium sulfonate) is thoroughly mixed with each soil sample. iii) Differential cultivars of peas (53) are grown for 30–40 days at 16–22 C in 10 cm plastic pots of the collected soil and then examined for wilt symptoms.

In heavily infested soils the susceptible cultivars are dead within 20–25 days. When symptoms are questionable, the 3rd and 4th nodes from each test plant are surface disinfested and assayed on acidified potato dextrose agar for *F. oxysporum* after 7 days incubation at room temperature. Through the use of this soil assay, growers can be advised as to the race or strain of wilt *Fusarium* dominant in a field and thereby which cultivars might be best to plant (54).

The occurrence and development of Fusarium wilt in western Washington has not followed any traceable evolutionary pattern, and therefore it is not known whether races 5 and 6 both evolved separately from race 1 or 2, though this remains quite possible. In the classical sense, new races or strains of a pathogen have developed due to, or following, the introduction of new cultivars of the host (50). The pea industry in western Washington began in 1924 and in the early 1930s race 1 resistant cultivars were needed to control wilt. This resistance was adequate until 1963. Between 1963 and 1976 the new races, 5 and 6, developed into serious economic factors limiting pea production. In addition, three more strains or races have been detected and are currently under study. The development of these new races and strains has occurred under a cropping practice in which primarily race 1 resistant cultivars have been grown. To date there is no evidence to indicate that a second race or strain of wilt will become dominant in a field initially infested with either race 5 or the new race 6. As an example, a race 5 wilt nursery was established in 1968 and has been planted to peas for 9 consecutive years. During this time the breeding lines tested have been changed progressively from a majority of susceptible to predominantly resistant types. However, peas resistant only to race 5 still survive 99–100% in this nursery. Whether this stability will remain for another 10 years is yet to be seen.

Because of the appearance of new races or strains of the pea wilt *Fusarium* in western Washington, a program has been initiated to incorporate vertical or horizontal resistance to wilt. The future of the pea-processing industry in western Washington depends on the development of a permanent and economically feasible control. This control may well be horizontal or vertical resistance integrated with cultural practices and fungicides.

Fusarium Wilt of Lentils

On a worldwide basis, the most serious Fusarium disease of lentil *(Lens esculenta)* appears to be Fusarium wilt cause by *Fusarium oxysporum* f. sp. *lentis* (38,75,102,108,127) and this discussion will be limited to this disease. Although the disease has not been important in the United States (131), Vasudeva and Srinivasan (123) reported that wilt of lentils is a serious constraint on production in India. Symptoms consist of leaf curling that begins on the lower leaves and extends upward, followed by a general collapse and death of the plant. The disease is reported to be most severe when soil moisture levels approximate 25% with a temperature range of 17–31 C.

Vasudeva and Srinivasan (123) reported that *F. oxysporum* f. sp. *lentis* is host specific and will not infect several other legumes tested. In culture the fungus is indistinguishable from other formae speciales of *F. oxysporum*. Macro- and microconidia are abundant, as are chlamydospores. Optimum growth *in vitro* is 27–30 C and test isolates of *F. oxysporum* f. sp. *lentis* can be classified into eight different groups as determined by morphological and cultural characteristics (46).

Resistance to *F. oxysporum* f. sp. *lentis* exists in cultivars and breeding lines in India (72) and Russia (75). However, reports on the genetics of resistance or the presence of biotypes or races of *F. oxysporum* f. sp. *lentis* were not found. Kamaiyan and Nene (67) reported that some cultivars of lentils are more resistant in the seedling stage, while others are more resistant in the mature plant stage.

Mehrotra and Claudius (90) found that pre-soaking lentil seeds in an 80 ppm concentration of Zn and Mn salts reduced Fusarium wilt. Also, late-sown (November-December) cultivars wilted less than earlier-sown cultivars due to soil temperature differences. Claudius and Mehrotra (37) found that root exudates from 21-day-old lentil plants were inhibitory to spore germination of *F. oxysporum* f. sp. *lentis*. This inhibitory root exudate from 21-day-old plants was thought to partially explain why older plants are more resistant than seedlings when artifically inoculated with the fungus.

Conclusions

Availability of good sources of simply inherited resistance to known races of *Fusarium oxysporum* causing wilt in peas and lentils indicates that these diseases are likely to be held in check through breeding. Because resistance to the cortical rots, caused by *F. solani,* in presently available sources, is based upon multiple genetic factors that affect disease severity both directly and indirectly, the development of cultivars resistant to these diseases is tedious and difficult. The task is most difficult with regard to snap (green) beans and green peas, which require precise refinements in horticultural and processing characteristics. Nevertheless, while better sources of resistance are sought, good progress is being made in establishing available resistance in acceptable beans and peas of all classes. This resistance, used in conjunction with appropriate cultural practices, promises to provide dependable, adequate controls.

The picture may further improve as techniques are developed for better utilization of ecological principles in cultural practices. The discovery and utilization of the principles of "Fusarium suppression" as it occurs naturally in certain soils, and the development of effective systemic chemicals also hold promise as means of disease control in the future.

Literature Cited

1. Abumiya, H., and K. K. Kitazawa. 1963. Studies on the root rot of bean caused by *Fusarium solani* f. *phaseoli.* IV On the growth habit of the causal organism in soil. Res. Bull. Hokkado Nat. Exp. Stn. 81:43–48.
2. Armstrong, G. M., and J. K. Armstrong. 1963. Fusarium wilt of bean in South Carolina and some host relations of the bean *Fusarium.* Plant Dis. Rep. 47:1088–1091.
3. Armstrong, G. M., and J. K. Armstrong. 1964. Pathogenicity of isolates of the bean wilt *Fusarium* from England and the United States. Plant Dis. Rep. 48:846–847.
4. Armstrong, G. M., and J. K. Armstrong. 1974. Races of *Fusarium oxysporum* f. *pisi;* causal agents of wilt of peas. Phytopathology 64:849–857.
5. Azam, H. A., and W. A. Frazier. 1958. Inheritance of resistance to Fusarium root rot in *Phaseolis vulgaris* L., and *Phaseolus coccineus* L. Ann. Rep. Bean Imp. Coop. No. 1:9–10.
6. Basu, P. K., N. J. Brown, R. Crete, C. O. Gourley, H. W. Johnston, H. S. Pepin, and W. L. Seaman. 1976. Yield loss conversion factors for Fusarium root rot of pea. Can. Plant Dis. Survey 56:25–32.
7. Beute, M. K., and J. L. Lockwood. 1968. Mechanism of increased root rot in virus-infected peas. Phytopathology 58:1643–1651.
8. Bisby, G. R. 1918. A Fusarium disease of garden peas in Minnesota. Phytopathology 8:77.
9. Bolton, A. T., V. W. Nutall, and L. H. Lyal. 1966. A new race of *Fusarium oxysporum* f. *pisi.* Can. J. Plant Sci. 46:343–347.
10. Bolton, A. T., and A. G. Donaldson. 1972. Variability in *Fusarium solani* f. *pisi* and *F. oxysporum* f. *pisi.* Can. J. Plant Sci. 52:189–196.
11. Boomstra, A. G., and F. A. Bliss. 1975. New sources of Fusarium root rot resistance in beans. Ann. Rep. Bean Imp. Coop. 18:16–18.
12. Bravo, A., D. H. Wallace, and R. E. Wilkinson. 1969. Inheritance of resistance to Fusarium root rot in beans. Phytopathology 59:1930–1933.
13. Burke, D. W. 1962. Preconditioning and placement of organic materials in the control of bean root rot. Phytopathology 52:727 (Abstr.).
14. Burke, D. W. 1964. Time of planting in relation to disease incidence and yield of beans in Central Washington. Plant. Dis. Rep. 48:789–793.
15. Burke, D. W. 1965. Plant spacing and Fusarium root rot of beans. Phytopathology 55:757–759.
16. Burke, D. W. 1965. Fusarium root rot of beans and behavior of the pathogen in different soils. Phytopathology 55:1122–1126.
17. Burke, D. W. 1965. The near immobility of *Fusarium solani* f. *phaseoli* in soils. Phytopathology 55:1188–1190.
18. Burke, D. W. 1966. Predisposition of bean plants to Fusarium root rot. Phytopathology 56:872 (Abstr.).
19. Burke, D. W. 1968. Root growth obstructions and Fusarium root rot of beans. Phytopathology 58:1575–1576.
20. Burke, D. W. 1975. New red and pink beans resistant to Fusarium root rot. Ann. Rep. Bean Imp. Coop. 18:19–20.
21. Burke, D. W., and A. W. Barker. 1966. Importance of lateral roots in Fusarium root rot of beans. Phytopathology 56:293–294.
22. Burke, D. W., D. J. Hagedorn, and J. E. Mitchell. 1970. Soil conditions and distribution of pathogens in relation to pea root rot in Wisconsin soils. Phytopathology 60:403–406.

23. Burke, D. W., L. D. Holmes, and A. W. Barker. 1972. Distribution of *Fusarium solani* f. sp. *phaseoli* and bean roots in relation to tillage and soil compaction. Phytopathology 62:550–554.
24. Burke, D. W., and J. M. Kraft. 1974. Responses of beans and peas to root pathogens accumulated during monoculture of each crop species. Phytopathology 64:546–549.
25. Burke, D. W., D. E. Miller, L. D. Holmes, and A. W. Barker. 1972. Counteracting bean root rot by loosening the soil. Phytopathology 62:306–309.
26. Burke, D. W., and C. E. Nelson. 1965. Effects of row and plant spacings on yields of dry beans. Wash. Agric. Exp. Stn. Tech. Bull. 664. 6 p.
27. Burke, D. W., and C. E. Nelson. 1967. Response of field beans to nitrogen fertilizers on *Fusarium*-infested and non-infested land. Wash. Agric. Exp. Stn. Bull. 687. 5 p.
28. Burke, D. W., and C. E. Nelson. 1968. NH_4^+ vs NO_3^- fertilization of dry field beans on *Fusarium*-infested land. Wash. Agric. Exp. Stn. Circ. 490. 8 p.
29. Burkholder, W. H. 1916. Some root diseases of the bean. Phytopathology 6:104 (Abstr.).
30. Burkholder, W. H. 1919. The dry root rot of the bean. N. Y. (Cornell) Agric. Exp. Stn. Mem. 29:999–1033.
31. Butler, E. J., and S. G. Jones. 1949. Plant pathology. MacMillan and Co. Ltd., London, England. 979 p.
32. Buxton, E. W. 1955. Fusarium diseases of peas. Trans. Brit. Mycol. Soc. 38:309–319.
33. Buxton, E. W., and D. A. Perry. 1959. Pathogenic interactions between *Fusarium oxysporum* and *Fusarium solani* on peas. Trans. Brit. Mycol. Soc. 43:378–387.
34. Bywater, J. 1959. Infection of peas by *Fusarium solani* var. *martii* and the spread of the pathogen. Trans. Brit. Mycol. Soc. 42:201–212.
35. Christensen, J. A., and L. A. Hadwiger. 1973. Induction of pisatin formation in the pea foot region by pathogenic and nonpathogenic clones of *Fusarium solani*. Phytopathology 63:784–790.
36. Christou, T., and W. C. Snyder. 1962. Penetration and host parasite relationships of *Fusarium solani* f. *phaseoli* in the bean plant. Phytopathology 52:219–226.
37. Claudius, G. R., and R. S. Mehrotra. 1973. Root exudates from lentil [*Lens culinaris* (Medik.)] seedlings in relation to wilt diseases. Plant Soil 38:315–320.
38. Claudius, G. R., and R. S. Mehrotra. 1973. Wilt and root rot diseases of lentil *(L. esculenta)* at Sagar. Indian Phytopathol. 26:268–273.
39. Cook, R. J., and N. T. Flentje. 1967. Chlamydospore germination and germling survival of *Fusarium solani* f. *pisi* in soil as affected by soil water and pea seed exudation. Phytopathology 57:178–182.
40. Cook, R. J., E. J. Ford, and W. C. Snyder. 1967. Mating types, sex, dissemination, and possible sources of clones of *Hypomyces (Fusarium) solani* f. *pisi* in South Australia. Australian J. Agric. Res. 19:253–259.
41. Cook, R. J., and W. C. Snyder. 1965. Influence of host exudates on growth and survival of germlings of *Fusarium solani* f. *phaseoli* in soil. Phytopathology 55:1021–1025.
42. Cook, R. J., and R. D. Watson, (ed.) 1969. Nature of the influence of crop residues on fungus-induced root diseases. Wash. Agric. Exp. Stn. Bull. 716. 32 p.
43. Cruickshank, I. A. M. 1962. Studies on phytoalexins. IV. The anti-microbial spectrum of pisatin. Australian J. Biol. Sci. 15:147–159.
44. Cruickshank, I. A. M., and D. R. Perrin. 1961. Studies on phytoalexins. III. The isolation, assay and general properties of a phytoalexin from *Pisum sativum* L. Australian J. Biol. Sci. 14:336–348.
45. Cruickshank, I. A. M., and D. R. Perrin. 1963. Studies on phytoalexins. IV. Pisatin: the effect of some factors on its formation in *Pisum sativum* L. and the significance of pisatin in disease resistance. Australian J. Biol. Sci. 16:111–128.
46. Dhingra, O. D., S. C. Agrawal, M. N. Khare, and L. S. Kushwaha. 1974. Temperature requirements of eight strains of *Fusarium oxysporum* f. *lentis* causing wilt of lentil. Indian Phytopathol. 24:408–410.
47. Dorn, S. 1974. The role of isomarticin, a toxin of *Fusarium martii* var. *pisi*, in the pathogenesis of stem and root rot of peas. Phytopathol. Z. 81:193–239.
48. Enchandi, E. 1967. Yellows of beans (*Phaseolus vulgaris* L.) provoked by *Fusarium oxysporum* f. *phaseoli* (Sp.). Turrialba 17:409–410.
49. Farley, J. D., and J. L. Lockwood. 1964. Increased susceptibility to root rots in virus-infected peas. Phytopathology 54:1279–1280.
50. Flor, H. H. 1971. Current status of the gene-for-gene concept. Annu. Rev. Phytopathol. 9:275–296.
51. Hagedorn, D. J. 1960. Testing commercial pea varieties for reaction to Fusarium root rot, *Fusarium solani* f. *pisi*. Phytopathology 50:637 (Abstr.).
52. Haglund, W. A. 1968. An atypical Fusarium wilt of peas in northwest Washington. Western Wash. Hort. Assoc. Proc. 3–5. Jan. 1968. Puyallup, WA.
53. Haglund, W. A. 1974. Race concept in *Fusarium oxysporum* f. *pisi*. Pisum Newsletter 6:20–21.
54. Haglund, W. A. 1976. Sampling for Fusarium wilt of peas. Western Wash. Hort. Proc. 42–43. Jan. 1976. Olympia, WA.
55. Haglund, W. A., and J. M. Kraft. 1970. *Fusarium oxysporum* f. *pisi* race 5. Phytopathology 60:1861–1862.
56. Hare, W. W., J. G. Walker, and E. J. Delwich. 1949. Inheritance of a gene for near wilt resistance in the garden pea. J. Agric. Res. 78:239–250.
57. Harter, L. L. 1929. A Fusarium disease of beans. Phytopathology 19:84 (Abstr.).

58. Hildreth, R. C. 1958. Genetic variation and variability of *Fusarium solani* f. *pisi* and *Fusarium oxysporum* f. *pisi* (race 2). Diss. Abstr. 18:11961.
59. Hubbeling, N. 1974. Testing for resistance to wilt and near wilt of peas caused by race 1 and race 2 of *Fusarium oxysporum* f. *pisi*. Overdruk UIT: Mededeling Fakulteit Landbouw-wetenschappen Gent. 29:991–1000.
60. Huber, D. M. 1963. Investigations on root rot of beans caused by *Fusarium solani* f. *phaseoli*. Ph.D. Thesis. Mich. State Univ. 97 p.
61. Huber, D. M., and A. L. Anderson. 1962. Interrelation of bacterial necrosis to crop rotation, isolation frequency, and bean root rot. Phytopathology 52:737 (Abstr.).
62. Huber, D. M., R. D. Watson, and G. W. Steiner. 1965. Crop residues, nitrogen, and plant diseases. Soil Sci. 100:302–308.
63. Hutton, D. G., R. E. Wilkinson, and W. F. Mai. 1973. Effect of two plant parasitic nematodes on Fusarium dry root rot of beans. Phytopathology 63:749–751.
64. Jones, F. R. 1923. Stem and root rot of peas in the United States caused by species of *Fusarium*. J. Agric. Res. 26:459–476.
65. Jones, F. R., and M. B. Linford. 1925. Pea disease survey in Wisconsin. Wis. Agric. Exp. Stn. Bull. 64. 30 p.
66. Kadow, K. J., and L. K. Jones. 1932. Fusarium wilt of peas with special reference to dissemination. Wash. Agric. Exp. Stn. Bull. 272. 30 p.
67. Kamaiyan, J., and Y. L. Nene. 1975. Note on the effect of sowing dates on the reaction of twelve lentil varieties to wilt disease. Madras Agric. J. 62:240–242.
68. Keenan, J. G., H. D. Moore, N. Oshima, and L. E. Jenkins. 1974. Effect of bean root rot on dryland pinto bean production in southwestern Colorado. Plant Dis. Rep. 58:890–892.
69. Kendrick. J. B., and W. C. Snyder. 1942. Fusarium yellows of beans. Phytopathology 32:1010–1014.
70. Kern, H., and S. Naef-Roth. 1965. Formation of phytotoxic pigments by *Fusarium* spp. of the martiella group (Ge). Phytopathol. Z. 53:45–64.
71. Kerr, A. 1963. The root rot-Fusarium wilt complex of peas. Australian J. Biol. Sci. 16:55–59.
72. Khare, M. N., and H. C. Sharma. 1970. Field screening of lentil varieties against Fusarium wilt. Mysore J. Agric. Sci. 4:354–357.
73. King, T. H., H. G. Johnson, H. Bissonnette, and W. A. Haglund. 1960. Development of lines of *Pisum sativum* resistant to Fusarium root rot and wilt. Hort. Sci. 75:510–516.
74. Knavel, D. E. 1967. Studies on resistance to Fusarium root rot, *Fusarium solani* f. *pisi* (F. R. Jones), in *Pisum sativum* L. Proc. Amer. Soc. Hort. Sci. 90:260–267.
75. Kovacikoua, E., and A. Suchanek. 1974. Odolnost Kolekce Cocky K Fusariovemv Vadnuti V. Podminkach Prirozene Infekce. [The resistance of the lentil collection to Fusarium wilt under natural infection conditions.] Shov. UVTI - Ochr. Rostl. 10:59–68.
76. Kraft, J. M. 1974. The influence of seedling exudates on the resistance of peas to Fusarium and Pythium root rot. Phytopathology 64:190–193.
77. Kraft, J. M. 1975. A rapid technique for evaluating pea lines for resistance to Fusarium root rot. Plant Dis. Rep. 59:1007–1011.
78. Kraft, J. M. 1977. The role of delphinidin and sugars in the resistance of pea seedlins to Fusarium root rot. Phytopathology 67:1057–1061.
79. Kraft, J. M., and J. W. Berry. 1972. Artificial infestation of large field plots with *Fusarium solani* f. sp. *pisi*. Plant Dis. Rep. 56:398–400.
80. Kraft, J. M., and D. W. Burke. 1974. Behavior of *Fusarium solani* f. sp. *pisi* and *Fusarium solani* f. sp. *phaseoli* individually and in combination on peas and beans. Plant Dis. Rep. 58:500–504.
81. Kraft, J. M., F. J. Muehlbauer, R. J. Cook, and F. M. Entemann. 1974. The reappearance of common wilt of peas in eastern Washington. Plant Dis. Rep. 58:62–64.
82. Kraft, J. M., and D. D. Roberts. 1969. Influence of soil, water, and temperature on the pea root rot complex caused by *Pythium ultimum* and *Fusarium solani* f. sp. *pisi*. Phytopathology 59:149–152.
83. Kraft, J. M., and D. D. Roberts. 1970. Resistance in peas to Fusarium and Pythium root rot. Phytopathology 60:1814–1817.
84. Linford, M. B. 1928. A Fusarium wilt of peas in Wisconsin. Wis. Agric. Exp. Stn. Bull. 85. 44 p.
85. Lockwood, J. L., and J. C. Ballard. 1960. Evaluation of pea introductions for resistance to Aphanomyces and Fusarium root rots. Mich Agric. Exp. Stn. Quart. Bull. 42:704–713.
86. Maloy, O. C., and W. H. Burkholder. 1959. Some effects of crop rotation on the Fusarium root rot of bean. Phytopathology 49:583–587.
87. Matthews, S., and R. Whitbread. 1968. Factors influencing pre-emergence mortality in peas. I. An association between seed exudates and the incidence of pre-emergence mortality in wrinkle-seeded peas. Plant Pathol. 17:11–17.
88. Matuo, T., and W. C. Snyder. 1972. Host virulence and the *Hypomyces* stage of *Fusarium solani* f. sp. *pisi*. Phytopathology 62:731–735.
89. Matuo, T., and W. C. Snyder. 1973. Use of morphology and mating populations in the identification of formae speciales in *Fusarium solani*. Phytopathology 63:562–565.
90. Mehrotra, R. S., and G. R. Claudius. 1973. Effect of chemical amendments and foliar applications on lentil wilt. Plant Soil 39:695–698.
91. Miller, D. E., and D. W. Burke. 1974. Influence of soil bulk density and water potential on Fusarium root rot of beans. Phytopathology 64:526–529.
92. Miller, D. E., and D. W. Burke. 1975. Effect of soil aeration on Fusarium root rot of beans. Phytopathology 65:519–523.

93. Miller, D. E., and D. W. Burke. 1977. Effect of temporary excessive wetting on aeration and Fusarium root rot of beans. Plant. Dis. Rep. 61:175–179.
94. Muehlbauer, F. J., and J. M. Kraft. 1973. Evidence of heritable resistance to *Fusarium solani* f. sp. *pisi* and *Pythium ultimum* in peas. Crop Sci. 13:34–36.
95. Muller, K. O. 1961. The phytoalexin concept and its methodological significance. Recent Adv. Bot. 1:396–400.
96. Nash, S. M., T. Christou, and W. C. Snyder. 1961. Existence of *Fusarium solani* f. *phaseoli* as chlamydospores in soil. Phytopathology 51:308–312.
97. Nash, S. M., and W. C. Snyder. 1962. Quantitative estimations by plate counts of propagules of the bean root rot *Fusarium* in field soils. Phytopathology 52:567–572.
98. Nash, S. M., and W. C. Snyder. 1964. Dissemination of the root rot *Fusarium* with bean seed. Phytopathology 54:880 (Abstr.).
99. Nash, S. M., and W. C. Snyder. 1967. Comparative ability of pathogenic and saprophytic Fusaria to colonize primary lesions. Phytopathology 57:293–296.
100. Natti, J. J. 1963. Influence of cabbage-bean cropping sequence on root rot and yield of dry beans. Ann. Rep. Bean Imp. Coop. 6:24.
101. Natti, J. J., and D. C. Crosier. 1971. Seed and soil treatments for control of bean root rot. Plant Dis. Rep. 55:483–486.
102. Padwick, G. W. 1941. Report of the Imperial Mycologist. Sci. Rep. Agric. Res. Inst. New Delhi 1939–40:94–101. (Rev. Appl. Mycol. 21:1–2.)
103. Papavizas, G. C. 1963. Microbial antagonism in bean rhizosphere as affected by oat straw and supplemental nitrogen. Phytopathology 53:1430–1435.
104. Papavizas, G. C. 1967. Evaluation of various media and anti-microbial agents for isolation of *Fusarium* from soil. Phytopathology 57:848–852.
105. Perry, D. A. 1959. Studies on the mechanism underlying the reduction of pea wilt by *Fusarium solani* f. *pisi*. Trans. Brit. Mycol. Soc. 42:388–396.
106. Pierre, R. E. 1966. Histopathology and phytoalexin induction in beans resistant and susceptible to *Fusarium* and *Thielaviopsis*. Diss. Abstr. Sec. B 27(5):1353B.
107. Pierre, R. E., and R. E. Wilkinson. 1970. Histopathological relationship of *Fusarium* and *Thielaviopsis* with beans. Phytopathology 60:821–824.
108. Prissyajnyuk, A. A. 1931. [Contribution to the diseases of field crops in the lower Volga region]: Plant. Prot., Leningrad 7:323–337.
109. Pueppke, S. G., and H. D. Van Etten. 1974. Pisatin accumulation and lesion development in peas infected with *Aphanomyces euteiches*, *Fusarium solani* f. sp. *pisi*, or *Rhizoctonia solani*. Phytopathology 64:1433–1440.
110. Reichle, R. E., W. C. Snyder, and T. Matuo. 1964. *Hypomyces* stage of *Fusarium solani* f. *pisi*. Nature 203:664–665.
111. Reinking, O. A. 1942. Distribution and relative importance of various fungi associated with pea root rot in commercial pea growing areas in New York. N. Y. (Cornell) Agric. Exp. Stn. Bull. 264. 43 p.
112. Rintelen, J. 1973. Influence of weeds on the infection of peas and flax by soil-borne Fusaria. III. Experiments on competition between pathogenic and apathogenic isolates of *Fusarium solani* and *Fusarium oxysporum* in infection of peas. Z. Pflanzenkr Pflanzenschutz 80:466–470.
113. Schreuder, J. C. 1951. Een oderzoek over de Amerikaanse vaatziekte van de erwten in Nederland, Rijdschr. Plantenziekten. 57:175–206.
114. Schroeder, W. T. 1953. Root rots, wilts and blights of peas, p. 401–408. *In* A. Stefferud (ed.), Plant Diseases, the Yearbook of Agriculture 1953, U. S. Dept. Agric., Washington, D. C.
115. Schroth, M. N., and F. F. Hendrix. 1962. Influence of nonsusceptible plants on the survival of *Fusarium solani* f. *phaseoli* in soil. Phytopathology 52:906–909.
116. Schroth, M. N., and W. C. Snyder. 1961. Effect of host exudates on chlamydospore germination of the bean root rot fungus, *Fusarium solani* f. *phaseoli*. Phytopathology 51:389–393.
117. Short, G. E., and M. L. Lacy. 1976. Carbohydrate exudation from pea seeds. Effect of cultivars, seed age, seed color and temperature. Phytopathology 66:182–187.
118. Silbernagel, M. J., and W. J. Zaumeyer. 1973. Beans, p. 253–269. *In* R. R. Nelson (ed.), Breeding plants for disease resistance: Concepts and applications, Pa. State Univ. Press, University Park, PA.
119. Snyder, W. C. 1933. A new vascular Fusarium disease of peas. Science 77:327.
120. Snyder, W. C., M. N. Schroth, and T. Christou. 1959. Effect of plant residues on root rot of bean. Phytopathology 49:755–756.
121. Steadman, J. R., E. D. Kerr, and R. F. Mumm. 1975. Root rot of bean in Nebraska: Primary pathogen and yield loss appraisal. Plant Dis. Rep. 59:305–308.
122. Toussoun, T. A., and Z. A. Patrick. 1963. Effect of phytotoxic substances from decomposing plant residues on root rot of bean. Phytopathology 53:265–270.
123. Vasudeva, R. S., and K. V. Srinivasam. 1952. Studies on the wilt disease of lentil (*Lens esculenta* Moench). Indian Phytopathol. 5:23–32.
124. Wade, B. L. 1929. The inheritance of Fusarium wilt resistance in canning peas. Wisc. Agric. Exp. Stn. Res. Bull. 97. 32 p.
125. Wade, B. L., W. J. Zaumeyer, and L. L. Harter. 1938. Variety studies in relation to Fusarium wilt of peas. U. S. Dept. Agric. Cir. 473. 26 p.
126. Walker, J. C. 1931. Resistance to Fusarium wilt in

garden, canning, and field peas. Wisc. Agric. Exp. Stn. Res. Bull. 107:1–15.
127. Walker, J. C. 1952. Disease of vegetable crops. McGraw-Hill Book Company. 525 p.
128. Walker, J. C., and W. W. Hare. 1943. Pea diseases in Wisconsin. Wisc. Agric. Res. Bull. 145:20–22.
129. Wallace, D. H., and R. E. Wilkinson. 1965. Breeding for Fusarium root rot resistance in beans. Phytopathology 55:1227–1231.
130. Weinke, K. E. 1962. The influence of nitrogen on root disease of bean caused by *Fusarium solani* f. *phaseoli*. Phytopathology 52:757 (Abstr.).
131. Wilson, V. E., and J. Brandsberg. 1965. Fungi isolated from diseased lentil seedlings in 1963–64. Plant Dis. Rep. 49:660–662.
132. Worf, G. L., and D. J. Hagedorn. 1962. Interactions of two pea Fusaria in soil and associated host responses. Phytopathology 52:1126–1132.
133. Wyse, D. L., W. F. Meggitt, and D. Penner. 1976. Herbicide root rot interactions in Navy Bean. Weed Sci. 24:16–21.
134. Zaumeyer, W. J. 1962. Pea diseases. U. S. Dept. Agric. Handbook 228. 30 p.
135. Zaumeyer, W. J., and H. R. Thomas. 1957. Bean diseases and methods for their control. U. S. Dept Agric. Tech. Bull. 868. 255 p.

15 *Fusarium*-Incited Diseases of Tomato and Potato and Their Control*

John Paul Jones and S. S. Woltz

Tomato Wilt

Fusarium wilt, incited by *Fusarium oxysporum* f. sp. *lycopersici,* was first described by Massee (66) in 1895. The disease is of world-wide importance, having been reported in at least 32 countries. In southern locations the disease is destructive in the field, but in northern areas it is limited by temperature primarily to glasshouse crops (90).

Tomato wilt is a warm weather disease, most prevalent on acid, sandy soils (92). Soil and air temperatures of 28 C are optimum for disease; too warm (34 C) or too cool (17–20 C) soils retard wilt development (19). If soil temperatures are optimum but air temperatures below optimum, the fungus will extend into the lower parts of the stem, but the plants will not exhibit external symptoms.

In general, factors favoring wilt development are: soil and air temperatures of 28 C, soil moisture optimum for plant growth, plants preconditioned with low nitrogen and phosphorus and high potassium, low soil pH, short day length, and low light intensity (37, 50, 54, 76, 77, 91). Conditions, in general, that predispose plants to disease may retard post-inoculation development of wilt. For example, wilt development is favored by dry soils prior to inoculation and by soil-moisture approaching field capacity after inoculation (20).

Dissemination of *F. oxysporum* f. sp. *lycopersici* is via seed, tomato stakes, soil, and infected transplants (11, 30, 55). Long distance spread is through seed and transplants; local dissemination is by transplants, tomato stakes, wind- and water-borne infested soil, and farm machinery.

Wilt once was the most common and destructive disease of tomato. However, Bohn and Tucker (9, 10) in 1939–40 reported the discovery of high-level resistance to wilt in Missouri Accession 150 (*Lycopersicon pimpinellifolium* P.I. 79532). This resistance was governed by a single dominant gene (I) which soon was incorporated into commercial cultivars (74) that under field conditions were nearly "immune" to Fusarium wilt.

Shortly after the discovery of the I gene, Alexander of the Ohio Agricultural Experiment Station secured breeding lines with the I gene from Tucker. Alexander made crosses and tested the progeny for resistance to wilt using a *Fusarium* isolate obtained from diseased plants found in a northern Ohio glasshouse. To his surprise, all plants succumbed to the disease. Alexander thereupon informed Tucker that the Missouri lines were not resistant to Fusarium wilt. Tucker replied that the material indeed was resistant and questioned Alexander's ability to make tomato crosses properly (L. J. Alexander, *per-*

*Florida Agricultural Experiment Stations Journal Series No. 747.

sonal communication). To resolve the differences in results, the men exchanged *Fusarium* isolates and breeding lines. They then jointly discovered that Alexander's isolate was a new pathogenic form which they designated as race 2 (3). All tomato cultivars up to 1969 were susceptible to race 2, including all cultivars with the I gene.

Although race 2 was first reported in 1945, it did not become widespread or of concern until after 1961 when Stall (81) reported its occurrence in Florida. Rapidly thereafter, it was discovered in several states of the United States and in several other countries including Australia, Brazil, Great Britain, Israel, Mexico, Morocco, and the Netherlands (38, 60, 62, 73, 87).

Because Alexander (1, 2) had the foresight to screen introductions of *Lycopersicon esculentum* and wild species as early as 1955, Stall and Walter (82) in 1965 were able to rapidly locate resistance to both race 1 and race 2 in accession P.I. 126915 (*L. esculentum* x *L. pimpinellifolium* hybrid). The resistance in P.I. 126915, according to Cirulli and Alexander (18), was regulated by two single dominant genes, one conferring resistance to race 1 and one conferring resistant to race 2. They suggested that the symbol I-2 represent the resistance gene to race 2. The gene for resistance to race 1 in P.I. 126915 was left undesignated because they were unable to determine whether it was the same as the I gene (gene for resistance to race 1 found in Missouri Accession 160). If two single dominant genes are involved, they are apparently very closely linked, because repeated attempts by Crill and Jones failed to separate them *(unpublished data)*. There are no known breeding lines or cultivars susceptible to race 1 and resistant to race 2. If a line is resistant to race 2, it has been by experience resistant to race 1.

Currently, several tomato breeding programs throughout the world are, in part, concerned with the development of cultivars resistant to both races of the pathogen. Numerous cultivars and hybrids are commercially available with resistance to both races. One of these, Walter (83), developed primarily by J. M. Walter and R. E. Stall of the University of Florida, is widely used and has saved the Florida and Mexican fresh market tomato industries from considerable loss and expense that would have been incurred with susceptible cultivars.

A third race of *F. oxysporum* f. sp. *lycopersici* supposedly inciting typical wilt symptoms on tomato was reported in 1966 in Brazil (88). However, the report was tentative and has not yet been verified.

The Committee on Genetic Vulnerability of Major Crops (22) reported that tomato was vulnerable to Fusarium wilt because of the wide-spread dependence on the I and I-2 genes for resistance. However, *F. oxysporum* f. sp. *lycopersici* has been quite stable in regard to the evolution of new pathogenic races. Cultivars with the I gene completely controlled wilt for 20 years before race 2 first caused significant crop reductions in Florida. Moreover, the cultivar Walter, 8 years after its introduction, remains resistant to all known verified wilt races, despite the fact that nearly 100% of the fresh market tomato acreage in Mexico and Florida since 1970 has been planted to Walter or other cultivars with the same resistance genes.

A great number of polygenic tolerant tomato cultivars were released from 1909 (34) to 1950, with the most widely used being Marglobe (75) and Rutgers (12). However, when environmental conditions, especially temperature, were favorable, a large percentage of the plants developed wilt symptoms and many died. Later when race 2 evolved in Florida, the use of the polygenic tolerant cultivar Homestead 24 (tolerant to race 2, resistant to race 1) did not prevent the establishment or spread of the pathogen, nor did it prevent significant field losses (27). Thus far the tomato has been far more vulnerable to losses caused by Fusarium wilt when polygenic tolerant rather than monogenic resistant (vertical resistance) cultivars were utilized.

Moreover, pathogen spread is curtailed with monogenic resistant cultivars, whereas polygenic tolerant cultivars encourage dissemination of the pathogen. Polygenic tolerance mechanisms should be utilized only when monogenic resistance is not available or when the utilization of monogenic resistance mechanisms is not feasible (26).

Root-knot nematode *(Meloidogyne incognita)* infection reportedly reduces the resistance conferred by the I or I-2 gene (13, 21, 39, 47). It was even suggested that breeding for monogenic resistance be discontinued because of the root-knot nematode effect on breaking resistance to Fusarium wilt of tomato (78). However, there are several reports that root-knot nematode infection did not reduce monogenic resistance (8, 43, 53, 61). Furthermore, there is no field evidence of I gene cultivars succumbing to wilt incited by race 1 or I-2 gene cultivars succumbing to race 1 or 2, despite the fact that these cultivars are commonly grown in fields highly infested with either or both Fusarium races as well as the root-knot nematode. Thus the utilization of monogenic resistance for control of Fusarium wilt incited by race 1 or 2 has been and should remain successful, even in areas where root-knot nematode field populations are sufficient to cause severe root galling.

In areas where there are no horticulturally adapted monogenic resistant cultivars, soil fumigation can greatly alleviate disease losses (17, 25, 50, 69, 70, 72, 86, 99, 100). Several of the better soil fumigants are DD-MENCS (methyl isothiocyanate + chlorinated C_3 hydrocarbons), chloropicrin, chloropicrin + chlorinated C_3 hydrocarbons or ethylene dibromide, methyl bromide, and methyl bromide + chloropicrin mixtures.

All of the broad-spectrum soil fumigants should be injected into the soil no deeper than 8 inches and no shallower than 5 inches (71). The injection streams should be 8 to 10 inches apart for DD-MENCS, chloropicrin, and chloropicrin + nematicide mixtures, and 10 to 12 inches apart for methyl bromide and methyl bromide + chloropicrin.

The soil to be fumigated should be of seedbed tilth so that the fumigant vapors can readily permeate it. All plant debris should be well rotted because pathogens in nondecomposed tissue are not destroyed by soil fumigants. Soil moisture sufficient to permit seed germination should be maintained at least 2 weeks prior to fumigation to break the dormancy of pathogen resting bodies and weed seeds. Adequate soil moisture at time of fumigation is essential to prevent premature loss of fumigant vapors. The soil should be compacted simultaneously with injection of the fumigant to help maintain moisture and contain the fumigant.

Mulches, such as polyethylene film, greatly improve the benefits of fumigation by retaining soil moisture and preventing subsequent loss of fumigant vapors (51). Mulches also prevent recontamination of fumigated soil, help confine root systems to treated soil, protect roots from mechanical injury, reduce fruit rots on ground or bush crops, provide weed control, and probably keep the major part of the rhizosphere less supportive of *F. oxysporum* because of low fertility (to be discussed later). Ideally, mulches should be applied immediately after fumigation.

A waiting period between planting and fumigating is necessary to permit fumigant vapors to dissipate and to avoid crop injury. The length of the waiting period depends on the fumigant, the weather, and the method of planting. The average period is 2 weeks; however, if the soil temperature is low or if heavy rains seal the soil, preventing vapor loss, the waiting period may extend to 3 or 4 weeks, especially when chloropicrin is used. Seeded crops are less likely to be injured than transplants. If transplants are used, they should be pathogen free.

Although field fumigation is expensive, costs can be considerably reduced and disease control maintained, by strip fumigation (50, 70). Moreover, Overman and Jones have preliminary evidence that the volume of treated soil per acre can be greatly decreased with the use of certain fumigants without decreased wilt control.

Fears have been expressed that recontamination of pasteurized soil will cause greater disease losses than would have been realized had the soil not been treated. These fears may be exaggerated; Jones and Crill (49) demonstrated that *Verticillium albo-atrum*-inoculated tomato seedlings could be transplanted into fumigated soil with yield losses not exceeding those incurred by transplanting healthy seedlings into nonfumigated soil infested with *V. albo-atrum*. Moreover, in experimental plots at the University of Florida Agricultural Research and Education Center-Bradenton, for over 15 years soil fumigation has, without exception, resulted in yield increases sufficient to cover the cost of fumigation and usually in sizeable net profits. If recontamination did occur during these years, it did not cause serious losses.

Steam treatments, especially in glasshouses, result in excellent control of Fusarium wilt and some of the relatively inexpensive contact, persistent nematicides reduce the occurrence of wilt and considerably increase yields (52, 65).

Benomyl in greenhouse experiments on young plants has provided control of Fusarium wilt of tomato (7, 85). Additionally, complete control of wilt of pot chrysanthemum and aster was obtained when benomyl was used in conjunction with lime and nitrate-nitrogen soil amendments (32,33). However, benomyl may injure tomato seedlings (7), and in the field the chemical does not seem to be as practical for wilt control as the broad-spectrum soil fumigants.

The original objective of steam "sterilization" in the greenhouse was eradication of soil-borne pathogens. However, the objective was not realized, since commonly 20% of the tomato plants would die from Fusarium wilt despite the use of steam. Currently, the objective of soil pasteurization, whether with broad-spectrum soil fumigants, steam, or nematicides, is not eradication of soil-borne pathogens, but the reduction of the populations sufficient to delay infection long enough that fruit yields are not adversely affected. Generally, in warm climates, *Fusarium* populations (as well as other pathogen populations) increase to relatively high levels within 3 months after fumigation. Consequently, by harvest many tomato plants exhibit slight to moderate wilt symptoms. The longer the season the greater the yield losses. However, with determinate-vined plants with a concentrated fruit set where fruit are harvested only one to three times, yields are not significantly decreased.

Fusarium wilt of tomato (and several other crops) also can be effectively controlled by adjustments in soil fertility and cultural practices leading to a systematic starvation of the pathogen (54, 56, 57, 93, 94, 95, 96). In soils that are low or deficient in native nutrients, fertilization and liming can be manipulated to retard the growth of *F. oxysporum* f. sp. *lycopersici* and to reduce disease incidence and severity. *Fusarium* requires most of the same nutrients as tomato plants; however, there are important differences. The tomato, with its extensive root system, penetrates a large portion of the topsoil, whereas *Fusarium* propagules are limited to a very local microenvironment. Consequently, if the soil infestation level is low, the rate of *Fusarium* proliferation can be greatly inhibited by using localized fertilizer placement to nourish the tomato plant without making fertilizer nutrients available throughout the topsoil. The system developed at the University of Florida Agricultural Research and Education Center-Bradenton retards development of wilt by reducing the growth, sporulation, and virulence of *Fusarium* by creating nutritional deficiencies, especially of micronutrients.

Fusarium oxysporum f. sp. *lycopersici* has a relatively high requirement for micronutrients. Deficiencies of copper, iron, manganese, molybdenum, and zinc reduce growth and sporulation. Moreover, *Fusarium* grown in liquid cultures devoid of molybdenum or zinc is not as virulent as *Fusarium* grown on optimal levels of these two micronutrients. The response of *Fusarium* to manganese, iron, and zinc is very pronounced. Increasing amounts of these micronutrients above those usually found in soil solutions for tomato culture are increasingly beneficial to growth and sporulation of the pathogen, whereas concentrations below average inhibit growth and spore production.

Liming *F. oxysporum* f. sp. *lycopersici*-infested field and glasshouse soil to pH 7.0–7.5 greatly limits the availability of micronutrients and consistently decreases the occurrence of wilt in naturally low-pH soils. However, when high-pH soil is further amended with lignosulfonate metal complexes of zinc and manganese or iron and manganese (these metal complexes are available for plant growth at high soil pH values), the beneficial effect of pH elevation is reversed. Moreover, amending infested field soil with gypsum ($CaSO_4$) does not increase soil pH and does not reduce the occurrence of wilt. The calcium content of soil amended with gypsum is as great as that amended with hydrated lime, and the calcium content of plants grown in soils amended with gypsum is as great as that of plants grown in soil amended with hydrated lime. Consequently, the beneficial effects of liming are not caused by an increased soil or tissue calcium content, but rather by the unavailability of micronutrients created by the high soil pH, which in turn limits the growth, sporulation, and virulence of the pathogen.

Increases in soil pH limit not only the availability of micronutrients but also of other elements essential for *F. oxysporum,* especially phosphorus and magnesium. Supplemental applications of superphosphate (in field plots) above the amount required for growth of tomato greatly increase occurrence of wilt in soils of pH 6.0. At pH 7.0 or 7.5 supplemental applications do not increase wilt occurrence because at these pH values phosphorus becomes unavailable. Increasing amounts of available magnesium supplied above deficiency levels increase stepwise the growth, sporulation, and virulence of *F. oxysporum.*

Nitrogen source greatly affects development of Fusarium wilt: nitrate-nitrogen (NO_3-N) inhibits, whereas ammonium-nitrogen (NH_4-N) encourages wilt development. The effect of nitrogen source appears to be threefold: first, NO_3-N increases and NH_4-N decreases soil pH; second, *F. oxysporum* f. sp. *lycopersici* grown on NH_4-N is far more virulent than *Fusarium* grown on NO_3-N; and third, tomato seedlings grown on NO_3-N are preconditioned more resistant to Fusarium wilt than seedlings grown on NH_4-N. The predominant factor seems to be the effect of nitrogen source on soil pH. However, NH_4-N applied to soil weekly may reverse the usual disease-preventative effect of high soil pH without a concomitant decrease in soil pH.

Liming inhibits development of Fusarium wilt not only because of its effect on the availability of micronutrients, but also because of its effect on soil microflora. Actinomycete and bacterial populations are favored by high soil pH values (89). Certain of these microorganisms are antagonistic to *F. oxysporum* f. sp. *lycopersici,* preventing spore germination and vegetative growth by means of toxic compounds. Often zones of *Fusarium* inhibition extend beyond the boundaries of the inhibiting organism. Bacteria and actinomycetes also compete with *Fusarium* for organic and inorganic nutrients in the soil solution.

Fusarium oxysporum f. sp. *lycopersici* races 1 and 2 can utilize several compounds as carbon sources including sugars, starch, cellulose, pectin, protein, and amino acids (S. S. Woltz and J. P. Jones, *unpublished data*). Of these, cellulose and pectin support a slower growth rate. However, *Fusarium* cultured on a pectin medium is more virulent than *Fusarium* grown on a dextrose medium. The lack of a readily available carbon source in the soil greatly retards growth of the pathogen. Thus a fallow period to bring about decomposition of organic matter residues will inhibit *Fusarium* growth and subsequent disease development. Fallowing also destroys weeds and grasses whose roots harbour the pathogen (59).

Davet et al. (28) reported that a cool (18–29 C) temperature strain of *F. oxysporum* f. sp. *lycopersici* race 1 existed in Morocco that produced severe wilt symptoms, especially if affected plants were irrigated with NaCl, $NaNO_3$, or MgCl. Calcium nitrate and calcium sulfate protected against the pathogen because, they hypothesized, Na or Mg ions replaced the Ca ions of the polygalacturonate chains of the cell membrane. This was consistent with the work of Edgington and Walker (29), who found that disease declined with an increase in calcium. The latter noted that the effect of boron was dependent on the calcium supply, and suggested that the effect of these elements on cell wall development be studied. Corden (24) also found that severity of wilt was increased by a calcium deficiency and suggested that calcium inhibited the activity of the polygalacturonase produced by the pathogen. However, Corden theorized that excess calcium would not give practical control of Fusarium wilt. Jones and Woltz (54, 56) found that calcium sulfate amendments failed to affect wilt development in pot experiments or in the field, supporting Corden's contention that calcium will not give practical field control. Possibly, measures that control wilt under unfavorable environmental conditions, such as cool weather, may be less effective under conditions highly favorable for wilt development.

Beckman et al. (6) found that tylose development in susceptible cultivars (compared to resistant cultivars) is delayed 2 days and complete occlusion is delayed 7 days after inoculation with *F. oxysporum* f. sp. *lycopersici.* They concluded that the sealing-off process by tylose development is a general type of resistance in higher plants against vascular pathogens. Cook implied a year later (23) that tomato wilt could be controlled by depriving infected plants of water, thereby reducing or ceasing transpiration. Since *F. oxysporum* microspores are carried in the transpiration stream, the upward spread of the pathogen would cease or be slowed sufficiently to permit time for the host occlusion reaction to seal it off. However, in the field, manipulation of soil water is not practical. Moreover, by the time symptoms are definitely evident, the pathogen probably has invaded and ramified throughout many plants. It has been our observation that once symptoms are apparent, withdrawal of water accelerates disease development.

Plants can be preconditioned for resistance or susceptibility by the use of certain amino acids or analogues of amino acids (58). Ethionine is very effective in protecting against wilt in the greenhouse. Although the fungus invades and colonizes the stem, symptoms do not develop. Preconditioning with ethionine, however, does not protect seedlings transplanted to the field (Jones and Woltz, *unpublished data*).

Summary of Control Measures

1. Use resistant cultivars where available.
2. Use soil fumigants if available and if economically feasible.
3. Manipulate soil fertility in such a way to decrease the growth, sporulation, and virulence of the pathogen.
   A. Add lime amendments to obtain soil pH of at least 7.0.
   B. Avoid use of soil amendments of micronutrients.
   C. Avoid excessive use of phosphorus and magnesium soil amendments.
   D. Use nitrate rather than ammonium-nitrogen.
   E. Apply fertilizers as bands close to tomato roots, do not broadcast apply.
   F. Permit soil to be fallow prior to planting.
4. Avoid the use of diseased transplants or infested seed.
5. Prevent dissemination of the pathogen by eliminating movement of infested soil into disease-free areas.

It is strategically sound to control as many factors as possible since under field conditions the complete control of a single factor can not be assured.

The system of two wilt races of *F. oxysporum* f. sp. *lycopersici* and two single dominant genes makes the tomato-Fusarium wilt system attractive to investigations by physiologists, biochemists, and pathologists delving into the nature of host resistance and pathogen variability. Reports on this aspect of tomato Fusarium wilt are too numerous and complex to be discussed here. For an excellent review and evaluation of this type of work see Walker's monograph on Fusarium wilt of tomato (90).

Other *Fusarium*-Incited Diseases of Tomato

Fusarium root rot, incited by *F. oxysporum* f. sp. *lycopersici*, appears to be the second most damaging *Fusarium*-incited disease of tomato. The disease was first reported in Japan and has become increasingly prevalent. Recently it has been noted in Japan, Canada, and the United States (Arizona, California, Florida, and Ohio) (31, 35, 46, 80, 98).

Fusarium root rot, similar to the cool weather Fusarium wilt reported by Davet, is favored by 10 to 20 C temperatures. Diseased plants exhibit root and crown lesions. Tap roots often are killed and lesions form on the lateral roots. Adventitious roots frequently form, especially if soil is placed around the stems of infected plants. Nevertheless, the root system is greatly reduced, and as the temperature increases, causing concomitant increase in water demand, infected plants wilt and may die. Vascular browning usually develops but does not extend more than 20 cm above soil line.

Yamakawa and Yosui (97) located resistance to Fusarium root rot in their breeding lines IRB 301–30 and 31 (*Lycopersicon esculentum* x *L. peruvianum* P.I. 126944 crosses) and in a cultivar from the Netherlands. The resistance in the IRB breeding lines is governed by a single dominant gene which Yamakawa and Yosui designated as J_3. Tomato breeders in Japan, Canada, Ohio, California, and Florida are attempting to incorporate the J_3 gene into commercially acceptable cultivars.

Farley et al. (36) initially reported failure of steam and chemical treatments to control Fusarium root rot in greenhouses in Ohio, and Sonoda (80) reported that a chloropicrin-methyl bromide treatment failed to control the disease in the field in Florida. However, apparently both researchers encountered recontamination immediately after soil pasteurization. Farley later (*personal communication,* August, 1976) indicated that steam sterilization was effective if the greenhouse interior were treated with formaldehyde. Canadian workers (*personal communication,* August, 1976) reported control with soil fumigants if sources of recontamination, such as compost piles of infected plants, were eliminated.

Sonoda (80) found relatively little disease where tomato was preceded by three crops of pepper. Disease was much more prevalent on a tomato crop preceded by three consecutive tomato crops.

In summary, control of Fusarium root rot should be through the use of resistant cultivars when available, fumigation or steam pasteurization, and crop rotation.

Fusarium solani was reported by Joffe and Palti (48) to cause a wilt of tomato in Israel. No other symptoms or information concerning the disease or organism is known. *Fusarium solani* also causes a foot rot of tomato (11). The symptoms (reported by W. Pont, Dept. Prim. Industries, Kamerunga Hort. Res. Stn., Queensland) are very similar to those incited by *F. oxysporum* f. sp. *lycopersici.* A dry rot of the tap root occurs, with vascular browning extending 10 to 20 cm up the stem. A spotting of the apical leaves develops, in contrast to the *F. oxysporum* root rot. No control measures are known, although soil fumigation or steam pasteurization should provide excellent control.

Various species of *Fusarium* cause rots of tomato fruit in the field and in post-harvest storage. Fruit matured during warm, wet weather are especially susceptible. Fusarium fruit rot, whether in the field, in the ripening room, or in transit, should be well controlled by field applications of foliar fungicides. High concentrations of ethylene may decrease Fusarium rots in ripening rooms and in transit (63).

Fusarium-Incited Diseases of Potato

A number of *Fusarium* species are associated with potato "wilts," stem rots, and tuber rots. Some of these, according to the classification of Booth (11) are: *Fusarium oxyspo-*

rum f. sp. *tuberosi* (Syn: *F. oxysporum* f. 1, *F. oxysporum, F. oxysporum* f. sp. *solani, F. oxysporum* var. *solani*); *Fusarium solani; Fusarium solani* f. sp. *eumartii* (Syn: *F. eumartii, F. solani* var. *eumartii*); *Fusarium solani* var. *coeruleum* (Syn: *F. solani* f. sp. *radicicola, F. coeruleum, F. radicicola*); *Fusarium avenaceum* (Syn: *F. roseum, F. roseum* 'Avenaceum'); *Fusarium sambicinum* (Syn: *F. roseum, F. roseum* 'Sambucinum'); and *Fusarium trichothecioides* (Syn: *F. roseum*).

Fusarium oxysporum f. sp. *tuberosi* and *F. solani* f. sp. *eumartii* are reported to cause "wilts" of potato. However, neither pathogen should be considered as a true vascular pathogen. The former primarily causes a stem rot and the latter principally a root rot and occasionally a stem rot.

Fuarium oxysporum f. sp. *tuberosi* was first reported by Smith and Swingle (79) in 1904 as causing a wilt of potato. Since then the disease has been reported throughout the world. Inoculation results indicate that *F. oxysporum* f. sp. *tuberosi* is a weak pathogen, that infection is highly dependent on environmental conditions, and that conditions favorable for infection are adverse for growth of the potato plant (40, 41).

Fusarium oxysporum f. sp. *tuberosi* (64), *F. solani* (42), and *F. avenaceum* (68) cause rosetting and chlorosis of the vine following rot of the lower stem. Root and seed piece infection from the soil is common and rot of new tubers may occur.

Fusarium solani f. sp. *eumartii* was first reported by Carpenter (16) in 1915 to cause a field and storage rot of tubers. Haskell (44) a year later reported that the pathogen caused a wilt of the potato plant and stem-end rot of the tubers. Since 1916 the disease has been reported from many countries.

Fusarium solani f. sp. *eumartii* is much more virulent than *F. oxysporum* f. sp. *tuberosi* (41, 68). Inoculation with the former pathogen almost invariably results in the development of typical field symptoms. These symptoms, following infection through the roots, first appear as yellowing and necrosis of the interveinal areas of the young leaves. Necrosis of the pith of the upper stem occurs, and the lower stem develops vascular discoloration. Root hairs are destroyed and the cortex of roots slough off. Extensive rot of the stem-end and vascular discoloration of the new tuber occur (41, 45, 84).

In contrast to the above disease syndrome, when invasion occurs through the seed piece, severe rotting of the underground stem occurs, resulting in a rapid wilt of the infected plant.

The "wilts" and stem rots are normally controlled by crop rotation, sanitation, and the use of seed piece treatment (41, 45). Soil fumigation is possible.

Fusarium solani var. *coeruleum, F. sambucinum,* and *F. trichothecioides* incite dry rot of tubers, one of the major diseases of potato in storage, transit, or handling (84). Tuber rot is world-wide in distribution and typically is a dry rot in which affected tissue becomes hard and chalky. In early stages the rot may be moist and cheesy; later the affected tissue becomes hard and leathery. Cavities containing white threads of the fungus form in the rotted tissue. The cavity walls are tinted salmon pink or blue. In the later stages of decay, compact mycelial tufts form on the surface of the tubers.

Bacterial soft rot frequently is associated with dry rot, causing a rot more typical of bacterial soft rot. Dry rot also may infrequently follow late blight (caused by *Phytophthora infestans*) or frost injury.

The fungi that cause dry rot are soil-borne and invade through wounds caused by digging and handling operations. Immature tubers are more susceptible to rot than mature tubers. In storage, tubers become increasingly susceptible as storage time lengthens (14, 67).

The cultivars Hunter, Keswick, Pontiac, Warba, and Netted Gem are susceptible to rot incited by *F. solani* var. *coeruleum,* whereas Kennebec, Sebago, Green Mountain, Cherokee, Merrimac, Ontario, Early Gem, and Menominee are resistant. Sebago, Keswick,

Chippewa, and Pontiac are susceptible to rot caused by *F. sambucinum*, whereas Cherokee, Houma, Irish Cobbler, Hunter, Warba, Merrimac, and Netted Gem are resistant (4, 15).

Control of dry rot can be obtained by the avoidance of mechanical injury to tubers. Superficial moisture should be removed from the tubers as soon as possible, but sufficiently high humidity and temperature must be maintained to promote suberization and wound healing. Seed piece treatment may be of benefit (5).

Literature Cited

1. Alexander, L. J. 1959. Progress report of National Screening Committee for Disease Resistance in Tomato for 1954–1957. Plant Dis. Rep. 43: 55–65.
2. Alexander, L. J., and M. M. Hoover. 1955. Disease resistance in wild species of tomato. Ohio Agric. Exp. Stn. Res. Bull. 752. 76 p.
3. Alexander, L. J., and C. M. Tucker. 1945. Physiological specialization in the tomato wilt fungus *Fusarium oxysporum* f. *lycopersici*. J. Agric. Res. 70: 303–313.
4. Ayers, G. W. 1956. The resistance of potato varieties to storage decay caused by *Fusarium sambucinum* f. 6 and *Fusarium coeruleum*. Amer. Potato J. 33: 249–254.
5. Ayers, G. W., and D. B. Robinson. 1956. Control of Fusarium dry rot of potatoes by seed treatment. Amer. Potato J. 33: 1–5.
6. Beckman, C. H., D. M. Elgersma, and W. E. MacHardy. 1972. The localization of Fusarial infections in the vascular tissue of single-dominant-gene resistant tomatoes. Phytophatology 62: 1256–1260.
7. Biehn, W. L. 1973. Curative action of foliar sprays of acidified benomyl suspensions against Fusarium wilt of tomato. Plant Dis. Rep. 57: 37–38.
8. Binder, E., and M. T. Hutchenson. 1959. Further studies concerning the effect of the root-knot nematode *Meloidogyne incognita acrita* on the susceptibility of the Chesapeake tomato to Fusarium wilt. Plant Dis. Rep. 43: 972–978.
9. Bohn, G. W., and C. M. Tucker. 1939. Immunity to Fusarium wilt in the tomato. Science 89: 603–604.
10. Bohn, G. W., and C. M. Tucker. 1940. Studies on Fusarium wilt of the tomato. I. Immunity in *Lycopersicon pimpinellifolium* and its inheritance in hybrids. Mo. Agric. Exp. Stn. Res. Bull. 311. 82 p.
11. Booth, C. 1971. The genus *Fusarium*. Commonwealth Mycol. Inst., Kew, Surrey, England. 237 p.
12. Boswell, V. R. 1937. Improvement and genetics of tomatoes, peppers, and eggplant. U.S. Dept Agric. Yearbook 1937: 176–206.
13. Bowman, P., and J. R. Bloom. 1966. Breaking the resistance of tomato varieties to Fusarium wilt by *Meloidogyne incognita*. Phytopathology 56: 871.
14. Boyd, A. E. W. 1952. Dry-rot disease of the potato. V. Seasonal and local variations in tuber susceptibility. Ann. Appl. Biol. 39: 330–338.
15. Boyd, A. E. W. 1952. Dry-rot disease of the potato. VI. Varietal difference in tuber susceptibility obtained by injection and riddle-abrasion methods. Ann. Appl. Biol. 39: 339–350.
16. Carpenter, C. W. 1915. Some potato tuber rots caused by species of *Fusarium*. J. Agric. Res. 5: 183–210.
17. Ciccarone, A. 1955. Control tests on Fusarium disease of tomato, with particular reference to the use of methyl bromide as a soil fumigant. (Trans. title, in Italian). Tecn. Agric. 1955: 11–12.
18. Cirulli, M., and L. J. Alexander. 1966. A comparison of pathogenic isolates of *Fusarium oxysporum* f. *lycopersici* and different sources of resistance in tomato. Phytopathology 56: 1301–1304.
19. Clayton, E. E. 1923. The relation of temperature to the Fusarium wilt of the tomato. Amer. J. Bot. 10: 71–88.
20. Clayton, E. E. 1923. The relation of soil moisture to the Fusarium wilt of the tomato. Amer. J. Bot. 10: 133–147.
21. Cohn, E., and G. Minz. 1960. Nematodes and resistance of tomato to Fusarium wilt. (Trans. title, in Hebrew). Hassadeh 40: 1347–1349.
22. Committee on Genetic Vulnerability of Major Crops, 1972. Genetic vulnerability of major crops. National Academy of Science, Washington, D. C. 307 p.
23. Cook, R. J. 1973. Influence of low plant and soil water potentials on diseases caused by soil-borne fungi. Phytopathology 63: 451–458.
24. Corden, C. E. 1965. Influence of calcium nutrition on Fusarium wilt of tomato and polygalacturonase activity. Phytopathology 55: 222–224.
25. Cox, R. S. 1963. Control of Fusarium wilt, root-rot, and weeds on trellis-grown tomatoes in south Florida. Proc. Fla. State Hort. Soc. 76: 131–134.
26. Crill, P., J. P. Jones, D. S. Burgis, and S. S. Woltz. 1972. Controlling Fusarium wilt of tomato with resistant varieties. Plant Dis. Rep. 56: 695–699.
27. Crill, P., J. P. Jones, and D. S. Burgis. 1973. Failure of "Horizontal Resistance" to control Fusarium wilt of tomato. Plant Dis. Rep. 57: 119–121.

28. Davet, P., C. M. Messiaen, and P. Rieuf. 1966. Interpretation of winter manifestation of Fusarium wilt of tomato in North Africa, favored by irrigation water salts. (Trans. title, in French). Proc. First Cong. Mediterr. Phytopathol. Union: 407–416.
29. Edgington, L. V., and J. C. Walker. 1958. Influence of calcium and boron nutrition on development of Fusarium wilt of tomato. Phytopathology 48: 324–326.
30. Elliott, J. A., and R. E. Crawford. 1922. The spread of tomato wilt by infected seed. Phytopathology 12: 428–434.
31. Endo, E. M., and J. V. Leary. 1974. Biology and control of the Fusarium crown rot disease of tomato. Fresh Market Tomato Res. Univ. Calif. Veg. Crops Series 171: 20–22.
32. Engelhard, A. W. 1975. Aster Fusarium wilt: complete symptom control with an integrated fungicide-NO_3-pH control system. Proc. Amer. Phytopathol. Soc. 2: 62. (Abstr.).
33. Englehard, A. W., and S. S. Woltz. 1973. Fusarium wilt of chrysanthemum: complete control of symptoms with an integrated fungicide-lime-nitrogen regime. Phytopathology 63: 1256–1259.
34.. Essary, S. H. 1920. Report of the botanist. Tenn. Agric. Exp. Stn. Rept. 1919–20: 15–16.
35. Farley, J. 1975. Identified: New tomato root and crown rot disease. Amer. Veg. Grower 23: 30.
36. Farley, J., G. Oakes, and C. Jaberg. 1975. A new greenhouse tomato root-rot disease caused by *Fusarium oxysporum:* A preliminary report. Ohio Agric. Res. Dev. Center Res. Summary 82: 27–29.
37. Foster, R. E., and J. C. Walker. 1947. Predisposition of tomato to Fusarium wilt. J. Agric. Res. 74: 165–185.
38. Gabe, H. L., and B. C. Knight. 1973. The occurrence of a second race of the tomato Fusarium wilt in the greenhouse. Proc. 7th Brit. Insecticide Fungicide Conf. 1013–1018.
39. Goode, M. J., and J. M. McGuire. 1967. Relationship of root-knot nematode to pathogenic variability in *Fusarium oxysporum* f. sp. *lycopersici.* Phytopathology 57: 812.
40. Goss, R. W. 1923. Relation of environment and other factors to potato wilt caused by *Fusarium oxysporum.* Univ. Nebr. Agric. Exp. Stn. Res. Bull. 23. 83 p.
41. Goss, R. W. 1936. Fusarium wilts of potato, their differentiation, and the effect of environment upon their occurrence. Amer. Potato J. 13: 171–180.
42. Goss, R. W. 1940. A dry rot of potato stems caused by *Fusarium solani.* Phytopathology 30: 160–165.
43. Harrison, A. L., and P. A. Young. 1941. Effect of root-knot nematode on tomato wilt. Phytopathology 31: 749–752.
44. Haskell, R. J. 1916. Potato wilt and tuber rot caused by *Fusarium eumartii.* Phytopathology 6: 321–327.
45. Hodgson, W. A., D. D. Pond, and J. Munro. 1974. Diseases and pests of potatoes. Canada Dept. Agric. Publ. 1492. 69 p.
46. Jarvis, W. R, H. J. Thorpe, and B. H. MacNeill. 1975. A foot and root rot disease of tomato caused by *Fusarium oxysporum.* Can. Plant Dis. Survey 55: 25–26.
47. Jenkins, W. R., and B. W. Coursen. 1957. The effect of root-knot nematodes, *Meloidogyne incognita acrita* and *M. hapla,* on Fusarium wilt of tomato. Plant Dis. Rep. 41: 182–186.
48. Joffe, A. Z. and J. Palti. 1965. Species of *Fusarium* found associated with wilting of tomato varieties resistant to *F. oxysporum* in Israel. Plant Dis. Rep. 49: 741.
49. Jones, J. P., and P. Crill. 1975. Reaction of resistant, tolerant, and susceptible tomato varieties to Verticillium wilt. Plant Dis. Rep. 59: 3–6.
50. Jones, J. P., and A. J. Overman. 1971. Control of Fusarium wilt of tomato with lime and soil fumigants. Phytopathology 61: 1415–1417.
51. Jones, J. P., and A. J. Overman. 1972. The effect of mulching on the efficacy of DD-MENCS for control of Fusarium wilt of tomato. Plant Dis. Rep. 56: 953–956.
52. Jones, J. P., and A. J. Overman. 1976. Tomato wilts, nematodes, and yields as affected by soil reaction and a persistent contact nematicide. Plant Dis. Rep. 60: 913–917.
53. Jones, J. P., A. J. Overman, and P. Crill. 1976. Failure of root-knot nematode to affect Fusarium wilt resistance of tomato. Phytopathology 66: 1339–1341.
54. Jones, J. P., and S. S. Woltz. 1967. Fusarium wilt (race 2) of tomato: effect of lime and micronutrient soil amendments on disease development. Plant Dis. Rep. 51: 645–648.
55. Jones, J. P., and S. S. Woltz. 1968. Field control of Fusarium wilt (race 2) of tomato by liming and stake disinfestation. Proc. Fla. State Hort. Soc. 81: 187–191.
56. Jones, J. P., and S. S. Woltz. 1969. Fusarium wilt (race 2) of tomato: calcium, pH, and micronutrient effects on disease development. Plant Dis. Rep. 53: 276–279.
57. Jones, J. P., and S. S. Woltz. 1970. Fusarium wilt of tomato: interaction of soil liming and micronutrients on disease development. Phytopathology 60: 812–813.
58. Jones, J. P. and S. S. Woltz. 1972. Effect of amino acids on development of Fusarium wilt of resistant and susceptible tomato cultivars. Proc. Fla. State Hort. Soc. 85: 148–151.
59. Katan, J. 1971. Symptomless carriers of the tomato Fusarium wilt pathogen. Phytopathology 61: 1213–1217.
60. Katan, J., and I. Wahl. 1966. Occurrence in Israel of new dangerous isolates of the tomato Fusarium wilt pathogen. Proc. First Cong. Mediterr. Phytopathol. Union: 425–430.
61. Kawamura, T., and K. Hirano. 1967. Studies of the complex disease caused by root-knot nematode

and Fusarium wilt fungus in tomato seedlings. (Trans. title, in Japanese). Tech. Bull. Fac. Hort., Chiba Univ., Japan 15: 7–19.
62. Laterot, H., and P. Pecaut. 1966. Presence of race 2 of *Fusarium oxysporum* f. sp. *lycopersici* in tomato crops in Morocco. (Trans. title, in French). Proc. First Cong. Mediterr. Phytopathol. Union: 421–433.
63. Lockhart, C. L. 1970. Suppression by ethylene of *Fusarium oxysporum* growth in culture and rots of tomato in controlled atmosphere storage. Can. J. Sci. 50: 347–349.
64. MacMillan, H. G. 1919. Fusarium blight of potatoes under irrigation. J. Agric. Res. 16: 279–303.
65. Magee, C. J. 1931. Steam sterilization of soils with special reference to glasshouses. Agric. Gaz. N. S. Wales 42: 428–432.
66. Masee, G. 1895. The "sleepy disease" of tomatoes. Gard. Chron. Ser. 3, 17: 707–708.
67. McKee, R. K. 1954. Dry-rot disease of the potato. VIII. A study of the pathogenicity of *Fusarium coeruleum* (Lib.) Sacc. and *Fusarium avenaceum* (Fr.) Sacc. Ann. Appl. Biol. 41: 417–434.
68. McLean, J. G., and J. C. Walker. 1941. A comparison of *Fusarium avenaceum*, *F. oxysporum*, and *F. solani* var. *eumartii* in relation to potato wilt in Wisconsin. J. Agric. Res. 63: 495–525.
69. Miller, P. M., and G. S. Taylor. 1973. Fusarium wilt of tomato influenced by pervoius soil pasteurization with DD-MENCS or steam. Plant Dis. Rep. 57: 267–269.
70. Overman, A. J., J. P. Jones, and C. M. Geraldson. 1965. Relation of nematodes, diseases, and fertility to tomato production on old land. Proc. Fla. State Hort. Soc. 78: 136–142.
71. Overman, A. J., J. P. Jones, and C. M. Geraldson. 1971. Soil preparation for tomatoes on old sand land. Gulf Coast Stn. Mimeo Rept. GCS 71–6. 4 p.
72. Perotta, G., and G. Cartia. 1965. Greenhouse trials against tracheomycoses of tomato. (Trans. title, in Italian). Tec. Agr. Catania 17: 1–15.
73. Peterson, R. A. 1973. Occurrence of race 2 of Fusarium wilt of tomato in Queensland. Queensl. J. Agric. Anim. Sci. 30: 323–326.
74. Porte, W. S., and H. B. Walker. 1941. The Pan America tomato, a new red variety highly resistant to Fusarium wilt. U. S. Dept. Agric. Circ. 611. 4 p.
75. Pritchard, F. J. 1927. Marglobe tomato. Market Grower J. 40: 104–108.
76. Scott, I. T. 1924. The influence of hydrogen ion concentration on the growth of *Fusarium lycopersici* and on tomato wilt. Mo. Agric. Exp. Stn. Res. Bull. 64. 32 p.
77. Sherwood, E. C. 1923. Hydrogen ion concentration as related to the Fusarium wilt of tomato seedlings. Amer. J. Bot. 10: 537–553.
78. Sidhu, G., and J. M. Webster. 1974. Genetics of resistance in the tomato to root-knot-wilt-fungus complex. J. Hered. 65: 153–156.
79. Smith, E. F., and D. B. Swingle. 1904. The dry rot of potatoes due to *Fusarium oxysporum*. U. S. Dept. Agric Bur. Plant Indus. Bull. 55. 64 p.
80. Sonoda, R. M. 1976. The occurrence of a Fusarium root rot of tomatoes in south Florida. Plant Dis. Rep. 60: 271–274.
81. Stall, R. E. 1961. Development of Fusarium wilt on resistant varieties of tomato caused by a strain different from race 1 isolates of *Fusarium oxyporum* f. *lycopersici*. Plant Dis. Rep. 45: 12–15.
82. Stall, R. E., and J. M. Walter. 1965. Selection and inheritance of resistance in tomato to isolates of races 1 and 2 of the Fusarium wilt organism. Phytopathology 55: 1213–1215.
83. Strobel, J. W., N. C. Hayslip, D. S. Burgis, and P. H. Everett. 1969. Walter, a determinant tomato resistant to races 1 and 2 of the Fusarium wilt pathogen. Fla. Agric. Exp. Stn. Circ. S-202. 9 p.
84. Talburt, W. F., and O. Smith. 1959. Potato processing. Avi. Publ. Co., Westport, Conn. 475 p.
85. Thanassoulopoulos, C. C., C. N. Giannopolitis, and G. T. Kitsos. 1970. Control of Fusarium wilt of tomato and watermelon with benomyl. Plant Dis. Rep. 54: 561–564.
86. Tobolsky, I., and I. Wahl. 1963. Effectiveness of various soil treatments for the control of Fusarium wilt on tomatoes in the greenhouse. Plant Dis. Rep. 47: 301–305.
87. Tokeshi, H., and F. Galli. 1966. Variability of *Fusarium oxysporum* f. *lycopersici* (Wr.) Snyd. and Hans. in Sao Paulo. (Trans. title, in Portuguese). Anais. Esc. sup. Agr. "Luiz Queiroz" 23: 195–209.
88. Tokeshi, H., F. Galli, and C. Kurozawa. 1966. A new race of tomato Fusarium in Sao Paulo. (Trans. title, in Portuguese). Anais. Esc. sup. Agr. "Luiz Queiroz" 23: 217–227.
89. Waksman, S. A. 1927. Principles of soil microbiology. The Williams and Wilkins Co., Baltimore, Md. 897 p.
90. Walker, J. C. 1971. Fusarium wilt of tomato. The American Phytopathological Soc. Monograph No. 6. 56 p.
91. Walker, J. C., and R. E. Foster. 1946. Plant nutrition in relation to disease development. III. Fusarium wilt of tomato. Amer. J. Bot. 33: 259–264.
92. Walter, J. M. 1967. Heriditary resistance to disease in tomato. Annu. Rev. Phytopathol. 5: 131–162.
93. Woltz, S. S., and J. P. Jones. 1968. Micronutrient effects on the in vitro growth and pathogenicity of *Fusarium oxysporum* f. sp. *lycopersici*. Phytopathology 58: 336–338.
94. Woltz, S. S., and J. P. Jones. 1973. Tomato Fusarium wilt control by adjustments in soil fertility. Proc. Fla. State Hort. Soc. 86: 157–159.
95. Woltz, S. S., and J. P. Jones. 1973. Tomato Fusarium wilt control by adjustments in soil fertility: a systematic approach to pathogen starvation. Agr. Res. Ed. Center, Bradenton Res. Rept. GC1973-7. 4 p.

96. Woltz, S. S., and J. P. Jones. 1973. Interactions in source of nitrogen fertilizer and liming procedure in the control of Fusarium wilt of tomato. HortScience 8: 137–138.
97. Yamakawa, K., and N. Nagata. 1975. Three tomato lines obtained by use of chronic gamma radiation with combined resistance to TMV and Fusarium race J-3. Inst. Rald. Breeding Tech. News No. 16.
98. Yamamoto, I., H. Komado, K. Kuniyasu, M. Saito, and A. Ezuka. 1974. A new race of *Fusarium oxysporum* f. sp. *lycopersici* inducing root rot of tomato. Proc. Kansai Plant Protection Soc. 16: 17–29.
99. Young, P. A. 1940. Soil fumigation with chloropicrin and carbon bisulphide to control rootknot nematode and wilt. Phytopathology 30: 860–865.
100. Young, P. A. 1962. Control of Fusarium wilt of tomato with Vorlex. Plant Dis. Rep. 46: 151.

16 Fusarium Wilt of Muskmelon

P. Mas, P. M. Molot, and Georgette Risser

Symptoms of the Disease

The Fusarium wilt fungus can attack muskmelon at any age, even before the plants are sprouted, but especially when the fruit are maturing. The symptoms on muskmelon are the same as those observed in the other wilt diseases caused by *Fusarium*, especially those on watermelon.

In France, on diseased plants, either a slow wilting accompanied by a progressive yellowing, or a sudden wilting without prior yellowing occurs (28). The first case is the more common one: the veins of some leaves turn yellow on one side, then later these leaves become completely yellow, thicken, and become brittle. At this stage, the leaves have a slight odor of violets. A longitudinal brown necrotic streak then appears on the stems from which gum exudes. In the final stage, the fungus sporulates in the necrotic zone and forms pinkish-colored sporodochia. In the second case the plants show a sudden wilting without yellowing or odor. The tips of the stems are generally attacked first and shrivel; wilting then progresses toward the base of the plant. In both cases a transverse section of the stem shows clear brown vascular discoloration corresponding to the vessels that have been attacked. Vascular discoloration is more predominant in those plants with yellowing symptoms.

Influence of the Environment

External factors can play an important role in the development and the severity of Fusarium wilt of muskmelon (30, 42) but it is difficult to separate the influence of each factor.

The influence of temperature on the pathogenic capabilites of *Fusarium oxysporum* f. sp. *melonis* is not identical with the influence of temperature on mycelial growth in vitro. On every isolate tested, the optimum temperature for pathogenesis is lower than that for growth, which is 26 C (35). The most severe symptoms are observed between 18 and 22 C. At high temperatures (30 C) the plant is infected without showing symptoms. Isolations reveal the presence of the parasite in the vessels and the transfer of plants to 22 C brings about sudden death. Contrary to most of the Fusarium wilts, wilt of muskmelon is a disease that occurs in cool soils and early in the season, since the most severe symptoms are found at temperatures betwen 18 and 20 C (17, 32). Insufficient illumination and short day lengths increase the severity of vascular wilt due to *Fusarium* (8, 11, 30).

Insofar as the moisture of the soil is concerned, Miller (32) has noted that the frequency of the disease is higher under dry conditions. In artificial inoculation tests we have also noticed that symptoms occur more quickly and are more intense when the relative humidity of the air is between 50 and 65%, and that they rarely occur when it goes beyond 80 or 90%.

Mineral nutrition also influences Fusarium wilt of muskmelon. The addition of nitrogen increases the severity of the disease (48, 51), either by increasing susceptiblity of the host (51) or by increasing virulence of the fungus. In vitro, the addition of nitrate to the growth medium of *Fusarium* enhances sporulation and the production of fusaric acid (16). On the other hand, the addition of potassium or calcium decreases disease severity (51); in vitro these two elements have an unfavorable effect on the production of fusaric acid (16); under these conditions potassium also has an inhibitory effect on the synthesis and activity of the enzymes of *Fusarium* (12). The addition of cellulose to the growth medium of the host reduces the intensity of the symptoms (13), perhaps by modifying the saprophytic behavior of the fungus (14).

Materials and Methods

Isolation of the parasite The presence of *F. oxysporum* f. sp. *melonis* in the soil can be determined by the use of a method described by Bouhot and Rouxel (5). Fusarium can be isolated from diseased plants by taking fragments of vascular tissue or sections of stem and plating them on a special nutrient agar medium containing antibiotics and tetrachloronitrobenzene (37). The use of these media is helpful but it is not indispensable for isolation of the fungus from diseased plants.

Conservation The isolates of *F. oxysporum* f. sp. *melonis* can be kept by repeated subculture on potato dextrose agar (PDA) or similar media, but these conditions are not the best, and it seems preferable to keep the isolates in sand or vermiculite soaked with a nutrient solution. After the medium is colonized by the parasite, it is allowed to dry slowly and the *Fusarium* is thus kept in the form of chlamydospores. Use of these media makes it possible to extend the time interval between subcultures to one year without problems.

Production of inoculum In order to obtain suspensions of microconidia (from 10^7 to 10^8 per milliliter), one can grow *F. oxysporum* f. sp. *melonis* on a Knop liquid medium with malt and saccharose (24). The fungus is grown for 7 days at 25 C in shake culture with a light period of 15 hr per day. Prior to use the culture is filtered through a nylon mesh and adjusted to the concentration desired (in general 10^7 microconidia/ml). One can also grow *F. oxysporum* f. sp. *melonis* on a solid substrate, for example vermiculite moistened with a nutrient solution (25). This inoculum then is utilized in infesting the soil.

Root inoculation The muskmelons are sown on sand and transplanted when they are at the two expanded cotyledons stage. The roots are washed, partially cut, and dipped in the spore suspension for a few minutes; the seedlings are then quickly replanted in sterilized sand. This rather brutal technique (since wounding of the roots is involved) has the advantage of assuring the penetration of a great amount of the inoculum (24).

Soil infestation The addition of inoculum to the soil can be accomplished either prior to planting, by incorporating infested vermiculite (0.5% by volume), or after planting, by injecting a suspension of microconidia with a syringe into the soil near the roots. Inoculation can also be done by dipping the tray of seedlings, when the plants have reached the two expanded cotyledons stage, in dilute liquid inoculum. These two techniques avoid replanting of seedlings and come closer to natural conditions.

Methods of control Numerous authors have demonstrated the remarkable capacity of survival in soil of *F. oxysporum* f. sp. *melonis* in the form of chlamydospores, in the organic matter, or on the roots of various plants which are symptomless carriers (1, 10, 20, 40, 47). Therefore, control through crop rotation appears to be difficult or impossible.

Disinfection or fumigation of the soil Disinfection of soil is feasible only for early plantings where the financial return is important. It is necessary to use products with strong action such as steam, steam-air mixture, and fumigants, but technical difficulties in the application of these materials impeded the utilization of these treatments. A mixture of chloropicrin (50 ml/m^2) and of methylbromide (75g/m^2) gives promising results (25), but following the application of the fumigants the risk of recolonization of the soil by the pathogen is serious.

Therapeutic products Products with a therapeutic action, such as benomyl and thiophanate, can be used to prevent infection in early stages of plant growth. The hope that these compounds could be used in an effective preventive control program has not been realized (23, 49), and applications giving a curative action require very high dosages which are close to the threshold of toxicity (50).

Grafting Grafting of melon on *Benincasa cerifera* was first mentioned by Louvet and co-workers (18, 19). This stock has an excellent compatibility with muskmelons and resistance to penetration by all known races of *Fusarium* (10). However, it is susceptible to attack by *Pyrenochaeta lycopersici, Verticillium albo-atrum,* and *Phomopsis sclerotioides,* albeit to a lesser degree than many cultivars of muskmelon. Some producers in France and in Italy use grafting for control of this disease.

Use of resistant cultivars The first work on resistance of melon to *Fusarium* began in 1932 (7). Presently, three types of resistance have been demonstrated and utilized in practice (see section on resistance and its genetic control).

Field resistance exists in the American cultivars of the type Delicious 51. The first cultivar of this type, Golden Gopher, brought out by the experiment station at St. Paul, Minnesota, was commercially used in 1941(6). Other cultivars were selected later and included Iroquois in 1944, Delicious 51 in 1951, and Harvest Queen in 1954 (52).

Resistance due to specific resistance genes *Fom1* and *Fom2* has been described in French studies (31, 43, 46). The simultaneous utilization of these two genes gives an excellent resistance to the pathotypes 0, 1, and 2. Doublon and Orlinabel were the first cultivars having the gene *Fom1;* cultivars such as Printadou, carrying the genes *Fom1* and *Fom2,* were released by the station of the Amelioration des plantes maraicheres in Montfavet, France, in 1965 and 1974. American cultivars, such as Persian Small Type or Chaca No 1, have the gene *Fom1*. Numerous cultivars grown in the Near East and in Asia have the gene *Fom2*.

The so-called horizontal resistance or non-specific resistance was found in certain Asiatic cultivars (e.g., Kogane Nashi Makuwa or Ogon 9) by Risser and Rode (45). This polygenic resistance is being introduced into the European cultivars.

Host-Parasite Interaction

Most studies, especially in the USA, Canada, India, and France, have dealt with the interaction muskmelon-*F. oxysporum* f. sp. *melonis;* unfortunately, these studies were often done on different material and with no coordination between workers, which makes synthesis difficult. We will deal with i) resistance of the muskmelon to *Fusarium* and its genetic control, ii) biochemical mechanisms of resistance, and iii) phenomena of cross protection between physiologic races.

Resistance of the melon to *Fusarium* and its genetic control The earliest work in the USA and in Canada touched on the resistance of the cultivars Golden Gopher, Iroquois and Delicious 51. This resistance is a field resistance and it is not expressed in the seedling stage, at low temperatures, at high concentrations of inoculum, or against very aggressive isolates (20, 39, 47). The fungus penetrates resistant plants just as readily as susceptible plants but develops less rapidly in the vessels of resistant plants (41).

Mortensen (38) studied the inheritance of resistance of the cultivar Iroquois and attributed resistance to a dominant gene R and two complementary genes A and B which acted in the absence of R. This study needs to be confirmed because of the highly fluctuating reaction of resistant plants of this type and because the use of a commercial cultivar as a parent, perhpas not homozygous for the genes for resistance, makes interpretation of the results difficult.

As early as 1952, Eide et al. (9) thought that various pathotypes existed in *F. oxysporum* f. sp. *melonis,* but they were unable to demonstrate the presence of physiologic races. The existence of physiologic races was finally shown in 1965, using a set of differential hosts (44). This work, confirmed by Banihashemi (2) and later Risser and Mas (44) and Risser (43), demonstrated the existence of a specific resistance system caused by two independent genes. Thus, four physiologic races (Table 16-1) are now defined (46).

Resistance is expressed very clearly at the seedling stage and under very severe conditions of infection such as dipping the seedlings in a suspension of spores of the pathogen before replanting. The mechanism of action of this resistance is not known, but it is believed that the genes *Fom1* and *Fom2* control a phenomenon of host-parasite interaction recognition, because a host infected with a mixture of compatible and incompatible physiologic races shows a resistant reaction (see section on cross protection below). Perhaps further research will allow us to link the field resistance of Delicious 51 with this specific resistance mechanism, for although Delicious 51 is as sensitive as Charentais T to the 4 races at the seedling stage, the hybrid F_1 Charentais T x Delicious 51 is partially resistant to the races 0 and 2 (G. Risser, *unpublished data*).

A third type of partial resistance has been shown in muskmelons from the Far East. This resistance is not specific and acts against weakly pathogenic clones or under mild conditions of infection. Its genetic control is polygenic (43).

Biochemical mechanisms of resistance A study of the biochemical mechanisms following infection was done by Bhaskaran and Prasad (3) on the susceptible cultivar of *Cucumis melo,* Delta Gold, and on *Cucumis callosus,* a species resistant to *F. oxysporum* f. sp. *melonis.* The resistant tissues are characterized by high total phenol and orthodiphenol content. After infection polyphenoloxydase increases and this increase is more intense in the resistant cultivar than in the susceptible cultivar. The polyphenoloxydase acts principally at the expense of the orthodiphenols. On the other hand, susceptible tissues contain high amounts of total nitrogen and amino nitrogen, and the difference between *Cucumis melo* and *C. callosus* increases following infection. Moreover, contamination induces an important decrease in glucides (reducing sugars in the resistant cultivar, non-reducing sugars in the susceptible cultivar). The increase in the polyphenoloxydase and peroxydase

Table 16-1. Classification of *Fusarium oxysporum* f. sp. *melonis* races according to differential host cultivars of *Cucumis melo*

	Differential Hosts and Resistance Genes		
Races of *Fusarium oxysporum melonis*	Charentais T.	Doublon *(Fom1)*	LJ 17187 *(Fom2)*
Race 0	S	R	R
Race 1	S	S	R
Race 2	S	R	S
Race 1-2	S	S	S

S = susceptible; R = resistant

activity has been confirmed in the case of infected plants and in melons at the adult stage (21). In addition, cucurbitacine, a substance that existed prior to infection and which has fungistatic properties, might play a role in the resistance to *Fusarium* (4, 15). Unfortunately these studies were done on very different hosts, making it impossible to be certain that the differences observed are due solely to genes controlling resistance or susceptibility, or that the mechanism of resistance of *C. callosus* is identical to that of *C. melo*.

Cross protection between physiologic races of *Fusarium oxysporum* f. sp. *melonis* When a plant is inoculated successively or simultaneously with two isolates belonging to two different races (the first one being nonpathogenic, the second one being pathogenic), an attenuation or even a complete suppression of the pathogenic effect is observed. Thus the presence of spores of race 0 or race 1 in melon LJ 17.187 causes a defense reaction to the race 1–2 which is shown by the absence of symptoms (22, 29). The phenomenon of cross protection in muskmelon against Fusarium wilt is an interesting model for the study of host-parasite interactions. We will briefly mention the optimum conditions which will bring about cross protection and then try to explain the mechanisms involved.

Conditions necessary for cross protection Conditions that have been determined to be most favorable to the expression of the cross protection phenomenon are: i) choice of the genetic material (29) (the cross protection reactions are more obvious in the line LJ 17.187 containing the gene *Fom2* than in the line Doublon containing the gene *Fom1*); ii) the modes of application of the two races (we will see further that a simultaneous application is always more effective than a staggered application); iii) the influence of temperature (an alternation of 22 C during the day and 18 C during the night); iv) the influence of humidity; and v) the fertility of the soil.

Among all the factors which bring about good protection to the muskmelon, one of the most important is the concentration of microconidia. There should be at least as many microconidia from the nonpathogenic race as from the pathogenic race. Protection is increased as the ratio of the respective concentrations increases. This condition is not sufficient of itself, for if such inoculum is diluted (the ratio of the two components remaining the same), there is an increase in severity of the symptoms. Therefore two factors are important: the value of the ratio of the incompatible race over the compatible race, and the absolute quantity of microconidia used (27).

Hypothesis of the mechanisms involved in the phenomenon of cross protection The two hypotheses that have led the way in this research are competition between races and defense reactions of the plant. The phenomena of competition between races (for example, nonpathogenic race 0 and pathogenic race 1–2 on LJ 17.187) were studied in vitro and in vivo.

When grown together on a nutrient agar substrate, the various physiologic races have a normal growth and the mycelia mingle without the slightest indication of antagonism. In addition, when grown in liquid culture in contact with one another but without a mixing of their microconidia (separated by a filter membrane), they keep their pathogenic capabilities. Therefore, in vitro, there is no demonstration of a competetive effect.

To test the effects of competition in vivo seedlings are inoculated by dipping the roots in a mixture of equal quantities of microconidia belonging to the two physiologic races. Under these conditions we have studied penetration of the inoculum into the plant, and multiplication of the races in the tissues.

Effects of competition during penetration The two races are grown on a liquid medium, one in the presence of ^{32}P, the other in the present of ^{45}Zn. After washing, the marked microconidia are mixed in equal quantity for each of the races and placed in contact with the roots of the melon seedling. The radioactive tracings of the root tissues show that penetration is very rapid, and initially the microconidia of the nonpathogenic race are more numerous than those of the pathogenic race (34). Therefore, it appears that there was a selection of the microconidia at the time of their penetration into the vascular system of the plant. However, the predominance of the nonpathogenic race over the pathogenic race is not sufficient to explain the protective effect obtained, for sufficient pathogenic microconidia enter the plant to cause symptoms.

Effect of competition during the multiplication of the pathogen in host tissues In order to study the multiplication of *Fusarium* in infected host tissues, a method for the evaluation of the quantity of the pathogen has been developed (37). This is done by grinding the tissues of the plant and plating out the dilution onto a specific nutrient medium. Small colonies appear that are easy to count; if the isolates have been carefully chosen, it is possible to recognize those belonging to the pathogenic race and those belonging to the nonpathogenic race.

Since the greatest portion of the inoculum remains at the root level, the dosages of the *Fusarium* were determined at this level in the course of time. Ten days after inoculation there is practically no multiplication of the inoculum (perhaps because of the rooting of the seedlings). The nonpathogenic race keeps its numerical advantage over the pathogenic race. But from 10 days on, the amount of *Fusarium* increases very rapidly, the rate of increase being a function of the respective proportions of the microconidia of the two races in the inoculum.

If at the time of inoculation the inoculum contains the same amount of both races, after 10 days the number of propagules of the pathogenic race within the tissues of the host increases and surpasses that of the nonpathogenic race (which remains practically unchanged). We should note, however, that the multiplication of the pathogenic race is significantly lower than that observed in control plants inoculated only with this race.

If the inoculum contains 10 times more microconidia of the nonpathogenic race, one again observes within the tissues an increase of the pathogenic race, but in this case it is compensated for by a much more intense multiplication of the nonpathogenic race.

It would seem that the most important point in the phenomenon of cross protection is the quantity of the nonpathogenic race present in the tissues of the host after 10 days. This quantity seems to be strictly conditioned by the composition of the inoculum used for infection.

Defense reactions of the plant Regardless of the physiologic race or of the melon cultivar used, the microconidia of *Fusarium* are capable of penetrating, establishing themselves, and remaining in the vascular system of the muskmelon. Under these conditions the presence of microconidia or hyphae of a nonpathogenic race induces a more or less transitory resistance in the plant to the pathogenic race.

It may be hypothesized that while in contact with the host, the nonpathogenic races of the melon *Fusarium* either possess or secrete an inducing factor causing a defense reaction in muskmelons possessing a resistance gene and preventing the proliferation of the parasite either by formation of a mechanical barrier or by the synthesis of fungistatic substances. Under these conditions pathogenic races, which normally lack inducing factors, would be stopped during their development. Unfortunately, at the present time the existence of barriers appears to be uncertain and the presence of substances of the phytoalexin type in infected tissues has never been shown. However, the hypothesis of a defense reaction in the plant should not be set aside. Various experimental tests, such as the effect of temperature (36) and the effect of filtrates of the germinating microconidia (26, 33) on cross protection indirectly support this idea.

Effect of temperature on cross protection Instead of inoculating simultaneously with the two physiologic races of *Fusarium*, one first inoculates with one race (usually the nonpathogenic race), then a few days later with the other race: the longer the interval between the two inoculations, the more susceptible the muskmelon. Moreover, by means of this method, protection is always less than in the case of simultaneous inoculation by a mixture of both races. This would indicate that the plant reacts very early to the infection and that this action diminishes with time.

However, there is the possibility of delaying the reaction of the plant with temperature during the interval separating the two inoculations (35). Thus, when muskmelons are first inoculated by a nonpathogenic race, then put at different temperatures (from 18 C to 30 C) for 4 days, finally put at 22 C and soon after infected by a pathogenic race, there are differences in symptom expression according to temperatures: if the incubation occurs at 18 C there is little disease, but if it is at 30 C symptoms are much more severe. One can suppose that at 18 C the defense reaction of the plant is blocked and will be effective only on return to 22 C, at the time of the second inoculation. On the other hand, at 25 C it would be immediate but its effects, no doubt ephemeral and localized, would be lessened 4 days later during the second inoculation.

Effect of the filtrates of germinating spore medium on cross protection An attenuation of symptoms is also observed, but to a lesser degree and for a shorter period, when the nonpathogenic microconidia are replaced by the filtrate of the medium where they have germinated (26). The filtrates have the characteristic of protecting melons that do not have genes for resistance against any pathogenic race. Their efficacity depends more on the isolate than on the physiologic race of *Fusarium* that was used in their manufacture.

In these active filtrates, a dialyzable, thermostable substance was present and was identified as fusaric acid, but other factors not yet isolated must also be present. Fusaric acid has no direct effect on the pathogen, but it is capable of inducing a defense reaction in the host (33).

Conclusions

The studies on cross protection indicate that there is no simple interaction between two physiologic races and no simple interaction between pathogen and host but that the two processes occur simultaneously. Thus, the efficacy of cross protection depends on several factors which interact with the respective concentrations of the microconidia of the two races, inoculum amount, length of the interval separating the two inoculations, temperature during this interval, etc. All these phenomena suggest that there are at least two mechanisms at play.

Literature Cited

1. Armstrong, G. M., and J. K. Armstrong. 1948. Nonsusceptible hosts as carriers of wilt Fusaria. Phytopathology 38:808–826.
2. Banihashemi, Z. 1968. The biology and ecology of *Fusarium oxysporum* f. sp. *melonis* in soil and the root zones of host and nonhost plants. Ph.D. Thesis, Michigan State Univ., East Lansing. 114 p.
3. Bhaskaran, R., and N. N. Prasad. 1971. Certain biochemical changes in two *Cucumis* species in response to Fusarium infection. Phytopathol. Mediterr. 10:238–243.
4. Bhaskaran, R., and N. N. Prasad. 1972. Role of preformed inhibitors in *Cucumis* sp. in wilt resistance. Phytopathol. Z. 73:75–77.
5. Bouhot, D., and F. Rouxel. 1971. Technique d'analyse selective et quantitative des *Fusarium oxysporum* et *Fusarium solani* dans le sol. Mode d'emploi. Ann. Phytopathol. 3:251–254.
6. Currence, T. M., C. J. Eide, and J. B. Leach. 1941. The Golden Gopher muskmelon. Market Gr. J. 68:14–16.
7. Currence, T. M., and J. G. Leach. 1934. Progress in developing muskmelon strains resistant to Fusarium. Proc. Amer. Soc. Hort. Sci. 32:481–482.
8. Davis, D. 1963. Investigations on physiology of selective pathogenicity in *Fusarium oxysporum* in test tube culture. Phytopathology 53:133–139.

9. Eide, C. J., and A. Nakila. 1952. An isolate of Fusarium with unusual virulence for the resistant Iroquois muskmelon. Phytopathology 42: 465. (Abstr.).
10. El-Mahjoub, M. 1971. Quelque aspects des rapports entre les parasites vasculaires et leurs hôtes: fusariose du melon et verticilliose de l'oeillet. Thèse de 3ème Cycle. Univ. Rennes, France. 108 p.
11. Foster, R. E., and J. C. Walker. 1947. Predisposition of tomato for Fusarium wilt. J. Agric. Res. 74:165–185.
12. Habibullah, V. M., and N. N. Prasad. 1976. Effect of potassium on growth and production of pectinolytic and cellulolytic enzymes of *Fusarium oxysporum* f. *melonis*. Indian J. Exp. Biol. 14:733–734.
13. Kannaiyan, S., and N. N. Prasad. 1973. Effect of organic nutrients on incidence of muskmelon wilt. Annamalai Univ. Agric. Res. Annu. 4/5:62–64.
14. Kannaiyan, S., and N. N. Prasad. 1975. Effect of certain amendments on the saprophytic activity of muskmelon wilt pathogen in soil. Indian Phytopathol. 28:385–386.
15. Kesayan, R., and N. N. Prasad. 1974. Correlation between crude cucurbitacin content in certain muskmelon varieties and Fusarium wilt incidence. Indian J. Exp. Biol. 12:476–477.
16. Kesayan, R., and N. N. Prasad. 1975. Effect of certain carbon and nitrogen sources on in vitro production of fusaric acid by muskmelon wilt pathogen. Indian Phytopathol. 28:28–32.
17. Leach, J. G., and T. M. Currence. 1938. Fusarium wilt of muskmelon in Minnesota. Minn. Agric. Exp. Stn. Bull. 129.
18. Louvet, J., and C. Lemaitre. 1961. L'utilisation des melons greffes pour lutter contre la fusariose. Rev. Hort. 2239:8–10.
19. Louvet, J., and J. Peyriere. 1962. Interêt du greffage du melon sur *Benincasa cerifera*. Compt. Rend. XVI Congr. Int. Hort., Brussels II:167–171.
20. McKeen, C. D. 1951. Investigation of Fusarium wilt of muskmelons and watermelons in southwestern Ontario. Sci. Agric. 31:413–423.
21. Maraite, H. 1973. Changes in polyphenoloxidases and peroxidases in muskmelon infected by *Fusarium oxysporum* f. sp. *melonis*. Physiol. Plant Pathol. 3:29–49.
22. Mas, P. 1967. Protection du melon contre la fusariose par infection préalable de la plantule avec d'autres souches de *Fusarium*. Compt. Rend. Acad. Agric. Fr. 53:1034–1040.
23. Mas, P. 1971. Résultats préliminaires de l'étude de l'action du méthylthiophanate sur le Fusarium du melon. 60–64. *In* Proceedings symposium thiophanates, March 22–24, 1971, Marseille, France. 120 p.
24. Mas, P. 1973. Recherches sur les fusarioses. XI. Pénétration du *Fusarium oxysporum* f. sp. *melonis* dans les plantules de melons sensibles et résistants selon deux techniques d'inoculation. Ann. Phytopathol. 5:213–218.
25. Mas, P., and D. Bouhot. 1974. Essai d'efficacité de quelques fumigants contre la fusariose du melon en serre *(Fusarium oxysporum* f. *melonis)*. Phytiatrie-Phytopharmacie 23:249–258.
26. Mas, P., and P. M. Molot. 1974. Atténuation de la sensibilité du melon au *Fusarium oxysporum* f. sp. *melonis*. 1. Rôle des filtrats de milieux de germination de spores. Ann. Phytopathol. 6:237–244.
27. Mas, P., and P. M. Molot. 1977. Influence de la concentration de l'inoculum en microconidies sur la prémunition du melon contre *Fusarium oxysporum* f. sp. *melonis*. Ann. Phytopathol. 9:71–75.
28. Mas, P., and G. Risser. 1966. Caracterisation, symptomes et virulence de diverses races de *Fusarium oxysporum* Schl. f. sp. *melonis* Sn. et Hans. I. Congr. Union Phytopathol. Mediterr. Bari:503–508.
29. Mas, P., G. Risser, and J. C. Rode. 1969. Phénomènes de prémunition entre formes specialisées ou races du *Fusarium oxysporum* chez *Cucumis melo*. Ann. Phytopathol. 1:213–216.
30. Messiaen, C. M., and P. Mas. 1969. Recherches sur les fusarioses. IV. Mise au point sur l'activité parasitaire du *Fusarium oxysporum* et sur divers facteurs rendant les plantes plus ou moins sensibles aux fusarioses vasculaires. Ann. Phytopathol. 1:401–426.
31. Messaien, C. M., G. Risser, and P. Pecaut. 1962. Etude des plantes resistantes au *Fusarium oxysporum* f. sp. *melonis* dans la varieté de melon Cantaloup Charentais. Ann. Amerlior. Plantes 12:157–164.
32. Miller, J. J. 1945. Studies on the Fusarium of muskmelon wilt. II. Infection studies concerning the host of the organism and the effect of environment on disease incidence. Can. J. Res. C 23:166–187.
33. Molot, P. M., and P. Mas. 1974. Attenuation de la sensibilité du melon au *Fusarium oxysporum* f. sp. *melonis*. 2. Rôle de l'acide fusarique. Ann. Phytopathol. 6:245–253.
34. Molot, P. M., and P. Mas. 1974. Utilisation du marquage radioactif dans l'étude de la pénétration des spores de *Fusarium oxysporum* f. *melonis* chez le melon contaminé par un melange de deux races physiologiques. Compt. Rend. Acad. Sci., Paris 278D;3327–3329.
35. Molot, P. M., and P. Mas. 1975. Influence de la température sur la croissance mycélienne et le pouvoir pathogène des quatre races physiologiques de *Fusarium oxysporum* f. sp. *melonis*. Ann. Phytopathol. 7:115–121.
36. Molot, P. M., and P. Mas. 1975. Influence de la température sur l'efficacité de la prémunition du melon contre *Fusarium oxysporum* f. sp. *melonis* par une race incompatible. Ann. Phytopathol. 7:175–178.
37. Molot, P. M., P. Mas, and H. Ferrière. 1979. Etude de la prémunition à l'aide d'une technique de dosage biologique du *Fusarium oxysporum* f. sp. *melonis* dans les tissus de melon. Ann. Phytopathol. 11:209–222.

38. Mortensen, J. A. 1959. The inheritance of Fusarium resistance in muskmelon. Diss. Abstr. 19:220.
39. Munger, H. N. 1954. Two new varieties from Cornell: Delicious 51, an early Fusarium resistant muskmelon, and the Valnorth tomato. Farm Res. 20(1):89.
40. Owen, J. H. 1956. Cucumber wilt, caused by *Fusarium oxysporum* f. *cucumerinum* n. f. Phytopathology 46:153–157.
41. Reid, J. 1958. Studies on the Fusaria which cause wilt in melons. I. The occurrence and distribution of races of the muskmelon and watermelon Fusaria and a histological study of the colonization of muskmelon plants susceptible or resistant to Fusarium wilt. Can. J. Bot. 36:393–410
42. Reid, J. 1958. Studies on the Fusaria which cause wilt in melons. II. The effect of light, nutrition, and various chemicals on the sporulation of certain fusarial isolates and preliminary investigations on the etiology of wilting on the muskmelon Fusarium. Can. J. Bot. 36:507–537.
43. Risser, G. 1973. Etude de l'hérédité de la résistance du melon *(Cucumis melo)* aux races 1 et 2 de *Fusarium oxysporum* f. *melonis*. Ann. Amelior. Plantes 23:259–263.
44. Risser, G., and P. Mas. 1965. Mise en évidence de plusieurs races de *Fusarium oxysporum* f. *melonis*. Ann. Amelior. Plantes 15:405–408.
45. Risser, G., and J. C. Rode. 1973. Breeding for resistance to *Fusarium oxysporum* f. *melonis*, p. 37–39. *In* La sélection du melon, Proceedings Symposium EUCARPIA, June 19–22, 1973, Montfavet-Avignon, France. 82 p.
46. Risser, G., Z. Banihashemi, and D. W. Davis. 1976. A proposed nomenclature of *Fusarium oxysporum* f. sp. *melonis* races and resistance genes in *Cucumis melo*. Phytopathology 66:1105–1106.
47. Rodriguez, R. A. 1960. The nature of resistance in muskmelons to Fusarium wilt. Ph.D. Thesis, Univ. Minn., St. Paul. 87 p.
48. Stoddard, D. L. 1947. Nitrogen, potassium, and calcium in relation to Fusarium wilt of muskmelon. Phytopathology 37:875–884.
49. Vigouroux, A., and P. Mas. 1970. Etude de l'action de benomyl sur deux agents de trachéomycose en culture légumière (résumé). 7th Int. Congr. Plant Prot., 21–25 Sept., Paris.
50. Wensley, R. N., and C. M. Huang. 1970. Control of Fusarium wilt of muskmelon and other effects of benomyl soil drenches. Can. J. Microbiol. 16:615–620.
51. Wensley, R. N., and C. D. McKeen. 1965. Some relationships between plant nutrition, fungal populations and incidence of Fusarium wilt of muskmelon. Can. J. Microbiol. 11:581–594.
52. Whitaker, T. W., and G. N. Davis. 1962. Cucurbits: botany, cultivation and utilisation. Leonard Hill Ltd., London. 249 p.

17 Disease Caused by *Fusarium* in Forest Nurseries

W.J. Bloomberg

Hartig, in 1892, recognized *Fusarium* as a pathogen of interest to German forest nurserymen and noted its broad host range and types of damage (30). It was reported as a seed-destroying fungus in Canada in 1901 (21). In the United States, first reports crediting damping-off ability to *Fusarium* (24,32) were later qualified by recognition of considerable pathogenic variation in the genus (34). During the 1930s, the importance of *Fusarium* as a nursery pathogen was established in Eastern Canada and Holland (68, 91). Later, it was implicated in "conifer sickness" in British nurseries (113).

The Disease

Disease types Hartley's forest nursery disease types (34) have been well documented for *Fusarium* (7, 17, 18, 61, 63, 69, 91, 116). In order of occurrence, they are seed rot, pre-emergence damping-off or germination failure, post-emergence damping-off or Fusariosis, top damping-off or cotyledon blight, root rot or late damping-off. Wilting appears to be a secondary symptom accompanied or preceded by root rot and not a true tracheomycosis. All disease types do not necessarily occur in a nursery, or to the same extent, but all are usually confined to the first growing season. *Fusarium* also causes cankers and blights of poplar coppice shoots, sets, and cuttings (4,56); soft rot of elm cuttings (79); and storage rot of several broadleaf species (114). Reduction in seedling growth as well as mortality has been attributed to *Fusarium* (86, 113). In addition, symptomless infections occur with, and sometimes predominate over those with typical symptoms (6, 37, 102, 104).

Fusarium complexes with other fungi cause several nursery diseases. Black root rot, which stunts or kills southern pines, has some root symptoms attributable to *Fusarium* spp., and others attributable to *Sclerotium bataticola* (22, 36). *Fusarium* spp. were associated with *Pythium* in Sitka spruce stunting in British nurseries (18, 113). In corky root disease of Douglas fir, *F. oxysporum* precedes *Cylindrocarpon radicicola* in young seedlings (13) but does not produce symptoms by itself(7).

Disease distribution For brevity, the term "damping-off" includes top, pre-, and post-emergence damping-off. Forest nursery diseases caused by *Fusarium* have been reported in all continents and in climatic regions from temperate to tropical. *Fusarium* is the main cause of all disease types in Quebec (69), and is the fungus predominantly associated with damping-off and root rot in the Canadian maritime provinces (112), with damping-off in central Canada (20, 99, 103, 104), and with root rot in British Columbia (6, 11, 78). In the United States, it was first recorded as a damping-off organism in California, Colorado, Kansas, and the Great Lakes states (31). Tint (93) acquired pathogenic isolates from nurseries throughout the northeastern states. Root rot is especially important in southern, southeastern, western and Pacific northwestern states (42, 84, 115). Black root rot occurs throughout the southern states (23, 40).

In Europe, *Fusarium* is the most frequent cause of damping-off in Hungary (25), Poland (45), and Czechoslovakia (43). It is the fungus most frequently associated with damping-off and root rot in South Australia (102, 106), East and West Africa (37, 66), and Cuba (50). It is the most important cause of damping-off and root rot in Japan (63, 117) and one of the main causes in northeastern China (116). All conifers in Canadian prairie province nurseries (103) and all southern hardwoods (76) are attacked by *Fusarium.* Spruce appears to be more resistant to damping-off and root rot than pine (52, 69), but equally susceptible to black root rot (74). Pine species differ in susceptibility (36, 50, 57, 74). Douglas fir is very susceptible to all Fusarium diseases (42, 74). *Cupressus, Cedrus,* and *Chamaecyparis* are relatively resistant (57, 74).

The Pathogen

Frequency Frequency of *Fusarium* spp. in forest nurseries appears to vary by location (Table 17-1). Frequency may change with time; *F. oxysporum* comprised 20 and 90% of all Fusarium isolates at a Canadian nursery sampled in 1941 and 1965, respectively (3, 6); it may also change during the growing season (18). There is little evidence for distribution of *Fusarium* spp. by host species or disease types.

Pathogenicity Pathogenicity of *Fusarium* spp. is not necessarily correlated with frequency of occurrence. Some North American tests showed that *F. sporotrichioides* was most pathogenic but least frequent (34, 72). *Fusarium monoliforme* was the most pathogenic of the common species. *Fusarium solani* was widespread but less pathogenic. In other tests, 19 *Fusarium* spp. were pathogenic to eight conifers (93). *Fusarium poae* caused most pre-emergence and *F. avenaceum* most post-emergence damping-off. There was considerable variation among conifers in their susceptibility to each *Fusarium* sp. In Quebec, *F. solani* was the most pathogenic to red pine, *F. sporotrichioides* and *F. monoliforme* were less so (68). In Japan, *F. oxysporum, F. solani, F. monoliforme, F. roseum,* and *F. lateritium* were all pathogenic to several conifers (63, 117). In Europe, *F. oxysporum, F. sporotrichioides,* and *F. equiseti* were pathogenic to pine species (28); *Fusarium avenaceum, F. solani,* and *F. sambucinum* also had strong pathogenic potential (49). In Britain, *F. oxysporum* caused greater and earlier disease in Sitka spruce than *F. roseum* (18). In Africa, pathogenicity of *F. oxysporum* and *F. solani* was greater than that of *F. monoliforme* and correlated well with frequency of isolation (39).

Table 17-1. Relative frequency of *Fusarium* spp. in diseased nursery seedlings

Location									
British Columbia (3)[b]		Quebec (68)		Japan (63)		E. Africa (37)		Hungary (25)	
	%[a]		%[a]		%[a]		%[a]		%[a]
F. avenaceum	55	*F. solani*	34	*F. oxysporum*	88	*F. oxysporum*	32	*F. oxysporum*	45
F. sambucinum	23	*F. equiseti*	20	*F. solani*	5	*F. solani*	32	*F. equiseti*	36
F. oxysporum	19	*F. sambucinum*	13	*F. roseum*	4	*F. moniliforme*	24	*F. sporotrichioides*	12
F. solani	3	*F. oxysporum* var. redolens	12	*F. lateritium*	4	*F. equiseti*	8	*F. solani*	5

[a]Per cent of most frequent identified *Fusarium* isolates.
[b]Reference no. in literature cited.

Races and strains Pathogenic variation in *Fusarium* may occur as races or strains. Two culturally and morphologically different strains of *F. oxysporum* differed in their ability to kill more and older pine and spruce seedlings (29). A forma specialis and two races of *F. oxysporum* were identified by their relative pathogenicity to several conifer and non-conifer hosts (63). *Fusarium oxysporum* isolates from various disease types showed sufficient differences in pathogenicity and disease curves to justify separation into strains (7, 14). Isolates from various conifer hosts were more pathogenic to conifers than those from broadleaf and herbaceous hosts and showed sufficient differential pathogenicity to be considered as separate races (52).

Damage

Relative importance Generally, European forest pathologists classify *Fusarium* as a primary pathogen, equalling or exceeding *Pythium* or *Rhizoctonia* in damage caused. In some North American evaluations, symptomless Fusarium infections and the presence of other pathogens indicate lesser importance (31, 102, 107); in other opinions, absence or scarcity of other pathogens, severe damage, and frequent isolation of *Fusarium* imply strong pathogenicity (5, 42, 69, 115).

Frequency of damage Some nurseries have incurred frequent, heavy losses from damping-off or root rot (5, 26, 42). In others, these diseases occur every year but severity varies (8, 78). Occasionally, epidemics occurs over a period of years (102). Top damping-off is generally less frequent (31, 69). Nurseries may vary in the frequency of different disease types (31, 69, 78).

Severity Pre- and post-emergence damping-off caused by *Fusarium* occasionally wipe out entire sowings (26, 37, 69), and 20–50% losses are common (49, 66, 116). Root disease losses from *Fusarium* alone may average 10–30% (1, 8, 80, 116), but those associated with *Fusarium* complexes may be much higher (13, 35, 40, 61, 113). Losses from each disease type are not necessarily related (7). Almost complete loss of stored maple seed (21) and seedlings (100, 114) was associated with *Fusarium.* Nearly 8% of containerized Douglas-fir and ponderosa pine seedlings exhibited root rot symptoms (51).

Disease development Post-emergence damping-off disease curves in Quebec nurseries showed well-defined peaks, tapering off by mid-July (69). Start, slope, and skew of curves varied by year and seedling species, depending on climatic factors affecting germination time and infection conditions. Post-emergence damping-off in Finnish and Polish nurseries peaked at 3–4 (29), and 4–8 weeks (45), respectively. Root rot disease curves in British Columbia nurseries peaked in late August and tapered off in November, but seedlings continued to die throughout the winter (8). In Britain, root rot reached a maximum in early August and tapered off rapidly (18).

Disease Factors

Temperature Correlations between temperature and damping-off have been noted in nurseries where *Fusarium* is a major pathogen (31, 69). Root rot severity was greatest during years with the hottest summers, in nurseries with the most bright sunshine during early summer, in the hottest parts of the beds (8, 115), and in dark-colored seedling containers (51). Severe root rot in South Australian nurseries was attributed to the hot, dry climate (106). Experimental evidence supports field observations: as temperature increased, severity or pre- and post-emergence damping-off increased, disease onset advanced, and disease curves rose more steeply (28, 40, 81, 95). Root rot occurred at 27 C or higher (40, 81). It was increased by a regime comprising 24 C for more than 6 hr/day during the first month but not by lower temperature during this period (8). Black root rot was greatest at 26–35 C (40, 75). Fusarium isolates from diseased seedlings behaved as high-temperature fungi, with growth peaking at 25–35 C (45, 77, 95). The fungus grew approximately 1 mm/day in Douglas-fir roots at 25 C (10). The infection process was not affected between 10–30 C (9). Slight increases in storage temperature greatly enhanced molding of broadleaf seedlings (114) and reduced germinability in pine and larch seeds infected with *Fusarium* (46).

Moisture Increased soil moisture has been related to both increased and decreased disease. Cooling of soil by irrigation or precipitation during hot weather reduced root rot (31, 115), but wet soils in spring favored damping-off as soon as they warmed up (69, 108), while saturation conditions favored root rot due to *F. roseum* (80). Low soil moisture *per se* does not appear to affect root rot (10) but may favor cankering by *F. lateritium* in poplar shoots or cuttings by reducing bark moisture (4). Desiccation of stored broadleaf seedlings enhanced growth of Fusarium molds (114).

pH Tolerance of Fusarium isolates to a broad pH spectrum (45, 77, 94) and frequency of *Fusarium* on seedlings in acid and alkaline nursery soils (6, 48, 103) argue against the importance of this factor. Inconsistency in control of Fusarium disease by soil acidification has generally discredited this method.

Light Although shading reduced damping-off and root rot in nurseries (8, 69), it is difficult to separate its effects on light, temperature, and moisture (95). With other factors held constant, reduced light intensity increased damping-off, probably because of increased seedling succulence and fungus growth (95). Effect of light intensity on susceptibility to Fusarium attack varied with seedling species (90). Very heavy shade in nurseries increased seedling mortality associated with weak pathogens including some Fusaria (97, 98).

Nutrients Losses from Fusarium diseases are often lower in less-fertile nurseries (33, 78) and are increased by nitrogenous manures (31). In sand cultures inoculated with *Fusarium* spp., increased N promoted seedling succulence and damping-off (94). By contrast, increase Ca and P decreased both. Other macro-elements had less well-defined effects, generally reducing damping-off when deficient. Nursery experiments have corroborated in vitro studies. Root rot of Douglas fir increased with addition of N (82) and more so with the sulphate than the nitrate ammonium salt (108). Timing of nitrogen application had no effect on total seedling mortality (82). Addition of N alone or in combination with P or K increased black root rot but P or K without N had no effect on either black root rot (75) or root rot (82). Addition of superphosphate at sowing reduced damping-off (49).

Soil type Early opinions that soils low in organic matter were less conducive to damping-off (33, 69) have been refuted by experience that sandy soils may support high levels of Fusarium diseases (41, 66) and that soils naturally rich in or amended with organic materials have less disease (3, 16, 17, 20, 83, 87, 115). Organic matter effects have been attributed to reduction of pH (3), substances inhibitory to *Fusarium* survival (96), stimulation of antagonistic microflora (54, 87), reduction of available N (75), and temperature modification (115).

Vegetational history prior to nursery production has been recognized as a factor in Fusarium root diseases (11, 41). Some soils naturally high in antagonistic flora have low disease levels (8, 110).

Inoculum Source of *Fusarium* inoculum in forest nurseries has been ascribed to fragments of diseased roots from previous seedling crops (9, 51), seed-borne spores (27, 67,

73), and introduction of infested soil (101). Due to cultivation, inoculum is generally uniformly distributed to about 20-cm soil depth (9, 49, 71) but causes greater disease at shallow depths (86). It may spread around nurseries by wind, irrigation, or machinery (101).

Soil population There is evidence for the importance of soil microbial populations in Fusarium diseases. Healthy and diseased seedlings had qualitatively and quantitatively different rhizosphere populations (43, 92). Fungal populations of disease-prone soils had lower levels of antagonism to *Fusarium* than disease-free soils (59). Populations recolonizing disinfected soil had different proportions of fungus genera and were correlated with disease incidence but not *Fusarium* counts (5). Several common soil fungi depressed growth of *F. culmorum* (85). However, in paired tests with over 100 fungi, bacteria, and actinomycetes, *Fusarium* was relatively insensitive to antagonism (105).

Interest in specific organisms has centered on *Trichoderma* spp., *Mycelium radicis atrovirens,* and actinomycetes. Disease was reduced in pine seedlings grown in nursery soil inoculated with *Trichoderma* (62) and in sterile soil inoculated with *Fusarium* and *Trichoderma* as compared to that with *Fusarium* alone (91). *Mycelium radicis atrovirens* was consistently present in roots of healthy but not of diseased seedlings (6, 59). Some strains of the fungus reduced disease when introduced into soil together with *F. oxysporum* (60). Actinomycete populations of nursery soil appeared to be negatively correlated with root rot (8). About 10% of actinomycete strains showed in vitro antagonism to a *Fusarium* causing damping-off (110). However, soil inoculation with actinomycetes showed no reduction of damping-off (111) or root rot (115).

Inoculation of soil with the mycorrhizal fungus *Laccarius laccata* reduced root rot and stunting in Douglas fir (82), provided that roots made contact with it before *F. oxysporum* (86). The effect did not appear to be mycorrhizal. Nitrogen stimulation of root rot was partly offset by adding spores to the soil. Successful control of damping-off was achieved by "bacterized" compost (47). Rhizosphere of healthy seedlings had nearly six times more spore-forming bacteria antagonistic to *F. culmorum* than diseased seedlings; a bacillus applied to soil or pine seeds considerably delayed damping-off (92).

Abundance of parasitic nematodes associated with damping-off has suggested that they enhance disease by creating infection courts in roots or hypocotyls (11, 39, 55, 109). Disease caused by *Fusarium* inoculum combined with nematodes or mechanical wounding exceeded that caused by *Fusarium* alone (39, 55).

Host factors Poor-quality seed significantly increased damping-off (88). Susceptibility to damping-off and root rot varied by provenances (8, 52) and was greatest in very young seedlings (7, 8, 29, 45, 86, 113). Root rot was not related to mean seedling growth but may be related to the growth rate of individual seedlings (86). Vigor of cuttings expressed as bark turgidity was inversely related to attack by *F. lateritium* (4).

Control

Chemical treatment of soil For reasons discussed under the section on pH, soil acidification by inorganic chemicals has proved unreliable in controlling damping-off and root rot (17, 32, 115). Soil fumigation has been more successful against Fusarium diseases than other chemical treatments. Formalin is a moderately effective fumigant (24, 33, 49) but has weak action against black root rot (58). Evidence of *Fusarium* tolerance to it makes application questionable (18). Methyl bromide and chloropicrin have been the most effective fumigants against damping-off and root rot (5, 84, 115), but effects on seedling growth have been noted with both (11, 115). Trizone, a combination of methyl bromide, chloropicrin, and propargyl bromide, was as effective as methyl bromide and more beneficial to seedling growth (36, 88). Timing and dosage of both chemicals had significant effects on disease control and seedling growth. Other, general purpose fumigants, especially Mylone, Vorlex, and Vapam, also exerted some disease control (5, 23, 82). In the presence of abundant parasitic nematodes (11, 35, 58), or sometimes in their absence (23), nematicidal fumigants DD and ethylene dibromide reduced root rot and black root rot, respectively.

Soil fungicides Although a wide variety of fungicides show in vitro activity against *Fusarium* (2, 64), they have generally provided less control in soil than fumigants (84). Thiram and captan have been most consistent in controlling damping-off (2). Treatment timing was critical in controlling root rot with captan soil drench (15).

Seed fungicides Captan and thiram are the most commonly used fungicidal seed dressings to prevent losses in *Fusarium*-infested nurseries, and have given moderate to good control of damping-off (11, 19, 88, 115) but not root rot (11, 44). However, phytotoxicity may outweigh disease control benefits, especially on weak or damaged seed or at high dosage rates (12, 89). Benomyl was effective and less phytotoxic (53). Rhizoctol (10% methylarsinic sulphide, Bayer, West Germany) was the most effective seed dressing against damping-off of pine in East Africa (38). Fungicide application to seedlings prior to storage did not prevent molding (114).

Antibiotics Varied success has been reported in controlling Fusarium diseases with antibiotics. Although actidione and other standard antibiotics, in vitro and in sterile sand tests, reduced *Fusarium* growth and damping-off, they were ineffective in nursery soils (70, 111). In the USSR, good control with streptomycin, penicillin and other antibiotics (65), especially as seed treatments, has been reported.

Cultural control There is considerable evidence that the addition of organic matter reduces all types of Fusarium losses. Peat and fresh sawdust incorporated into the soil appear to be most beneficial (3, 16, 17, 20, 58, 66, 75, 115). Careful fertilizer scheduling is critical in control of Fusarium diseases. Delayed application of N until 2 months after sowing (75), the fall (115), or even the second growing season (108) has been recommended. Applications of P and K before sowing (108) improved seedling resistance (49, 75).

Heavy irrigation was effective in controlling root rots (75, 115), provided there was free drainage; otherwise waterlogging favored the disease (80). Raised beds improved drainage and reduced damping-off (69). Shading was beneficial (8, 69), provided it was not so dense as to reduce seedling vigor (98).

Crop rotation has reduced root rot (115), but fallowing was ineffective against black root rot (58). Reduction of inoculum (9) and minimizing spread (101) are essential in keeping disease levels to a minimum. Fall sowing resulted in less root rot and black root rot (8, 58), but more damping-off (69, 71) than spring sowing. Avoiding dense seedling stands is also recommended (20, 69, 113).

Literature Cited

1. British Columbia Forest Service. 1950. Reforestation. Forest Nurseries, p. 23–24. *In* B.C. For. Serv. Annu. Rep. 1960.
2. Belcher, J., and L.W. Carlson. 1968. Seed-treatment fungicides for control of conifer damping-off: laboratory and greenhouse tests, 1967. Can. Plant Dis. Surv. 48: 47–52.
3. Bier, J.E. 1942. Laboratory experiments on the control of damping-off in Douglas-fir using commercial peat as the planting medium. Can. Dep. Agric. Dom. For. Pathol. Service (Mimeo. Rep.) 18 p.
4. Bier, J.E. 1959. The relation of bark moisture to the development of canker diseases caused by

native facultative parasites. 2. Fusarium canker on black cottonwood. Can. J. Bot. 37: 781–788.
5. Bloomberg, W.J. 1965. The effect of chemical sterilization on the fungus population of soil in relation to root disease of Douglas-fir seedlings. For. Chron. 41: 182–187.
6. Bloomberg, W.J. 1966. The occurrence of endophytic fungi in Douglas-fir seedlings and seed. Can. J. Bot. 44: 413–420.
7. Bloomberg, W.J. 1971. Diseases of Douglas-fir seedlings caused by *Fusarium oxysporum*. Phytopathology 61: 467–470.
8. Bloomberg, W.J. 1973. Fusarium root rot of Douglas-fir seedlings. Phytopathology 63: 337–341.
9. Bloomberg, W.J. 1976. Distribution and pathogenicity of *Fusarium oxysporum* in a forest nursery soil. Phytopathology 66: 1090–1092.
10. Bloomberg, W.J. 1976. Stimulation of Douglas-fir seedling root growth, damping-off, and root rot. Can. For. Service Rep. BC–X–127. 43 p.
11. Bloomberg, W.J., and W.R. Orchard. 1969. Chemical control of root disease of Douglas-fir seedlings in relation to fungus and nematode populations. Ann. Appl. Biol. 64: 239–244.
12. Bloomberg, W.J., and J. Trelawny. 1970. Effect of thiram on germination of Douglas-fir seed. Phytopathology 60: 1111–1116.
13. Bloomberg, W.J., and J.R. Sutherland. 1971. Phenology and fungus-nematode relations of corky root disease of Douglas-fir. Ann. Appl. Biol. 69: 265–276.
14. Bloomberg, W.J., and W. Lock. 1972. Strain differences in *Fusarium oxysporum* causing diseases of Douglas-fir seedlings. Phytopathology 62: 481–485.
15. Bloomberg, W.J., and W. Lock. 1974. Importance of treatment timing in the control of Fusarium root rot of Douglas-fir seedlings. Phytopathology 64: 1153–1154.
16. Buckland, D.C. 1942. Further experiments in the control of damping-off in Douglas-fir using commercial peat as the planting medium. Can. Dep. Agric. Dom. For. Pathol. Service (Mimeo. Rep.) 12 p.
17. Buckland, D.C. 1947. Investigations on the control of Fusarium top blight of Douglas-fir seedlings. Can. Dep. Agric. Sci. Service (Mimeo Rep.) 21 p.
18. Buxton, E.W., I. Sinha, and V. Ward. 1962. Soil-borne diseases of Sitka spruce seedlings in a forest nursery. Trans. Brit. Mycol. Soc. 45: 433–448.
19. Cockerill, J. 1955. The use of thiram as a control for damping-off of red pine. Can. Dep. Agric. Bi-M. Prog. Rep. 11(4): 1.
20. Cockerill, J., and J.D. Cafley. 1951. Damping-off of red pine seedlings at Orono, Ont. Can. Dep. Agric. Bi-M. Prog. Rep. 7(5): 1–2.
21. Fletcher, J. 1902. Maple seed blight (*Fusarium* sp.), p. 259. *In* Rep. Entomologist and Botanist, Can. Dep. Agric.
22. Foster, A.A. 1959. Nursery diseases of southern pines. U.S. Dep. Agric. For. Service Pest Leafl. 32. 7 p.
23. Foster, A.A. 1961. Control of black root rot of pine seedlings by soil fumigation in the nursery. Ga. For. Res. Council Rep. No. 8, 5 p.
24. Gifford, C.M. 1911. The damping off of coniferous seedlings. Vt. Agric. Exp. Stn. Bull. 157: 140–171.
25. Hangyal, W. 1971. *Fusarium*-arten, als erreger der umfallkrankheit an sämlingen der gemeinen kiefer und schwarzkiefer. Erdeszeti Kut. 67: 167–180.
26. Hangyal, W. 1973. Microflora examinations on Scots and Black pine seeds. Erdeszeti Kut. 69: 171–179.
27. Hangyal, T. 1974. Lucfenyötüalmos fenyövetés fertözési kísérleteinek eredményei. Erdo 23: 406–408.
28. Hangyalne Balul, W. 1975. Fenyö csemetedölést okozó néhány gombafaj patogenitásának vizsgálata. Erdo 24: 65–68.
29. Hanioja, P. 1969. Taimipoltteen aiheuttajista metsäntutkimuslaitoksen Punkaharjun Koeaseman Taimitarhoissa. Comm. Inst. For. Fenn. 69. 21 p.
30. Hartig, R. 1892. Ein neuer keimlingspilz. Fortstl. Naturw. Z. Jahrg. 1: 432–436.
31. Hartley, C. 1921. Damping-off in forest nurseries. U.S. Dep. Agric. Bull. 934. 100 p.
32. Hartley, C., and T.C. Merrill. 1914. Preliminary test of disinfectants in controlling damping-off in various nursery soils. Phytopathology 4: 89–92.
33. Hartley, C., and R. G. Pierce. 1917. The control of damping-off of coniferous seedlings. U.S. Dep. Agric. Bull. 453. 32 p.
34. Hartley, C., T.C. Merrill, and A.S. Rhoads. 1918. Seedling diseases of conifers. J. Agric. Res. 15: 521–558.
35. Henry, B.W. 1951. Ehylene dibromide controls a root rot at the W.W. Ashe Nursery. U.S. For. Service Tree Planters Notes 7: 2–4.
36. Henry, B.W. 1953. A root rot of southern pine nursery seedlings and its control by soil fumigation. Phytopathology 43: 81–88.
37. Hocking, D. 1968. Fungi associated with damped-off and healthy pine seedlings and with seed in East African pine nurseries. Trans. Brit. Mycol. Soc. 51: 221–226.
38. Hocking, D., and A.A. Jaffer. 1969. Damping-off in pine nurseries: fungicidal control by seed pelleting. Commonwealth For. Rev. 48: 355–363.
39. Hocking, D., E.C. Setliff, and A.A. Jaffer. 1968. Potential pathogenicities of fungi associated with damped-off pine seedlings in East African pine nurseries. Trans. Brit. Mycol. Soc. 51: 227–232.
40. Hodges, C.S. 1962. Black root rot of pine seedlings. Phytopathology 52: 210–219.
41. Hodges, C.S. 1975. Black root rot of pine, p. 14–16. *In* U.S. Dep. Agric. For. Service, Agric. Handbook, 470. 125 p.
42. Holtzmann, O.V. 1955. Organisms causing damping-off of coniferous seedlings and their control. Ph.D. Thesis, Wash. State Coll., Pullman. 59 p.

43. Jancarik, V. 1967. The relationship of rhizosphere and soil fungi to seedlings of forest tree species. Commun. Inst. For. Cech. 5: 35–45.
44. Johnson, D.W., and R.D. Harvey. 1975. Seed-protectant fungicides for control of Douglas-fir and ponderosa pine seedling root rots. U.S. For. Service Tree Planters Notes 26: 3–5.
45. Kozlowska, C. 1962. Badania nad biologia *Fusarium oxysporum* (Schl.) S et H. I. Próby jego zwalczania. Prace Inst. Bad. Lesn. 246: 3–91.
46. Kozlowska, C. 1968. Investigations on pathogenic fungi on seeds of forest trees pine, larch, oak, birch. For. Res. Inst. Warsaw Report on Proj. No. E21–FS–21, U.S. Dep. Agric. Grant No. FG-PO–153. 118 p.
47. Krasilnikov, N.A., and F.A. Raznitsina. 1946. A bacterial method of controlling damping-off of Scots pine seedlings caused by *Fusarium*. (In Russian) Agrobiologiya 5/6: 109–121.
48. Kubikova, J. 1968. *Fusarium oxysporum* (Schlecht.) Snyd. et Hans.—a dominant fungus species on the root-surface of woody plant seedlings. Plant Soil. 28: 306–312.
49. Kulakova, N.M. 1964. Preventing damping-off of Scots pine seedlings. (In Russian) Lesn. Khoz. 17: 41–44.
50. Labrada, A. 1973. Damping-off en los viveros de pinos en Cuba. Baracoa 3: 34–44.
51. Landis, T.D. 1976. Fusarium root disease of containerized tree seedlings. U.S. Dept. Agric. For. Service, Biological Evaluation R2–76–16. 6 p.
52. Lang, K.J. Experimente mit erregern der umfallkrankheit. II. Einzel- und mischinfektionen mit umfallerregern und Trichoderma viride Pers. ex. Fr. an Konifersämlinger verschiedener herkunft. Eur. J. For. Pathol. 6: 46–56.
53. Lock, W., J.R. Sutherland, and L.J. Sluggett. 1975. Fungicide treatment of seeds for damping-off in British Columbia nurseries. U.S. For. Service Tree Planters Notes 26: 16–28.
54. Lu, K.C. 1968. Effect of organic amendments on soil microflora in relation to Fusarium root rot of Ponderosa pine seedlings, p. 40–45. *In* Proc. West. For. Nurs. Council, 1968.
55. Macara, A.M. 1972. Nematodes and associated fungi found in forest nurseries, p. 321–326. *In* Act. III Congr. Univ. Fitopathol. Mediterr. 1972.
56. Magnani, G. 1954. Alcuni casi di deperimento di pioppelle in vivaio. Cellulosa e Carta 5: 14–15.
57. Magnani, G. 1975. Sulla suscettibilita di alcune specie di conifere alle morie in semenzaio. Cellulosa e Carta 26: 19–25.
58. Maki, T.E., and B.W. Henry. 1951. Root-rot control and soil improvement at the W.W. Ashe nursery. Southern For. Exp. Stn. Occasional Pap. 119. 23 p.
59. Manka, K. 1970. Parasitäre sämlingskrankheiten der forstbäume und die bodenpilze. Zentralbl. Bakteriol. parasitenkd. infektionkr. Hyg.(Zweite Abt.)124: 450–459.
60. Manka, K., and M. Gierczak. 1972. Mozliwosc uzycia grzyba Mycelium radicis atrovirens Melin do biologicznego zwalczania zgorzeli siewek sosny zwyczajnej. (In Polish, English summary) Pr. Kom. Nau,. Roln. Lesn. 34: 111–119.
61. Marks, G.C. 1965. The pathological histology of root rot associated with late damping-off in *Pinus lambertiana*. Australian For. 29: 30–38.
62. Mattis, G.Ya., and A.P. Badanov. 1973. Biological method of control of *Fusarium* on seedlings. (In Romanian) Khozyaistvo 9: 58–60.
63. Matuo, T. and O. Chiba. 1966. Species and formae speciales of Fusaria causing damping-off and root rot of coniferous seedlings in Japan. Ann. Phytopathol. Soc. Jap. 32: 14–22.
64. Mocanu, V. 1972. Contributii la studiul bolii "fuzarioza" plantulelor de pin şi molid. Măsuri de prevenire şi combatere. Rev. Padurilor 87: 535–540.
65. Molotkova, N.D., and V.G. Yakovlev. 1974. Application of trichothecin and phytobacteriomycin to control lodging of conifer seedlings (In Russian). Mikol. Fitopathol. 8: 55–59.
66. Ofosu-Asiedu, A., and A. Otoo. 1972. Effect of organic matter on the pathogenicity of three damping-off fungi. For. Prod. Res. Inst. Ghana, Tech. Newsletter 6(3/4): 11–15.
67. Pawuk, W.H., and J.P. Barnett. 1974. Root rot and damping-off of container-grown southern pine seedlings, p. 173–176. *In* Tinus, R.W., W.I. Stein, and W.E. Balmer (ed.), Proc. N. Amer. Containerized Forest Tree Seedling Sympos.
68. Pomerleau, R. 1934. The fungi responsible for seedling blight of conifers at the Berthierville forest nursery, p. 58–61. *In* Rep. Quebec Soc. Plant Protect. 1932/34.
69. Pomerleau, R. 1942. Etudes sur la fonte des semis de conifères. Rev. Trimest. Can. 28: 127–153.
70. Pomerleau, R. 1963. Essais sur l'efficacité de quelques antibiotiques contre la fonte des semis de conifères. Phytoprotection 44: 116–119.
71. Rathbun, A.E. 1918. The fungous flora of pine seedbeds. Phytopathology 8: 469–483.
72. Rathbun, A.E. 1922. Root rot of pine seedlings. Phytopathology 12: 213–220.
73. Rosnev, B. 1975. Lodging of seedlings caused by *Fusarium* species. (In Russian) Gorsko Stopanstvo 31: 32–36.
74. Rowan, S.J. 1960. The susceptibility of twenty-three tree species to black root rot. Plant Dis. Rep. 44: 646–647.
75. Rowan, S.J. 1971. Soil fertilization, fumigation, and temperature affect severity of black root rot of slash pine. Phytopathology 61: 184–187.
76. Rowan, S.J., T.H. Filer, and W.R. Phelps. 1972. Nursery diseases of southern hardwoods. U.S. Dep. Agric. For. Service For. Pest Leaflet 137. 7 p.
77. Salisbury, P.J. 1952. The effect of temperature and hydrogen-ion concentration on growth of certain cultures of *Fusarium oxysporum* isolated from

Douglas-fir seedlings. Can. Dep. Agric. Div. For. Biol. (Mimeo Rep.) 7 p.

78. Salisbury, P.J. 1954. A review of damping-off of Douglas-fir seedlings in British Columbia nurseries. For. Chron. 30: 407–410.
79. Schreiber, L.R. 1967. A soft rot of Elm root cuttings caused by *Fusarium solani*. Phytopathology 57: 920–921.
80. Schwalm, J.M. 1973. Delaware nursery root rot in seedlings triggered by wet weather and poor drainage. U.S. For. Service Tree Planters Notes 24: 18.
81. Shea, K.R., and J.H. Rediske. 1961. Pathological aspects of germination and survival of Douglas-fir in controlled environment. Weyerhaeuser Co. For. Res. Note 41. 8 p.
82. Sinclair, W.A., D.P. Cowles, and S.M. Hee. 1975. Fusarium root rot of Douglas-fir seedlings: suppression by soil fumigation, fertility management, and inoculation with spores of the fungal symbiont *Laccaria laccata*. For. Sci. 21: 390–399.
83. Smith, R.S. 1967. Decline of *Fusarium oxysporum* in the roots of *Pinus lambertiana* seedlings transplanted into forest soils. Phytopathology 57: 1265.
84. Smith, R.S. 1975. Fusarium root disease, p. 9–10. *In* Forest Nursery Diseases in the United States, U.S. Dep. Agric., Agric. Handbook 470. 125 p.
85. Sokolov, D.V., and V.I. Scedrova. 1968. Trial of antibiotic substances produced by Polyporaceae and certain other fungi for control of Fusariosis (damping-off) of Scots pine. (In Russian) Nauchn. Tr. Leningr. Lesotekh. Akad. 110: 109–113.
86. Stack, R.W., and W.A. Sinclair. 1975. Protection of Douglas-fir seedlings against Fusarium root rot by a mycorrhizal fungus in the absence of mycorrhiza formation. Phytopathology 65: 468–472.
87. Stankova-Opocenska, E., and J. Lehovec. 1971. Vzcházení, rÅust, zdravotní stav a mikrobní osídlení kořenÅu okrasných jehličin, pěstovaných v jílovita-rašelinovém substrátu. Acta PrÅuhonic. 24: 119–140.
88. Sutherland, J.R. 1967. Field tests for control of red pine seedling diseases. Phytoprotection 48: 58–67.
89. Sutherland, J.R., W. Lock, and L.J. Sluggett. 1975. Damping-off in British Columbia nurseries. Can. For. Service Rep. BC–X–125. 20 p.
90. Taher, M.M., and R.C. Cooke. 1975. Shade-induced damping-off in conifer seedlings. I. Effects of reduced light intensity on infection by necrotrophic fungi. New Phytol. 75: 567–572.
91. Ten Houten, J.G. 1939. Kiemplantenziekten van coniferen. Drukkerij J. van Boekhoven, Amsterdam. 128 p.
92. Timonin, M.I. 1966. Rhizosphere effect of healthy and diseased lodgepole pine seedlings. Can. J. Microbiol. 12: 531–537.
93. Tint, H. 1945. Studies in the Fusarium damping-off of conifers. I. The comparative virulence of certain Fusaria. Phytopathology 35: 421–439.
94. Tint, H. 1945. Studies in the Fusarium damping-off of conifers. II. Relation of age of host, pH, and some nutritional factors to the pathogenicity of *Fusarium*. Phytopathology 35: 440–457.
95. Tint, H. 1945. Studies in the Fusarium damping-off of conifers. III. Relation of temperature and sunlight to the pathogenicity of *Fusarium*. Phytopathology 35: 498–510.
96. Toussoun, T.A., W. Menzinger, and R.S. Smith. 1969. Role of conifer litter in ecology of *Fusarium:* Stimulation of germination in soil. Phytopathology 59: 1396–1399.
97. Vaartaja, O. 1952. Forest humus quality and light conditions as factors influencing damping-off. Phytopathology 42: 501–506.
98. Vaartaja, O. 1953. Disease caused by weak parasites associated with shade in seedbeds. Can. Dep. Agric. Bi-M. Prog. Rep. 9(2): 3–4.
99. Vaartaja, O. 1953. Seedling diseases of conifers in Saskatchewan. Can. Dep. Agric. Bi-M. Prog. Rep. 9(5): 2.
100. Vaartaja, O. 1955. Storage molding of maple seedlings. Can. Dep. Agric. Bi-M. Prog. Rep. 11(1): 3.
101. Vaartaja, O. 1967. Reinfestation of sterilized seedbeds by fungi. Can. J. Microbiol. 13: 771–776.
102. Vaartaja, O. 1967. Damping-off pathogens in South Australian nurseries. Phytopathology 57: 765–768.
103. Vaartaja, O., and W.H. Cram. 1956. Damping-off pathogens of conifers and of caragana in Saskatchewan. Phytopathology 46: 391–397.
104. Vaartaja, O., and A.W. Hill. 1965. Fungi isolated from damping-off of conifers in Ontario, 1964. Can. Dep. Agric. Bi-M. Prog. Rep. 21(4): 2–3.
105. Vaartaja, O., and P.J. Salisbury. 1965. Mutual effects in vitro of micro-organisms isolated from tree seedlings, nursery soil and forests. For. Sci. 11: 160–168.
106. Vaartaja, O., and M. Bumbieris. 1967. Organisms associated with root rots of conifers in South Australian nurseries. Plant Dis. Rep. 51: 473–476.
107. Vaartaja, O., W.H. Cram, and G.A. Morgan. 1961. Damping-off etiology especially in forest nurseries. Phytopathology 51: 35–42.
108. Van den Driessche, R. 1963. Nursery experiments with Douglas-fir. Commonwealth For. Rev. 42: 242–254.
109. Van Weerdt, L.G., A.P. Martinez, and R.P. Esser. 1960. Pine seedlings in Florida State nurseries, Sta. Board Florida Lab. Notes (Mimeo Rep.), 4 p.
110. Veselinovic, N., and M. Peno. 1971. Zastupljenost mikroorganizama antagonista prema *Fusarium oxysporum* var. *orthoceras* soj 12 u raznim tipovima zemljista. Sumarstvo 24: 29–36.
111. Vörös, J. 1954. Antibotikumok és antagonista mikroogranizmusok felhasnálása fenyöcsemetedölés ellen. Növèntermelés 3: 115–122.
112. Wall, R.E. 1974. Recent conifer disease problems in forest nurseries in the Maritime provinces. Can. Plant Dis. Curv. 54: 116–118.

113. Warcup, J.H. 1952. Effect of partial sterilization by steam or formalin on damping-off of Sitka spruce. Trans. Brit. Mycol. Soc. 35: 248–262.
114. Wilner, J., and O. Vaartaja. 1958. Prevention of injury to tree seedlings during cellar storage. For. Chron. 34: 132–138.
115. Wright, E., G.M. Harvey, and C.A. Bigelow. 1963. Tests to control Fusarium root rot of Pondrosa pine in the Pacific Northwest. U.S. For. Service Tree Planters Notes 59: 15–20.
116. Wu, Y-S., Y. Kao, S-F. Ku, C.-N Liu, and S-F. Chong. 1963. The damping-off of coniferous seedlings. I. Symptoms, causal organisms, and cultural practices in relation to disease development. (In Chinese, English summary) Phytophyl. Sin. 2: 179–186.
117. Yamamoto, M., C-Y. Hong, and H. Hashimoto. 1965. Studies on the damping-off of pine seedlings. (In Japanese, English summary) J. Jap. For. Soc. 47: 30–34.

18 Pitch Canker of Southern Pines

L.D. Dwinell, E.G. Kuhlman, and G.M. Blakeslee

Pitch canker was first described in 1946 (20) and since 1974 has caused severe damage to pines in the southern United States (10, 14, 15, 29, 30). This canker disease is caused by a species of *Fusarium* and research during the past few years has dramatically altered forest pathologists' concept of pitch canker. This article summarizes many of these new reports.

Hosts and Geographical Distribution

Pitch canker is currently causing severe damage to slash *(Pinus elliottii* var. *elliottii),* loblolly *(P. taeda),* Virginia *(P. virginiana)* and shortleaf *(P. echinata)* pines (10, 14, 15, 29, 30). The disease also has been reported on South Florida slash pine *(P. elliottii* var. *densa)* (4), eastern white *(P. strobus)* (1), longleaf *(P. palustris),* Table-Mountain *(P. pungens),* and pitch *(P. rigida)* (21) pines in the United States and on *P. occidentalis* in Haiti (21). Monterey *(P. radiata)* (18) and Scotch *(P. sylvestris)* pines (21), which are not native to the southeastern United States, have also been found to be hosts by artificial inoculation. Until a few years ago, loblolly pine was not considered to be a host of the pitch canker fungus (21), but in 1971 a strain pathogenic to loblolly pine was found in eastern North Carolina (19). More recently it has been found that the pitch canker fungus from several pine hosts can infect loblolly pine (10, 11, 15). No strains of the fungus with pathogenicity limited to loblolly pine have been discovered.

The disease is now known to occur on pines from Virginia to southern Florida and west to Louisiana. Current outbreaks have been most severe on planted slash pine in Florida and on loblolly, slash, and Virginia pines in seed orchards in Alabama, Georgia, Mississippi, and North Carolina. A 1976 disease survey revealed pitch canker to be present in nearly all Florida counties but a serious problem only in the central and southern counties (29). The disease was also found in 22 of 25 slash pine seed orchards in Florida. On loblolly pine, the disease has caused considerable damage at the J. P. Weyerhaeuser Seed Orchard (Washington, North Carolina) and McNair Seed Orchard (McNair, Mississippi) (15). Pitch canker has also been observed in other seed orchards in the southern United States. Current research deals with the elucidation of outbreaks in Florida and in pine seed orchards throughout the southeastern United States.

Symptoms

There are two primary symptoms of pitch canker disease (16). The classic symptom is bleeding, resinous cankers on the trunk or large branches (3, 20). The canker retains its bark and is usually sunken (3). On some pine species, the wood beneath the canker is deeply pitch-soaked. Species such as slash, South Florida slash, longleaf, Virginia, and shortleaf pines may show pitch-soaking to the pith; however, bole cankers on loblolly pine, which appear to be uncommon, usually contain slight wedges of pitch-soaked tissue (15). This pitch-soaking of the underlying wood is a diagnostic character useful in separating pitch canker from other canker diseases of pines (3). The production of pitch in some pine species is so copious that attempts were made between 1947 and 1954 to make commercial use of the fungus to stimulate gum flow for naval stores production (9, 17).

The other symptom is shoot dieback in the upper crown, which is the predominant symptom of pitch canker on loblolly pine in seed orchards and on slash pine in plantations in Florida (10, 14, 15, 30). Shoot dieback, which results from cankers developing primarily on the summer flushes, is noticeable in the fall, when the fully developed needles turn yellow to reddish brown. Dieback of terminal and lateral branches within the crown can become severe. Studies conducted in one 20-year-old plantation of slash pine typical of areas of pitch canker epidemics in Florida revealed that more than 25% of the live crown was destroyed in 20% of the trees (Blakeslee, *unpublished*). This high frequency of cankers per tree and extensive dieback has not been reported in previous outbreaks (4, 20, 24, 31). Developing cankers are usually pitch-soaked, and resin is often exuded. Cankers on slash pine are usually more resinous than those on loblolly pine. Once a canker girdles a stem, all needles distal to the girdle become dehydrated and die. Spread down the stem appears to be arrested by nodes. Cankers continue to develop during the winter and spring. In the spring, expanding buds and shoots die because the girdled shoots are unable to supply water. The new growth may be fully expanded before being killed by the infection on the older tissue (14, 15).

Dead shoots may remain in the crown for several years and serve as indicators of the disease. Needles which remain on the killed shoots eventually turn gray to grayish brown. Witches'-broom develops in some trees as adventitious buds form in response to repeated infections and dieback (15).

The Causal Agent

In 1945, pitch flowing from bark and soaking the wood was the primary symptom observed on Virginia pines near Asheville, North Carolina. The disease was called pitch canker, and a species of *Fusarium* was indicated as the causal agent (20). This was the first report of a *Fusarium* species causing a disease of pines. The causal agent produced abundant microconidia, few macroconidia, and no chlamydospores, and was thought to belong in the section Liseola (20). However, in 1949, the fungus was named *F. lateritium* f. sp. *pini* (23), placing it in a distinctly different section of a very large and diverse genus. Recent isolations made from pitch cankers on several pine species consistently yield a fungus which has been identified as *F. moniliforme* var. *subglutinans,* a member of the section Liseola (10, 11, 15, 23, 30). A key characteristic of the section Lateritium is a lack of microconidia (8); since the pitch canker fungus produces abundant microconidia (16, 20, 23, 33), it cannot be placed in Lateritium. Furthermore, the microconidia are formed on polyphialides as well as on simple phialides (23); such formations, according to Booth (8) and Nirenberg (27), also fit the section Liseola. The section Lateritium has only simple phialides. The production of microconidia in false heads distinguishes *F. moniliforme* var. *subglutinans* from *F. moniliforme,* in which the microconidia are formed in chains (8). In culture, the pine pitch canker fungus generally fits the description provided by Booth (8, 23). The optimum temperature for growth is 24 C (16, 23). Strains of the fungus from slash pine grew significantly better at 32 C than those from loblolly, shortleaf, and Virginia pines (23). The fungus is heterothallic, and the perfect stage, *Gibberella fujikuroi* var. *subglutinans,* has formed following matings of pine isolates with single ascospore isolates from India (23). Morphologically and genetically, it appears that the fungus originally designated *F. lateritium* f. sp. *pini* (33) should be reclassified as *F. moniliforme* var. *subglutinans* (23).

When strains of *F. moniliforme* var. *subglutinans* were originally isolated from cankers on slash and loblolly pines, they were considered to be secondary invaders. However, when the same species was subsequently isolated from typical pitch cankers on longleaf, Virginia, South Florida slash, and shortleaf pines, pathogenicity studies were initiated (10, 11). The fungus was able to cause cankering and shoot mortality in Virginia, shortleaf, slash, and loblolly pines (10, 11, 23). These pine species appeared to vary in their susceptibility to *F. moniliforme* var. *subglutinans* (11). The fungus also shows evidence of variation in virulence (11). Schmidt (30) has also demonstrated pathogenicity of *F. monoliforme* var. *subglutinans* on slash pine. Isolates of *F. lateritium* from slash and loblolly pine were not pathogenic (11, 23). On the basis of the morphology and pathogenicity of the fungus, we believe that we are dealing with the same fungus which was described in 1949 (23) as the cause of pitch canker.

In 1959, it was suggested that pitch canker may have been introduced into the United States from Haiti (3). However, *F. moniliforme* var. *subglutinans* occurs on a wide range of hosts over a broad geographical area (8) and is common in the southeastern United States (P. E. Nelson, *personal communication*). The disease is probably endemic to the southern United States. Recent research also indicates that strains of the fungus from other crops are pathogenic to pine. For example, strains from gladiolus corms have been found to cause shoot mortality in one-year-old slash and loblolly pine seedlings (2, 13). The possible interrelationships of these current outbreaks of pitch canker disease with other agricultural and horticultural crop diseases are not known.

According to the literature, *F. moniliforme* var. *subglutinans* is soil-borne (8). Strains of this fungus pathogenic to pine have been recovered from plantation soils in Florida and soils from seed orchards in Georgia, Mississippi, and North Carolina. The fungus was recovered only from soils of a young slash pine plantation and two seed orchards where there was a ground cover of grass. In older plantations with a layer of needles, the fungus was present at negligible levels (12). Wind-borne dust containing the pine pitch canker fungus may serve as a source of inoculum for the spread of the pathogen.

Although Hepting (18) reported fruiting of the asexual stage of the pitch canker pathogen on artificially inoculated *P. radiata* seedlings, the occurrence of sporodochia on natural cankers is considered rare (3) and has been reported only once from shortleaf and Virginia pines in North Carolina (34). During the current pitch canker outbreaks in Florida, the salmon-pink sporodochia of *F. moniliforme* var. *subglutinans* have been found in abundance on diseased branches of slash pine (7). The sporodochia form primarily in fasicle scars exposed when needles are dislodged from the diseased branch. This field-produced inoculum has good viability and 1.5-year-old slash pines inoculated with the spores consistently express typical symptoms of pitch canker. The sporodochia have been observed in all seasons of the year, and the spores are readily spread by wind or splashing rain.

Geographical and Clonal Variation

There is considerable variation among pines in susceptibility to pitch canker. In three Coastal Plains seed orchards, pitch canker incidence is markedly higher in the Piedmont seed source of loblolly pines. At the McNair Seed Orchard, it was fairly uniform in distribution over all geographical sources of loblolly pine (15). In Florida, the incidence of pitch canker on slash pine is highest in plantations. It appears that the seed source was primarily from south Georgia. The incidence is significantly lower in natural stands of slash pine.

A marked clonal variation in disease susceptibility has been noted in both slash and loblolly pines in seed orchards (15, 29). At the McNair Seed Orchard, only one clone has a history of no pitch canker infeciton. At the J.P. Weyerhaeuser Seed Orchard, even in the hard-hit Piedmont section, six clones have no damage and four others have only minimal damage (15). In a 1976 survey of slash pine seed orchards in Florida, pitch canker was found in 22 of 25 orchards. Susceptibility to pitch canker was found to vary according to clones (5, 29).

Insects and Wounding

Fusarium moniliforme var. *subglutinans* probably could not attack an unwounded host plant. Insect wounds have been suggested as suitable infection courts (3, 21, 25, 26). The cyclic nature of disease outbreaks in Florida (29, 31) on slash pine suggests that conflicting relationships exist between pitch canker and feeding injury of the pine tip moth *(Rhyacionia subtropica)* (25, 26). The deodar weevil *(Pissodes nemorensis)* may be contributing to the incidence of the disease in north Florida (32). Recent studies conducted in areas of pitch canker outbreaks have shown a high degree of association of the pathogen and *P. nemorensis,* both during the weevil's developmental stages within brood trees and during emergence of adult weevils (6). In loblolly pine seed orchards, the needle midge (*Contarinia* sp.) is common (28) and causes a needle wound that is colonized by *F. moniliforme* var. *subglutinans* (15).

Pitch canker is also associated with mechanical wounding. In slash pine seed orchards, bole cankers are often associated with injuries caused by mechanical shakers used for cone harvest (14). Infection frequently occurs at cone scars on loblolly pine when the cones are removed from the branches during harvest (15). Broken branches caused by cattle grazing in young slash pine plantations also serve as infection courts (Dwinell, *unpublished*).

The pitch canker fungus also infects rust galls caused by *Cronartium quercuum* f. sp. *fusiforme* (3, 15, 22). The fungus has been isolated from pitch-cankered fusiform rust galls on loblolly pines in Florida and Georgia (15) and on slash and longleaf pines in South Carolina (22).

Stress

Various stress factors have been suggested as predisposing trees to infection. Drought may be associated with the disease in Florida (32). Neither moisture deficiency nor early frost, however, appears to be involved in the outbreaks of pitch canker in the loblolly pine seed orchards (15). Cultural practices such as fertilization or insecticide applications may be involved. Late summer applications of ammonium nitrate to promote flowering may increase susceptibility by providing a more succulent host or by prolonging growth in the fall. In a few cases in Florida, the incidence of pitch canker on slash pine increased following applications of fertilizer (35), although the disease is also severe in unfertilized stands.

Damage

The rate of development of a pitch canker in slash pine plantations and loblolly pine seed orchards is extremely rapid. At the McNair Seed Orchard, for example, only 18% of the ramets were known to be affected in August 1975, but by December 45% of the trees had moderate to severe shoot dieback. A similar pattern was noted at the J. P. Weyerhaeuser Seed Orchard (15). In Florida, vast acreages of slash pine developed pitch canker in only a few months. One area in southern Volusia County, Florida, for example, had relatively little pitch canker in 1975. However, during the fall and winter of 1976, some 20,000 acres became severely affected (14). Infection levels exceeded 90% in many stands (Blakeslee, *unpublished*).

Trees which exhibit shoot dieback often recover (15, 24, 30, 31). Loblolly pines in seed orchards appear to recover faster than slash pines in plantations. Thus, primary economic impacts are reduced growth, malformed stems, and decreased production of cones.

It is difficult to estimate accurately the severity of the damage caused by outbreaks of pitch canker occurring since 1975. On the basis of a limited outbreak in 1968–70, Schmidt and Underhill (31) estimated that growth losses and mortality would reduce yields in this area by 5.4 cords per acre at a rotation age of 25 years. On the basis of the current outbreak in permanent plots, Earl Underhill *(personal communication)* predicts a reduction in growth of 72% during the first year of infection and an initial loss of 0.90 cords per acre. Although most infected areas have sustained relatively little mortality as a result of pitch canker (29), a recent study in Florida found nearly 25% mortality 2 years after a heavy infection (Blakeslee, *unpublished*). The mortality was in response to the severe infection and dieback of the crown and was not attributable to such factors as girdling caused by main stem cankers or insect attacks. In a 12-month period beginning March 1976, two companies had to cut prematurely over 2,000 acres of slash pine because of high incidence of pitch canker (14).

In seed orchards, the loss is primarily in cone production. In 1975, a record 3,428 bushels of cones were harvested at the McNair Seed Orchard. A year later, production was only 467 bushels, largely because of the pitch canker problem (15).

Control

A proven disease control program is not yet available for pitch canker. In plantations, salvage harvesting of severely diseased, merchantable stands can limit further losses to growth reduction and mortality. Timely thinning or other improvement cutting can rid a stand of infected trees that would otherwise die or become stagheaded culls. Systematic removal of infected trees reduces fire hazard and inoculum sources while providing more growing space for better trees.

Southern pines vary in their susceptibility to pitch canker (5, 14). The use of native or local seed sources may substantially reduce the incidence of disease (14). Pockets of high disease incidence in the Southeast have been traced to off-site seed sources. In the future, trees selected for disease resistance should be available for stocking high-hazard sites.

In the seed orchard, management practices should be aimed at minimizing wounding of the trees. Tree shakers used for cone harvest should be properly adjusted and operated by trained personnel. Trees with bole cankers should be rogued. On loblolly pine, the cones should be removed by means other than twisting and tearing. Wounding of the tree base and exposed anchor roots can be avoided by careful mowing.

Chemical control is economically feasible in seed orchards where individual trees have high monetary value. There is evidence that insecticides may be useful in reducing wounds caused by insects when applied during the last flushes of growth in late summer and early fall (Dwinell, *unpublished*).

Systemic fungicides such as the benzimidazoles may be effective for protection. The pitch canker fungus is highly sensitive to benomyl in vitro. However, spray tests in Florida failed to give significant control. Other methods of applying systemic fungicides are being evaluated.

Summary

Pitch canker, a disease of southern pines native to the United States, is causing severe damage to planted slash pine in Florida and to several species of pines in seed orchards throughout the South. Primary symptoms of this disease are resin exudation, pitch-soaked cankers, and shoot dieback developing principally during the fall and winter. Infection usually results in reduced growth and malformed stems. Mortality resulting from the shoot dieback is generally low. The greatest economic impact in seed orchards is decreased production of cones. The causal organism is considered to be *Fusarium moniliforme* var. *subglutinans* because of its asexual morphology and sexual compatability with single ascospore isolates of *Gibberella fujikuroi* var. *subglutinans.* Pathogenicity of this fungus has been confirmed. Strains from non-pine hosts have also been found to be pathogenic to slash and loblolly pines. The fungus produces sporodochia on slash pine in nature and is generally considered to be airborne. Insects are probably involved in the etiology and may contribute to the cyclic nature of the disease. Controls have not yet been developed, although the possibility of selection and breeding of resistant pines is promising.

Literature Cited

1. Artman, J.E. 1973. Eastern white pine—a new host for *Fusarium lateritium* f. *pini.* Plant Dis. Rep. 57:182–184.
2. Barrows, J.B., and L.D. Dwinell. 1978. Decay of gladiolus corms caused by the pine pitch canker fungus, *Fusarium moniliforme* var. *subglutinans.* Phytopathol. News 12:174 (Abstr).
3. Berry, C.R., and G.H. Hepting. 1959. Pitch canker of southern pines. USDA For. Serv. For. Pest Leafl. 35. 3 p.
4. Bethune, J.E., and G.H. Hepting. 1963. Pitch canker damage to South Florida slash pine. J. For. 61:517–522.
5. Blakeslee, G.M., and D.L. Rockwood. 1978. Variation in resistance of slash pine to pitch canker caused by *Fusarium moniliforme* var. *subglutinans.* Phytopathol. News 12:207 (Abstr.).
6. Blakeslee, G.M., S.W. Oak, W. Gregory, and C.S. Moses. 1978. Natural association of *Fusarium moniliforme* var. *subglutinans* with *Pissodes nemorensis.* Phytopathol. News 12:208 (Abstr.).
7. Blakeslee, G.H., S. Kratka, R.A. Schmidt, and C.S. Moses. 1978. Sporodochia of the pitch canker fungus *(Fusarium moniliforme* var. *subglutinans)* as found on diseased slash pine in Florida. Plant Dis. Rep. 62:656–657.
8. Booth, C. 1971. The genus *Fusarium.* Commonw. Mycol. Inst., Kew, Surrey, England. 237 p.
9. Clapper, R.B. 1954. Stimulation of pine oleoresin flow by fungus inoculation. Econ. Bot. 8:269–284.
10. Dwinell, L.D. 1976. A shoot dieback of loblolly pine in seed orchards. Proc. Amer. Phytopathol. Soc. 3:335 (Abstr.).
11. Dwinell, L.D. 1978. Susceptibility of southern pines to infection by *Fusarium moniliforme* var. *subglutinans.* Plant Dis. Rep. 62:108–111.
12. Dwinell, L.D., and J.B. Barrows. 1978. Recovery of the pine pitch canker fungus from pine plantation and seed orchard soil. Phytopathol. News 12:207 (Abstr.).
13. Dwinell, L.D., and P.E. Nelson. 1978. Susceptibility of slash and loblolly pines to strains of *Fusarium moniliforme* and its variety *subglutinans.* Phytopathol. News 12:207 (Abstr.).
14. Dwinell, L.D., and W.R. Phelps. 1977. Pitch canker of slash pine in Florida. J. For. 75:488–489.
15. Dwinell, L.D., R.L. Ryan, and E.G. Kuhlman. 1977. Pitch canker of loblolly pine in seed orchards, p. 130–137. *In* Proc. 14th South. For. Tree Improv. Conf. (Gainesville, Florida, June 1977).
16. Fakir, G.A. 1973. Linear growth rate and morphology of four isolates of *Fusarium lateritium* f. *pini* Hepting. Bangladesh J. Bot. 2:67–72.
17. Hepting, G.H. 1954. Gum flow and pitch soak in Virginia pine following Fusarium inoculation. Southeast. For. Exp. Stn. Pap. 40. 8 p.
18. Hepting, G.H. 1961. *Pinus radiata* susceptible to pitch canker. Plant Dis. Rep. 45:889–890.
19. Hepting, G.H. 1971. Diseases of forest and shade trees of the United States. USDA For. Serv. Agric. Handb. 386. 658 p.
20. Hepting, G.H., and E.R. Roth. 1946. Pitch canker, a new disease of some southern pines. J. For. 44:742–744.
21. Hepting, G.H., and E.R. Roth. 1953. Host relations and spread of the pine pitch canker disease. Phytopathology 43:475 (Abstr.).
22. Kraus Sch., H., and W. Witcher. 1977. Survey of pine pitch canker in South Carolina. Plant Dis. Rep. 61:976–978.
23. Kuhlman, E.G., L.D. Dwinell, P.E. Nelson, and C. Booth. 1978. Characterization of the *Fusarium* causing pitch canker of southern pines. Mycologia 70:1131–1143.
24. Laird, P.O., and C.W. Chellman. 1972. An evaluation of a pitch canker outbreak in a Florida slash pine plantation. USDA For. Serv., State & Priv. For., Southeast Area, Rep. 73-1-16. 7 p.
25. McGraw, J.R., R.C. Wilkinson, R.A. Schmidt, and E.M. Underhill. 1970. Tip moths and pitch

canker in Florida. Inst. Food & Agric. Sci., Univ. Fla. Gainesville. Prog. Rep. 76–1. 2 p.

26. Matthews, F.R. 1962. Pitch canker-tip moth damage association on slash pine seedlings. J. For. 60:825–826.
27. Nirenberg, H. 1976. Untersuchungen uber die morphologische und biologisches Differencierung in der *Fusarium*–Sektion Liseola. Mitt. Biol. Bundesanstalt Land.-und Forst. Berlin-Dahlem 169:1–117.
28. Overgaard, N.A., H.N. Wallace, C. Stun, and G.D. Hedet. 1976. Needle midge (Diptera: Cecidomagiidae) damage to loblolly pines in the Erambert Federal Seed Orchard, Mississippi. USDA For. Serv., State & Priv. For., Southeast. Area, Rep. 76–213. 11 p.
29. Phelphs, W.R., and C.W. Chellman. 1976. Evaluation of "pitch canker" in Florida slash pine plantations and seed orchards, 1976. USDA For. Serv., State & Priv. For., Southeast. Area. 22 p.
30. Schmidt, R.A. 1976. Pitch canker in Florida: History, current status, and future research. Recent developments in forestry research. Univ. Fla. Gainesville. Resour. Rep. 3. p. 54–57.
31. Schmidt, R.A., and E.M. Underhill. 1974. Incidence and impact of pitch canker in slash pine plantations in Florida. Plant Dis. Rep. 58:451–454.
32. Schmidt, R.A., R.C. Wilkinson, C.S. Moses, and F.S. Broerman. 1976. Drought and weevils associated with severe incidence of pitch canker in Volusia County, Florida. Inst. Food & Agric. Sci., Univ. Fla. Gainesville. Prog. Rep. 76–2. 4 p.
33. Snyder, W.C., E.R. Toole, and G.H. Hepting. 1949. Fusaria associated with mimosa wilt, sumac wilt, and pine pitch canker. J. Agric. Res. 78:365–382.
34. Stegall, W.A., Jr. 1966. Fruiting of *Fusarium lateritium pini* on naturally infected pine pitch cankers. Plant Dis. Rep. 50:476–477.
35. Wilkinson, R.C., E.M. Underhill, J.R. McGraw, W.L. Pritchett, and R.A. Schmidt. 1977. Pitch canker incidence and fertilizer-insecticide treatment. Inst. Food & Agric. Sci., Univ. Fla. Gainesville. Prog. Rep. 77–1. 4 p.

19 The Fusarium Diseases of Turfgrass

Patricia L. Sanders and Herbert Cole, Jr.

Pathogenic species of *Fusarium* are commonly found in soil, surviving primarily as chlamydospores or as saprophytes on organic debris (69). In established stands of turfgrass, any given tiller is probably closely associated with propagules of these pathogens during most of its development. The three diseases of turfgrass commonly attributed to *Fusarium* spp., Fusarium patch, silver-top, and Fusarium blight, are found on stressed, slowly growing grass or on plants that have been wounded. Fusarium patch is most severe when grass is dormant, and, as such, constitutes a disease of a rather passive host. When the silver-top symptom is incited by *F. poae*, the fungus apparently must be introduced in a wound just distal to the terminal node. Fusarium blight similarly occurs on turfgrass which is under stress, typically moisture and/or heat stress. Thus, it would appear that *Fusarium* spp. are not particularly aggressive pathogens on turfgrass, requiring rather specific and narrow environmental conditions for pathogenesis to occur. When these conditions are met, however, the *Fusarium*-incited diseases of turfgrass can be devastating.

Details of the development of Fusarium patch have been understood for over 50 years. Fusarium blight and silver-top, however, have been the subjects of much controversy since they were first described. Fusarium blight, in particular, remains an enigma, and much of the current knowledge about the disease is open to question. Field development of Fusarium blight is not well-understood, particularly the factors responsible for the field symptoms. In the case of silver-top disease, controversy has revolved around the identification of the causal agent(s). At this time, all researchers do not agree on the identity of a causal agent.

Fusarium Patch

Introduction

Fusarium patch of turfgrass (pink snow mold, Fusarium snow mold) is a significant disease problem on amenity and recreational turf in areas where temperatures frequently fall to 0 C. It is an important problem in the maintenance of fine turf in the northern U.S. and Canada, and is considered to be one of the major diseases of turfgrass in the Pacific northwest areas of the U.S. and Canada. It occurs during the late fall, winter, and early spring in these areas, and is called snow mold because of its association with snow. This name is misleading because the presence of snow is not necessary for the development of this disease.

Fusarium patch of turfgrass has been recognized since the mid-1800s. Some of the earliest studies of the disease and its causal organism were done by Fries (25), Unger (82), Porkorny (53), Fuckel (26), and Sorauer (70, 71, 72). In 1933, Dahl (23) published a complete, detailed study of Fusarium snow mold of turfgrass, and most of this information is still pertinent.

Disease Description

The *Fusarium* spp. which incite this disease grow on turfgrass in prolonged cool, wet weather or under snow cover, and the symptoms are usually apparent in autumn before snow fall and/or as soon as the snow melts. The disease appears as irregular circular patches of dead, matted grass, from 5 cm to 0.5 m or more in diameter. In severe infections, these patches may coalesce, covering large areas. The patches are a bleached gray-white color, and may appear slimy when wet. When conditions are favorable for development, white aerial mycelium may be visible on the infection centers. With exposure to light, this mycelium may be a faint pinkish color. When disease incidence is mild, grass may recover, but under severe disease conditions, the pathogen invades and kills the entire plant, leaving circular patches of dead turfgrass.

Fusarium patch is distinguishable from Typhula snow mold by the faint pink color of the aerial mycelium, absence of the brown sclerotia characteristic of *Typhula,* and the presence of large numbers of crescent-shaped macroconidia characteristic of Fusaria.

The disease occurs on most genera of grass used in amenity and recreational turf situations in northern and temperate regions. Host genera include bentgrass (*Agrostis* spp.), Bermudagrass (*Cynodon* sp.), fescues (*Festuca* spp.), velvetgrass (*Holcus* sp.), ryegrasses (*Lolium* spp.), and bluegrasses (*Poa* spp.) (17).

Causal Organism(s)

Fusarium patch is incited chiefly by *F. nivale,* but Schaffnit (56) was able to produce snow-mold-like symptoms by inoculating turfgrass with *F. roseum* 'Culmorum.' Fulton (27) also found isolates of *F. roseum* which were most virulent at 10 C on seedling bluegrass. Sanders and Cole (*unpublished data*) have isolated only *F. roseum* 'Culmorum' from snow mold outbreaks, and *F. tricinctum* has been implicated on rare occasions (P. E. Nelson, *unpublished data*). Thus, it would appear that, while the preponderance of Fusarium patch is incited by *F. nivale,* isolates of other *Fusarium* spp. may, on occasion, be involved.

In laboratory studies (56, 78), the temperature range for growth of *F. nivale* on culture media is 18–22 C. This may vary with the isolate, however, for an optimum range of 5–10 C is reported by Toussoun and Nelson (77). In greenhouse tests on seedling turfgrass (23, 39) and field studies (23, 42), the temperature range for optimum disease development on the host is 0–5 C. To explain the activity of *F. nivale* under snow cover during winter, Dahl (23) placed thermograph bulbs just beneath the surface of the turf, under the snow cover. The temperature remained almost constant at 0 C, regardless of the air temperature, which fell to a low of −30 C.

In laboratory studies *F. nivale* grew best at high water potentials and growth diminished markedly with water stress greater than −6 bars (10). This finding is supported by the fact that the pathogen flourishes in the humid atmosphere under snow and in cool, rainy weather.

Infection and Disease Development

Fusarium nivale apparently is able to invade either above- or below-ground parts of the grass plant as the primary infection site. In greenhouse studies, Dahl (23) inoculated seedling bluegrass and fescue, and mature, field-grown creeping bentgrass by placing mycelia of test isolates grown on sterilized grass clippings on the foliage of test grass species. Infections developed on all plants in moist chambers at 0–5 C. In most cases, infection of the leaves had taken place within 2 days after inoculation. Histological studies of infected leaves showed that the mycelium developed along the outside of the leaf, penetrated stomatal openings, and moved into the intercellular spaces. Growth of the pathogen in the tissue was intercellular and was followed by collapse of the cells. The fungus then filled the space occupied by the collapsed cells and penetrated the vascular system. In severely affected leaves, sporodochia developed through the stomatal openings.

In greenhouse inoculation studies, Holmes and Channon (40) introduced mycelial macerates or spore suspensions of *F. nivale* into holes 2.5 cm deep in the soil between perennial ryegrass seedlings. The inoculated seedlings were subjected to four successive 12-hr periods at 0 C. Spore suspensions failed to incite infection, but mycelial macerates introduced into the soil caused considerable reduction in root number and length, and concomitant losses in shoot height and weight. Seven days after inoculation, many of the infected seedlings had collapsed with necrotic leaves. Holmes and Channon report that reasons for the failure of the spore inocula to incite disease are not clear, but they speculate that the lack of infectivity may be related to spore concentration or experimental environment.

High levels of nitrogen and late fall nitrogen fertilization increased the severity of injury by *F. nivale* (8, 23, 32, 44, 81). This increase in disease severity may vary with the grass species and nitrogen source. Brauen (8) superimposed various sulfur treatments across nitrogen applications and found that applications of 112 and 224 kg/ha annually of sulfur (93% wettable) prevented the disease increase that was coincident with the application of nitrogen as urea. No Fusarium patch was observed at any time during this study in the plots treated with high amounts of sulfur.

The use of straw mulch and the accumulation of a heavy thatch layer have also been linked with increased snow mold damage (17). The mulch and thatch layers may provide the insulation necessary to maintain winter temperature in the thatch at the 0–5 C optimum for growth of the pathogen, even in the absence of snow cover.

Control

The control of Fusarium patch is complicated by the fact that there may be snow-mold-inciting fungi other than *F. nivale* causing disease in the same place at the same time. It is important to know the composition of the pathogen population, since control measures for Fusarium patch may not be effective against all fungi involved in a mixed snow mold infection.

Management practices aimed at reducing predisposition of turfgrass to Fusarium patch include avoiding late fall nitrogen application, straw mulch, thatch accumulation, and snow buildup in areas with a history of the disease. Apart from these practices, spring and late fall application of fungicides are the most prevalent method of controlling Fusarium patch. For many years, organic and inorganic mercury compounds and cadmium salts were the mainstays of chemical control of this disease (34, 36). At the time this review was prepared, mercury fungicides were registered by the U.S. Environmental Protection Agency for the control of snow mold on fine turf, but their use for this purpose has been banned by certain states. Pentachloronitrobenzene (PCNB) and dithiocarbamate fungicides are effective when applied on regular 10–14 day schedules. Because it may be impossible to adhere to such rigorous application schedules, the control emphasis since 1967 has been on the more persistent systemic benzimidazole fungicides (33).

Results of field trials for chemical control of Fusarium patch (35, 87) showed that satisfactory control of the disease could be obtained with a diverse group of chemicals, including benzimidazoles, thiophanates, mercurials, dithiocarbamates, PCNB, and two experimental chemicals which give promise of excellent control of Fusarium patch. Perhaps the wisest course at this time is a program alternating the several classes of chemicals which are presently available for control of Fusarium patch. This allows the turf manager greater flexibility in schedules and materials, and reduces the possibility of the buildup of benzimidazole-resistant components of the pathogen population.

Fusarium Blight

Introduction

Fusarium blight is a severe mid-summer disease of Kentucky bluegrasses (*Poa pratensis*) on home lawns, golf course fairways, and other recreational areas. The use of the bluegrass cultivar Merion is widespread on such areas because of its resistance to other diseases, vigor of growth and recovery, and excellent play characteristics. Merion is particularly susceptible to Fusarium blight, and, because of its widespread use, loss of recreational and amenity turf to this disease can be severe in mid-summer. Fusarium blight is also an important concern of the sod production industry, and is causing increasing problems for these enterprises.

The disease was first described by Couch and Bedford (19), who observed it on Merion Kentucky bluegrass in southeastern Pennsylvania. During 1960, 1961, and 1963, the disease became epiphytotic on bluegrass and bentgrass stands in Ohio, New York, New Jersey, Delaware, Maryland, and the District of Columbia (19). Bean (3) observed a similar disease on Merion bluegrass lawns in the Washington, D.C., area. Subirats and Self (75) described Fusarium blight on centipede grass in Alabama, and Bell (6) reported a *Fusarium*-incited disease of zoysia.

Since Couch and Bedford (19) first described Fusarium blight, a great deal of research effort has been expended in attempts to understand and control this disease. Much of the research to date has led to contradictory conclusions, and control of Fusarium blight is still a difficult problem. In spite of intensive study, many aspects of Fusarium blight remain unexplained, including the role that Fusaria play in the development of the field symptoms. At this time, it is not possible to say with confidence what biotic and environmental factors may be involved in the etiology of Fusarium blight.

Disease Description

Couch and Bedford (19) reported that disease symptoms initially appeared in the field as scattered light green circular patches 5–15 cm in diameter. In later stages of disease development the patches were 0.3—1 m in diameter, light tan in color, and often contained center tufts of unaffected grass, producing a characteristic "frog-eye" appearance. In addition to "frog-eyes," affected areas also appeared as crescents, streaks, and circles. Couch (18) postulated that both a leaf-blighting and a root rot phase may occur.

Bean (3, 4) reported that Fusarium blight infection centers were small (dollar spot size) at the outset and increased in size as the disease progressed. In his observations, "frog-eyes" were rare, and the disease in the field was confined primarily to the crown area. Endo and Colbaugh (24) described Fusarium blight of Kentucky bluegrass in California as being confined to the stem base and crown. "Frog-eyes" were not observed, and all plants within a blighted area were killed.

Fulton (29), describing Fusarium blight in Pennsylvania, reported that the first symptoms noted were wilt and darkening of scattered areas of turf in the heat of the day.

Within 48 hr, these areas took on a permanently wilted, gray-green appearance, and then collapsed and bleached to a tan color. "Frog-eyes" were common, and the size of the affected area did not increase from the time the darkened patch was observed to the time the plants collapsed and turned tan. In severe infections, these patches coalesced to form large areas of blighted turf.

Leaf lesions in the field are reported to have light tan centers surrounded by a light brown margin of infected tissue (3). Leaf lesions produced by inoculation of greenhouse-grown bluegrass seedlings with *Fusarium* (19, 29) are initially dark green, fading to tan, sometimes with a red-brown margin.

Leaves, crowns, and roots of Merion Kentucky bluegrass from field infection centers of Fusarium blight were examined histologically (55). Specimens from the "frog-eye," dead ring, and periphery of the dead ring exhibited varying degrees of vascular plugging in roots and crowns. A coarse, septate mycelium atypical of *Fusarium* in morphology and staining reaction was noted in the cortex of the crowns and in the cortex, perivascular fibers, endodermis, and xylem vessels of roots from the "frog-eye" and the dead ring. The fungus was most prevalent in specimens from the dead ring, and was never found in crowns from symptomless grass.

Causal Organisms

Couch and Bedford (19) identified the Fusarium blight pathogens as *Fusarium roseum* f. sp. *cerealis* 'Culmorum' and *Fusarium tricinctum* f. sp. *poae*. Following an intensive survey of Fusarium blight in the Washington, D.C., area, Bean (3) found that *F. roseum* f. sp. *cerealis* 'Culmorum' could be readily isolated from crowns and leaf lesions, but not from root tissue. Subirats and Self (75) isolated *F. roseum* 'Culmorum' from a leaf blight and crown rot of centipede grass, and Endo and Colbaugh (24) found it associated with Fusarium blight of bluegrass in California. Sanders et al. (55) isolated *F. roseum* 'Acuminatum,' *F. roseum* 'Crookwell,' and *F. roseum* 'Equiseti' from crowns of bluegrass exhibiting symptoms of Fusarium blight in Pennsylvania.

Agriculture Handbook 165 (1) lists *F. scirpi* var. *acuminatum* as associated with rot of secondary roots of Kentucky bluegrass and creeping bentgrass and *F. culmorum* as associated with root rot and snow mold of colonial bentgrasses. Couch (17) stated that *F. roseum* f. sp. *cerealis* was the cause of root rot of turfgrasses, including bentgrass, bluegrass, and fescue. Cole et al. (13) isolated *F. roseum* and *F. tricinctum* from 12 of 22 seed lots of the Kentucky bluegrass cultivars Delta, Merion, Park, and Newport.

Couch and Bedford (19) concluded that *F. roseum* and *F. tricinctum* were the primary incitants of Fusarium blight, based on (i) consistency of isolation over a broad geographic area for several growing seasons, (ii) the high degree of pathogenicity shown by these isolates on greenhouse-grown seedling grass, and (iii) the lack of pathogenicity or inconsistency of isolation of other microorganisms. Koch's postulates have never been completed for Fusarium blight because attempts to produce the field symptoms on mature bluegrass with consistency by inoculation or environmental manipulation have been largely unsuccessful. Pathogenicity studies with Fusaria have usually been done by foliar inoculation of immature, greenhouse-grown grasses (13, 19, 22, 29, 45). Occasionally, greenhouse inoculation of grasses with Fusaria have been attempted by placement of the fungal propagules in the root zone (4, 6, 55). In all but one instance (29) greenhouse studies have not resulted in disease development that paralleled development in the field. Without a reliable inoculation procedure, it is not possible to determine with certainty whether the Fusaria that are regularly isolated from bluegrass showing symptoms of Fusarium blight are indeed the causal pathogens. A study by Smiley et al. (62), which attempted to determine the relationship between the numbers of Fusaria isolated from bluegrass areas and the incidence of Fusarium blight, revealed a negative correlation between these two parameters.

Host and Cultivar Susceptibility

Fusarium blight is not a serious problem on most cool season Kentucky bluegrasses in areas where summer stress conditions are not severe. It is in the long, hot summers of the transition zone that the disease can assume epiphytotic proportions. Funk (30) observed that very few of the best bluegrass cultivars have the qualities which allow production of an attractive, compact, dense, disease-free turf in severe summer environments. He therefore suggests that susceptibility to Fusarium blight is the result of an interaction between the turfgrass cultivar and the environment under which the grass is grown. Turgeon (79) and Funk (30) have presented data on the relative susceptibility of many bluegrass cultivars and blends to Fusarium blight. Turgeon and Meyer (80) have reported that the incidence of Fusarium blight varied with bluegrass cultivar, mowing height, and fertilization level. In their studies, the order of cultivar susceptibility, from highest to lowest, was: Kenblue, Fylking, Pennstar, Merion, and Nugget. Funk (30) observed that cultivars adapted to higher mowing were severely blighted when clipped closely.

Fulton (29) reported that, in greenhouse studies, Merion was more susceptible than Fylking or Delta. Couch and Bedford (19), in growth chamber studies with 6-week-old seedlings of Highland colonial bentgrass, Merion Kentucky bluegrass, and Pennlawn creeping red fescue, reported that Highland was most susceptible, followed by Merion, and then Pennlawn.

Disease Development

Bean (4) postulated that chlamydospores in soil provide the initial inoculum for early-season infection. Couch (18) suggested that the source of initial inoculum was dormant mycelium in infected grass plants and debris.

Funk (30) presented a list of factors which might predispose turfgrass to Fusarium blight. These included high temperature, high humidity, recurring drought stress, reduced air circulation, excessive nitrogen, dense lush growth, thatch, close mowing, nematodes, and other diseases. Partyka (51) states that, since the *Fusarium* pathogens were present in most turf areas, infection and disease development were related to stress conditions, primarily drought and high temperature. He noted that anything restricting roots, such as compaction, shortage of nutrients, thatch development, nematodes, or insects, could predispose turf to drought stress and Fusarium blight development. He further stated that anything impeding water penetration, such as thatch, slope, or compacted soil, may induce water stress conditions.

Bean (4) concluded that severity of Fusarium blight was directly proportional to intensity and duration of sunlight. The disease was never seen in heavy shade. Pure stands of Merion were most severely infected, especially in sunny, sloping locations.

Most research into the effect of temperature indicates that Fusarium blight incidence and severity increases as summer temperatures increase. Studies employing artificial inoculation of greenhouse-grown seedling grass showed that disease was most severe at temperatures between 27 and 33 C, and that severity increased as temperatures increased (22, 24, 29, 75). In the field, Bean (3) observed that the disease was active only during the warmest part of the summer, and Fulton (29) states that temperatures of 27 C or above are necessary in the 7 to 14 days prior to the development of Fusarium blight symptoms. However, a survey of weather records for 1966 and for 1973–1978 (63) indicated that there was no apparent relationship between disease outbreaks and summer temperatures.

Most workers report drought stress as an important factor in the etiology of Fusarium blight (4, 19, 22, 24, 30, 51, 76, 83). Bean (4) postulated that moisture stress was essential for the disease to occur. In greenhouse experiments with artificially inoculated seedling grass, employing soils with moisture equivalents of 20–80%, Troll (76) found infection most severe on plants growing on soil with the lowest moisture equivalents. Cutwright and Harrison (22) studied the effect of environmental factors on Fusarium infection of greenhouse-grown grass. Disease severity was greatest under low irrigation at all inoculum levels. Couch and Bedford (19) found no significant differences in disease incidence between any of the *Fusarium* isolate/soil moisture regimes investigated in the greenhouse. In a field irrigation plot experiment (24), as the amount of irrigation increased from marginal to abundant, Fusarium blight decreased from 40% of the plot area affected to less than 1%. Troll (76) observed that Fusarium blight was severe in Massachusetts during the moisture stress summer years of 1960–1966, but was far less severe during the wet summer of 1967; a Merion bluegrass home lawn was symptomless in 1967, but symptoms occurred again in 1968, a year with a dry summer. Fulton (29) observed that outbreaks of Fusarium blight in Pennsylvania are almost always preceded by a period of warm, wet weather followed by dry weather. A survey of weather data in New York (63) indicated, however, that Fusarium blight outbreaks always occurred soon after a major rain.

Fusarium blight appears to be most prevalent on older, established bluegrass areas. Bean (4) observed that a stand must be 3 years old or more before the disease will occur, and Fulton (29) reported that Fusarium blight in Pennsylvania was common on turf stands older than 5 years and was seldom seen on stands less than 2 years old. Cutwright and Harrison (21) observed Fusarium blight on stands of Merion that were less than 2 years old. Sanders and Cole (*unpublished data*) have observed the disease on stands of bluegrass less than 1 year old.

There is a commonly held belief among those who work with turf that thatch accumulation predisposes turfgrass to Fusarium blight, but there is little research evidence to support this hypothesis. Bean (4), Turgeon (79), and Smiley and Craven (58) found no relationship between thatch depth and Fusarium blight severity. Turgeon (79) observed no relationship between the thatching tendency of a bluegrass cultivar and its susceptibility to Fusarium blight. On a sodded turf area, Fusarium blight was most severe where the thatch layer was thinnest (58). Smiley et al. (62) suggested that disease incidence may be related to the thatch decomposition rate. Turgeon (79) and Funk (30) have observed coincident increase in thatch and Fusarium blight severity on field plots treated with tricalcium arsenate.

There appears to be a consistent relationship between severity of Fusarium blight and level of nitrogen fertilization. Endo and Colbaugh (24) found that the disease was most severe in the field on plants receiving high levels of nitrogen and low amounts of water. Turgeon (79) reported that certain bluegrass cultivars which had little disease at low levels of nitrogen (1 kg/93 m^2) were severely blighted at high levels (3–4 kg/93 m^2). Bean (4), however, observed no correlation, in the Washington, D.C., area, between soil nitrogen and disease severity. Greenhouse studies (19, 22) showed increased susceptibility to Fusarium infection when grass was grown under high levels of nitrogen.

Couch and Bedford (19) reported the results of a greenhouse investigation of the influence of normal and 0.1 normal levels of calcium on the susceptibility of foliage of seedling fescue, bluegrass, and bentgrass. In this study, disease incidence was significantly higher in plant groups grown under 0.1 X calcium. Bean (3) was unable to correlate disease severity in the field with low calcium levels. Applications of 1 kg hydrated lime/month/93 m^2 were found to significantly increase incidence of Fusarium blight (Cole and Sanders, *unpublished data*).

Partyka (51) has raised the possibility that predisposing root organisms may be involved under certain conditions. He questioned whether organisms such as *Pythium* or *Rhizocto-*

nia may be present at low levels of activity early in the growing season and may be capable of weakening the turf so that Fusaria become established readily under favorable conditions.

Vargas and Laughlin (86) investigated the role of *Tylenchorhynchus dubius,* a migratory ectoparasitic stunt nematode, in Fusarium blight development on Merion Kentucky bluegrass. A survey by the authors indicated that Merion plants growing in blighted areas had poorly developed root systems and were associated with high soil populations of *T. dubius* (200–1500/100 cc soil). In two greenhouse experiments, seedling Merion bluegrass was grown in a sterile sand:loam:peat mix artificially infested with either *F. roseum, T. dubius,* or a combination of *F. roseum* and *T. dubius*. Based on the results of these experiments, Vargas and Laughlin (86) and Vargas (83) suggested that Fusarium blight may involve an interaction between *F. roseum* or *F. tricinctum* and *T. dubius,* or that *T. dubius* alone may be responsible. Couch and Bedford (19), however, were unable to associate high levels of parasitic nematodes with incidence of Fusarium blight. Cole observed that levels of parasitic nematodes were low on two Pennsylvania golf courses with histories of severe Fusarium blight. Vargas et al. (84) and Smiley et al. (64) have reported field control of Fusarium blight with nematicide application. Couch (20) obtained no disease suppression in field trials employing four nematicides.

There is a lack of consistency in results of research done at varying places and times to determine the effect of specific environmental stresses on the development of Fusarium blight. Recently, Smiley (*personal communication*) suggested that, since response to stress at the cellular level is similar regardless of the stress, Fusarium blight may be triggered by stress in general, rather than a particular stress or stresses.

Fusarium Blight on Newly Sodded Areas

The use of sod to establish and repair lawns and recreational areas has become increasingly prevalent. Newly installed sod can be destroyed rapidly by Fusarium blight, with severe infections occurring in sod that has been established only 1–2 years. Purchasers of such sod believe that they have been sold "diseased grass." The question raised in such situations is whether the nature of the sod-growing site or the installation site was responsible for the disease outbreak. Bean (4) reported that approximately the same number of *Fusarium* propagules were recovered from a sod-growing site, the blighted areas in the installation site, and the blight-free areas in the installation site. Thus it would appear that potentially pathogenic Fusaria occur with approximately equal frequency in both the sod-growing and the installation sites. There may be stress factors in the transportation of sod, as well as in the environment of the installation site, which can predispose newly laid sod to Fusarium blight. Partyka (51) suggested that the drying or heating of sod in transit may allow Fusaria subsequently to cause disease. He also observed that sod laid down on dry soil or not watered adequately could be so stressed. He further stated that another source of stress was poor permeation of water or capillary action at the sod/soil (clay) interface, which might result in a poor root system and resultant stress. Bean (3) reported that when turf was removed from *Fusarium*-blighted areas and these areas resodded with 1-year-old healthy sod, the newly sodded areas were as heavily blighted as the original turf by the following summer. Cole found that when *Fusarium*-blighted sod was transferred to a blight-free site, the Fusarium blight symptoms abated and did not return.

Control

Because of the regular association of Fusarium blight with hot, dry environments, Bean (4) recommended heavy watering for disease suppression. There is research evidence (24) that adequate irrigation suppresses Fusarium blight; unfortunately, such treatment also encourages invasion of quality bluegrass turf by annual bluegrass (*Poa annua*) and other weed grasses.

Funk (30) pointed out that Kentucky bluegrass cultivars originating in areas of cool, moist summers were often severely damaged by the disease, and suggested that Fusarium blight resistance is to be found in "southern turf-type" bluegrass cultivars which are heat and drought tolerant. He stated that these cultivars should do well in the southern transition zone of bluegrass adaptation where Fusarium blight is most damaging. He also suggested the overseeding of turf-type ryegrasses into blighted bluegrass stands, adding the caution that improved resistance to *Pythium* and *Rhizoctonia* was needed for good summer performance of ryegrass in the humid summer heat stress region.

Turgeon (79) suggested, in addition to the "plant-oriented" control approach of resistant Kentucky bluegrass cultivars, an "environment-oriented" approach of avoiding excessive nitrogen fertilization during spring, providing adequate moisture for turfgrass survival during stress periods through irrigation, performing appropriate cultivation practices to control thatch and alleviate soil compaction, and applying effective fungicides.

In the period from 1966 through 1976, many chemical control experiments involved the contact fungicides, such as mancozeb (Dithane M-45), thiramorganic mercury (Tersan OM), anilazine (Dyrene), Difolatan, chlorothalonil (Daconil), and others (3, 5, 19). Although a partial level of control was obtained in a few instances, the level of disease suppression was not satisfactory for practical use. For example, Bean et al. (5) showed that a single application of hydrated lime (11 kg/93 m^2) reduced Fusarium blight four fold, and Tersan OM applied in weekly sprays provided significant disease suppression. On the other hand, Difolatan, Dyrene, Dithane M-45, Daconil 2787, and Panogen turf fungicide provided no significant benefit. In spite of favorable small plot results, however, neither hydrated lime nor Tersan OM were successful on a commercial scale. Applications of hydrated lime (1 kg/month/93 m^2) were found by Cole and Sanders (*unpublished data*) to increase significantly the incidence of Fusarium blight.

Soil fumigation to control Fusarium blight was tested by Cutwright and Harrison (21). Sod was stripped from a test area that had been severely blighted, replicated plots were fumigated with three soil fumigants, and the area was seeded with Merion Kentucky bluegrass 2 weeks after fumigation. Only a few scattered *Fusarium*-blighted areas developed in the experimental area during the following summer. The disease ratings for 1968, however, indicated that none of the treatments effectively controlled the disease over even a 2-year period.

The introduction of systemic fungicides provided the first effective chemical treatment. Cutwright and Harrison (21) reported excellent control of Fusarium blight with weekly sprays of benomyl. Muse (46) reported control of Fusarium blight with thiophanate methyl and benomyl applied as drenches, and Fulton et al. (28) obtained suppression of the disease with pre-symptom application of foliar sprays of triarimol and benomyl, followed by a drenching water irrigation. Vargas and Laughlin (85) reported excellent control of the disease with regular preventive applications of benomyl as a drench (foliar sprays were not effective), but no control with Dithane M-45 or thiabendazole, regardless of mode of application. In the ensuing years, others (9, 14, 15, 31, 39, 49, 65, 66, 85) have obtained successful control of Fusarium blight through preventive and curative applications of benzimidazoles, triarimol (EL 273), and fenarimol (EL 222). However, control failures were reported with benzimidazole-derivative fungicides due to apparent pathogen resistance to the fungicides (12, 67).

During the 1977 and 1978 growing seasons, excellent control of Fusarium blight was obtained with the experimental fungicides, BC 6447 (11, 43, 47), CGA 64251 (43, 47, 59), DPX 4424 (59), and RP 26019 (11, 59, 64, 68). Studies with triadimefon (BC 6447) (54) and iprodione (RP 26019) (60, 61) demonstrated no in vitro fungitoxicity to *Fusarium* spp.

Iprodione was found to increase sporulation of *Fusarium* spp. in vitro and to increase the number of Fusarium propagules in treated field soil (60, 61).

Discussion

The literature on Fusarium blight is extensive, which attests both to the tenacity of turf researchers and the elusive nature of the problem. Many aspects of Fusarium blight remain unexplained, and information is still lacking in several critical areas. Much of the experimental work has been done in moist chambers with seedlings, while in the field the disease is most damaging on mature bluegrass under stress. Relating growth chamber experiments on seedlings to field situations involving mature stands is precarious. Although potentially pathogenic *Fusarium* spp. are regularly isolated from turfgrass showing symptoms of Fusarium blight, no one has been able to induce the field "frog-eye" or "dead ring" symptom by inoculating turfgrass with these Fusaria. The experimental fungicide triadimefon provided complete suppression of the field symptoms of Fusarium blight, but exhibited a lack of in vitro fungitoxicity to *Fusarium* spp. isolated from diseased turfgrass (54). Fungitoxicity to these pathgens could not be demonstrated in bioassays of Kentucky bluegrass which had been treated with triadimefon, but fungitoxicity to other turfgrass pathogens was readily demonstrated with both techniques. Thus, one can reach the conclusion that the designated pathogen(s) have not incited the field symptom, and the chemical which suppresses the field symptom does not control the pathogen(s). More likely these results indicate that *Fusarium* spp. are pathogenic within an unexplained biotic or environmental complex which produces the field symptoms of Fusarium blight.

The literature indicates that the majority of the Fusarium blight outbreaks are associated with moisture stress. Work (16, 50) on plant water stress and soil water potential and the effects of these factors on colonization of soil organic matter and host plants by *Fusarium* spp. may be relevant to understanding Fusarium blight. Foot rot of dryland wheat, incited by *F. roseum* f. sp. *cerealis* 'Culmorum,' was demonstrated to be most severe under high nitrogen regimes and in seasons of drought stress. *Fusarium* spp. were able to colonize organic matter under low soil moisture conditions that inhibited the activity of antagonistic microorganisms. In a like manner, the fungi were able to colonize wheat plants rapidly under internal moisture stress. High nitrogen fertilization increased moisture stress within wheat plants, accentuating colonization by *Fusarium* spp. This increased internal water stress probably was not due to a direct nutritional effect, but rather to indirect effects produced by larger root systems, which increased water extraction from the soil, and greater leaf areas, which increased transpiration. The "frog-eye" or "dead ring" field symptom and the sharp delineation of the infection center borders often occur over a very short distance, an abrupt rather than a gradual transition from healthy grass to severely blighted grass. One tiller may be dead and the next one apparently healthy, producing a distinct "dead spot" on the turf. If water stress is critical in disease development, then why, over a very short distance, should one tiller be under water stress and an adjacent one apparently free from such stress? This symptom has never been observed on turfgrass grown in a greenhouse or a growth chamber.

Soil moisture availability, as affected by proliferation of soil Basidiomycetes, may influence the pattern of disease development in the field, especially the "frog-eye" symptom. Shantz and Piemeisel (57) reported that in fairy ring development, soil moisture levels in the actual ring portion were lower and water percolation rates were reduced compared with the ring interior or exterior. "Hard dry spot" associated with fungal proliferation has long been observed on bentgrass greens by golf course superintendents. Although other factors may contribute to soil and plant moisture stress from place to place in a turf area, the radial expansion pattern of Basidiomycete fungi in soil and the hydrophobic nature of

Basidiomycete mycelium and by-products (37) may present one explanation for the "dead ring" and "frog-eye" symptom.

Growth chamber studies indicate that bentgrass is most susceptible to Fusarium blight, while field observations suggest that the disease is primarily a problem on bluegrass. Traditionally, bentgrass is grown under adequate, often excessive, irrigation, especially during mid-summer, but bluegrass fairways are traditionally kept "dry" to minimize invasion by undesirable grasses. This difference in irrigation practices may explain the absence of Fusarium blight in most bentgrass areas.

Fusarium blight, and other soil-borne diseases, are increasingly becoming a serious threat to turfgrass, particularly at a time when the soil microflora is being modified by the use of long-lasting systemic fungicides. Elucidation of the etiology of Fusarium blight is essential to the development of soundly based control practices, and may lead to the development of research protocols which can be utilized in the study of other soil-borne diseases of turfgrass.

Silver-Top

Introduction

Silver-top (white-heads, white top) is a disease of many wild and cultivated grasses in which the flowers on some or all panicles are sterile, and the affected panicle is white or pale green. Silver-top is one of the most serious problems in turfgrass grown for seed production. Outbreaks of silver-top have been known to cause up to 85% yield reduction, and complete loss of seed crops has been reported in stands over 3 years old (41).

Disease Description

Symptoms appear soon after head emergence, usually before the seed head has fully expanded. The seed head and the stem above the terminal node are bleached and dead, while the flag leaf and sheath, the stem, and the remainder of the plant below the terminal node show no symptoms (7). The panicle can be pulled easily from the flag leaf sheath, breaking about 3 mm above the terminal node and revealing a shriveled or necrotic area about 13–16 mm in length on the stem above the region of the break. Measurements indicate that there are no differences in panicle length or weight between healthy and diseased panicles, but seed weights and germination studies reveal that silver-topped heads produce practically no normal seeds. This would appear to indicate that injury occurs after the head has emerged, but before pollination (73).

The disease is ubiquitous on wild and cultivated grasses in Europe, Canada, and the U.S., and is of economic importance where fescue, Kentucky blue, bent, and other grasses are being grown for seed (41). Kentucky bluegrass (*Poa pratensis*) and red fescue (*Festuca rubra*) are particularly susceptible (7).

Causal Organisms and Disease Development

According to Stewart and Hodgkiss (74), the disease was first described in 1875 by Professor J. H. Comstock in his "Syllabus of a Course of Lectures." Since the initial description by Comstock, who reported that the symptoms were caused by thrips, the disease has been variously attributed to species of thrips, maggots, aphids, mites, true bugs, and to the fungus, *Fusarium poae.*

In 1891, Osborn (48) noted insect punctures in great abundance on affected plants in

Canada, and concluded that the silver-top injury could be produced by sucking insects, such as species of *Homoptera* or *Hemiptera.* While the injury caused by feeding of sucking insects just above the terminal node was more conspicuous, i.e., silver-top, these insects caused similar injury in older plant parts, but the resultant shrivelling was confined to the few cells immediately surrounding the feeding puncture. In those instances where the insect beak was thrust into the succulent area above the terminal node, the effect was to kill the cells of the area through which the sap nourishing the terminal internode passed, resulting in death of the plant parts distal to the injury.

In 1908, Stewart and Hodgkiss (74) reported the occurrence of the fungus, *Sporotrichum poae* (*F. poae*), and the mite *Pediculopsis graminum* (*Siteroptes graminum*), on silver-topped grass in New York State. They concluded that the association of the fungus and the mite with the silver-top symptom was not consistent, since the majority of affected culms which they examined were free from *Sporotrichum* (*Fusarium*). Further investigation revealed that the *Sporotrichum* (*Fusarium*) and the mite found on affected grass were identical with the *Sporotrichum* (*Fusarium*) and mite which cause bud-rot of carnation.

Keil (41) isolated *F. poae* from grasses with silver-top symptoms, inoculated healthy grass with the fungus, obtained silver-top symptoms, and was able to reisolate the fungus. He also obtained silver-top symptoms by allowing the mite, *Siteroptes graminum,* which was contaminated with *F. poae,* to feed on healthy grass plants. According to Peterson and Vea (52), J. M. Scholl also produced silver-top symptoms on healthy grass by inoculating turfgrass with *F. poae* isolated from affected plants. In both of these studies the affected culms showed necrosis or decay at the inoculation site, distinct from the shriveled culm which was observed in the field in Wisconsin and Minnesota.

Hardison (38) presented evidence that the mite, *S. graminum,* and the fungus, *F. poae,* were not the causal agents of silver-top in Oregon. He found that *S. graminum* was absent on 95% of the affected plants which he examined, and the disease was controlled by dichlorodiphenyltrichloroethane (DDT), which he did not consider an effective miticide. He isolated *F. poae* from less than 20% of the affected plants which he cultured. Based on the fact that *F. poae* was increasingly isolated as the season and the infections progressed, he concluded that *F. poae* was a secondary invader of dead stem tissue, rather than the primary cause of the silver-top symptom. Berkenkamp's work (7) suggested that *F. poae* was not the primary cause of silver-top in Canada, since the fungus was isolated from less than half of the affected stems cultured.

Starks and Thurston (73) investigated virus, genetic characteristic, and mechanical injury as causes of silver-top. No virus symptoms were obtained with mite transmission, mechanical inoculation, or transmission by dodder bridges. Affected plants sprayed with insecticides and kept in screened cages through the winter developed no silver-topped heads the following year. This result suggested that silver-top was not a genetic characteristic. A silver-top symptom, not entirely characteristic of the field symptom, was produced by mechanical injury of the stem above the terminal node.

Arnott and Bergis (2) noted minute punctures opposite the areas of injury in affected plants, and observed a high incidence of various species of plant bugs in fields where grass showed silver-top symptoms. Peterson and Vea (52), showed that a capsus bug, *Capsus simulans,* was an important factor in silver-top of bluegrass in Minnesota. Small numbers of capsus bugs were coincident with extensive injury in bluegrass seed fields, and, in cage experiments, a single capsus bug produced silver-top on healthy grass after only a 10-min feeding period. They concluded that *Capsus simulans* secreted a toxic material injected into the plant during feeding, which caused the stem injury and the symptom of silver-top.

Discussion

In reviewing the studies on silver-top of grasses to date, one is struck by the differing conclusions of the various investigators. The ascribed causal agents appear to vary as the environment or the symptom description varies. If silver-top is considered simply to be a devastating symptom on grasses which are grown for seed, regardless of the type of injury which produces it, a more consistent picture emerges. To quote Osborn (48), "It is evident, upon examination of the subject, that the same appearance of the grass may be produced by very different agents, provided they attack the same point in the stem. Any injury to the juicy base of the terminal node that cuts off the flow of the sap to the head during a certain stage of its growth must produce the withering and whitening so conspicuous in affected fields. Starting with this premise it is reasonable to conclude that the trouble *may* result from a number of different agents, and such, I believe, to be actually the case. . . ."

Control

Whether the silver-top symptom is caused by mechanical injury by feeding insects, toxic material injected by feeding insects, or *F. poae* inoculated by feeding mites, control measures are similar. In all cases, the requisite control is the destruction of the insect/mite or the overwintering eggs. Burning of grass fields after harvest is a common method for controlling silver-top. Burning destroys overwintering eggs and usually makes further control unnecessary (52). Air pollution legislation may shortly make this method of control impossible, leaving the use of contact and systemic insecticides and/or miticides as the only recourse.

Literature Cited

1. Anon. 1960. Index of plant diseases in the United States. USDA Agric. Handbook No. 165. U.S. Government Printing Office, Washington, D.C. 531 p.
2. Arnott, D. A., and I. Bergis. 1967. Causal agents of silver top and other types of damage to grass seed crops. Can. Entomol. 99:660–670.
3. Bean, G. A. 1966. Observations on Fusarium blight of turfgrasses. Plant Dis. Rep. 50:942–945.
4. Bean, G. A. 1969. The role of moisture and crop debris in the development of Fusarium blight of Kentucky bluegrass. Phytopathology 59:479–481.
5. Bean, G. A., R. N. Cook, and A. E. Rabbit. 1967. Chemical control of Fusarium blight of turfgrass. Plant Dis. Rep. 51:839–841.
6. Bell, A. A. 1967. Fungi associated with root and crown rots of *Zoysia japonica*. Plant Dis. Rep. 51:11–14.
7. Berkenkamp, B., and J. Meeres. 1975. Observations on silvertop of grasses in Alberta. Can. Plant Dis. Survey 55:83–84.
8. Brauen, S. E., R. L. Goss, C. J. Gould, and S. P. Orton. 1976. The effects of sulphur in combinations with nitrogen, phosphorus and potassium on colour and Fusarium patch disease of Agrostis putting green turf. J. Sports Turf Res. Inst. 51:83–91.
9. Brown, G. E. 1976. EL222 evaluated for control of Fusarium blight. Amer. Phytopathol. Soc. Fung. Nemat. Test Rep. 31:157.
10. Bruehl, G. W., and B. Cunfer. 1971. Physiologic and environmental factors that affect the severity of snow mold of wheat. Phytopathology 61:792–799.
11. Burpee, L. L., H. Cole, Jr., and P. L. Sanders. 1977. Control of benzimidazole-tolerant *Fusarium* spp. on annual and Kentucky bluegrass. Proc. Amer. Phytopathol. Soc. 3:302 (Abstr.).
12. Cole, H., Jr. 1976. Factors affecting Fusarium blight development. Weeds, Trees, and Turf, July, p. 35–37.
13. Cole, H., S. W. Braverman, and J. Duich. 1968. Fusaria and other fungi from seeds and seedlings of Merion and other turf-type bluegrasses. Phytopathology 58:1415–1419.
14. Cole, H., C. W. Goldberg, and J. M. Duich. 1973. Fusarium blight recovery following curative treatments with systemic fungicides. Amer. Phytopathol. Soc. Fung. Nemat. Test Rep. 28:122.
15. Cole, H., L. B. Massie, J. M. Duich, and T. C. Ryker. 1971. Fusarium blight control with Benlate. Amer. Phytopathol. Soc. Fung. Nemat. Test Rep. 26:123.
16. Cook, R. J., and R. I. Papendick. 1972. Influence

of water potential of soils and plants on root disease. Annu. Rev. Phytopathol. 10:349–374.
17. Couch, H. B. 1962. Diseases of turfgrasses. Reinhold Publishing Co., New York, N.Y. 289 p.
18. Couch, H. B. 1976. Fusarium blight of turfgrasses—an overview. Weeds, Trees, and Turf, July, p. 9, 34–35.
19. Couch, H. B., and E. R. Bedford. 1966. Fusarium blight of turfgrasses. Phytopathology 56:781–786.
20. Couch, H. B., J. M. Garber, and J. A. Fox. 1979. Relative effectiveness of fungicides and nematicides in Fusarium blight control. Golf Business, July, p. 12–16.
21. Cutright, N. J., and M. B. Harrison. 1970. Chemical control of Fusarium blight of 'Merion' Kentucky bluegrass turf. Plant Dis. Rep. 54:771–773.
22. Cutright, N. J., and M. B. Harrison. 1970. Some environmental factors affecting Fusarium blight of 'Merion' Kentucky bluegrass. Plant Dis. Rep. 54:1018–1020.
23. Dahl, A. S. 1934. Snowmold of Turfgrasses as caused by *Fusarium nivale.* Phytopathology 24:197–214.
24. Endo, R. M., and P. F. Colbaugh. 1974. Fusarium blight of Kentucky bluegrass in California. Proc. 2nd Int. Turfgrass Res. Conf., p. 325–326.
25. Fries, E. 1825. Systema orbis vegetabilis. Part 1:317–318. London.
26. Fuckel, L. 1869. Symbolae mycologicae. P. 142, 370. Wiesbaden.
27. Fulton, D. E. 1972. Etiological factors associated with Fusarium blight of turfgrass. M.S. Thesis. Pa. State Univ., University Park, Pa. 72 p.
28. Fulton, D., H. Cole, C. Goldberg, and L. B. Massie. 1972. Fusarium blight control with fungicides. Amer. Phytopathol. Soc. Fung. Nemat. Test Rep. 27:136–137.
29. Fulton, D. E., H. Cole, Jr., and P. E. Nelson. 1974. Fusarium blight symptoms on seedling and mature Merion Kentucky bluegrass plants inoculated with *Fusarium roseum* and *Fusarium tricinctum.* Phytopathology 64:354–357.
30. Funk, C. R. 1975. Developing genetic resistance to Fusarium blight. Weeds, Trees, and Turf, July, p. 41–44.
31. Goldberg, C. W., H. Cole, Jr., L. L. Burpee, and J. M. Duich. 1974. Fusarium blight control under minimal irrigation bluegrass culture. Amer. Phytopathol. Soc. Fung. Nemat. Test Rep. 29:118–119.
32. Goss, R. L., and C. J. Gould. 1968. Some interrelationships between fertility levels and Fusarium patch disease of turfgrasses. J. Sports Turf Res. Inst. 44:19.
33. Gould, C. J. and R. L. Goss. 1971. Comparison of light-frequent, heavy-infrequent, and alternating applications of fungicides to control Fusarium patch. Proc. 2nd Int. Turfgrass Conf., p. 339–343.
34. Gould, C. J., R. L. Goss, and V. L. Miller. 1961. Fungicidal tests for control of Fusarium patch disease of turf. Plant Dis. Rep. 45:112–118.
35. Gould, C. J., R. L. Goss, and W. E. Vassey. 1976. New fungicides for the control of *Fusarium nivale.* J. Sports Turf Res. Inst. 51:67–73.
36. Gould, C. J., V. L. Miller, and R. L. Goss. 1965. New experimental and commercial fungicides for control of Fusarium patch disease of bentgrass turf. Plant Dis. Rep. 49:923–927.
37. Griffin, D. M. 1969. Soil water in the ecology of fungi. Annu. Rev. Phytopathol. 7:289–310.
38. Hardison, J. R. 1959. Evidence against *Fusarium poae* and *Siteroptes graminum* as causal agents of silver top of grasses. Mycologia 51:712–728.
39. Hoadley, B., and L. T. Palmer. 1973. Fusarium blight control in Nebraska. Amer. Phytopathol. Soc. Fung. Nemat. Test Rep. 28:122–123.
40. Holmes, S. J. I., and A. G. Channon. 1975. Glasshouse studies on the effect of low temperature on infection of perennial ryegrass seedlings by *Fusarium nivale.* Ann. Appl. Biol. 79:43–48.
41. Keil, H. L. 1946. "White-heads" of grasses. Ph. D. Thesis. Pa. State Univ., University Park, Pa. 37 p.
42. Lebeau, J. B. 1964. Control of snow mold by regulating winter soil temperature. Phytopathology 54:693–696.
43. Loughner, D., P. L. Sanders, F. W. Nutter, Jr., and H. Cole, Jr. 1979. Suppression of Fusarium blight symptoms on Merion Kentucky bluegrass with single and double applications of fungicide, 1978. Amer. Phytopathol. Soc. Fung. Nemat. Test Rep. 34:144.
44. Madison, J. H., L. J. Peterson, and T. K. Hodges. 1960. Pink snow mold on bentgrass as affected by irrigation and fertilizer. Agron. J. 52:591–592.
45. Meyer, W. A., and F. H. Berns. 1976. Techniques for rapid determination of Fusarium blight susceptibility in Kentucky bluegrass selections. Weeds, Trees, and Turf, July, p. 44–46.
46. Muse, R. R. 1971. Chemical control of Fusarium blight of 'Merion' Kentucky bluegrass. Plant Dis. Rep. 55:333–335.
47. Nutter, F. W., Jr., P. L. Sanders, D. Loughner, and H. Cole, Jr. 1979. Control of Fusarium blight with fungicides under fairway conditions in mixed Kentucky bluegrass—annual bluegrass, 1978. Amer. Phytopathol. Soc. Fung. Nemat. Test Rep. 34:145.
48. Osborn, H. 1891. Silver top in grass and the insects which may produce it. Can. Entomol. 23:93–96.
49. Palmer, L. T. 1974. Fusarium blight control. Amer. Phytopathol. Soc. Fung. Nemat. Test Rep. 29:118.
50. Papendick, R. I., and R. J. Cook. 1974. Plant water stress and development of Fusarium foot rot in wheat subjected to different cultural practices. Phytopathology 64:358–363.
51. Partyka, R. E. 1976. Factors affecting Fusarium blight in Kentucky bluegrass. Weeds, Trees, and Turf, July, p. 37–38.
52. Peterson, A. G. and E. V. Vea. 1971. Silver top of bluegrass in Minnesota. J. Econ. Entomol. 64:247–252.
53. Pokorny, A. 1865. Notiz uber das diessjahrige massenhafte Auftreten des Schneeschimmels (*La-*

nosa nivalis Fr.) im Wiener Stadtpark. Verhandl. k. k. Zoologisch-bot. Gesell. in Wien 15:281–286.

54. Sanders, P. L., L. L. Burpee, H. Cole, Jr., and J. M. Duich. 1978. Uptake, translocation, and efficacy of triadimefon in control of turfgrass pathogens. Phytopathology 68:1482–1487.
55. Sanders, P. L., B. W. Pennypacker, and H. Cole, Jr. 1980. Histological study of 'Merion' Kentucky bluegrass showing symptoms of Fusarium blight. Phytopathology 70:468. (Abstr.).
56. Schaffnit, E. 1913. Der Schneeschimmel und die ubrigen durch *Fusarium nivale* Ces. hervorgerufenen Krankheitserscheinungen des Getreides. Landw. Jahrb. 43:521–648.
57. Shantz, H. L., and R. L. Piemeisel. 1917. Fungus fairy rings in eastern Colorado and their effect on vegetation. J. Agric. Res. 11:191–245.
58. Smiley, R. W., and M. M. Craven. 1977. Control of benzimidazole-tolerant *Fusarium roseum* on Kentucky bluegrass. Plant Dis. Rep. 61:484–488.
59. Smiley, R. W., and M. M. Craven. 1979. Fungicides for controlling Fusarium blight of Kentucky bluegrass, 1978. Amer. Phytopathol. Soc. Fung. Nemat. Test Rep. 34:146.
60. Smiley, R. W., and M. M. Craven. 1979. *Fusarium* species in soil, thatch, and crowns of *Poa pratensis* turfgrass treated with fungicides. Soil Biol. Biochem. 11:355–363.
61. Smiley, R. W., and M. M. Craven. 1979. In vitro effects of Fusarium blight-controlling fungicides on pathogens of *Poa pratensis*. Soil Biol. Biochem. 11:365–370.
62. Smiley, R. W., M. M. Craven, and J. A. Bruhn. 1980. Fusarium blight and physical, chemical, and microbial properties of Kentucky bluegrass sod. Plant Dis. 64:60–62.
63. Smiley, R. W., M. M. Craven, and R. C. O'Knefski. 1980. Fusarium blight of Kentucky bluegrass, in relation to precipitation and to redox potential of thatch. Agron. J. 72: (in press).
64. Smiley, R. W., M. M. Craven, and D. C. Thompson. 1978. Helminthosporium leaf spot and Fusarium blight control on Kentucky bluegrass, 1977. Amer. Phytopathol. Soc. Fung. Nemat. Test Rep. 33:145.
65. Smiley, R. W., and M. T. Cinque. 1976. Fungicides for controlling Fusarium blight of turfgrass. Amer. Phytopathol. Soc. Fung. Nemat. Test Rep. 31:155–156.
66. Smiley, R. W., and M. B. Harrison. 1975. Split vs. single applications for Fusarium blight control. Amer. Phytopathol. Soc. Fung. Nemat. Test Rep. 30:123–124.
67. Smiley, R. W., and R. J. Howard. 1976. Tolerance to benzimidazole-derivative fungicides by *Fusarium roseum* on Kentucky bluegrass turf. Plant Dis. Rep. 60:91–93.
68. Smiley, R. W., and W. Nass. 1976. Fungicides for controlling Fusarium blight of turfgrass. Amer. Phytopathol. Soc. Fung. Nemat. Test Rep. 31:155.
69. Smith, S. N. 1970. The significance of populations of pathogenic Fusaria in soil, p. 28–30. *In* T. A. Toussoun, R. V. Bega, and P. E. Nelson, (ed.), Root diseases and soil-borne pathogens, Univ. Calif. Press, Berkeley.
70. Sorauer, P. 1901. Der Schneeschimmel. Zeitschr. Pflanzenkr. 11:217–228.
71. Sorauer, P. 1901. Uber Den Scheeschimmel. Mitt. Deut. Landw. Gesell. 16:93–95.
72. Sorauer, P. 1903. Uber Forstbeschadigungen am Getreide und damit in Verbindung stehende Pilakrankheiten. Landw. Jahrb. 32:1–68.
73. Starks, K. J., and R. Thurston. 1962. Silver top of bluegrass. J. Econ. Entomol. 55:865–867.
74. Stewart, F. C., and H. E. Hodgkiss. 1908. The Sporotrichum bud-rot of carnations and the silver top of June grass. N.Y. (Geneva) Agric. Exp. Stn. Tech. Bull. 7. 119 p.
75. Subirats, F. J., and R. L. Self. 1972. Fusarium blight of centipede grass. Plant Dis. Rep. 56:42–44.
76. Troll, J. 1969. Fusarium blight and Merion bluegrass. U.S. Golf Assoc. Green Section Record 7(1):7–9.
77. Toussoun, T. A., and P. E. Nelson. 1976. A pictorial guide to the identification of *Fusarium* species. (2 ed.). Pa. State Univ., University Park, Pa. 43 p.
78. Tu, C. 1929. Physiologic specialization in *Fusarium* spp. causing headblight of small grains. Phytopathology 19:143–154.
79. Turgeon, A. J. 1975. Effects of cultural practices on Fusarium blight incidence in Kentucky bluegrass. Weeds, Trees, and Turf, July, p. 38–40.
80. Turgeon, A. J., and W. A. Meyer. 1974. Effects of mowing height and fertilization level on disease incidence in five Kentucky bluegrasses. Plant Dis. Rep. 58:514–516.
81. Tyson, J. 1936. Snow mold injury to bentgrasses. Mich. Quarterly Bull. 19(2):87–92.
82. Unger, F. 1844. Uber *Lanosa nivalis*. Frs. Bot. Zeit. 2:569–575.
83. Vargas, J. M., Jr. 1975. The role of nematodes in the development of Fusarium blight. Weeds, Trees, and Turf, July, p. 40–41.
84. Vargas, J. M., Jr., R. Detweiler, C. W. Laughlin, and S. Worrall. 1973. 1973 turfgrass fungicide research report. Plant Dis. Rep. 25, Michigan State Univ., East Lansing, p. 21–30.
85. Vargas, J. M., Jr., and C. W. Laughlin. 1971. Benomyl for the control of Fusarium blight of 'Merion' Kentucky bluegrass. Plant Dis. Rep. 55:167–170.
86. Vargas, J. M., Jr., and C. W. Laughlin. 1972. The role of *Tylenchorhynchus dubius* in the development of Fusarium blight of Merion Kentucky bluegrass. Phytopathology 62:1311–1314.
87. Warren, C. G., L. L. Burpee, and H. Cole, Jr. 1977. Fusarium snow mold control with fungicides: 1975–1976. Amer. Phytopathol. Soc. Fung. Nemat. Test Rep. 32:159.

20 Fusarium Infections in Human and Veterinary Medicine

Gerbert Rebell

Knowledge of the capacity of *Fusarium* species to cause opportunistic infections in man and animals has developed rather recently, and the number of reports of such infections is increasing due to awareness of their occurrence. The generally greater number of opportunistic infections reported is the result of improved medical care of severely ill, immune-deficient, and heavily medicated patients, and improved laboratory diagnosis of mycotic disease. The highlights of Fusarium infections in man are (i) mycotic keratitis (corneal ulcers), of which *Fusarium* species, especially *F. solani,* are the most common cause; (ii) the inclusion of *Fusarium* species among those fungi which, in patients with compromised host resistance, exhibit a tendency to invade blood vessels, leading to fatal haemotogenous dissemination of the infection; (iii) the occurrence of *F. oxysporum* as a cause of superficial, white onychomycosis (nail infection); and (iv) the occurrence of white grain mycetoma due to *Fusarium* species. Much less data is available on infections in animals other than man, with the exception of marine turtles.

Eye Infections

Fusarium species, and in particular *F. solani,* are the most frequent cause of mycotic infections of the cornea (keratomycosis, mycotic ulcerative keratitis) (Fig. 20-1A). There are numerous other illustrations of typical ocular lesions in the medical literature (22, 38, 54). These infections often follow very minor abrasions of the cornea, caused by a variety of materials, such as dust, vegetable and animal matter, and metal. In one case in Miami, the source of *F. solani* causing a corneal infection in a 9-year-old boy was infested sawdust, into which the boy had tumbled while playing on exercise apparatus in a public school yard. *Fusarium solani* has also been found in mascara used by a patient suffering from *F. solani* keratitis (65).

Clinical disease Fusarium keratitis is usually seen in apparently healthy patients. The course of infection is progressive, leading to perforation of the cornea and blindness of the infected eye within a few weeks. There is, at this time, no evidence that in human cases Fusarium keratitis may sometimes resolve without treatment or without damage to the eye. Anti-inflammatory steroids, which are sometimes prescribed by a physician who does not recognize the lesion as being mycotic, increase the severity of the infection. Early treatment of the infection with antifungal medication is usually effective. There is, therefore, great importance in the rapid identification of corneal lesions as being of fungal origin.

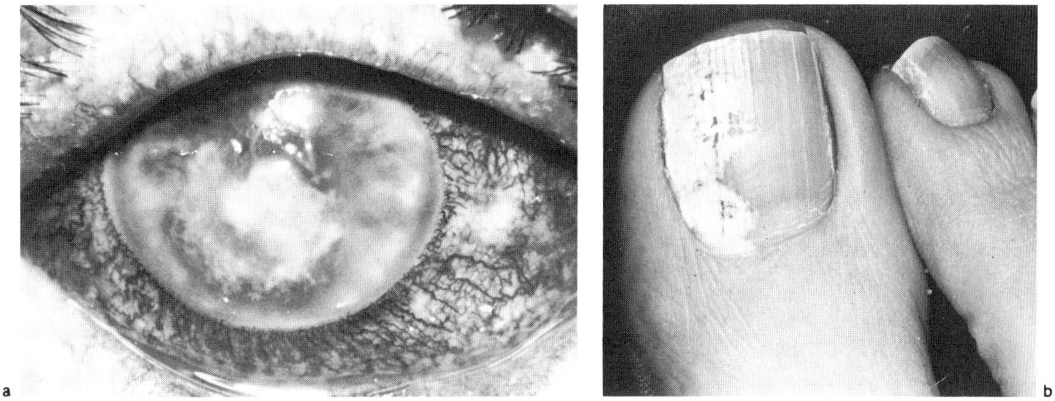

Fig. 20-1. Infections by two *Fusarium* species occurring simultaneously in a 46-year-old man: (A) Keratomycosis caused by *F. solani*. Note the "hyphate" feathery edge typical of a mycotic ulcer. (B) Superficial white onychomycosis of the toe nails caused by *F. oxysporum*.

Diagnosis The diagnosis of fungal keratitis is accomplished by means of stained smears and cultures from scrapings obtained from the infected cornea by the ophthalmologist, usually with a platinum spatula which can be sterilized in an alcohol flame, and by using the slit lamp microscope (56, 66). Smears may be stained by the Gram, Giemsa, and Grocott methenamine silver methods, of which the last is the most reliable (22, 24) (Fig. 20-2A). Cultures are made by inoculating scrapings of the cornea on blood agar or on Sabouraud's agar containing 50 μg/ml gentamicin sulfate. Initial growth of the fungus is usually evident within 24 hr, at which time a tape touch with a lactophenol cotton blue preparation, viewed under the microscope, will reveal the presence of Fusarium conidia (Fig. 20-3, 20-4).

Treatment Fusarium keratitis requires treatment with topical antifungal drugs, suitable for use in the eye. Of these, a 5% aqueous suspension of the polyene antibiotic natamycin (pimaricin) has been the only drug with extensively documented successful use (23, 37, 51). Imidazole compounds, such as clotrimazole, miconazole, and econazole, have been used and seem to be potentially as good as, or (in instances when the fungus is growing deep in the cornea) perhaps better than natamycin (33, 35, 57). Publication of treatment results with available imidazole compounds is much needed to complete the account of treatment alternatives. All of the above compounds are still classed by the Food and Drug Administration as experimental when used topically in the eye in the United States, although natamycin has been made available for use without an Investigator's New Drug permit. Other polyene antibiotics, such as nystatin and amphotericin B, which were used before natamycin was introduced, appear to have been less successful in the treatment of keratitis (36), although when diluted to 0.15%, to reduce the irritating properties of the bile salts included as solubilizing agents, amphotericin B appears to be effective (67). In vitro, isolates of *F. solani* from keratitis are usually sensitive to 2.5 ug/ml natamycin, 5 μg/ml nystatin, and 1.25 μg/ml amphotericin B, at 37 C, the endpoints varying with the conditions of the tests. Additional features of natamycin, which is a polyene of smaller molecular size than nystatin and amphotericin B, may account for its possibly greater clinical efficacy.

Treatment failures with natamycin occur in up to 10% of the cases. These failures are thought, in some cases, to be due to poor penetration of natamycin into the cornea, particularly when the fungus is growing deep in the cornea. In other cases, however, there is deterioration of the lesion after the fungus has been rendered non-viable by antifungal treatment, as tested by cultures inoculated with the surgically removed cornea. This latter

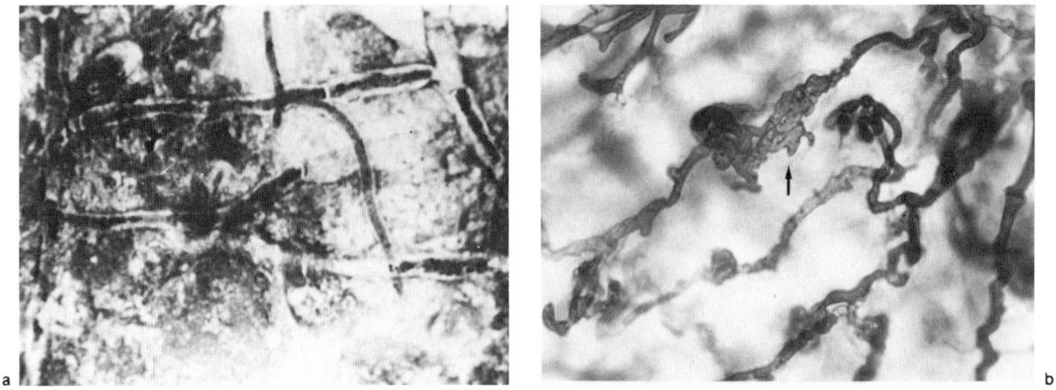

Fig. 20-2. *Fusarium* in infected tissue scrapings: (A) *F. oxysporum* in a scraping from the cornea stained with Giemsa stain. (B) *F. oxysporum* in scraping from toe nail stained with periodic acid Shiff stain. Arrow points to the "eroding frond" and possible "perforating organ" associated with invasion of the nail plate. (Photo supplied by N. Zaias.)

type of treatment failure may be due to hypersensitivity or immune phenomena initiated by fungus antigens. Tests to date have not revealed fungal collagenase activity in vitro or other factors that account for the pathogenicity of *F. solani* (or other fungi) in the apparently healthy cornea following abrasion or trauma. However, collagenase activity in the cornea does occur in experimental infections with *F. solani* in rabbits, as it does following alkali burns of the cornea. The role of trichothecene compounds or other mycotoxins produced by *Fusarium* has not been studied, although an extract from *F. moniliforme* was found to damage the rabbit cornea by Dudley and Chick (14). Corneal transplants (penetrating keratoplasty), while not recommended as a substitute for treatment with antifungal compounds in a non-perforating keratitis, are effective in saving vision when perforation has occurred or clinically appears to be imminent (19, 23, 54).

Complications Fungal endophthalmitis, infection within the eye, follows as a serious consequence of the growth of the fungus through the cornea into the anterior chamber and iris and lens of the eye, or of perforation of the ulcer into the eye. These complications leading to mycotic endophthalmitis require the intraocular instillation of 2.5 to 1.0 μg doses of amphotericin B into the vitreous humor and, possibly, as experience develops, that of other antimycotic drugs, such as the imidazole compounds (20, 46, 58). Natamycin and nystatin are not suitable for use within the eye. Treatment of fungus infection localized in the anterior chamber (between the cornea and lens) may be best approached through use of a penetrating topical antifungal agent, such as an imidazole compound (33), or by depot injection of antifungal compounds into the subconjunctival tissue supplying blood vessels to the corneal margin (limbus). Such externally applied treatment, however, is probably ineffective if the infection has reached the vitreous chamber behind the lens.

Primary endophthalmitis Mycotic endophthalmitis directly following penetrating trauma or surgery to a previously uninfected eye may occur, and has been reported for *F. oxysporum* following a thorn puncture (58), and *F. moniliforme, F. episphaeria,* and *F. solani,* following surgery and trauma (20).

In animals One might expect that mycotic keratitis would occur frequently in non-human vertebrates, but to date this has not been observed. There has been one reported infection of the cornea with *F. solani* in a horse in the southern part of the United States (49), and another case occurred in Miami, Florida, in which treatment with natamycin failed.

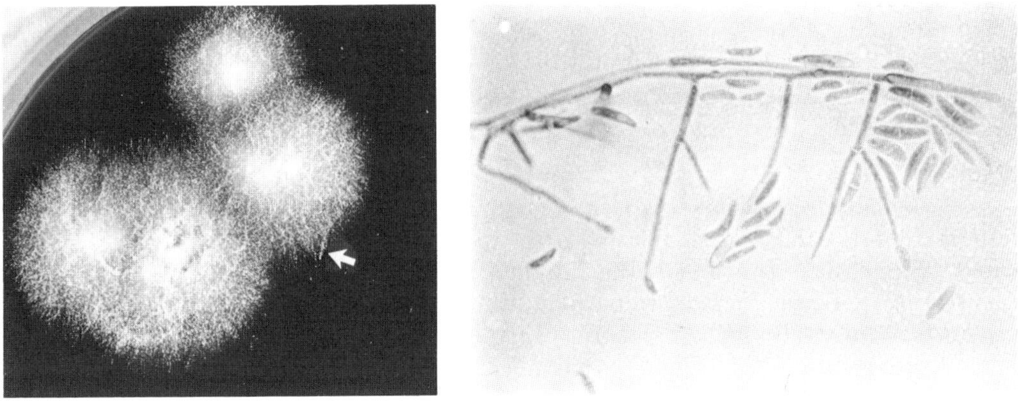

a b

Fig. 20-3. *Fusarium solani* keratomycosis: (A) Thallus outgrowth on blood agar at 48 hr. Arrow points to the "C" streak, the preferred method of inoculating corneal scrapings on the agar. (B) Hyphae and unbranched phialides and microconidia, seen characteristically in tape mount preparations made as soon as growth is detected (within 24 hr after inoculation).

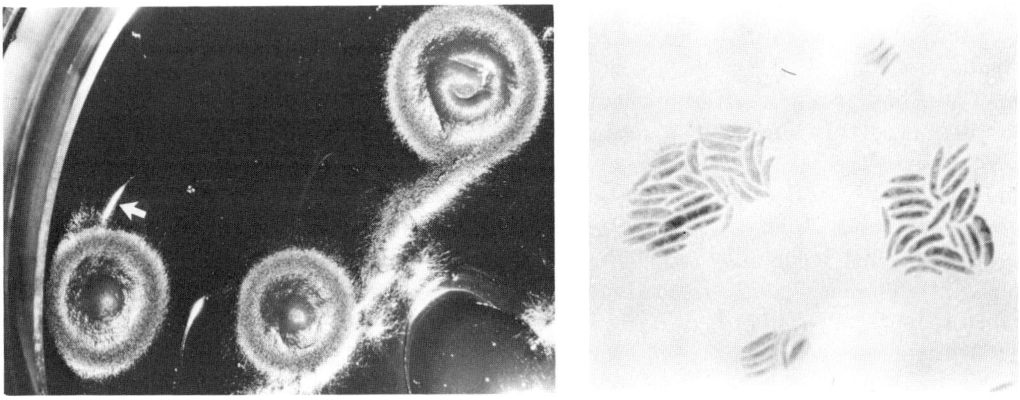

a b

Fig. 20-4. *Fusarium dimerum* keratomycosis: (A) Thallus outgrowth on blood agar at 96 hr. Arrow points to the "C" streak, the preferred method of inoculating corneal scrapings on the agar. (B) Conidia, characteristically without hyphae, seen in tape mount preparation made as soon as growth is detected (24–48 hr after inoculation).

Experimental infections with *F. solani* have been produced in the corneas of rabbits (21), owl monkeys, and guinea pigs (*unpublished data*), and in rats (11, 32). The majority of such infections undergo resolution within 2 weeks, with relatively little damage to the eye, unless the animal is treated with anti-inflammatory steriods. A satisfactory animal model, paralleling progressive *F. solani* keratitis in man, has therefore not been achieved.

Mycology Fungi in 37 genera are capable of producing exogenous (externally induced) mycotic keratitis. *Fusarium solani,* however, is clearly the most frequent cause, and is responsible for perhaps as many cases as the rest of the causative fungi combined. This is based on published and unpublished results in the United States and data from other countries (21), especially from tropical and subtropical regions, where the prevalence of mycotic keratitis is also highest. *Fusarium solani* keratitis has been reported in various parts of the United States (22, 30, 36, 37), Colombia, Argentina (5, 71, 73, 74), Peru, Japan (44), the Philippines (61, 62, 63), Australia, Malaysia (34), and Nigeria (28). In Miami, Florida, in the 10½-year period, January 1969 to July 1979, we have diagnosed 168 cases of mycotic keratitis, of which 110 (65.5%) were caused by *Fusarium;* of these, 101 (91.8%) were *F. solani.* The frequency of *F. solani* keratitis may correlate with the increased prevalence of this fungus in soils in tropical regions. Isolates from the cornea appear to share the special property of being able to grow reasonably well at 37 C (16, 37).

One of the interesting aspects of *F. solani* keratitis is that *F. solani* was only recently recognized as a cause of mycotic keratitis (5, 30, 34, 36; C. C. Olson, M. Okumoto, and C. Halde, *unpublished data),* leading us to the conjecture that some older reports of Cephalosporium keratitis were based on the misidentification of the microconidia of *F. solani* as *Cephalosporium* (7, 18, 50, 54). Some early reports of *F. oxysporum* keratitis may also have been based on misidentification of *F. solani* isolates (6, 27, 43, 45). There are also reports of Fusarium keratitis without speciation (26, 47, 48, 50, 51, 55, 60, 75).

After *F. solani, Fusarium* species with yellow to orange, more or less glabrous thalli are most frequently isolated from mycotic keratitis. These have been identified as *F. episphaeria, F. dimerum, F. nivale,* and atypical isolates of *F. roseum.* Of these, *F. dimerum* may occur the most frequently (25, 72). *Fusarium episphaeria* and *F. dimerum,* as defined above, have been reported from keratitis in California, Florida, Argentina, and the Philippines (23, 30, 62). There is one published report of *F. nivale* (52) from mycotic keratitis. *Fusarium oxysporum* and *F. moniliforme* (3, 35, 39) complete the list of species known to cause mycotic keratitis. Infections with these less-frequent causes are not clinically distinguishable from *F. solani* infections, and have been of at least equal severity. One case of keratitis caused by *Cylindrocarpon tonkinense* has been reported from Colombia, S.A. (42). Of the nine cases of Fusarium keratitis diagnosed in Miami which have not been *F. solani,* two were caused by *F. oxysporum,* one by *F. moniliforme,* and the remaining six were in the yellow-orange glabrous group and require further taxonomic work.

Infections of the Nails

In contrast to the predominant association of *F. solani* with mycotic keratitis, a type of nail infection called white superficial onychomycosis (or leukonychia mycotica) is caused by *F. oxysporum* (59, 69, 70) (Fig. 20-1B), of which it may be the most prevalent cause after the dermatophyte *Trichophyton mentagrophytes.* This infection, which is thought to require an enzymatic attack by the fungus on the horny keratin substance of the nail plate, is also caused by *Aspergillus terreus* and *Acremonium* sp. (probably *A. potroni).* Cultures from nails sometimes yield *F. solani* and other species of *Fusarium.* However, to date, these have not been found to cause nail infections of the progressive and chronic type caused by *F. oxysporum.*

Diagnosis of *F. oxysporum* nail infections is made by culture and by demonstrating hyphae in the nail substance. The cultures are preferably repeated and multiple, to eliminate the likelihood of Fusarium contamination rather than actual infection. The Hotchkiss-McManus periodic acid Shiff stain, as modified by Kligman, Mescon, and DeLamater (40), is useful because Fusarium hyphae in the nail are not as refractive as those of dermatophytes when viewed in KOH preparations.

Fusarium oxysporum nail infections are treated with imidazole compounds in dimethyl sulfoxide (DMSO) or acetone, 10% glutaraldehyde (64) and surgical evulsion of the affected nails (70).

Other Fusarium Infections Reported in Man

Keratitis, caused chiefly by *F. solani,* and superficial infections of the nail plate, which may be caused by *F. oxysporum,* are the two currently known, regularly occurring Fusarium infections in man. The pathogenic potential of *Fusarium* species as opportunistic agents of mycotic disease, however, is manifest also in primary endophthalmitis, and in other infections which have been reported; these will probably increase in reported prevalence in the future, as diagnosis improves, and as modern medical practice, prolonging the life of patients whose resistance is seriously compromised by disease and medication, continues. The species which have caused opportunistic mycotic infections reported to date are *F. solani, F. oxysporum, F. moniliforme,* and *F. roseum.*

Infections of burns There has been one case reported of fatal systemic mycotic invasion by an unidentified *Fusarium* species (possibly *F. oxysporum)* in a burned child, in which the fungus grew into the necrotic, burned skin and into blood vessels, leading to metastatic infections in the kidney, heart and brain (1). *Fusarium solani* and *F. roseum* have been known to infect the necrotic skin of burn victims (31, 53), and other case reports of Fusarium infection of burns, with systemic invasion, may be expected.

Infection of necrotic skin and the border of ulcers A tendency of *F. solani,* and possibly *F. oxysporum,* to invade necrotic and serum-wetted skin is manifest in reports by English, et al. (16, 17) of the growth of these organisms superficially in the skin around chronic venous statis ulcers of the legs, and in the necrotic skin of the feet of a diabetic patient (15). An isolate of *F. solani* from skin surrounding a leg ulcer grew relatively poorly at 37 C, as compared to strains from keratitis.

Inflammatory infections localized in the skin There have been two reports of localized inflammatory Fusarial infections in the skin. In one case a granulomatous lesion of the face was produced by *F. solani* (8). The lesion was removed surgically. The patient was later found to have defective leukocyte enzymes characteristic of chronic granulomatous disease. In a second case, pustular lesions were produced by *F. moniliforme* on the hand of an otherwise apparently healthy patient (13). These lesions healed without treatment, as did comparable lesions produced experimentally by inoculating mice with the fungus.

Disseminated infections in patients with leukemia and lymphoma There have been two reports of invasive Fusarium infections in patients with hemapoetic malignancies, leading to death. In one case symptoms included multiple lesions in the skin, orbital cellulitis, and clinical evidence of systemic infection by *F. solani,* including seizures indicative of CNS involvement; the patient was a 2½-year-old child with acute leukemia for 1½ years, who was receiving immunosuppressive therapy (12). In the second case there was infection of the skin and many internal organs by *F. moniliforme* and the yeast, *Candida tropicalis,* in a 32-year-old man with a 5-year history of malignant lymphoma (68).

Visceral infection with possible endogenous mycotoxicosis An initially healthy 66-year-old woman developed gastrointestinal symptoms and depressed bone marrow following in-

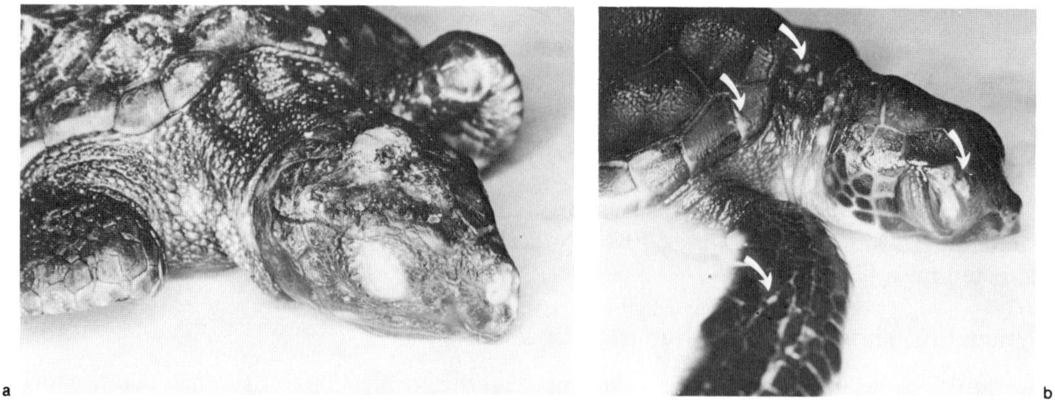

Fig. 20-5. Lesions containing *Fusarium solani* in marine turtles. (A) Infection of the eye and nares seen in a baby Atlantic Ridley Turtle (*Lepidochelys kempi*) collected on the Texas Gulf Coast. (B) Infection seen in a newly hatched Atlantic Green Turtle (*Chelonia m. mydas*), hatched and reared in confinement at the Cayman Turtle Farm, Grand Cayman, British West Indies. Arrows point to some of the lesions located on the head, neck, flippers, and between the scutes.

fluenza immunization. The symptoms were suggestive of alimentary toxic aleukia resulting from Fusarium toxin. In addition, the patient exhibited muscle weakness of the type associated with the Eaton-Lambert syndrome (29). The patient expired, and the autopsy revealed granuloma in the esophageous, liver, spleen, and cecum containing a dense growth of *F. oxysporum*, which had also invaded the surrounding tissue.

Mycetoma and osteomyelitis (penetrating wound infection) There are two known cases of white grain mycetoma caused by *Fusarium*. One of these was caused by *F. solani* (4), and the cause of the second, initially reported to be *Cephalosporium* (2) has, according to Ajello, been reidentified as *F. moniliforme*. These cases call into question other reports of mycetoma caused by unspeciated *Cephalosporium*.

Bourguignon et al. have reported a case of osteomylitis of the tibia caused by *F. oxysporum* in a 7-year-old boy who received a penetrating wound on the knee with a thorn (10). The patient was successfully treated by surgical debridement and systemic amphotericin B.

The isolation of *Fusarium* from urine There is a report of the isolation of *Fusarium* spp. from the urine of ten adult female patients and one boy without symptoms or with mild cystitis (41). It is impossible, however, to entirely rule out the possibility of contamination of the catheterized specimens. *Fusarium* has been isolated from the urine of patients with systemic Fusarium infection disseminated to the kidney.

Fusarium Infection in Marine Turtles

Except for mycotic keratitis in horses, and a single report of intestinal ulcers caused by *F. solani* in a captive tortoise (9), there is an absence of reports of Fusarial infections in animals other than man. The single exception is the common occurrence of *F. solani* in lesions of marine turtles (Fig. 20-5A). The underlying cause of these lesions is not known, but they are apparently manifestations of the capacity of *F. solani* to invade necrotic animal tissue. The necrosis itself is sometimes suspected of being herpetic (viral), and sometimes traumatic, although proof is lacking. There is a high recovery of *F. solani* from calcareous beach sands mixed with decomposed marine and other vegetation in Florida and the Caribbean Islands, which may account for these infections. In a unique situation at the Lerner Marine Laboratory, formerly maintained by the American

Museum of Natural History on Bimini, Bahamas, there occurred a high incidence of *F. solani* infection of the shells and skin of baby Loggerhead Turtles; it was traced to *Fusarium*-infected rubberized horse hair matting used as a filter in the onshore marine well, dug in the sand, which supplied water to the wet lab facility (Fig. 20-6). At the Cayman (Green) Turtle Farm on Grand Cayman, British West Indies, there is a high incidence of proven herpetic skin lesions, as well as traumatic lesions, which are never associated with the presence of *F. solani,* although the fungus is abundant on the beach sand at the farm, where some of the turtle eggs are laid and hatched. The turtles are not in direct contact with this sand after hatching; like those at the Lerner Marine Laboratory, they are raised in concrete tanks supplied with flowing sea water. At the Cayman Turtle Farm this water is pumped directly from the sea, without filtration through sand. There are, however, occasional newly hatched turtles with characteristic elevated lesions which are pathonomic for the presence of *F. solani* (Fig. 20-5B). These appear to be primary Fusarium infections, and occur in less than 0.01% of all turtles hatched at the farm. Pulmonary and visceral mycotic infections sometimes occur in baby turtles at the farm, but these are never associated with *Fusarium,* the principal causative agents being *Paecilomyces lilacinus* and *Scolecobasidium.* One may surmise from this that either the *F. solani* is not capable of producing systemic infection, or does not reach the lungs and viscera by whatever route of infection occurs in the case of those fungi responsible for pulmonary and visceral lesions.

a

Fig. 20-6. *Fusarium solani* infection in baby Atlantic Loggerhead Turtle (*Caretta c. caretta*) reared in confinement at the Lerner Marine Laboratory, Bimini, Bahamas: (A) Lesions in growing portions of the shell scutes of affected turtle. (B) *F. solani* growing on horse hair from sea well (unstained). Arrow points to frondose hyphal tip, indicative of keratinophilic attack of fungus on the hair. (C) *F. solani* in turtle lesion growing in cornified layer of growing turtle scute stained with GMS.

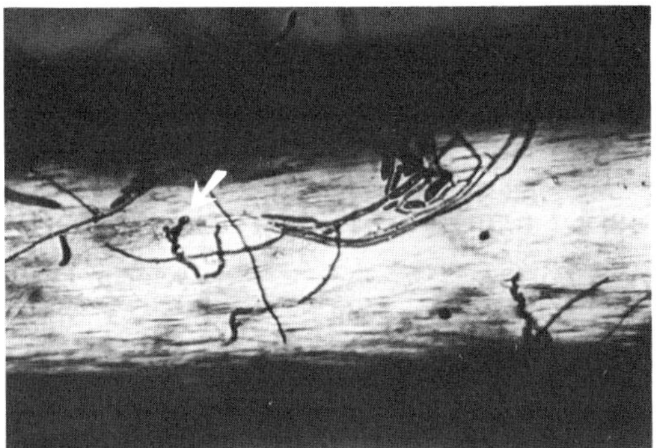

b

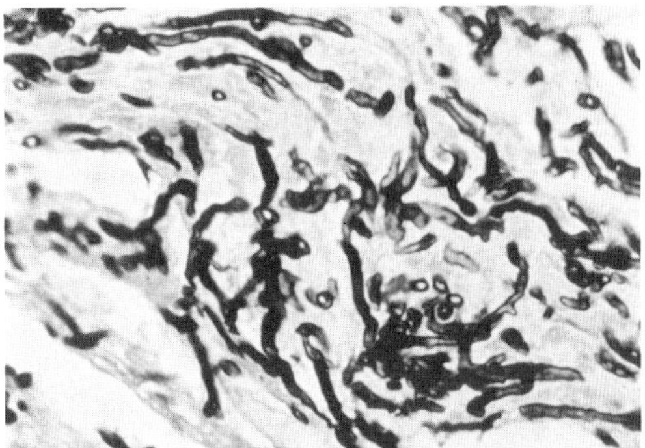

c

Literature Cited

1. Abramowsky, C. R., D. Quinn, W. D. Bradford, and N. F. Conant. 1974. Systemic infection by *Fusarium* in a burned child. The emergence of a saprophytic strain. J. Pediatrics 84:561–564.
2. Albereci, F., and L. Morganti, 1877. Sul primo caso de micetoma gel tiele da *Acremonium sp.* osservato in Italia. Giorn. Malettie Instettive Parassitarie. 12:1–4.
3. Anderson, B., S. S. Roberts, C. Gonzalez, and E. W. Chick. 1959. Mycotic ulcerative keratitis. Arch. Ophthalmol. 62:169–179.
4. Anderson, B., Jr., and E. W. Chick. 1963. Mycokeratitis treated with amphotericin B and mechanical debridement. Southern Med. J. 56:220–224.
5. Arrechea, A. de, R. C. Zapater, E. Storero, and V. H. Guevara. 1971. Quertomicosis por *Fusarium solani.* Arch. Oftalmol. Hisp. Amer. 46:123–127.
6. Barsky, D. 1959. Keratomycosis, a report of six cases. Arch. Ophthalmol. 61:547–552.
7. Bedell, A. J. 1947. *Cephalosporium* keratitis. Amer. J. Ophthalmol. 30:997–1000.
8. Benjamin, R. P., J. L. Callaway, and N. F. Conant. 1970. Facial granuloma associated with Fusarium infection. Arch. Dermatol. 101:598–600.
9. Borst, G. H. A., C. Vrolge, F. G. Paelma, P. Zwart, W. J. Strik, and L. C. Peters. 1972. Pathological findings on animals in the Royal Zoological Gardens of the Rotterdam Zoo during the years 1963, 1964 and 1965. Acta Zool. Pathol. Antverpiensia 56:3–19.
10. Bourguignon, R. L., A. F. Walsh, J. C. Flynn, C. Baro, and E. Spinos. 1976. *Fusarium* sp. osteomyelitis. J. Bone Joint Surg. 58:722–723.
11. Burda, C. D., and E. Fisher, Jr. 1959. The use of cortisone in establishing experimental fungal keratitis in rats. A preliminary report. Amer. J. Ophthalmol. 48:330–335.
12. Cho, C. T., T. S. Vats, J. T. Lowman, J. W. Brandsberg, and F. E. Tosh. 1973. *Fusarium solani* infection during treatment for acute leukemia. J. Pediatrics 83:1028–1031.
13. Collins, M. S., and M. G. Rinaldi. 1977. Cutaneous infection in man caused by *Fusarium moniliforme.* Sabouraudia 15:151–160.
14. Dudley, M. A., and E. W. Chick. 1964. Corneal lesions produced in rabbits by an extract of *Fusarium moniliforme.* Arch Ophthalmol. 72:346–350.
15. English, M. P. 1968. Invasion of the skin by filamentous, non-dermatophyte fungi. Brit. J. Dermatol. 80:282–286.
16. English, M. P. 1972. Observations on strains of *Fusarium solani* and *F. oxysporum* and *Candida parapsilosis* from ulcerated legs. Sabouraudia 10:35–42.
17. English, M. P., R. J. Smith, and R. R. M. Harmoan. 1971. The fungal flora of ulcerated legs. Brit. J. Dermatol. 84:567–581.
18. Focosi, M. 1932. Keratomycosis due to *Cephalosporium.* Bull. Ocul. 11:250–264.
19. Forster, R. K. 1975. Therapeutic surgery in medical treatment failures of keratitis. Brit. J. Ophthalmol. 59:366–371.
20. Forster, R. K. 1978. Endophthalmitis, chap. 24. *In* T. D. Duane (ed.), Clinical ophthalmology, Harper & Row, Hagerstown, Md.
21. Forster, R. K., and G. Rebell. 1975. Animal model of *Fusarium solani* keratitis. Amer. J. Ophthalmol. 79:510–515.
22. Forster, R. K., and G. Rebell. 1975. The diagnosis and management of keratomycosis; 1. Cause and diagnosis. Arch. Ophthalmol. 93:975–978.
23. Forster, R. K., and G. Rebell. 1975. The diagnosis and management of keratomycosis; 2. Medical and surgical management. Arch. Ophthalmol. 93:-1134–1136.
24. Forster, R. K., M. G. Wirta, M. Solis, and G. Rebell. 1979. Methenamine-silver-stained corneal scrapings in keratomycosis. Amer. J. Ophthalmol. 82:261–265.
25. Garcia, N. P., E. Escani, and R. Zapater. 1972. Queratomycosis por *Fusarium dimerum.* Arch. Oftalmol., Buenos Aires 47:332–334.
26. Gillespie, F. D. 1963. Fungus corneal ulcer. Amer. J. Ophthalmol. 56:823–825.
27. Gingrich, W. D. 1962. Keratomycosis. J. Amer. Med. Assoc. 179:602–608.
28. Gugnani, H. C., R. S. S. Talwar, A. N. U. Njoku-Obi, and H. Kodilenye. 1976. Mycotic kereatitis in Nigeria; a study of 21 cases. Brit. J. Ophthalmol. 60:607–613.
29. Gutman, L., S. M. Chou, and R. S. Pore. 1975. Fusariosis, myasthenic syndrome, and aplastic anemia. Neurology 25:922–926.
30. Halde, C., and M. Okumoto. 1966. Ocular mycoses: a study of 82 cases. Proc. 20th Int. Congr. Ophthalmol., Munich. Neth. Excerpta Medica. Int. Congr. Ser. 146:705–712.
31. Holzegel, K., and H. J. Kempf. 1964. Fusariummykose auf der Haut eines Verbrannten. Dermatol. Monatsschr. 150:651–658.
32. Ishibishashi, Y. 1978. Experimental fungal keratitis due to *Fusarium.* Improvement of inoculation technique and re-examination of animal model. Acta Soc. Ophthalmol. Japan 82:643–651.
33. Jones, B. R. 1975. Principles in the management of oculomycosis. Amer. J. Ophthalmol. 79:719–751.
34. Jones, B. R., D. B. Jones, and S. M. Lim. 1970. Corneal ulcer and intraocular infection due to *Fusarium solani.* Trans. Ophthalmol. Soc. U.K. 84:757–779.
35. Jones, D. B. 1976. Fungal keratitis, chap. 21. *In* T. D. Duane (ed.), Clinical ophthalmology (Vol. 4), Harper & Row, Hagerstown, Md.

36. Jones, D. B., R. Sexton, and G. Rebell. 1970. Mycotic keratitis in South Florida. A review of 39 cases. Trans. Ophthalmol. Soc. U.K. 84:781–797.
37. Jones, D. B., R. K. Forster, and G. Rebell. 1972. *Fusarium solani* keratitis treated with natamycin (pimaricin); eighteen consecutive cases. Arch. Ophthalmol. 88:147–154.
38. Kaufman, H. E., and R. M. Wood. 1965. Mycotic keratitis. Amer. Ophthalmol. 59:992–1000.
39. Kidd, G. H., and F. T. Wold. 1973. Dimorphism in a pathogenic *Fusarium*. Mycologia 65:1371–1375.
40. Kligman, A. M., H. Mescon, and E. D. DeLamater. 1951. The Hotchkiss-McManus stain for the histopathologic diagnosis of fungus disease. Amer. J. Clin. Pathol. 21:88–91.
41. Lazarus, J. A., and L. H. Schwarz. 1948. Infection of urinary bladder with unusual fungus strain (*Fusarium*). Urologic Cutaneous Rev. 52:185–189.
42. Laverde, C. L., L. H. Moncada, A. Restrepo, and C. L. Vera. 1973. Mycotic keratitis: 5 cases caused by unusual fungi. Sabouraudia 11:119–123.
43. Lynn, J. R. 1964. *Fusarium* keratitis treated with cycloheximide. Amer. J. Ophthalmol. 58:637–641.
44. Matsumoto, T., and N. Soejima. 1976. Keratomycosis. Mykosen 19:217–222.
45. Mikami, R., and G. N. Stemmermann. 1968. Keratomycosis caused by *Fusarium oxysporum*. Amer. J. Clin. Pathol. 29:257–262.
46. Miller, J. S., G. Rebell, R. C. Magoon, S. M. Kulvin, and R. K. Forster. 1979. Intravitreal antimycotic therapy and the cure of endophthalmitis caused by a *Paecilomyces lilacinus* contaminated pseudophakos. Ophthalmic Surg. 9:54–63.
47. Ming, Y. N., and T. F. Yu. 1966. Identification of a *Fusarium* species, isolated from a corneal ulcer. Acta Microbiol. Sin. 12:1180–1186.
48. Miranda, H., W. Fernandez, and V. Sauchez. 1969. Quertomicosis: presentacion de tres cases causados por *Fusarium* sp. Mycopathol. Mycol. Appl. 37:179–185.
49. Mitchell, J. S., and M. H. Attleberger. 1973. *Fusarium* keratomycosis in the horse. Vet. Med. Small Anim. Clin. 68:1257–1260.
50. Naumann, G., W. R. Green, and L. E. Zimmerman. 1967. Mycotic keratitis, a histopathologic study of 73 cases. Amer. J. Ophthalmol. 64:668–682.
51. Newmark, E., A. C. Ellison, and H. E. Kaufman. 1970. Pimaricin therapy of *Cephalosporium* and *Fusarium* keratitis. Amer. J. Ophthalmol. 69:458–466.
52. Perez, M. 1966. *Fusarium nivale* as a cause of corneal mycosis. Klin. Oczna. 36:609–612.
53. Peterson, J. E., and T. J. Baker. 1959. An isolation of *Fusarium roseum* from human burns. Mycologia 51:453–456.
54. Polack, F. M., H. E. Kaufman, and E. Newmark. 1971. Keratomycosis, medical and surgical treatment. Arch. Ophthalmol. 85:410–416.
55. Putanna, S. T. 1967. Primary keratomycosis. J. All-India Ophthalmol. Soc. 17:171–200.
56. Rebell, G., and R. K. Forster. 1974. Fungi of keratomycosis, p. 482–490. *In* F. H. Lennette, J. P. Spaulding and J. P. Truant (ed.), Manual of clinical microbiology, Amer. Soc. Microbiol., Washington, D.C.
57. Rippon, J. J. 1977. Mycotic infections of the eye and ear, p. 475. *In* Medical mycology, the pathogenic fungi and the pathogenic actinomycetes, W. B. Saunders, Philadelphia, Pa.
58. Rowsey, J. J., D. L. Teaceers, J. A. Smith, D. L. Newson, and J. Rodriquez. 1979. *Fusarium oxysporum* endophthalmitis. Arch. Ophthalmol. 97:103–105.
59. Rush-Munro, F. M., H. Black, and J. M. Dingley. 1971. Onychomycosis caused by *Fusarium oxysporum*. Australian J. Dermatol. 12:18–29.
60. Sigtenhorst, M. L., and W. L. Gingrich. 1957. Bacteriologic studies of keratitis. Southern Med. J. 50:346–350.
61. Salceda, S. R. 1973. Keratomycosis with emphasis on the diagnostic and therapeutic value of anterior keratectomy. Trans. Asia-Pacific Acad. Ophthalmol. *In* Trans. Ophthalmol. Soc. New Zealand 25:202–212.
62. Salceda, S. R. 1976. Fungi and the human eye. Kaliklasan 5:143–174.
63. Salceda, S. R., M. J. Valenton, L-Fc. Nievera, and R. M. Abendanio. 1974. Corneal ulcer: A study of 110 cases. J. Philippine Med. Assoc. 49:493–508.
64. Suringa, D. W. R. 1970. Treatment of superficial onychomycosis with topically applied glutaraldehyde. Arch. Dermatol. 102:163–167.
65. Wilson, L. A., J. W. Kuehne, S. W. Hall, and D. G. Ahearn. 1971. Microbial contamination in ocular cosmetics. Amer. J. Ophthalmol. 71:1298–1302.
66. Wilson, L. A., and R. B. Sexton. 1976. Laboratory aids in diagnosis, chap. 1. *In* T. B. Duane (ed.), Clinical ophthalmology (Vol. 4), Harper & Row, Hagerstown, Md.
67. Wood, T. O., and W. Williford. 1976. Treatment of keratomycosis with amphotericin B (0.15%). Amer. J. Ophthalmol. 81:847–849.
68. Young, N. A., K. J. Kwon-Chung, T. T. Kubota, A. E. Jennings, and R. Fisher. 1978. Disseminated infection by *Fusarium moniliforme* during treatment for malignant lymphoma. J. Clin. Microbiol. 7:589–594.
69. Zaias, N. 1966. Superficial white onychomycosis. Sabouraudia 5:99–103.
70. Zaias, N. 1971. Onychomycosis. Arch. Dermatol. 105:263–274.
71. Zapater, R. C. 1971. Keratomycoses por *Fusarium*. Proc. Fifth Congr. Int. Soc. Human Anim. Mycol., Paris, 187–188.
72. Zapater, R. C., A. Arrechea, and V. H. Guevara. 1972. Quertomicosis por *Fusarium dimerum*. Sabouraudia 10:274–275.

73. Zapater, R. C., and A. Arrechea. 1975. Mycotic keratitis by *Fusarium*. A review and report of two cases. Ophthalmologica 170:1–12.
74. Zapater, R. C., M. A. Brunzini, E. V. Albesi, and C. A. Silicarto. 1976. El genero *Fusarium* como agente etiologico de micosis oculares: presentacion de 7 cases. Arch. Oftalmol. Buenos Aires 51:279–286.
75. Zimmerman, L. E. 1962. Ocular mycoses, p. 1151–1160. *In* R. D. Baker and E. W. Chick (ed.), International symposium on opportunistic fungus infections (Durham, N. C.), Williams Wilkins Co., Baltimore, Md.

II The Fungus *Fusarium*: Ecology

Prologue

T. A. Toussoun

Plant pathologists were aware of the importance of ecology in the overall understanding of the biology of microorganisms for a good many years before ecology attained its present notoriety. Thus, after the war, when the study of soil-borne fungus diseases was taken up in earnest, the ecology of these fungi was the principal focus of interest. This work was concentrated in England, the western U.S.A., and Australia for a number of years. *Fusarium* was a leading organism studied, being of the first rank among these fungi. I cannot emphasize enough the pioneering nature of these investigations—it was akin to the discovery of a new world. There are some excellent reviews of this subject, among which are the books of S. D. Garrett of Cambridge University; *The Ecology of Soil-borne Pathogens: Prelude to Biological Control,* edited by K. F. Baker and W. C. Snyder; *Ecology of Fungi,* by D. M. Griffin, which emphasizes the physical aspects of soil ecology; and *Biological Control of Plant Pathogens,* by K. F. Baker and R. J. Cook. This list is merely an introduction to a subject whose vitality is still increasing in research stations throughout the world.

The contributions in this book will bring the reader up to date on several of the main aspects of the ecology of *Fusarium.* Thus Cook addresses himself to water as it affects *Fusarium* growth, sporulation, and survival in soil, as well as pathogenesis. This aspect of soil ecology came rather late in the course of events and to a very large extent was due to the work of D. M. Griffin. Jim Cook and his colleagues, among whom we should mention R. I. Pappendick, have for the past 14 years emphasized the role of water in the life of *Fusarium* in soil, and therefore the treatment of this subject in this book is quite authoritative. R. Baker deals with competition for nutrients, a subject which, in contrast to water, was the first of the soil influences studied on *Fusarium.* No one is better qualified to discuss this than Tex Baker; when Tex is not on moon or space biology projects, he does terrestrial computations on *Fusarium* chlamydospores, which incidentally is the subject dealt with by Schippers and van Eck in this book. For many Fusaria the chlamydospore is the critical soil-borne propagule. Noticed by Burkholder during the early years of this century, it was not until very much later that its importance in the life cycle was realized. This also brought about a fundamental change in thinking: soil fungi by and large are quiescent in the soil until activated by nutrients; if these are from a host, then the pathogen is predisposed to attack. It is therefore crucial to understand what affects the formation, germination, and renewal of these spores. Today biological control seeks to intervene in the cycle of these chlamydospores and, by extension, other propagules. Naturally occurring biological control in the form of suppressive soils is the subject of the paper by Louvet, Alabouvette, and Rouxel, who base their work on soils in southern France which are consistently free of Fusarium wilt despite the fact that susceptible crops are grown year after year. This work deserves to be closely followed, for their experiments are most interesting and constructive, as is their current appraisal of the reason for the suppressiveness. Where most other studies have focused on the antagonistic activities of lytic bacteria and actinomycetes, Louvet and his colleagues believe suppressiveness is caused by saprophytic clones of *F. oxysporum* and *F. solani* native to the soil. This is most logical, for there are no closer kin to *F. oxysporum* pathogens than *F. oxysporum* saprophytes and thus the latter would compete most effectively for the same ecological niches under the same circumstances; *F. solani* has been shown in other cases to effectively block the establishment in the host of the isolates of *F. oxysporum* that cause wilt diseases.

Stoner's paper takes us to non-cultivated soils, and the reader will realize, upon reading this good summation, that *Fusarium* really is an omnipresent soil fungus. I have left the first paper, that of Burgess, to the last because he touches on the ecology of *Fusarium* in a manner quite different from that of all others. Heretofore ecological studies have centered on a few clones or a particular pathogen in a restricted locality. Burgess deals with whole *Fusarium* populations in continents. He also emphasizes the importance of air-dispersal even in those forms without perfect states. This is most welcome, for up to now *Fusarium* conidia have been thought to be water-borne only, and chlamydospores to be immobilized in plant residues in soil. Burgess points out that this does not fit the facts. On the contrary, air-dispersal of these spores is a common, important, and totally overlooked method of travel.

21 General Ecology of the Fusaria

L.W. Burgess

The Fusaria are widely distributed in soil, on subterranean and aerial plant parts, plant debris, and other organic substrates (4,16,53,55). They are common in tropical and temperate regions and are also found in desert, alpine, and arctic areas, where harsh climatic conditions prevail (4,6,11,15,16,22,24,25,28,51,53,54,55). Many species occur abundantly in fertile cultivated and rangeland soils but are relatively uncommon in forest soils (28,41,45). A few representatives are regarded as entomogenous fungi (4,15,16) and several others are reported to be fungal parasites (4). The Fusaria are often regarded as soil-borne fungi because of their abundance in soil and their frequent association with plant roots, either as parasites or saprophytes. However, many have active or passive means of dispersal in the atmosphere and are common colonizers of aerial plant parts, where they may result in diseases of considerable economic importance (3,4,11,12,20,26, 29,35,39,53). Some of these air-borne Fusaria are rarely encountered in isolations from soil or roots. The widespread distribution of most *Fusarium* populations can probably be attributed to two key factors, their ability to colonize a wide range of substrates and their efficient mechanisms for dispersal in time and/or space. These two factors also contribute to their ability to readily adapt to new ecological niches created by man.

In this article I propose to discuss the propositions summarised above in three parts: modes of existence, geographic distribution, and the influence of man's activities. The nomenclature used is according to the nine-species system of Snyder & Hansen as illustrated by Toussoun and Nelson (46).

Modes of Existence

An analysis of the behaviour of species and infra-species populations of the Fusaria indicates that there are three basic modes of existence. Two of these can be designated by the traditional terms, soil-borne and air-borne, while a third is included for those Fusaria which are common in soil and have efficient mechanisms for air dispersal. Survey data and routine isolations do, however, indicate that many Fusaria can occur outside their commonly accepted habitats. Abnormal weather conditions and man's activities probably account for such occurrences. I will define each mode of existence and then discuss the ecology of Fusaria which are typical representatives of each mode.

The majority of the Fusaria are normally found in or on soil, where they exist as colonizers of living plant parts or plant residues within the soil or adjacent to the soil surface, normally within 100 cm (4,6,11,14,24,28,33,41,51,52). In the absence of a suitable substrate, these fungi persist as resistant (dormant) hyphae in plant residues colonized parasitically or saprophytically, or as discrete propagules such as chlamydospores and resistant conidia. Traditionally these populations have been regarded as soil-borne fungi, although soil-based would seem a more appropriate term. Typical representatives are *Fusarium oxysporum, F. solani, F. roseum* 'Culmorum,' *F. roseum* 'Graminearum' Group 1 (12), and *F. roseum* 'Equiseti.'

A number of Fusaria normally exist as primary or secondary colonizers of aerial plant parts (4,38,53). These fit the traditional definition of air-borne fungi. They each have at least one efficient method of air-dispersal, viz. passive wind-borne or rain-splash dispersal or active dispersal by ascospores. In the absence of suitable substrates they persist as resistant hyphae or conidia, or as immature perithecia in dead twigs and branches attached to healthy branches in the plant canopy or in residues on the soil surface (4,39,52,53). This mode of existence is characteristic of such populations as *F. rigidiusculum, F. lateritium,* and *F. roseum* 'Graminearum' Group 2.

I have included a third mode of existence, additional to the traditional soil-borne and air-borne categories, to focus attention on those populations which are generally regarded as soil-borne fungi but which are also adapted to dispersal in the atmosphere and are common colonizers of aerial plant parts. *Fusarium moniliforme* and *F. roseum* 'Semitectum' are good examples.

Soil-borne Fusaria Although these Fusaria are abundant in soil and commonly associated with roots and the lower parts of plants, they are also encountered on aerial plant parts. Their presence on aerial parts is not surprising and is not different from the behaviour of many other soil-based genera. It can be attributed to their passive dispersal in the atmosphere (17,19,35,36), their ability to utilise a wide range of substrates, and in some instances, abnormal environmental conditions (12). The fact that these fungi can be dispersed in the atmosphere is often overlooked and is not well documented (19,20,35, 36). The soil-borne Fusaria often form sporodochia on substrates on or just above the soil surface, and the spores in these fruiting bodies can be dispersed by rain splash directly or after they are washed onto the soil surface (17,19). Conidia in sporodochia are probably not suited to dispersal by wind in dry weather because they are usually cemented to the substrate (19). It has also been suggested that air dispersal of these Fusaria may be associated with dust particles (8,20,35), and Ooka and Kommedahl (35) have presented evidence of long-distance dispersal in dust storms over several hundred kilometres. It is probable that these fungi will also be common in the atmosphere where soil is being cultivated or crops are being harvested. The role of insects in aerial dispersal of soil-borne Fusaria has not been studied in detail but is probably significant in some circumstances (4,8). Since these fungi are present in the atmosphere, some of their inocula will be deposited on aerial plant surfaces where they may remain inactive or, given suitable conditions, germinate and colonize the substratum or part thereof. Such colonization may be parasitic or saprophytic in nature.

Fusarium oxysporum is one of the most common soil-borne Fusaria (4,14,16,24,25,28, 51,52,53). Some infra-specific populations of this highly variable fungus are very pathogenic, causing wilts and crown and root rots, while others appear to be vigorous saprophytes which are common colonizers of senescent or damaged plant tissue (2,4,21,23,25, 53). It is one of the most common Fusaria isolated from roots and crowns and is active under a wide range of environmental conditions (4,6,21,23,24,25,53). This species readily forms chlamydospores and so is well adapted to long-term survival in soil (4,31). *Fusarium oxysporum* can also persist as hyphae in organic residues (4). Thus pathogenic members of this species are particularly difficult to eliminate from infested soil by rotation or long fallow. Elimination of some of the wilt Fusaria is rendered more difficult by their ability to colonize non-susceptible species (1,2,18). Although *F. oxysporum* has been isolated from the air (20,36) it is not well adapted to air dispersal. Its presence in the atmosphere can probably be attributed to wind-blown soil or organic debris (20,35). Abundant sporodochia are often formed on the stems of plants affected by Fusarium wilt and these would provide a source of inoculum for localised dispersal by rain splash. *Fusarium oxysporum* is not known to form perithecia. I am not aware of any reports of its association with localised diseases of aerial plant parts and none are listed by Booth (4). Air dispersal may, however, be important in the ecology of the fungus in some situations. For example Rowe et al. (36) have produced evidence that air-borne microconidia of *F. oxysporum* f. sp. *radicis-lycopersici*, which causes crown and root rot of tomato (21), can bring about re-colonization of steamed soil in greenhouses. The sources of the air-borne microconidia were presumed to be infested tomato-plant residues outside the greenhouses.

In temperate areas *F. solani* also exists as a common soil-borne fungus (4,14,23,28, 51,52), which includes both saprophytes and parasites (40). Some parasitic members are important root and crown rot pathogens (4,7,8,53), while others have been implicated as canker-causing organisms of several hardwood trees (4). In the absence of suitable substrates the fungus survives primarily as chlamydospores, the importance of which was first suggested by Burkholder (8) and later confirmed by Nash et al. (32) and others (7,31). Data from studies by Schroth and Hendrix (37) suggest that survival of *F. solani* f. sp. *phaseoli* in agricultural land is enhanced by temporary supplies of nutrients in diffusates from non-susceptible plants and crop residues which enable the fungus to vegetate and form new chlamydospores. It is highly probable that the survival of other members of *F. solani* is also enhanced by diffusates from some plant and crop residues. *Fusarium solani* has occasionally been isolated from air in temperate areas, where its presence is probably a result of wind-blown soil or organic debris (20,35) or rain-splashed conidia from sporodochia (17), but perithecia are rarely encountered in temperate areas, where *F. solani* is generally not a common colonizer of aerial plant parts (4). In contrast, in tropical areas this fungus is not only abundant in soil but is also commonly isolated from aerial plant parts either as a superficial colonizer or in association with twig die-back, cankers, and leaf spots (53; L. W. Burgess, *unpublished data*). I suggest that in these areas *F. solani* is probably a common member of the air spora and thus would be deposited regularly on aerial plant parts. Air-borne inocula could originate from two sources. First, the combination of high rainfall and sporodochial formation would provide an abundant source of rain-splashed conidia. Second, the fungus commonly forms fertile perithecia in the tropics on aerial plant parts and residues (4); these would enable active dispersal in the atmosphere by ascospores. *Fusarium solani* provides an excellent example of how the environment can alter the pattern of existence of a particular *Fusarium*.

Fusarium roseum 'Equiseti' is also extremely common in and on soil, where it exists as a vigorous saprophyte and survives as chlamydospores or mycelial fragments either free in the soil matrix or in organic residues in or on soil (31; L. W. Burgess, *unpublished data*). It is widely distributed, being most abundant in warm-temperate and sub-tropical areas (4,22). Although it normally exists as a saprophyte, *F. roseum* 'Equiseti' has occasionally been implicated in disease problems (4,22). It is a common colonizer of senescing wheat and maize tissues, from which it can be readily isolated (6,11). Since it is extremely common in soil, it is not surprising that it has been isolated from the air, where it is thought to be associated with wind-blown soil particles (20; P. E. Nelson, *personal communication*). The fungus also forms conidia in sporodochia on plant residues from which dispersal could occur by rain splash. Dispersal by rain-splash or wind-blown soil would account for its ability to colonize aerial plant parts, particularly those close to the soil surface.

Some populations of *F. roseum* 'Equiseti' may actually be adapted to wind dispersal. Many isolates of this fungus from semi-arid regions of western New South Wales (Australia) form abundant microconidia from polyphialides in culture (L. W. Burgess, P. E. Nelson, and T. A. Toussoun, *unpublished data*). If the fungus produces these spores under natural conditions on soil-surface residues, they would facilitate wind-dispersal of the fungus.

Air-borne Fusaria These Fusaria are colonizers of aerial plant parts and are adapted to dispersal in the atmosphere by passive or active means. They are rarely encountered in soil. Many representatives of this group are important pathogens causing, for example, twig blights and cankers, stem cankers, collar rots, and inflorescence blights. Rain-splash of conidia constitutes the main means of passive dispersal, although insects probably have a role in some situations. Active dispersal of ascospores provides a major means of dispersal for those members which form abundant perithecia in nature. These fungi survive on infested plant parts that may either remain attached to the plant or fall onto the soil surface. They are rarely isolated from soil, although I have regularly isolated *F. roseum* 'Graminearum' Group 2 (*Gibberella zeae*) from debris in maize soils where the fungus was presumably incorporated by cultivation.

The population of *F. lateritium* which causes twig blight and dieback of ornamental mulberry trees in the Sydney area of New South Wales, Australia, is a typical air-borne *Fusarium*. This fungus is active parasitically in wet conditions in spring, with ascospores providing the main source of primary inoculum. Sporodochia and perithecia develop on dead twigs over summer, and the fungus then spreads by rain-splash of condia as well as by ascospores. The fungus overseasons as perithecia on dead twigs in the tree canopy or on the soil surface. On occasions I have also found viable conidia in the vicinity of perithecia on dead twigs at the end of winter, which climatically is quite mild. Infection of twigs is more severe in wet springs in trees where perithecia are abundant on dead infested twigs within the tree canopy. In these situations the proximity of the inoculum source to the fresh substrate (young twigs) probably accounts for severe infection. I have not isolated the fungus from soil under affected trees.

Fusarium lateritium 'Stilboides,' the fungus responsible for Storey's bark disease, scaly bark, and collar rot of coffee (38), is closely related to the *F. lateritium* on mulberry and is also adapted to dispersal in the atmosphere. However, the formation of perithecia by *F. lateritium* 'Stilboides' in nature has not been observed, and detailed studies in Malawi by Siddiqi and Corbett (39) indicate that the fungus is spread mainly by rain-splashed conidia. The conidia are formed in sporodochia which develop abundantly in association with lesions. The fungus persists through the dry season in Malawi in old lesions as mycelium which starts sporulating with the onset of wet weather. Conidia of *F. lateritium* 'Stilboides' are not dispersed by wind (39). The same authors occasionally isolated the fungus from the integument of certain insects and found that the fungus survived very poorly in soil.

Fusarium rigidiusculum, a tropical species, has commonly been isolated from aerial plant parts colonized either saprophytically or parasitically (4,16,53) and appears to be well adapted to dispersal in the atmosphere. Unfortunately there are no reports of studies on the dispersal of this species. It forms fertile perithecia in nature (4), and in culture it produces both macroconidia in sporodochia and microconidia in chains. Thus it would appear to be adapted for both active dispersal by ascospores and passive dispersal by rain-splashed or wind-blown conidia. The fungus does not form chlamydospores in culture and there is no evidence that it occurs in soil.

Fusaria in subterranean and aerial habitats These Fusaria are not only common colonizers of subterranean plant parts and residues in soil but also of aerial plant parts and residues on the soil surface. Traditionally they have been regarded as soil-borne fungi.

A typical representative of this category is *F. moniliforme,* a species which is widely distributed in tropical and temperate regions (4,11,16,35,53). It is common in soil and has been isolated from the roots of members of the Gramineae, Leguminoseae, and other families. *Fusarium moniliforme* is also commonly encountered as a primary or secondary colonizer of aerial plant parts. It is, for example, responsible for stalk rots of maize and sorghum, head rots, stem and top rot of sugar cane ('pokkah boeng' disease), and rot of banana flowers and fruit (4,53). It has been commonly isolated from cultivated and rangeland soils (14,24,25,27,51,52,54). Apparently the fungus survives as mycelium in plant residues in soil or on the soil surface (31,34). Chlamydospore formation by *F. moniliforme* has not been observed in culture (4) or soil (31,34).

Fusarium moniliforme is well adapted for dispersal in the atmosphere in dry or wet weather as it forms macroconidia in sporodochia and chains of microconidia in dry powdery masses. Ooka and Kommedahl (35), in studies on its dispersal in maize fields, concluded that conidia can be dislodged and dispersed by wind or by rain-splash but they did not identify the type of conidia involved. Presumably the microconidia are predominantly involved in wind dispersal, whereas both microconidia and macroconidia could be dispersed by rain-splash. Ooka and Kommedahl (35) also suggested that the fungus could be dispersed in wind-blown soil and fragments of infested crop residues. Although this fungus is reported to produce perithecia on dead plant material (4), there is no evidence that ascospores have a significant role in the dispersal of the fungus (53; L. W. Burgess, *unpublished data*).

Fusarium moniliforme 'Subglutinans' can also be included in this category. It has regularly been isolated from soil (51,52) and is a common colonizer of aerial plant parts (3,11,26). It has, for example, long been recognised as a cause of cob and grain rot of maize (4,10). Recent studies have also shown it to be the cause of pitch canker of southern pines in the United States (26) and fruit rot of pineapple in Brazil (3). Infection of aerial plant parts presumably results from conidia dispersed in the air. However, *F. moniliforme* 'Subglutinans' is known to produce perithecia in nature (4), so ascospores may aid dispersal in some environments. There have been no studies on the mode of dispersal of the conidia produced by this fungus, but it does form abundant macroconidia in sporodochia and microconidia in false-heads. Rain-splash of these spores would provide an efficient means of localised air dispersal.

Fusarium roseum 'Semitectum,' another *Fusarium* which is common in soil, is also adapted to dispersal in the atmosphere and the colonization of aerial plant parts. It is widely distributed and is particularly common in tropical and sub-tropical areas (4), where it is generally regarded as a secondary colonizer of plant tissue. It has commonly been isolated from soil (14,51,52), where presumably it survives as mycelium in residues or as chlamydospores. Although chlamydospore formation has not been demonstrated in soil, it has been observed in culture (4). Lukezic and Kaiser (29) studied the dispersal of *F. roseum* 'Semitectum' in a Honduras banana plantation (these authors referred to the fungus by the name *F. roseum* 'Gibbosum,' but it has since been confirmed as *F. roseum* 'Semitectum' [F. L. Lukezic, *personal communication*]). They found that conidia formed on aerial conidiophores were easily dislodged by wind and dispersed in the atmosphere with wind speeds as low as 2.4 kph. Maximum spore concentration in air was found to occur in well-defined peaks in the afternoon on rainless days. The peaks occurred following a period of light rain. On one occasion the peak occurred during heavy rain, indicating rain-splash dispersal. In laboratory tests Lukezic and Kaiser (29) also found that conidia formed in sporodochia on banana fruit could not be dislodged by air blasts. Such conidia would be subject to rain-splash dispersal under natural conditions.

Although the traditional ecological categories, soil-borne and air-borne, can be applied to the majority of the Fusaria, the evidence presented indicates that some populations can colonize substrates outside their accepted habitat. There are usually simple explanations for such occurrences and the Fusaria should not be credited with unusual or unique qualities because of their ubiquitous distribution.

Variation with respect to habitat In defining the mode of existence of a particular *Fusarium* it must be borne in mind that it may consist of two or more morphologically similar populations which differ in respect to their modes of existence and physiological attributes. For example, two populations designated Groups 1 and 2 have been shown to exist within *F. roseum* 'Graminearum' (12). They are distinguished by the presence or absence of perithecia in culture. Representatives of these two populations form similar macroconidia in culture but produce slightly different colonies, both in respect to morphology and growth rate, on potato-dextrose agar. Members of Group 1 do not form perithecia in culture, do so very rarely in nature, and are normally associated with crown and root rots of cereals and grasses (6,12). In contrast, members of Group 2 form perithecia abundantly in culture and in nature and are normally associated with diseases of aerial plant parts, both as primary and secondary colonizers (12). They are responsible for diseases such as head scab of wheat, barley, and oats; stalk and cob rot of maize and pearl millet; and stub-dieback of carnations (11,12).

Members of Group 1 survive as hyphae in residue in and on soil for at least one to two years under certain environmental conditions (5,50). They do not produce chlamydospores in soil (50). Presumably infection of plants by members of Group 1 is a result of direct contact between the plant parts and infested residue. However, the fungus can form abundant sporodochia on leaf-sheath tissue of cereals and grasses affected by crown rot under wet conditions. It is thought that rain-splash dispersal of these spores provides the source of inoculum which results in occasional infection of the inflorescences of cereals and grasses (head scab) by members of Group 1 (12).

Members of Group 2 survive as perithecia and hyphae on and in residues, respectively (52). Under suitable conditions the perithecia actively discharge ascospores, which are thought to act as the main source of inoculum for infection of aerial plant parts. A limited amount of infection probably results from rain-splashed macroconidia or by direct contact with infested residue.

Although the two populations probably occur in most areas, one is usually more abundant because it is favoured by the prevailing environmental conditions and/or the presence of particular hosts (12). Thus members of Group 1 are usually more common in drier areas, possibly because dry conditions favour the survival of this fungus in residue (5). Crown rot of wheat and other grasses caused by members of Group 1 is more severe when infected plants are affected by dry conditions. This factor may also favour an increase in the incidence of the fungus. However, it is known that symptomless wheat plants can be extensively colonized by members of Group 1 under optimal soil moisture conditions (44).

Members of Group 2 tend to be more common in areas where it is warm and humid for at least part of the year when suitable substrates are available (11,12). Warm humid conditions favour the formation of perithecia and ascospore discharge (47) and hence infection and colonization of aerial plant parts.

Seedling blight occurs in both types of environment. In humid regions it is most probably the result of seed-borne inoculum of Group 2 on or in seed from heads affected by scab or rot. In dry areas seedling death is less common but can occur in drought periods and is normally caused by soil-borne inoculum of Group 1.

When abnormal environmental conditions prevail, isolates from both groups will be found occupying the same ecological niche. For example, members of both groups have occasionally been isolated from adjacent wheat heads affected by head scab. This has been noted in the northern wheat belt of New South Wales, Australia, when wet conditions prevailed around flowering time in crops which contained plants which were extensively colonized by members of Group 1 at the base and where crop residues contained perithecia of Group 2. Ascospores probably constituted the inoculum which resulted in head scab caused by Group 2, whereas the scab caused by Group 1 was probably the result of infection by rain-splashed macroconidia.

It is important to distinguish such populations as Groups 1 and 2 of *F. roseum* 'Graminearum.' Both are important plant pathogens and the selection of control measures is, to a degree, governed by their respective modes of existence. Such populations may also exist within other species of the Fusaria (e.g., *F. lateritium, F. solani*).

Geographic Distribution

Caution must be exercised in generalising about the distribution and relative abundance of the Fusaria in different climatic zones for several reasons. First, there have been remarkably few intensive and systematic surveys of the Fusaria associated with a particular substrate or with soil from different climatic zones. Second, it is difficult to compare the data from surveys made by different workers because techniques vary considerably. The range and frequency of isolation of Fusaria from soil is reflected in the bias imposed by isolation methods and the time of sampling (25,28,48,51). Similarly, the spectrum of Fusaria isolated and their frequency of isolation from plant material also reflects the choice of techniques (11,25). Third, the interpretation of data obtained by different workers is further complicated by the use of different taxonomic systems and confusion as to the nature and boundaries of certain populations. Finally, the objective measurement of the abundance of the Fusaria is difficult because there are no individuals analogous to those of higher plants and animals, which can be simply and accurately counted. The frequency of isolation from soil or plant material is the parameter normally adopted as a measure of relative abundance. However, frequency of isolation of a *Fusarium* does not necessarily reflect the actual biomass of the fungus present in soil or plants because colonies of the Fusaria on soil or plant isolation plates may originate from mycelia or spores (discrete or in clumps). Colonies (mycelia) of the Fusaria are indeterminate in size, and spore formation varies between species. Thus the interpretation of isolation results requires considerable caution. Correlating differences in isolation frequencies with particular environmental factors such as temperature, rainfall, substrate, or soil type is complicated because these factors are interrelated.

Influence of temperature and rainfall on distribution In spite of the above limitations, it is possible to make some generalisations about the distribution of the Fusaria in respect to temperature and rainfall. The majority of these fungi are widely distributed, occurring in a variety of climatic zones (4,16). However, some Fusaria do appear to be restricted to certain climatic zones and are rarely encountered in other areas. For example, at least two members, *F. rigidiusculum* and *F. roseum* 'Longipes,' are relatively common in tropical and sub-tropical regions but have rarely been encountered in temperate areas (4,16,53; L. W. Burgess, *unpublished data*). In contrast *F. roseum* 'Sambucinum' is common in cold-temperate and alpine soils, whereas it is uncommon in the tropics (16,23,24,28; L. W. Burgess, *unpublished data*). Similarly, members of *F. nivale* associated with grasses have been isolated from many cold-temperate and temperate areas but not from the tropics (4,16,41,53).

Survey data indicate that many of the Fusaria which are widely distributed are more common in some areas. *Fusarium roseum* 'Equiseti' and *F. roseum* 'Acuminatum' are widely distributed, but the former is usually isolated more frequently from warm-temperate and sub-tropical areas, whereas the latter is isolated more frequently from temperate and cold-temperate areas (4,6,11,14,15,16,22,23,24,51,52). In my experience *F. roseum* 'Equiseti' is also more common in drier areas than in areas of high rainfall, and it is abundant in debris from semi-arid rangeland soils of eastern Australia (L. W. Burgess, *unpublished data*).

Fusarium moniliforme and *F. moniliforme* 'Subglutinans' are also widely distributed, but the results of systematic suverys involving isolations from plants, soil, and seed indicate that *F. moniliforme* is isolated more frequently than *F. moniliforme* 'Subglutinans' from subtropical and temperate areas. Conversely, *F. moniliforme* 'Subglutinans' is usually isolated more frequently from temperate and cool-temperate areas (4,6,11,24,30,51,53).

Although *F. solani* is widespread in tropical and temperate areas, it tends to be more common in warmer regions and appears to have an affinity for areas of high rainfall or irrigated soils (4,6,14,15,16,23,28,41,51,53; L. W. Burgess, *unpublished data*).

Variation with respect to climate Within some populations of the Fusaria certain isolates tend to be more common in particular climatic regions. For example, isolates of *F. roseum* 'Semitectum' which produce very long spores are more frequently isolated from tropical than temperate areas of eastern Australia (L. W. Burgess, *unpublished data*). Isolates of *F. roseum* 'Equiseti' which form very long macroconidia and abundant microconidia have been isolated more frequently from semi-arid rangeland soils of western New South Wales than from higher rainfall areas of eastern New South Wales, Australia (L. W. Burgess, *unpublished data*). Average temperatures in the semi-arid areas are significantly higher than in the wetter areas. The influence of climate on the relative abundance of the two populations within *F. roseum* 'Graminearum,' Groups 1 and 2 has been discussed above.

Influence of Man's Activities

Survey data and experimental results indicate that the relative abundance, and range, of Fusaria isolated from soil can be related to climate (24,28,51), vegetation (41), soil moisture (27,43), soil type (42,45), and fertility (41,49). The frequency of isolation and range of Fusaria isolated from plants can also be correlated with environmental factors (4,6,11,23,24,41,53). Thus it is not surprising that man's activities influence the relative abundance and the range of Fusaria present at a particular location. His activities affect the Fusaria through changes in the nature and availability of substrates, their influence on dispersal in time and space, and through modification of the environment.

Changes in the nature and availability of substrate occur whenever a pasture or crop rotation is altered. Snyder and Nash (41), for example, found that *F. roseum* 'Culmorum' was extremely common in long-term cereal plots at Rothamsted in England, whereas it was quite rare in long-term plots planted to broad-leaf crops in the same locality. All plots were cultivated. The use of resistant cultivars of a particular host crop may also lead to changes in the population of the relevant pathogen. Such changes have been well documented in respect to the wilt Fusaria (2).

The influence of cultivation on Fusaria in undisturbed grassland or forest soils is difficult to assess because it is usually accompanied by a dramatic change in the nature and availability of substrate and the use of fertilizers. Comparisons of Fusaria in cultivated and adjacent undisturbed soils indicate that the cultivation and cropping of soils does have a significant effect on the abundance of Fusaria in soil, and in some instances on the range of populations which can be recovered (28,33,45,54). For example, Windels and Kommedahl (54) found that the recovery of Fusaria from undisturbed prairie soils differed significantly from the recovery from similar soils planted to maize. Their results indicated that cultivation and maize production resulted in a significant increase in the abundance of *F. roseum* and *F. moniliforme,* whereas other species were generally not affected.

Changes in the abundance of Fusaria which follow the clearing of forests or woodlands can be quite dramatic. This is not surprising as survey data indicate that the Fusaria are common in cultivated soils but much less common in forest and woodland soils (28,41,45). An appropriate example is provided by the influence of cultivation, cropping, and pasture production on the Atherton Plateau in north Queensland, Australia. This plateau is in the tropics and prior to settlement in the last century was covered by rainforest. The Fusaria are common in the cultivated and pasture soils of the plateau. *Fusarium roseum* 'Graminearum' Group 2 is extremely common as a colonizer of maize stalks, senescing grass stems, and plant residues on the soil surface. However, this fungus has not been isolated from soil or plant samples from remnants of the rainforest on the plateau. Indeed, the recovery of Fusaria generally from these rainforest soils was quite low (L. W. Burgess, *unpublished data*). Warren and Kommedahl (49) have shown that the retention as opposed to the removal of residues can also affect the level of various Fusaria in soil.

The dispersal of the Fusaria in space can be affected by man's activities, thus changing the distribution patterns of these fungi. Dispersal can be enhanced through the transport of contaminated seed (4,10,13), soil (4), propagating material (4), and dust caused by cultivation (8,20,35). Such effects will probably have a negligible impact on the occurrence of the cosmopolitan Fusaria but could be very important in the successful dissemination of some pathogens which are not widely distributed.

The survival (dispersal in time) of Fusaria in crop residues will depend on whether these residues are burnt, retained on the soil surface, or incorporated. Unfortunately there is a lack of information on this aspect of survival. It is known that the retention of maize residues colonized by *F. roseum* 'Graminearum' Group 2(12) (*Gibberella zeae*) on the soil surface enhances the survival of this fungus as perithecia (L. W. Burgess, *unpublished data*). If such residues are incorporated, the perithecia are mechanically disrupted, decomposed, or cannot release their ascospores into the atmosphere. Thus the retention of residues infested with this fungus on the soil surface should enhance its dispersal and hence its ability to colonize new substrates. There is a need for detailed quantitative studies of the effect of residues on survival, per se, of the Fusaria in relation to their abundance and disease severity.

The survival of the Fusaria as chlamydospores may be influenced by the choice of crop species, as discussed earlier (37). The persistence of the wilt Fusaria in non-susceptible species will also be influenced by man's choice of crops (1,2,18).

Modification of the physical environment by man can also have significant effects on the ecology of the Fusaria. The introduction of irrigation to an arid area, for example, permits a change in the nature and the availability of substrates and alters the physical environment, both factors which can affect the occurrence of the Fusaria. Irrigation of some semi-arid inland areas of New South Wales, Australia, has permitted maize production, and *F. roseum* 'Graminearum' Group 2 has become established in these maize fields despite the fact that it has not been recovered from adjacent undisturbed rangeland which has not been irrigated (L. W. Burgess, *unpublished data*).

The use of fertilizers, especially nitrogen, has also been correlated with frequency of isolation of the Fusaria from soil and plants (41,49). Studies by Butler (9) also suggest that the nitrogen status of soil may influence saprophytic persistence. Further studies on the influence of soil nitrogen on survival are warranted.

The above examples provide a brief summary of some of man's activities which can have a profound influence on the distribution and abundance of the Fusaria and which in some instances can result in their establishment in areas where the environment would normally limit their activity.

Conclusions

Some generalisations can be made concerning the modes of existence and geographic distribution of the Fusaria. However, there is an urgent need for more systematic surveys of the Fusaria associated with a particular substrate (crop) or soils from a variety of contiguous climatic areas differing in respect to rainfall, temperature, and other environmental factors. There is also a need for standardisation of the techniques adopted for surveys so that the results of different workers can be compared. Such surveys not only provide information on the ecology of the Fusaria but can also provide an abundance of isolates for detailed studies on the characteristics of particular populations.

Literature Cited

1. Armstrong, G.M., and J.K. Armstrong. 1948. Nonsusceptible hosts as carriers of wilt Fusaria. Phytopathology 38: 808–826.
2. Armstrong, G.M., and J.K. Armstrong. 1975. Reflections on the wilt Fusaria. Annu. Rev. Phytopathol. 13: 95–103.
3. Bolkan, H.A., J.C. Dianese, and F.P. Cupertino. 1979. Pineapple flowers as principal infection sites for *Fusarium moniliforme* var. *subglutinans*. Plant Dis. Rep. 63: 655–657.
4. Booth, C. 1971. The genus *Fusarium*. Commonwealth Mycological Institute, Kew, Surrey, England. 237 p.
5. Burgess, L.W., and D.M. Griffin, 1968. The recovery of *Gibberella zeae* from wheat straws. Australian J. Exp. Agric. Anim. Husb. 8: 364–370.
6. Burgess, L.W., A.H. Wearing, and T.A. Toussoun. 1975. Surveys of Fusaria associated with crown rot of wheat in eastern Australia. Australian J. Agric. Res. 26: 791–799.
7. Burke, D. W. 1965. Fusarium root rot of beans and behaviour of the pathogen in different soils. Phytopathology 55: 1122–1126.
8. Burkholder, W.H. 1919. The dry root-rot of the bean. Cornell Univ. Agric. Expt. Sta. Mem. 26: 999–1033.
9. Butler, F.C. 1959. Saprophytic behaviour of some cereal root-rot fungi. 4. Saprophytic survival in soils of high and low fertility. Ann. Appl. Biol. 47: 28–36.
10. Edwards, E.T. 1936. Maize seed selection and disease control. The problem of internal seedborne infection. N.S.W. Dep. Agric. Gazette 47: 303–306.
11. Francis, R.G., and L.W. Burgess. 1975. Surveys of Fusaria and other fungi associated with stalk rot of maize in eastern Australia. Australian J. Agric. Res. 26: 801–807.
12. Francis, R.G., and L.W. Burgess. 1977. Characteristics of two populations of *Fusarium roseum* 'Graminearum' in eastern Australia. Trans. Brit. Mycol. Soc. 68: 421–427.
13. Gordon, W.L. 1952. The occurrence of *Fusarium* species in Canada. II. Prevalence and taxonomy of *Fusarium* species in cereal seed. Can. J. Bot. 30: 209–251.
14. Gordon, W.L. 1956. The occurrence of *Fusarium* species in Canada. V. Taxonomy and geographic distribution of *Fusarium* species in soil. Can. J. Bot. 34: 833–846.
15. Gordon, W.L. 1956. The taxonomy and habitats of the *Fusarium* species in Trinidad, B.W.I. Can. J. Bot. 34: 847–864.
16. Gordon, W.L. 1960. The taxonomy and habitats of *Fusarium* species from tropical and temperate regions. Can. J. Bot. 38: 643–658.
17. Gregory, P.J., E.J. Guthrie, and M.E. Bunce. 1959. Experiments in splash dispersal of fungus spores. J. Gen. Microbiol. 20: 328–354.
18. Hendrix Jr., F.F., and L.W. Nielson. 1958. Invasion and infection of crops other than the forma suscept by *Fusarium oxysporum* f. *batatas* and other formae. Phytopathology 48: 224–228.
19. Hirst, J.M. 1965. The dispersal of soil microorganisms, p. 69–81. *In* K.F. Baker and W.C. Snyder (eds.), Ecology of soil-borne plant pathogens, Univ. Calif. Press, Berkeley.
20. Horst, R.K., P.E. Nelson, and T.A. Toussoun. 1970. Aerobiology of *Fusarium* spp. associated with stem rot of *Dianthus caryophyllus*. Phytopathology 60: 1296 (Abstr.).
21. Jarvis, W.R., and R.A. Shoemaker. 1978. Taxonomic status of *Fusarium oxysporum* causing foot and root rot of tomato. Phytopathology 68: 1679–1680.
22. Joffe, A.Z., and J. Palti. 1967. *Fusarium equiseti* (Cda.) Sacc. in Israel. Israel J. Bot. 16: 1–18.
23. Kommedahl, T., and E.I. Siggeirsson. 1973. Prevalence of *Fusarium* species in roots and soil of grassland in Iceland. Res. Inst. Nedhri As. Hveragerdhi, Iceland. Bull. 14. 27 p.
24. Kommedahl, T., C.E. Windels, and D.S. Lang. 1975. Comparison of Fusarium populations in grasslands of Minnesota and Iceland. Mycologia 67: 38–44.
25. Kreutzer, W.A. 1972. *Fusarium* spp. as colonists and potential pathogens in root zones of grassland plants. Phytopathology 62: 1066–1070.
26. Kuhlman, E.G., L.D. Dwinell, P.E. Nelson, and C. Booth. 1978. Characterization of the *Fusarium* causing pitch canker of southern pines. Mycologia 70: 1131–1143.
27. Lim, G. 1967. Fusarium populations in rice field soils. Phytopathology 57: 1152–1153.
28. Lim, G., and C.H. Chew. 1970. *Fusarium* in Singapore soils. Plant Soil 33: 673–677.
29. Lukezic, F.L., and W.J. Kaiser. 1966. Aerobiology of *Fusarium roseum* 'Gibbosum' associated with crown rot of boxed bananas. Phytopathology 56: 545–548.
30. Marasas, W.F.O., N.P.J. Kriek, V.M. Wiggins, P.S. Steyn, D.K. Towers, and T.J. Hastie. 1979. Incidence, geographic distribution, and toxigenicity of *Fusarium* species in South African corn. Phytopathology 69: 1181–1185.
31. Messiaen, C.M., P. Mas, A. Beyries, and H. Vendran. 1965. Recherches sur l'écologie des champignons parasites dans le sol. IV. Lyse mycélienne et formes de conservation dans le sol chez les *Fusarium*. Ann. Epiphyties 16: 107–128.
32. Nash, S.M., T. Christou, and W.C. Snyder. 1961. Existence of *Fusarium solani* f. *phaseoli* as chlamydospores in soil. Phytopathology 51: 308–312.
33. Nash, S.M., and W.C. Snyder. 1965. Quantitative and qualitative comparisons of Fusarium popula-

tions in cultivated fields and non-cultivated parent soils. Can. J. Bot. 43: 939–945.
34. Nyvall, R.F., and T. Kommedahl. 1968. Individual thickened hyphae as survival structures of *Fusarium moniliforme* in corn. Phytopathology 58: 1704–1707.
35. Ooka, J.J., and T. Kommedahl. 1977. Wind and rain dispersal of *Fusarium moniliforme* in corn fields. Phytopathology 67: 1023–1026.
36. Rowe, R.C., J.D. Farley, and D.L. Coplin. 1977. Airborne spore dispersal and recolonization of steamed soil by *Fusarium oxysporum* in tomato greenhouses. Phytopathology 67: 1513–1517.
37. Schroth, M.N., and F.F. Hendrix Jr. 1962. Influence of nonsusceptible plants on the survival of *Fusarium solani* f. *phaseoli* in soil. Phytopathology 52: 906–909.
38. Siddiqi, M.A., and D.C.M. Corbett. 1963. Coffee bark diseases in Nyasaland. I. Pathogenicity, description, and identity of the causal organism. Trans. Brit. Mycol. Soc. 46: 91–101.
39. Siddiqi, M.A., and D.C.M. Corbett. 1968. Coffee bark diseases in Malawi. II. Properties of the causal organism and conditions favouring the disease. Trans. Brit. Mycol. Soc. 51: 129–135.
40. Snyder, W.C., and H.N. Hansen. 1941. The species concept in *Fusarium* with reference to section Martiella. Amer. J. Bot. 28: 738–742.
41. Snyder, W.C., and S.M. Nash. 1968. Relative incidence of Fusarium pathogens of cereals in rotation plots at Rothamsted. Trans. Brit. Mycol. Soc. 51: 417–425.
42. Stotsky, C., and R.T. Martin. 1963. Soil mineralogy in relation to the spread of Fusarium wilt of banana in Central America. Plant Soil 18: 317–337.
43. Stover, R.H. 1953. The effect of soil moisture on *Fusarium* species. Can. J. Bot. 31: 693–697.
44. Taylor, P.W.J., and L.W. Burgess. 1980. The influence of water stress on colonization of wheat by *Fusarium roseum* 'Graminearum' Group 1 in relation to symptom development. Australian Plant Pathol. Soc. Newsletter 9: 34 (Abstr.).
45. Toussoun, T.A. 1975. Fusarium-suppressive soils, p. 145–151. *In* G.W. Bruehl (ed.), Biology and control of soil-borne plant pathogens, Amer. Phytopathol. Soc., St. Paul.
46. Toussoun, T.A., and P.E. Nelson. 1976. A pictorial guide to the identification of *Fusarium* species (2nd Ed.). Pa. State Univ. Press, University Park, Pa.
47. Tschanz, A.T., R.K. Horst, and P.E. Nelson. 1976. The effect of environment on sexual reproduction of *Gibberella zeae*. Mycologia 68: 327–340.
48. Warcup, J.H. 1965. Growth and reproduction of soil microorganisms in relation to substrate, p. 52–68. *In* K.F. Baker and W.C. Snyder (ed.), Ecology of soil-borne plant pathogens, Univ. Calif. Press, Berkeley.
49. Warren, H.L., and T. Kommedahl. 1973. Fertilisation and wheat refuse effects on *Fusarium* species associated with wheat roots in Minnesota. Phytopathology 63: 103–108.
50. Wearing, A.H., and L.W. Burgess, 1977. Distribution of *Fusarium roseum* 'Graminearum' Group 1 and its mode of survival in eastern Australian wheat belt soils. Trans. Brit. Mycol. Soc. 69: 429–442.
51. Wearing, A.H., and L.W. Burgess. 1977. The relative frequency of isolation of Fusaria from wheat soils in the eastern Australian wheat belt. Australian Plant Pathol. Soc. Newsletter 6: 26–28.
52. Wearing, A.H., and L.W. Burgess. 1978. Distribution and mode of survival of *Fusarium roseum* 'Graminearum' Group 2 in maize soils of eastern Australia. Trans. Brit. Mycol. Soc. 70: 480–486.
53. Wellman, F.L. 1972. Tropical American plant disease. Scarecrow Press, Inc., Metuchen, N.J. 989 p.
54. Windels, C.E., and T. Kommedahl. 1974. Population differences in indigenous *Fusarium* species by corn culture of prairie soil. Amer. J. Bot. 61: 141–145.
55. Wollenweber, H.W., and O.A. Reinking. 1935. Die Fusarien. Paul Parey, Berlin. 355 p.

22 Water Relations in the Biology of *Fusarium*

R. James Cook

Introduction

In general, *Fusarium* species are most active and survive best in dry soils (50). Accordingly, Fusarium diseases are commonly more important under dry rather than wet conditions (21,25,39), but important exceptions exist; vascular-wilts are most important under conditions of ample moisture available for growth of the host (21), and leaf and head blights of cereals require high humidity (chapter 4 above). Nevertheless, whether in soil, in plant tissues, or on aerial plant parts, the governing principles in water/fungus relationships are the same. This chapter treats the principles of water relations in the biology of *Fusarium* under: (i) physical principles; (ii) water relations in growth, survival, and reproduction of *Fusarium;* and (iii) water relations in development of disease.

Physical Principles

Water Potential

Water, whether in soil, plant tissue, or air, is best described (49) in terms of its potential energy (water potential). Pure, free, liquid water is arbitrarily assigned a water potential of zero. Water with solutes dissolved (osmotic) or absorbed by adhesion or cohesion to the surfaces of soil particles or cell wall material (matric) is less than pure or free and thus has negative potential energy. Water in air space at 100% relative humidity (RH) (the equilibrium relative humidity of pure, free water) has zero water potential. Water in air at less than 100% RH has negative potential energy. Water potential may be expressed as Joules/kg, bars, or Pascals (100 Joules/kg = 1 bar = 10^5 Pascals). In this paper, I use bars to express water potential. Numerous reviews on the subject of water potential applied to mycology and plant pathology are available (2,3,14,17,19,21,22,30,31,34,35,36,37,40,44,46,47).

The flow of water through the soil-plant-air continuum is a function of (i) the water potential gradient along the continuum, (ii) the amount of water available for flow, and (iii) the various resistances to flow along the continuum (44). The water potential of soil is mostly the sum of matric and osmotic components due, respectively, to adsorption to the soil particles and to the salts in the soil solution. The water potential of a cell, whether fungal or plant, is due largely to osmotic and turgor components.

Fungal cells probably exist mostly in equilibrium, or nearly so, with their environment, whether substrate or soil. This is the same as saying that cells or hyphae of a fungus like

Fusarium cannot remove water from the space around them at a rate sufficient to lower the water potential of that surrounding space. [Note: This principle was the basis for the conclusion by Clark (10) that unlike higher plants, microorganisms do not compete for water.] When in soil or crop refuse, the chlamydospores and hyphae of *Fusarium* are under the influence of the soil or refuse water potential. Refuse buried in soil will probably be in equilibrium with (at the same temperature and water potential as) the soil. However, whereas the main component of soil water potential is matric, the main component of refuse water potential probably is osmotic. When in the living host, Fusarium hyphae are under the influence of the plant water potential. The water potential of parenchyma tissues is largely osmotic, but that of xylem is largely a negative turgor and has only a small osmotic component (44). When in agar media, *Fusarium* functions under the influence of the water potential of the medium which is largely osmotic.

Water Content

As a soil becomes progressively wetter, the soil water potential approaches zero and becomes less important to *Fusarium*. Instead, the amount of air-filled pore space becomes the important consideration (33,35). The percentage of a soil volume as air-filled pore space ranges from 0 in soil saturated with water to 30–50% in soil at field capacity (−0.3 bar), depending on the soil type.

The poor soil aeration associated with excessively wet soils includes: (i) reduced availability of oxygen due to the slower rate of oxygen diffusion through the soil (32,33,35); (ii) a greater volume of microsites (35,38) with low redox potential due to lack of oxygen and loss of nitrates through denitrification; and (iii) increased activity of facultative and strict anaerobes and hence, increased production of ethylene, nitrous oxide, carbon dioxide, and other products associated with anaerobiosis. Ethylene production, with implications for both *Fusarium* and host (15) is maximal at saturation, greatly slowed at −5 bars, and virtually prevented at −10 to −15 bars (24).

The amount of soil water also influences availability of nutrients to *Fusarium* (19). For example, the amount of exudate released from pea seeds during their germination is directly proportional to soil water content; the lower the soil water content, the less the amount exudate (18,41). Chlamydospore germination by *F. solani* f. sp. *pisi* was proportional, in turn, to the amount of pea seed exudate (15). Fusarium spores are capable of germination at water potentials down to −80 bars or lower (20,51); the limit to germination by availability of host exudates in drier soils provides a mechanism whereby the *Fusarium* does not respond (in spite of suitable water potential) unless soil moisture is also suitable for growth of the host (17).

Water Relations in Growth, Survival, and Reproduction

The effects of water potential on Fusarium structures can be expressed in order of decreasing water potential as (i) optimal for growth and/or reproduction, (ii) minimal for growth (the fungus becomes dormant), and (iii) detrimental or even lethal to the fungus during dormancy.

The optimal and minimal osmotic potentials for spore germination and hyphal growth of specific Fusaria have been studied on agar media (6,16,23,42,51) or soil systems (23) amended with KCl or other solutes to lower water potential. In general, *Fusarium* species grow best between −5 and −25 bars and not at all below −100 to −120 bars (Fig. 22-1). *Fusarium nivale* seems to be an exception in that it grows maximally on agar media with KCl at −1 to −2 bars and grows little or not at all below about −50 bars (6). Ordinary laboratory media such as potato dextrose agar have osmotic potentials in the range of −1.5 and −3.0 bars.

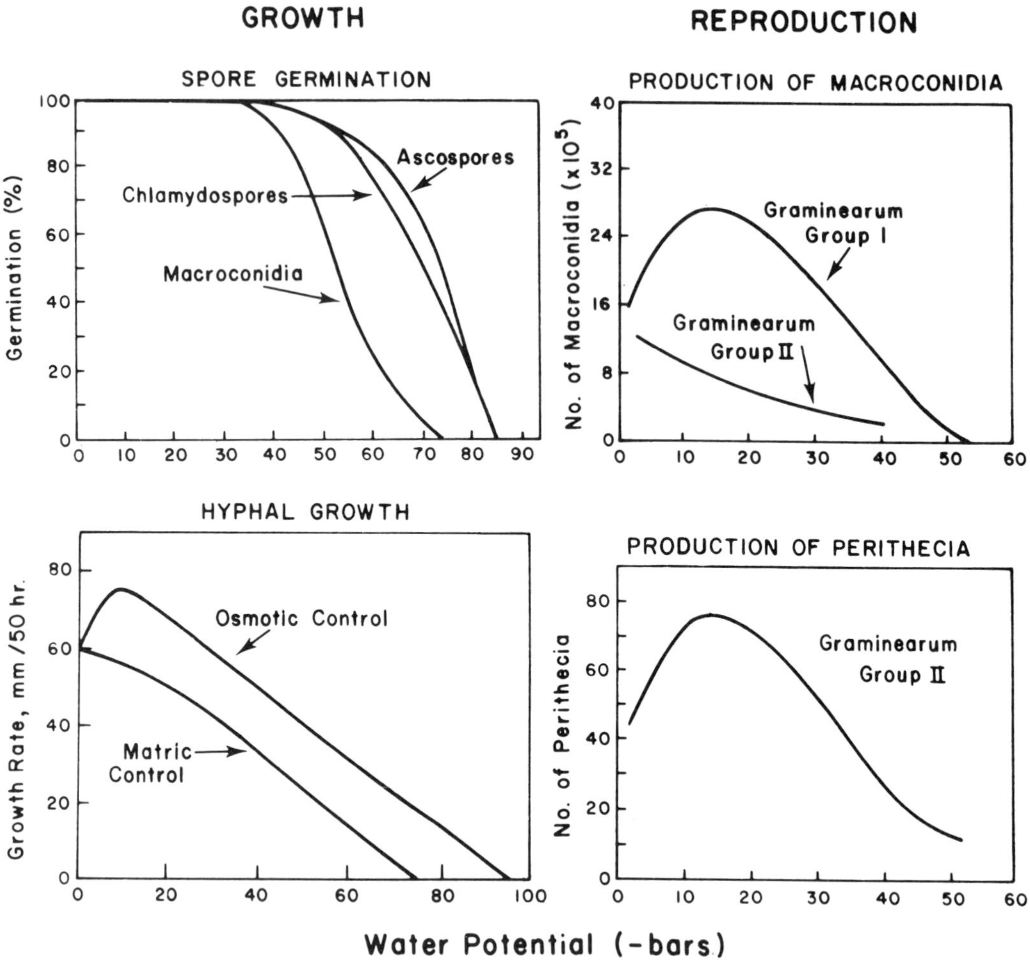

Fig. 22-1. Influence of water potential on growth and reproduction in *Fusarium roseum*, as illustrated by: data of Sung and Cook (51) for germination of macroconidia and chlamydospores of Culmorum and ascospores of Graminearum Group II; data of Cook et al. (23) for hyphal growth of Culmorum in soil adjusted to different water potentials with salts (osmotic) or with different water contents (matric); and by data of Sung and Cook (51) for production of macroconidia and perithecia in Graminearum Groups I and II.

The stimulation in growth rate of hyphae with slightly lowered water potentials (−5 to −15 bars) apparently holds for osmotic potential but not for matric potential (Fig. 22-1). Thus, hyphal growth of *F. roseum* 'Culmorum' was increased in soil over the range of −0.3 to −15 bars osmotic potential (soils wetted to uniform water content using KCl of different molalities and hence, osmotic potentials), but was progressively slower as the matric potential was lowered below −0.3 bars (soils simply adjusted to different water contents and hence, matric potentials) (23). The germination of conidia, chlamydospores, and ascospores of *F. roseum* likewise showed no evidence of stimulation with reduced osmotic potentials (51). Instead, germination of all three spore types was uniformly maximal between −1.4 and −20 bars, then progressively slower with each increment drop in osmotic potential between −20 and about −80 bars, the lower limit (Fig. 22-1, 51).

We can assume that for uptake of water, the water potential inside the fungus cell must be as low as or probably lower than that of the environment around the cell or hypha (1,14). The hypha also must maintain a turgor pressure if growth is to continue. Osmoregulation is the process whereby cell osmotic potentials are raised or lowered as necessary to maintain cell turgor (4). The slower growth rates with progressively lower matric potentials may be the result of the additional energy that must be expended by the fungal cell in osmoregulation (4). The progressively better growth in response to salts in the range of −5 to −15 bars may result from ion uptake by the cell, thereby resulting in cell osmotic potentials more ideal for cell functions and maintenance of turgor. The use of ions from the medium rather than constitutive osmotica is possibly the easiest means to osmoregulation and maintenance of turgor for a fungal hypha. Spores contain a rich supply of food reserves that undoubtedly contribute in osmoregulation and maintenance of turgor; this might explain why spores of *F. roseum* germinated uniformly rapidly (exhibited no evidence of improved germination with salts) over the range of ~0 to −20 bars (51).

For some *Fusarium* species, the higher the temperature in the range of 15 to 35 C, the lower the optimal water potential for maximal growth. Conversely, the lower the water potential, the higher the temperature at which the *Fusarium* can grow (15,41). For example, *F. roseum* 'Graminearum' grows maximally at about −10 to −15 bars at 20 C, and at −55 bars at 35 C (16). Previous work (25) indicated that Graminearum grows at 33 C, but not at 35 C, but that work was based on growth on ordinary nutrient agar media which rarely are drier than −1.5 to −3.0 bars when fresh. Likewise, for *F. oxysporum* f. sp. *lycopersici,* as the temperature was increased to 35 C, this fungus grew better in drier media (42). The shift in optimal temperature with drier conditions may be an adaptive mechanism in these fungi since, in general, dry conditions and high temperatures occur simultaneously. It seems that these Fusaria are not only adapted to a drier habitat with higher temperature, but may even prefer this combination of conditions.

The influence of water potential on reproduction has been studied for *Fusarium roseum* 'Culmorum,' 'Avenaceum,' 'Graminearum' Group I, and 'Graminearum' Group II (51). Production of macroconidia by the isolates of Culmorum, Avenaceum, and Graminearum Group I was maximal at about −15 bars and nil in the range of −60 to −80 bars (Fig. 22-1). Between about −80 and −100 bars, these isolates produced only mycelium. Production of macroconidia by the isolates of Graminearum Group II was maximal at −1.4 bars (water agar with no salts added) and progressively less as the substrate water potential was lowered with either KCl or NaCl below −1.4 bars (Fig. 22-1). The isolates of Graminearum Group II produced perithecia in culture in contrast to those of Culmorum, Avenaceum, and Graminearum Group I, which did not. Perithecial production was stimulated by reductions in osmotic potential, was maximal at about −15 bars, and was nil at −60 bars and below (Fig. 22-1). Thus, as the osmotic potential was lowered, the isolates of Graminearum Group II not only produced progressively fewer macroconidia, they made perithecia instead.

The isolates of Culmorum, Avenaceum, and Graminearum Group I used in the studies by Sung and Cook (51) were all from Washington State and were all known root pathogens. In contrast, the isolates of Graminearum Group II were from Pennsylvania and were known causes of carnation stem rot and corn stalk rot. Diseases of wheat and lentils in Washington State are favored by dry soil conditions, whereas the Pennsylvania isolates are normally operative on above-ground plant parts in a humid climate. The respective isolates apparently are equipped to sporulate maximally in the range of substrate water potential under which they normally must function. The results with Graminearum Group II suggest further that these isolates produce sporodochia under conditions of free moisture (~0 bars) on the plant surface, and shift to perithecial production as the plant tissues dry.

Chlamydospores of *F. solani* f. sp. *phaseoli* allowed to equilibrate with soil at about 50% RH (~750 bars) for more than 90 days remain alive, but subsequently require 48 hr to germinate rather than the usual 4–6 hr (3). The extended lag phase is needed for rehydration by the chlamydospores. Possibly, longer incubation of these spores at such low water potentials would kill them. Rapidity of drying is also important; Sitton and

Cook (48) showed that rapid drying (rapid equilibration with very low water potentials) killed the chlamydospores of *F. roseum* 'Graminearum' but not those of 'Culmorum.'

Incubation on wet soil also can be detrimental to survival of Fusarium spores (50). This indirect effect probably is caused by the absence of oxygen, the action of soil microorganisms, or both.

The indirect effects of water potential have been studied most in terms of interaction of *Fusarium* with soil bacteria. *Fusarium* generally can grow at water potentials down to about −100 bars, but soil bacteria are active in soil only down to −10 or −15 bars (19,20). Thus, below −15 bars *Fusarium* escapes soil bacteria. This can explain why colonization of wheat straws by Culmorum in natural soil was greatest at −80 to −90 bars (13), the minimal water potential for growth of the fungus (23). Presumably, the least amount of growth by antagonists of Culmorum occurred at this water potential.

Moisture is also one of the more important variables determining whether *Fusarium* can remain as an occupant of host refuse, or will be replaced in the refuse by saprophytes. Burgess and Griffin (8) showed for *F. roseum* 'Graminearum' (=*G. zeae*) that the more favorable the conditions for microbial activity (e.g., 35 C and 100% RH=0 bars) the shorter the life-span of Graminearum in the straws. At 25 C, survival was poorest between −200 to −350 bars where Penicillia were still active and replaced the presumably dormant Graminearum. At still lower water potentials, even competitors were dormant. In general, a pioneer occupant of host tissues (in this case, Graminearum in wheat straw) is most successful in maintaining possession of that tissue under conditions favorable for its own metabolism (5).

Water Relations in the Production of Plant Disease

This section deals with the influence of water potential on *Fusarium* inside the host plant and on the host-pathogen interaction. For the purpose of this discussion, the section is divided into i) stalk, culm, stem, root, and foot rots; ii) foliar and head blights; iii) vascular wilts; and iv) diseases of storage organs.

Stalk, culm, stem, foot and root rots This category of Fusarium diseases includes root and foot rot of wheat caused by *F. roseum* 'Culmorum' and *F. roseum* 'Graminearum,' stalk rot of sorghum caused by *F. moniliforme,* stalk rot of corn caused by *F. roseum* 'Graminearum,' and the root and basal stem rots of sweet potatoes (39), beans, and peas caused by formae speciales of *F. solani.* All are favored by dry soil conditions—and probably plant water stress.

The onset of severe dryland foot rot of winter wheat infected with Culmorum in the Pacific Northwest relates to the development of internal plant water potentials of about −32 to −35 bars or lower, beginning at heading or slightly later (45). Plants with water potentials between −15 and −25 bars (typical of wheat with ample available soil moisture) may become infected but do not develop severe foot rot (14). Less pathogenesis by the *Fusarium* in wheat with the higher plant water potentials probably relates to a form of general resistance in the host that stops functioning when plant water potentials drop below −32 to −35 bars (14). The precise, threshold, plant-water potential favorable to pathogenesis is not known, and may be different for young plants compared to older plants.

The practices used to prevent plant water stress in wheat, and thus to reduce Fusarium foot rot, are i) reduce the amount of nitrogen applied to the crop and thereby reduce the leaf area; ii) seed late, which keeps plants smaller and also conserves the finite quantity of soil water; and iii) grow cultivars of winter wheat with the ability to avoid stress due to better roots or less loss of water through transpiration. All of these factors contribute to higher (wetter) plant water potentials.

Data are not available on the relationship between host water potential and the development of root, stem, or stalk rots of beans, peas, sorghum, or corn infected with the respective Fusaria. There seems little doubt, however, that these diseases also are favored by low plant water potentials because in each case, practices that reduce water stress also reduce disease severity. Fusarium root and stem rot of beans and peas can be reduced by subsoiling to loosen the soil and permit deeper root growth and hence greater water uptake (9). The Fusarium stalk rot of sorghum can be reduced in western Nebraska by use of minimum or no-tillage (27); this makes 12–15 cm more water available to the sorghum crop and keeps the soil slightly cooler longer into the growing season. As for wheat, the development of plant water stress favorable for pathogenesis is probably the limiting factor to disease in these crops during periods of ample moisture.

Fusarium root rot of beans and probably peas may also be favored by temporary soil flooding (43). Flooding is another form of stress on the host, in this case, stress due to low oxygen. The development of more severe disease when the supply of soil oxygen is reduced probably is not much different than what happens with low plant water potentials; a certain level of general resistance functions against the pathogen so long as the host is fully functional physiologically, but ceases to function with stress due to an inadequate oxygen supply to the host.

Foliar and head blights Infection of leaves and floral parts is favored by rains and high humidity. A spore or hypha of *Fusarium* has the same water-potential requirements above-ground as below-ground, but whereas −90 bars or wetter is common in soil, the equivalent water potential can be achieved and maintained above-ground only with frequent rains and/or high relative humidity. (Reminder: at 20 C the equilibrium relative humidity for −90 bars water potential is about 94%; and at −120 bars, about 92%.) The minimal water potential can thus be achieved on leaf or flower surfaces only if the relative humidity is sufficiently high (92–94%) for a sustained period. The water potential is not likely to remain above this level for long with bright sun or a drying breeze. The presence of free water (0 bars), although higher than needed for growth and infection by *Fusarium*, helps maintain the required water potential on the plant surface for some time after a rain.

Once inside the host, stress associated with low plant water potentials probably is just as important for foliar or head blight as for stalk, stem, or root rots. Thus, the severity of Fusarium blight of turfgrass (chapter 19 above) is directly related to plant water stress. As with beans, corn, and sorghum (discussed above), the actual leaf water potentials present in water-stressed turf are unknown, but probably are well within the range required by the *Fusarium* for growth.

Vascular wilts Water affects the occurrence and severity of vascular wilts during: i) infection; ii) growth or movement in the host stem; and iii) production of wilt symptoms.

Fusarium oxysporum is probably not much different from *F. roseum* or *F. solani* in that while in soil or in the root zone, growth and survival are favored by the drier conditions (51), presumably because there is less antagonism from other soil microorganisms. Once in the host, *F. oxysporum* depends in many cases on movement as microconidia in the xylem. Conditions that slow transpiration, e.g., dry soil or a low evaporative demand, presumably also slow the transport of *F. oxysporum* as conidia up the stem. *Fusarium oxysporum* may also grow as hyphae up the stem, but this process is less efficient compared with movement as microconidia in the transpirational stream (26). Tomato has been reported (12) to become "resistant" to wilt when the soil is dry, probably for this reason (21).

The two most common theories proposed to explain wilting are i) a toxin causes the membranes of the leaf cells to lose their semi-permeability and thus their turgor; and ii) plugging of the xylem elements increases the resistance to water flow to the leaves, thereby causing loss of turgor. The theory that a toxin causes wilt due to loss of semi-permeability by cell membranes is still largely unproven (1,26). On the other hand, evidence continues to accumulate in support of the theory that wilting is due to increased resistance to water flow in the infected stems. Duniway (28,29) showed that the resistance to water flow in the main stem of tomato infected with *F. oxysporum* f. sp. *lycopersici* cannot account for wilting. In contrast, resistance to water flow approaches infinity in the leaf petioles and veins of infected tomato. Apparently, the critical sites of increased resistance to water flow are in the leaf and petiole xylem and not in the stem xylem.

The plugging theory can explain the sudden onset of wilting of infected plants following a hot day with a high evaporative demand. With a high evaporative demand, water flow to the leaf cells cannot keep pace with loss to the atmosphere and thus wilting occurs.

Diseases of storage organs The diseases of storage organs caused by *Fusarium* generally are favored by dry environments, e.g., "dry" rot of potatoes caused by *F. solani* 'Coeruleum' or, in some cases, *F. roseum* 'Sambucimum,' and the corm and bulb rots of gladiolus, tulips, and daffodils caused by *F. oxysporum*. The osmotic pressure (osmotic component of the water potential) of potato tuber tissues is about −7 to −8 bars (22), which is ideal for growth of Fusaria. Probably the occurrence of a Fusarium rot of potatoes, squash, bulbs, or other storage organs is not a function of water potential of the tissues (which is fairly ideal and constant under many storage conditions), but rather relates to factors such as the presence of water films over the organ, supply of oxygen into the tissues, competition from bacteria on the tissues, and other indirect effects (22). More work is needed to clarify the direct and indirect effects of water and water potential on Fusarium diseases of storage organs.

The water in cereal grains is usually expressed as a percentage of dry weight of the grain, or as water activity (a_w) (11). Water activity is the equilibrium relative humidity multiplied by 0.01. Thus, at 20 C, 0.94 a_w is equal to about −90 bars water potentials, and 0.92 a_w about −120 bars. *Fusarium* is subject to the same water potential requirements in seed as in any other medium. So long as the minimum water potential is satisfied, *Fusarium* in a seed theoretically will make growth provided no other environmental factor is limiting.

There are no data on the relationship of water potential to mycotoxin production by *Fusarium,* but since water potential affects production of antibiotics in microorganisms (7,52), there is little doubt that it also affects production of mycotoxins. There is also the possibility that the water potential ideal for mycotoxin production will be different from that most suitable for growth.

Literature Cited

1. Adebayo, A. A., and R. F. Harris. 1971. Fungal growth responses to osmotic as compared to matric water potential. Soil Sci. Soc. Amer. Proc. 35:465–469.
2. Ayres, P. G. 1978. Water relations of diseased plants, p. 1–60. *In* T. T. Kozlowski (ed.), Water deficits and plant growth V: Water and plant disease, Academic Press, New York, San Francisco, and London. 323 p.
3. Baker, K. F., and R. J. Cook. 1974. Biological control of plant pathogens. W. H. Freeman and Co., San Francisco. 433 p.
4. Brown, A. D. 1976. Microbial water stress. Bacteriol. Rev. 40:803–846.
5. Bruehl, G. W. 1975. Systems and mechanisms of residue possession by pioneer fungal colonists, p. 77–83. *In* G. W. Bruehl (ed.), Biology and control of soil-borne plant pathogens, Amer. Phytopathol. Soc., St. Paul, Minnesota. 216 p.
6. Bruehl, G. W., and B. Cunfer. 1971. Physiologic and environmental factors that affect the severity of snow mold of wheat. Phytopathology 61:792–799.
7. Bruehl, G. W., B. Cunfer, and M. Toiviainen.

1972. Influence of water potential on growth, antibiotic production, and survival of *Cephalosporium gramineum*. Can. J. Plant Sci. 52:714–423.
8. Burgess, L. W., and D. M. Griffin. 1968. The recovery of *Gibberella zeae* from wheat straws. Austr. J. Exp. Agric. Anim. Husb. 8:364–370.
9. Burke, D. W., D. E. Miller, L. D. Holmes, and A. W. Barker. 1972. Counteracting bean root rot by loosening the soil. Phytopathology 62:306–309.
10. Clark, F. E. 1965. The concept of competition in microbial ecology, p. 339–345. *In* K. F. Baker and W. C. Snyder (ed.), Ecology of soil-borne plant pathogens, Univ. Calif. Press, Berkeley, Los Angeles and London. 571 p.
11. Christensen, C. M. 1978. Moisture and seed decay, p. 199–221. *In* T. T. Kozlowski (ed.), Water deficits and plant growth V: Water and plant disease, Academic Press, New York, San Francisco, and London. 323 p.
12. Clayton, E. E. 1923. The relation of soil moisture to the Fusarium wilt of tomato. Amer. J. Bot. 10:133–147.
13. Cook, R. J. 1970. Factors affecting saprophytic colonization of wheat straw by *Fusarium roseum* f. sp. *cerealis* 'Culmorum.' Phytopathology 60:1672–1676.
14. Cook, R. J. 1973. Influence of low plant and soil water potentials on diseases caused by soilborne fungi. Phytopathology 63:451–458.
15. Cook, R. J. 1977. Management of the associated microbiota, p. 145–166. *In* J. G. Horsfall and E. B. Cowling (ed.), Plant disease, an advanced treatise, Vol. I, How disease is managed, Academic Press, New York and London. 463 p.
16. Cook, R. J., and A. A. Christen. 1976. Growth of cereal root-rot fungi as affected by temperature-water potential interactions. Phytopathology 66:193–197.
17. Cook, R. J., and D. M. Duniway. 1980. Water relations in the life-cycles of soilborne plant pathogens. *In* J. F. Parr and W. R. Gardner (ed.), The concept of water potential applied to soil mirobiology and biochemistry, Amer. Soc. Agron. Madison, Wisc.
18. Cook, R. J., and N. T. Flentje. 1967. Chlamydospore germination and germling survival of *Fusarium solani* f. *pisi* in soil as affected by soil water and pea seed exudation. Phytopathology 57:178–182.
19. Cook, R. J., and R. I. Papendick. 1970. Effect of soil water on microbial growth, antagonism, and nutrient availability in relation to soil-borne fungal diseases of plants, p. 81–88. *In* T. A. Toussoun, R. V. Bega, and P. E. Nelson (ed.), Root diseases and soil-borne pathogens, Univ. Calif. Press. Berkeley. 255 p.
20. Cook, R. J., and R. I. Papendick. 1970. Soil water potential as a factor in the ecology of *Fusarium roseum* f. sp. *cerealis* 'Culmorum' Plant Soil 32:131–145.
21. Cook, R. J., and R. I. Papendick. 1972. Influence of water potential of soils and plants on root disease. Annu. Rev. Phytopathol. 10:349–372.
22. Cook, R. J., and R. I. Papendick. 1978. The role of water potential in microbial growth and development of plant disease with special reference to postharvest pathology. HortScience 13:559–564.
23. Cook, R. J., R. I. Papendick, and D. M. Griffin. 1972. Growth of two root-rot fungi as affected by osmotic and matric water potentials. Soil Sci. Soc. Amer. Proc. 36:78–82.
24. Cook, R. J., and A. M. Smith. 1977. Influence of water potential on production of ethylene in soil. Can. J. Microbiol. 23:811–817.
25. Dickson, J. G. 1923. Influence of soil temperature and moisture on the development of the seedling-blight of wheat and corn caused by *Gibberella saubinetii*. J. Agric. Res. 23:837–870.
26. Dimond, A. E. 1970. Biophysics and biochemistry of the vascular wilt syndrome. Annu. Rev. Phytopathol. 8:301–322.
27. Doupnik, B., Jr., M. G. Boosalis, G. Wicks, and D. Smika. 1975. Eco-fallow reduces stalk rot in grain sorghum. Phytopathology 65:1021–1022.
28. Duniway. J. M. 1971. Resistance to water movement in tomato plants infected with *Fusarium*. Nature (London) 230:252–253.
29. Duniway, J. M. 1971. Water relations of Fusarium wilt in tomato. Physiological Plant Pathol. 1:537–546.
30. Duniway, J. M. 1973. Pathogen-induced changes in host water relations. Phytopathology 63:458–466.
31. Duniway, J. M. 1979. Water relations of water molds. Annu. Rev. Phytopathol. 17:431–460.
32. Griffin, D. M. 1963. Soil moisture and the ecology of soil fungi. Biol. Rev. 38:141–166.
33. Griffin, D. M. 1968. A theoretical study relating the concentration and diffusion of oxygen to the biology of organisms in soil. New Phytol. 67:561–577.
34. Griffin, D. M. 1969. Soil water in the ecology of fungi. Annu. Rev. Phytopathol. 7:289–310.
35. Griffin, D. M. 1972. Ecology of soil fungi. Chapman and Hall. London 193 p.
36. Griffin, D. M. 1977. Water potential and wood-decay fungi. Annu. Rev. Phytopathol. 15:319–329.
37. Griffin, D. M. 1978. Effect of soil moisture on survival and spread of pathogens, p. 175–197. *In* T. T. Kozlowski (ed.), Water deficits and plant growth V: Water and plant disease, Academic Press, New York, San Francisco, and London. 323 p.
38. Greenwood, D. J., and D. Goodman. 1965. Oxygen diffusion and aerobic respiration in columns of fine crumbs. J. Sci. Food Agric. 16:152–160.
39. Harter, L. L., and W. A. Whitney. 1927. Relation of soil temperature and soil moisture to the infection of sweet potatoes by the stem-rot organism. J. Agric. Res. 34:435–441.
40. Kozlowski, T. T. (ed.). 1978. Water deficits and plant growth. Vol. V. Water and plant disease.

Academic Press. New York, San Francisco, and London. 323 p.
41. Kerr, A. 1964. The influence of soil moisture on infection of peas by *Pythium ultimum.* Austr. J. Biol. Sci. 17:676–685.
42. Manandhar, J. B., and G. W. Bruehl. 1973. *In vitro* interaction of *Fusarium* and *Verticillium* wilt fungi with water, pH, and temperature. Phytopathology 63:413–419.
43. Miller, D. E., and D. W. Burke. 1975. Effect of soil aeration on Fusarium root rot of beans. Phytopathology 65:519–523.
44. Papendick, R. I., and G. S. Campbell. 1975. Water potential in the rhizosphere and plant and methods of measurement and experimental control, p. 39–49. *In* G. W. Bruehl (ed.), Biology and control of soil-borne plant pathogens, Amer. Phytopathol. Soc., St. Paul, Minnesota. 216 p.
45. Papendick, R. I., and R. J. Cook. 1974. Plant water stress and development of Fusarium foot rot in wheat subjected to different cultural practices. Phytopathology 64:358–363.
46. Schoeneweiss, D. F. 1975. Predisposition, stress, and plant disease. Annu. Rev. Phytopathol. 13:193–211.
47. Schoeneweiss, D. F. 1978. Water stress as a predisposing factor in plant disease, p. 61–100. *In* T. T. Kozlowski (ed.), Water deficits and plant growth V: Water and plant disease, Academic Press, New York, San Francisco, and London. 323 p.
48. Sitton, J. W., and R. J. Cook. 1981. Comparative morphology and survival ability of chlamydospores of *Fusarium roseum* 'Culmorum' and 'Graminearum.' Phytopathology 71:85–90.
49. Slatyer, R. O. 1967. Plant water relationships. Academic Press, New York. 366 p.
50. Stover, R. H. 1953. The effect of soil moisture on *Fusarium* species. Can. J. Bot. 31:693–697.
51. Sung, J. M., and R. J. Cook. 1981. Effect of water potential on reproduction and spore germination of *Fusarium roseum* 'Graminearum,' 'Culmorum,' and 'Avenaceum.' Phytopathology 71:499–504.
52. Wong, T. W. 1972. Effect of soil water on bacterial movement and Streptomycete-fungal antagonism. Ph.D. Thesis, Univ. Sydney, Sydney, Australia. 167 p.

23 Ecology of the Fungus, *Fusarium:* Competition

Ralph Baker

The term *competition* will be used in the narrower sense of active demand in excess of the immediate supply of material and condition on the part of two or more organisms (15, 16, 33). As such it is one of the three mechanisms of antagonism (34), the others being antibiosis and exploitation.

In exploring where competition is operating in the life cycle and activities of the Fusaria, certain principles can be elaborated. A sizeable number of soil microorganisms are dormant at any given time. The primary, but certainly not the sole reason for this, is the lack of suitable and available energy sources. Thus, competition among soil organisms is one factor imposing dormancy and the associated advantages of survival.

Clark (15) effectively reduced the occurrence of competition in soil to the phenomena associated with energy sources: ". . . basically competition among microorganisms is for a substrate, in the specific form and under the specific conditions in which that substrate is presented."

The chronic starvation confronting soil microorganisms is probably the major factor contributing to the typical countertendency of soils to significant change in biological characteristics. This biological "buffering capacity," then, is in large part a manifestation of competition. Given this antagonistic background, it is not surprising that the strategy for using competition in biological control is applied to soils at periods in the life cycle of the pathogen when energy sources, in the course of events, become available. The idea is to deprive the pathogen of essential nutrients at a point where (usually) saprophytic activity is essential for success in induction of disease. It follows that the key portions of the life cycle of Fusaria where competition may operate occur when the pathogen is carrying on metabolic activities in germination, penetration of a host, saprophytic thallus production, and sporulation.

Hypothetically, any of these activities may be diminished or completely inhibited if an essential nutritional factor in the substrate is limiting. A review (2) indicated that candidates suggested as potentially being in short supply in soil under certain conditions are carbon, nitrogen, and vitamins. Currently, vitamins are not considered among candidates for limiting factors leading to biological control.

Bruehl (10) beautifully characterized the results of escape from intense competition with examples in the genus *Fusarium*. He calls this "radiate evolution." For example, the wild forms of *Fusarium oxysporum* (having a probable common origin) became physiologically distinct into diverse formae speciales with little effective competition among them. Again, populations of *Fusarium solani* f. sp. *phaseoli* and *pisi* may coexist in soil, both using root exudates of their respective nonsuscepts (28). *Fusarium solani* f. sp. *pisi* even invaded beans but incited no damage. Circumvention of competition is a valid strategy for survival, too; we should not be surprised if the soil-borne Fusaria have developed cunning methods to avoid it when necessary.

Competition in Germination and Penetration

The substrates of importance for ingress of a given soil-borne pathogen are associated with host exudates. This is the place where competition can occur. Elements most likely to be limiting in the rhizosphere or rhizoplane turn out to be those nutrients most critical in germination and penetration by Fusaria, namely, carbon and nitrogen.

The requirements for germination of Fusarium propagules are treated in detail elsewhere in this volume (Griffin, chapter 30). The nutritional relationships of all *Fusarium* spp. have not received the careful study afforded by Griffin and his coworkers, but there are general principles to apply. At inoculum densities found in soil (24, 25) propagule germination percentages of the Fusaria are quite low. This is due to soil fungistasis. Dormancy may be overcome by the addition of available carbon (19). Typically, this element is supplied in soil by exudates from hosts or other plants (20) or by addition of organic matter. Nitrogen compounds in soil are also required for germination (23). This element is not usually limiting in nonamended soil, however, and may also be present in exudates or in organic matter. The type of nitrogenous compound also influences germination. Ammonium induced greater germination of conidia of *F. solani* f. sp. *phaseoli* than nitrate nitrogen, although there was no difference in effects of these types on *Fusarium roseum* (13).

The importance of nutrition in penetration was emphasized by Toussoun (37). In bean root rot, the pathogen typically forms a thallus on the host surface and then penetrates host tissue (14). Toussoun et al. (38) found penetration of hypocotyl tissue abundant in solutions containing nitrogen but sparse in glucose solutions. Most infection by *F. solani* f. sp. *phaseoli* occurred when both carbon and nitrogen were supplied. Development of the thallus on hypocotyls was enhanced and symptom expression increased particularly by the ammonium form of nitrogen (32, 39).

These basic considerations regarding requirements of the pathogen for successful penetration and infection formed the basis for an understanding of the mechanism of biological control of bean root rot with soil amendments with wide C/N ratios such as barley straw (35), cellulose, and glucose (32). In soil amended with cellulose, nitrogen immobilization occurred within a week after application at established amendment ratios at and above 25:1. Competition for nitrogen by the general soil microflora thus precluded successful germination and penetration of the host by the pathogen.

Other work suggests competition for carbon rather than nitrogen. Lignin added to soil had no effect on disease development of bean root rot (31). The same was true for chitin; however, a small but significant amount of control was observed when these two amendments were combined in the system. The addition of nitrate nitrogen increased the amount of control. Clearly, competition for nitrogen was not the mechanism of biological control. Addition of glucose to the lignin-chitin system nullified this control, suggesting carbon competition as the mechanism. In use of the relatively complex carbon compounds in chitin and lignin, the soil microflora should also metabolize any simpler compounds such as those exuded in the infection court of the host. Such exudates would be rapidly metabolized by the microflora, now in higher populations because of the presence of the organic amendments. Thus the pathogen is deprived of essential carbon.

Attempts were made, using model systems, to test this hypothesis (5). Glucose was added to soil systems containing various organic amendments (added previously) to simulate the introduction of a carbon compound from host exudates. Added glucose was no longer detected 3–4 days after incorporation in chitin-lignin amended soils, but in raw soil, it was detectable even after 12 days had elapsed. The same relative trend was noticed with glutamic acid added to soil to simulate a root exudate compound containing both nitrogen and carbon.

In these studies (5) utilization of glucose introduced into cellulose-amended soil was delayed, differences in complete utilization being approximately 8 days. This raises the possibility of simple, normally available carbon sources persisting for a relatively long period in soil when there is a deficiency of nitrogen (imposed by competition). The possibility exists also that breakdown products of complex organic amendments like chitin or lignin (e.g., ammonia) are fungistatic or fungicidal to a soil-borne pathogen.

Competition also may be involved in the spread of Fusarium wilt of banana. Stotzky and Martin (36) suggested that the absence or limited activity of *F. oxysporum* f. sp. *cubense* in soils containing montmorillonite clays is related to bacterial antagonism in the soils encouraged by the more stable pH of these soils.

Biological control from competition predicts little or no effect on viability of populations of the pathogen in the soil before and after treatment. Propagules should not germinate, but they would not be killed, as would be the case if the mechanism involved a fungicidal antibiotic (1). This was the case in biological control of bean root rot; inoculum densities in the form of chlamydospores of *F. solani* f. sp. *phaseoli* were statistically the same before and after application of cellulose to the soil resulting in biological control (4).

For what types of Fusarium diseases is biological control through competition most likely to be achieved? The known instances in which this mechanism has been detected are in systems where the host infection court is nonmotile (1). This is explicable in terms of the amounts of inoculum encountered by infection courts (2, 3). In the Fusarium wilt diseases, penetration typically occurs through the moving root tip. Thus, these moving infection courts encounter much more inoculum than fixed ones (such as hypocotyls). Competition typically does not immediately reduce inoculum density. Even if competition reduces germination in soil inoculum to very low levels, enough propagules can be encountered by these moving infection courts to make up the difference. This may explain why Guy and Baker (26) were unable to obtain biological control for Fusarium wilt of pea with a cellulose amendment.

Competition also must operate over relatively short periods since limiting factors are typically immobilized temporarily. Thus competition might protect an infection court if the host tissue was only temporarily susceptible (e.g., pathogens inciting damping-off of germinating seeds or seedlings). If the host is still susceptible to infection after competition factors cease to operate, obviously ingress of the pathogen is only delayed.

Competition in Model Systems

Marshall and Alexander (30) observed that a variety of soil bacteria induced suppression of *F. oxysporum* f. sp. *cubense* in liquid medium or sterile soil systems. The competitive effect was overcome by additions of inorganic nitrogen and thus they ascribed the suppression to biologically induced nitrogen deficiency. Later Finstein and Alexander (21) implicated carbon as the main limitation to development of the same pathogen in similar model systems with deficiencies of nitrogen and phosphorus demonstrated after the shortage of carbon was relieved. Modeling of competition using *Fusarium* was approached in a different way by Lindsey (29). He used soil microbiological sampling tubes (SMST) which, in effect, introduced a substrate into soil containing two Fusaria with different competitive saprophytic activities. While cultivars of *F. roseum* may not have as much ability to colonize all substrates in soil (18) as previously thought (22), this species probably has more competitive saprophytic ability than *F. solani* f. sp. *phaseoli* (12). In Lindsey's (29) system, *F. roseum* dominated the isolates obtained from SMST when the two species were incubated together in previously sterilized soil. *Fusarium solani* was recovered only when it had been placed in soil originally at 50 times the inoculum density of *F. roseum. Fusarium solani* competed effectively for substrates in the SMST, but only when sufficient nitrogen was available. Thus, the observed phenomena in this model system could best be explained in terms of competition for nitrogen.

Byther (13), using the same pair of fungi, found conidia of *F. roseum* germinated in vitro much better than those of *F. solani* at low levels of nitrogen and carbon, especially at pH levels above 6.5. This difference in germination efficiency of *F. solani* could only be overcome by adding carbon as glucose. Also, conidia of *F. solani* germinated much better in the presence of ammonium than nitrate compounds, whereas germination of conidia of *F. roseum* was indifferent to the form of nitrogen at low nutrient levels. In nitrogen-impoverished soil, chlamydospore germination of the bean *Fusarium* was also higher when supplemented with the ammonium rather than the nitrate form of nitrogen. Such observations may provide partial explanations for the observed influences of various forms of nitrogen on certain plant diseases (27).

Competition in Other Stages of the Life Cycle

Detailed studies of mechanisms of competition in the Fusaria have been confined, for the most part, to those events leading to breeching of host barriers. Less attention has been directed to the basic reasons for possession of substrates once they are invaded. Bruehl (9) describes and gives implications related to the phenomenon in which Fusaria may colonize tissues of roots and/or stems of living plants. Obvious responses by the host in the form of symptoms may or may not be present, but such prior establishment of the organism is often so strong that other fungi are virtually excluded when the plant dies. Again, if clean, glossy wheat straw is introduced into soil infested with chlamydospores of *F. roseum* 'Culmorum,' competitive saprophytic colonization of the substrate may occur (17). The same pathogen is unable to colonize old gray straw already occupied by weak parasites or saprophytes. Thus, occupancy of wheat refuse under field conditions by *F. roseum* 'Culmorum' was due to parasitism (18), but the vigor in competitive use of foods in wheat straw must be a prime factor in the ability of the fungus to use a substrate and eventually achieve long-term survival (8). In such cases, the level of antagonism may be the major factor determining successful saprophytism, and among mechanisms, competition is likely to predominate (11). Indeed, subsequent sporulation, growth, and inoculum production (6, 7) in debris may be influenced by the nutrient status of the substrate. Plant pathologists have yet to take full advantage of these "weak links" in the life cycle of many Fusaria by initiating measures leading to competition for nutrients essential to the activities of pathogens.

Literature Cited

1. Baker, R. 1968. Mechanisms of biological control of soil-borne pathogens. Annu. Rev. Phytopathol. 6:263–294.
2. Baker, R. 1969. Fungal populations and incidence of disease, p. 8–11. *In* R. J. Cook and R. D. Watson (ed.), Nature of the influence of crop residues on fungus-induced root diseases, Washington Agric. Exp. Stn. Bull. 716.
3. Baker, R. 1970. Use of population studies in research on plant pathogens in soil, p. 11–15. *In* T. A. Toussoun, R. V. Bega, and P. E. Nelson (ed.), Root diseases and soil-borne pathogens, Univ. Calif. Press, Berkeley and Los Angeles.
4. Baker, R., and Shirley M. Nash. 1965. Ecology of plant pathogens in soil. VI. Inoculum density of *Fusarium solani* f. sp. *phaseoli* in bean rhizosphere as affected by cellulose and supplementary nitrogen. Phytopathology 55:1381–1382.
5. Benson, D. M., and R. Baker. 1970. Rhizosphere competition in model soil systems. Phytopathology 60:1058–1061.
6. Booth, R. H., and G. S. Taylor. 1976. Fusarium diseases of cereals. X. Straw debris as a source of inoculum for infection of wheat by *Fusarium nivale* in the field. Trans. Brit. Mycol. Soc. 66:71–75.
7. Booth, R. H., and G. S. Taylor. 1976. Fusarium

diseases of cereals. XI. Growth and saprophytic activity of *Fusarium nivale* in soil. Trans. Brit. Mycol. Soc. 66:77–83.
8. Bruehl, G. W. 1969. Factors affecting the persistence of fungi in soil, p. 11–14. *In* R. J. Cook and R. D. Watson (ed.), Nature of the influence of crop residues on fungus-induced root diseases, Washington Agric. Exp. Stn. Bull. 716.
9. Bruehl, G. W. 1975. Systems and mechanisms of residue possession by pioneer fungal colonists, p. 77–83. *In* G. W. Bruehl (ed.), Biology and control of soil-borne plant pathogens, Amer. Phytopathol. Soc., St. Paul, Minnesota.
10. Bruehl, G. W. 1976. Management of food resources by fungal colonists of cultivated soils. Annu. Rev. Phytopathol. 6:247–264.
11. Burgess, L. W., and D. M. Griffin. 1967. Competitive saprophytic colonization of wheat straw. Ann. Appl. Biol. 60:137–142.
12. Burke, D. W. 1965. The near immobility of *Fusarium solani* f. sp. *phaseoli* in natural soil. Phytopathology 55:1188–1190.
13. Byther, R. 1965. Ecology of plant pathogens in soil. V. Inorganic nitrogen utilization as a factor of competitive saprophytic ability of *Fusarium roseum* and *F. solani.* Phytopathology 55:852–858.
14. Christou, T., and W. C. Snyder. 1962. Penetration and host parasite relationships of *Fusarium solani* f. *phaseoli* in the bean plant. Phytopathology 52:219–226.
15. Clark, F. E. 1965. The concept of competition in microbial ecology, p. 339–345. *In* K. F. Baker and W. C. Snyder (ed.), Ecology of soil-borne plant pathogens, Univ. Calif. Press, Berkeley and Los Angeles.
16. Clements, F. E., and V. E. Shelford. 1939. Bio-ecology. John Wiley and Sons, New York. 425 p.
17. Cook, R. J. 1970. Factors affecting saprophytic colonization of wheat straw by *Fusarium roseum* f. sp. *cerealis* 'Culmorum.' Phytopathology 60:1672–1676.
18. Cook, R. J., and G. W. Bruehl. 1968. Relative significance of parasitism versus saprophytism in colonization of wheat straw by *Fusarium roseum* 'Culmorum' in the field. Phytopathology 58:306–308.
19. Cook, R. J., and M. N. Schroth. 1965. Carbon and nitrogen compounds and germination of chlamydospores of *Fusarium solani* f. *phaseoli.* Phytopathology 55:254–256.
20. Cook, R. J., and W. C. Snyder. 1965. Influence of host exudates on growth and survival of germlings of *Fusarium solani* f. *phaseoli* in soil. Phytopathology 55:1021–1025.
21. Finstein, M. S., and M. Alexander. 1962. Competition for carbon and nitrogen between *Fusarium* and Bacteria. Soil Sci. 94:334–339.
22. Garrett, S. D. 1956. Biology of root-infecting fungi. Cambridge Univ. Press. 293 p.
23. Griffin, G. J. 1964. Long term influence of soil amendments on germination of conidia. Can. J. Microbiol. 10:605–612.
24. Griffin, G. J. 1970. Carbon and nitrogen requirements for macroconidial germination of *Fusarium solani:* dependence on conidial density. Can. J. Microbiol. 16:733–740.
25. Griffin, G. J. 1974. Soil fungistasis: fungus spore germination in soil at spore densities corresponding to natural population levels. Can. J. Microbiol. 20:751–754.
26. Guy, S. O., and R. Baker. 1977. Inoculum potential in relation to biological control of Fusarium wilt of peas. Phytopathology 67:72–78.
27. Henis, Y. 1976. Effect of mineral nutrients on soil-borne pathogens and host resistance. International Potash Institute, 12th Colloquim:17–28.
28. Kraft, J. M., and D. W. Burke. 1974. Behavior of *Fusarium solani* f. sp. *pisi* and *F. solani* f. sp. *phaseoli* individually and in combinations on peas and beans. Plant Dis. Rep. 58:500–504.
29. Lindsey, D. 1965. Ecology of plant pathogens in soil. III. Competition between soil fungi. Phytopathology 55:104–110.
30. Marshall, K. C., and M. Alexander. 1960. Competition between soil bacteria and *Fusarium.* Plant Soil 12:143–153.
31. Maurer, C. L., and R. Baker. 1964. Ecology of plant pathogens in soil. I. Influence of chitin and lignin amendments on development of bean root rot. Phytopathology 54:1425–1426.
32. Maurer, C. L., and R. Baker. 1965. Ecology of plant pathogens in soil. II. Influence of glucose, cellulose, and inorganic nitrogen amendments on development of bean root rot. Phytopathology 55:69–72.
33. Milne, A. 1961. Definition of competition among animals, p. 40–61. *In* F. L. Milthorpe (ed.), Mechanisms in biological competition, Symp. Soc. Exptl. Biol. 15, Cambridge Univ. Press, London.
34. Park, D. 1960. Antagonism—the background of soil fungi, p. 148–159. *In* D. Parkinson and J. S. Waid (ed.), The ecology of soil fungi, Liverpool Univ. Press, Liverpool.
35. Snyder, W. C., M. N. Schroth, and T. Christou. 1959. Effect of plant residues on root rot of bean. Phytopathology 49:755–756.
36. Stotzky, G., and R. T. Martin. 1963. Soil mineralogy in relation to the spread of Fusarium wilt of banana in Central America. Plant Soil 18:317–338.
37. Toussoun, T. A. 1970. Nutrition and pathogenesis of *Fusarium solani* f. sp. *phaseoli,* p. 95–98. *In* T. A. Toussoun, R. V. Bega, and P. E. Nelson (ed.), Root diseases and soil-borne pathogens, Univ. Calif. Press, Berkeley and Los Angeles.
38. Toussoun, T. A., Shirley M. Nash, and W. C. Snyder. 1960. The effect of nitrogen sources and glucose on the pathogenesis of *Fusarium solani* f. *phaseoli.* Phytopathology 50:137–140.
39. Weinke, K. E. 1962. Influence of nitrogen on the root disease of bean caused by *Fusarium solani* f. *phaseoli.* Ph.D. Thesis, Univ. Calif., Berkeley.

24 Formation and Survival of Chlamydospores in *Fusarium*

B. Schippers and W.H. van Eck

For many Fusaria, survival in soil depends on chlamydospores that have the ability to withstand adverse environmental conditions (43). Selective media for isolation of Fusaria from soil (5, 32, 44), direct observation methods using soil smears and membrane filters (43, 48, 57, 72), electronmicroscopy (48, 72), and autoradiography (26) all have been used to study chlamydospore formation and survival in soil. Most attention has been paid to chlamydospores of the plant pathogenic species, namely *Fusarium solani, F. oxysporum,* and *F. roseum.* Three major phases in the life cycle of chlamydospores can be distinguished: formation, dormancy and germination. This paper discusses the origin of chlamydospores in field soil, the factors that induce chlamydospore formation, the ultrastructure of their formation, their lysis during dormancy, and the germination of chlamydospores as far as it affects their formation and longevity. It also discusses the consequences of cultural practices for their formation and survival. Finally, a definition of the term chlamydospore for Fusaria is proposed.

Chlamydospore Formation in Field Soil

Chlamydospore formation in pathogenic Fusaria commonly takes place in hyphae in the infected and decaying host tissue (11, 43, 71). Nash et al. (43) observed that the site in the cortex where chlamydospores are formed affects their morphology and physiology; those formed on the surface of stems and roots are mainly large structures rich in oil droplets with thick warty walls, while those formed in the intercellular space of the outer cortex are small with smooth, thinner walls. Chlamydospores may also be formed abundantly from macroconidia that originate from sporodochia on lesions at the soil level (11, 43); when washed into soil, the macroconidia convert quickly to chlamydospores that are formed either at germ tubes or by transformation of the macroconidial cells (19, 42) (Fig. 24-1). The importance of asexual sporulation in contributing to survival in soil may depend on the host. In the Pacific Northwest U.S.A., oats infected with *F. roseum* f. sp. *cerealis* 'Culmorum' support greater asexual sporulation and thus contribute to higher propagule densities under field conditions than does infected wheat (12). Macroconidia formed during active saprophytic growth of mycelium following addition of organic substances to infested field soil also may be converted to chlamydospores and thus result in more chlamydospores per unit of soil than originally present (49).

Finally, chlamydospores can be produced in germ tubes of chlamydospores stimulated to germinate in soil in response to a temporary supply of nutrients provided by diffusates from seeds, living host or non-host plants, or crop residues in their vicinity. Such germination is short-lived but is commonly of sufficient duration to permit formation of the replacement chlamydospore. This "recycling" process is probably very important in accounting for the longevity of some Fusaria in soil (33). Such germination is a non-specific response to carbon and nitrogen sources in soil and rhizosphere and may even increase the number of chlamydospores (13, 49, 59, 62, 79). On the other hand, if available C and N in the nutrients, together with inorganic nitrogen in soil, exceeds the minimum requirements of the pathogen for chlamydospore germination, the soil microflora also stimulated by the nutrients may cause lysis of the germ tube before new chlamydospores have been formed (13).

Although chlamydospores are commonly formed from hyphae or conidia in infected and decaying host tissue in the field, macroconidia cultivated on rich agar media are the propagules used in most model experiments in the laboratory to study the factors that induce chlamydospore formation. Many biotic and abiotic factors in soil can modify the induction of chlamydospore formation, e.g., temperature (4, 46), CO_2 content (45), and pH (21, 57). Energy depletion of inoculum and metabolites of the soil microflora, however, are considered to be the major factors responsible for the induction of chlamydospore formation and are discussed in the following paragraphs.

Induction of Chlamydospore Formation by Energy Depletion and Inorganic Salts

In most studies absence or low levels of organic carbon sources appear to be a prerequisite for chlamydospore formation from macroconidia or germlings of *Fusarium* (21, 22, 23, 27, 30, 57, 66). Convincing evidence of energy deprivation as the primary stimulus for chlamydospore formation from macroconidia was presented by Meyers and Cook (40), who obtained rapid and synchronous chlamydospore formation after sudden withdrawal of exogenous sucrose from the germlings of three clones each of *F. solani* f. sp. *phaseoli* and *F. solani* f. sp. *pisi* in liquid culture. The importance of energy depletion as a stimulant for chlamydospore formation was confirmed by Hsu and Lockwood (30). They demonstrated that for several clones of formae speciales of *F. solani, F. oxysporum,* and *F. roseum,* chlamydospore formation from germlings on membrane filters leached with distilled water was comparable to that on membrane filters on natural soil.

Qureshi and Page (54), however, reported that addition of either organic or inorganic carbon sources was needed to obtain chlamydospores formed from mycelium of *F. oxysporum* in a two-salt medium. Stevenson and Becker (66) could not confirm this dependence of mycelium of *F. oxysporum* on exogenous carbon. They suggested the dependence of chlamydospore formation on exogenous carbon as found by Qureshi and Page (54) to be due to the starved condition of the mycelium used. The reduced chlamydospore formation in macroconidia at very low nutrient levels in soil extract observed by Alexander et al. (3) and Goyal et al. (20) was also ascribed to insufficient energy of the inoculum. In macroconidia of *F. solani* f. sp. *pisi* "low" in lipid content, chlamydospre formation in soil was reduced in comparison to that in macronconidia "high" in lipid content (76). Chlamydospore formation thus seems to depend on the nutrient status of the inoculum. The nutrient status of the inoculum apparently determines whether deprivation or addition of nutrients stimulates chlamydospore formation. Under natural conditions starved inoculum may be more common than in most experiments reported, where "well-fed" macroconidia produced on rich agar media are often used as inoculum. Chlamydospore formation in certain circumstances in the field therefore could very well be supported by carbohydrates released from decaying plant tissue or from roots.

In considering the significance of the nutrient status of the soil in chlamydospore formation, the carbon dioxide concentration of the soil should also be taken into account: carbon dioxide at concentrations commonly found in the soil atmosphere has been demonstrated to be fixed by mycelium of *F. oxysporum* f. sp. *cubense* (67), to stimulate the multiplication of this fungus, and to inhibit the formation of chlamydospores (45).

The possible role of inorganic salts in stimulating chlamydospore formation in conidia was studied by Hsu and Lockwood (30). Germlings of macroconidia of *F. solani* f. sp. *phaseoli* on membranes floating on water readily formed chlamydospores, in contrast to non-germinated macroconidia, which, in distilled or phosphate-buffered water, did not produce chlamydospores unless inorganic salts were added (30). Of several weak salt solutions, Na_2SO_4 solution was the most comparable to soil extracts and soil in stimulating numbers of chlamydospores to form. Na_2SO_4 did not increase chlamydospore production from germlings on membranes on water. The environmental conditions most favourable for chlamydospore formation apparently differ according to whether the macroconidium is germinated or not germinated. The beneficial effect of inorganic salts, e.g., $MgCl_2$, on chlamydospore formation from macroconidia of *F. solani* f. sp. *radicicola* and *F. solani* 'Coeruleum' had also been noticed by Griffin (22). The close resemblance of chlamydospore formation in weak salt solutions to that on soil and in soil extracts led Hsu and Lockwood (30) to conclude that an environment deficient in energy, but with an appropriate weak salt solution, may be all that is required for chlamydospore formation.

In contrast to the effects of carbon sources, the reported effects of nitrogen on chlamydospore formation are contradictory. At low macroconidial densities in salt solutions, inorganic nitrogen did not inhibit chlamydospore formation in *F. solani* f. sp. *radicicola* and *F. solani* f. sp. *phaseoli* (25). Schippers (55) and Schippers and Old (58) found NH_4Cl and nitrate, but especially the former, to inhibit chlamydospore formation in *F. solani* f. sp. *cucurbitae* in liquid culture as well as in soil at relatively high conidial densities. Conidial densities appear to be a complicating factor in experimental systems used to study chlamydospore formation (24). At high conidial densities, macroconidia in *F. sulphureum* did not germinate, but every conidium converted into a chlamydospore. At low conidial densities, the conidia germinated but did not convert into chlamydospores (61). The failure of macronidia of *F. sulphureum* to germinate at high densities apparently resulted from the presence of a self-inhibitor. Inhibition of spore germination at high conidial densities with consequences for chlamydospore formation has also been related to self-inhibitors by Griffin (23; see also ch. 30 in this volume). Self-inhibitors also may explain the reduced formation of chlamydospores from macroconidia of *F. solani* f. sp. *cucurbitae* at higher densities in soil (48). Macroconidia of *F. sulphureum* from older colonies converted more readily into chlamydospores than those harvested from younger ones (61), suggesting that age of macroconidia is also important.

Induction of Chlamydospore Formation by Microbial Metabolites

A single staling factor detected in agar and liquid cultures of *F. oxysporum* was shown to inhibit the growth of this *Fusarium* and at higher concentrations to induce chlamydospores to form in the mycelium (51, 53). The same substance from *F. oxysporum* also induced chlamydospore formation in *F. coeruleum* (52). The chemical nature of the staling agent is unknown, and no information exists on its formation in heavily infected decaying host tissues where it might be responsible for the abundant chlamydospore formation (11).

While chlamydospore formation by members of *F. oxysporum* is possibly insured by the ability to produce chlamydospore-inducing substances, it has been shown that *F. solani* f. sp. *phaseoli* is unable to produce such a substance (14). Chlamydospore formation in *F. solani* f. sp. *phaseoli* depends instead on external factors such as metabolites of other soil micro-organisms and nutrient depletion in hyphae or conidia. The stimulatory effect of soil bacteria on chlamydospore production by *F. solani* and *F. roseum* was demonstrated by Venkat Ram (78) and Park (50) respectively. Sterilized soil extracts also were found to contain inducing factors for chlamydospore formation in *F. solani* 'Coeruleum' (22) and *F. solani* f. sp. *phaseoli* (2, 7, 34). The factor inducing chlamydospore formation in *F. solani* f. sp. *phaseoli* was not recoverable from all soils tested (3). Extracts of soil samples collected at various times during the year differed in their ability to induce chlamydospore formation (15). Based on differences in response it was supposed that different chlamydospore-inducing substances originate from *Protaminobacter, Arthrobacter* and *Bacillus*. Differences in chlamydospore formation from macroconidia of clones of *F. solani* f. sp. *phaseoli* were considered to be specific responses to the different inducing bacterial isolates from soil tested (18). The authors therefore suppose that changes in numbers and types of bacteria in soil may affect the ability of specific clones to produce chlamydospores (16).

The chlamydospore-inducing factor in sterile soil extracts, probably an organic compound (6), could be counteracted by high nutrient levels that stimulated germination of the macroconidia and mycelial growth. Increased production of chlamydospores in sterile inorganic salt solutions by additions of autoclaved or irradiated soil was also ascribed to compounds, probably of biological origin (40). Meyers and Cook (40) suppose that such compounds may act as stimulants for chlamydospore formation additional to stimulation by energy depletion. Based on their experiences, Ford et al. (17) consider chlamydospore formation in *F. solani* f. sp. *phaseoli* to be a function of the genetic constitution of the fungal clone involved, the kind and amount of inducing substances present in soil, and the nutrient level of the soil. In contrast, Hsu and Lockwood (30) conclude from their experiments that the chlamydospore-inducing effects of sterile soil extracts and bacterial isolates can be ascribed to energy depletion and inorganic salts rather than to specific organic compounds of microbial origin. The possible role of microbial metabolites and of inorganic salts in the induction of chlamydospore formation in soil is still rather obscure and needs further exploration.

Ultrastructure of Cell Walls of Chlamydospores

Alexander (2) examined the ultrastructure of chlamydospores of *F. solani* and *F. oxysporum* formed in sterile soil extracts. During the conversion of hyphal or conidial cells into chlamydospores, new cell wall material was deposited adjacent to the original hyphal or conidial wall. This agrees with recent observations on chlamydospore formation in *F. culmorum* (10), *F. oxysporum* (66), and *F. solani* (77) in sterile cultures, and in *F. solani* in natural soil (48, 73). Two layers are recognized in chlamydospore walls of *F. culmorum* (10), *F. oxysporum* (66), and *F. solani* (77), the original conidial or hyphal wall plus the newly deposited wall layer. During chlamydospore maturation in sterile cultures, the original macroconidial wall of *F. culmorum* and *F. solani* f. sp. *cucurbitae* gradually disappears (10, 76), which sometimes results in the loss of the entire conidial wall (76). During chlamydospore formation in *F. solani* f. sp. *cucurbitae* in soil, the outer cell wall layers slough off (48). In this pattern of chlamydospore wall formation, the original hyphal or conidial wall and the newly deposited wall layer can be considered as primary and secondary walls, respectively. This pattern of chlamydospore wall formation is in contrast with recent observations of Schneider et al. (60, 61) on chlamydospore formation in *F. sulphureum* in sterile cultures. During chlamydospore formation in *F. sulphureum* there is no accretion of new cell wall material adjacent to the conidial wall; instead, both the outer and inner layers of the macroconidial wall increase in thickness (30% and 250%, respectively), forming together the new thick chlamydospore wall (60). Schneider and Wardrop (62) observed that in chlamydospore walls of *F. sulphureum,* the chitin component forms one continuous layer of randomly orientated microfibrils. In chlamydospores of *F. sulphureum,* therefore, they only distinguish a primary wall (62).

The question arises whether the accretion of new cell wall material adjacent to the original hyphal or macroconidial wall during chlamydospore formation in *F. culmorum, F. oxysporum,* and *F. solani* (10, 48, 66, 73, 77) justifies the distinction of a secondary wall. To preclude discrepancies on presence or absence of different wall layers in chlamydospores due to dissimilar use of the term "layer," Griffith (28) suggested that in the absence of definite changes in microfibrillar orientation within the cell wall, the apparent differentiated lamellae seen in the walls should be referred to as "zones" rather than "layers." The distinction of "layers" in chlamydospore walls of *F. culmorum, F. oxysporum,* and *F. solani* (10, 48, 66, 73, 77) is based on heavy-metal-stained ultrathin sections and involves mainly the organization of the matrix material (glucans and proteins), as the chitin fibrils of the walls do not react with the heavy metal stains and hence are not revealed by electron microscopy. For chlamydospores of *F. culmorum, F. oxysporum,* and *F. solani,* no information exists on the microfibrillar organization of the chlamydospore wall. The disappearance of the original macroconidial wall in chlamydospores of *F. culmorum* and *F. solani* (10, 77), however, favours the conclusion that a secondary wall is deposited during chlamydospore formation. Information on the microfibrillar structure of cell walls can be obtained by removing the matrix material from the wall by hydrolysis, leaving the microfibrils which then can be examined by electron microscopy. Using this procedure, Schneider and Wardrop (62) observed that the microfibrils in chlamydospore walls of *F. sulphureum* formed in sterile culture are randomly orientated in one continuous layer. This favours the conclusion that in these chlamydospores only a primary wall is present. The possibility remains, however, that the natural configuration of the microfibrils is lost during the drastic procedure of isolation of the microfibrils from the walls.

The present information suggests that in different *Fusarium* species chlamydospore wall formation occurs acording to two basically different patterns. A comparative study of the different *Fusarium* species is necessary to clarify these contradictory conclusions on the ultrastructure of chlamydospore wall formation. Standardization of the fixation procedures ($KMnO_4$ versus Aldehyde and OsO_4 fixation) and evaluation of the procedure of isolation and examination of microfibrils from walls is needed.

Chlamydospores formed in sterile cultures may not be identical in their ultrastructure to those formed in soil. During chlamydospore formation in soil, new electron-dense material accumulates in and on the original macroconidial wall, giving the chlamydospore a verrucose (warty) appearance (73; Fig. 24-2, 24-3), as can also be seen with the light microscope (Fig. 24-1). The warts on the chlamydospores of *F. solani* were only formed in natural soil and in sterile soil extracts but not in sterile shake cultures (2, 73, 77). These warts disintegrate in soil, a process in which soil micro-organisms seem to be involved (Fig. 24-4) and that finally may leave only the newly deposited wall layer behind which then seems to be the main cell wall of chlamydospores in soil (73). The wartiness of the chlamydospores varies among the formae speciales of *F. solani* (73).

The ultrastructure of chlamydospores and chlamydospore wall formation in *Fusarium* in relation to persistance of chlamydospores in soil should preferentially be studied in chlamydospores formed in natural soil. Chlamydospores formed in sterile culture have the disadvantage of lacking the warts, while their survival capacity in soil depends on the culture medium used (57).

Chemical Composition of Chlamydospore Wall

Van Eck (74) and Schneider et al. (60) investigated the chemical composition of both macroconidial and chlamydospore walls of *F. solani* f. sp. *cucurbitae* and *F. sulphureum,* respectively. Their results are similar, showing that the walls mainly consist of glucosamine, glucose, mannose, galactose, amino acids, and glucuronic acids, and that there are different proportions of these compounds in the two spore types. The cell walls of both

1 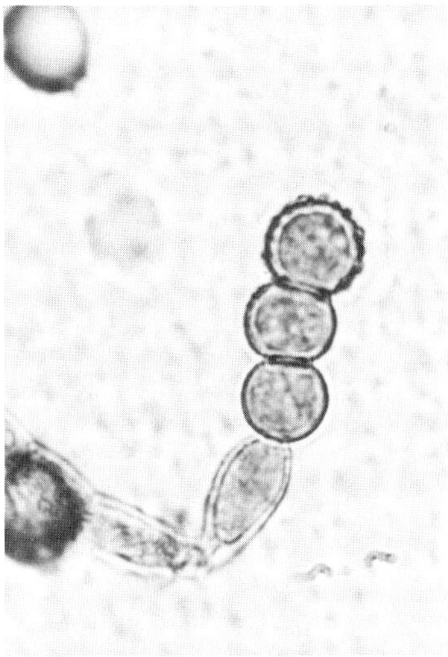

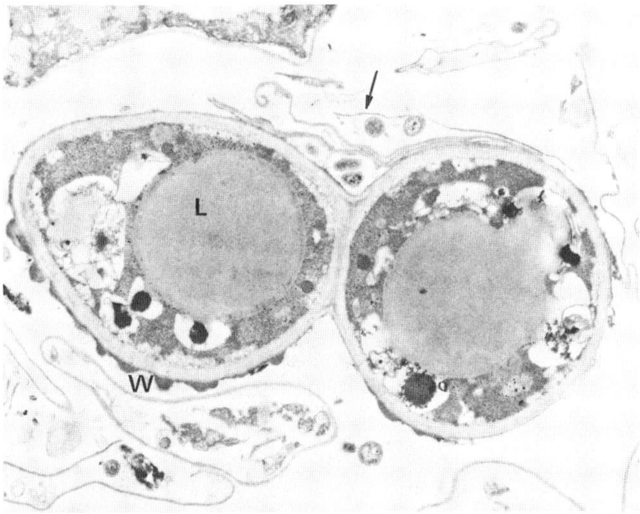

2

3 

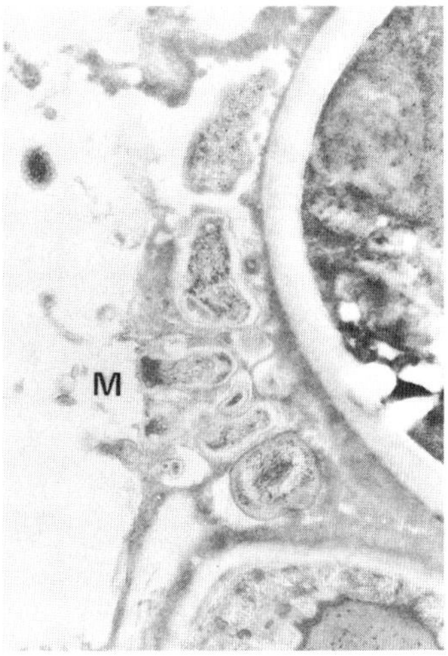

4

Fig.24-1. A chain of chlamydospore cells of *Fusarium solani* f. sp. *cucurbitae* produced from a germinated macroconidium incubated in soil on a Nuclepore membrane filter. Light microscopy.

Fig.24-2. Chlamydospore of *Fusarium solani* f. sp. *pisi* from soil. Note the large lipid bodies (L) and the electron-dense warts (W) on the chlamydospore cell wall. Sloughing off of the outer cell wall layers is associated with micro-organisms (arrow). Transmission electron microscopy.

Fig.24-3. Warty chlamydospore of *Fusarium solani* f. sp. *pisi* formed on a Nuclepore membrane filter in soil. Scanning electron microscopy.

Fig.24-4. Micro-organisms (M) invading warts on chlamydospore walls of *Fusarium solani* f. sp. *phaseoli* in soil. Transmission electron microscopy.

macroconidia and chlamydospores of *F. solani* f. sp. *cucurbitae* are readily solubilized by β(1-3) glucanase, whereas chitinase only affects the septa. Possibly other enzymes in the glucanase may be partly responsible for the solubilization of the cell wall material.

During chlamydospore formation, electron-dense material, which is resistant to β(1-3) glucanase, is deposited in the macroconidial cell wall layer that later becomes the outer layer of the chlamydospore cell wall. It is not known whether this material has a melanin-like nature. Melanin is known to enhance resistance of fungal cell walls to microbial lysis (35).

Survival of Chlamydospores in Soil

The significance of chlamydospores for survival of pathogenic Fusaria in agricultural soils varies widely among species and formae speciales and apparently depends on soil characteristics and climate factors. Sitton and Cook (64) observed that chlamydospores from Graminearum are much more sensitive to rapid drying and high temperatures than those of Culmorum. This may explain why Graminerum is rarely recovered from the dryland areas of the state of Washington (USA). Such a difference in persistence in soil between chlamydospores that are morphologically similar has been documented by Nash and Alexander for *F. solani* f. sp. *cucurbitae* and *F. solani* f. sp. *phaseoli* (42). Alexander (2) obtained evidence with electron microscopy that chlamydospore walls of *F. solani* f. sp. *cucurbitae* formed in soil extracts were thinner than those of the bean *Fusarium* and therefore might be less suited to survival. Van Eck (73) observed no differences in wall thickness between chlamydospores of these two fungi formed in soil, except for wartiness of the outer layer. The warts, however, are rapidly degraded by soil micro-organisms and do not contribute to persistence of the chlamydospores in soil (Fig. 24-4). Instead, he noticed that chlamydospores of *F. solani* f. sp. *cucurbitae* contained relatively little stored lipids compared with chlamydospores of the f. sp. *phaseoli* and *pisi*. To test the relevance of stored lipids to persistence (Fig. 24-2), macroconidia of *F. solani* f. sp. *pisi* were produced on a basal medium containing 2% and 1/64% sucrose, respectively. Chlamydospores formed in soil from these macroconidia had 50% and 25% of their cell contents occupied by lipid bodies, respectively, yet they did not differ in persistence in soil over the first two months after their formation (76).

Do soil micro-organisms that degrade cell walls of living chlamydospores play an important role in destruction of chlamydospores? Old and Schippers (48) noticed with electron microscopy that sloughing off of outer cell wall layers during chlamydospore formation in soil was closely correlated with the presence of soil micro-organisms (Fig. 24-2). Van Eck and Schippers (77), however, observed identical processes during chlamydospore formation of *F. solani* f. sp. *cucurbitae* in sterile cultures. Warts on the chlamydospores in soil were readily invaded by soil micro-organisms (Fig. 24-4); perforations of the walls, on the other hand, were seldom observed (73). Although the isolated chlamydospore cell wall, except for the electron-dense layer, is readily solubilized by β(1-3) glucanase, in soils amended with chitin and laminarin [β(1-3) glucan], no perforations or cell wall degradations were observed. Lysis of chlamydospores in these soils was nevertheless enhanced (75). It was concluded therefore, that under experimental conditions, lysis of chlamydospores is an autolytic rather than a heterolytic process. Reduction in the population density of *Fusarium* in soil after amendment of the soil with chitin to stimulate a chitinoclastic microflora (9, 31, 41, 56) cannot be ascribed to heterolytic degradation of the chlamydospore wall. It also cannot be explained by increased competition for nutrients, as increased lysis of chlamydospores did not correlate with increase in total numbers of micro-organisms in soil. It is more likely that the increased lysis of chlamydospores in chitin-amended soils (31, 41, 74) is due to the toxic effect of ammonia being released during chitin degradation (58, 81).

Cultural Practices and Their Consequences for Formation and Survival of Chlamydospores

Efforts to reduce diseases caused by Fusaria by soil amendments with crop residues or organic compounds have been successful incidentally. Success is correlated to inhibition of chlamydospore germination in *F. solani* f. sp. *phaseoli* (38). Crop residue decomposition products may have the ability to decrease propagule numbers by stimulating chlamydospore germination followed by germ tube lysis (70) or by direct killing of the chlamydospore (81). Zakaria et al. (81) gave strong evidence that reduction in population density of *F. oxysporum* and *F. solani* was due to killing of chlamydospores by ammonia released by degradation of oilseed meal amendments. These observations strongly suggest that the increased lysis observed in chitin-amended soils (31, 41, 74) is also due to the toxic effect of ammonia being released during chitin degradation (58). In other studies with soil amendments, numbers of chlamydospores of *F. solani* f. sp. *phaseoli* even increased, due to stimulation of sporulation or to formation of replacement chlamydospores (1, 36, 37, 39, 49). Control of populations of chlamydospore-forming Fusaria based on incorporation of particular residues appears to be highly difficult to stabilize in the field (see chapter 14 of this volume).

"*Fusarium*-suppressive soils" have been described (68) in which pathogenic Fusaria either do not establish or establish slowly. The phenomenon is based on the inability to produce chlamydospores in these soils (8, 69) or on formation of chlamydospores of inferior quality (in terms of size, wall thickness, and stored nutrients) which are therefore unable to survive for long periods (68; see also the next chapter of this volume). The inability of chlamydospores to establish, germinate, or survive in suppressive soils could well be caused by the specific elements of the microflora and needs further exploration (16, 65, 75).

The Term Chlamydospore in *Fusarium*

Based on an elaborate treatise on the origin, structure, and function of chlamydospores in fungi, Griffiths (29) suggested a definition of the term *chlamydospore* as follows:

> *Chlamydospore:* (Greek = mantle)—a viable, asexually produced accessory spore resulting from the structural modification of a vegetative hyphal segment(s) possessing an inner secondary wall, usually impregnated with hydrophobic material, and whose function is primarily perennation and not dissemination.

In *Fusarium,* however, chlamydospore formation may take place in both vegetative hyphal segments and in conidial cells (11, 43). General agreement exists on the accretion of new cell wall material during chlamydospore formation. Whether this results in a differentiate secondary wall formation (10, 66, 77) or not (60) is subject to controversy. A "double wall" (primary plus secondary wall) often used for the recognition of chlamydospores is also a questionable criterion because primary wall material may disappear quickly (48, 73, 77). On the basis of present knowledge we suggest defining the term chlamydospore for Fusaria as follows:

> *Chlamydospore:* —a viable, asexually produced accessory spore resulting from the structural modification of a vegetative hyphal segment(s) or conidial cell possessing a thick wall mainly consisting of newly synthesized cell wall material; its function is primarily survival in soil.

According to this definition the "changed macroconidia" in *F. roseum* Graminearum described by Nyvall (46) and also the thick-walled hyphae such as described by Nyvall and Kommedahl (47) have to be considered as chlamydospores. Any thick-walled cell of a *Fusarium* that has survival value, whether or not double-walled, is a chlamydospore.

Literature Cited

1. Adams, P.B., J.A. Lewis, and G.C. Papavizas. 1968. Survival of root-infecting fungi in soil. IX. Mechanism of control of Fusarium root rot of bean with spent coffee grounds. Phytopathology 58: 1603–1608.
2. Alexander, J.V. 1964. A study of the ultrastructure in the development and germination of chlamydospores of *Fusarium*. Ph. D. Thesis, Univ. Calif. (Berkeley).
3. Alexander, J.V., J.A. Bourret, A.H. Gold, and W.C. Snyder. 1966. Induction of chlamydospore formation by *Fusarium solani* in sterile soil extracts. Phytopathology 56: 353–354.
4. Banihashemi, Z., and D.J. de Zeeuw. 1973. The effect of soil temperature on survival of *Fusarium oxysporum* f. *melonis* (Leach and Currence) Snyder and Hansen. Plant Soil 38: 465–468.
5. Bouhot, D., and F. Rouxel. 1971. Recherches sur l'écologie des champignons parasites dan le sol. IV. Nouvelles mises au point concernant l'analyse sélective et quantitative des *Fusarium oxysporum* et *Fusarium solani* dans le sol. Ann. Phytopathol. 3: 171–188.
6. Bourret, J.A. 1965. Physiology of chlamydospore formation and survival in *Fusarium*. Ph. D. Thesis. Univ. Calif. (Berkeley).
7. Bourret, J.A., A.H. Gold, and W.C. Snyder. 1965. Inhibitory effect of CO_2 on chlamydospore formation in *Fusarium solani* f. *phaseoli*. Phytopathology 55: 105 (Abstr.).
8. Burke, D.W. 1965. Fusarium root rot of beans and the behavior of the pathogen in different soils. Phytopathology 55: 1122–1126.
9. Buxton, E.W., O. Khalifa, and V. Ward. 1965. Effect of soil amendment with chitin on pea wilt caused by *Fusarium oxysporum* f. *pisi*. Ann. Appl. Biol. 55: 83–88.
10. Campbell, W.P., and D.A. Griffith. 1974. Development of endoconidial chlamydospores in *Fusarium culmorum*. Trans. Brit. Mycol. Soc. 63: 221–228.
11. Christou, T., and W.C. Snyder. 1962. Penetration and host-parasite relationships of *Fusarium solani* f. *phaseoli* in the bean plant. Phytopathology 52: 219–226.
12. Cook, R.J. 1968. Influence of oats on soil-borne populations of *Fusarium roseum* f. sp. *cerealis* 'Culmorum.' Phytopathology 58: 957–960.
13. Cook, R.J., and W.C. Snyder. 1965. Influence of host exudates on growth and survival of germlings of *Fusarium solani* f. *phaseoli* in soil. Phytopathology 55: 1021–1025.
14. Ford, E.J. 1969. Production of chlamydospore-inducing substances by *Fusarium oxysporum*. Phytopathology 59: 1026.
15. Ford, E.J., A.H. Gold, and W.C. Snyder. 1970. Soil substances inducing chlamydospore formation by *Fusarium solani*. Phytopathology 60: 124–128.
16. Ford, E.J., A.H. Gold, and W.C. Snyder. 1970. Induction of chlamydospore formation in *Fusarium solani* by soil bacteria. Phytopathology 60:479–484.
17. Ford, E.J., A.H. Gold, and W.C. Snyder. 1970. Interaction of carbon nutrition and soil substances in chlamydospore formation by *Fusarium*. Phytopathology 60: 1732–1737.
18. Ford, E.J., and E.E. Trujillo. 1967. Bacterial stimulation of chlamydospore production in *Fusarium solani* f. sp. *phaseoli*. Phytopathology 57: 811 (Abstr.).
19. French, E.R., and L.W. Nielson. 1966. Production of macroconidia of *Fusarium oxysporum* and their conversion to chlamydospores. Phytopathology 56: 1322–1323.
20. Goyal, J.P., H. Maraite, and J.A. Meyer. 1973. Abundant production of chlamydospores by *Fusarium oxysporum* f. sp. *melonis* in soil extract with glucose. Neth. J. Plant Pathol. 79: 162–164.
21. Griffin, G.J. 1964. Influence of carbon and nitrogen nutrition on chlamydospore formation by *Fusarium solani* f. *radicicola*. Phytopathology 54: 894 (Abstr.).
22. Griffin, G.J. 1965. Chlamydospore formation in *Fusarium solani* 'Coeruleum.' Phytopathology 55: 1060 (Abstr.).
23. Griffin, G.J. 1970. Carbon and nitrogen requirements for macroconidial germination of *Fusarium solani:* dependence on conidial density. Can. J. Microbiol. 16: 733–740.
24. Griffin, G.J. 1970. Exogenous carbon and nitrogen requirements for chlamydospore germination by *Fusarium solani:* dependence on spore density. Can. J. Microbiol. 16: 1366–1368.
25. Griffin, G.J. 1976. Roles of low pH, carbon and inorganic nitrogen source use in chlamydospore formation by *Fusarium solani*. Can. J. Microbiol. 22: 1381–1389.
26. Griffin, G.J., and R.H. Ford. 1974. Soil fungistasis: fungus spore germination in soil at spore densities corresponding to natural population levels. Can. J. Microbiol. 20: 751–754.
27. Griffin, G.J., and T. Pass. 1969. Behaviour of *Fusarium roseum* 'Sambucinum' under carbon starvation conditions in relation to survival in soil. Can. J. Microbiol. 15: 117–126.
28. Griffith, D.A. 1973. Fine structure of the chlamydospore wall in *Fusarium oxysporum*. Trans.Brit. Mycol. Soc. 61: 1–6.
29. Griffith, D.A. 1974. The origin, structure and function of chlamydospores in fungi. Nova Hedwigia 25: 503–547.
30. Hsu, S.C., and J.L. Lockwood. 1973. Chlamydospore formation by *Fusarium* in sterile salt solutions. Phytopathology 63: 597–601.
31. Khalifa, O. 1965. Biological control of Fusarium wilt of peas by organic soil amendments. Appl. Biol. 56: 129–137.
32. Komada, H. 1975. Development of a selective

medium for quantitative isolation of *Fusarium oxysporum* from natural soil. Rev. Plant Protection. Res. 8: 114–125.
33. Kraft, J.M., F.J. Muehlbauer, R.J. Cook, and F.M. Entemann. 1974. The reappearance of common wilt of peas in eastern Washington. Plant Dis. Rep. 58: 62–64.
34. Krikun, J., and R.E. Wilkinson. 1963. Production of chlamydospores in sterile soil extracts sterilized by passage through different types of filters. Phytopathology 53: 880 (Abstr.).
35. Kuo, M.J., and M. Alexander. 1967. Inhibition of the lysis of fungi by melanins. J. Bacteriol. 94: 624–629.
36. Lewis, J.A., and G.C. Papavizas. 1968. Survival of root-infecting fungi in soil. VII. Decomposition of tannins and lignins in soils and their effects on Fusarium root rot of bean. Phytopathol. Z. 63: 124–134.
37. Lewis, J.A., and G.C. Papavizas. 1975. Survival and multiplication of soil-borne plant pathogens as affected by plant tissue amendments, p. 84–89. *In* G.W. Bruehl (ed.), Biology and control of soil-borne plant pathogens, Amer. Phytopathol. Soc., St. Paul, Minn.
38. Lewis, J.A., and G.C. Papavizas. 1977. Effect of plant residues on chlamydospore germination of *Fusarium solani* f. sp. *phaseoli* and on Fusarium root rot of beans. Phytopathology 67: 925–929.
39. Maurer, C.H., and R. Baker. 1965. Ecology of plant pathogens in soil. II. Influence of glucose, cellulose and inorganic nitrogen amendments on development of bean root rot. Phytopathology 55: 69–73.
40. Meyers, J.A., and R.J. Cook. 1972. Induction of chlamydospore formation in *Fusarium solani* by abrupt removal of the organic carbon substrate. Phytopathology 62: 1148–1153.
41. Mitchell, R., and M. Alexander. 1963. Lysis of soil fungi by bacteria. Can. J. Microbiol. 9: 169–177.
42. Nash, S.M., and J.V. Alexander. 1965. Comparative survival of *Fusarium solani* f. *cucurbitae* and *F. solani* f. *phaseoli* in soil. Phytopathology 55: 963–966.
43. Nash, S.M., T. Christou, and W.C. Snyder. 1961. Existence of *Fusarium solani* f. *phaseoli* as chlamydospores in soil. Phytopathology 51: 308–312.
44. Nash, S.M., and W.C. Snyder. 1962. Quantitative estimations by plate counts of propagules of the bean root rot Fusarium in field soils. Phytopathology 52: 567–572.
45. Newcombe, M. 1960. Some effects of water and anaerobic conditions on *Fusarium oxysporum* f. *cubense* in soil. Trans. Brit. Mycol. Soc. 43: 51–59.
46. Nyvall, R.F. 1970. Chlamydospores of *Fusarium roseum* 'Graminearum' as survival structures. Phytopathology 60: 1175–1177.
47. Nyvall, R.F., and T. Kommedahl. 1966. Thickened hyphae as a survival mechanism in *Fusarium moniliforme*. Phytopathology 56: 893 (Abstr.).
48. Old, K.M., and B. Schippers. 1973. Electron microscopical studies of chlamydospores of *Fusarium solani* f. *cucurbitae* formed in natural soil. Soil Biol. Biochem. 5: 613–620.
49. Papavizas, G.C., P.B. Adams, and J.A. Lewis. 1968. Survival of root-infecting fungi. V. Saprophytic multiplication of *Fusarium solani* f. sp. *phaseoli* in soil. Phytopathology 58: 414–420.
50. Park, D. 1956. Effect of substrate on a microbial antagonism with reference to soil condition. Trans. Brit. Mycol. Soc. 39: 239–259.
51. Park, D. 1961. Morphogenesis, fungistasis, and cultural staling in *Fusarium oxysporum* Snyd. and Hans. Trans. Brit. Mycol. Soc. 44: 377–399.
52. Park, D. 1963. Evidence for a common fungal growth regulator. Trans. Brit. Mycol. Soc. 46: 541–548.
53. Pratt, C.A. 1924. The staling of fungal cultures; I. Genetical and chemical investigation of staling by *Fusarium*. Ann. Bot. 38: 563–599.
54. Qureshi, A.A., and O.T. Page. 1970. Observations on chlamydospore production by *Fusarium* in a two-salt solution. Can. J. Microbiol. 16: 29–32.
55. Schippers, B. 1972. Reduced chlamydospore formation and lysis of macroconidia of *Fusarium solani* f. *cucurbitae* in nitrogen-amended soil. Neth. J. Plant Pathol. 78: 189–197.
56. Schippers, B., and W.M.M.M. De Weijer. 1972. Chlamydospore formation and lysis of macroconidia of *Fusarium solani* f. *cucurbitae* in chitin-amended soil. Neth. J. Plant Pathol. 78: 45–54.
57. Schippers, B., and K.M. Old. 1974. Factors affecting chlamydospore formation by *Fusarium solani* f. *cucurbitae* in pure culture. Soil Biol. Biochem. 6: 153–160.
58. Schippers, B., and L.C. Palm. 1973. Ammonia, a fungistatic volatile in chitin-amended soil. Neth. J. Pl. Path. 79: 279–281.
59. Schippers, B., and J.S. Voetberg. 1969. Germination of chlamydospores of *Fusarium oxysporum* f. sp. *pisi* race 1 in the rhizosphere and penetration of the pathogen into roots of a susceptible and a resistant pea cultivar. Neth. J. Plant. Pathol. 75: 241–258.
60. Schneider, E.F., L.R. Barran, P.J. Wood, and I.R. Siddiqui. 1977. Cell wall of *Fusarium sulphureun* II. Chemical composition of the conidial and chlamydospore walls. Can. J. Microbiol. 23: 763–769.
61. Schneider, E.F., and W.L. Seaman. 1974. Development of conidial chlamydospores of *Fusarium sulphureum* in distilled water. Can. J. Microbiol. 20: 247–254.
62. Schneider, E.F., and A.B. Wardrop. 1979. Ultrastructural studies on the cell walls in *Fusarium sulphureum*. Can J. Microbiol. 25: 75–85.
63. Schroth, M.N., and F.F. Hendrix, Jr. 1962. Influence of nonsusceptible plants on the survival of *Fusarium solani* f. *phaseoli* in soil. Phytopathology 52: 906–909.
64. Sitton, J.W., and R.J. Cook. 1976. Chlamydospores of *Fusarium roseum* 'Culmorum' and *F.*

roseum 'Graminearum.' Fusarium Notes FAO, Plant Pathology Newsletter, 2: 3–4.
65. Smith, S.N. 1977. Comparison of germination of pathogenic *Fusarium oxysporum* chlamydospores in host rhizosphere soils conducive and suppressive to wilts. Phytopathology 67: 502–510.
66. Stevenson, I.L., and S.A.W.E. Becker. 1972. The fine structure and development of chlamydospores of *Fusarium oxysporum*. Can. J. Microbiol. 18: 997–1002.
67. Stover, R.H., and S.R. Freiberg. 1958. Effect of carbon dioxide on multiplication of *Fusarium* in soil. Nature 181: 788–789.
68. Toussoun, T.A. 1975. Fusarium-suppressive soils, p. 145–151. *In* G.W. Bruehl (ed.), Biology and control of soil-borne plant pathogens, Amer. Phytopathol. Soc., St. Paul, Minn.
69. Toussoun, T.A., W. Menzinger, and R.S. Smith, Jr. 1969. Role of conifer litter in ecology of *Fusarium:* Stimulation of germination in soil. Phytopathology 59: 1396–1399.
70. Toussoun, T.A., Z.A. Patrick, and W.C. Snyder. 1963. Influence of crop residue decomposition products on the germination of *Fusarium solani* f. *phaseoli* chlamydospores in soil. Nature 197: 1314–1316.
71. Trujillo, E.E., and W.C. Snyder. 1963. Uneven distribution of *Fusarium oxysporum* f. *cubense* in Honduras soils. Phytopathology 53: 167–170.
72. Van Eck, W.H. 1976. Suitability of membrane-filter techniques to study the ultrastructure of *Fusarium solani* in soil. Can. J. Microbiol. 22: 1628–1633.
73. Van Eck, W.H. 1976. Ultrastructure of forming and dormant chlamydospores of *Fusarium solani* in soil. Can. J. Microbiol. 22: 1634–1642.
74. Van Eck, W.H. 1978. Chemistry of cell walls of *Fusarium solani* and the resistance of spores to microbial lysis. Soil Biol. Biochem. 10:155–157.
75. Van Eck, W.H. 1978. Autolysis of chlamydospores of *Fusarium solani* f. sp. *cucurbitae* in chitin and laminarin amended soils. Soil Biol. Biochem. 10: 89–92.
76. Van Eck, W.H. 1978. Lipid body content and persistence of chlamydospores of *Fusarium solani* in soil. Can J. Microbiol. 24:65–69.
77. Van Eck, W.H., and B. Schippers. 1976. Ultrastructure of developing chlamydospores of *Fusarium solani* f. *cucurbitae* in vitro. Soil Biol. Biochem. 8: 1–6.
78. Venkat Ram, C.S. 1952. Soil bacteria and chlamdospore formation in *Fusarium solani.* Nature 170: 899.
79. Whalley, W.M., and G.S. Taylor. 1976. Germination of chlanydospores of physiologic races of *Fusarium oxysporum* f. *pisi* in soil adjacent to susceptible and resistant pea cultivars. Trans. Brit. Mycol. Soc. 66: 7–13.
80. Zakaria, M.A., and J.L. Lockwood. 1980. Reduction in Fusarium populations in soil by oilseed amendments. Phytopathology 70: 240–243.
81. Zakaria, M.A., J.L. Lockwood, and A.B. Filonow. 1980. Reduction in Fusarium population density in soil by volatile degradation products of oilseed meal amendments. Phytopathology 70: 495–499.

25 Microbiological Suppressiveness of Some Soils to Fusarium Wilts*

J. Louvet, C. Alabouvette and F. Rouxel

The microbiological activity of certain soils can prevent the establishment of plant pathogens or can inhibit their pathogenic activities. The properties of these soils have been designated in various ways, and are summarized by Baker and Cook (4, p. 62) as "resistant, long-life, immune, intolerant, antagonistic, suppressive," and by others as "unfavorable" (42), or "disease-free" (1). The expressions most commonly used by English authors, which we will use in this paper, are "suppressive soils," and their opposite, "conducive soils." The expression "pathogen-suppressive soil" sometimes employed does not necessarily mean the elimination of the pathogen from the soil, but rather the absence or suppression of the disease when susceptible plants are cultivated in that soil; the correct expression is therefore, "disease-suppressive soil," or better yet according to Burke (10) "disease-reducing soil," which implies a quantitative phenomenon. In French, we prefer "sol résistant au développement d'une maladie" (soil resistant to the development of a disease), pointing out that this "resistance" is of a microbiological nature (23). Soil is thus considered to be a living entity which manifests a certain "resistance" to infestation and to the expression of the pathogenic capacities of an organism.

The existence of suppressive soils has been known for some time in the cases of *Gaeumannomyces graminis* var. *tritici* (13), *Fusarium roseum* f. sp. *cerealis* 'Culmorum' (4, p. 62), *Verticillium albo-artrum* (32), *Fusarium solani* f. sp. *phaseoli* (10), *Rhizoctonia solani* (5), and *Phytophthora cinnamomi* (8). Studies on the Fusarium wilt suppressive soils have been summarized by Toussoun (45). These studies deal with soils suppressive to Fusarium wilts of banana (36, 43, 44); of legumes such as peas (48), muskmelon (23, 49) and sweet potato (41); and of fiber crops such as flax (47), and cotton (1).

These studies show that the mechanisms of suppressiveness are varied and complex; they are of great interest because, on the one hand, they lead to a better understanding of the relationships existing between populations of saprophytic and pathogenic microorganisms, and on the other hand, they lead to the possibility of utilizing this natural resistance in the control of soil-borne diseases.

Evidence for Soil Suppressiveness

The absence of a soil-borne disease in a crop of susceptible plants does not necessarily mean that the soil is suppressive. In certain cases, the pathogen may not have been

*Accepted for publication in 1977.

introduced into the region. In other cases, the soil may be infested but climatic and cultural factors are unfavorable for the development of the disease. These possibilities may be eliminated if one observes, after artificial infestation of the soil and cultivation of highly susceptible plants under conditions highly conducive to the disease, that the plants remain healthy or are only lightly attacked in comparison to others grown under the same conditions in other soils.

When one has thus shown that the absence of the disease is due solely to the specific properties of the soil, one must endeavour to determine if it is tied to phenomena which take place at the level of the susceptible host tissues or in the soil itself.The prior colonization of susceptible plants by non-pathogenic fungi can induce mechanisms of the resistance against *Fusarium oxysporum* pathogens, a phenomenon of cross protection (26). Thus, Mas (25) obtained cross protection (pre-immunizing effect) against Fusarium wilt by growing muskmelons in a compost artificially infested with non-pathogenic Fusaria. On the other hand, Louvet et al. (23) observed that the absence of the disease on muskmelons grown under normal conditions in suppressive soils was not due to a resistance acquired by these plants, because they were quickly attacked if transplanted into infested, conducive soil. In addition, cross protection was effective for only 10–20 days, while muskmelons grown in suppressive soils remained healthy for the entire growing period. It is obvious, therefore, that different biological phenomena are involved. Burke (10) also observed that when he transplanted beans from a root rot (*Fusarium solani* f. sp. *phaseoli)* suppressive soil or from a conducive soil, the disease occurred regardless of the soil of origin. Therefore, in the context of the studies made to date on Fusarium disease suppressive soils, the observed phenomena have no relation to a resistance acquired by the plants as a consequence of their growth in these soils.

On the other hand, Pope and Jackson (33), following experiments on take-all decline (*Gaeumannomyces graminis*) reported that "The transplanting experiments show that the roots of seedlings grown for 5 days in decline soil before transplanting to non-decline soil suffer significantly less infection than seedlings grown continuously in non-decline soil or for 5 days in non-decline soil before transplanting to decline soil."

It is likely, therefore, that soils exist in which disease suppressiveness is due to phenomena which take place in the soil itself, apart from the host plant, while in others, the phenomena that are responsible for this suppressiveness take place in the rhizosphere or in the rhizoplane of cultivated plants. In all the cases suppressiveness is basically microbiological, that is to say, particular microorganisms act directly on the pathogen, preventing its installation and/or its development in the soil or inhibiting its pathogenic activities

Table 25-1. Percent of diseased muskmelons in relation to inoculum density of *Fusarium oxysporum* f. sp. *melonis* and soil suppressiveness

Chlamydospores/g soil[a]	Soils		
	A	B	C
	%	%	%
0	0	0	0
10	0	0	0
100	0	0	45
200	0	0	40
400	0	0	33
800	0	0	45
1600	0	33	92
3200	0	33	92

[a]Formed according to a method proposed by Locke and Colhoun (21).

toward the host. This corresponds to the two categories established by Baker and Cook (4, p. 61): "soils where pathogens fail to establish" and "soils where pathogens become established but fail to produce disease." Thus in order to prove disease suppressiveness of a soil, in the sense in which we have defined it, it is necessary to show that the observed phenomena are due to soil microorganisms other than the pathogen. The simplest procedure consists of studying the effects of biocidal agents applied to soil. When the absence of a disease is due simply to the unfavorable direct action of certain physical or chemical properties of the soil on the pathogen, autoclaving of the soil does not render it conducive. This is the case with the soils of Guadaloupe suppressive to bacterial wilt *(Pseudomonas solanacearum)* studied by Bereau and Messiaen (7). Generally heat treatment, gamma irradiation, or fumigation, which reduce the populations of saprophytic microorganisms, diminish or negate the specific properties of microbiologically suppressive soils (8, 10, 17, 23, 38).

Characteristics of Fusarium Wilt Suppressive Soils

Having demonstrated the microbiological nature of suppressiveness, it is of interest to determine its principal characteristics by observing the disease level in plant populations grown in conducive and suppressive soils, either under normal cultural conditions or following various treatments.

Wilt Suppressive Potential

Stover (44) reported unpublished work by Volk, who observed the spread of Fusarium wilt of banana in different soils of Central America in 1930 and distinguished "non-resistant," "semi-resistant," and "resistant soils." Similarly, Wensley and McKeen (49) classed Canadian soils as a function of their wilt potential with respect to the muskmelon.

The best way of comparing the "wilt-suppressive potential" of different soils is to artificially infest them with increasing doses of inoculum and to grow susceptible plants in them under controlled conditions. Disease severity is then an indicator of the level of suppressiveness. In fact, the fundamental characteristic of suppressive soils is to give evidence of an inoculum potential (6, 22) that is extremely weak even in the presence of a high inoculum density.

Thus, we have compared the "receptivity" of a soil (A) from the Chateaurenard region of the Durance Valley to that of two other soils (B, C) by infesting them with chlamydospores in amounts varying from 10 to 3,200 per gram of soil (Table 25-1). The results show that soil A has a remarkable suppressiveness which was not overcome even by the application of inoculum doses much greater than those found in plantings prior to cropping (50–500 per gram). This enables us to understand why no disease has ever been observed in fields of that region. Conversely, soil C is very conducive, since the percentage of plants attacked is already high at inoculum concentrations equal to those found in soil at planting time. Soil B is suppressive, but less than soil A, its suppressiveness being overcome by inoculum doses above 1,000 chlamydospores per gram. These results recall those of Smith and Snyder (41) on *Fusarium oxysporum* f. sp. *batatas* and those of Burke (10) on *F. solani* f. sp. *phaseoli,* but the disease suppressive potential of the soils they studied was relatively weak. Indeed, it was manifested only at low inoculum concentrations.

It is necessary to have complementary studies to define quantitatively the level of suppressiveness of soils and to classify these as a function of their conduciveness to certain diseases.

Specificity of Suppressiveness

When one observes that a soil is suppressive to a Fusarium wilt, one can question whether it is suppressive to all Fusarium wilts and to all other soil-borne diseases. An interesting example of the suppressiveness of certain soils to a group of Fusarium wilts was mentioned by Baker and Cook (4, p. 64) as follows: "Soils of the coastal area near Castroville, California, are conducive to Fusarium wilts of peas, crucifers, and other wilt-susceptible crops grown there, but soils in the Salinas and San Joaquin valleys are suppressive to Fusarium wilts of tomato, peas, sweet potatoes, or other wilt-susceptible crops grown in those areas. Smith and Snyder (41, 42) showed that soils suppressive to one forma specialis of *F. oxysporum* may be suppressive to other formae speciales, but conversely, wilt-conducive soils are favorable to many different formae speciales." Toussoun (45) added that "In the Salinas Valley . . . Fusarium wilts are absent but other Fusarium diseases like bean root rot are prevalent."

We have also shown (J. Louvet et al., *unpublished data*) that the suppressiveness of alluvial soils in the Chateaurenard region is exerted simultaneously on the Fusarium wilts of muskmelon, tomato, cucumber, and carnation, but does not operate on bean-rot *(F. solani* f. sp. *phaseoli)* or on varied soil-borne diseases caused by *Verticillium dahliae* and *Phomopsis sclerotioides* on Cucurbitaceae, *Phytophthora* spp. and *Phomopsis lycopersici* on Cucurbitaceae and Solanaceae, and *Macrophomina phaseolina* on sunflower.

In cases studied to date therefore, the suppressiveness of certain soils has precise qualitative limits and operates specifically against the Fusarium wilts, which implies special mechanisms to which the *Fusarium oxysporum* wilt pathogens would be sensitive.

Origin and Evolution of the Suppressiveness of a Soil

If suppressiveness exists in a soil prior to cultivation or before susceptible plants are planted, it is then spoken of as native suppressiveness. Such suppressiveness can be observed at the first planting and increases upon successive cultivation of the susceptible host, contrary to what takes place in conducive soil, where monoculture generally causes a progressive aggravation of disease due to increase of the inoculum. This is what has been observed by Smith and Snyder (41) for Fusarium wilt of sweet potato, by Louvet et al. (23) for Fusarium wilt of muskmelon, and by Tu et al. (47) for Fusarium wilt of flax.

In the case of other diseases, suppressiveness does not seem to exist prior to cultivation but is acquired by the soils in the course of successive and continuous cultivation of susceptible plants. Mitchell (30) has qualified this type of suppressiveness as adaptive. Thus, following Fellows and Ficke (13), various researchers have observed that the phenomenon of take-all decline was linked to cereal monoculture. For example, Lemaire and Coppenet (20) determined that after 1–2 years of severe disease, wheat monoculture brings about a reduction of the disease so that after 4–5 years there is only negligible effect on yield. The interruption of this monoculture by the cultivation of a dicot causes the recurrence of the disease 2 years later. The other cited cases of such an acquired suppressiveness are potato scab (27) and bean root rot (10). No suppressiveness of this type seems to have been reported for the Fusarium wilts.

It is to be noted that the suppressiveness of a soil is not a static property; it is in effect the consequence of the activity of specific microbial populations whose development can be increased or decreased, as shown by transmission experiments.

Transmissibility of Suppressiveness from Soil to Soil

The best way to show transmissibility of suppressiveness from soil to soil is to mix different proportions of suppressive and conducive soils, and to compare the disease-

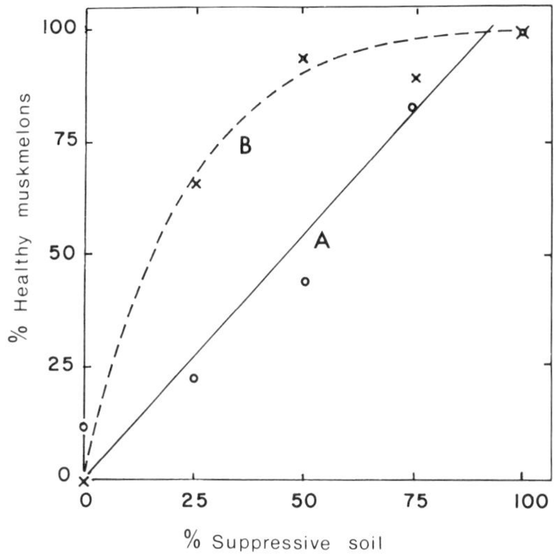

Fig.25-1. Suppressiveness to Fusarium wilt of mixtures of suppressive and conducive soils, untreated (A), or heat treated at 100 C for 30 min (B). Re-infestation after mixing. (23)

suppressive potentials in the mixes thus obtained. Thus Fig. 25-1 shows that the suppressiveness to Fusarium wilt of muskmelon is transmitted differently depending on the type of mixture. When the conducive soil has not been previously treated, the suppressiveness obtained is directly proportional to the concentration of the suppressive soil (line A), just as if this was a question of a simple mixture of inert elements which do not react with one another. On the other hand, in mixtures made with conducive soil that have been heat treated, such a straight line relationship does not exist and the suppressiveness obtained is always superior to that of mixtures obtained with non-treated soil (curve B). This difference coincides with a greater abundance in the treated soil of free ecologic niches which can be colonized by microorganisms coming from the suppressive soil. One can increase this colonization and thereby increase the suppressiveness of the soil by lengthening the incubation period, that is to say the contact time between the two soils before their infestation. Thus, an incubation period of 30 days on mixtures of 10% suppressive soil and 90% conducive, heat-treated soil resulted in 58% healthy plants as compared to 25% healthy plants in similar but nonincubated mixtures.

With certain soils, it would be also possible to obtain a suppressiveness which would be graphed by a curve situated below line A (Fig. 25-1). This would result from a weak competitive saprophytic ability of the microflora responsible for suppressiveness or from an inability of the soil to acquire this suppressiveness.

In the experiments just cited, the mixtures to transfer suppressiveness were made prior to the introduction of the *Fusarium* pathogen. But, as shown in Fig. 25-2, it is equally possible to transmit suppressiveness in previously infested soil. These results derive from phenomena of the same type as those observed by Shipton et al. (38). They show the great aptitude of the microorganisms responsible for suppressiveness to colonize substrates, even in competition with the native microflora.

It is important to control temperature and soil moisture during the incubation phase because these two factors can influence the microbiological equilibrium of the mixture and thus modify its wilt suppressive potential. The addition of nutritive substances to the mixture may also help the transmission of suppressiveness. Thus, Menzies (27), by the addition of 1% alfalfa meal, reduced from 50% to 10% the amount of suppressive soil necessary for an efficient transmission of suppressiveness to potato scab. In the same way, by comparing the effect of different substances such as glucose, cellulose, peptone, starch, and gelatin on the transmission of suppressiveness in a mixture of 1% soil from the

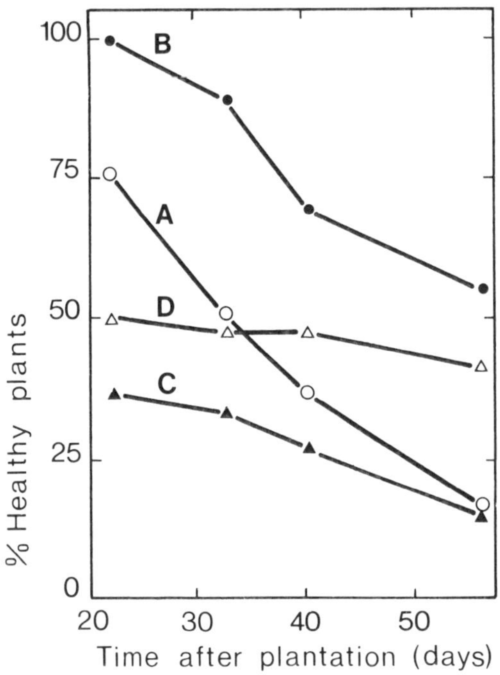

Fig.25-2. Transmission of suppressiveness to previously infested conducive soils. A = conducive soil infested with *Fusarium oxysporum* f. sp. *melonis*. B = 10% suppressive soil + 90% soil A. C = conducive soil infested with *F.oxysporum* f. sp. *lycopersici*. D = 10% suppressive soil + 90% soil B.

Chateaurenard region and 99% of a heat-treated conducive soil, we obtain the best results by addition of starch.

Preliminary experiments have shown that the suppressiveness acquired by these mixtures lasted for more than a year and could be transmitted in turn to other soils. This, however, does not mean that the characteristics of the inital suppressiveness are completely transmitted. In particular, the physical and chemical properties of the mixtures are clearly different from those of the originally suppressive soils and it is possible that they play a role in the acquisition, development, and conservation of suppressiveness.

Physical and Chemical Characteristics of Suppressive Soils

Burke (10) was not able "to transfer the rot-suppressive properties of the resistant soil to other soils by transfer of soil microorganisms," even with the addition of nutritive substances or by autoclaving the receptor soil. According to Burke (10), this "suggests that rot suppression depends upon the predominance of both physical and microbiological components of the resistant soil and that neither component alone is sufficient to produce the effect." A test whereby suppressiveness was transferred back to the original soil which had been rendered conducive by autoclaving, would have indicated if certain non-living factors present in that suppressive soil played an important role.

It is on soils suppressive to Fusarium wilt of banana that the most important studies were aimed at trying to establish correlations between suppressiveness and the physical and chemical characteristics of the soils. Stotzky (43) concludes that "analysis of soils . . . showed no correlations between the rate of disease spread and their chemical (i.e., pH, cation exchange capacity, total N, organic matter, extractable K, Mg, Ca, P, Mn, Fe, Al, and B) or physical (i.e., sand-, silt-, and clay-sized particles, water-holding capacity, ⅓ bar water content) properties, as determined by standard soil-testing procedures. A high correlation, however, was apparent between the rate of disease spread and the clay mineralogy of the soils : a particular species of clay mineral was present in almost all soils in which disease spread was slow, and was absent in almost all soils in which it was rapid. The distinguishing clay mineral . . . tended toward montmorillonite in the montmorillonite-vermiculite sequence." We have also noted (23) that the suppressive soils we studied contained 5% montmorillonite.

Following numerous other observations, one can conclude that, generally speaking, suppressiveness to Fusarium wilts is manifested in clay soils (36, 47, 48, 49). The lack of precision in this important area points to the urgent need for research to establish if there really is a constant correlation between the particular physical and chemical characteristics of soils and their suppressiveness to Fusarium wilts.

Microbiological Phenomena Involved in Suppressiveness

In previous paragraphs the principal general characteristics of suppressive soils have been determined by the observation of the state of health of susceptible plants. In order to understand the active microbiological mechanisms, one must examine the behaviour of the *F. oxysporum* populations responsible for the wilts and that of the populations of the microorganisms responsible for suppressiveness. Pathogens and saprophytes constitute populations whose specific properties and interactions determine the wilt-suppressive potential of these soils.

Behavior of *Fusarium oxysporum* Clones in Suppressive and Conducive soils

Various workers have attempted to assay populations of *F. oxysporum* responsible for wilts and populations of saprophytic Fusaria in suppressive and in conducive soils. Even though the counts were obtained under varying conditions and with different techniques, the results are generally in agreement. They allow for a determination of trends in the qualitative and quantitative evolution of *Fusarium* populations.

Reinking (35) indicated that 10 years after the planting of susceptible bananas, *F. oxysporum* f. sp. *cubense* could be isolated from conducive soils, whilst it was not easy to do so from suppressive soils. But, as Toussoun (45) observed, these results are questionable because the pathogenic clones were differentiated from the saprophytic clones solely on the basis of morphological criteria. Wensley and McKeen (49) did not find differences between levels of populations of *F. oxysporum* f. sp. *melonis* in the suppressive soils and in the conducive soils, but they did observe that the populations of *F. oxysporum* saprophytes were higher in the suppressive soil. Nash et al. (31) and Smith and Snyder (41,42) reported that during the cultivation of vegetables in suppressive soils, "although neither the pathogenic formae of *F. oxysporum* nor the wilt diseases are found in these soils, saprophytic *F. oxysporum* establishes abundantly and in a variety of clonal types." In the presence of susceptible plants, the populations of *F. oxysporum* pathogens introduced artificially remained constant in the suppressive soil, while they doubled in the conducive soil. In the absence of plants, the addition of nutritive substances caused a more marked lowering of the introduced pathogenic populations of *F. oxysporum* and increase in the native populations of *F. oxysporum* in the suppressive soils than in the conducive soils. Moreover, in the course of studies during which soils were kept out of doors for 2 years after infestation with the special forms *raphani, cucumerinum,* and *lycopersici,* Komada and Ezuka (18) observed that in certain soils the number of propagules of *F. oxysporum* pathogens diminished more than in others, while populations of *F. oxysporum* saprophytes remained practically constant. Louvet et al. (23) found, following the cultivation of susceptible muskmelons in soils artifically infested with propagules varying in number from 100 to 800 per g of soil, that the populations of *F. oxysporum* f. sp. *melonis* established themselves at levels below 500 in suppressive soil, whilst they were above 1,000 in conducive soil. On the other hand, further work (J. Louvet et al., *unpublished data*) has shown that in the absence of cultivation, the total population of *Fusarium* saprophytes, estimated by dilution plate technique, was very high in suppressive soils, reaching 20–40% of total fungal population, whereas in the majority of cultivated soils in temperate regions, the percentages observed were generally below 10–15%

We thus note between soils suppressive and conducive to wilts differences in the equilibria of populations of *Fusarium* spp. In the suppressive soils, the *F. oxysporum* pathogens seem to become established with difficulty, while the populations of saprophytic *Fusarium* species are more varied and reach higher levels than in the conducive soils. The following observations on the formation of chlamydospores from macroconidia, their germination, and the formation of secondary chlamydospores, were made in order to establish the manner in which these equilibrium modifications took place. They confirm and partially explain the results which just have been presented.

By placing macroconidia of *F. oxysporum* f. sp. *melonis* in aqueous non-sterile soil extracts, we have observed (J. Louvet et al., *unpublished data*) that chlamydospores are formed in lesser numbers in extracts of suppressive soils; they are poorly formed, thin walled, and therefore less apt to persist. These results are quite similar to those obtained by Burke (10) with *F. solani* f. sp. *phaseoli.* In addition, we have been able to confirm with the specialized forms *melonis* (3 races), *lycopersici, dianthi,* and *albedinis,* the observations of Smith and Snyder (42), Smith (40), and Tu et al. (47) made on the special forms *batatas, lycopersici, cubense,* and *lini.* These works show that in suppressive soil, the germination of chlamydospores of these *F. oxysporum* pathogens is not as good as in conducive soil, the germ tubes are shorter, the development of mycelium is less vigorous, and the secondary chlamydospores are formed in lesser numbers. Finally, one can also observe that the competitive saprophytic ability of the *F. oxysporum* pathogens is expressed less well in suppressive soil than in conducive soil. Thus, by applying the method of Rao (34), we have obtained saprophytic colonization ratings of 54 in suppressive soils and 85 in conducive soil.

These observations recall those that were made on the absence of *Fusarium* sp. in the soils of conifer forests in California in which the lysis of germ tubes occurred prior to the formation of secondary chlamydospores (39, 45, 46).

It is interesting to note that the majority of the observations which have just been mentioned were made in the absence of plants, which once again confirms that the factors which determine suppressiveness to Fusarium wilts act in the mass of the soils themselves. The particular behaviour of the Fusaria in these soils results from fungistatic and lytic phenomena which are conditioned by the activities of certain groups of microorganisms.

Other Soil Microorganisms Implicated in Suppressiveness

The comparative study of the behaviour of the *F. oxysporum* pathogens in suppressive and conducive soils has also allowed for certain observations on the behaviour of the saprophytic *Fusarium* populations. But other soil microorganisms, such as fungi, bacteria, and actinomycetes, can play a part in the phenomenon of suppressiveness to Fusarium wilts. The best way of determining the respective action of these organisms is to use soil treatments which selectively eliminate, reduce, inhibit, or stimulate a part of the microflora and then estimate the effect of these treatments on the increase or decrease of suppressiveness. The selective factors that can be used are temperature, humidity, ionizing radiations, sonication, pH, antibiotics, and various chemical substances of specific activity.

It is the effect of treatments at different temperatures which has been the most thoroughly studied, particularly following Baker et al. (5). These authors have shown that the percentage of fungal species, bacteria, and actinomycetes killed by steam air treatments applied during 30 min increased particularly between 55 and 100 C, and that the antagonism of the soils to *Rhizoctonia solani* diminished accordingly.

Rouxel et al. (37) studied the effect of aerated steam treatments for 30 min on the

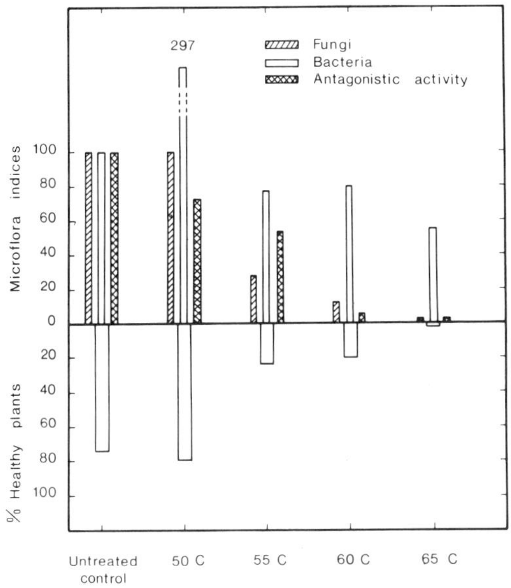

Fig.25-3. Effects of differential heat treatments applied for 30 min to a soil suppressive to Fusarium wilt of muskmelon. Density and activity of microbial populations (indices 100 = $12x10^3$ fungi, $94x10^5$ bacteria/g soil and activity of antagonists in untreated control). Percentage of healthy plants 3 weeks after planting in treated soils artificially reinfested with 350 chlamydospores/g. (37)

microflora and the suppressiveness of a soil to Fusarium wilt of muskmelon. Figure 25-3 shows that the minimum temperature necessary to strongly reduce suppressiveness is between 50 and 55 C. This is the same temperature at which a clear-cut break in the fungal population is observed, while the bacterial population remains abundant even at higher temperatures and when the population of the antagonists to *Fusarium* diminishes gradually. The disappearance of the major portion of the suppressiveness is thus correlated with the reduction in numbers of fungi. Further experiments (J. Louvet et al., *unpublished data*) were done to determine the capacity of the most common isolates to re-establish suppressiveness in soil which had been sterilized. Among the species of *Fusarium, Cephalosporium, Gliocladium,* and *Myrothecium* very sensitive to the 55 C treatment, it is the Fusaria, and in particular *F. oxysporum* and *F. solani,* which re-established the highest level of suppressiveness. On the other hand, an *Aspergillus* sp. which withstood the 55 C treatment and composed 50% of the residual fungal flora, had no effect. These results support the hypothesis that fungal saprophytes, particularly native *Fusarium* saprophytes which are abundant in this soil, play a role in the suppressiveness of Fusarium wilt of muskmelon.

The application of differential soil heat treatments to other types of suppressive soils has led to linking suppressiveness with *Pseudomonas fluorescens,* killed at 60 C, in the case of take-all decline(11), or with spore-forming *Bacillus* sp. or Actinomycetes, which survive 60 C but are killed at 100 C, in the case of suppressiveness to *Phytophthora cinnamomi* (9).

Other observations and experiments have likewise led to bacterial activity as the basis for the suppressiveness of certain soils to Fusarium wilt. Thus Stotzky (43) affirms that "Basically, the presence of montmorillonite apparently maintains the micro-habitats favorable for the development of bacteria, which, in turn, probably controls fungal proliferation." His opinion is reinforced by Baker and Cook (4, p. 65) who state that "clay soils would remain favorably moist for bacteria over longer periods and thus relatively unfavorable to Fusaria." Likewise Smith and Snyder (42) have observed greater increase in bacterial populations in suppressive soils than in conducive soils, and Smith (40) states that "Evidence from several countries indicates that one factor which characterizes Fusarium wilt suppressive soils is the increase in *Arthrobacter* on *Fusarium* germlings." Finally, the experiments of Tu et al. (47) suggest that the decrease of Fusarium wilt of flax observed under certain conditions is due to bacteria present either in the suppressive soils or on stems of diseased plants.

Komada and Ezuka (14) have shown that in those soils most unfavorable to the preservation of the *F. oxysporum* pathogens, the levels of the populations of antagonistic actinomycetes were highest. Arjunarao (2,3) came to the same conclusion from studies in which he compared wilt-free and wilt-sick soils to *F. oxysporum* f. sp. *vasinfectum.* He showed also that the addition of these actinomycetes to a conducive soil partially controlled this disease of cotton. These results recalled the observations of Messiaen et al. (28), who noted that actinomycetes were the soil microorganisms most often producing lysis in *Fusarium.*

In the present state of our knowledge we can thus suppose that, according to the situation, different microorganisms play, separately or in association, a primary role in the phenomena of suppressiveness of soils to Fusarium wilts.

General Mechanisms for Suppressiveness

As in all cases of biological control of soil-borne pathogens, the studies made to date do not permit us to have a clear idea of the microbiological mechanisms which cause the suppressiveness of soil. Thus, for example, even though take-all decline has been very well studied, opposing theories still exist (11). They are based either on a loss of pathogenicity of *Gaeumannomyces* (19), or on a microbial antagonism which operates by antibiosis or by nutrient competition.

No one has proposed to explain the phenomena observed in Fusarium wilt suppressive soils by a diminution of the pathogenicity of the inoculum. On the other hand, evidence furnished to prove the hypothesis that antagonistic microbes intervene by the action of toxic metabolites is not very convincing. The role of competition for nutrients in these phenomena of suppressiveness was assessed in the course of studies dealing with the influence of carbon and nitrogen sources on the behaviour of both the *F. oxysporum* pathogens and the saprophytic microbial populations. Thus, Marshall and Alexander (24) concluded from their experiments on soils suppressive to *F. oxysporum* f. sp. *cubense* that "It is suggested that competition for nutrients is a significant means of ecological control among members of the soil microflora and, together with competitive interactions for space and oxygen, may be the major factors governing biological control of soil-borne fungi." In addition, Smith and Snyder(42) have affirmed that the bacteria which develop most rapidly in a suppressive soil as a consequence of slight additions of glucose and asparagine "must be considered active as nutrient competitors in this soil rather than antagonists . . . and may, at least in part, account for a depletion of nutrients before the comparatively slow-starting *(F. oxysporum)* pathogens have grown much." On the other hand, they had observed that at higher levels of nutrients the chlamydospores of the pathogen germinated well. We have likewise observed (J. Louvet et al., *unpublished data*) that after the addition of glucose (2% wt/wt) in a suppressive soil, the germination percentage of *F. oxysporum* f. sp. *melonis* chlamydospores doubled and the muskmelon plants were attacked. These facts corroborate the hypothesis that in these cases, phenomena of competition for nutrients must determine, at least in part, the fungistatic activity which intervenes in the suppressiveness of soils to Fusarium wilts.

It may well be that competition occurs between pathogenic and saprophytic clones of *Fusarium* both for ecological possession of niches in the soil and for the development by the pathogens of thalli whose vigor would determine their capacity to infect host plants. Such a hypothesis based on the antagonism between closely allied organisms from the systematic point of view as well as from the ecologic point of view has already been proposed in the cases of muskmelon Fusarium wilt (29) and of take-all decline (12).

The fact that suppressive soils are much more favorable to the establishment of *Fusarium* saprophytes than to *Fusarium* pathogens has been expressed in a particularly clear manner by Toussoun (45). He states that "The suppressive soil . . . was wilt-pathogen suppressive, [but] it was not suppressive to the saprophytes native to it." A certain number of facts brought forth in the present paper support this statement and differentiate the behaviour of Fusaria in suppressive and conducive soils: i) The populations of *F. oxysporum* pathogens are less abundant in suppressive soils than in conducive soils, while the reverse is observed for the *Fusarium* saprophytes; ii) the chlamydospores of the *Fusarium* pathogens form and survive less well in suppressive soils, germinate less rapidly and less easily, and the thalli produced are less capable of saprophytic colonization and formation of secondary chlamydospores; and iii) heat treatments which destroy the *Fusarium* saprophytes also remove suppressiveness to Fusarium wilts, and that suppressiveness can be partially re-established by artificial re-infestation with saprophytic clones of *F. oxysporum* and *F. solani* isolated from a suppressive soil.

The rapidity of the development of certain clones of *Fusarium* saprophytes in suppressive soils would give them a relative competitive advantage, permitting them to better utilize substrates and therefore allowing them to occupy numerous ecologic sites. The *F. oxysporum* pathogens, being in competition for the same sites, would be at a disadvantage in getting established in the soil and in the rhizosphere, and as a consequence their capacity to infect roots of host plants would be reduced.

Baker and Cook (4, p. 172) have suggested that the differences in behaviour in soil between pathogens and saprophytes could be related to some of their physiologic characters. They state that "Parasites may be less competitive because of their specialized ability to derive food from living tissue. It seems generally true that the organisms most apt to succeed in mixed populations competing for the same food supply will be those that are no more specialized than is required in the particular situation."

In addition, Ford et al. (14, 15, 16) showed that certain soil bacteria produced chlamydospore-inducing substances which act in a specific manner on each of the clones of *Fusarium*, depending on their genetic constitution. The presence and the activity of these bacteria in certain soils and not in others, and the fact that the substances that they produce can condition in such a precise manner the development and survival of Fusaria, could have a bearing on the differing behaviour of saprophytic and parasitic clones in suppressive and conducive soils.

These hypotheses do not exclude the fact that the implicated mechanisms could be different depending on the categories of wilt suppressive soils, and that several of them could work simultaneously in each case.

Fusarium Wilt Suppressive Soils and Biological Control

A better understanding of both the mechanisms at work and the pathogens and saprophytes that are implicated should permit a utilization of Fusarium wilt suppressive soils in biological control. The difficulties inherent in the fight against this type of disease justify the development of research in this direction.

First of all, it would be useful to classify soils as a function of their natural suppressiveness, or conversely of their conduciveness, to Fusarium wilts. Conducive soils are certainly much more abundant than suppressive soils, but, as is emphasized by Toussoun (45), "Fusarium suppressive soils may be more prevalent than is generally realized. This picture is probably obscured, as it is in other areas, by the widespread use of resistant varieties." There are considerable differences of receptivity between soils. Thus, it has been observed that soon after the first planting of susceptible plants in certain virgin soils,

Fusarium wilts caused damage; these diseases became widespread in the second growing period. On the other hand, these diseases have never been observed in certain soils in areas that are geographically or climatically adjoining. It would thus be helpful to classify the different cultivated soils as a function of these characters. For this to be done would require measurement of the levels of suppressive potential based on an evaluation of either the behaviour of *F. oxysporum* pathogens in the soil, or disease ratings in correlation with increasing inoculum densities. This would permit growers to know in advance the risk of the occurence of Fusarium wilts in the different soil types they cultivate.

It is also important to preserve and increase naturally occurring suppressiveness. Since its basis is microbiological, it is evident, and has been proven (23), that the application of certain biocidal treatments to suppressive soils is to be avoided. We also know that certain cultural practices, in particular organic amendments and crop rotation sequences, cause quantitative and qualitative upheavals in the microbiological activity of soils. These modifications in equilibrim have effects on the receptivity of soils which merit thorough studies from both theoretical and practical standpoints.

The transmission to conducive soils of factors responsible for suppressiveness would be an elegant method of biological control, but its use requires, first of all, a better knowledge of these factors, their mode of action and conditions of efficacy. One can search for either a prevention against infestation by the *F. oxysporum* pathogens, or an eradication by biological disinfestation. Stotzky (43) considers the possibility of lowering the receptivity of soils by amending them with montmorillonite. But is generally through the addition of suppressive soils to conducive soils that the transfer of suppressiveness has been most frequently attempted. The most clear-cut results have been obtained when the conducive soils first received a treatment which destroyed a portion of their microflora, in order to facilitate a controlled recolonization by microorganisms that are specific for suppressiveness. One could likewise improve this colonization by selectively increasing the biological agents responsible for suppressiveness in a "transfer medium" or by adding to the receiving soil nutrients favorable for growth of these microorganisms.

However, these are only paths to be explored while solving a number of problems in science, techniques, and economics.

Conclusion

The two most important examples of soil-borne disease suppressiveness, take-all decline and Fusarium wilt suppressive soils, are fundamentally different in nature. In the first case, the observed phenomenon is strictly linked to cereal monoculture and occurs in very diverse soils. In the other case, Fusarium wilt suppressiveness is a particular and specific characteristic of certain soils and exists independently of crop-rotation sequences.

Even though the mechanisms involved are quite different, one can note certain analogies between "resistance" or "susceptibility" of the soils to the development of diseases, in particular to Fusarium wilts, and resistance or susceptibility of plants to these diseases. The absence of the disease can be linked in both cases either to a total inhospitability toward the pathogen (the phenomenon of "immunity"), or to a sheltering under conditions which do not permit it to manifest its pathogenic activities ("tolerance"). In general, a narrow specificity links each *F. oxysporum* pathogen to its host plant and determines the existence of special forms and races as well as horizontal and vertical resistance. In all cases studied to date, one may consider that the Fusarium wilt suppressive soils possess a "horizontal resistance," since it is manifested toward several special forms and races. Just as in resistance in plants, "resistance" of soils is transmissible. From the point of view of control, genetic resistance of plants to diseases has been effectively transferred to numerous individuals by application of the laws that control the transmission of this hereditary character and by knowledge of the genes that are the fundamental physical support. This progress has been achieved in spite of a lack of knowledge of a large part of the intermediary physiological mechanisms. In the same way, when we better know the microorganisms which are the basis for soil "resistance" and the factors most favorable for their installation and activity, we will be able to transmit more effectively the "resistance" of one soil to another. One could then, perhaps, as in resistance of plants, characterize the different types of "resistance" of soils as "dominant or recessive," "constitutional or induced," etc.

Acknowledgements

The authors wish to thank T.A. Toussoun and P.E. Nelson for translating and reviewing the manuscript.

Literature Cited

1. Arjunarao, V. 1971. Biological control of cotton wilt. I. Soil fungistasis and antibiosis in cotton fields. Proc. Indian Acad. Sci. Sect. B 73 : 265–272.
2. Arjunarao, V. 1971. Biological control of cotton wilt. II. In vitro effects of antagonists on the pathogen *Fusarium vasinfectum*. Proc. Indian Acad. Sci. Sect B 74 : 16–28.
3. Arjunarao, V. 1971. Biological control of cotton wilt. III. In vivo effect of antagonists on the pathogen *Fusarium vasinfectum*. Proc. Indian Acad. Sci. Sect B 74 : 53–62.
4. Baker, K.F., and R.J. Cook. 1974. Biological control of plant pathogens. W.H. Freeman Co., San Francisco. 433 p.
5. Baker, K.F., N.T. Flentje, C.M. Olsen, and H.M. Stretton, 1967. Effect of antagonists on growth and survival of *Rhizoctonia solani* in soil. Phytopathology 57 : 591–597.
6. Baker, R. 1968. Mechanisms of biological control of soil-borne pathogens. Annu. Rev. Phytopathol. 6 : 263–294.
7. Bereau, M., and C.M. Messiaen. 1975. Réceptivité comparée des sols à l'infestation par *Pseudomonas solanacearum*. Ann. Phytopathol. 7 : 191–193.
8. Broadbent, P., and K.F. Baker. 1974. Behaviour of *Phytopthora cinnamomi* in soils suppressive and conducive to root rot. Australian J. Agric. Res. 25 : 121–137.
9. Broadbent P., and K.F. Baker. 1975. Soils suppressive to Phytophthora root rot in Eastern Austalia, p. 152–157. *In* G.W. Bruehl (ed.), Biology and control of soil-borne plant pathogens, Amer. Phytopathol. Soc., St. Paul, Minnesota.
10. Burke, D.W. 1965. Fusarium root rot of beans and the behaviour of the pathogen in different soils. Phytopathology 55 : 1122–1126.
11. Cook, R.J., and A.D. Rovira. 1976. The role of bacteria in the biological control of *Gaeumannomyces graminis* by suppressive soils. Soil Biol. Biochem. 8 : 269–273.
12. Deacon, J.W. 1976. Biological control of the take-all fungus, *Gaeumannomyces graminis*, by *Phialophora radicicola* and similar fungi. Soil Biol. Biochem. 8 : 275–283.
13. Fellows, H., and C.H. Ficke. 1939. Soil infestation by *Ophiobolus graminis* and its spread. J. Agric. Res. 58 : 505–519.
14. Ford, E.J., A.H. Gold, and W.C. Snyder. 1970. Soil substances inducing chlamydospore formation by *Fusarium*. Phytopathology 60 : 124–128.
15. Ford, E.J., A.H. Gold, and W.C. Snyder. 1970. Induction of chlamydospore formation in *Fusarium solani* by soil bacteria. Phytopathology 60 : 479–484.
16. Ford, E.J., A.H. Gold, and W.C. Snyder. 1970. Interaction of carbon and nutrition and soil substances in chlamydospore formation by *Fusarium*. Phytopathology 60 : 1732–1737.
17. Gerlach, M. 1968. Introduction of *Ophiobolus graminis* into new polders and its decline. Neth. J. Plant Pathol. 74 : 1–97.
18. Komada, H., and A. Ezuka. 1970. Ecological study of Fusarium diseases of vegetable crops. I. Survival of pathogenic Fusaria in different soil types, p. 1–6 (English summary.) Res. Progr. Rept. Tokai-Kinki Nat. Agric. Exp. Stn. 6.
19. Lapierre, H., J.M. Lemaire, B. Jouan, and G. Molin. 1970. Mise en évidence de particules virales associées à une perte de pathogenicité chez le Piétin-échaudage des céréales, *Ophiobolus graminis*. Compt. Rend. Acad. Sci. (Paris), Série D, 271 : 1833–1836.
20. Lemaire, J.M., and M. Coppenet. 1968. Influence de la succession céréalière sur les fluctuations de la gravité du Piétin-échaudage *(Ophiobolus graminis)*. Ann. Epiphyties 19 : 589–599.

21. Locke, T., and J. Colhoun. 1974. Contributions to a method of testing oil palm seedlings for resistance to *Fusarium oxysporum* f. sp. *elaeidis*. Phytopathol. Z. 79 : 77–92.
22. Louvet, J. 1973. Les perspectives de lutte biologique contre les champignons parasites des organes souterrains des plantes, p. 48–58. *In* Perspectives de lutte biologique contre les champignons parasites des plantes cultivées et des tissus ligneux, Proceed. International Symposium. Station Féd. Rech. Agron., Lausanne, Switzerland.
23. Louvet, J., F. Rouxel, and C. Alabouvette. 1976. Recherches sur la résistance des sols aux maladies. I. Mise en évidence de la nature microbiologique de la résistance d'un sol au développement de la Fusariose vasculaire du melon. Ann. Phytopathol. 8 : 425–436.
24. Marshall, K.C., and M. Alexander. 1960. Competition between soil bacteria and *Fusarium*. Plant Soil 12 : 143–153.
25. Mas, P, 1967. Protection du melon contre la Fusariose par infection préalable de la plantule avec d'autres souches de *Fusarium*. Compt. Rend. Acad. Agric. Fr., Paris : 1034–1040.
26. Matta, A. 1971. Microbial penetration and immunization of uncongenial host plants Annu. Rev. Phytopathol. 9 : 387–410.
27. Menzies, J.D. 1959. Occurrence and transfer of a biological factor in soil that suppresses potato scab. Phytopathology 49 : 648–652.
28. Messiaen, C.M., P. Mas, A. Beyries, and H. Vendran. 1965. Lyse meycélienne et formes de conservation dans le sol chez les *Fusarium*. Ann. Epiphyties 16 : 107–128.
29. Meyer, J.A., and H. Maraite. 1971. Multiple infection and symptom mitigation in vascular wilt diseases. Trans. Brit. Mycol. Soc. 57 : 371–377.
30. Mitchell, J.E. 1973. The mechanisms of biological control of plant diseases. Soil Biol. Biochem. 5 : 721–728.
31. Nash, S. M., and W.C. Snyder. 1965. Quantitative and qualitative comparison of Fusarium populations in cultivated fields and noncultivated parent soils. Can. J. Bot. 43 : 939–945.
32. Nelson, R. 1950. Verticillium wilt of peppermint. Mich. Agric. Exp. Stn. Tech. Bull. 221 : 1–259.
33. Pope, A.M.S., and R.M. Jackson. 1973. Effects of wheatfield soil on inocula of *Gaeumannomyces graminis* var. *tritici* in relation to take-all decline. Soil Biol. Biochem. 5 : 881–899.
34. Rao, A.S. 1959. A comparative study of competitive saprophytic ability in twelve root-infecting fungi by an agar plate method. Trans. Brit. Mycol. Soc. 42 : 97–111.
35. Reinking, O.A., 1935. Soil and Fusarium diseases. Zent. Bakt. Parasitenk. Infekt. 91 : 243–255.
36. Reinking, O.A., and M.M. Manns. 1933. Parasitic and other Fusaria counted in tropical soils. Z.Parasitenk. 6 : 23–75.
37. Rouxel, F., C. Alabouvette, and J. Louvet. 1977. Recherches sur la résistance des sols aux maladies. II. Incidence de traitements thermiques sur la résistance microbiologique d'un sol à la Fusariose vasculaire du melon. Ann. Phytopathol. 9 : 183–192.
38. Shipton, P.J., R.J. Cook, and J.W. Sitton. 1973. Occurrence and transfer of a biological factor in soil that suppresses take-all of wheat in eastern Washington. Phytopathology 63 : 511–517.
39. Smith, R.S., Jr. 1967. Decline of *Fusarium oxysporum* in the roots of *Pinus lambertiana* seedlings transplanted into forest soils. Phytopathology 57 : 1265.
40. Smith, S.N. 1975. Association of *Arthrobacter* with banana wilt Fusaria in suppressive soils. Proc. Amer. Phytopathol. Soc. 2 : 19 (Abstr.).
41. Smith, S.N., and W.C. Snyder. 1971. Relationship of inoculum density and soil types to severity of Fusarium wilt of sweet potato. Phytopathology 61 : 1049–1051.
42. Smith, S.N., and W.C. Snyder. 1972. Germination of *Fusarium oxysporum* chlamydospores in soils favorable and unfavorable to wilt establishment. Phytopathology 62 : 273–277.
43. Stotzky, G. 1973. Techniques to study interactions between microorganismes and clay minerals in vivo and in vitro. Bull. Ecol. Res. Comm., Stockholm 17 : 17–28.
44. Stover, R.H. 1962. Fusarial wilt (Panama disease) of bananas and other Musa species. Commonwealth Mycol. Inst., Phytopathol. Paper 4. 117 p.
45. Toussoun, T.A. 1975. Fusarium-suppressive soils, p. 145–151. *In* G.W. Bruehl (ed.), Biology and control of soil-borne plant pathogens, Amer. Phytopathol. Soc., St. Paul, Minnesota.
46. Toussoun, T.A., W. Mentzinger, and R.S. Smith, Jr. 1969. Role of conifer litter in ecology of *Fusarium* : stimulation of germination in soil. Phytopathology 59 : 1396–1399.
47. Tu, C.C., Y.H. Cheng, and M. Chen. 1975. Flax Fusarium wilt-suppressive soil in Taiwan. Plant Prot. Bull., Taiwan 17 : 390–399.
48. Walker, J.C., and W.C. Snyder. 1933. Pea wilt and root rots. Wis. Agric. Exp. Stn. Bull. 424. 16 p.
49. Wensley, R.H., and C.D. McKeen. 1963. Populations of *Fusarium oxysporum* f. *melonis* and their relation to the wilt potential of two soils. Can. J. Microbiol. 9 : 237–249.

Literature on Fusarium wilt suppressive soils published after acceptance of this paper (1977)

Alabouvette, C., F. Rouxel, and J. Louvet. 1977. Recherches sur la résistance des sols aux maladies. III. Effets du rayonnement γ sur la microflore d'un sol et sa résistance à la Fusariose vasculaire du Melon. Ann. Phytopathol. 9 : 467–471.

Alabouvette, C., F. Rouxel, and J. Louvet. 1979.

Characteristics of Fusarium wilt suppressive soils and prospects for their utilization in biological control. *In* Schippers, B., and W. Gams (eds.) Soil-borne plant pathogens 165–182. Acad. Press, London.

Alabouvette, C., F. Rouxel, and J. Louvet. 1980. Recherches sur la résistance des sols aux maladies. VI. Mise en évidence de la spécificité de la résistance d'un sol vis-à-vis des Fusarioses vasculaires. Ann. Phytopathol. 12 : 11–19.

Alabouvette, C., F. Rouxel, and J. Louvet. 1980. Recherches sur la résistance des sols aux maladies. VII. Etude comparative de la germination des chlamydospores de *Fusarium oxysporum* et de *Fusarium solani* au contact de sols résistant et sensible aux Fusarioses vasculaires. Ann. Phytopathol. 12 : 21–30.

Alabouvette, C., R. Tramier, and D. Grouet. 1980. Recherches sur la résistance des sols aux maladies. VIII. Perspectives d'utilisation de la résistance des sols pour lutter contre les Fusarioses vasculaires. Ann. Phytopathol. 12 : 83–93.

Alabouvette, C., Y. Couteaudier, and J. Louvet. 1981. Comparaison de la réceptivité de différents sols et substrats de culture aux Fusarioses vasculaires. Agronomie (in press).

Garibaldi, A., G. Pergola, and C. Dalla Guda. 1980. Sulla presenza in Italia di terrani repressivi nei rignardi della Fusarioso del garafano. Atti Giornate Fitopatologiche, 457–464.

Rouxel, F., C. Alabouvette, and J. Louvet. 1979. Recherches sur la résistance des sols aux maladies. IV. Mise en évidence du rôle des Fusarium autochtones dans la résistance d'un sol à la Fusariose vasculaire du Melon. Ann. Phytopathol. 11 : 199–207.

Sher, F.M., and R. Baker. 1980. Mechanism of biological control in a Fusarium-suppressive soil. Phytopathology 70 : 412–417.

Smith, S.N. 1977. Comparison of germination of pathogenic *Fusarium oxysporum* chlamydospores in host rhizosphere soils conducive and suppressive to wilts. Phytopathology 67 : 502–510.

Tramier, R., J.C. Pionnat, A. Bettachini, and C. Antonini. 1979. Recherches sur la rèsistance des sols aux maladies. V. Evolution de la Fusariose de l'Oeillet en fonction des substrats de culture. Ann. Phytopathol. 11 : 476–481.

Tu, C.C., Y.C. Cheng, Y.C. Chang. 1978. Antagonistic effect of some bacteria from Fusarium wilt-suppressive soil and their effect on the control of flax wilt in the field. Journ. Agr. Res. China 27 : 245–258.

26 Ecology of *Fusarium* in Noncultivated Soils

Martin F. Stoner

Introduction

Fusarium species occur in noncultivated soils around the world, yet most of the literature on *Fusarium* pertains to cultivated soils (4,13,30,66). This emphasis on cultivated soils together with isolated reports on the absence or low levels of *Fusarium* from certain wildland areas has fostered the impression that *Fusarium* is rare or unimportant in noncultivated soils. A better understanding of *Fusarium* in noncultivated soils is necessary as we become more involved in the management of native ecosystems. The species concept of Snyder and Hansen (94) is followed here to simplify treatment of diverse information.

All species of *Fusarium* have been found in noncultivated soils. Information on the occurence of particular pathogenic forms, cultivars, or clones is too limited to reveal the extent or any patterns of distribution. Among the Fusaria, *F. oxysporum, F. solani,* and *F. roseum* are the most widespread and predominant species in noncultivated soils. Of these species, *F. oxysporum* is most cosmopolitan, as is true in cultivated soils (66). The ubiquity of these species is supported by their many biological forms, wide range of hosts and substrates, competitive saprophytic ability, and ability to form chlamydospores (4,42, 58,59,65,71). Other species of *Fusarium* have more restricted distributions determined by substrates or other requirements.

Fusarium species in noncultivated soils are associated with roots and organic matter such as host debris rather than being widespread throughout the mineral soil (6,43,44,46, 47,67,91,97,98). Therefore, the largest populations of *Fusarium* are often in the uppermost horizon, and usually in the top 5–15 cm of soil (21,33,86,97,98). *Fusarium oxysporum,* which can survive and reproduce effectively as a parasite or as a pioneer saprophyte on organic debris (59,65,88), tends to predominate in humic soils or layers. Some cultivars of *F. roseum* are protected within colonized plant remains (42). *Fusarium nivale* and *F. rigidiuscula* are active primarily on aerial plant parts or litter and are reported infrequently from soil except as survival stages (25,27,43,81). *Fusarium nivale* is known to survive in leaf debris in soil throughout the world in the temperate zone. It has been found in leaf remains in soils of grassy areas within chaparral communities of the San Gabriel foothills of southern California (M. Stoner, *unpublished data*).

In many but not all instances, populations of Fusaria are smaller in noncultivated than in cultivated soils (49,59,101; T. Matuo, Japan, *personal communication*);but small populations are not indicative of lack of importance in native ecosystems. In noncultivated soils, these and other resident fungi have various ecological niches (65,85) affecting the structure and function of the ecosystems. Since the roles of these fungi are not restricted to parasitism and disease and can involve nutrient cycling and other functions in noncultivated soils, it is conceivable that large populations are not prerequisite to significant ecological impact. Indeed, the presence of *Fusarium* at relatively low levels in some noncultivated soil ecosystems such as forests points to ecological roles different from those in cultivated soils.

Little information is available regarding the host-specificity of Fusaria or the occurrence or activity of plant pathogenic forms of *Fusarium* in noncultivated soils. *Fusarium solani* f. sp. *pisi* was found in wind-blown soil in the Mallee Scrub area of South Australia, where propagules probably came from pea fields in the vicinity (9). *Fusarium oxysporum* f. sp. *vasinfectum* was absent or occurred in "relatively small" populations in noncultivated soils of Tanzania (17). Kreutzer (43) considered some isolates of *F. roseum* obtained from Colorado grasslands to be potentially pathogenic to commercial cereals.

The noncultivated habitats of *Fusarium* include forests and other woodlands; scrub communities; savannahs; prairies; pastures and other grasslands; deserts; swamps; littoral zones; and other areas from cold alpine and boreal zones to seacoast and tropical climates. Pastures are included here because, although they may represent disturbed areas (90), they usually are not subject to the same disturbances (59,90,101) associated with cultivation, and the mycoflora of such soils is similar to that of noncultivated grasslands. Soils of closed forests and woodlands or swampy areas generally have qualitatively and quantitatively limited *Fusarium* populations (8,48,49,55,74,84,85,88,93), whereas the largest populations together with the greatest species diversity of Fusaria have been found in soils of mesic grasslands (6,18,63,85,90) and nonforested communities or open woodlands with understories of abundant grasses and other herbs (3,63,85). Some soils are conducive to Fusaria, while others are suppressive (59,92); such soil qualities are attributable to vegetation, plant exudates, and microbial factors.

Detection of Fusaria in Noncultivated Soils

The limited knowledge on *Fusarium* in noncultivated soils is due largely to our emphasis on agricultural situations and to failures in detection due to selectivity of isolation media and methods. Many general floristic studies of noncultivated soils have employed techniques that are unlikely to detect *Fusarium* even where it occurs. The tendency for selective occurrence of *Fusarium* in the root zones or remains of plant associates (41,44,46,47,85,97), together with relatively small populations, may contribute to the difficulty of detection in noncultivated soils. A more pronounced localization of Fusaria in noncultivated as compared to cultivated soils could be due to greater spacing of host plants, lack of severe, frequent disturbance of the soil structure that disperses propagules and substrata, a different system of niches, and sharply stratified habitats (85). Therefore, it is essential to assure detection by managing both the selectivity of isolation techniques (39,85,94,96) and analytical approaches to particular habitats (83,84,85).

Occurrence in Specific Soil Ecosystems

Forests

All *Fusarium* species except *F. nivale* have been reported to occur in forest soils. *Fusarium oxysporum* and *F. solani* are noted most often, and reports collectively suggest that the latter species may be more widespread than the former in forests, particularly in the tropics. *Fusarium roseum* is uncommon in closed forests (13). The population density of Fusaria in general is relatively low in forest soils, especially in closed forests, as compared to grasslands and other open communities. Ishii (34; *personal communication)* believes that *Fusarium* is a relatively minor component of various forest soils "all over Japan."

Temperate and subtropical forests *Fusarium* has been found in soils of hardwood forests in England (77), Denmark (36), India (72), Japan (T. Matuo, *personal communication*), South Africa (20), and the U.S.A. (7,29,51,54,62). *Fusarium oxysporum* (20,29,77) and *F. solani* (29,51,54,62,72) have been reported most frequently. *Fusarium roseum* (29,77), *F. moniliforme* (20,54), *F. tricinctum* (54,72), and *F. episphaeria* (29) are less frequent and, where present, usually sparse. Morrow (56) noted the presence of *Fusarium* in pine-oak soils in Texas. *Fusarium oxysporum, F. roseum, F. tricinctum* (75), and other *Fusarium* species (73) were found in Canadian boreal forests. *Fusarium roseum* and *F. episphaeria* were found in mixed hardwood-conifer forests in Japan (T. Matuo, *personal communication)*. Thornton (90) found no Fusaria in mixed *Podocarpus-Beilschmiedia* forest in New Zealand.

Fusarium apparently has more limited occurrence in coniferous than in hardwood forests. Occurrence in forests of pine, cedar, fir, and other trees has been noted in Canada (2,99), Japan (T. Matuo, *personal communication*), and the U.S.A. (8,102). Ellis (24) noted the absence of *Fusarium* from pinewoods; but Wright and Bollen (102) isolated the genus frequently from soils in a 25-year-old Douglas-fir *(Pseudotsuga)* forest in Oregon, and Widden and Parkinson (99) reported *Fusarium* from Canadian pine forests. In a study of a 40-year-old *Pinus radiata* plantation and an adjacent pasture site 50 feet away, both with the same soil type in a high rainfall area near Mt. Gambier in South Australia, R. J. Cook *(personal communication)* found that *Fusarium* populations were nil in the pine soil but were 5–10,000 propagules/g in the pasture. *Fusarium oxysporum* (2), *F. solani* (2,99), *F. roseum* (2; T.Matuo, *personal communication*), and *F. episphaeria* (8) are mentioned in isolated reports on coniferous forests.

Smith (76) noted that *F. oxysporum* failed to establish in native pine soils in California. Experiments suggested strongly that this might have been due to combined effects of stimulated spore germination and subsequent germ tube lysis caused by leachates from pine litter (93). These relatively dry forest soils support pure pine stands and a very thick litter layer, conditions not common to all coniferous forests; therefore, the applicability of this work to other areas is unclear.

Tropical forests *Fusarium* occurs in closed hardwood forests in Costa Rica (26), Hawaii (81,85), India (72), Malaysia (33), Panama (26), and Trinidad (31). *Fusarium solani* (26,31,33,72,81,85) apparently is most common, although *F. oxysporum* (31,33,81,85) and *F. moniliforme* (26,33,85) have been found frequently. *Fusarium tricinctum* has been reported from India (72) and Malaysia (33), and *F. lateritium* and *F. episphaeria* from Malaysia (33). Based on its associations elsewhere, the scarcity of *F. roseum* in closed forests may be attributable to the general absence of grasses and/or the warmer climate in tropical areas. In extensive studies of Hawaiian soils (81,85; M. Stoner *unpublished data*), *F. rigidiuscula* has been found only at one location in an old, dense kipuka forest (Kipuka Puaulu) in deep rich soils that supported *Sapindus* and other hardwoods.

Hee (33) noted that primary (undisturbed) jungle in Malaysia had fewer species and smaller populations of *Fusarium* than secondary (regrown) jungle. Lim and Chew (49) found no *Fusarium* in primary forests of Singapore. In the Pekan Nanas area of Malaysia (33), populations of *F. solani* and *F. oxysporum* in secondary forests rivaled those of cultivated soils. Hee (33) attributed this situation tentatively to relatively high moisture and organic content in jungle soils.

Fusarium occurs in open tropical hardwood forests of Tanzania (17; D. L. Ebbels, *unpublished data*), Honduras (86), and Hawaii (M. Stoner, *unpublished data;* 85). Only *F. solani* is known to occur in all three places, while *F. oxysporum* is reported from Hawaii, and *F. episphaeria* and *F. roseum* from Tanzania.

Grasslands

Fusarium oxysporum, F. solani, and *F. roseum* are the most frequently reported species from grasslands in general and from individual types of grassland: native grassland, prairie, pasture, and savannah (L. Burgess, *personal communication;* D. Ebbels, *unpublished data;* M. Stoner and G. E. Baker, *unpublished data;* 3,17,19,21,22,25,31, 40,42,43,45,47,48,50,52,57,59,78,85,86,90,91,97,100). At least two of these three species were found in each grassland studied. *Fusarium roseum* is strongly and regularly represented in grasslands. Ebbels (17) found large populations of *F. solani* in calcareous soils of the Shinyanga district of Tanzania. Infrequent reports of *F. nivale* and *F. rigidiuscula* in noncultivated soils are due apparently to their absence from areas, selective residency in aerial plant parts, or to poor isolation techniques.

Fusarium species are "prominent and characteristic" members of native grassland and prairie communities (63). *Fusarium oxysporum* and *F. roseum* are common (3,19,21,22,31, 41,47,57). *Fusarium solani* has been found in prairies of Manitoba, Canada (3); the U.S.A., where it is very common in some soils (43,100,101); and from Iceland (42) and Ceylon (57). There have been a few reports of *Fusarium tricinctum* in prairie soils of Canada (3) and the U.S.A. (63); *F. moniliforme* in the U.S.A. (25,100); and *F. episphaeria* in the U.S.A. (100) and in Iceland, where it is common (40). In a study of contiguous prairies extending along an environmental gradient in Wisconsin, Orpurt and Curtis (63) found two unidentified *Fusarium* species in wet prairie; *F. oxysporum* and *F. solani* in mesic prairie; and *F. tricinctum* in dry prairie. *Fusarium nivale* and *F. rigidiuscula* have been found in both prairies and abandoned agricultural fields in central Oklahoma (25).

Mueller-Dombois and Perera (57) used the differential distributions of *F. solani, F. roseum,* and other fungi to distinguish montane grassland (patana) zones of Ceylon; *F. oxysporum* occurred in all zones. *Fusarium oxysporum* is widespread in African savannah communities (21,22,45,50); *F. solani,* in the Transvaal (22) and Honduras (86); and *F. roseum,* in the Transvaal (22). Lanneau et al. (45) found only *F. oxysporum* in savannah and cultivated soils in Katanga Province, Republic of the Congo (now Zaire). *Fusarium oxysporum* is common in soils of the MacRitchie Reservoir catchment area in Singapore (47).

Evidence suggests a differential distribution of *F. roseum* cultivars in grasslands according to geographic location and climatic zones within regions (3,25,30,78,89,97,100). Kommedahl and Siggeirsson (41) found *F. roseum* 'Sambucinum' to be predominant in Iceland. In another study (100), *F. roseum* 'Equiseti' was more common than Graminearum or Avenaceum in certain prairies. Associations with certain plants apparently are responsible for the distribution pattern of *F. roseum* cultivars and *F. oxysporum* in tussock-grassland soils of New Zealand (89).

Fusarium populations of pastures resemble those of undisturbed grasslands, as indicated by reports from Australia (19), England (52), the continental U.S.A. (64), Hawaii (85; G. E. Baker and M. Stoner, *unpublished data*), New Zealand (90,91), and Tanzania (18). Variations in the species composition and size of *Fusarium* populations in pastures as opposed to native grasslands do not appear to follow particular patterns. The species-selective factors in pastures probably include soil types and plant hosts and, in some cases, the effects of fertilizers (32,90). In addition to the major grassland-prairie species, *F. moniliforme* was found in Tanzania (18); *F. tricinctum,* in Hawaii (G. E. Baker, *unpublished data);* and *F. episphaeria,* in England (52). McKenzie (52) found *F. solani* to be more abundant in pasture soils than in land under cereal culture.

Deserts

Fusarium has been found in noncultivated desert soils of central Iraq (1), Israel (5,38), the French Sahara (60), and the southwestern U.S.A. (16,70,78). Ranzoni (70) reported the widespread distribution of *Fusarium* in the Sonoran Desert, noting the occurrence of *F. oxysporum, F. solani, F. tricinctum, F. roseum, F. lateritium,* and *F. nivale.* Al-Doory et al. (1) listed the same species for Iraq, with the exceptions that *F. lateritium* was absent and *F. moniliforme* was present. *Fusarium moniliforme, F. oxysporum, F. roseum,* and *F. solani* occur in desert soils of the Sodom and Be'er Sheva' regions of Israel (38). *Fusarium oxysporum* and *F. roseum* have been found in desert soils of northern South Australia (J. Warcup, *personal communication*). Durrell and Shields (16) noted that *Fusarium roseum* occurred in 40–80% of their samples from the Frenchman and Yucca Flats areas of Nevada, and was common in soils within atomic bomb tests sites.

In a study of fungi in cool-desert soils (mostly sandy loams or sands) of northern Arizona and southern Utah, States (78) associated certain *Fusaria* with six of seven sampled plant communities. *Fusarium moniliforme* was found in grassland, pinyon-juniper woodland, and sagebrush shrub and ponderosa pine forest communities; *F. nivale* was limited to one of three grassland sites and to sand shrub communities; *F. oxysporum* was found in all communities except the forests; *F. roseum* 'Equiseti' was found at grassland, blackbrush shrub, sand dune shrub, and sagebrush shrub sites, while *F. roseum* 'Gibbosum' occurred in all communities; and *F. solani* occurred in grasslands, one of five pinyon-juniper woodlands, most blackbrush shrub sites, and was very common in all sand dune shrub soils. *Fusarium oxysporum* and *F. roseum* occurred with high frequency in samples of grassland soils. *Fusarium* was not found at a mixed conifer forest site.

Littoral Habitats

Populations in mangrove swamps include *F. oxysporum* (most common), *F. solani, F. tricinctum, F. roseum, F. episphaeria,* and *F. lateritium* in Hawaii (44,79; G. E. Baker, *unpublished data*); *F. solani* in Singapore (49); and *Fusarium* species in Africa (87). Tidal salt marshes and mud flats in England support *F. oxysporum* (23) and *F. roseum* (23,68). Steele (79) isolated *Fusarium* from beach sands of the Line and Phoenix Islands in the Pacific. Dunn (15) recorded *F. lateritium, F. moniliforme, F. roseum,* and *F. solani* from Hawaiian beach sand. *Fusarium solani* was widespread, although not as common as on Eniwetok Atoll in the Marshall Islands. *Fusarium solani* was the most common species in all littoral sand zones of Eniwetok, although its frequency was low in the anaerobic zone. It retained viability in seawater and fresh water for at least 4 weeks; made optimal growth at 25–30 C; grew under anaerobic conditions; and was killed by incubation with sulfide as might occur in some submerged sand layers. *Fusarium moniliforme* was widespread on Oahu, Hawaii. *Fusarium tricinctum* has been found infrequently in Hawaiian beach sands (G. E. Baker, *unpublished data). Fusarium solani* in Hawaii (M. Stoner, *unpublished data*) and *F. episphaeria* in Singapore (49) have been associated with soils supporting native beachside vegetation.

Miscellaneous Records

Fusarium oxysporum has been found in Tundra soils (13), low moors of Denmark (36), bottom mud dredged from a depth of 200 fathoms near Eva Beach, Oahu (G. E. Baker, *personal communication*), and in the brown "slimes" (thin, moist layers of mineral colloids and organic components including root fragments) on walls of lava caves on the island of Hawaii (M. Stoner and F. G. Howarth, *unpublished data*). *Fusarium solani* was found in Nile River silt in Sudan (61) and mamane (*Sophora*) grassland soils of leeward Hawaii Island (M. Stoner, *unpublished data*). *Fusarium tricinctum* was found in soil of a grass-fern community in Bora Bora and, with *F. episphaeria,* in bottom mud of Flathead Lake in Montana (G. E. Baker, *unpublished data*). *Fusarium roseum* occurs in recently glaciated soil near Glacier Bay, Alaska (10). Unidentified Fusaria are recorded from tundra soils (14), Danish beech mor soil (37), bush soil of South Australia (12), alpine soil in Wyoming (69), and myrtaceous heathland in Australia (53). *Fusarium* apparently is absent from many ericaceous *(Calluna)* heathlands in the British Isles (66,74,88).

Factors Determining Occurrence and Distribution

Higher Plant Communities

Except in extreme situations, it is difficult to correlate the distribution of *Fusarium* species with individual soil factors such as pH, moisture, and organic matter. Distribution is governed most directly by associated plants or plant communities, which in turn reflect a concert of environmental determinnants (6,35,61,81,85,89,90,93,97). Since noncultivated soils usually support mixed plant communities, relationships between specific higher plants and particular species of *Fusarium* are not readily apparent, and knowledge in this area is limited. No *Fusarium* species is known to be limited to, or always associated with, any particular plant or plant community. Although most studies of Fusaria in noncultivated soils have not been concerned with the hosts of reported fungi, our general knowledge of *Fusarium* indicates that fungus-host associations are likely and that occurrence in these soils may depend on suitable hosts. Nour (61) detected *F. solani* and *F. roseum* in Nile River silts only after establishment of "gerf" vegetation.

Ecological Succession

Fusarium can be a member of pioneer, transitional, and climax communities. Its greatest diversity and populations tend to occur in well-established communities where organic matter has accumulated and supportive hosts are present. In a study of British maritime dunes, Brown (6) found that *Fusarium* occurred primarily in intermediate to climax communities of dune grass and dune heath. Mallik and Rice (51) considered *Fusarium* as characteristic of transitional to climax areas in bottomland hardwood forest in Oklahoma. On the island of Hawaii, *Fusarium* is virtually absent in fragile soil ecosystems (82) of mature-pioneer and transitional *Metrosideros* ('ōhi'a)-*Cibotium* (tree fern) rain forests (84,85). This lack of *Fusarium* probably is the result of changes in plant communities and edaphic conditions associated with ecological transitions from open areas to closed forests and from rocky or sandy to muck soils. In contrast, several species of *Fusarium* with small to large populations occur in mesic grasslands, open montane forests, and mountain parkland areas on the island of Hawaii. This diversity may indicate a relatively stable or climax character in some of these communities (M. Stoner, *unpublished data;* 85,95). Although more research is needed, I believe that *Fusarium* and other genera can be used as valuable indicators of the stages of ecological succession and community development.

Spatial Distribution within Ecosystems

Stoner et al. (81,84,85) studied the distribution of soil-borne fungi along an altitudinal gradient (transect) with noncultivated, acid soils that extended from rain forest at 1197 m (3920 ft) elevation uphill through mesic forest, savannah, and scrub to an alpine desert at 3050 m (10,000 ft) on the east flank of Mauna Loa volcano, on the island of Hawaii. *Fusarium* species occurred differentially within distinct plant community zones along the environmental gradient. The occurrence of individual Fusaria corresponded closely to one or more zones on the gradient. Distribution was not correlated clearly or commonly with individual factors such as soil acidity or organic matter content. *Fusarium* was absent from the low-elevation *Metrosideros*-tree fern rain forests with wet muck soils. *Fusarium* occurred in the higher, more mesic to cool, dry zones of the gradient and in some mesic sites at lower, warmer elevations; and populations were highest in the mesic montane zones. *Fusarium solani* appeared only in the mesic, lower montane closed and open forests in old kipuka areas (Kipuka Puaulu and Kipuka Ki, old islands of vegetation surrounded by recent lava flows) with deep, rich soils and diverse vegetation at 1220–1279 m (4000–4200 ft) elevation. *Fusarium rigidiuscula* was found only at Kipuka Puaulu. *Fusarium oxysporum* had the widest distribution over the mesic center of the transect (1200–2040 m, 4000–6700 ft), with its highest populations in a broad, central zone of savannah and mountain parkland vegetation including grasses, hardwood trees and shrubs on shallow to moderately deep soils. Populations of *F. oxysporum* tapered sharply at either end of the mesic zone, with sporadic distribution and limited numbers in an open, scrub forest at 1200 m (4000 ft) elevation and at the upper edges of the mountain parkland (2040 m, 6700 ft). *Fusarium lateritium* was limited to dry, cool sub-alpine scrub-forest areas at 2130 m (7000 ft) to 2440 m (8000 ft, treeline) with shallow soils scattered among outcroppings of lava. Both *F. oxysporum* and *F. solani* had relatively large populations (1000 propagules/g soil) in open forest but low numbers (less than 100/g) in adjacent closed forest areas of Kipuka Ki and Kipuka Puaulu. Open areas supported abundant grasses and other herbs, while the understory in closed forests was made up largely of scattered dicotyledonous shrubs and herbs. At higher, mesic elevations on the gradient, areas with abundant grasses also had relatively large populations of *F. oxysporum*. The association of *F. oxysporum* and grasses has been noted elsewhere (66,90,97). The occurrence of much larger populations of *F. solani* below 912m (3000 ft) in a savannah, and its apparent absence above 1220 m (4000 ft), indicate a possible requirement for warmer temperatures, at least within this locality in Hawaii.

Cultivation

Nash and Snyder (59) noted the marked effects of cultivation on the size and species-cultivar composition of *Fusarium* in related soils of adjacent noncultivated and agricultural land in California. They reported the presence of *F. oxysporum* (100 propagules/g dry soil), *F. solani* (common; unspecified numbers), *F. roseum* (300/g), and *F. episphaeria* (300/g) from grassy woodlands and scrub-grassland communities. In the Castroville area, total Fusaria were estimated at 1500–3700/g in noncultivated soils compared to 11,000–27,000/g in cultivated areas. *Fusarium episphaeria* and some cultivars of *F. roseum* were equally numerous in both noncultivated and cultivated soils.

Microbial Interactions

Fusarium species exist as integral members of well-defined and spatially limited soil-fungal communities (84,85). Stability in the species composition and distribution of these fungal groups allied with plant communities indicates a high level of niche differentiation among the Fusaria and other microbes that has been determined in part by microbial interactions (84,85). The regular occurrence of certain species of *Fusarium* in particular plant-fungus communities as exemplified by some grasslands (41,85,101) may be in part a result of interspecific competition among Fusaria for a limited number of niches in these environments.

Soil Environment

Organic matter Very wet, highly organic soils are not conducive to *Fusarium*. Park (66) found that *F. oxysporum* did not occur in certain soils with over 23% organic matter.

Fusarium is rare in acid peat soils of the Alakai Swamp, Kauai, Hawaii (M. Stoner, *unpublished data*), and in other rain forest soils where populations are sporadic and small: *Fusarium moniliforme* (rare, windward Mauna Kea, Hawaii Island, 40 propagules/g dry soil); *F. solani* (trace, Alakai Swamp; homothallic strain occasionally encountered, less than 200/g, windward rain forests, Hawaii Island). Glynn and Kavanagh (28), Bisby et al. (3) and Dale (11) reported the absence or rarity of *Fusarium* in acid peats. Kommedahl (40), however, noted numerous propagules in acid peats; and *Fusarium* is common in some alkaline peaty soils (80). *Fusarium* can colonize some peat or peaty soils providing they are suitably enriched (28) or otherwise altered by cultivation.

pH *Fusarium* occurs in diverse noncultivated soils ranging from acidic (8,40,42,54,75,85) to neutral and alkaline (2,22,23,29,42,70). While the reported range is approximately pH 4.0–9.6, most records fall between 4.5 and 8.0. In a study of undisturbed grasslands in England, Warcup (97) noted a differential distribution of members of *F. roseum,* apparently according to pH: *F. roseum* 'Sambucinum' was common on alkaline to neutral calcareous soils, whereas *F. roseum* 'Acuminatum' was associated with most acid soils. *F. oxysporum* has been found over a wide range of pH (70,85,97). In Iceland, propagule counts of Fusaria are generally higher in acid as opposed to basic grassland soils (40). In contrast, Brown (6) found *F. oxysporum* and *F. roseum* mostly in alkaline, maritime sand dunes; *F. roseum* occurred in acid dunes in one area only. Within the extremes, the probable influence of pH is through effects on the nutrition and structure of plant communities that support the presence of *Fusarium*. *Fusarium* may be more abundant in some acid soils because of reduced bacterial competition (41); however, the abundance of both bacteria and *Fusarium* in some acid soils (85) indicates that the role of pH is more complex.

Temperature All species of *Fusarium* have been reported from temperate to tropical areas, and some have ranges extending into very cold climatic zones. Limited data suggest that *F. lateritium* and *F. rigidiuscula* tend to occur in warmer temperate and tropical areas. *Fusarium solani* is cited frequently as being more common in warm, cultivated soils (13); but in noncultivated soils, *F. solani* is widely distributed in temperate as well as tropical climates (13,17,18,49), although it has not been reported from extreme alpine or boreal climates (69,73,75). In certain regions, *F. solani* shows a preference for, or tolerance of, warmer habitats (15,85).

Kommedahl (40) found *Fusarium* to be widespread in grassland soils of Iceland, but it was absent from the coldest soils (5–6 C at the time of sampling). *Fusarium roseum,* which predominates in grassland soils of Iceland, may benefit from low temperatures. Since the fungus survives largely in host debris, slower decomposition of protective substrata in cold soils may compensate for the weak competitive saprophytic ability of predominate Fusaria (42).

Moisture *Fusarium* species tend to be qualitatively and quantitatively more prevalent where soil moisture is abundant (29,33,40,62,75,85). However, *Fusarium* is sparse or absent in very wet or flooded environments such as open bogs and conifer swamps in Wisconsin (8) and muck soils of rain forests in Hawaii (84,85). The matter of soil moisture must be considered in connection with plant communities. *Fusarium* can be well represented in areas with very limited or temporary soil moisture and/or high evaporation rates, provided an adapted plant community is established there (5,60,70,78). Species such as *F. solani, F. oxysporum,* and *F. roseum* that have abilities as primary colonizers, strong competitive saprophytic ability, or capability to form chlamydospores are better adapted to habitats with limited water.

Literature Cited

1. Al-Doory, N., M. K. Tolba, and H. Al-Ani. 1959. On the fungal flora of Iraqi soils. II. Central Iraq. Mycologia 51:429–439.
2. Bhatt, G. C., 1970. The soil microfungi of white cedar forests in Ontario. Can. J. Bot. 48:333–339.
3. Bisby, G. R., M. I. Timonin, and N. James. 1935. Fungi isolated from soil profiles in Manitoba. Can. J. Res., Sect. C. 13:47–65.
4. Booth, C. 1971. The genus *Fusarium*. Commonwealth Mycological Inst., Kew, Surrey, England. 237 p.
5. Borut, S. 1960. An ecological and physiological study of soil fungi of the northern Negev (Israel). Bull. Res. Council Israel 8:65–80.
6. Brown, J. C. 1958. Soil fungi of some British sand dunes in relation to soil type and succession. J. Ecol. 46:641–664.
7. Christensen, M., W. F. Whittingham, and R. O. Novak. 1962. The soil microflora of wet-mesic forests in southern Wisconsin. Mycologia 54:374–388.
8. Christensen, M., and W. F. Whittingham. 1965. The soil microfungi of open bogs and conifer swamps in Wisconsin. Mycologia 57:882–896.
9. Cook R. J., E. J. Ford, and W. C. Snyder. 1968. Mating types, sex, dissemination and possible sources of clones of *Hypomyces (Fusarium) solani* f. *pisi* in South Australia. Austral. J. Agric. Res. 19:253–259.
10. Cooke, W. B., and C. B. Lawrence. 1959. Soil mould fungi isolated from recently glaciated soils in southeastern Alaska. J. Ecol. 47:529–549.
11. Dale, E. 1914. On the fungi of soil. II. Fungi from chalky soil, uncultivated peat, and the "black earth" of the reclaimed fenland. Ann. Mycol., Berl. 12:33–62.
12. Dixon, D. 1928. The microorganisms of cultivated and bush soils. Austral. J. Exp. Biol. Med Sci. 5:223–232.
13. Domsch, K. H., and W. Gams. 1972. Fungi in agricultural soils. Halstead Press Division, John Wiley and Sons., Inc., New York. 290 p.
14. Dowding, P., and P. Widden. 1974. Some relationships between fungi and their environment in Tundra regions, p. 123–150. *In* A. J. Holding, O. W. Heal, S. F. Maclean, Jr., and P. W. Flanagan, (ed.), Soil organisms and decomposition in tundra, International Biological Program Tundra Biome Steering Committee, Stockholm.
15. Dunn, P. H. 1973. The ecology of filamentous heterotrophic microorganisms in subtropical and tropical marine psammen habitats. Ph.D. Thesis, Univ Hawaii, Honolulu. 226 p.
16. Durrell, L. W., and L. M. Shields. 1960. Fungi isolated in culture from soils of the Nevada test site. Mycologia 52:636–641.
17. Ebbels, D. L. 1973. Plant pathology, p. 30–37. *In* Cotton research report, 1970–71, Cotton Research Corp., London.
18. Ebbels, D. L. 1974. Plant pathology, p. 24–32. *In* Cotton research report, 1971–72, Cotton Research Corp., London.
19. Edmunds, R. L., and W. A. Heather. 1973. Root diseases in pine nurseries in the Australian Capital Territory. Plant Dis. Rep. 57:1058–1062.
20. Eicker, A. 1969. Microfungi from surface soil of forest communities in Zululand. Trans. Brit. Mycol. Soc. 53:381–392.
21. Eicker, A. 1970. Vertical distribution of fungi in Zululand soils. Trans. Brit. Mycol. Soc. 55:45–57.
22. Eicker, A. 1974. The mycoflora of an alkaline soil of the open-savannah of the Transvaal. Trans. Brit. Mycol. Soc. 63:281–288.
23. Elliott, J. S. B. 1930. The soil fungi of the Dovey Salt Marshes. Ann. Appl. Biol. 17:284–305.
24. Ellis, M. 1940. Some fungi isolated from pinewood soil. Trans. Brit. Mycol. Soc. 24:87–97.
25. England, C. M., and E. L. Rice. 1957. A comparison of the soil in a tall-grass prairie and of an abandoned field in central Oklahoma. Bot. Gaz. 118:186–190.
26. Farrow, W. M. 1954. Tropical soil fungi. Mycologia 46:632–646.
27. Ford, E. J., J. A. Bourret, and W. C. Snyder. 1967. Biologic specialization in *Calonectria (Fusarium) rigidiuscula* in relation to green point gall of cacao. Phytopathology 57:710–712.
28. Glynn, A. N., and T. Kavanagh. 1973. Survival and inoculum potential of *Fusarium oxysporum* f. sp. *lycopersici* in sphagnum and fen peats. Irish J. Agric. Res. 12:273–278.
29. Gochenaur, S. E., and W. F. Whittingham. 1967. Mycoecology of willow and cottonwood lowland communities in southern Wisconsin. I. Soil microfungi in the willow-cottonwood forests. Mycopath. Mycol. Appl. 33:125–139.
30. Gordon, W. L. 1956. The occurrence of *Fusarium* species in Canada. V. Taxonomy and geographic distribution of *Fusarium* species in soil. Can. J. Bot. 34:833–846.
31. Gordon, W. L. 1956. The taxonomy and habitats of the *Fusarium* species in Trinidad, B. W. I. Can. J. Bot. 34:847–864.
32. Guillemat, J., and J. Montegut. 1960. The effect of mineral fertilizers on some soil fungi, p. 98–106. *In* D. Parkinson and J. S. Waid, (ed.), The ecology of soil fungi, Liverpool Univ. Press.
33. Hee, L. W. 1971. *Fusarium* species from west Malaysian soils: a study of their distribution, biology and pathogenicity. M. Agr. Sci. Thesis. Univ. Malaya, Kuala Lumpur. 105 p.
34. Ishii, H. 1974. Fungal flora, their geographical and ecological distribution in the forest soils of Japan. Soil Microorganisms 15:12–20. (in Japanese).
35. Jackson, R. M. 1960. Soil fungistasis and the rhizosphere, p. 168–176. *In* D. Parkinson and J. S. Waid, (ed.), The ecology of soil fungi, Liverpool Univ. Press.

36. Jensen, H. L. 1931. The fungus flora of the soil. Soil Sci. 31:123–158.
37. Jensen, V. 1963. Studies on the microflora of Danish beech forest soils. Zentr. Bakteriol. Parasitent, Abt. II. 117:167–179.
38. Joffe, A. Z., and J. Palti. 1977. Species of *Fusarium* found in uncultivated desert-type soils in Israel. Phytoparasitica 5(2):119–121.
39. Komada, H. 1976. A new selective medium for isolating *Fusarium* from natural soil. Proc. Amer. Phytopathol. Soc. 3:76.
40. Kommedahl, T. 1972. *Fusarium* species in Icelandic soils in relation to winterkilling of grasses. Res. Inst. Nedhri Ás, Hveragerdhi, Iceland. Bull. 11. 19 p.
41. Kommedahl, T., and E. I. Siggeirsson. 1973. Prevalence of *Fusarium* species in roots and soil of grassland in Iceland. Res. Inst. Nedhri Ás, Hveragdhi, Iceland. Bull. 14. 27 p.
42. Kommedahl, T., C. E. Windels, and D. S. Lang. 1975. Comparisons of *Fusarium* populations in grasslands of Minnesota and Iceland. Mycologia 67:38–44.
43. Kreutzer, W. A. 1972. *Fusarium* spp. as colonists and potential pathogens in root zones of grassland plants. Phytopathology 62:1066–1070.
44. Kubíková, J. 1968. *Fusarium oxysporum* (Schlecht.) Snyd. et Hans.—a dominant fungus species of the root-surface of woody plant seedlings. Plant Soil 28:306–312.
45. Lanneau, C., J. A. Meyer, and P. Staner. 1967. Contribution à l'étude de la flore fongique des sols du Katanga. Bull. Seances (Bruxelles) 1967-3:540–545.
46. Lee, B. K. H., and G. E. Baker. 1973. Fungi associated with the roots of red mangrove, *Rhizophora mangle*. Mycologia 65:894–906.
47. Lim, G. 1969. Some observations on soil and root-surface mycoflora. Plant Soil 31:143–148.
48. Lim, G. 1974. Distribution of *Fusarium* in some British soils. Mycopath. Mycol. Appl. 52:231–237.
49. Lim, G., and C. H. Chew. 1970. *Fusarium* in Singapore soils. Plant Soil 33:673–677.
50. Locke, T. 1972. A study of vascular wilt disease of oil palm seedling. Ph.D. Thesis. Univ. Manchester. 120 p.
51. Mallik, M. A. B., and E. L. Rice. 1966. Relation between soil fungi and seed plants in three successional forest communities in Oklahoma. Bot. Gaz. 127:120–127.
52. McKenzie, F. M. 1972. Survival of *Fusarium culmorum* (W. G. Sm.) Sacc. in the soil. Ph.D. Thesis. Univ. Manchester. 102 p.
53. McLennan, E. I., and S. C. Ducker. 1954. The ecology of the soil fungi of an Australian heathland. Austral. J. Bot. 2:220–245.
54. Miller, J. H., J. E. Giddens, and A. A. Foster. 1957. A survey of the fungi of forest and cultivated soils of Georgia. Mycologia 49:779–808.
55. Morral, R. A. A., and T. C. Vanterpool. 1968. The soil microfungi of upland boreal forest at Candle Lake, Saskatchewan. Mycologia 60:642–654.
56. Morrow, M. B. 1932. The soil fungi of a pine forest. Mycologia 24:398–402.
57. Mueller-Dombois, D., and M. Perera. 1971. Ecological differentiation and soil fungal distribution in the montane grasslands of Ceylon. Ceylon J. Sci. (Biol. Sci.) 9:1–41.
58. Nash, S. M., T. Christou, and W. C. Snyder. 1961. Existence of *Fusarium solani* f. *phaseoli* as chlamydospores in soil. Phytopathology 51:308–312.
59. Nash, S. M., and W. C. Snyder. 1965. Quantitative and qualitative comparisons of *Fusarium* populations in cultivated fields and noncultivated parent soils. Can. J. Bot. 43:939–945.
60. Nicot, J. 1960. Some characteristics of the microflora in desert sands, p. 94–97. *In* D. Parkinson and J. S. Waid, (ed.), The ecology of soil fungi, Liverpool Univ. Press.
61. Nour, M. A. 1956. A preliminary survey of fungi in some Sudan soils. Trans. Brit. Mycol. Soc. 39:357–360.
62. Novak, R. O., and W. F. Whittingham. 1968. Soil and litter microfungi of a maple-elm-ash floodplain community. Mycologia 60:776–787.
63. Orpurt, P. A., and J. T. Curtis. 1957. Soil microfungi in relation to the prairie continuum in Wisconsin. Ecology. 38:628–637.
64. Paine, F. S. 1927. Studies on the fungus flora of virgin soils. Mycologia 19:248–267.
65. Park, D. 1959. Some aspects of the biology of *Fusarium oxysporum* Schl. in soil. Ann. Bot. 23:35–49.
66. Park, D. 1963. The presence of *Fusarium oxysporum* in soils. Trans. Brit. Mycol. Soc. 46:444–448.
67. Parkinson, D., and W. B. Kendrick. 1960. Investigations of soil microhabitats, p. 22–28. *In* D. Parkinson and J. S. Waid, (ed.), The ecology of soil fungi, Liverpool Univ. Press.
68. Pugh, C. J. F. 1960. The fungal flora of tidal mud flats, p. 202–208. *In* D. Parkinson and J. S. Waid, (ed.), The ecology of soil fungi, Liverpool Univ. Press.
69. Rall, G. 1965. Soil fungi from the alpine zone of the Medicine Bow Mountains, Wyoming. Mycologia 57:872–881.
70. Ranzoni, F. V. 1968. Fungi isolated in culture from soils of the Sonoran Desert. Mycologia 60:356–371.
71. Rao, A.S. 1959. A comparative study of competitive saprophytic ability in twelve root-infecting fungi by an agar plate method. Trans. Brit. Mycol. Soc. 42:97–111.
72. Rao, P. 1970. Studies on soil fungi. III. Seasonal variation and distribution of microfungi in some soils of Andhra Pradesh (India). Mycopath. Mycol. Appl. 40:277–298.
73. Reddy, T. K. R., and R. Knowles. 1965. The fungal flora of a boreal forest raw humus. Can. J. Microbiol. 11:837–843.
74. Sewell, G. W. F. 1959. The ecology of fungi in Calluna-heathland soils. New Phytol. 58:5–15.

75. Singh, P. 1976. Some fungi in the forest soils of Newfoundland. Mycologia 68:881–890.
76. Smith, R. S., Jr. 1967. Decline of *Fusarium oxysporum* in the roots of *Pinus lambertiana* seedlings transplanted into forest soils. Phytopathology 57:1265.
77. Snyder, W. C., and S. M. Smith. 1968. Relative incidence of Fusarium pathogens of cereals in rotation plots at Rothamsted. Trans. Brit. Mycol. Soc. 51:417–425.
78. States, J. S. 1978. Soil fungi of cool-desert plant communities in northern Arizona and southern Utah. J. Arizona-Nevada Acad. Sci. 13:13–17.
79. Steele, C. W. 1967. Fungus populations in marine waters and coastal sands of the Hawaiian, Line, and Phoenix Islands. Pacific Sci. 21:317–331.
80. Stenton, H. 1953. The soil fungi of Wicken Fen. Trans. Brit. Mycol. Soc. 36:304–314.
81. Stoner, M. F. 1974. Ecology of *Fusarium* species in noncultivated soils of Hawaii. Proc. Amer. Phytopathol. Soc. 1:102 (Abstr.).
82. Stoner, M. F. 1976. Proposed theory on ohia forest decline in Hawaii: a precipitant phenomenon related to soil conditions and island maturation. Proc. Amer. Phytopathol. Soc. 3:215 (Abstr.).
83. Stoner, M. F. 1976. Reference fungus method for ecological comparisons of soils. Proc. Amer. Phytopathol. Soc. 3:276–277 (Abstr.).
84. Stoner M. F., and G. E. Baker. 1980. Soil and leaf fungi. *In* D. Mueller-Dombois et al. (ed.), Island ecosystems: biological organization in selected Hawaiian communities, Dowden, Hutchinson and Ross, Inc., Pennsylvania. (In press).
85. Stoner, M. F., D. K. Stoner, and G. E. Baker. 1975. Ecology of fungi in wildland soils along the Mauna Loa Transect. United States/International Biological Program Tech. Rep. No. 75. Univ. Hawaii, Honolulu. 102 p.
86. Stotzky, G., R. D. Goos, and M. I. Timonin. 1962. Microbial changes occurring in soil as a result of storage. Plant Soil 16:1–18.
87. Swart, H. J. 1958. An investigation of the mycoflora in the soil of some mangrove swamps. Acta Bot. Neerl. 7:741–768.
88. Thornton, R. H. 1956. Fungi occurring in mixed oakwood and heath soil profiles. Trans. Brit. Mycol. Soc. 39:485–494.
89. Thornton, R. H. 1958. Biological studies of some tussock-grassland soils. New Zealand J. Agric. Res. 1:922–938.
90. Thornton, R. H. 1960. Growth of fungi in some forest and grassland soils, p. 84–91. *In* D. Parkinson and J. S. Waid, (ed.), The ecology of soil fungi, Liverpool Univ. Press.
91. Thornton, R. H. 1965. Studies of fungi in pasture soils. I. Fungi associated with live roots. New Zealand J. Agric. Res. 8:417–449.
92. Toussoun, T. A. 1975. Fusarium-suppressive soils, p. 145–151. *In* G. W. Bruehl, (ed.), Biology and control of soil-borne plant pathogens, Amer. Phytopathol. Soc., St. Paul, Minn.
93. Toussoun, T. A., W. Menzinger, and R. S. Smith, Jr. 1969. Role of conifer litter in ecology of *Fusarium;* stimulation of germination in soil. Phytopathology 59:1396–1399.
94. Toussoun, T. A., and P. E. Nelson. 1976. *Fusarium,* a pictorial guide to the identification of *Fusarium* species according to the taxonomic system of Snyder and Hansen. (2nd ed.). Pennsylvania State Univ. Press, Univ. Park, Pa. 43 p.
95. Tresner, H. S., M. P. Backus, and J. T. Curtis. 1954. Soil microfungi in relationship to the hardwood forest continuum in southern Wisconsin. Mycologia 46:314–333.
96. Tsao, P. H. 1970. Selective media for isolation of pathogenic fungi. Annu. Rev. Phytopathol. 8:157–186.
97. Warcup, J. H. 1951. The ecology of soil fungi. Trans. Brit. Mycol. Soc. 34:376–399.
98. Warcup, J. H. 1957. Studies on the occurrence and activity of fungi in a wheat-field soil. Trans. Brit. Mycol. Soc. 40:237–262.
99. Widden, P., and D. Parkinson. 1973. Fungi from Canadian coniferous forest soils. Can. J. Bot. 51:2275–2290.
100. Windels, C. E., and T. Kommedahl. 1971. Comparisons of *Fusarium* spp. and populations in cultivated and noncultivated soils. Phytopathology 61:1026 (Abstr.).
101. Windels, C. E., and T. Kommedahl. 1974. Population differences in indigenous *Fusarium* species by corn culture of prairie soils. Amer. J. Bot. 61:141–145.
102. Wright, W., and W. B. Bollen. 1961. Microflora of Douglas-fir forest soil. Ecology 42:828–835.

III The Fungus *Fusarium:* Genetics and Cytology

Prologue

T. A. Toussoun

Fusarium has always been considered a highly variable organism. One might even say that *Fusarium* and variation are synonymous. The literature on fungi, particularly that of the first half of this century, is replete with the saltants, variants, and mutants of *Fusarium.* In their book *Introductory Mycology,* C. J. Alexopoulos and C. W. Mimns state: "Few mycologists attempt to identify form species of *Fusarium* because of the great variability in this group, a variability that makes identification uncertain for all but the few specialists." As a student working on *Hypomyces solani* f. sp. *cucurbitae,* now called *Nectria haematococca* f. sp. *cucurbitae,* I soon had my share of mutants which appeared spontaneously on potato dextrose agar media. The extent of this variation was truly remarkable, ranging from changes in growth rate and pigmentation, to forms that produced non-ostiolate perithecia (cleistothecia?), and others with genetic blocks that prevented the formation of protoperithecia, or had variations in the development of protoperithecia, the fertility of perithecia, and their color.

Whence comes this variability which is apparently accelerated by the laboratory environment but is by no means restricted to it? What is its nature, and how is it transmitted? How can this variability be integrated with the often marked host specificity? We know amazingly little about the genetics and cytology of *Fusarium.* The reader will be forewarned of this by noting that there are only three contributions to this section of the book. The first paper is a review by Puhalla of what is known of the cytology and genetics of the asexual and sexual forms of *Fusarium.* One can clearly see the many unexplored avenues of research into such crucial areas as the genetics of pathogenicity and host specificity, heterokaryosis, parasexuality, and extra-chromosomal inheritance, as well as the possibilities in the use of modern genetic techniques in the elucidation of the variability of the genus. The second paper by Daboussi-Bareyre and Parisot exemplifies one of these findings and should stimulate others to confirm them. Finally, Bouhot's contribution deals with the pathogenic potential of *F. oxysporum.* Surely this is one of the most intriguing aspects of this fungus and one for which it would seem to be ideally suited to serve as a model. And yet, this contribution is, to my knowledge, unique in this area. Bouhot and his colleagues are alone in investigating the origin of races in *Fusarium.* It would seem that within the complex of forms attacking cucurbits there exist genes governing pathogenicity to watermelon and muskmelon, and that mutations occur from one to the other. Apparently, formae speciales form genetic "families," and mutations between these do not occur. For example, forms that do not attack cucurbits do not mutate to forms that attack cucurbits, nor do saprophytes mutate to become pathogens. This work of Bouhot and his colleagues is extremely thought provoking and hopefully will stimulate more research in this area.

It is difficult to understand the paucity of genetic work with *Fusarium*, particularly in view of the pioneering studies of H. N. Hansen. He was, in my view, one of the outstanding investigators of variation in fungi. His accomplishments have not received the recognition that is their due, probably because he was not a prolific writer. He developed a rapid and reliable single-spore culturing technique which proved invaluable in analyzing fungi for variability. It was the foundation for and led directly to the studies culminating in the revision of *Fusarium* taxonomy known as the Snyder & Hansen classification. Hansen and his colleagues and students developed the detailed mechanisms of heterocaryosis and explained their significance in the biology and the asexual variability of fungi. It was they who established the universality of this phenomenon in fungi and who showed that the dissociation of the heterokaryotic condition during conidium formation played a major role in the variation of imperfect fungi. Hansen laid the basis for the further studies of B.O. Dodge, C.C. Lindegren, G.W. Beadle, and G. Pontecorvo. Hansen's work on the genetics of *Hypomyces solani (= Nectria haematococca)* f. sp. *cucurbitae* led him to studies on the origin and mechanism of sexuality and to distinguishing the two mechanisms he called compatibility heterothallism and sex heterothallism, subjects which also deserve greater attention than they have received to date.

Surely such auspicious beginning should not go long unfulfilled!

27 Genetic Considerations of the Genus *Fusarium*

John E. Puhalla

In this chapter I will review those areas of research in *Fusarium* species that could be called genetic. Despite the difficulties of asexual genetic analysis, the bulk of this research has centered on the asexual forms. This emphasis on asexual forms may be due to the fact that many of them are extremely important plant pathogens. During the last 40 years, however, the sexual stages for several species of *Fusarium* have been found, and some genetic studies of them have been initiated. Genetic analyses cannot be separated from considerations of morphology, physiology, and cytology. Therefore, in this review I will also consider such areas where appropriate. Finally, I will discuss the usefulness of genetic methodology in other phases of *Fusarium* research.

Nuclear Studies of *Fusarium*

The two major asexual spore forms of the Fusaria are microconidia and macroconidia. The microconidia are unicellular and uninucleate (15, 36); the more conspicuous macroconidia are multicellular, but each cell has only one nucleus. All nuclei of one macroconidium, however, are mitotic descendants of the same progenitor nucleus and are therefore genetically identical (26).

Sexual stages of *Fusarium* are ascomycetous; the sexual spore is the ascospore. Some perfect states have two-celled ascospores and are now assigned to the genus *Nectria*. (Early workers placed these types in the genus *Hypomyces*, but more recent studies do not support this view [10]. I will, however, retain the scientific name used by the early workers when I am discussing their findings.) El-Ani (35) showed that each cell of a two-celled ascospore is uninucleate, and both nuclei are genetically identical. Other perfect states, such as *Gibberella*, form multicellular ascospores. I could find no reports, however, on their nuclear condition.

Reports on the nuclear state of the cells of the vegetative hyphae are conflicting. Dickinson (26) found that nearly all hyphal cells of *Fusarium oxysporum* f. sp. *vasinfectum* had only one nucleus. Schneider (87) found this to be true for *F. avenaceum*, as did Punithalingam (81) for *F. culmorum* and Hoffman (56) for *F. solani*. Yet other workers have reported multinucleate cells in some *Fusarium* species (4, 52, 64). Hoffman (56) examined several *Fusarium* species and concluded that the number of nuclei per cell was species dependent. Thus, cells of *F. oxysporum* and *F. solani* were essentially uninucleate, whereas those of *F. culmorum* and *F. graminearum* were mainly multinucleate. Most workers found that it was the younger cells that were multinucleate; older cells were mainly uninucleate.

These conflicting reports could be due in part to the species examined, the age of the cells observed, and the methods of preparing the cells for examination. In some studies fixed and stained cells were examined; in others, phase microscopy was used. Nutrition may also be a critical factor, because Hoffman (52) found that nuclear number depended on pH and C:N ratio of the growth medium as well as on temperature.

Under certain conditions young hyphal cells of *Fusarium* can undoubtedly be multinucleate. Aist (3) found that in such cells the nuclei divided synchronously. Division was rapid and was followed by septum formation. The daughter nuclei moved at random into the newly forming cells. In these same studies Aist also found several nuclei both in the tip cell and in some cells behind it. Because older cells are mainly uninucleate, subsequent septum formation must occur in these cells without further nuclear divisions.

Both light and electron microscopic studies of somatic nuclear division revealed major differences from mitosis in higher organisms (4, 5). The nuclear membrane remained intact throughout the entire division. Start of division was signaled by disappearance of the nucleolus. The chromosomes condensed to deeply staining bodies. Aist and Williams (4) could find no evidence that the chromosomes were all attached to a thread or ring-like structure, as had been reported in other fungi (38, 47). There was a spindle but not a true metaphase plate. The spindle consisted of microfibrils that extended from the chromosomes to the centriole, which was in turn very closely associated with the nuclear membrane. The microfibril-centriole complex may mediate not only chromosome separation at anaphase but also the movement of the daughter nuclei through the cytoplasm.

Meiosis was studied in depth by Hirsch (50) and El-Ani (35) in *Hypomyces solani* f. sp. *cucurbitae* (now called *Nectria haematococca)*, the sexual stage of one member of the species *F. solani.* The various phases of meiosis were clearly defined and were comparable to those in other organisms. At late diakinesis of prophase I, four separate, condensed bivalent chromosomes were clearly visible. El-Ani concluded that in this fungus the haploid number of chromosomes (n) = 4. One chromosome was long, two were intermediate in length and one was quite short.

Chromosome counts from meiotic division figures have been made for several sexual species of *Fusarium,* and they are listed in Table 27-1. Chromosome counts based on somatic division figures are much more difficult because the chromosomes at this stage are smaller and the division figures are less clear. Aist and Williams (4) counted at least four centromeres during division in *F. oxysporum* and concluded that n = 4. Punithalingam (81) suggested that n = 8 from his somatic nuclear studies of *F. culmorum.*

Studies on the DNA content of the Fusaria are rare. Kumari and associates (65) found that the nuclear DNA content of microconidia of *F. oxysporum* f. sp. *melonis* was comparable to that of other fungi. No base composition determinations of the DNA have been published to my knowledge.

Table 27-1. Haploid chromosome numbers (n) of sexual forms of Fusarium

Species	n[a]	References
Hypomyces solani f. sp. *cucurbitae*[b]	4	32, 49
Hypomyces solani (homothallic)	6	50
Gibberella roseum	6	51
Gibberella lateritium	6	51
Gibberella cyanogena	4	57
Gibberella cyanea	4	34, 57
Gibberella fujikuroi	4	57
Gibberella stilboides	4	57
Gibberella zeae	4	57
Calonectria rigidiuscula	7	51
Calonectria nivalis	7	57

[a] Counts made on fixed and stained chromosomes during meiotic division.
[b] = *Nectria haematococca.*

Variation in *Fusarium* Cultures

A culture of *Fusarium* freshly isolated from nature might remain unchanged through repeated agar subculture, or it might gradually change its morphology, or it might suddenly produce sectors of growth radically different in appearance from the parent. This morphological variability has caused no end of controversy in classifying species of *Fusarium*. Variants of one species were sometimes morphologically indistinguishable from members of another species (14). Species determinations were often based on cultures that had changed morphologically from the original isolate. Brown (14) noted that such cultures were also frequently avirulent and thus contended that they did not represent the situation in nature. Clearly, an investigation of variation in *Fusarium* is essential to an understanding of the genus.

Studies of morphological variation in *Fusarium* are conflicting and often confusing, but certain generalizations about this variation can be made: (i) Not all isolates are variable, and some isolates are more variable than others. (ii) The changes are not continuous but occur in discrete steps. A morphological change, however, may be so small that it is overlooked by all but the most astute observer. The altered morphology may appear as sectors or patches in the parent colony, or it may arise among the spore progeny of an apparently normal colony. (iii) The morphological changes are usually limited to loss of aerial mycelium, an increase in macroconidial production, and sometimes an increase in pigmentation. (iv) Such variants are mostly stable and can be perpetuated by hyphal tip or monoconidial transfer.

Variability in agar cultures may or may not reflect what occurs in nature. Early workers found that variants arose more frequently on certain media than others; Richard's medium enhanced variation, whereas soil or soil extract agar cultures rarely produced variants (13, 76). Several researchers have reported that strains isolated directly from nature were more uniform, usually having abundant aerial mycelia and few macroconidia (7, 23, 77, 100, 101). Miller (75) proposed, therefore, that each *Fusarium* species had a characteristic "wild type" and that spontaneous morphological variation in nature was rare or strongly selected against. In some species, however, several morphological types have been isolated directly from nature (16, 69, 82, 93). Nonetheless it does seem that the extreme morphological variability in agar cultures does not occur among natural isolates.

Spontaneous variation in the nutrition of *Fusarium* cultures has rarely been assayed directly. Prasad (79) did find a tightly growing variant of *F. solani* f. sp. *cucurbitae* that required biotin for normal growth. Georgopoulos (41) recovered variants of this same fungus that were resistant to chloronitrobenzene compounds. Such variants were found on a medium containing the chloronitrobenzenes, but presumably they were spontaneous mutations.

Natural isolates of certain plant-pathogenic species of *Fusarium* showed significant differences in pathogenicity. Such isolates fell into groups whose members could attack only certain plant species. Snyder and Hansen (91) called these groups "formae speciales." Often the formae speciales could be further subdivided into "physiological races" that attacked only certain cultivars of a particular plant species. In general, isolates of the different formae speciales were morphologically indistinguishable. The trait of host specificity was extremely stable in culture; spontaneous changes in host range were rarely, if ever, found. One exception is the report of Armstrong and Armstrong (6), who found that some variants of *F. oxysporum* f. sp. *tracheiphilum* attacked only cowpea, whereas the parent culture had attacked both cowpea and soybean.

Some morphological variants had altered virulence but this was invariably a change in aggressiveness; that is, the variant caused greater or less disease to the same extent in all plant hosts tested (7, 46). Unlike morphological variation, such changes in pathogenicity were reversible (39). A variant less aggressive than the parent could produce secondary variants with increased aggressiveness (39). Some workers claimed that loss of aerial mycelium was associated with decreased aggressiveness, but there were frequent exceptions. In general, there is no stringent correlation between morphological type and pathogenicity.

Most reports of variation in *Fusarium* are woefully lacking in quantitative data; and even if such data were presented, they might still be difficult to interpret because of differences in experimental procedure. Thus, some workers scored variants as sectors in colonies; others scored aberrant types among conidial progeny. The two sources of variation need not be comparable. Sector variants are probably mainly suppressive to the parent strain (74). Moreover, Cormack (23) found that the type of spore (macroconidia vs. microconidia) and the source of the spore (single conidiophores vs. sporodochia) profoundly affected the frequency of variants. Only macrospores from sporodochia yielded significant numbers of variants. Sporodochia normally develop late in freshly isolated cultures, and Miller (75) has even proposed that such sporodochia are patches of mutant tissues. His view is not shared by most other workers. The fact that some strains are more unstable than others suggests that the frequency of variation also is under genetic control.

Procedures have been proposed to prevent variability of *Fusarium* stock cultures. Maintenance on sterile soil or soil extract agar, low-temperature storage, and stock transfer by a single hyphal tip or microconidium have all been suggested to reduce variation. These procedures are effective, and they may provide additional clues as to the basis of variation.

The genetic and molecular bases for spontaneous variability in *Fusarium* are still unknown. Early studies on variation were largely descriptive. Even more recent studies remained descriptive, although in the meantime rather sophisticated genetic techniques have become available. Certain of these observations are, however, indicative. Variants were readily found in single spore (and therefore presumably homokaryotic) cultures. The stability of spontaneous variants through vegetative transfer suggested they were true genetic mutants. As early as 1932, Dickinson (26) used a heterokaryon analysis to show that a morphological variant of *F. fructigenum* was due to a lesion in the nucleus. Later in 1954, Buxton (15) used a similar analysis to demonstrate the nuclear genetic basis of morphological variants in *F. oxysporum* f. sp. *gladioli.* Several morphological variants in *Hypomyces solani* f. sp. *cucurbitae* and *H. ipomoeae* were found to behave like single nuclear gene mutations in matings (29, 90). I will discuss the details of these studies later.

The high frequency of morphological variants has been offered by some workers as evidence against their being true nuclear mutations. High frequencies of directed, relatively non-random changes are characteristic of cytopasmic mutation in some fungi. Singh (88) induced "ropy" mutants in *F. oxysporum* with acriflavine, a chemical known to cause cytoplasmic variants in other fungi. Ropy types are common spontaneous variants in *Fusarium.* No studies have been published, however, that firmly establish the cytoplasmic or nuclear bases for spontaneous variants.

Specific treatments are known that greatly enhance variation in cultures of *Fusarium.* The high frequency of variants on Richard's medium has already been mentioned. In 1945 Miller (74) stated that ultraviolet light (UV) irradiation greatly increased morphological variants, but he presented no data. Several years later in 1956 Buxton (17) treated conidia of *F. oxysporum* f. sp. *pisi* with UV and recovered 0.1–0.3% morphological variants. He also recovered an equal number of variants, termed auxotrophs, that had acquired specific nutritional requirements. Ultraviolet light was also effective in inducing variants in *F. oxysporum* that were resistant to the fungicide dodine (66) and variants of *F. moniliforme* with increased gibberellin production (60). Variants have also been induced with such chemicals as nitrous acid, ethylmethanesulfonate and nitrosoguanidine (11, 86, 98), all of which are known to be mutagenic in most organisms. Bouhot used nitrosoguanidine to change the forma specialis or physiological race designation of isolates of *F. oxysporum.* Details of his studies are presented in this volume (12). Dimock (27) claimed that high concentrations of zinc salts induced morphological variation in *F. oxysporum.* The fungicides benomyl (24) and fernasan (1) have been implicated as mutagens. Such findings could be extremely important because of the widespread use of these fungicides.

There are claims that the host plant or its metabolites can alter *Fusarium* isolates. Follin and Laville (39) reported that a pionnotal variant of *F. oxysporum* f. sp. *cubense* changed to a sporodochial type during one "passage" through the banana host plant. The possibility of cross contamination, however, could not be eliminated. Bolton and Davidson (9) inoculated *F. oxysporum* f. sp. *pisi,* the wilt *Fusarium,* into peas and recovered forms that caused root rot, the characteristic symptom of *F. solani* f. sp. *pisi.* Most workers regard these two species as quite distinct, although Buxton and Ward (20), Hildreth (48), and Massey (68) considered them to overlap. Armstrong et al. (7) found no alteration in the pathogenicity of isolates of *F. oxysporum* f. sp. *vasinfectum* passed through cotton plants. In 1958 Buxton (18) reported a most extraordinary experiment. Conidia of Race 1 of *F. oxysporum* f. sp. *pisi,* inhibited by root exudate of the resistant pea cultivar Alaska, were incubated for 2 weeks in the exudate and then allowed to grow in a medium free of exudate. Some of the progeny were then able to attack Alaska. This change in virulence was stable in vegetative culture. The implications of these results are profound. If Fusaria can indeed acquire new host specificities so readily, this fact must be carefully considered by both plant pathologists and breeders. I wish to emphasize, however, that this study by Buxton was unique and incomplete and must be verified.

Asexual Genetic Analysis

Heterokaryosis and parasexuality are the main tools for genetic analyses of asexual fungi. Heterokaryosis can arise in a fungus when two genetically different hyphae fuse and their cytoplasms intermix. Dickinson (26) made the first definitive observations on hyphal fusion in *Fusarium.* He found that in agar cultures most fusions took place between hyphae that arose as side branches some distance back from the colony front. Dickinson defined two fusion types, "real" and "attempted." In a real fusion two side branches grew toward each other, touched tip to tip, and the wall between them dissolved away allowing cytoplasmic passage between them. Often one side branch was nothing more than a short peg, but fusion was always tip to tip. One cell could produce several side branches, each of which could fuse. In attempted fusions the contact was tip to side; that is, only one hypha was active. The cell wall at point of contact remained intact.

Buxton (15) reported fusions between germ tubes of macroconidia, a phenomenon Dickinson did not see. Later Schneider (87) reported that such fusions did not occur on laboratory media but would occur on glass slides coated with albumin or dilute malt agar. Buxton had indeed made his observations on hyphae growing in water on glass slides. Recently Mesterhazy (73) claimed that fusions formed between macroconidia still in the sporodochium. A definite chemical environment may be essential for fusions. Robinson (85) found that succinic acid induced branching in subapical regions of *F. oxysporum.* Other chemicals may be attractants for the approaching side branches. Temperature is also important; Dickinson found that high temperatures depressed fusion.

Dickinson (26) paired hyphal segments from a normal culture of *F. fructigenum* with those of a morphological variant. The hyphae fused and their cytoplasms intermingled, but the nuclei remained intact. The fusion cells were cut out and cultured. From some of them, both morphological types were recovered among the conidial progeny. Dickinson concluded that such segregation reflected the partitioning of the nuclei into the uninucleate spores and that the determinants for the morphological differences resided in the nucleus. He regarded the morphological variant as a true nuclear mutation.

Genetic studies like those of Dickinson were not again pursued until 1954, when Buxton (15) paired two morphological variants of *F. oxysporum* f. sp. *gladioli.* Growth from these pairings was morphologically distinct from either variant. He took single hyphal tips from this growth, cultured them and examined their conidial progeny. Both variant morphologies were recovered from some of these single tip cultures. Buxton assumed that the variants were due to true nuclear gene mutations and therefore concluded that the segregating tips were heterokaryotic. Since he found that only tip cells were multinucleate, he reasoned that heterokaryosis was confined to them.

In 1956 Buxton (17) used UV-induced auxotrophs of *F. oxysporum* f. sp. *pisi* to "force" heterokaryons; that is, he paired complementary auxotrophs on a medium which supported only non-auxotrophic growth. On this medium only the fusion product, the heterokaryon, would grow. He reported forced heterokaryons between isolates of Race 1 from widely separated geographic areas and also heterokaryons between Race 1 and Race 2 of this pathogen. In the latter heterokaryons nearly all conidia were auxotrophic, like one or the other parent; but a small proportion (3 in 10^8) were prototrophic, that is, they were able to grow without supplementation. Colonies from these prototrophic conidia produced mainly prototrophic conidia that had a diameter 1.25× greater than that of the parent auxotrophic conidia. The implication was that the prototrophic conidia were diploid. Around 4% of the conidia of the putative diploids were auxotrophic, but often their auxotrophic requirements were in new, non-parental combinations. Buxton contended that these recombinants arose through a parasexual cycle. The pathogenicity of these progeny also had recombined. Whereas Race 1 attacked only pea cultivar A and Race 2 attacked both cultivars A and B, some auxotrophic progeny were totally avirulent and one attacked cultivars A and C. Cultivar C was not attacked by either Race 1 or Race 2.

Buxton's next published genetic studies switched to the banana wilt fungus *F. oxysporum* f. sp. *cubense* (19). He paired complementary auxotrophs of four strains showing different degrees of virulence and different symptomatology. He reported low levels of diploidy among conidia of these heterokaryons and demonstrated recombination both for auxotrophy and for pathogenicity. His data were scant, however, and he made no attempt to quantify his results.

Tuveson and Garber (94) claimed to have recovered heterokaryons between Race 1 and Race 2 of *F. oxysporum* f. sp. *pisi,* but direct proof of heterokaryosis was lacking. Both parental auxotrophic types were recovered among conidia of these heterokaryons, but often the ratios of the two were very disparate. Later (95) they recovered more equal ratios by supplementing the pairing medium with the requirements of the minority component. They also found a small proportion of prototrophic conidia (5–40 in 10^7). Such conidia developed into cultures with mostly prototrophic conidia. The few auxotrophic conidia were sometimes recombinant, but an exhaustive parasexual analysis was not made.

Garber and colleagues (40) tried to force heterokaryons between different formae speciales of *F. oxysporum.* They paired complementary auxotrophs of the formae speciales *pisi, niveum, lycopersici* and *conglutinans.* These formae speciales also had distinctive morphologies. At first the pairings produced only a thin growth, but later vigorously growing sectors appeared that usually resembled one of the component strains in morphology. Conidia from these sectors formed colonies on unsupplemented medium. Garber and coworkers offered these findings as evidence for heterokaryon formation between formae speciales. Their data were very incomplete, however, and their conclusions were based on very circumstantial evidence.

Dhillon and coworkers (25) attempted to force heterokaryons between different strains of *Gibberella fujikuroi (F. moniliforme)* and between *G. fujikuroi* and *F. oxysporum.* They obtained limited prototrophic growth, but their data did not confirm heterokaryosis because cross feeding could not be ruled out.

The cytology and asexual genetics of the aster wilt fungus *F. oxysporum* f. sp. *callistephi* were studied in a series of papers by Hoffmann (52, 53, 54, 55). He found that most hyphal cells were uninucleate and that anastomoses occurred exclusively between older hyphal cells. He concluded, therefore, that heterokaryosis did not arise in the tip cells as claimed by Buxton (15), but rather must be confined to those areas of the colony back from the edge. Most of his forced pairings of complementary auxotrophs did not grow on unsupplemented medium. In the few that did, the initial growth was thin, but later rapidly growing sectors appeared. Such sectors appeared sooner in pairings between auxotrophs from the same wild type than in those involving two different wild types. Conidia from the sectors were usually prototrophic but the same size as the parent conidia. A few sectors also yielded auxotrophic conidia, some with the growth requirements of both heterokaryon components. Hoffmann concluded that the sectors were actually already haploid recombinants from a parasexual process. Because the heterokaryon and diploid phase were not detected, they were considered to be extremely unstable and transitory.

Recently there have been reports of heterokaryosis and parasexuality in wilt Fusaria from cotton plants (2) and from tomato (86). In both cases complementary auxotrophs were paired and evidence for heterokaryosis was presented. In the tomato wilt both heterokaryon components were obtained from single hyphal tips. In the cotton wilt, recombinant haploids were recovered. In neither study were diploids detected directly.

The recovery of recombinants from pairings of complementary auxotrophs provides convincing evidence that at least transient heterokaryons can form in *Fusarium.* In most cases, however, a stable heterokaryotic culture did not form. Therefore, the nuclear condition of the heterokaryons and the mechanism of its formation have not been determined. Isolates of *Fusarium,* however, have many features in common with the wilt fungi *Verticillium dahliae* and *V. albo-atrum,* in which the nature of the heterokaryon has been established (80, 96). In these fungi the hyphal cells are also uninucleate, and fusions occur only between older hyphal cells. Heterokaryosis is confined to the fusion cells, but these cells do not proliferate further. The products of complementation formed within fusion cells are sufficient to support the growth of the remaining non-fused, homokaryotic cells. The heterokaryon is, therefore, a mosaic of heterokaryotic and homokaryotic hyphal cells.

The heterokaryon of *V. dahliae* is relatively stable and can be maintained indefinitely at 24 C (80). At 30 C, on the other hand, the heterokaryon ceases growth. This is because hyphal fusion decreases sharply at high temperatures, and without continued fusions the heterokaryon cannot be maintained. Hyphal fusion in *Fusarium* is also temperature sensitive (26). Perhaps stable heterokaryons of *Fusarium* could be obtained from pairings at rather low incubation temperatures. Lack of stable heterokaryosis may also be due to the inability of the heterokaryotic fusion cells to support the homokaryotic regions.

Diploids have been isolated only in *F. oxysporum* f. sp. *pisi.* In most cases the diploids were so unstable, their presence was detected only indirectly, i.e., by the recovery of recombinant haploids. In *V. dahliae* vegetative diploids were also unstable but could be maintained by hyphal tip transfer and high-temperature incubation (80). Perhaps under similar incubation temperatures the diploids of *Fusarium* could also be found and maintained.

The recovery of recombinant haploids also suggests a parasexual mechanism in *Fusarium* like that described in *Aspergillus* by Pontecorvo (78). In *Aspergillus* this parasexual cycle consists of heterozygous diploid formation from forced heterokaryons, somatic genetic recombination in the diploid nuclei, and return to the haploid stage by a non-meiotic mechanism, probably involving aneuploid intermediates. No workers have ever demonstrated somatic genetic recombination at the diploid level in *Fusarium*. The haploidization process has also not been characterized. All stages of the parasexual cycle in *Aspergillus* occur only rarely. In contrast, in *Fusarium* recombinants readily form from auxotroph pairings. Such a high frequency of recombination, however, has also been found in *V. albo-atrum,* where most steps of the parasexual cycle have been demonstrated (45). A clear demonstration of parasexuality in *Fusarium,* therefore, is still lacking.

Sexual Genetic Analysis

Like the asexual forms of *Fusarium,* many of the sexual forms are also quite variable, and some are plant pathogens. Of the nine species defined by Snyder and Hansen (92), all but two contain isolates with perfect states (90). In all cases the female gametangium is a protoperithecium, and any nucleated vegetative propagule can serve as the male (28). When the protoperithecia are fertilized, they develop into perithecia containing four- or eight-spored asci. These sexual forms are amenable to such procedures as controlled parentage crosses (89) and unordered tetrad (ascus) analysis (33, 41). Although sexual genetic analysis is far easier than asexual analysis, few such sexual isolates have been studied genetically, and most of them have been in the species *F. solani.*

In 1973 Matuo and Snyder (71) reported that six of ten formae speciales of *F. solani* had been induced to fruit. The fruiting structures in all six formae speciales were indistinguishable, and all were relegated to the species *Hypomyces solani.* All six fruiting formae speciales were heterothallic; that is, fertile perithecia developed only when two isolates of opposite mating type were paired. Two mating types, designated + and − (or A and a), were found. Many isolates were hermaphroditic, being able to function as both male and female, but isolates which were strictly unisexual male or female were also found in each of the six formae speciales. In spite of the fact that all six formae speciales were assigned to the same species, isolates from one forma specialis would not mate with isolates from another. Isolates within a forma specialis, however, usually mated with each other if they were of the opposite mating types. In the forma specialis *cucurbitae,* however, two distinct mating groups were found. Snyder and Matuo, therefore, called these seven groups "mating populations" (MP).

The first published genetic studies of sexual forms of *F. solani* were conducted by Dimock (28,29,30,31). He examined the fungus *Hypomyces ipomoeae* which is now classified as *H. solani* f. sp. *batatas.* This forma specialis includes isolates pathogenic to sweet potato and belongs to MP II of Matuo and Snyder. Dimock determined that the fungus was heterothallic and that the two mating types were controlled by two alleles at a single nuclear gene locus (28). He then crossed two spontaneous morphological variants, called *diffusa* and *rosa,* each with the normal type. About half the ascospore progeny of each cross were mutant and half were normal in morphology (29). Although his data were scanty, Dimock concluded that each variant was due to a single mutation in the nucleus (Table 27-2).

Later Dimock examined several other morphological variants in *H. ipomoeae.* One, called *aborta,* was recovered from a culture grown on a medium containing nearly toxic concentrations of a zinc salt (30). The variant was red and formed few conidia and no protoperithecia. When *aborta* was crossed with the normal strain, only four-spored asci developed and nearly all the ascospores formed normal colonies. Dimock concluded that the *aborta* trait was due to a single nuclear gene mutation and that nuclei containing this mutation could not form ascospores in the ascus. From some of the progeny of these crosses he also found other variant types. In a later paper (31) he chose seven of these variants that were morphologically distinct from each other and from the normal type. He crossed each to the normal type and concluded from progeny analyses that six of the

Table 27-2. Genetic mutants found in sexual species of *Fusarium*

Species	Mutant (Symbol)	Mutation type[a]	Rfs.
Hypomyces ipomoeae	*Diffusa (D)*	SNG	29
	Rosa (R)	SNG	29
	Aborta (1)	SNG	30
	Vinifera	SNG?	31
	Alba	SNG?	31
	Convoluta	SNG?	31
	Reverta	SNG?	31
	Restricta	SNG?	31
Hypomyces solani f. sp. *cucurbitae*	Loss of maleness (*m*)	SNG	44
	Loss of femaleness (*c*)	SNG	44
	White perithecia (*w*)	SNG	89
	Ascospore color (*h*)	SNG	42
	Sterile perithecia (*stp-4*)	SNG	42
	Button morphology (*b-55*)	SNG	42
	Tolerance to chlorinated nitrobenzenes		
	(*cnb-1,-2,-3*)	SNG	41
	(*cnb-4,-5*)	SNG	43
	Resistance to dodine (*dod-1,-2,-3,-4*)	SNG	61, 62
	Reduced virulence	SNG	63
	Conidial color	P	90
Hypomyces solani f. sp. *pisi*	Sporulating colonial (*sc*)	SNG	98
	Spider (*s*)	SNG	97
Hypomyces solani (*saprophyte*)	Dwarf colonies (*N*)	?	59
	Fluffy growth (*E*)	?	59

[a] Symbols: SNG = single nuclear gene, P = polygenes, ? mutation is heritable but its nature is not known.

seven were due to single gene mutations in the nucleus. His data, however, were very scanty. Moreover, all six variants were unstable and frequently reverted to a normal morphology. These findings are summarized in Table 27-2.

Since the time of Dimock's work most studies of *F. solani* have centered on the forma specialis *cucurbitae* that attacks cucurbits. In fact, this forma specialis has received more genetic study than any other sexual form of *Fusarium*. The sexual stage of this fungus, called *Hypomyces solani* by some researchers (90) and *Nectria haematococca* by others (23, 62), does not occur in nature, but it can be obtained in the laboratory. Apparently its absence in nature is due to the geographical separation of the opposite mating types. Two physiological races are defined in this forma specialis, and each behaves like a separate MP. Only the genetics of Race 1 (MP I) have been studied.

In 1946 Hansen and Snyder (44) reported the frequent occurrence of female unisexual types in *H. solani* f. sp. *cucurbitae,* race 1. These types could still form protoperithecia and resembled the hermaphrodite morphologically, but they could not function as a male. Less frequently, male unisexual types were found. They could be distinguished from the hermaphrodite by their increased mycelial growth and their inability to form sporodochia or protoperithecia. When either unisexual form was crossed to the hermaphrodite, unisexuality segregated in the progeny in a 1:1 ratio, that is, like a single nuclear gene mutation. A cross between a male unisexual and a female unisexual yielded not only the parental types but also the original hermaphroditic type and a new type called neuter because it could function neither as a male nor as a female. In these respects neuter forms behave exactly like asexual *Fusarium* isolates.

The genetic basis for unisexuality was contested for many years. Hirsch (50) claimed that the parental hermaphrodite had four chromosomes, one large, two medium-sized and one small. The female unisexual form arose when one of the medium-sized chromosomes was lost; the male arose when the other medium-sized chromosome was lost. The neuter form lacked both medium-sized chromosomes. Later, El-Ani (33, 34), analyzing unordered asci of unisexual crosses, concluded that chromosomal loss was not involved. Rather, female unisexuality was due to mutation at a nuclear gene locus called *m*, and male unisexuality was due to a mutation at a second linked locus called *c*. In their recent review Snyder and colleagues (90) accepted the viewpoint of El-Ani. Such unisexual mutants impose mating restrictions on this fungus which have been termed sex heterothallism (90).

Recently Bistis and Georgopoulos (8) examined the sexual development of *H. solani* f. sp. *cucurbitae* in more detail. They found that macroconidia in the proximity of protoperithecia of opposite mating type underwent marked morphological changes. The protoplasm of the spore was redistributed so that certain cells were engorged with protoplasm; papillae frequently developed on these cells. Subsequently, long, slender hyphae, called trichogynes, developed from the protoperithecia, grew in a semi-directed manner to the altered macroconidia, and fused with them. Microconidia did not alter visibly under these conditions, but fusions between them and the trichogynes were sometimes seen. If the conidia carried the *m* gene and were thus unable to function as males, all the above processes, including fusion, still occurred. Bistis and colleagues concluded that spore differentiation, trichogyne development and fusion were controlled by the mating-type locus. The *m* gene locus might control some later process in sexual development such as passage of the male nuclei through the trichogyne or subsequent nuclear fusion.

The species *H. solani* f. sp. *cucurbitae* is quite variable (79), and spontaneous morphological variants have been recovered among both natural isolates and laboratory cultures. The first trait studied extensively was perithecial color. Snyder (89) found isolates from nature with either red or white protoperithecia. When he crossed the two, he recovered fertile perthecia. The perithecia were always the same color as the protoperithecial, or female, parent. In other words, perithecial color was maternally influenced. However, half the ascospore progeny of these perithecia produced white protoperithecia, and half produced red. He concluded that perithecial color was determined by a single nuclear gene locus. Perithecial color segregated independently of mating type.

Several other spontaneous variants of *H. solani* f. sp. *cucurbitae* have also been shown to be true gene mutants. With the exception of conidial color, all mutations were in single nuclear genes (Table 27-2). Some instances of linkage have been reported. The *cnb*-2 and *cnb*-4 loci lie on opposite sides of the mating type locus, which in turn shows linkage to the centromere (42,43). The *c* and *m* loci lie close to, but on opposite sides of, a second centromere. The *stp*-4 and *w* loci show tight linkage with each other and also some centromere linkage. Georgopoulos (42) concluded that these various associations represented three different linkage groups.

Kappas and Georgopoulos (61, 62) published the first reports of induced mutation in a sexual form of *Fusarium*. They treated *H. solani* f. sp. *cucurbitae* with ultraviolet and X-ray irradiation and recovered 13 mutants that were resistant to the fungicide dodine. All 13 mutants were due to single lesions at any one of four unlinked nuclear gene loci (Table 27-2). The loci *dod*-1 and *dod*-2 were in the same linkage group as the mating type locus. The *dod*-4 locus was in the same linkage group as the *w* locus. Some of the dodine-resistant mutants were less virulent than their normal progenitors (63). One such avirulent mutant was subjected to unordered tetrad analysis. The reduction in virulence was found not to be due to the *dod* mutation but rather to mutation at another unlinked locus.

Very recently Bistis and Georgopoulos (8) recovered auxotrophic mutants from *H. solani* f. sp. *cucurbitae* irradiated with ultraviolet light. To my knowledge, this is the first published report of auxotrophs in a sexual form of *Fusarium*. No genetic analyses of these auxotrophs, however, were reported.

Hypomyces solani f. sp. *pisi* has also received some genetic study. Matuo and Snyder (70) isolated strains of this forma specialis from pea, ginseng, and mulberry. Isolates from one host would not always attack the other hosts, and some isolates were saprophytic. Yet all were classed in this forma specialis because they were interfertile and belonged to MP VI. This classification poses a dilemma: if the forma specialis *pisi* is defined on the basis of host range, i.e., the ability to attack pea, then isolates that are avirulent to pea or are saprophytic do not belong in this forma. All are, however, clearly genetically similar. As more is learned about the sexual forms of *Fusarium,* such dilemmas should become quite common.

Van Etten and Kolmark (98) used the mutagen nitrosoguanidine to induce morphological mutants in *Nectria haematococca.* One of the induced variants studied by Van Etten and Kolmark produced tightly growing, heavily sporulating colonies. It behaved like a single nuclear gene mutation in crosses (Table 27-2). A second induced mutant called *spider* (*s*) that was also found in this study has recently been shown to be due to a single gene mutation (97).

Some isolates of *H. solani* are homothallic, that is, monosporic cultures produce fertile perithecia. Such homothallic types are invariably saprophytic. The relationship between heterothallic and homothallic types is still unknown. Genetically they may be quite distinct; Hirsch (50) has reported different *n* chromosome numbers for the two types (Table 27-1). Hwang (59) examined a saprophytic, homothallic isolate of *H. solani* found on banana. The wild type N frequently produced dwarf colony types called B; and B forms, in turn, could give rise to E forms having greater aerial mycelium and sporulation Changes in the opposite direction were not seen. Because all three types fruited and produced ascospore progeny all like the respective parent, Hwang concluded that the B and E types were true genetic mutations. However, since the fungus is homothallic, controlled parent crosses were not made, and the exact nature of the genetic lesions was not determined.

A few other sexual states of *Fusarium* have received limited genetic study. The species *Calonectria rigidiuscula,* the perfect stage of *F. rigidiusculum,* also contains both homothallic and heterothallic isolates (37, 84). Like *H. solani,* the homothallic forms of this species are non-pathogenic. The heterothallic, pathogenic isolates produce four- or eight-spored asci, whereas the homothallis types have exclusively four-spored asci. Cook (21) described *Gibberella avenacea* as the sexual stage of *Fusarium roseum* f. sp. *cerealis* 'Avenaceum.' He noted that wild-type perithecia were blue, but that both white and flesh-colored types occurred in nature. No genetic analysis of perithecial color was made. Hsieh and coworkers (58) surveyed the species *G. moniliformis (=G. fujikuori)* which is the sexual stage of *F. moniliforme.* They found that the fungus was normally a heterothallic hermaphrodite, but male and female unisexual variants also occurred. Three mating populations like those in *H. solani* were also indicated. Initial studies of *G. avenacea* and *G. moniliformis* showed that they were genetically homogeneous. *Calonectria nivalis,* the sexual stage of *F. nivale,* was also reported to be quite uniform (22).

Genetics as a Tool in *Fusarium* Research

Many facets of the basic biology of the Fusaria—their physiology, taxonomy, variability and pathogenicity—are still unknown. In the last 20 years genetic techniques have been developed in other fungi and bacteria which should be applicable to *Fusarium* and which should shed light on these unknown areas. Such techniques, however, have rarely been exploited in *Fusarium.* I will discuss some of the exceptions here.

In 1958 McDonnell (72) used UV-induced mutants to explore the disease physiology of *F. oxysporum* f. sp. *lycopersici*. The mutants grew poorly on a medium containing pectin or sodium polypectate as the carbon source, and one of them was found to lack certain pectolytic enzymes, including endopolygalacturonase. This enzyme has been considered by many workers to be vital to symptom expression of wilt diseases. The mutant lacking enzymatic activity did not cause wilt or death like the parent, but it still produced such symptoms as epinasty, leaf yellowing, and vascular occlusion. Later Mann (67) also induced mutants in this fungus that could not utilize pectin. Of seven such mutants some produced little or no disease, but others were about as virulent as the parent. She concluded that pectolytic activity of the fungus was not essential for disease expression.

Auxotrophy in a pathogen may lead to a loss or diminution of pathogenicity. Proper nutrition is certainly necessary for the host and pathogen in disease development, but this effect of auxotrophy may be indirect. In other words, those pathways affected by the auxotrophic lesion may not be directly involved in pathogenesis. However, Rao and Shanmugasundarum (83) induced two auxotrophs in the cotton wilt Fusarium that may have affected the pathways of pathogenesis directly. One auxotroph required paraminobenzoic acid *(pab)*, the other required nicotinic acid *(nic)*. The two mutants were grown in liquid culture, and the culture liquid was extracted with petroleum ether. Cotton plant cuttings, when placed in the extract of the *nic* mutant, wilted more severely than those in extract of wild type. The extract of the *pab* mutant did not wilt the cuttings. Moreover, the *nic* mutant produced more fusaric acid than the wilt type. Fusaric acid may be an important phytotoxin in Fusarium wilt disease.

Recently Van Etten and Smith (99) showed that *F. solani* f. sp. *phaseoli* could detoxify the substance phaseollin, a phytoalexin produced by bean plants in response to infection by this fungus. This detoxification may explain why this fungus is not affected by phaseollin. Detoxification involves conversion of phaseollin to 1a-hydroxyphaseollone. Such a reaction is enzymatic and therefore surely under genetic control. No genetic studies of this system have been made.

Mutants carrying blocks in the normal sexual development of *Fusarium* have been found. They include the unisexual male and female types, the neuter types and variants with sterile perithecia (*stp*). Such mutants may permit an elucidation of the steps in sexual development in *Fusarium*. Other mutants of this type, however, are needed to clarify all stages in this development.

In my opinion, future genetic studies of the genus *Fusarium* should center on the sexual forms. Rather sophisticated genetic techniques, in particular mutant induction and gene mapping, are already available for these studies. Many of the most important features of the asexual forms—extreme variability, pathogenicity, and host specificity—also occur in some sexual states. Even such processes as heterokaryosis and parasexuality might be found in such isolates. Sexual forms could therefore serve as model systems for the entire genus.

Literature Cited

1. Abdalla, M. H. 1975. Sectoring of fungal colonies induced by low concentrations of fernasan. Mycopathologia 56:39–40.
2. Ahamed, N. M. M., and E. R. B. Shanmugasundaram. 1972. Biochemical and genetical studies in host-parasite relationship. Effect of heterokaryosis on the degree of virulence of nutritional mutants of *Fusarium vasinfectum*. Phytopathol. Z. 75:349–359.
3. Aist, J. R. 1968. The mitotic apparatus in fungi, *Ceratocystis fagacearum* and *Fusarium oxysporum*. J. Cell Biol. 40:120–135.
4. Aist, J. R., and P. H. Williams. 1972. Ultrastructure and time course of mitosis in the fungus *Fusarium oxysporum*. J. Cell Biol. 55:368–389.
5. Aist, J. R., and C. L. Wilson. 1968. Interpretation of nuclear division figures in vegetative hyphae of fungi. Phytopathology 58:876–877.

6. Armstrong, G. M., and J. K. Armstrong. 1950. Biological races of the *Fusarium* causing wilt of cowpeas and soybeans. Phytopathology 40:181–193.
7. Armstrong, G. M., J. D. MacLachlan, and R. Weindling. 1940. Variation in pathogenicity and cultural characteristics of the cotton-wilt organism, *Fusarium vasinfectum*. Phytopathology 30:515–520.
8. Bistis, G. N., and G. S. Georgopoulos. 1979. Some aspects of sexual reproduction in *Nectria haematococca* var. *cucurbitae*. Mycologia 71:127–143.
9. Bolton, A. T., and A. G. Donaldson. 1972. Variability in *Fusarium solani* f. *pisi* and *F. oxysporum* f. *pisi*. Can. J. Plant Sci. 52:189–196.
10. Booth, C. 1971. The genus *Fusarium*. Commonw. Mycol. Inst., Kew, Surrey, England. 237 p.
11. Bouhot, D. 1970. Variations induites du pouvoir pathogène chez *Fusarium oxysporum* f. sp. *melonis*. Ann. Acad. Sci. fenn. A., IV. 168:25–27.
12. Bouhot, D. 1980. Some aspects of the pathogenic potentialities in formae speciales and races of *Fusarium oxysporum* on Cucurbitaceae. Chap. 27 of this volume.
13. Brown, W. 1926. Studies in the genus *Fusarium*. IV. On the occurrence of saltations. Ann. Bot. 40:223–243.
14. Brown, W. 1928. Studies in the genus *Fusarium*. VI. General description of strains, together with a discussion of the principles at present adopted in the classification of *Fusarium*. Ann. Bot. 42:285–304.
15. Buxton, E. W. 1954. Heterocaryosis and variability in *Fusarium oxysporum* f. *gladioli* (Snyder & Hansen). J. Gen. Microbiol. 10:71–84.
16. Buxton, E. W. 1955. The taxonomy and variation in culture of *Fusarium oxysporum* from gladiolus. Trans. Brit. Mycol. Soc. 38:202–212.
17. Buxton, E. W. 1956. Heterokaryosis and parasexual recombination in pathogenic strains of *Fusarium oxysporum*. J. Gen. Microbiol. 15:133–139.
18. Buxton, E. W. 1958. A change of pathogenic race in *Fusarium oxysporum* f. *pisi* induced by root exudate from a resistant host. Nature 181:1222–1224.
19. Buxton, E. W. 1962. Parasexual recombination in the banana-wilt Fusarium. Trans. Brit. Mycol. Soc. 45:274–279.
20. Buxton, E. W., and V. Ward. 1962. Genetic relationships between pathogenic strains of *Fusarium oxysporum, Fusarium solani* and an isolate of *Nectria haematococca*. Trans. Brit. Mycol. Soc. 45:261–273.
21. Cook, R. J. 1967. *Gibberella avenacea* sp. n., perfect stage of *Fusarium roseum* f. sp. *cerealis* "Avenaceum." Phytopathology 57:732–736.
22. Cook, R. J., and G. W. Bruehl. 1966. *Calonectria nivalis,* perfect stage of *Fusarium nivale,* occurs in the field in North America. Phytopathology 56:1100–1101.
23. Cormack, M. W. 1951. Variation in the cultural characteristics and pathogenicity of *Fusarium avenaceum* and *F. arthrosporioides*. Can. J. Bot. 29:32–45.
24. Dassenoy, B., and J. A. Meyer. 1973. Mutagenic effect of benomyl on *Fusarium oxysporum*. Mut. Res. 21:119–120.
25. Dhillon, T. S., E. D. Garber, and E. G. Wyttenbach. 1961. Genetics of phytopathogenic fungi. VI. Heterocaryons involving *Gibberella fujikuroi* and formae of *Fusarium oxysporum*. Can. J. Bot. 39:785–792.
26. Dickinson, S. 1932. The nature of saltation in *Fusarium* and *Helminthosporium*. Minn. Agric. Exp. Stn. Bull. 88. 42 p.
27. Dimock, A. W. 1936. Variation in a species of *Fusarium* induced by high concentrations of zinc salts. Zentralbl. Bakteriol., Abt. 2, 95:341–347.
28. Dimock, A. W. 1937. Observations on sexual relations in *Hypomyces ipomoeae*. Mycologia 29:116–127.
29. Dimock, A. W. 1937. Hybridization experiments with natural variants of *Hypomyces ipomoeae*. Bull. Torrey Bot. Club 64:499–507.
30. Dimock, A. W. 1937. Hybridization studies on a zinc-induced variant of *Hypomyces ipomoeae*. Mycologia 29:273–285.
31. Dimock, A. W. 1939. Studies on ascospore variants of *Hypomyces ipomoeae*. Mycologia 31:709–727.
32. El-Ani, A. S. 1954. Chromosomes of *Hypomyces solani* f. *cucurbitae*. Science 120:323–324.
33. El-Ani, A. S. 1954. The genetics of sex in *Hypomyces solani* f. *cucurbitae*. Amer. J. Bot. 41:110–113.
34. El-Ani, A. S. 1956. Cytogenetics of sex in *Gibberella cyanogena* (Desm.) Sacc. Science 123:850.
35. El-Ani, A. S. 1956. Ascus development and nuclear behavior in *Hypomyces solani* f. *cucurbitae*. Amer. J. Bot. 43:769–778.
36. El-Ani, A. S. 1968. The cytogenetics of the conidium in *Microsporum gypseum* and of pleomorphism and the dual phenomenon in fungi. Mycologia 60:999–1015.
37. Ford, E. J., J. A. Bourret, and W. C. Snyder. 1967. Biologic specialization in *Calonectria (Fusarium) rigidiuscula* in relation to green point gall of cocoa. Phytopathology 57:710–712.
38. Fjeld, A., and M. M. Laane. 1970. The nuclear division and parasexual cycle in *Penicillium*. Genetica 41:517–524.
39. Follin, J. C., and E. Laville. 1966. Variations chez la *Fusarium oxysporum* f. *cubense* (Agent causal de la maladie de Panama du bannanier). Fruits 21:261–268.
40. Garber, E. C., E. G. Wyttenbach, and T. S. Dhillon. 1961. Genetics of phytopathogenic fungi. V. Heterocaryons involving formae of *Fusarium oxysporum*. Amer. J. Bot. 48:325–329.
41. Georgopoulos, S. G. 1963. Tolerance to chlorinated nitrobenzenes in *Hypomyces solani* f. *cucurbitae* and its mode of inheritance. Phytopathology 53:1086–1093.

42. Georgopoulos, S. G. 1963. Genetic markers and linkage relationships in *Hypomyces solani* f. *cucurbitae*. Can. J. Bot. 41:649–659.
43. Georgopoulos, S. G., and N. J. Panopoulos. 1966. The relative mutability of the cnb loci in *Hypomyces*. Can. J. Genet. Cytol. 8:347–349.
44. Hansen, H. N., and W. C. Snyder. 1946. Inheritance of sex in fungi. Proc. Nat. Acad. Sci. U.S. 32:272–273.
45. Hastie, A. C. 1967. Mitotic recombination in conidiophores of *Verticillium albo-atrum*. Nature 214:249–252.
46. Haymaker, H. H. 1928. Pathogenicity of two strains of the tomato-wilt fungus, *Fusarium lycopersici* Sacc. J. Agric. Res. 36:675–695.
47. Heale, J. B., A. Gafoor, and K. C. Rajasingham. 1968. Nuclear division in conidia and hyphae of *Verticillium albo-atrum*. Can. J. Genet. Cytol. 10:321–340.
48. Hildreth, R. C. 1958. Genetic variation and variability of *Fusarium solani* f. *pisi* and *F. oxysporum* f. *pisi* Race 2. Diss. Abst. 18:1196.
49. Hirsch, H. E. 1947. Cytological phenomena and sex in *Hypomyces solani* f. *cucurbitae*. Proc. Nat. Acad. Sci., Wash. 33:268–270.
50. Hirsch, H. E. 1949. The cytogenetics of sex in *Hypomyces solani* f. *cucurbitae*. Amer. J. Bot. 36:113–121.
51. Hirsch, H. E., W. C. Snyder, and H. N. Hansen. 1949. Chromosome numbers in the Hypocreaceae. Mycologia 41:411–415.
52. Hoffmann, G. M. 1964. Untersuchungen über die Kernverhältnisse bei *Fusarium oxysporum* f. *callistephi*. Arch. Mikrobiol. 49:51–63.
53. Hoffmann, G. M. 1966. Untersuchungen über die Heterokaryosebildung und den Parasexualcyclus bei *Fusarium oxysporum*. I. Anastomosenbildung im Mycel and Kernverhältnisse bei der Conidienentwicklung. Arch. Mikrobiol. 53:336–347.
54. Hoffmann, G. M. 1966. Untersuchungen über die Heterokaryosebildung und den Parasexualcyclus bei *Fusarium*. II. Gewinnung und Identifizierung auxotrophen Mutanten. Arch. Mikrobiol. 53:348–357.
55. Hoffmann, G. M. 1967. Untersuchungen über die Heterokaryosebildung und den Parasexualcyclus bei *Fusarium oxysporum*. III. Paarungsversuche mit auxotrophen Mutanten von *Fusarium oxysporum* f. *callistephi*. Arch. Mikrobiol. 56:40–59.
56. Hoffmann, G. M. 1968. Kernverhältnisse bei pflanzenpathogenen imperfekten Pilzen, insbesondere Arten der Gattung *Fusarium*. Zentralbl. Bakteriol., Abt. 2, 122:405–519.
57. Howson, W. T., R. C. McGinnis, and W. L. Gordon. 1963. Cytological studies on the perfect stages of some species of *Fusarium*. Can J. Genet. Cytol. 5:60–64.
58. Hsieh, W. H., S. N. Smith, and W. C. Snyder. 1977. Mating groups in *Fusarium moniliforme*. Phytopathology 67:1041–1043.
59. Hwang, S. 1948. Variability and perithecium production in a homothallic form of the fungus *Hypomyces solani*. Farlowia 3:315–326.
60. Imshenetski, A. A., and O. M. Ulianowa. 1962. Experimental variability in *Fusarium moniliforme* Sheld. leading to the formation of gibberellins. Nature 195:62–63.
61. Kappas, A., and S. G. Georgopoulos. 1968. Radiation-induced resistance to dodine in *Hypomyces*. Experientia 24:181–182.
62. Kappas, A., and S. G. Georgopoulos. 1970. Genetic analysis of dodine resistance in *Nectria haematococca* (Syn. *Hypomyces solani*). Genetics 66:617–622.
63. Kappas, A., and S. G. Georgopoulos. 1971. Independent inheritance of avirulence and dodine resistance in *Nectria haematococca* var. *cucurbitae*. Phytopathology 61:1093–1094.
64. Koenig, R., and F. L. Howard. 1962. Nuclear division and septum formation in hyphal tips of *Fusarium oxysporum*. Amer. J. Bot. 49:666 (Abstr.).
65. Kumari, L., J. R. Decallonne, and J. A. Meyer. 1975. Deoxyribonucleic acid metabolism and nuclear division during spore germination in *Fusarium oxysporum*. J. Gen. Microbiol. 88:245–252.
66. MacNeil, B. H., and J. V. Sabanayagam. 1968. The induction of dodine tolerance in *Fusarium oxysporum* f. *melonis:* a technique applicable to the study of the bionomics of soil-borne fungal pathogens. Can. J. Microbiol. 14:1262–1263.
67. Mann, B. 1962. Role of pectic enzymes in the Fusarium wilt syndrome of tomato. Trans. Brit. Mycol. Soc. 45:169–178.
68. Massey, L. M. 1926. Fusarium rot of Gladiolus corms. Phytopathology 16:509–523.
69. Mathur, B. L., and N. Prasad. 1964. Variation in *Fusarium oxysporum* f. *cumini* in nature. Indian J. Agric. Sci. 34:273–277.
70. Matuo, T., and W. C. Snyder. 1972. Host virulence and the Hypomyces stage of *Fusarium solani* f. sp. *pisi*. Phytopathology 62:731–735.
71. Matuo, T., and W. C. Snyder. 1973. Use of morphology and mating populations in the identification of formae speciales in *Fusarium solani*. Phytopathology 63:562–565.
72. McDonnell, K. 1958. Absence of pectolytic enzymes in a pathogenic strain of *Fusarium oxysporum* f. *lycopersici*. Nature 182:1025–1026.
73. Mesterhazy, A. 1973. The morphology of an undescribed form of anastomosis in *Fusarium*. Mycologia 65:916–919.
74. Miller, J. J. 1945. Studies on the Fusarium of muskmelon wilt. I. Pathogenic and cultural studies with particular reference to the cause and nature of variation in the causal organism. Can. J. Res. C. 23:16–43.
75. Miller, J. J. 1946. Cultural and taxonomic studies on certain Fusaria. I. Mutation in culture. Can. J. Res. C. 24:188–212.
76. Miller J. J., L. W. Koch, and A. A. Hildebrand. 1947. A comparison of cultural methods for the

maintenance of çertain economic fungi. Sci. Agric. 27:74–80.
77. Oswald, J. W. 1949. Cultural variation, taxonomy and pathogenicity of *Fusarium* species associated with cereal root rots. Phytopathology 39:359–376.
78. Pontecorvo, G. 1956. The parasexual cycle in fungi. Annu. Rev. Microbiol. 10:393–400.
79. Prasad, N. 1949. Variability of the cucurbit root-rot fungus, *Fusarium (Hypomyces) solani* f. *cucurbitae.* Phytopathology 39:133–141.
80. Puhalla, J.E., and J.E. Mayfield. 1974. The mechanism of heterokaryotic growth in *Verticillium dahliae.* Genetics 76:411–422.
81. Punithalingam, E. 1972. Cytology of *Fusarium culmorum.* Trans. Brit. Mycol. Soc. 58:225–230.
82. Rai, J. N., and R. P. Singh. 1973. Fusarial wilt of Brassica juncea. Indian Phytopathol. 26:225–232.
83. Rao, K. R., and E. R. B. Shanmugansundarum. 1966. Biochemical genetic studies on host-parasite relationship. Pathogenicity of two mutants of *Fusarium vasinfectum* Atk. on cotton plants. Experientia 22:138–139.
84. Reichle, R. E., and W. C. Snyder. 1964. Heterothallism and ascospore number in *Calonectria rigidiuscula.* Phytopathology 54:1297–1299.
85. Robinson, P. M. 1972. Isolation and characterization of a branching factor produced by *Fusarium oxysporum.* Trans. Brit. Mycol. Soc. 59:320–322.
86. Sanchez, L. E., J. V. Leary, and R. M. Endo. 1976. Heterokaryosis in *Fusarium oxysporum* f. sp. *lycopersici.* J. Gen. Microbiol. 93:219–226.
87. Schneider, R. 1958. Untersuchungen über Variabilität und Taxonomie von *Fusarium avenaceum* (Fr.) Sacc. Phytopathol. Z. 32:95–126.
88. Singh, U. P. 1973. Effect of acriflavine on UV-induced mutants of Fusarium species. Mycopathol. Mycol. Appl. 50:183–193.
89. Snyder, W. C. 1940. White perithecia and the taxonomy of *Hypomyces ipomoeae.* Mycologia 32:646–648.
90. Snyder, W. C., S. G. Georgopoulos, R. K. Webster, and S.N. Smith. 1975. Sexuality and genetic behavior in the fungus *Hypomyces (Fusarium) solani* f. sp. *cucurbitae.* Hilgardia 43:161–185.
91. Snyder, W. C., and H. N. Hansen. 1940. The species concept in *Fusarium.* Amer. J. Bot. 27:64–67.
92. Snyder, W. C., and H. N. Hansen. 1954. Variation and speciation in the genus *Fusarium.* Ann. N.Y. Acad. Sci. 60:16–23.
93. Subramanian, C. V. 1951. Is there a "wild-type" in the genus *Fusarium?* Proc. Nat. Inst. Sci., India 17:403–411.
94. Tuveson, R. W., and E. D. Garber. 1959. Genetics of phytopathogenic fungi. II. The parasexual cycle in *Fusarium oxysporum* f. *pisi.* Bot. Gaz. 121:74–80.
95. Tuveson, R. W., and E. D. Garber. 1961. Genetics of phytopathogenic fungi. IV. Experimentally induced alterations in nuclear ratios of heterocaryons of *Fusarium oxysporum* f. *pisi.* Genetics 46:485–492.
96. Typas, M. A., and J. B. Heale. 1976. Heterokaryosis and the role of cytoplasmic inheritance in dark resting structure formation in *Verticillium* spp. Mol. Gen. Genet. 146:17–26.
97. Van Etten, H. D. 1978. Identification of additional habitats of *Nectria haematococca* mating population VI. Phytopathology 68:1552–1556.
98. Van Etten, H. D., and H. G. Kølmark. 1977. Modifying the growth habit of the filamentous fungus *Fusarium solani* to facilitate replica plating procedures. Can. J. Bot. 55:848–851.
99. Van Etten, H. D., and D. A. Smith. 1975. Accumulation of anti-fungal isoflavonoids and 1a-hydroxyphaseollone, a phaseollin metabolite, in bean tissue infected with *Fusarium solani* f. sp. *phaseoli.* Physiol. Plant Pathol. 5:225–237.
100. Waite, B. H., and R. H. Stover. 1960. Studies on Fusarium wilt of bananas. VI. Variability and the cultivar concept in *Fusarium oxysporum* f. *cubense.* Can. J. Bot. 38:985–994.
101. Wellman, F. L., and D. J. Blaisdell. 1941. Pathogenic and cultural variation among single-spore isolates from strains of the tomato-wilt Fusarium. Phytopathology 31:103–120.

28 Nucleocytoplasmic Interactions Implicated in Differentiation in *Nectria haematococca*

Marie-Josée Daboussi-Bareyre and Denise Parisot

Introduction

The presence of extrachromosomal heredity factors has been shown in the majority of fungi used in genetic research. The existence of these factors has been deduced from criteria such as non-mendelian segregation during sexual reproduction, somatic segregation, and the spreading of a different phenotype from one mycelium to another following hyphal anastomosis (25). However, the positive identification of the molecular nature of the cytoplasmically transmitted factors would lead to a definite conclusion (7). The non-mendelian inherited characters in *Saccharomyces cerevisiae* (19) and in *Neurospora crassa* (33) are due to alterations of the mitochondrial DNA. Virus-like particles containing double-stranded RNA are found consistently associated with the "killer" factor in yeasts (34) and *Ustilago* (26). However in the majority of cases such as senescence in *Podospora anserina* (31) and *Aspergillus glaucus* (25) or the formation of the "red" sectors of *A. nidulans* (25), the molecular nature of the determinants which are distributed independently of the nuclear chromosomes is not known.

These extrachromosomal factors are not autonomous: we know for example that in *S. cerevisiae,* nuclear mutations cause the disappearance of the virus-like particles and of the "killer" character associated with them (34). Similarly the biogenesis of mitochondria is controlled by nuclear as well as organelle genes (33). Nucleocytoplasmic interactions have also been found for factors whose physical nature is not known; for example, the formation of the cytoplasmic incompatibility factor S in *P. anserina* is strictly dependent on the presence of the nuclear gene *S* (2). The presence of such relationships between nuclear genes and extrachromosomal factors and the possibility of biochemical and genetical analyses in fungi (11) make these questions of cytoplasmic heredity of particular interest in view of the problems concerning the occurence and the stability of differentiated states in the eukaryotic organisms.

Numerous species of fungi have the capability of varying their morphological or physiological characters; some of these changes have been attributed to a mutation or to a heterokaryotic situation which may be followed by mitotic recombination; some others show certain peculiarities which suggest that a nuclear phenomenon is not responsible for these modifications but rather that they have a cytoplasmic origin. This is particularly the case with the sudden changes affecting the morphology and the growth of *Pestalozzia annulata* (10), *Curvularia pallescens* (13), *Hypomyces ipomoeae* (4), and *Nectria haematococca* (27). In these organisms, when cultures from a single uninucleate cell are placed under uniform growth conditions, a mycelial type occurs which is substituted for the

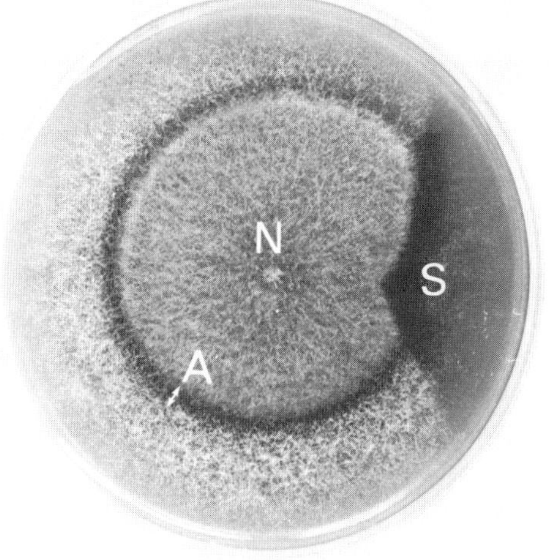

Fig. 28-1. *Wild* culture of *Nectria haematococca* showing the 3 mycelial types: normal (N), ring (A) and sector (S).

original type; the new type appears suddenly and spreads to the adjoining filaments. In the last two species, the sexual stage permits the analysis of interactions between nuclear genes and cytoplasmic factors. The study of two transmittable differentiations in *N. haematococca* is the subject of this article.

Characteristics of the Two Differentiations in *Nectria haematococca*

When grown on a potato dextrose agar medium containing 2% glucose at 26 C, *wild* thalli of a homothallic isolate of *N. haematococca,* started from uninucleate spores, produce aerial mycelium which is dense and white, later becoming grayish and spotted with violet. These thalli grow 4–4.5 mm per day. Two morphological modifications appear randomly in the submarginal region of these cultures as small brown spots in which hyphal elongation is reduced. The first develops as an arc 2–3 mm in width; this arc does not increase in width and spreads rapidly from either side of its point of origin, soon encircling the whole thallus with a dark ring (Fig. 28-1). The second differentiation appears as a wedge-shaped brownish sector (Fig. 28-1). These two differentiations can be observed on the same thallus, provided they appear at points fairly well separated on the margin; they do not invade one another and each limits the extension of the other.

Once initiated, the ring and the sector progressively invade the margin of the culture at a speed which determines their shape: this speed is respectively 20 (in the case of the ring) and 2 times (in the case of the sector) the growth rate of the hyphae. These differentiated areas differ from the normal mycelium originating directly from spores by three morphologic characteristics: fewer aerial hyphae, a decrease in the linear growth rate of the hyphae, and diffusion of a brown pigment into the medium (6, 27).

Stability of the Differentiated States

Massive inocula (1 mm x 1 mm) taken from differentiated areas immediately produce differentiated subcultures. However, normal subcultures can be obtained under the influence of various factors, reducing, at least momentarily, the metabolic activity of the cells (aging of the mycelium, exhaustion of the substrate, presence of inhibitors of synthesis such as actinomycine D and cycloheximide) (6, 12). Apparently this is due to an inhibition of the expression of these differentiations, since the newly formed mycelium can regain the capability of differentiation as soon as the constraints are eliminated.

The reversion to the normal state is observed when inocula from the differentiated areas are very small. Thus, 50% of small hyphal fragments (containing 3 to 5 cells) taken in a sector which is 24 hr old, and 97% of fragments obtained from rings of the same age, regenerate normal thalli. With the other small fragments (respectively 50% and 3%) a new phenomenon appears: a retarded expression of the differentiations. The differentiated phenotypes, ring or sector, appear only after a delay of approximately 24 hr, prior to which the young thalli express the normal phenotype. An analogous phenomenon has been observed during germination of the microconidia of *Podospora anserina* carrying the cytoplasmic factor S: the incompatibility reactions appear only after a growth period of a few hours to 2–3 days (2). The hypothesis that a threshold concentration of cytoplasmic factors is necessary may be the reason for the retarded expression of the differentiations in *Nectria haematococca.* Arguments supporting this hypothesis have been obtained in the study of the conditional mutant *727,* which will be dealt with in the second part of this article.

The fragmentation of hyphae into nucleocytoplasmic units, capable of regenerating thalli which can transmit the potentiality contained at the time of transfer, is equivalent to a sampling of the cytoplasm (3). The presence of hyphal fragments which do not regenerate a differentiated mycelium shows that the cytoplasm is heterogenous. The ring and sector zones are thereby shown to be mosaics of normal cells and differentiated cells.

A more critical sampling of the cytoplasm can be done by means of microconidia obtained from 24-hr-old differentiated zones: 95% of the conidia from young sectors give rise to normal thalli; the remaining 5% give rise to thalli where the phenotypic expression of the differentiation is delayed as with hyphal fragments obtained from sectors or rings. It seems, therefore, that the transmission of differentiation depends on the amount of cytoplasm used in the transfer; the greater the amount of cytoplasm present in the transfered material, the greater the chances of the differentiation being transmitted. However, isolated spores do not give the same type of sampling as hyphal fragments. Conidia are reproductive structures and during their formation the cytoplasm and/or organelles included within the spore may be somewhat selected. Ascospores never transmit these differentiations. This may be due to the ontogeny of these spores (22) and/or the length of the process (about 20 days), since it is known that the old differentiated zones give rise to a higher proportion of normal thalli than young sectors or rings. This is the result of a decreasing proportion of differentiated cells due to proliferation of normal mycelium coming from either the normal cells of the mosaic or germinating conidia. Therefore, the ability to transmit the differentiation is reduced (28). It is important to remember that the normal thalli which come from the "dedifferentiation" process have not completely lost their capacity to differentiate; the ring and the sector always reappear after more or less time.

Evidence for a Cytoplasmic Origin of the Contagious Character of these Differentiations

The differentiated areas have a speed of radial extension less than that of the normal mycelium, and yet they expand. This invasive condition is due to the infection of the normal cytoplasm by the differentiated cytoplasm. This infection does not take place through the substrate; any barrier which maintains continuity of the agar substrate but prevents anastomoses between adjoining filaments stops the tangential extension of the differentiation which takes place only through anastomoses. On the other hand, these

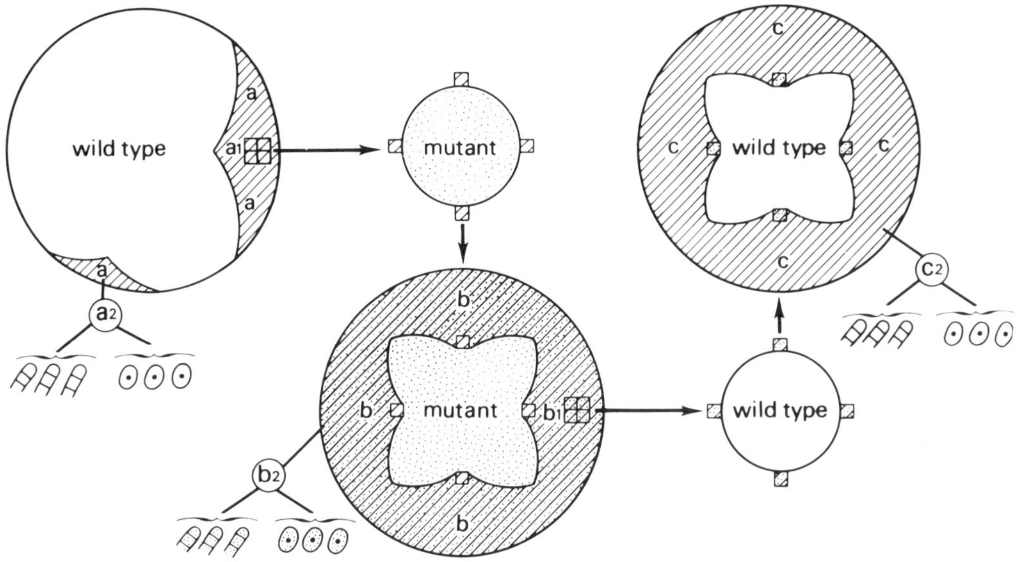

Fig. 28-2. Scheme showing the transmission of the sector differentiation in the absence of heterokaryosis. Reciprocal transmission experiments between two genetically different cultures: *wild* type and a morphological mutant.

a) Sector differentiation which appeared spontaneously in the *wild* culture
  a_1: massive inoculum
  a_2: conidia and hyphal fragments which gave rise to *wild* cultures
b) Sector differentiations induced on the mutant
  b_1: massive inoculum
  b_2: conidia and hyphal fragments which gave rise to *mutant* cultures
c) Sector differentiations induced on the *wild* culture
  c_2: conidia and hyphal fragments which gave rise to *wild* cultures

differentiations can be induced experimentally within normal thalli by putting large inocula (4 mm^2 agar blocks taken from rings or sectors of suitable age) alongside the growing margin of a young mycelium. The development of differentiations is similar in all respects to those which appear spontaneously. The same result is obtained when the receptor thallus and the donor inoculum come from two genetically different clones (Fig. 28-2); conidia or hyphal fragments obtained from induced differentiated zones express only the genotype of the receptor clone. This suggests that there is no migration of nuclei from the donor inoculum to the receptor normal hyphae and that the transmissible factors are of a cytoplasmic nature (6). Since *N. haematococca* is homothallic, confirmation of the extrachromosomal nature of the determinants of the differentiations cannot be obtained by studying reciprocal crosses. The major argument in favor of the cytoplasmic nature of these determinants is the association of the differentiated phenotype with the nuclei of the normal thallus following heterokaryosis localized at the point of anastomosis, which is the first stage of the infection. It is this very same infection test which allows us to show that each of these differentiations is due to a specific factor; in fact, all the inocula taken from the margin of sectors and brought to the margin of normal thalli produce only sectors in the latter. In the same manner all inocula obtained from the ring only induce rings in those thalli that receive them. The hyphae of *N. haematococca* can thus contain two types of factors transmitted independently of nuclear markers (29).

The cytoplasmic transmission of these two differentiations does not mean that the nucleus plays no role in the phenomenon. The role of the nucleus is shown through the mutation of specific nuclear genes leading to the disappearance or the modification of these differentiations.

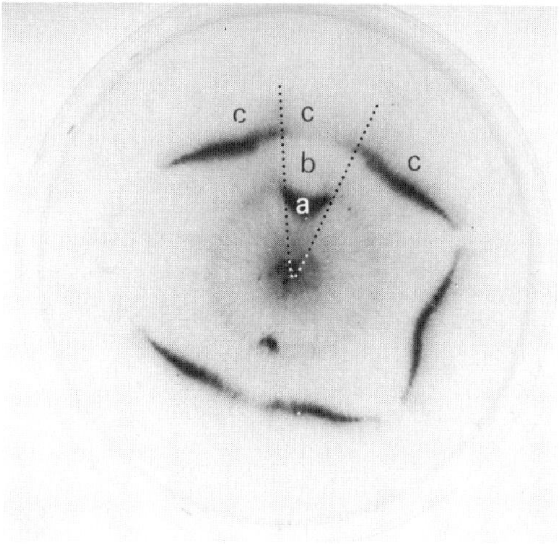

Fig. 28-3. Influence of temperature on the expression of the sector differentiation.
a) Sector initiated at 19 C
b) Normal zone grown at 26 C
c) Expression and propagation of the sector along the same radii as (a) upon return to 19C

Genetic Control of the Differentiated States

In order to show that a phenomenon of differentiation is controlled by nuclear genes one has to obtain mutants which normally do not produce the phenomenon being studied. Then one determines through genetic analysis which loci have mutated and what are the consequences of each of these mutations on the process of differentiation.

Search for Mutants

Microconidia of the *wild* clone (normal phenotype) suspended in distilled water were treated with concentrations of 100, 150 and 200 μg of N-methyl-N′-nitro-N-nitrosoguanidine per milliliter or exposed to 4500 ergs/mm^2 of UV light at 2537 Å (16). A few days after the treatment, thalli from the surviving spores were inoculated individually with massive inocula taken from rings and sectors (Fig. 28-2). We retained mutants incapable of developing rings or sectors after these inoculations. They can be grouped into three categories depending upon whether one, the other, or both differentiations are affected.

Properties of the Mutant Clones

Mutations simultaneously preventing the two differentiations Five mutants of this type have been obtained (1, 16). They differed morphologically from the *wild* clone by producing more aerial mycelium, lacking pigmentation, and suppressing the sexual stage through self fertilization. At 26 C they did not show either one of these two differentiations, spontaneously or following experimental inoculation. However, three of them were capable of differentiating at temperatures other than 26 C (1).

The effect of temperature on the expression of these differentiations was carefully studied in mutant *727* (14). This mutant shows the differentiations only between 14 and 19 C. Experiments in which *727* thalli carrying sectors were placed successively at 19 C, then 26 C, and back to 19 C gave information on the fate of sector factor in the mycelium at 26 C. A sector initiated at 19 C (Fig. 28-3a) stopped its propagation when the thallus was transferred to 26 C (Fig. 28-3b); the corresponding area grew with a normal phenotype. If the culture was brought back to 19 C, the sector started to spread again from its two edges, being shifted along two radii without any noticeable lateral extension during the 26 C period (Fig. 28-3c). The same situation was observed if zone (a) was removed after growth at 26 C. These experiments showed that the determinant persisted at 26 C, since a differentiated zone reappeared as soon as it was brought back to 19 C. The determinant present at 26 C in zone (b) did not give rise to the characteristic symptoms of the differentiation. Finally, even though the anastomoses appeared to be functional, the lateral propagation of determinant was strongly reduced or even suppressed at 26 C. At this temperature everything goes along as if the sector factor migrated passively towards the apices.

A subculture taken from the margin of a *727* sector incubated at 19 C gave rise to a differentiated thallus with an extension rate slower than that of a normal control thallus at the same temperature. Thus the synthesis of the determinant was accompanied by a slowing-down of growth like in the *wild* type. An identical subculture brought to 26 C generated a thallus with a normal phenotype and grew twice as fast as the differentiated control at 19 C and at the same speed as a normal *727* culture at 26 C.

Infection tests showed that this 26 C thallus retains the determinant. One can suppose that the normal phenotype is due to a strong reduction of synthesis of the determinant at 26 C. If this is the case, one should be able to follow the kinetics of disappearance of the determinant by evaluating at various times the quantity present in thalli growing at 26 C. The presence or absence of determinant in various regions of the thallus was checked by using the infection test.

The percentage of infectious inocula taken from differentiated thalli grown for 1–9 days at 19 C remained constant—around 95% —while the number of infectious inocula taken from thalli with the normal phenotype decreased with time at 26 C. This decrease reflected the dilution of the determinant according to the length of exposure to the restrictive conditions. In addition, the distribution of the determinant in the thallus grown at 26 C depended on the zone considered: the infectious inocula were proportionately more numerous in the periphery than in the center of the cultures, suggesting again that the determinant is pulled passively towards the growing margin.

The increasing rarity of the determinant at 26 C could be due to dilution, destruction, or to both processes. In order to determine which was the case, dilution was reduced as much as possible by using a nystatin substrate which stops the growth. Differentiated cultures were incubated 5 days at 19 C and then transferred on nystatin medium either at 19 C or at 26 C. Every day for 11 days inocula were taken from these nystatin-stopped thalli and tested for their infectious ability. These experiments showed that in the absence of growth of the thalli, the percentage of infectious inocula did not greatly vary during incubation at 26 C. Further, the similar evolution of these percentages at 19 and 26 C indicated that there was no selective destruction of the infectious factor at 26 C. The diminution of the quantity of determinant observed at 26 C can therefore be attributed to dilution. At 26 C the determinant is neither synthesized nor destroyed; the infectious factor present in the initial inoculum is gradually diluted as the thallus grows until a threshold is reached below which the inocula are no more infectious.

The observations made on the sectorial differentiation in this mutant indicate that it does not occur at 26 C because the synthesis of the corresponding infectious factor is stopped, and not because it is masked at this temperature. When there is differentiation (*wild* thalli, *727* thalli at 19 C), the morphological alterations appear as a result of the synthesis of infectious factor and not as a consequence of its presence in the cells (14).

Mutations affecting only the ring differentiation The ten mutants isolated can be classified into two groups. Eight isolates, a_1, a_2, ... a_8, are similar in morphology to the *wild* thalli and show the sector differentiation, but are incapable of ring differentiation spontaneously or through experimental infection, regardless of incubation temperature. Two other isolates, *58* and *105*, are red pigmented and never show the ring differentiation even though they possess the cytoplasmic factor that causes it. Inocula taken anywhere in a pigmented *58* thallus and placed at the margin of suitable receptor thalli all induce the ring in the receptor. Thus, the synthesis of the infectious ring factor occurs not only in the margin of *58* thalli but within the whole mycelium as well, including the sectors. Contrary to the *wild* type, this factor is not exhibited by the appearance of a ring and the slowing-down of growth but simply by the diffusion of a red pigment. The transmissible character and the pigmentation are perpetuated by continual mass transfers, but do not appear immediately after the germination of microconidia or ascospores. Three to four days must elapse before the appearance of the first red spots, which spread over the entire culture in the next 24 hr. Any inoculum taken from the red zone induces ring differentiation in the appropriate receptor, while any inoculum taken from a region that is still white does not. This red pigmentation is the expression of the synthesis of the ring factor in these mutants. Rings obtained from a *wild* receptor following infection by *58* or *105* mycelium are identical to those which appear spontaneously in the receptor. Therefore it is not possible to determine if the mutations affect the ring factor or a cellular constituent necessary for its expression (15).

Mutation affecting only the sector differentiation Only one mutant, *789,* has been isolated that is red pigmented, forms the ring as *wild* type, but never produces the sector. The sector factor is nevertheless present like the ring factor in clones *58* and *105* (1,16).

Genetic Analysis of the Mutants

***mutant* x *wild* crosses** Mutants were crossed with the *wild* clone in order to determine the number of loci modified in each isolate. The genetic analysis is done with random ascospores because the less than 5% hybrid perithecia cannot be distinguished from those produced by the homothallic *wild* isolate and each ascus rarely gives 8 germinating ascospores. Several hybrid perithecia from each cross were examined and the segregation of markers noted for each in a sample of approximately 300 ascospores.

For every cross studied we have found only ascospores giving rise to thalli similar to the two parents in hybrid perithecia. The proportions are generally mendelian, with about 50% of the ascospores giving rise to *wild* cultures and 50% giving rise to mutant cultures (Table 28-1). However, we have found an anomalous type during these analyses. In certain crosses between the *wild* type and a self-sterile mutant, we obtained hybrid perithecia with an excess of *wild* type ascospores, suggesting that the perithecia of *Nectria haematococca* could result from multiple fertilizations (5). This analysis leads to the simple conclusion that the 1 : 1 segregation means that these mutants differ from the *wild* type only by a single gene. The mutation of a single gene leading to the simultaneous absence of the ring and the sector (see above) suggests a common stage in the development of these two differentiations.

Table 28-1. Progeny of crosses between the *wild* culture and mutant cultures *58, 727*

		Number of ascospores giving rise to	
Crosses	Perithecia	*wild* thalli	*mutant* thalli
wild × *58*	1	231	218
	2	209	179
	3	205	202
wild × *727*	1	113	125
	2	274	261
	3[a]	391	154

[a]this perithecium showed an excess of *wild* spores.

***mutant* x *mutant* crosses** These crosses were performed to determine the relationships between the genes altered by the mutations previously described. Since most of the mutants are either self-sterile or less sterile than the *wild* type, the percentage of hybrid perithecia is larger than in the cross *mutant* x *wild*; it may be 100% in certain crosses between two self-sterile mutants. From these crosses we concluded that: (i) The five mutations which prevent the formation of both the sector and ring correspond to the alteration of four genes, two of which are linked. (ii) The 10 mutations affecting ring formation are localized on the same chromosome in a region whose length is inferior to one unit of recombination, because in crosses using two mutants of this type, not a single thallus developed the ring from 200 ascospores that were tested; this locus, called "A," is independent of the first four. (iii) The mutation affecting sector formation is controlled by a gene independent of the other genes described above. These mutagenic experiments, followed by genetic analysis, show that numerous genes are implicated in these two differentiations. The mutations obtained occur in various regions of the genome except those affecting specifically the ring which are all localized in a chromosomal region called *locus A*.

Nucleocytoplasmic Interactions in Heterokaryons

Formation of heterokaryons following fusion of cells is one method used for studying the influence of cytoplasm on the nuclear genome and vice-versa. It allows us to observe the behavior of a nucleus in the presence of cytoplasmic products resulting from the activity of another type of nucleus differing from the first either through mutations or by its gene expression pattern (23). This is also a method to test if mutations localized in the same chromosomal region affect one gene or two adjacent genes (7). These two reasons led us to analyze at first the relationships between the three allelic forms found at *locus A*.

Formation of Heterokaryons in *Nectria haematococca*

Heterokaryosis is not spontaneous in *N. haematococca*. One must force nuclei of different origins to cohabit in the same cytoplasm so that the association has a considerable selective advantage over each homokaryon. This is brought about by creating on a minimal medium a balanced heterokaryon between two cultures bearing non-allelic auxotrophic mutations in addition to the mutations to be studied (8, 9).

The arginine-requiring mutant nuclei *arg154* and *arg13* were introduced respectively into the clones *wild* and a_1; the clone *58* was made auxotrophic for lysine by incorporating the mutation *lys255* into it. These auxotrophic markers did not change the differentiation pattern of the strains, but the mycelia of double mutants were smoother than those of the corresponding simple prototrophic mutants (17).

The pairings were carried out according to the experimental scheme shown in Fig. 28-4A. The mycelium which arose at the junction of the two partners was carefully ground up, and apical as well as intercalary hyphal fragments having at least 3 cells were plated out on various media. Some were placed on minimal medium, in order to determine the proportion of the prototroph cells which contained both nuclear types, and others were placed on a complete medium, in order to estimate the proportions of the two auxotrophic partners. The different nuclear types present in the prototrophic cultures coming from isolated fragments were selected and analyzed by making cultures from uninucleate microconidia.

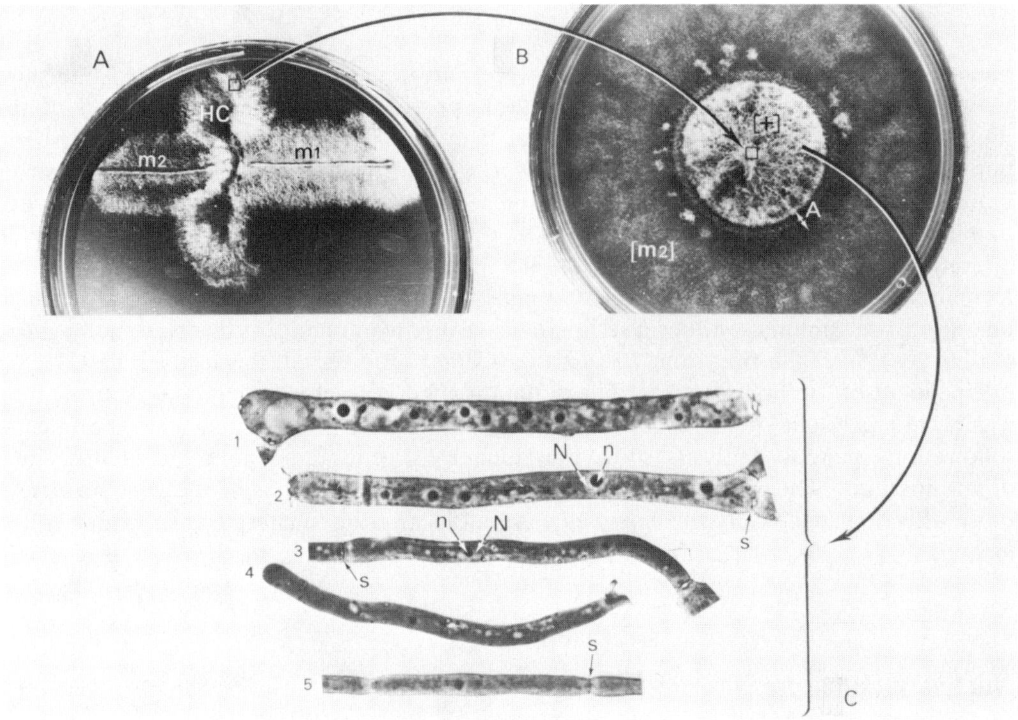

Fig. 28-4. A) Formation of the heterokaryotic mycelium (HC) during confrontation of cultures m1 and m2.
m1: cultures having allele *58*$^{+}$ and requiring arginine: *arg154(58*$^{+}$*)*
m2: culture having allele *58* and requiring lysine: *lys255 (58)*
The cultures m1 and m2 are grown on a cellophane film and placed on a complete medium for 48 hr. Then when the cultures are in contact with each other, the cellophane is transferred to a minimal medium which only allows growth of the heterokaryotic mycelium.
B) Heterokaryotic culture *lys255(58)/arg154 (+)* with a *wild* phenotype and showing a spontaneous ring (A) beyond which are only nuclei m2.
C) Examples of cellular types found in the heterokaryotic mycelium; cells with 8 nuclei (1), 5 nuclei (2), 2 nuclei (3), 1 nucleus (5) or uninucleate apex (4). N = nucleolus, s = septum.

Study of the Heterokaryons *lys255 (58)/arg154 (+)* and *lys255 (58)/arg13 (a_1)*

The mycelium coming from these two pairings yielded about 1 % of heterokaryotic hyphal fragments (Table 28-2). Examination of the heterokaryotic mycelium under the phase contrast microscope (Fig. 28-4C) showed multinucleate cells containing 2 to 10 nuclei, a situation favorable for the establishment of heterokaryosis. One also observed lines of uninucleate cells suggesting that the heterokaryotic association is limited to only a few cells.

Table 28-2. Estimation of the different cellular types in prototrophic mycelium arising from two pairings

	pairing *lys255 (58)/arg154 (+)*	pairing *lys255 (58)/agr13 (a_1)*
Homothallic hyphal fragments *(lys$^-$)* on complete medium	65 %	20 %
Homokaryotic hyphal fragments *(arg$^-$)* on complete medium	35 %	80 %
Heterokaryotic hyphal fragments on minimal medium	0.5%	1.5%

In the combination *lys255 (58)/arg154 (+)*, the heterokaryotic mycelium had the same morphology as the *wild* type. In spite of the preponderance of *lys255 (58)* nuclei in this mycelium (Table 28-2) the differentiated state associated with the functioning of gene *58* was never exhibited and the corresponding cytoplasmic ring factor was not synthesized. This suggests that the activity of allele *58* is repressed in presence of the allele (+) and therefore the latter is dominant.

At any time during growth, these heterokaryotic cultures could form rings, which is a property of the *wild* type. When these heterokaryons developed on a complete medium, the mycelium growing beyond the ring was a shorn phenotype characteristic of the auxotroph mutant associated with allele *58* (Fig. 28-4B). Subcultures and conidia taken from the zone exterior to the ring and placed on different media showed a requirement for lysine. The nuclei *arg154 (+)*, present in 35% of the hyphal fragments before the appearance of the ring (Table 28-2), were missing in the hyphae growing beyond this ring. Therefore, once the ring forms, the heterokaryotic association does not exist any more. That means that the cells containing nuclei bearing the *wild* type allele [*arg154 (+)* homokaryotes and *arg154 (+)/lys255 (58)* heterokaryotes] no longer participated in growth after having expressed the differentiation. Only the homokaryotic *58* cells allowed the thallus to grow further, because the synthesis of the infectious ring factor in these cells does not have the same effects on growth as in the *wild* type (15).

The *lys255 (58)/arg13* (a_1) heterokaryon had the properties of mutant *58*, exhibiting a red pigment and synthesing the infectious ring factor. This phenotype did not result from a preponderance of nuclei carrying the allele *58* (Table 28-2). The allele *58* therefore appeared to be capable of functioning as well in mixed cytoplasm with nuclei containing the allele a_1 as in cytoplasm which only contained nuclei *58*. Contrary to the allele *(+)*, the allele a_1 had no inhibitory activity towards the synthesis of the transmissible ring factor. The mutant phenotype of the heterokaryon indicated that there was no complementation between the alleles *58* and a_1. We can conclude that allele *58* is dominant over allele a_1 and the two mutations apparently affect the same gene.

Discussion of the Results of the Study of the Heterokaryons *lys255 (58)/arg154 (+) and lys255 (58)/arg13* (a_1)

In the heterokaryotic mycelium, the cells which possess the allele (+) and which expressed differentiation cease to divide. This result allows us to improve the conclusions of experiments on the fragmentation described earlier. In the *wild* type, the differentiated areas are mosaics of normal cells and differentiated cells characterized by a slower growth rate. This is clear because we now know that the cells which have shown differentiation do not divide. The growth of the differentiated zones is caused solely by those of the young apices which are still normal and susceptible to infection at any moment through contact with differentiated neighboring cells. The radial extension of the area is therefore not favored.

In the heterokaryotic mycelium *lys255 (58)/arg154 (+)* with a *wild* phenotype, the nuclei *58* do not become activated on the third day as in the case in the homokaryotic mycelium *58*, which arises from a microconidium (15) but much later. This fact indicates that the activity of allele (+) can repress the onset of the activity of the allele *58*. This repressive action of the allele (+) could be suppressed either by changing its activity in the course of its transition from a normal regime to a differentiated regime, or by the mutation a_1.

The absence of complementation observed in the heterokaryon *lys255 (58)/arg13* (a_1) between the mutations *58* and a_1 suggests that they affect the same functional unit *locus A*. One can explain their different effects on the expression of the ring differentiation by supposing that mutation a_1 stops the synthesis of a necessary product, while mutation *58* only brings about a minor modification to this product.

Conclusions

The *wild* clone of *N. haematococca* shows two distinctive differentiations, which are self-exclusive at the level of the cell, but which can coexist within a haploid thallus of monokaryotic origin. These differentiations can be indefinitely perpetuated by subculturing large fragments, but disappear in the course of asexual or sexual reproduction, only to reappear inevitably later in the course of growth. They can be transferred to a thallus that has not yet exhibited them by cytoplasmic contact with a differentiated transfer of suitable size and age. This transmission does not result from diffusion of substances within the medium, but is due to migration through anastomoses of specific determinants that propagate themselves independently of the nuclei. Expression of these differentiations also requires the structural integrity of genes detected through mutagenesis and genetic analysis. Apparently some of them interfere in an indirect manner with the synthesis of the determinants by acting on the cell metabolism. Other genes seem to be implicated more directly in the production of determinants. These are *locus A* for the transmissible ring factor and *locus 789* for transmissible sector factor.

As in the case of several other hereditary phenomena involving extra nuclear elements in fungi (7, 21), the physio-chemical nature of the determinants that induce these differentiations is not known. The expression of these inductors resembles symptoms of certain virus diseases in higher plants (18). However, infection cannot be obtained here by simply placing a few drops of a cell extract from a differentiated thallus on a *wild* thallus *(unpublished results)*. It is true that the virus particles shown in fungi are rarely transmitted to healthy isolates in a way other than by anastomoses with a virus-infected donor (30). However, attempts to show the presence of virus particles or of stable RNA associated with the differentiated states in *N. haematococca,* using some proven methods (20, 24, 32), have failed *(unpublished results)*. In the absence of proof regarding the viral nature of the cytoplasmic factors involved, we can suggest other hypotheses and propose, for example, that the ring and sector determinants are transitory cellular constituents. Within this context, the mechanisms which appear most reasonable to us to explain the characteristics of the differentiations are variations in the activity of certain genes in the course of development. If these genes function under a certain regime in the normal mycelium, then, in certain cells, this regime would be modified by the influence of an unknown primary stimulus. These genes would bring about the same change in the activity of neighboring genes by the diffusion of their product.

Acknowledgements

The authors thank Professor T.A. Toussoun and Dr. J.F. Lafay for the critical revision of the manuscript and its translation into English; Professors M. Bennoun-Picard and J. Chevaugeon, for their suggestions concerning the interpretation of the results; and C. Gerlinger and M. Maugin, for their excellent technical assistance.

Literature Cited

1. Bareyre, M.J., and D. Laillier-Rousseau. 1972. Propriétés de nouveaux mutants affectant les variations morphologiques chez le *Nectria haematococca.* Compt. Rend. Acad. Sci., Paris, série D, 274 : 3614–3615.
2. Beisson-Schecroun, J. 1962. Incompatibilité cellulaire et interactions nucléo-cytoplasmiques dans les phénomènes de barrage chez *Podospora anserina.* Ann. Génét. 4 : 1–50.
3. Belcour, L. 1976. Loss of a cytoplasmic determinant through formation of protoplasts in *Podospora anserina.* Neurospora Newsletter 23 : 26–27.
4. Boissonnet-Menes, M. 1969. Intervention du gènôme dans un phénomène extrachromosomique

chez l'*Hypomyces ipomoeae*. Compt. Rend. Acad. Sci., Paris, série D, 268 : 1593–1596.

5. Bouvier, J. 1969. Ordre d'action des génes et nombre de noyaux impliqués dans l'initiation et le développement du *Penicillium baarnense*. Compt. Rend. Acad. Sci., Paris, série D, 269 : 171–174.
6. Bouvier, J., and E. Laville. 1970. Origine et fonction des inducteurs d'états différenciés chez deux Ascomycètes. Physiol. Végétale 8 : 361–374.
7. Burnett, J.H. 1975. Mycogenetics. John Wiley and Sons, London. 375 p.
8. Buxton, E.W. 1962. Parasexual recombination in the banana wilt *Fusarium*. Trans. Brit. Mycol. Soc. 45 : 274–279.
9. Caten, C.E., and J.L. Jinks. 1966. Heterokaryosis: its significance in wild homothallic ascomycetes and fungi imperfecti. Trans. Brit. Mycol. Soc. 49 : 81–93.
10. Chevaugeon, J., and C. Lefort. 1960. Sur l'apparition régulière d'un "mutant" infectant chez un champignon du genre *Pestalozzia*. Compt. Rend. Acad. Sci., Paris, 250 : 2247–2249.
11. Chevaugeon, J. 1968. Etude expérimentale d'une étape du développement du *Pestalozzia annulata*. Ann. Sci. Natur., série Bot., 12è série, 9 : 417–432.
12. Chevaugeon, J. 1974. Stability of the differentiated state in *Pestalozzia annulata*. Trans. Brit. Mycol. Soc. 63 : 371–379.
13. Cuzin, F. 1961. Apparition régulière chez *Curvularia pallescens* d'une variation sectorielle contagieuse non transmissible par les thallospores. Compt. Rend. Acad. Sci., Paris, série D, 262 : 1656–1658.
14. Daboussi-Bareyre, M.J. 1976. Synthèse et migration de l'information morphogénétique chez le *Nectria haematococca* (Berk. et Br.) Wr. I : Etude chez un mutant conditionnel. Physiol. Végétale 14 : 517–532.
15. Daboussi-Bareyre, M.J. 1977. Synthèse et migration de l'information morphogénétique chez le *Nectria haematococca* (Berk. et Br.) Wr. II : Etude des modalités de l'expression. Physiol. Végétale 15 : 577–590.
16. Daboussi-Bareyre, M.J., D. Laillier–Rousseau, and D. Parisot. 1979. Contrôle génétique de deux états différenciés de *Nectria haematococca*. Can. J. Bot. 57 : 1161–1173.
17. Daboussi-Bareyre, M.J. 1980. Heterokaryosis in *Nectria haematococca:* Complementation between mutants affecting the expression of two differentiated states. J. Gen. Micriobiol. 116 : 425–433.
18. Diener, T.O. 1963. Physiology of virus-infected plants. Annu. Rev. Phytopathol. 1 : 197–218.
19. Dujon, B. 1975. Les fonctions génétiques mitochondriales chez *Saccharomyces cerevisiae*, p. 75–79. *In* S. Puiseux-Dao (ed.), Molecular biology of nucleocytoplasmic relationships, Elsevier, Amsterdam.
20. Dunkle, L.D. 1974. Double-stranded RNA mycovirus in *Periconia circinata*. Physiol. Plant Pathol. 4 : 107–116.
21. Esser, K., and W. Keller. 1976. Genes inhibiting senescence in the Ascomycete *Podospora anserina*. Mol. Gen. Genet. 144 : 107–110.
22. Hanlin, R.T. 1971. Morphology of *Nectria haematococca*. Amer J. Bot. 58 : 105–116.
23. Harris, H. 1971. Cell fusion—The Dunham Lectures. Clarendon Press, Oxford, p. 106–141.
24. Herring, A.J., and E.A. Bevan. 1974. Virus-like particles associated with the double-stranded RNA species found in killer and sensitive strains of the yeast *Saccharomyces cerevisiae*. J. Gen. Virol. 22 : 387–394.
25. Jinks, J.L. 1966. Mechanisms of inheritance. 4. Extranuclear inheritance, p. 619–660. *In* G.C. Ainsworth and A.S. Sussman (ed.), The fungal organism, vol. 2, Academic Press, New York.
26. Koltin, Y., and P.R. Day. 1976. Inheritance of killer phenotypes and double-stranded RNA in *Ustilago maydis*. Proc. Nat. Acad. Sci., USA, 73 : 594–598.
27. Laville, E. 1967. Sur une variation contagieuse de l'*Hypomyces haematococcus*. Compt. Rend. Acad. Sci., Paris, série D, 264 : 265–267.
28. Laville, E. 1967. Transmission et persistance de la variation annulaire chez l'*Hypomyces haematococcus*. Compt. Rend. Acad. Sci., Paris, série D, 264 : 904–906.
29. Laville, E. 1971. Etude des mécanismes de deux variations contagieuses chez le *Nectria haematococca* (Berk. et Br.) Wr. Thèse de Doctorat d'Etat, Université de Paris-Sud, Centre d'Orsay. 96 p.
30. Lemke, P.A., and C.H. Nash. 1974. Fungal viruses. Bacteriol. Rev. 38 : 29–56.
31. Marcou, D. 1961. Notion de longévité et nature cytoplasmique du déterminant de la sénescence chez quelques champignons. Ann. Sci. Natur., série Bot., 12è série, 11 : 653–764.
32. Morris, T.J., and E.M. Smith. 1977. Potato spindle tuber disease: procedures for the detection of viroid RNA and certification of disease-free potato tubers. Phytopathology. 67 : 145–150.
33. Watson, K. 1976. The biochemistry and biogenesis of mitochondria, p. 92–120. *In* J.E. Smith and D.R. Berry (ed.), The filamentous fungi, vol. 2, Edward Arnold, London.
34. Wickner, R.B. 1976. Killer of *Saccharomyces cerevisiae:* a double-stranded ribonucleic acid plasmid. Bacteriol. Rev. 40 : 757–773.

29 Some Aspects of the Pathogenic Potential in Formae Speciales and Races of *Fusarium oxysporum* on Cucurbitaceae

D. Bouhot

Introduction

Fusarium wilt of muskmelon has been known for a long time in France. The disease has been studied extensively and a summary of this work is presented by Mas et al. (9).

Georgette Risser's experiments on muskmelon breeding and selection have produced interesting resistant cultivars and have enabled us to determine precisely the number and the types of resistant genes in these cultivars (9). Thus, two independent genes, *Fom1* and *Fom2*, carried respectively by the muskemelon Doublon (D) and the muskemelon CM 17187 (CM), enable one to classify the races of *Fusarium oxysporum* f. sp. *melonis*. The cultivar susceptible to all races is Charentais T (C).

There presently exist, worldwide, four races of *F. oxysporum* f. sp. *melonis:* R_0, R_1, R_2 and $R_{1,2}$. Race $R_{1,2}$ is subdivided into isolates that cause wilting and isolates that cause necrotic yellowing. This new nomenclature, proposed by Risser et al. (12), is utilized in this paper.

Prior to 1964 no race of *F. oxysporum* f. sp. *melonis* was known in France, but the introduction of new resistant cultivars has made possible the study of these races. The rapidity with which these "new races" were observed is bothersome and seems to eliminate *a priori* the hypothesis which is often advanced that the *Fusarium* "adapts" to new plant cultivars. In order to understand and study this process, one normally must utilize the classic genetic method of making crosses between races and studying their segregation in succeeding generations. This method is difficult in *F. oxysporum* since these fungi have no known sexual state. For this reason chemical treatments were used to induce mutations at the sites of pathogenesis within the nuclei of the microconidia of *F. oxysporum* f. sp. *melonis* to try to understand better the genetic origin of races. In order to interpret the modifications of the pathogen characteristics within the mutants, H. H. Flor's gene-for-gene theory was used (7). The resistant genes in the differential muskmelon cultivars are used to show and classify the genes governing pathogenic behavior. Following the observations at the level of races, the question of the origin of formae speciales was also approached in the same manner. Through a similar process of induced mutation, the possibility of one forma specialis mutating to another has been verified. Our experiments on races and on formae speciales took place between 1965 and 1974 and were carried out with more than 4,000 mutants.

Studies on the Origin of Races in *F. oxysporum* f. sp. *melonis* (3,4,5)

Field Observations

It is not uncommon to observe attacks of Fusarium wilt the first year plants are grown in soils that have never previously grown muskmelons. Whence comes this parasite, since *F. oxysporum* f. sp. *melonis* is strictly specialized to the melon? Perhaps through a contamination of transplants or tools, or perhaps is it present at an inoculum level too low to be detected by laboratory analysis but detectable by a population of susceptible plants. Unfortunately it is impossible to answer this question directly.

Table 29-1 shows the history of muskmelon selections in France and with it the appearance of races of f. sp. *melonis*. Thus it is easy to note that resistant genes originating in other continents (Asia, America) and introduced in the new muskmelon cultivars are attacked by f. sp. *melonis* by the second year of planting of these cultivars. One would not think that these genes, which are absent from the European continent, could exercise a prior selective pressure on genes of the population of *F. oxysporum* f. sp. *melonis*. It is difficult to understand how natural and fleeting mutations could have systematically brought about the appearance of new genes for pathogenicity specially adapted to the resistant genes of the host. Race 1 is of particular interest. First noticed in 1965 through the introduction of the muskmelon cultivar Doublon, it was shown by soil analyses to be present all over France in 1965, when the Doublon cultivar had not yet been cultivated in all the country.

Laboratory Observations

In 1969 the laboratory collection of isolates of f. sp. *melonis* was tested on the three differential host cultivars (C, D, CM). This was not possible earlier since the gene *Fom1* of CM 17187 was not introduced in France until 1967. An isolate obtained from western France in 1957 was shown to be $R_{1,2}$ yellow type. In 1957 no race of f. sp. *melonis* was known in France. Race $R_{1,2}$ thus existed at least 12 years before muskmelon cultivars were selected for resistance to this race.

Table 29-1. History of melon selections in France and of the appearance of races in *Fusarium oxysporum* f. sp. *melonis*

	Differential host cultivars			
	Charentais T (C)	Doublon (D)	CM 17187 (CM)	
Geographic origin of cultivars	France	France	USA	
Nature of resistance gene	–	(*Fom1*)	(*Fom2*)	
Date when first cultivated in France	Prior to 1960, very old cultivar	1963	1963	1967
Date when disease was first observed	Before 1960	1964	1964[a]	1968
Corresponding races of f. sp. *melonis*	R_0	R_1	$R_{1,2}$ wilting	$R_{1,2}$ yellowing
Code for races	(C+D−CM−)	(C+D+CM−)	(C+D+CM+)	

[a] In reality it was the Asian cultivar Ogon, also possessing the resistance gene *Fom2*, which, when used in 1963–64, showed the presence of $R_{1,2}$ wilting type.

Induced Mutation Experiments

Each isolate of *F. oxysporum* is obtained and kept as single spore cultures. To induce mutations, microconidia are suspended in a solution of nitrosoguanidine, at a dose of 100 to 400 γ/ml, for one hr at 30 C. They are then washed, plated on potato dextrose agar, and 16 to 20 hr later the germinated microconidia are transferred to tubes for study. Two techniques are used for the inoculations: The first laboratory technique, using polyethylene bags as described by Bouhot and Rouxel (6), is easy to do and allows for a rapid testing of a large number of isolates. It favors virulence of the parasite over the resistance of the host. In the second technique, the inoculum is grown on vermiculite moistened with potato extract and introduced into soil in 15-cm-diam pots in the glasshouse. This technique is more "natural."

All the mutants are regularly tested with the first technique, and any appearing to be of interest are tested in the glasshouse by the second technique. Inoculations using the first technique are done in controlled chambers at 20 C with a photoperiod of 15 hr. Those in the glasshouse are done at 20 to 25 C. In order to better control variations in pathogenic behaviour (sensu Miles [11]), only those individuals which cause at least 60% mortality are considered pathogenic.

Morphologic Mutants

Morphologic mutants have been obtained for the races R_0 and R_1. They are characterized by the same pathogenic behaviour as the parents, the acquisition of a specific pigment in the culture (generally dark violet, dark red, or black), and a reduced growth rate. They are utilized as markers in ecological tests. Some of these have remained stable when kept in dry sterilized soil for more than 6 years.

Gibberellin-Producing Mutant

During mutation tests in 1969 on isolate 15 of *F. oxysporum* f. sp. *melonis* belonging to race R_0, the mutant 15-15 was found to produce an elongation of muskmelon seedlings. This mutant had lost the cultural characteristics typical of f. sp. *melonis,* but retained the capacity of penetrating the roots of the muskmelon. The gibberellin activity of the mutant was studied using as reference the English isolate *Gibberella fujikuroi* ACC 917. After the production in vitro of gibberellin was demonstrated (the amount determined by means of chromatography), the biologic activity of the eluates was tested on the dwarf pea cultivar Annonay.

Mutant 15-15 produces 100 times more gibberellin than the parental isolate and 15 to 10 times less than the reference isolate. The biologic test on pea was positive for the mutant and the control isolate and negative for the parental isolate. This mutant has thus acquired the property of synthesizing gibberellin to a high level, close to that of *G. fujikuroi.*

Pathogenic Mutants

Major changes in pathogenic behaviour appeared in the very first experiments and continued to appear for the next 3 years at the rate of two cycles of mutant production per year. Table 29-2 shows some of the different types of mutation observed in the three races of *F. oxysporum* f. sp. *melonis*. Two things are clear: i) through mutation, three characteristics of pathogenic behaviour [(C+), (D+), and (CM+)] were separated from each other and can be considered as independent though they could have been thought of as being linked in 1965–66 in the races R_0, R_1, and $R_{1,2}$ (Table 29-1). This independence is presently (in 1978) supported by the fact that in the muskmelon there also exist two independent

Table 29-2. An example of the frequency distribution of pathogenic characteristics obtained after a nitrosoguanidine treatment of the three French races of *Fusarium oxysporum* f. sp. *melonis*

Determination of races of f. sp. *melonis*				Parental isolates of *F. oxysporom* f. sp. *melonis* belonging to races known in France			
Races in the world	Genetic code determined by differential melon cultivars: C	D	CM[a]	Race R_0 isolate15	$Race_1$ isolate8	$Race_{1,2}$ wilting isolate7	$Race_{1,2}$ yellowing isolate 24
$R_{1,2}$	+	+	+[b]	2[c]	1	13	12
R_1	+	+	−	1	8	0	1
R_0	+	−	−	8	3	2	3
R_2	+	−	+	0	2	1	3
unknown	−	−	+	2	1	0	3
unknown	−	+	+	0	0	3	3
unknown	−	+	−	1	3	2	0
unknown	−	−	−	5	8	4	0
Total number of monoconidial progeny studied after treatment				19	26	25	25

[a] C = Charentais T, D = Doublon, CM = CM 17187 the differential melon cultivars
[b] + is given to progeny when the mortality rate caused by the progeny is equal to or greater than 60%
−is given to progeny when mortality rate is 0–50%
[c] number of microconidial progeny showing the left hand genetic code after treatment by nitrosoguanidine of microconidia of the four parental isolates.

resistant genes *Fom1* and *Fom2*. ii) Starting with the race R_0, it has been possible to obtain races R_1 and $R_{1,2}$; starting with R_1, races R_0, $R_{1,2}$, and R_2 were obtained; nearly all the races were obtained starting with $R_{1,2}$. Thus in 1966 a mutant R_2 was found in the laboratory in France, at a time when the race was still unknown there and when Banihashemi (2) was reporting its presence in the USA in Michigan. Another mutant (C−D+CM−), which doesn't conform to known races, was likewise obtained in the laboratory in 1965 and at the same time isolated in France in the Nantes region from the soil in a truck garden.

These two examples clearly show that certain new pathogenic characters, discovered through induced mutations, can later be found in the field, and nothing prevents one from believing that all the combinations found could naturally appear, for example (C−D+CM−) or (C−D−CM+). The introduction of new independent resistant genes in muskmelon further increases the chances of diversification within this spectrum of races; perhaps it will also allow a determination of the extent of the flexibility in pathogenic behaviour. On the other hand, it is only through the existence of this third resistance gene of the muskmelon that we will be able to increase our knowledge of the pathogenic potentialities of *F. oxysporum* f. sp. *melonis,* since no other process (chemical, serological, or immunoserological) has been able until now to supplant the differential cultivars of muskmelon in the classification of the pathogenic characteristics of *F. oxysporum* f. sp. *melonis.*

Stability of Pathogenic Mutants

The availability of isolates with pathogenic markers is useful to the pathologist and to the geneticist to the degree that these characters are stable over time. All of the mutants obtained in 1965 and 1966 have been kept in dry sterilized soil and have been regularly tested on muskmelon for more than 6 years, i.e., for seven successive inoculations on the three differential cultivars. The individuals which received the mutagenic treatment can be divided into two groups.

In the first group are those mutants with a stable pathogenic behaviour. Some have completely lost their pathogenic behaviour (C−D−CM−), others keep their parental characters [(C+D−CM−) or (C+D+CM−)]. In the first case, the mutation is effective at the pathogenic-behaviour level since the characters (C+) and (D+) have become (C−) and (D−); on the other hand, in the second case, one can assume that if the mutations have appeared they do not touch on the sites of pathogenicity.

In the second group are those mutants whose pathogenic behaviour has varied with time. In these, two types of variations have been seen. In certain mutants, pathogenicity, which had been profoundly modified, changed and became similar to that of the parental type. In other mutants, pathogenicity was modified through mutation but fluctuated considerably with time without any apparent or precise direction. For example, mutant No. 15 of R_1 has successively shown the following pathogenic behaviours: (C+D+CM−) = R_1, (C−D−CM−) = non virulent, (C+D+CM−) = R_1, (C+D+CM−), (C−D+CM−). Mutant No. 16 of R_0 has varied in the following manner: (C−D−CM−) = non virulent, (C+D+CM−) = R_1, (C+D+CM−), (C+D−CM−) = R_0, (C+D+CM−) = R_1.

The individuals retaining the parental type represent, in the mean, 35 to 50% of the mutants that were isolated, and the nonvirulent mutants represent 20 to 30%; those that are more or less stable but nevertheless interesting because of their new characters represent 20 to 30%.

Discussion

The instability of those mutants with new pathogenic capabilities leads one to believe that they may not exist in nature, but previous observations show that some of them were found in the field in France and in the USA. The induction of mutations at the pathogenic-behaviour level has shown that the pathogenic potentialities of f. sp. *melonis* are larger than those presently known through the use of the three differential host cultivars. In addition, f. sp. *melonis,* which contains these races, also has at least three genes affecting pathogenic behaviour. It is even possible that f. sp. *melonis* contains a greater number of genes or a greater combination of genes affecting pathogenicity than resistant genes in the host. This may explain why the creation of a resistant cultivar may lead to the appearance of a corresponding aggressive gene in the parasite, but it also casts a doubt on the efficacy of dominant monogenic resistance. Plant breeders have looked for and found resistant genes in the host and have studied its genetic determinants without taking into account the genetic potential of the parasite. They certainly are not the only ones to do this, since several phytopathologists have made similar studies, particularly with *F. oxysporum.* For all of these reasons, polygenic resistance should be a better tool, for it offers, in principle, a greater, more differentiated, and more polyvalent barrier to the agregate pathogenic potentialities of the parasite.

In ecological terms, several points can be made: i) The genes governing pathogenic behaviour (and corresponding races) of f. sp. *melonis* can be present in a soil well before susceptible cultivars of muskmelon are cultivated. ii) The parasite is capable of diversifying its pathogenic potential in the absence of the host, e.g., by mutation. iii) The host plant plays a double role in the process of the appearance of new races: by means of its resistant genes, it is a sensitive selective filter revealing within the pathogen the corresponding combination of genes affecting pathogenic behaviour; then, as was shown by Wensley and McKeen (13) and Banihashemi (2), the rhizosphere of the muskemelon plant rapidly increases the inoculum level at the end of the first crop of muskmelons so that the second crop is severely attacked.

The observations made here for the races of f. sp. *melonis* have already been seen in other pathogenic fungi. Zadoks (14) made an interesting synthesis of the subject when he explained the appearance of races, their disappearance, their evolution, and the attendant consequences for the plant breeder. Some of his conclusions are particularly apt and confirm our observations: "New physiologic races can be produced by mutation, hybridization, and heterokaryosis. . . . The genetics of the host-parasite relationships can be explained on the basis of the gene for gene hypothesis of Flor [7]. . . . Resistance breeding is an important agent in the appearance of a new race and in the evolution of pathogenicity. . . . [In the case of flax rust] races of the pathogen responsible for the breakdown of resistance were present at the time of introduction of the new variety. The variety selected and multiplied the race. . . ."

Zadoks (14) arrived at these conclusions from genetic studies on rusts of cereals and of flax [those of H.H. Flor (7) especially] and on late blight of potato, whose pathogens have a sexual state.

In spite of the absence of a sexual state, *F. oxysporum* behaves in a manner similar to *Puccinia* and *Phytophthora.* H.H. Flor's theory (7) allows one to progress a little in the understanding of the origin of physiologic races in the vascular wilt formae speciales of *oxysporum.* To do this it is necessary that the genetic makeup of the material be known precisely.

Studies on the Origin of Formae Speciales in *Fusarium oxysporum* Attacking Cucurbitaceae

Evidence in the preceding section shows that it is relatively easy to "fabricate" races, by induced mutation, with the f. sp. *melonis.* What takes place at the level of these special forms? Armstrong and Armstrong (1) present a synthesis of the work on classification, differentiation, enumeration, and validation of the numerous special forms of *F. oxysporum* observed in the world. The question we ask here is simple: is it possible by induced mutation to pass from one forma specialis to another? As with races, this is a way of studying the determinants of the formae speciales.

Methods

These are identical to those utilized for producing mutants at the race level: single spore isolates of *Fusarium,* mutation by means of nitrosoguanidine, inoculations of plants with laboratory techniques and in the glasshouse, and reisolation of the *Fusarium* from the stem of each plant showing wilt symptoms. Determination of the characteristic of f. sp. *melonis* is obtained with the muskmelon cultivar Charentais T, f. sp. *niveum* with the black-seeded watermelon, f. sp. *cucumerinum* with the cucumber cultivar Vert Long Maraicher.

Mutations from an unspecialized *F. oxysporum* An isolate of *F. oxysporum* isolated from roots of leek was kept in a laboratory as a single spore culture. One hundred microconidia obtained from this clone were inoculated into muskmelon. None of these showed the characteristics of f. sp. *melonis.* At the same time, 100 microcondia were treated with nitrosoguanidine and were also inoculated into muskmelon under the same conditions. Not one mutant acquired the character of f. sp. *melonis.*

Mutations arising from an f. sp. *gladioli* *Fusarium oxysporum* f. sp. *gladioli* is one of the least well defined and specialized of the formae speciales. It was interesting to see if it could acquire another pathogenic character. Isolate 53, obtained in 1967 from a plate rot of a gladiolus corm, is a typical f. sp. *gladioli* and nonpathogenic on muskmelon. In 1972, 100 microconidia of this isolate, which had previously been treated in nitrosoguanidine, were individually inoculated into muskmelon. Most of the mutants affected the development of the muskmelon seedlings. Three different effects were noted: i) inhibition of the root development of the seedlings where the roots approach the inoculum zone; ii) abnormal elongation of the stem of the seedlings, indicating the production of gibberellin by the mutants; iii) necrosis of the crown of the seedlings similar to that caused by *Rhizoctonia solani* and not typical of Fusarium wilt. Thus no mutant had acquired the characteristics of f. sp. *melonis*.

Mutations from an f. sp. *niveum* Due to the negative results obtained in the two prior efforts, we formulated the hypothesis that it might be easier to obtain transfers between formae speciales attacking Cucurbitaceae. In 1972, f. sp. *melonis, cucumerinum* and *niveum* were known and were widely distributed throughout the world; the f. sp. *luffae* and *lagenariae* were found only in Asia, respectively by Kawai et al. (8), and Matuo and Yamamoto (10). In France at that time, only f. sp. *melonis* was present. However, the race $R_{1,2}$ wilting type of f. sp. *melonis,* observed in France in 1964 and localized in the Marseille region, does attack watermelon. It therefore contains two characters of f. sp. *melonis* and *niveum* and is clearly distinct from the other races R_0, R_1, and $R_{1,2}$ yellowing type of f. sp. *melonis*. It is perhaps more of an f. sp. *niveum* than a f. sp. *melonis*. This example shows the possibility of coexistence of two characters of the forma specialis in the same isolate of *F. oxysporum* and explains why we advanced the hypothesis of "specialization on the Cucurbitaceae."

Isolate 1 of the f. sp. *niveum,* obtained from Morocco in 1965, has been kept in single spore culture and utilized for the following studies. Fifty-eight microconidia of isolate 1 obtained from this clone were inoculated into muskmelon and watermelon. All have shown the characteristics of f. sp. *niveum* but not the characteristics f. sp. *melonis*. Twice again, at 1-year intervals, microconidia of the parental isolate were treated with nitrosoguanidine and inoculated into muskmelon and watermelon. In 1971, mutant 1-5 of f. sp. *melonis* and in 1972 mutant 1-39b of f. sp. *niveum* (approximate mutation rate: 1%) behaved in the same manner: they acquired the characteristics of f. sp. *melonis* and lost the characteristics of f. sp. *niveum*. In addition the two mutants showed a remarkable stability of their new pathogenic capability for 2 years (four successive inoculations). Inoculated on the three differential muskmelon cultivars, they attacked only Charentais T, and they therefore belonged to the R_0 race of f. sp. *melonis*. These observations and experiments show that not only are the two characters f. sp. *niveum* and f. sp. *melonis* capable of coexisting in one clone, but also that a passage is possible by mutation between f. sp. *niveum* and f. sp. *melonis,* that is to say in the absence of the host plant.

Lineage of *F. oxysporum* f. sp. *cucumerinum* *Fusarium oxysporum* f. sp. *cucumerinum* has never been observed in France, but occurs in Belgium and in the Netherlands. It does not seem to pose a great threat to Europe at present. Prior to carrying out experiments on mutations in 1973, microconidia obtained from a single spore of isolate 56 of f. sp. *cucumerinum* originating from Naldwyck (Netherlands) in 1962, were inoculated into cucumber, muskmelon, and watermelon. Six months later the same was done with isolate 57 of f. sp. *cucumerinum* obtained from Dr. J.P. Jones at Bradenton, Florida, in 1973. The results are shown in Table 29-3. For isolate 56, the characteristics of f. sp. *cucumerinum, melonis,* and *niveum* are naturally present in the parental isolate and all the possible arrangements of these characteristics are present but in different frequencies. For isolate 57 of f. sp. *cucumerinum* the character of f. sp. *melonis* is practically absent, except for one single descendant in which the characteristics of the three formae speciales are present together. It is important to note that it is possible for these three characters to spontaneously coexist in the same microconidia. These two isolates already containing all the characteristics that can be observed are not good material to induce mutations which would show the possibilities of a passage from one forma specialis to another.

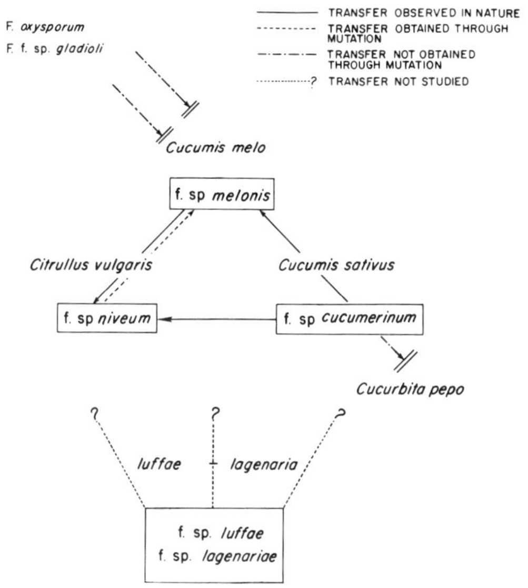

Fig. 29-1. Schematic diagram of the observed and induced mutations in three formae speciales of *Fusarium oxysporum* attacking the Cucurbitaceae.

Discussion

Figure 29-1 summarizes all the relationships observed in the field or brought about in the laboratory between the three formae speciales attacking the Cucurbitaceae. To date it has not been possible for an unspecialized *F. oxysporum* or an isolate of *F. oxysporum* f. sp. *gladioli* to acquire one of the three characteristics in f. sp. *cucurbitae.* This does not mean that this is impossible starting from another forma specialis.

The three characters, f. sp. *melonis, niveum,* and *cucumerinum,* apparently separated in the isolates that are recovered in the normal manner, can in reality coexist in one isolate. All of the possible combinations between them were observed: (*melonis* + *niveum*), (*melonis* + *cucumerinum*), (*niveum* + *cucumerinum*), (*melonis* + *niveum* + *cucumerinum*), showing a sort of preferential linkage between these characters, a linkage which does not exist within the f. sp. *gladioli* for example.

Table 29-3. Frequency distribution of the various f. sp. characteristics observed in the microconidial progeny of 2 isolates of f. sp. *cucumerinum*

Cucurbit species used in the determination of the three f. sp.			Parental isolate of f. sp. *cucumerinum*	
Cucumber for f. sp. *cucumerinum*	Melon for f. sp. *melonis*	Watermelon for f. sp. *niveum*	65	57
+[a]	−	−	52[b]	51
+	+	−	19	0
+	−	+	19	8
+	+	+	9	1
−	+	+	1	0
−	+	−	4	0
−	−	+	5	3
−	−	−	6	16
Total number of progeny studied after treatment			115	79

[a] + is given to progeny when the mortality rate caused by the progeny is equal to or greater than 60%
− is given to progeny when the mortality is 0–50%
[b] number of macroconidial progeny showing the left hand f. sp. characteristics after treatment by nitrosoguanidine of microconidia of two isolates (65 and 57) of f. sp. *cucumerinum*

One finds at the level of the formae speciales attacking the Cucurbitaceae the same properties of pathogenic characteristics as at the level of physiologic races, that is to say, the independence of the forma specialis characters and the possibility of linkage. The races are differentiated by the two muskmelon cultivars and the formae speciales are differentiated by the three species of Cucurbitaceae. There remains, however, an unknown with f. sp. *luffae* and *lagenariae,* described in Asia and not included in these experiments.

Likewise the changes f. sp. *niveum* toward f. sp. *cucumerinum* and f. sp. *melonis* toward f. sp. *cucumerinum* were not proven. Finally, it would be important to verify if a similar situation could exist in the Solanaceae or in other plant families. One should note, however, that the formae speciales attacking Cucurbitaceae are among those that are the most specialized, which might explain the family tie between them.

Squash (*Cucurbita pepo*) is not attacked by *F. oxysporum.* Experiments to obtain aggressive mutants on squash starting with a clone of an isolate of f. sp. *cucumerinum* have failed. The entity "cucurbitaceae," if it exists, would therefore have definite limits.

Upon reflection it is not astonishing to find, at the level of parasitic specialization in *F. oxysporum,* the same hierarchy as in the classification of species, genera, and families in plants. It is the logical consequence of H.H. Flor's theory (7) and of the necessity for the plant pathologist to define the characters of pathogenic behaviour starting solely from the responses of the host.

The idea of a "cucurbit identity" in *F. oxysporum* corresponds to the notion of the family of Cucurbitaceae in plants, and could be expressed by a "forma familialis" which would group together a certain number of formae speciales. But perhaps it is also proof of a precise determinism (one or several genes) which controls the character of the forma specialis just as the gene define the race.

Literature Cited

1. Armstrong, G.M., and J.K. Armstrong. 1980. Formae speciales and races in *Fusarium.* Chap. 34 of this volume.
2. Banihashemi, Z., 1968. The biology and ecology of *Fusarium oxysporum* f. sp. *melonis* in soil and the root zones of host and non host plants. Ph. D. Thesis Michigan State Univ., East Lansing. 114 p.
3. Bouhot, D. 1970. Variations induités du pouvoir pathogène chez *Fusarium oxysporum* f. sp. *melonis.* Ann. Acad. Sci. Fenn. A, IV Biologica, 168: 25–27.
4. Bouhot, D. 1973. Some studies on the origin of races and formae speciales in *Fusarium oxysporum* by use of nitrosoguanidine mutants. Abstr. Paper presented 2nd Int. Congr. Plant Pathol., Minneapolis, Minn.
5. Bouhot, D., and J. Louvet, 1971. Some observations and experiments on the origin of *Fusarium oxysporum* f. sp. *melonis* races in France. Int. Symp. Pathol. Wilt in Plants. Madras. 18–25 January.
6. Bouhot, D., and F. Rouxel. 1970. Deux techniques de déterminiation du pouvoir pathogène des *Fusarium oxysporum.* Ann. Phytopathol. 3: 591–594.
7. Flor, H.H. 1957. Genic systems in flax and the flax rust fungus. Robigo 4: 2–3.
8. Kawai, I., H. Suzuki, and K. Kawai. 1958. On the pathogenicity of wilt Fusarium of the cucurbitaceous plants and their forms. Shoznoka Agric. Exp. Stn. Bull. 3: 49–68. (Japanese, English summary).
9. Mas, P., P.M. Molot, and G. Risser. 1980. *Fusarium* wilt of muskmelon. Chap. 16 of this volume.
10. Matuo, T., and I. Yamamoto. 1957. On *Fusarium oxysporum* f. sp. *lagenariae* n. f. causing wilt of *lagenaria vulgaris* var. *hispida.* Trans. Mycol. Soc. Japan 2: 61–63.
11. Miles, A.A. 1955. The meaning of pathogenicity, p. 1–16. *In* J.W. Howie and A.J. O'Hea (ed.), Mechanisms of microbial pathogenicity, Fifth Symp. Soc. Gen. Microbiol, London, Cambridge Univ. Press.
12. Risser, G., Z. Banihashemi, and D.W. Davis. 1976. A proposed nomenclature of *Fusarium oxysporum* f. sp. *melonis* races and resistance genes in *Cucumis melo.* Phytopathology 66: 1105–1106.
13. Wensley, R.N., and C.D. McKeen. 1963. Populations of *Fusarium oxysporum* f. sp. *melonis* and their relations to the wilt potential of two soils. Can. J. Microbiol. 9: 237–249.
14. Zadoks, J.C. 1959. On the formation of physiologic races in plant parasites. Euphitica 8: 104–116.

IV The Fungus *Fusarium*: Physiology and Histopathology

Prologue

T. A. Toussoun

Part IV of this book deals with physiology of the pathogenic organism, a fascinating subject because it is really a study of the mechanisms of disease. What other area in plant pathology could be more to the point? With an organism as faceted as *Fusarium* the opportunities are almost limitless. I broached this field some time ago when working on the nutrition of some Fusaria. These laboratory studies showed that nitrogen sources had a striking effect on growth and sporulation, both asexual and sexual. In the latter case it affected almost everything: size, color, morphology, placement, and rate of maturation of perithecia, as well as the number of asci. Later, studies on pathogenesis in the greenhouse and in the field showed that nitrogen sources played a decisive role in this area, affecting every step of the process from propagule germination to penetration and invasion of the host root or hypocotyl. This student's work now seems rather antiquated and crude, but we still have a way to go. What is needed is the type of study which would elucidate why of two morphologically indistinguishable organisms, say *Fusarium oxysporum,* only one has the capability of invading certain tissues of a specific host; in other words we need an explanation of what some are pleased to call the parasitic habit. What seems to have held up progress are technical difficulties of analysis. The active components of parasitism and pathogenesis (let us call them metabolites) are produced in minute quantities at the cellular level. Furthermore, the contiguousness of the host and pathogen protoplasts and their interactions add to the difficulties of assay. This is perhaps the reason why plant physiologists or biochemists rarely are willing to tackle such programs. Nevertheless they must be tackled, utilizing the most modern techniques and armaments of micro-chemistry and micro-analysis. We must study the pathogen in the nutrient, physical, and chemical environments available to it in the host and replace the lima-bean-agar, the potato-dextrose-agar, and all the other agar media developed in the past century. The papers that follow may help give the impetus.

Beginning with the beginning, Griffin discusses the physiology of spore germination. One of the things that has inhibited this type of study has been our poor knowledge of the soil environment. As more became known, the interplay of the various factors was seen to result in responses quite different from thc simplistic in vitro germination experiments of the past. After all, how relevant are germination studies of macroconidia taken from a culture grown on potato-dextrose-agar and placed in a drop of water on a glass slide on a laboratory bench?

Jones and Woltz, working with wilts caused by *Fusarium oxysporum,* discuss the effects of nutrition on disease. Their ideas may lead to studies on disease control, not just symptom remission, through soil fertility management: feed the host and starve the pathogen. MacHardy and Beckman have a truly excellent treatment on infection and pathogenesis of the

Fusarium oxysporum wilts. This is no mere compilation of results but a critical analysis of these unique vascular parasites. For example, they point out that the upper root and hypocotyl are the critical areas in the development of systemic infections, where the struggle for dominance between pathogen and host is decided; that the mechanisms of host resistance are many and varied and that what is crucial is not how much but how fast a host responds to invasion; and that it is the amount of pathogen build-up and the extent of its distribution that distinguishes resistant and susceptible interactions. This is a very fine treatment of a subject which not too long ago engaged some of the best plant pathologists' minds at the Swiss Federal Institute of Technology in Zurich, The Connecticut Agricultural Experiment Station, and the University of Wisconsin—Madison, resulting in vigorous, stimulating debates. Later came the work carried out under the auspices of the United Fruit Company on banana—a plant whose size would seem to make it the ideal experimental tool for in vivo studies of the physiological processes operating in the course of these diseases. We have included the contribution of the Armstrongs here rather than with those papers dealing with taxonomy in Part VII, for it seemed that here they reinforce the diversity as well as the uniqueness of these fungi called *Fusarium oxysporum.*

Vesonder and Hesseltine deal with quite a different aspect—metabolites of *Fusarium* toxic to animals and to humans. These mycotoxins are yet another aspect of these multifaceted organisms. Finally Barbara Pennypacker contributes a paper on the neglected subject of pathological plant anatomy, wherein are discussed the cortical rot Fusaria and, once again, *Fusarium oxysporum* causing vascular wilts.

30 Physiology of Conidium and Chlamydospore Germination in *Fusarium*

G. J. Griffin

Conidium and chlamydospore germination are important early steps in the life cycle of Fusaria. Macroconidia formed on host tissues may be washed into the soil, where many are converted into chlamydospores, usually following macroconidium germination. The chlamydospore, formed either in soil or in host tissues, is the main survival structure of many Fusaria, and germination of this propagule in soil and in the rhizosphere of plants has been the subject of numerous ecological investigations. Knowledge of the physiology of the conidium and chlamydospore germination processes has been useful, and in some instances critical, in identifying the environmental factors limiting or stimulating germination of these spore forms. In this paper, I examine the conidium and chlamydospore germination processes, and nutritional, metabolic, and environmental considerations of germination in axenic culture. More is known about macroconidium germination than chlamydospore or microconidium germination.

The Conidium and Chlamydospore Germination Processes

Macroconidium germination by Fusaria can be a rapid process, completed after 4 to 7 hr of incubation in some instances (7, 14, 18). Slow macroconidium germination (over several days) may occur at high conidial densities in axenic culture (18). Chlamydospore germination appears to proceed almost as rapidly as macroconidium germination. For mixtures of *F. solani* macroconidia and chlamydospores, the former germinated earlier and suppressed chlamydospore germination (20). Wilson (45) indicated microconidia of *Fusarium oxysporum* f. sp. *cubense* germinated at a slower rate than macroconidia or chlamydospores in Warburg flasks, but these results may be due, in part, to the artificially high spore densities required for respiration studies. Macroconidium germination is typically characterized by the formation of one or two germ tubes (18), whereas microconidia and chlamydospores typically germinate by only one germ tube. Marchant (29) indicated that each macroconidium cell acts as a separate unit, but the foregoing tends not to support this.

According to Marchant (29, 30), walls of *F. culmorum* (= *F. roseum* 'Culmorum') macroconidia contain an outer mucilagenous sheath. Such a layer could account for the wettable nature of conidia, although other workers (40, 44) have not reported a mucilagenous sheath on conidia. Marchant and White (31) indicated that the sheath may function in nutrient uptake; it is not found on germ tubes. Kleinschuster and Baker (28) indicated that the outer spore surfaces of *F. solani* and *F. roseum* 'Avenaceum' macroconidia have carbohydrate-containing receptors for specific plant lectins. During germination at high spore density in a medium containing glucose and inorganic nitrogen, swelling of Culmorum macroconidia occurred, which was thought to be due in part to an increase in size of the mucilagenous layer (31). In contrast, little or no swelling was detected during exogenous-carbon independent macroconidium germination by *F. solani* (18). Wilson (45) reported microconidia of *F. oxysporum* doubled in size previous to germination, but no mention of chlamydospore or macroconidium swelling during germination was made. Possibly, conidium swelling is a response strictly to exogenous energy sources supplied to high spore densities.

Schneider and Wardrop (40) demonstrated that conidial and chlamydospore cell walls of *F. sulphureum* consist of randomly oriented microfibrils; X-ray diffraction analyses indicated the microfibrils are chitin. The microfibrils were considered to be embedded in a carbohydrate and protein matrix which constitutes the bulk of the wall (39, 40). During germination of Culmorum macroconidia, a new wall layer is deposited inside the macroconidium wall, according to Marchant (30). This is followed by lysis of an area of the conidium wall and emergence of the germ tube. In contrast, Stalhammar-Carlemalm (43) concluded that the germ tube wall extended from the inner layer of the conidial wall in *F. sambucinum* var. *coeruleum*. In conflict with previous work (30), Schneider and Wardrop (40) demonstrated that the hyphal apexes (and presumably germ tubes) of *F. sulphureum* and Culmorum are composed of chitin microfibrils. During germination of Culmorum and *F. solani* f. sp. *phaseoli* macroconidia, there appeared to be increases in numbers (29) and organization (36) of the ribosomes, respectively. Mitochondria appeared to become smaller and more numerous during *F. sambucinum* var. *coeruleum* conidial germination (43).

According to Griffiths (24), the chlamydospores of *F. oxysporum* have a thick microfibrillar wall that is laid down inside the hyphal wall during chlamydospore formation. During germination of *F. oxysporum* chlamydospores, cytoplasmic changes occur, including convolution of the cell membrane and abstriction of vesicles between the cell membrane and chlamydospore wall (25). A new wall layer is deposited between the cell membrane and the chlamydospore wall. Lysis of the chlamydospore wall is initiated prior to extension of the germ tube through the lysing area. Griffiths (25) observed no swelling of the chlamydospore during germination, which appears to be the case for chlamydospores of other *Fusarium* spp.

Nutritional and Energy Requirements for Conidial and Chlamydospore Germination

Spore density appears to be the single most important factor affecting the nutritional requirements for germination of conidia and chlamydospores in pure culture. Early research on the nutritional requirements for germination of macroconidia indicated that there were absolute requirements for exogenous carbon and nitrogen sources (11, 17, 23, 31, 42). These findings appear to have resulted mainly from using high conidial densities (10^5 to over 10^6 conidia/ml) in germination assays. Subsequent research has demonstrated that washed macroconidia are capable of rapid (within 7 hr) complete germination in the absence of exogenous organic carbon at low conidial densities (7, 18, 27). For *F. solani* macroconidia, full dependence on exogenous organic carbon was observed at $1–3 \times 10^5$ conidia/ml, while full dependence on both exogenous organic carbon and inorganic nitrogen was observed near 1×10^6 conidia/ml (18). Thus the lack of exogenous energy sources alone is not a factor limiting macroconidial germination in soil or other natural environments.

A somewhat different situation appears to exist for chlamydospore germination than for macroconidium germination. Washed *F. solani* chlamydospores germinated to a high degree in the absence of organic carbon in axenic culture, when the spore density was sufficiently reduced (19). However, washed chlamydospores had an exogenous carbon requirement for germination under the spore density conditions at which they were formed (19). Unlike conidia, chlamydospores are not readily disseminated (*F. solani* chlamydospores have an adhesive spore surface) to new environments where germination conditions may be more favorable than where the spores were formed. Chlamydospore germination in nature appears to be always dependent on exogenous energy sources (14, 17). Exogenous nitrogen was required for high or complete chlamydospore germination at high spore density in axenic culture (but not at low spore density) and in soil (14, 17, 19). The level of indigenous ammonium and nitrate is critical to assessments made in soil.

Under conditions of carbon dependence (1×10^4 chlamydospores/ml), glucose concentrations as low as 0.04 ng C/spore initiated chlamydospore germination for *F. solani* (20). Ethanol was stimulatory at 0.004 ng C/spore. While complete chlamydospore germination was supported by glucose alone, supplied at very high levels (200 ng C/spore), provision of exogenous nitrogen, as ammonium, reduced greatly the carbon requirement for complete germination (2.0 ng C plus 0.12 ng N/spore). Germ tube growth rate was similarly influenced by exogenous nitrogen.

At high spore density, a wide variety of organic carbon substrates (sugars, alcohols, organic acids and amino acids) can serve as energy sources and support partial or complete conidium or chlamydospore germination. In general, ethanol has been the most stimulatory carbon source tested thus far (6, 11, 20), with an exception for *F. roseum* (42). D-glucose, maltose, and sucrose have been good sources of carbon in most macroconidium or chlamydospore germination studies (11, 18, 19, 23, 31, 42); D-fructose, D-galactose, D-xylose, D-arabinose, and D-mannose have been good sources in several investigations (18, 19, 23, 31). Sugars that supported good macroconidium germination by *F. solani* f. sp. *phaseoli* were oxidized more rapidly and with less assimilation than those that did not (9). Organic acids such as citrate, malate, succinate, and acetate have been less satisfactory (11, 21, 31), although this may be due, in part, to decreased permeability of the spore to the tricarboxylic-acid-cycle compounds. Peptides and amino acids, such as asparagine (31), are likely to be good or excellent sources of both carbon and nitrogen for germination, but these compounds have not been examined to any extent as sole carbon and nitrogen sources in axenic culture. Recently, Barran (3) demonstrated that phenylalanine uptake by *F. sulphureum* conidia is effected by a general transport system for neutral and basic amino acids, and is partially repressed by sucrose. In soil, amino acids and peptides were highly stimulatory to chlamydospore germination (14, 17). DL-aspartate, glycine, L-alanine, DL-serine, and DL-threonine partially replaced the stimulatory effect of ethanol on macroconidium germination by *F. solani* f. sp. *phaseoli* at high conidial density (11). Carbon substrates supporting *F. solani* macroconidium germination also suported chlamydospore germination (18, 19, 21). Ammonium has been superior to nitrate as a nitrogen source, generally (6, 18, 31). Amino acid nitrogen (L-glutamate, L-asparagine, L-alanine, L-leucine, DL-aspartate, etc.) has been equal to ammonium, and no single amino acid appears to be superior to others (11, 18). Thus, Fusaria do not appear to be highly exacting, qualitatively, in the nutritional requirements for spore germination. For the *Fusarium* clones studied in axenic culture or soil, nutritional components in many plant root exudates are likely to be stimulatory to chlamydospores and conidia. If these propagules are present in the rhizoplane, and other factors (e.g., inhibitory exudates) are not limiting.

Role of Inhibitors in Conidial and Chlamydospore Germination

Park (34) demonstrated that in a dilute nutrient-agar medium, *F. oxysporum* chlamydospore and conidial germination occurred as part of a dynamic equilibrium between

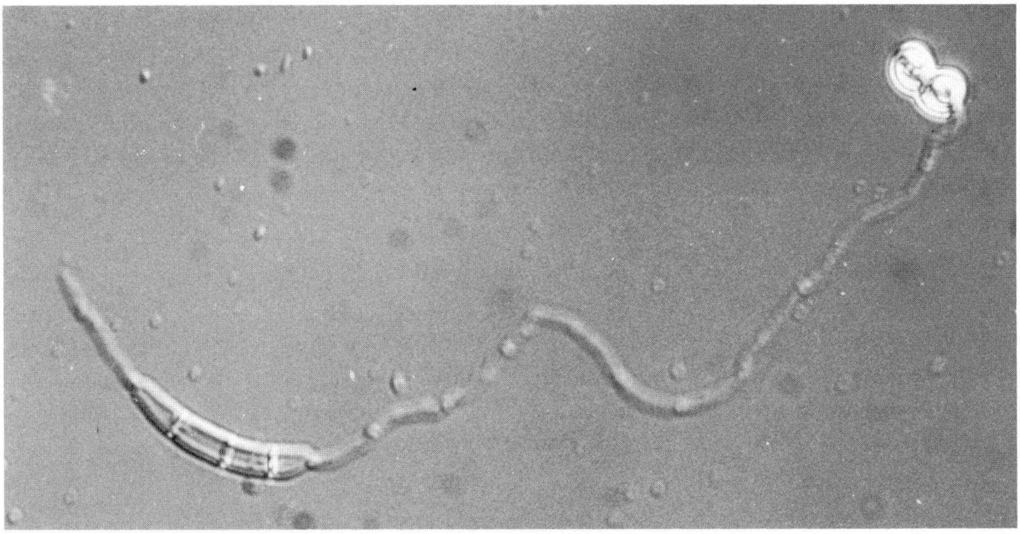

Fig. 30-1. *Fusarium solani* macroconidium germination and chlamydospore formation in an organic-carbon-free medium at 5 × 10^3 conidia per ml. Note that the macroconidium and germ tube (which passes over the chlamydospore) cytoplasm has been incorporated into the two-celled chlamydospore. Unstained interference-contrast photomicrograph.

growth, morphogenesis, lysis, and regrowth. He indicated that a labile substance accumulated in the growth medium was responsible for these events. Some compounds, such as acetaldehyde, have been isolated and identified as inhibitory to other fungi (37, 38). The influence of spore density on macroconidial and chlamydospore germination, described above, suggested to Griffin (18, 19) that an autoinhibitor, elaborated from endogenous spore reserves by *F. solani,* suppressed germination and retarded germ tube growth. From this view, conidium or chlamydospore density determines both the level of autoinhibitor in the medium, and the energy requirements for germination. Similar studies (16, 18, 20) indicated also that the autoinhibitor or some similar substance may induce the formation of chlamydospores on the tips of germ tubes of macroconidia in axenic culture, a morphogenesis that occurs also in soil and is recognized as a critical step in *Fusarium* survival. In pure culture, this macroconidium germination-chlamydospore formation process (Fig. 30-1) occurred for *F. solani* in the absence of exogenous organic carbon when the spore density was about 1 × 10^4 conidia/ml or lower; conidium germ tubes were shorter and chlamydospore formation more rapid when the spore density was closer to this value than at lower values (18), or, in separate tests, when the density of pregerminated conidia was increased (16). Ford et al. (15) postulated that a morphogen produced by Fusaria induces chlamydospore formation. It is possible that the substances postulated by the above workers are similar or identical. Fungistatic volatiles and other substances in soils may have similar effects on *Fusarium* conidium germination, and on chlamydospore formation and germination in soil, in addition to the postulated autoinhibitor and morphogen (15, 18, 21).

Endogenous Metabolism and Germination

The results discussed above indicate that endogenous substrates in the macroconidium are sufficient to support germination. *Fusarium* macroconidia, and especially chlamydospores, are known for having a high-lipid content (21, 29, 44). Lipid appears to be utilized during endogenous respiration by ungerminated macroconidia, and during exogenous carbon-dependent germination (at high conidial density) by Culmorum (32) and *F. solani* f. sp.

phaseoli (10) macroconidia. In macroconidia of the latter fungus, however, lipid materials accounted for only 37% of the loss in dry weight during endogenous respiration. Cultivation of this fungus on a high-glucose medium (4% glucose vs. 1% glucose) greatly increased the lipid content of macroconidia, but had little effect on reducing the carbon requirement for germination. The carbon requirement, under the conditions of this test, was quite high, however, and this may be responsible for the absence of a significant effect. Endogenous respiration rates, on an equivalent weight basis (QO_2), were approximately the same for high-lipid and low-lipid conidia, but on a per spore basis, a 29% increase in endogenous respiration rate was found for high-lipid conidia. Thus, the energy available for germination per spore would appear to be greater in high-lipid conidia, than in low-lipid conidia.

Ammonium ion, common in the environment of macroconidia and chlamydospores in nature, stimulated endogenous respiration and lowered the RQ (respiratory quotient) of macroconidia (10, 23, 42), but had no effect on germination in a balanced inorganic salt solution (18, 23). This effect of ammonium on respiration is thought to be due to incorporation of respiratory intermediates into amino acids, and, possibly, into protein (10). Ungerminated macroconidia of *F. solani* f. sp. *phaseoli* had a low capability to synthesize RNA and protein (13). Extracts prepared from macroconidia of this fungus were capable of incorporating leucine into polypeptide, but incorporation by germinated macroconidia was 10 times greater (36). Germination also increased the endogenous respiration of *F. solani* f. sp. *phaseoli* by seven-fold (10). Within the limited ranges tested, the endogenous respiration rate (QO^2) of ungerminated macroconidia was not affected by spore density, while exogenous glucose oxidation increased with decreasing spore density (9, 10).

In Culmorum macroconidia, lipid bodies were distributed around the perimeter of cells and became reduced with age (29). Griffin and Pass (23) found the endogenous respiration rate, endogenous reserves, and germinability of *F. roseum* 'Sambucinum' macroconidia decreased when incubated in the absence of exogenous organic carbon over a 9-month period at high conidial density. On a nutrient-agar medium, macroconidia surviving after this time synthesized cytoplasm, to fill the single surviving chlamydospore-like macroconidium cell, previous to germ tube emergence and, consequently, had a longer latent period and slower rate of germination than freshly harvested macroconidia. In contrast, chlamydospores of *F. solani* did not undergo similar changes, and 100% germinability of chlamydospores was observed after 12 months' incubation in the absence of an exogenous energy source (20). It is not known whether differences in endogenous respiration rates between chlamydosporic macroconidia and chlamydospores accounts for these results. However, the endogenous respiration rate of *Fusarium* macroconidia is much higher than that of constituitively dormant spores of microorganisms. Many chlamydospores appear able to survive desiccation (2, 25), and this may result in a reduced endogenous respiration rate, thus slowing down the exhaustion of endogenous reserves. The latent period for germination may be significantly lengthened by desiccation or low water potentials, however (2, 41).

Exogenous Carbon Metabolism and Germination

Data obtained by Cochrane et al. (12) suggested that both ethanol and acetate are oxidized by a combination of the tricarboxylic acid cycle and the glyoxylate cycle in *F. solani* f. sp. *phaseoli* macroconidia. The glyoxylate pathway was active also in Culmorum macroconidia (32). Glucose appeared to be metabolized via the Embden-Meyerhof pathway in *F. solani* f. sp. *phaseoli* (9). Acetate did not replace ethanol in supporting germination of this fungus at high spore density and, in addition, did not accelerate the production of CO^2 from exogenous glucose, as did ethanol (12). Ungerminated macroconidia oxidatively assimilated less acetate carbon than ethanol carbon. The greater reducing power of ethanol was not thought to be responsible for the effect of ethanol on germination, since increasing the level of glucose did not replace ethanol in germination for this fungus. However, Griffin (20) found increased glucose concentrations did duplicate the effect of ethanol on chlamydospore germination of a different clone of *F. solani*. Cochrane et al. (12) felt a key precursor of certain amino acids (acetaldehyde-diphosphothiamine) may have been responsible for the effect of ethanol on macroconidium germination. Ethanol depressed the endogenous respiration of *F. solani* f. sp. *phaseoli* macroconidia (12), while exogenous glucose did not affect endogenous respiration (9, 10).

More important than the effect of ethanol is the effect of exogenous inorganic nitrogen on exogenous carbon metabolism, because of its ecological significance and because of its apparent ability to reduce the energy requirement for germination (20). Ammonium, and the conidium germination process, had appreciable and similar effects on glucose metabolism in *F. solani* f. sp. *phaseoli* macroconidia (9). Glucose was respired more rapidly and completely in the presence of ammonium than in its absence. Amino acid synthesis and energy demands of protein synthesis may account for these responses to ammonium (9). In Culmorum macroconidial germination, there appeared to be a net increase in endogenous lipid utilization when ammonium was supplied with glucose (32). Along with synthesis of needed nitrogen-containing macromolecules, these stimulatory effects of ammonium would generate increased energy for the spore per carbon source molecule supplied, and may give rise to a lower exogenous energy requirement for germination. Similar metabolic effects of ammonium, and possibly of nitrate, may occur also during chlamydospore germination by Fusaria. Ammonium had no similar effect on ethanol oxidation by *F. solani* f. sp. *phaseoli* macroconidia (12).

Influence of the Physical and Chemical Enviroment on Germination

In addition to the effects of exogenous carbon and nitrogen substrates on germination, other physical and chemical factors often have a critical effect on conidium or chlamydospore germination. While macroconidia or chlamydospores may germinate in distilled water alone or in very dilute inorganic salt solutions (7, 19, 27), provision of a balanced inorganic salts medium provides a more suitable and natural environment for germination. Undesirable osmotic, pH, antagonistic ion, and cell membrane effects are less likely to occur when potassium and magnesium or calcium salts are present in the germination medium. For example, Byther (6) found KCl (or K_2HPO_4) and $MgSO_4$ stimulated macroconidium germination by *F. solani* f. sp. *phaseoli* in sodium phosphate buffer at pH 6.5. These effects of potassium and magnesium may be due, in part, to countering sodium ion toxicity and to stabilizing cell membrane function. Investigations demonstrating the stimulating effect of low concentrations of various osmotica (or decreased osmotic water potential) on mycelial growth of Fusaria (R. J. Cook, chapter 22 of this volume) have not been extended to include spore germination thus far. Armolik and Dickson (1) found *F. moniliforme* conidial germination less tolerant of low osmotic water potential than several Pencillia and Aspergilli (see also the section above on endogenous metabolism).

Although few precise studies on the influence of pH on conidium or chlamydospore germination have been made, most data indicate germination can occur over a wide range in pH (6, 7, 8). Chi and Hanson (8) found an isolate of *F. solani* from red clover germinated at high levels from pH 2.4 (final pH of medium = 2.7) to pH 9.0 (final pH of medium = 8.2). An optimum of pH 5 to 6 was found. In contrast, Byther (6) reported *F. solani* f. sp. *phaseoli* germinated poorly at pH 6.5 and above in solutions of low nutrient status. High germination was observed by Byther (6) and others (8, 11, 23) in the presence of high concentrations of carbon and nitrogen at pH 6.5 or 7.5. In addition to these modifying effects of nutrients, choice of buffer or physical nature of substrate may similarly modify response of macroconidia to pH.

As with many plant pathogens, Fusaria usually have an optimum temperature for conidium germination of 25 to 28 C (6, 23). Little or no germination typically occurs above 37 C, although only a few Fusaria have been examined. Macroconidia of low-temperature pathogens, such as *F. roseum* 'Sambucinum,' germinate readily at 6 C, but germination takes three times longer than at 25 C (23). Low temperature appeared to effect both the rate of germination and the latent period for germination. Both *F. solani* f. sp. *phaseoli* and *F. roseum* f. sp. *cerealis* macroconidia germinated optimally at 25 C, but germination by the latter fungus was affected less by reduced temperature (19 C) when inorganic nitrogen was supplied at high or low levels (6). Amino acid nitrogen sources may partially nullify the unfavorable effect of low or high temperatures on germination, as has been found with other fungi, but this has not been examined for Fusaria.

In many fungi, it is believed that CO_2 fixation is important during the early stages of spore germination, and that CO_2 is fixed into dicarboxylic acids of the Krebs cycle. Amino acids, purines and other key intermediates may then be synthesized and lead to incorporation into protein and nucleic acids. Fusaria appear to be tolerant to high levels of CO_2, but presently no uniform information exists on the role of CO_2 in spore germination by Fusaria. Bourret et al. (4) clearly demonstrated that increased levels of CO_2 (0–30%) stimulated chlamydospore germination by *F. solani* f. sp. *phaseoli* in the presence of glucose and ammonium in nonsterile soil, and CO_2 stimulated the growth of this fungus in sterile aqueous soil extracts (5). Part of the effect of high CO_2 levels on germination was attributed to a suppressive effect of CO_2 on competing soil microorganisms. In pure culture studies, removal of metabolic CO_2 from the germination medium accelerated macroconidium germination by *F. solani* f. sp. *phaseoli* in the early stages (5–6.5 hr) of germination (11). In contrast, macroconidial germination by another *F. solani* clone was 16–26% lower upon CO_2 removal under axenic conditions (G. J. Griffin, *unpublished data*). Germination was assayed after 7 hr.

Marchant and White (32) indicated that removal of CO_2 is inhibitory to conidium germination by Culmorum, but that increased CO_2 concentrations (not specified) inhibit macroconidium germination. Germinating macroconidia of this fungus fixed $^{14}CO_2$ into organic acids, glutamic acid, and glycerides, but the largest $^{14}CO_2$ activity was present in a compound tentatively identified as glutamine (32). $^{14}CO_2$ fixation increased up to 0.83% CO_2, the highest level tested. These workers suggested that CO_2 played a role in regulating chitin synthesis. The effect of CO_2 on germination does not appear to be due to a change in pH of the medium. The above findings, together with observations on *Fusarium* growth in inorganic media (20, 33, 35) and the inhibition of conidial germination by acidic soil volatiles (22), suggest further research is needed to clarify the role of CO_2 in spore germination of Fusaria.

Fusaria appear to have a tolerance to a wide range of O_2 concentrations. Formae speciales of *F. oxysporum* are able to grow at very low O_2 concentrations (0.01%), and *F. oxysporum* f. sp. *cubense* is able to grow under anaerobic conditions provided that yeast extract, MnO_2, nitrate, selenite, or ferric ions are present (26). Studies with *F. solani* f. sp. *phaseoli* indicated that maximum macroconidial germination ocurred at 3–4% O_2, under conditions of high spore density (11). It is likely that germination can proceed at low O_2 concentrations similar to those found for growth, especially at low spore densities.

Literature Cited

1. Armolik, N., and J. G. Dickson. 1956. Minimum humidity requirement for germination of conidia of fungi associated with storage of grain. Phytopathology 46:462–465.
2. Baker, K. F., and R. J. Cook. 1974. Biological control of plant pathogens. W. H. Freeman & Co. San Francisco. 433 p.
3. Barran, L. R. 1976. Transport of phenylalanine by conidia of *Fusarium sulphureum*. Can. J. Microbiol. 22:1390–1396.
4. Bourret, J. A., A. H. Gold, and W. C. Snyder. 1965. Inhibitory effect of CO_2 on chlamydospore formation in *Fusarium solani* f. *phaseoli*. Phytopathology 55:1052 (Abstr.).
5. Bourret, J. A., A. H. Gold, and W. C. Snyder. 1968. Effect of carbon dioxide on germination of chlamydospores of *Fusarium solani* f. sp. *phaseoli*. Phytopathology 58:710–711.
6. Byther, R. 1965. Ecology of plant pathogens in soil. V. Inorganic nitrogen utilization as a factor of competitive saprophytic ability of *Fusarium roseum* and *F. solani*. Phytopathology 55:852–858.
7. Cappellini, R. A., and J. L. Peterson. 1971. pH, nutrients and macrospore germination in *Gibberella zeae*. Mycologia 63:641–643.
8. Chi, C. C., and E. W. Hansen. 1964. Relation of temperature, pH, and nutrition to growth and sporulation of *Fusarium* spp. from red clover. Phytopathology 54:1053–1058.
9. Cochrane, V. W., S. J. Berry, F. G. Simon, J. C. Cochrane, C. B. Collins, J. A. Levy, and P. K. Holmes. 1963. Spore germination and carbon metabolism in *Fusarium solani*. III. Carbohydrate respiration in relation to germination. Plant Physiol. 38:533–540.
10. Cochrane, V. W., J. C. Cochrane, C. B. Collins, and F. G. Serafin. 1963. Spore germination and carbon metabolism in *Fusarium solani*. II. Endogenous respiration in relation to germination. Amer. J. Bot. 50:806–814.
11. Cochrane, J. C., V. W. Cochrane, F. G. Simon, and J. Spaeth. 1963. Spore germination and carbon metabolism in *Fusarium solani*. I. Requirements for spore germination. Phytopathology 53:1155–1160.
12. Cochrane, V. W., J. C. Cochrane, J. M. Vogel, and R. S. Coles, Jr. 1963. Spore germination and carbon metabolism in *Fusarium solani*. IV. Metabolism of ethanol and acetate. J. Bacteriol. 86:312–319.
13. Cochrane, J. C., T. A. Rado, and V. W. Cochrane. 1971. Synthesis of macromolecules and polyribosome formation in early stages of spore germination in *Fusarium solani*. J. Gen. Microbiol. 65:45–55.
14. Cook, R. J., and M. N. Schroth. 1965. Carbon and nitrogen compounds and germination of chlamydospores of *Fusarium solani* f. *phaseoli*. Phytopathology 55:254–256.
15. Ford, E. J., A. H. Gold, and W. C. Snyder. 1970. Interaction of carbon nutrition and soil substances in chlamydospore formation by *Fusarium*. Phytopathology 60:1732–1737.
16. Griffin, G. J. 1965. Chlamydospore formation in *Fusarium solani* 'Coeruleum.' Phytopathology 55:1060 (Abstr.).
17. Griffin, G. J. 1969. *Fusarium oxysporum* and *Aspergillus flavus* spore germination in the rhizosphere of peanut. Phytopathology 59:1214–1218.
18. Griffin, G. J. 1970. Carbon and nitrogen requirements for macroconidial germination of *Fusarium solani:* dependence on conidial density. Can. J. Microbiol. 16:733–740.
19. Griffin, G. J. 1970. Exogenous carbon and nitrogen requirements for chlamydospore germination by *Fusarium solani:* dependence on spore density. Can. J. Microbiol. 12:1366–1368.
20. Griffin, G. J. 193. Modification of the exogenous carbon and nitrogen requirements for chlamydospore germination of *Fusarium solani* by contact with soil. Can. J. Microbiol. 19:999–1005.
21. Griffin, G. J. 1976. Roles of low pH, carbon, and inorganic nitrogen source use in chlamydospore formation by *Fusarium solani*. Can. J. Microbiol. 22:1381–1389.
22. Griffin, G. J., T. S. Hora, and R. Baker. 1975. Soil fungistasis: elevation of the exogenous carbon and nitrogen requirements for spore germination by fungistatic volatiles in soil. Can. J. Microbiol. 21:1468–1475.
23. Griffin, G. J., and T. Pass. 1969. Behavior of *Fusarium roseum* 'Sambucinum' under carbon starvation conditions in relation to survival in soil. Can. J. Microbiol. 15:117–126.
24. Griffiths, D. A. 1973. Fine structure of the chlamydospore wall in *Fusarium oxysporum*. Trans. Brit. Mycol. Soc. 61:1–6.
25. Griffiths, D. A. 1973. Fine structure of chlamydospore germination in *Fusarium oxysporum*. Trans. Brit. Mycol. Soc. 61:7–12.
26. Gunner, H. B., and M. Alexander. 1964. Anaerobic growth of *Fusarium oxysporum*. J. Bact. 86:1309–1316.
27. Hsu, S. C., and J. L. Lockwood. 1973. Soil fungistasis: behavior of nutrient-independent spores and sclerotia in a model system. Phytopathology 63:334–337.
28. Kleinschuster, S. J., and R. Baker. 1974. Lectin-detectable differences in carbohydrate-containing surface moieties of macroconidia of *Fusarium roseum* 'Avenaceum' and *Fusarium solani*. Phytopathology 64:394–399.
29. Marchant, R. 1966. Fine structure and spore germination in *Fusarium culmorum*. Ann Bot. 30:441–445.
30. Marchant, R. 1966. Wall structure and spore germination in *Fusarium culmorum*. Ann. Bot. 30:821–830.

31. Marchant, R., and M. F. White. 1966. Spore swelling and germination in *Fusarium culmorum*. J. Gen. Microbiol. 42:237–244.
32. Marchant, R., and M. F. White. 1967. The carbon metabolism and swelling of *Fusarium culmorum* conidia. J. Gen. Microbiol. 48:65–77.
33. Mirocha, C. J., and J. E. de Vay. 1971. Growth of fungi on an inorganic medium. Can. J. Microbiol. 17:894–901.
34. Park, D. 1961. Morphogenesis, fungistasis and cultural staling in *Fusarium oxysporum* Snyder and Hansen. Trans. Brit. Mycol. Soc. 44:377–390.
35. Qureshi, A. A., and O. T. Page. 1970. Observations on chlamydospore production by *Fusarium* in a two-salt solution. Can. J. Microbiol. 16:29–32.
36. Rado, T. A., and V. W. Cochrane. 1971. Ribosomal competence and spore germination in *Fusarium solani*. J. Bacteriol. 106:301–304.
37. Robinson, P. M., and M. K. Garrett. 1969. Identification of volatile sporostatic factors from cultures of *Fusarium oxysporum*. Trans. Brit. Mycol. Soc. 52:293–299.
38. Robinson, P. M., and D. Park. 1966. Volatile inhibitors of spore germination produced by fungi. Trans. Brit. Mycol. Soc. 49:639–649.
39. Schneider, E. F., L. R. Barran, P. J. Wood, and I. R. Siddiqui. 1977. Cell wall of *Fusarium sulphureum*. II. Chemical composition of the conidial and chlamydospore walls. Can. J. Microbiol. 23:763–769.
40. Schneider, E. F., and A. B. Wardrop. 1979. Ultrastructure studies on the cell walls in *Fusarium sulphureum*. Can. J. Microbiol. 25:75–85.
41. Schneider, R. 1954. Untersuchungen über Feuchtigkeitsansprüche parasitischer Pilze. Phytopathol. Z. 21:63–78.
42. Sisler, H. D., and C. E. Cox. 1954. Effects of tetramethylthiuram disulfide on metabolism of *Fusarium roseum*. Amer. J. Bot. 41:338–345.
43. Stalhammar-Carlemalm, M. 1976. An electron microscope study of spore germination in *Fusarium sambucinum* var. *coeruleum*. Trans Brit. Mycol. Soc. 67:83–90.
44. Van Eck, W. H., and B. Schippers. 1976. Ultrastructure of developing chlamydospores of *Fusarium solani* f. *cucurbitae* in vitro. Soil Biol. Biochem. 8:1–6.
45. Wilson, E. M. 1960. Physiology of an isolate of *Fusarium oxysporum* f. *cubense*. Phytopathology 50:607–612.

31 Nutritional Requirements of *Fusarium oxysporum:* Basis for a Disease Control System*

S. S. Woltz and John Paul Jones

Fertilizers, Soil pH, and Disease

Variations in the incidence and severity of wilt diseases incited by *Fusarium oxysporum* have been produced by variations in fertilization and pH adjustment procedures (3, 4, 11, 16, 23, 28, 34, 45, 46, 47, 48). The most consistent observations related to liming practice, rate of nitrogen fertilization, and the chemical form in which nitrogen was applied. Liming soils to reduce acidity was frequently successful in reducing disease problems (7, 9, 14, 16, 17, 19, 20, 56, 57, 58, 59); when used with manure or large amounts of ammonium nitrogen, however, the lime treatment was ineffective (45, 56). Studies on the rate of nitrogen application generally showed more disease with higher rates of application (34, 57, 58, 59). The nitrate form of nitrogen was much less favorable to disease than the ammonium form (20, 56, 57, 58, 59); organic nitrogen favored disease more than inorganic nitrogen (45). Ammonium nitrogen became more favorable to disease as the nitrogen rate was increased, while nitrate nitrogen became increasingly unfavorable with increasing rate of application (56). *Fusarium oxysporum* cultured on ammonium nitrogen was more virulent than the same fresh weight of the organism cultured on nitrate nitrogen. Effects of nitrate and ammonium sources on disease were apparently related, in part at least, to soil pH effects; nitrate caused an elevation in soil pH while ammonium ion caused a reduction. Tomato plants grown in soil receiving nitrate plus lime had less disease than those grown in soil receiving ammonium nitrogen plus lime.

Raising soil pH toward or slightly above neutrality appears to be a common denominator in cultural control of Fusarium wilt, which commonly is a disease associated with acidic, sandy soils rather than heavier soils with higher pH values. Micronutrients apparently were deficient in respect to the requirements of *F. oxysporum,* since additions of iron, manganese, and particularly zinc significantly reversed the disease-retarding effects of liming (16, 17). Ammonium nitrogen applied in combination with sufficient lime to retain pH above neutrality reversed the effect of the lime even though pH remained at a high level (56). Rhizosphere pH was not measured but doubtless was lower than the pH of a general soil sample.

The inhibition of disease attributable to liming was reversed by soil applications of various combinations of iron, manganese, and zinc. The effects of these heavy metals appear to be subject to imbalance inter se (6, 13, 16, 17, 19, 24, 41, 42) as well as substitutive effects causing variations in activity of toxins (12, 26, 31, 38). The most effective element in terms of liming effect on nutrition appears to be zinc, which also

*Florida Agricultural Experiment Stations Journal Series No. 934.

needs to be present within the proper concentration range for toxin production (31). Micronutrient supply in the soil is a strategic weak point for control of *Fusarium,* since the fungus is more vulnerable to micronutrient deficiencies than the host plant. *Fusarium* is limited to locally available nutrients. Higher plants have a better micronutrient-extracting system because of the extensiveness of the root system and the solubilizing and transporting action of organic acids present in the root and excreted into the rhizosphere.

The effects of lime and the nitrate source of nitrogen in controlling *Fusarium* also are most likely related to lowering the concentrations of other nutrients in the soil solutions, namely, phosphorus, magnesium, sulfur, and possibly copper. Lacking these nutrients and iron, manganese, and zinc in adequate amounts, the Fusarium propagule is less likely to establish significant inoculum levels in the soil. The dependence of *Fusarium* on soil minerals is easily demonstrated by measuring growth of *Fusarium* on soil-agar plates made from fertilized and unfertilized soils. Virgin Myakka fine sand failed to support more than a trace of growth, while the same soil fertilized with 4-8-8 at 2000 pounds per acre grew *Fusarium* abundantly (S. S. Woltz and J. P. Jones, *unpublished data*). This method of evaluating soils has not been explored in detail, but the procedure represents a potential laboratory evaluation of the Fusarium support potential of various soils with and without supplemental fertilization.

Acid soil (pH 4.2) supported growth of *Fusarium* through the soil whereas a pH near neutrality prevented this growth (49). In vitro studies (27) indicated that this probably represents microbiological competition and antibiosis. Competitive effects of bacteria and actinomycetes, dependent on higher soil pH, have been demonstrated to lie in the competition for nutrients and to a lesser degree in antibiotic production. There is a gap here, however, between in vitro demonstrations and in situ action. It appears that a basic reason "fertile" non-acid soils do not support *F. oxysporum* and disease production is the competitive action of a healthy population of soil microflora. Also, the higher soil pH must frequently interfere with Fusarium nutrition in the heavier soils.

Soil amendment with nitrogen resulted in an increased incidence of Fusarium wilt of cotton, while phosphorus and potassium applied individually markedly reduced disease in comparison with an untreated control (34). Balanced fertilization with nitrogen, phosphorus, and potassium combined resulted in approximately the same incidence of wilt. Fertilization of wilt-sick soil in which flax was grown (23) resulted in a general reduction in incidence of Fusarium wilt compared with nonfertilized controls; this type of response occurred with ammonium nitrate, superphosphate, muriate of potash, lime, and sulfur when each was applied as a separate treatment. The effect of varied nitrogen and potassium nutrition was studied in solution culture for tomato relative to Fusarium wilt (45). Disease development increased in the following order: low nitrogen, high potassium, normal nutrition level of both, high nitrogen, and low potassium. This study indicated that high nitrogen and low potassium favored disease while low nitrogen and high potassium retarded it. Fusarium wilt (corm rot) of gladiolus generally has been worse with high levels of nitrogen nutrition (59). Adequate levels of calcium in terms of host plant requirements generally have been reported to favor wilt resistance (2, 5, 16, 22, 39). Boron deficiency increased the severity of Fusarium wilt (22).

Conflicting observations on fertilizer, pH, and plant nutrition effects are frequently encountered in the literature, which is not surprising in view of the intricacy of the Fusarium-host-soil complex. For direct applicability to disease control, a variation in fertilizer practices should not, on balance, economically reduce crop productivity. This narrows the list of available procedures. Some observed effects appear only to be related to the rate of plant growth and disease expression. Another consideration is that toxin production by *F. oxysporum* may be dependent on soil fertility and plant nutrition (6); a suppression of symptom expression may be related to toxin activity. The practical significance of disease control effects needs to be evaluated for specific objectives. Economic crop production may be adequate in one case while in another, plant material may need to be free of *Fusarium* for certification or propagation. Control usually is adequate if the disease only becomes a significant problem to plants after the normal harvest has been completed. The use of nutritionally oriented control of *F. oxysporum* in connection with other methods will be reviewed in a later section emphasizing the systems approach.

Fusarium Nutrition and Disease Control

Microorganism nutrition has received considerable attention as a tool in demonstrating the biologically essential elements (40). Pathogen studies have been undertaken to determine the response of the pathogen to variations in the chemical environment (10, 29, 35, 40, 51). Objectives of such studies were to delineate information about the effects of specific nutrients on growth, virulence, and survival of the pathogen in the biosphere.

In the case of *F. oxysporum* much attention has been given to carbon and organic nitrogen sources relative to in vivo and in vitro situations (15, 18, 32, 53). Carbon sources, in addition to providing the structural base for the organism, are also the customary energy sources, although other energy sources might be significant in vivo, namely, elemental sulfur, hydrogen, hydrogen sulfide, or ammonium compounds. This organism possesses a very highly developed versatility in utilization of compounds permitting growth and survival under many chemical-physical environments. Enzymatic adaptation is readily accomplished by *F. oxysporum* and facilitates growth and pathogenicity. Enzyme production is conservative, generally being formed as required; if not needed, such as when alternates are available, enzyme synthesis may be at very low levels. Exposure to nitrate and limitation to pectins or cellulose as carbon sources, for example, activates enzymes or enzyme synthesizing mechanisms. Availability of alternate nutrient sources frequently "short-circuits" enzyme synthesis. Glucose availability retards synthesis and activation of pectolytic enzymes. These features of *F. oxysporum* metabolism should, on further study, yield information relative to control procedures. Survival of chlamydospores depends on nutritional considerations in soil with regard to sugars and amino acids (37).

The stages of growth, decline, or resting of a Fusarium population in soil depend on the ecological balance and nutrient availability. The most critical limitation is chemical energy and the competitive use of the energy sources present in the soil. Other microorganisms (bacteria, actinomycetes, and other fungi) compete according to the magnitude of their genetically controlled reproductive capacity. Bacteria under favorable conditions are obviously most competitive. *Fusarium oxysporum* control by competitive microorganisms seems to be mainly due to their depletion of available carbon sources, but also to antibiotic production (27). Systems of Fusarium control should include an analysis of competition by soil microflora and methods of favoring competition.

Fusarium oxysporum is very capable as an autotroph, requiring only a carbon source, for structure and energy, and inorganic compounds. The list of essential nutrient elements currently includes carbon, hydrogen, oxygen, nitrogen, phosphorus, potassium, magnesium, sulfur, iron, manganese, molybdenum, and zinc (40). Copper has not been shown to be indispensable (40, 51). Chloride has not been shown to be essential to *F. oxysporum* but may benefit the disease-producing functions of the fungus. Chloride activates a number of pectolytic and amylolytic enzymes. The amounts required are of considerable magnitude and are not in the trace element category. Other elements will doubtless be added to this list with continued research and methodology development in purification of media ingredients and in preventing nutrient element contamination from culture vessels, air, cotton, and other sources.

Sources of carbon useful to *F. oxysporum* include most sugars, alcohols, pectins, cellulose, and amino acids, and a wide variety of additional carbon compounds. Nitrogen sources include amino acids, proteins, urea, and nitrate and ammonium compounds. The presence of specific amino acids in the combined culture of tomato and *F. oxysporum* resulted in deviation from normal *F. oxysporum* f. sp. *lycopersici* races 1 and 2 classification relative to tomato cultivar susceptibility (15, 18, 53). Tomato cultivars susceptible to wilt developed much greater contents of free amino acids in plant tissue than did resistant cultivars (33).

The classical work of R. A. Steinberg (40) led to some new concepts in disease control through soil fertility management (51). The principal themes pursued at the University of Florida (7, 8, 9, 16, 18, 20, 50, 51, 52, 53, 54, 55, 56, 57, 58, 59) have been that a promising approach to the control of this soil-borne disease includes: i) an evaluation of the nutritional requirements of *F. oxysporum;* ii) a study of the disease reactions to varied levels, sources, and combinations of nutrients; iii) an evaluation of effects upon disease by varied liming procedures; and iv) application of the information to control *F. oxysporum* diseases.

The main concept that was developed is that *F. oxysporum* is subject to a practical degree of control of virulence by soil fertility management procedures which limit nutrient availability to the fungus without adversely affecting the crop (9, 19, 20, 51, 55). To date, the simple procedures of liming sandy soils to the pH range of 6.0 to 7.5, together with the use of principally nitrate nitrogen have resulted in a significant degree of disease control for the horticultural crops aster, chrysanthemum, cucumber, gladiolus, tomato, and watermelon (7, 8, 9, 16, 17, 20, 55, 56, 57, 58, 59). It is likely that this work can be improved upon by using additional adjustments in soil fertility management. For example, fertilizer placement under plastic mulch at the soil surface reduces the nutrient content of the main volume of soil to the apparent detriment of *F. oxysporum* that is distributed throughout the soil, without damage to the crop. As another example, relatively low levels of a given nutrient could be tolerated in the soil if the crop received foliar nutritional sprays. Avoidance of broadcast fertilizer practices and excessively high nutrient applications will probably reduce *F. oxysporum* damage in sandy soils of low native fertility. These methods have not been thoroughly studied for feasibility but do illustrate a potential worth investigating in nutritionally oriented control of *F. oxysporum.*

Systems Approach

The systems approach to problem solving seems to be well adapted to the *F. oxysporum* disease because it needs to be analyzed systematically and inclusively at one point in order to provide for consideration of the many factors in the disease process. By viewing the system as a whole, research and control efforts can be conducted against a common background of a system so results may be compared and combined synthetically. Analysis of the overall system indicates the need for an interdisciplinary approach. The system proposed in the form of an equation for outlining and description is: *F. oxysporum* + host plant + soil + microflora + environment = disease. The principal feature in the systems approach is that the entire system should ultimately be charted. Objectives should be listed, as well as the order of priorities. Components and areas of the system should be subject to research in sequence to maximize productivity on a fundamental or practical basis. Priority should be decided on the basis of the importance of existing problems, the state of development of methodology, and probable ease of solution. When the system has been delineated, it should be published for reference and use by interested researchers. Steps in the use of a systems chart would include revision, expansion, and completion. The systems approach as envisioned here does not represent a significant deviation from conventional research but does require an overview of the problem area, including factors that are not thought to be of direct importance.

Objectives in the disease systems equation above are: i) to develop soil fertility adjustment programs that will minimize disease losses in crop production; ii) to combine these practices with other components of the system to further increase the level of disease control; iii) to provide special techniques for plant pathology research, e.g., plant breeding stock evaluation, where it is necessary to control the level of severity of disease to assure that resistance is evaluated quantitatively; and iv) to provide assistance in special plant culture programs, such as propagation, where production and maintenance of disease-free material is essential.

Components of the overall system include: 1) plant breeding, 2) exclusion, 3) pathogen eradication, 4) chemical fungicide control, 5) sanitation, and 6) soil fertility management. The last item is the component under consideration in the present paper, specifically, the role of soil fertility management in control of *F. oxysporum.*

Systems Factors for Fusarium Control

- I. Physical
  - A. Light
  - B. Temperature
  - C. Pressure
  - D. Electromagnetic radiation in addition to visible light
  - E. Soil structure
  - F. Water-holding capacity
- II. Chemical
  - A. Soil atmosphere: O_2, CO_2, H_2S, ethylene, methane, etc.
  - B. Soil moisture: free, capillary, and combined water.
  - C. *Fusarium oxysporum* essential nutrient elements for growth: C, H, O, N, P, K, Mg, S, Fe, Mn, Mo, Zn (Cu, Cl ?)
  - D. Higher plant essential elements in addition to those listed above in (C): Ca, B, Cu, Cl
  - E. Some elements essential to other life forms: F, Si, Se, Sn, V, Co, Li, Na, I
  - F. Elements significant by their presence in soil: Al, Pb, Hg, As, Cd, Br, Cr, Rb
- III. Biological
  - A. Host plants
  - B. Non-host plants
  - C. Bacteria
  - D. Actinomycetes
  - E. Animals
- IV. Cultural practices
  - A. General soil management: fertilization, liming, sulfur application, foliar nutritional sprays, mulches, herbicides, cover crops, weeds, fallowing, flooding, deep-plowing.
  - B. Crop management: cultivars, propagation material, cultivation, herbicides, soil fungicides, systemic fungicides, steaming, culture media.

A research program on one facet of the system should be undertaken with provision for relating to the system in the interest of applicability. The ultimate use of the research product should relate to the system to the extent that it accomplishes positive development that need not be redone in a different manner to be meaningful; increments of accomplishment should be useful in building solutions to the problem. The system component above involves too many contributing factors for disease to be successfully solved by mathematical modeling. Repeating and verification are essential. Research should be relevant to fundamental processes and applied uses to be used in the synthesis of coordinated solutions to the disease problem.

Methodology and Systems

Combined experience on the subject of *F. oxysporum* nutrition and disease has produced some observations on methodology requirements. The systems approach organization suggests that the applicability of results will be greatest when experimental conditions are established for environmental factors within limits that can be related to the final disease problem. This falls within the customary realm of plant pathology; however, clarification and emphasis are in order. The criteria below are intended to relate to the systems approach.

Solution cultures which may be shaken or stationary provide a method of culturing *F. oxysporum* useful in studying the essential chemical elements and compounds and in providing standardized inoculum. Study of essential compounds will involve purification of chemicals (35, 40, 51) and the use of contaminant-free containers such as polypropylene flasks covered with polypropylene beakers. Growth characteristics, abnormal metabolism, and cumulative effects of staling products need to be taken into account in avoiding holding cultures too long and in preventing unwanted anaerobiosis. Solution pH and salt concentrations should be kept in a normal range during periods of culture. Calcium carbonate and selected sources of nitrogen and phosphorus will prevent excessive acidity or alkalinity. Metabolism may be more representative in shaken cultures due to better gas exchange than in stationary cultures. If desired, filtered air may be forced through cultures. Stationary cultures are easy to use, however, and for micronutrient studies demonstrate deficiencies more easily, because of lower efficiency of extraction of nutrients from solutions, containers, and cotton plugs. Conclusions from this type of research should be incorporated in the systems approach to disease control by whatever methods are necessary.

Culture on solid media in petri dishes provides advantages in permitting a study of the growth characteristics of the colony and the development of sporulation structures and spores as compared with liquid cultures. Agar, soil-agar, and surface-sterilized living plant parts are useful to obtain information applicable to in vivo situations. Nutritional deficiencies are easily demonstrated in soil-agar plates leading to indirect evidence for features of a control system. For example, if acidic sandy soils of low native fertility are not fertilized or provided with nutrients, *F. oxysporum* growth is severely limited. Tests may be made to determine deficiencies in reference to the nutritional requirements of *Fusarium.* An analytical approach may thus be made relative to the native capacity of soils to support growth of *Fusarium,* sporulation, virulence (through inoculation tests), and survival as chlamydospores. Autoclaved soils produce quite different results than nonautoclaved soils because of competition by the soil microflora. In vivo soil culture, with native microflora present, must be incorporated into the studies. The survival and proliferation of a strain of *F. oxysporum* may be followed, after infestation, by direct microscopic examination, soil dilution plates, or by inoculating host plants as a bioassay.

The role of host nutrition is not easily extricated from the overall disease situation in most instances (1, 2, 5, 6, 21, 22, 25, 29, 30, 33, 36, 39, 43, 44), since soil or another substrate is usually the common meeting ground of *F. oxysporum* and host. Plant material may, however, be inoculated and held for disease development, representing host nutrition conditioning effects that began prior to inoculation; since nutrient uptake from the environment is interrupted, the condition of plant material samples remains essentially constant in regard to gain or loss of nutrients. Seedlings or plant propagation material representing varied nutritional backgrounds may be held for development of disease in direct response to pre-inoculation nutrition. If plants are grown under a given nutritional regime in solution culture before inoculation and then transferred to another regime after inoculation (22), effects may be ascribed to pre- and post-inoculation conditioning; interpretation may be complicated by secondary responses of host and *Fusarium* to the sequence of nutritional regimes. Foliar nutrition of host plants (1, 21) represents an approach to studying the effect of host nutrition on disease processes as well as a procedure for by-passing the nutrition of soil-borne *Fusarium*. Research is needed to isolate and measure the effects of host nutrition. It will be necessary to develop additional methodology to define roles of host nutrition in the disease response.

A range of experimental approaches is necessary in going from pure culture of *Fusarium* to greenhouse pot experiments, to field plot research and finally to application in the hands of growers. For example, grower methods of liming and fertilizing must be evaluated for disease control in order that research procedures can be applied to the field. Soil-infestation procedures should be evaluated to obtain a reasonable inoculum level in reference to the problem in the field. Pathogen population and added culture media should not unbalance the disease response. It may be necessary to establish suitable inoculum levels in preliminary experiments. The objective is to provide an adequate but not too rigorous test of a single component in a system. Varied inoculation procedures such as root wounding, and soil drenching should be examined for relevance to disease development in the field. The fate of *Fusarium* in the plant needs to be traced concurrently (30). Speed of development of disease and expression of symptoms should be within a normal range. If the rigorous methods of infestation or inoculation do not overcome nutritional disease control, then one need not search further. For example, the combined approaches to disease control by lime application plus nitrate-nitrogen use (20, 57, 59) plus fungicide supplementation (7, 9) have been successful in the face of rigorous methods of infestation and inoculation. A coordinated approach using all available adjustments to inhibit Fusarium disease processes should be the optimum.

Summary

Fusarium oxysporum diseases are affected by soil fertility practices. High levels of nitrogen fertilization frequently encourage disease development. High levels of potassium reduce severity of disease but appear to be related to a balance between potassium and nitrogen. If nitrogen is in surplus supply over potassium, disease development is more severe, and as potassium is supplied, disease development is inhibited. This information is only of limited application to disease control in the field since normal crop production requires adequate levels of both nitrogen and potassium; however, excessive rates of nitrogen should be avoided.

Relatively low levels of calcium appear more conducive to disease than normal levels; cation antagonism by competitive cations such as sodium adversely affects disease resistance of the host plant. Boron deficiency of host plants increases disease severity. When multiple nutrient deficiencies prevent a normal rate of plant growth, an applied nutrient that partially overcomes growth limitation may affect disease response. Increased growth permits symptom expression and toxin dilution. Specific disease responses may be related to toxin activation by heavy metals. Research results from normal plant growth are more directly applicable to disease control.

Fusarium oxysporum diseases were controlled significantly in six horticultural crops by limited adjustments in soil fertility. Sandy soils were limed uniformly throughout the soil volume to pH 6.5 to 7.5 which controlled Fusarium disease, in part at least, by limiting micronutrient availability. The effect on disease was probably related to other nutrient deficiencies as well as enhanced competition by bacteria and actinomycetes. Increasing rates of ammonium nitrogen caused increasing disease severity. Nitrate, when furnished as the principal nitrogen source, inhibited disease development but the inhibition was not as rate-dependent as for ammonium nitrogen. Fusarium inoculum cultured with ammonium nitrogen was more virulent than that grown with nitrate nitrogen.

Fusarium oxysporum requires 12 nutrient elements and a source of energy for normal growth as presently established. These are absolute requirements which, if unsatisfied, will limit growth, sporulation, and survival of the fungus. The disease-producing potential can be limited by limiting nutrient and energy availability. A systems approach procedure is outlined in this paper to orient research efforts for the control of diseases incited by *F. oxysporum*.

Literature Cited

1. Balasubramanian, A., and G. Rangaswami. 1969. Studies on the influence of foliar nutrient sprays on the root exudation pattern in four crop plants. Plant Soil 30:210–220.
2. Corden, M. E. 1965. Influence of calcium nutrition on Fusarium wilt of tomato and polygalacturonase activity. Phytopathology 55:222–224.
3. Couch, H. B., and J. R. Bloom. 1958. Influence of soil moisture, pH and nutrition on the alteration of disease proneness in plants. Trans. N. Y. Acad. Sci. 1958:432–437.
4. Davet, P., C. M. Messiaen, and P. Rieue. 1966. Interpretation des manifestations hivernales de la fusariose de la tomate en Afrique du nord, favorisées par la presence de sels dan les eaux d'irrigation. Rapport INRAT 1966:407–416.
5. Edgington, L. V., and J. C. Walker. 1958. Influence of calcium and boron nutrition on development of Fusarium wilt of tomato. Phytopathology 48:324–326.
6. Egli, T. A. 1969. Untersuchungen uber den Einfluss von Schwermetallen auf *Fusarium lycopersici* Sacc. und den Krankheitsverlauf der Tomatenwelke. Phytopathol. Z. 66:223–252.
7. Engelhard, A. W. 1975. Aster Fusarium Wilt: Complete symptom control with an integrated fungicide-NO_3-pH control system. Proc. Amer. Phytopathol. Soc. 2:62 (Abstr.).
8. Engelhard, A. W., and S. S. Woltz. 1971. Fusarium wilt of chrysanthemum: Symptomatology and cultivar reactions. Proc. Fla. State Hort. Soc. 84:351–354.
9. Engelhard, A. W., and S. S. Woltz. 1973. Fusarium wilt of chrysanthemum: Complete control of symptoms with an integrated fungicide-lime-nitrate regime. Phytopathology 63:1256–1259.
10. Hendrix, F. F., and T. A. Toussoun. 1964. Influence of nutrition on sporulation of the banana wilt and bean root rot Fusaria on agar media. Phytopathology 54:389–392.
11. Huber, D. M., and R. D. Watson. 1972. Nitrogen form and plant disease. Down to Earth 27:14–15.
12. Jhamaria, S. L. 1972. Nutritional requirement of *Fusarium oxysporum* f. *niveum*. Indian Phytopathol. 25:29–32.
13. Jones, J. P., and S. S. Woltz. 1967. Fusarium wilt (race 2) of tomato: Effect of lime and micronutrient soil amendments on disease development. Plant Dis. Rep. 8:645–648.
14. Jones, J. P., and S. S. Woltz. 1968. Field control of Fusarium wilt (race 2) of tomato by liming and stake disinfestation. Proc. Fla. State Hort. Soc. 81:187–191.
15. Jones, J. P., and S. S. Woltz. 1969. Effect of ethionine and methionine on the growth, sporulation, and virulence of *Fusarium oxysporum* f. sp. *lycopersici* race 2. Phytopathology 59:1464–1467.
16. Jones, J. P., and S. S. Woltz. 1969. Fusarium wilt (race 2) of tomato: Calcium, pH, and micronutrient effects on disease development. Plant Dis. Rep. 53:276–279.
17. Jones, J. P., and S. S. Woltz. 1970. Fusarium wilt of tomato: Interaction of soil liming and micronu-

trient amendments on disease development. Phytopathology 60:812–813.
18. Jones, J. P., and S. S. Woltz. 1972. Effect of amino acids on development of Fusarium wilt of resistant and susceptible tomato cultivars. Proc. Fla. State Hort. Soc. 85:148–151.
19. Jones, J. P., and S. S. Woltz. 1972. Effect of soil pH and micronutrient amendments on Verticillium and Fusarium wilt of tomato. Plant Dis. Rep. 56:151–153.
20. Jones, J. P., and S. S. Woltz. 1975. Effect of liming and nitrogen source on Fusarium wilt of cucumber and watermelon. Proc. Fla. State Hort. Soc. 88:200–203.
21. Kannaiyan, S., and N. N. Prasad. 1975. Influence of foliar nutrient sprays on the population of *Fusarium oxysporum* f. *melonis* and other soil microflora in the rhizosphere of muskmelon. Indian Phytopathol. 27:527–531.
22. Keane, E. M., and W. E. Sackston. 1970. Effects of boron and calcium nutrition of flax on Fusarium wilt. Can. J. Plant Sci. 50:415–422.
23. Kommedahl, T., J. J. Christensen, and R. A. Fredericksen. 1970. A half century of research in Minnesota on flax wilt caused by *Fusarium oxysporum*. Minn. Agric. Exp. Stn. Tech. Bull. 273. 35 p.
24. Kern, H., and S. Naef-Roth. 1966. Kupfer, Aluminium und Eisen als krankheitshemmende Faktoren bei der Fusskrankheit der Erbsen. Phytopathol. 57:289–297.
25. Louvet, J. 1972. Signification de l'étude qualitative and quantitative des champignons phytopathogènes telluriques. OEPP-EPPO Bul. 7:5–15.
26. Marshall, B. H., Jr. 1955. Some effects of inorganic nutrients on the growth and pathogenicity of five fungal pathogens of gladiolus. Phytopathology 45:676–680.
27. Marshall, K. C., and M. Alexander. 1960. Competition between soil bacteria and *Fusarium*. Plant Soil 12:143–148.
28. Moore, H., and C. Chupp. 1952. A physiological study of the Fusaria causing tomato, cabbage and muskmelon wilts. Mycologia 44:523–532.
29. Patil, S. S., and A. E. Dimond. 1968. Repression of polygalacturonase synthesis in *Fusarium oxysporum* f. sp. *lycopersici* by sugars and its effect on symptom reduction in infected tomato plants. Phytopathology 58:676–682.
30. Pennypacker, B. W., and P. E. Nelson. 1972. Histopathology of carnation infected with *Fusarium oxysporum* f. sp. *dianthi*. Phytopathology 62:1318–1326.
31. Prasad, Y. 1972. Zinc in the toxin metabolism of *Fusarium oxysporum* Schlecht. ex Fr. f. *lini* (Bolley) Snyd. & Hans. Indian J. Agric. Sci. 42:950–952.
32. Prasad, M., and S. K. Chaudhary. 1973. Variations in amino acids and sugars and their effect on growth and sporulation in *Fusarium oxysporum* f. *udum*. Phytopathol. Z. 78:147–151.
33. Raggi, V., A. Zazzerini, B. Barberini, F. Ferranti, and R. Draoli. 1974. Alterations in the composition of free and combined amino acids in susceptible and resistant tomato cultivars infected with *Fusarium oxysporum* f. *lycopersici*. Phytopathol. Z. 79:258–280.
34. Sadasivan, T. S. 1965. Effect of mineral nutrients on soil micro-organisms and plant disease, p. 460–469. *In* K. F. Baker and W. C. Snyder (ed.), Ecology of soil-borne plant pathogens, Univ. Calif. Press, Berkeley.
35. Saraswathi-Devi, L. 1958. Essentiality of trace elements to some soil fungi. J. Indian Bot. Soc. 37:509–517.
36. Schroeder, W. T., and J. C. Walker. 1942. Influence of controlled environment and nutrition on the resistance of garden pea to Fusarium wilt. J. Agric. Res. 65:221–248.
37. Smith, S. N., and W. C. Snyder. 1975. Persistence of *Fusarium oxysporum* f. sp. *vasinfectum* in fields in the absence of cotton. Phytopathology 65:190–196.
38. Soni, N. K., and K. M. Uyas. 1973. Respiratory and growth responses of *Fusarium oxysporum* Sacc. induced by various substances. Indian J. Exp. Biol. 11:217–219.
39. Standaert, J. Y., C. Myttenaere, and J. A. Meyer. 1973. Influence of sodium/calcium rations and ionic strength of the nutrient solution on Fusarium wilt of tomato. Plant Sci. Letters 1:413–420.
40. Steinberg, R. A. 1950. Growth on synthetic nutrient solutions of some fungi pathogenic to tobacco. Amer. J. Bot. 37:711–714.
41. Subramanian, D. 1956. Role of trace element chelation in the Fusarium wilt on cotton. Proc. Indian Acad. Sci., Sect. B, 43:302–307.
42. Sulochana, C. B. 1952. Soil conditions and root diseases. IV. The effect of micro-elements on the occurrence of bacteria, actinomycetes and fungi in soils. Proc. Indian Acad. Sci., Sect. B. 36: 19–33.
43. Sulochana, C. B. 1952. Soil conditions and root diseases. VII. Response of cotton plants to micro-element amendments and its relation to disease development. Proc. Indian Acad. Sci., Sect. B 36:234–242.
44. Timonin, M. I. 1940. The interaction of higher plants and soil micro-organisms. II. Study of the microbial population of the rhizosphere in relation to resistance of plants to soil-borne diseases. Can. J. Res. 18:444–455.
45. Walker, J. C. 1971. Fusarium wilt of tomato. Monograph 6, Amer. Phytopathol. Soc., Minneapolis, Minn. 56 p.
46. Walker, J. C., and R. E. Foster. 1946. Plant nutrition in relation to disease development. III. Fusarium wilt of tomato. Amer. J. Bot. 33:259–264.
47. Walker, J. C., and W. J. Hooker, 1945. Plant nutrition in relation to disease development. I. Cabbage yellows, Amer, J. Bot. 32:314–320.
48. Walker, J. C., and W. J. Hooker. 1945. Plant nutrition in relation to disease development. II. Cabbage clubroot. Amer. J. Bot. 32:487–490.

49. Wilson, I. M. 1946. Observations on wilt disease of flax. Trans. Brit. Mycol. Soc. 29:221–231.
50. Woltz, S. S. 1974. Gladiolus Fusarium disease: Assay of soilborne inoculum potential and cultivar susceptibility. Plant Dis. Rep. 58:184–187.
51. Woltz, S. S., and J. P. Jones. 1968. Micronutrient effects on the in vitro growth and pathogenicity of *Fusarium oxysporum* f. sp. *lycopersici*. Phytopathology 58:336–338.
52. Woltz, S. S., and J. P. Jones. 1971. Effect of varied iron, manganese and zinc nutrition on the in vitro growth of Race 2 *Fusarium oxysporum* f. sp. *lycopersici* and upon the wilting of tomato cuttings held in filtrates from cultures of the fungus. Proc. Fla. State Hort. Soc. 84:132–135.
53. Woltz, S. S., and J. P. Jones. 1970. Effects of twenty natural amino acids on pathogenesis of Homestead 24 tomato by *Fusarium oxysporum* f. sp. *lycopersici* race 1 and 2. Proc. Fla. State Hort. Soc. 83:175–179.
54. Woltz, S. S., and J. P. Jones. 1972. Response of Manapal tomato seedlings to variations in inorganic nutrition. Proc. Fla. State Hort. Soc. 85:175–177.
55. Woltz, S. S., and J. P. Jones. 1972. Tomato Fusarium wilt control by adjustments in soil fertility: A systematic approach to pathogen starvation. Bradenton AREC Univ. Fla. Res. Rep. GC1973-7. 4p.
56. Woltz, S. S., and J. P. Jones. 1973. Interactions in source of nitrogen fertilizer and liming procedure in the control of Fusarium wilt of tomato. HortScience 8:137–138.
57. Woltz, S. S., and J. P. Jones. 1973. Tomato Fusarium wilt control by adjustments in soil fertility. Proc. Fla. State Hort. Soc. 86:157–159.
58. Woltz, S. S., and A. W. Engelhard. 1973. Fusarium wilt of chrysanthemum: Effect of nitrogen source and lime on disease development. Phytopathology 63:155–157.
59. Woltz, S. S., and R. O. Magie. 1975. Gladiolus Fusarium disease reduction by soil fertility adjustments. Proc. Fla. State Hort. Soc. 88:559–562.

32 Metabolites of Fusarium

R. F. Vesonder and C. W. Hesseltine

Fusarium, a genus containing many common soil saprophytes as well as plant pathogens, is frequently found in cereal grains. Many species of *Fusarium* produce a number of secondary metabolites which elicit dissimiliar physiological and pharmacological responses in plants and animals. Recognized responses are: Mycotoxicoses (hemorrhagic diseases, alimentary toxic aleukia, estrogenism, and emesis-refusal response) in animals and humans on ingestion of grains infected with *Fusarium;* phytotoxicoses of plants; and initiation of the sexual stage in some fungi.

In the past decade, mycotoxins attained high research priority because of the effects of aflatoxins in turkey poults and their extreme carcinogenicity. Mycotoxins produced by the Fusaria are the 12, 13-epoxy-Δ^9-trichothecenes. *Fusarium* spp. elaborate more than half of the 37 trichothecenes which have been isolated. These metabolites are also produced by species of *Trichoderma, Trichothecium,* and *Stachybotrys.*

Many papers in the literature describe the toxic effect of trichothecenes in laboratory animals, and extrapolate these biological effects to man and farm animals. However, only the trichothecenes T-2 and vomitoxin have been isolated from grains molded by *Fusarium.* T-2 is associated with lethal toxicoses to dairy cattle, and vomitoxin produces a refusal-vomiting phenomenon in swine. An estrogenic toxin, zearalenone, is also commonly found in corn infected with *Fusarium* spp.

This chapter will list, in addition to the trichothecenes, specific groups of metabolites from *Fusarium,* based on biological activity. These groups are: Pigments, mycotoxins, antibiotics, phytotoxins, and phytoalexins. A miscellaneous group will include derivatives of zearalenone, di- and tri-terpenoids, along with various other ring-structured compounds whose biological activity has not been investigated. The following microbial products are excluded: Alicyclic compounds (steroids and gibberellins), aliphatic and related compounds (carboxylic acids, carotenoids), enzymes, simple nitrogen compounds.

The compounds listed in each biological category will include structure, physiochemical data, the reference source of structure, and biological activity. When a compound possesses various biological activities the compound will be cross-indexed.

Although there is extensive literature on *Fusarium* metabolites, we will list only key references, with several reviews and books. Our literature search included Chemical Abstracts and books on fungal metabolites (Turner, 1971; Miller, 1961; Shibata, Natori, and Udagawa, 1964; and Korszybski, Kowszyk-Gindifer, and Kurylowicz, 1967). Besides *Fusarium,* literature inquiry encompassed the perfect state names for *Fusarium,* including *Calonectria, Gibberella, Nectria,* and *Hypomyces.*

Our search indicated that some compounds were listed under more than one name; these are also reported. We were greatly aided by use of the *Dictionary of the Fungi*

(Ainsworth 1971). As one would expect, a number of toxins were reported for which structure has not yet been determined.

General References

1. Ainsworth, G. C. 1971. Ainsworth and Bisby's dictionary of the fungi, 6th ed. Commonwealth Mycol. Inst. Kew, Surrey, England. 663.
2. Brian, P. W. 1951. Antibiotics produced by fungi. Bot. Rev. 17:357–430.
3. Broadbent, D. 1966. Antibiotics produced by fungi. Bot. Rev. 32:219–242.
4. Korzybski, T., Z. Kowszyk-Gindifer, and W. Kurylowicz. 1967. Antibiotics, origin, nature, and properties. Vol. II, p. 1310–1325, 1402. Pergamon Press, London.
5. The Merck Index. 1968. 8th ed. Merck & Co., Rahway, New Jersey. 1713 p.
6. Miller, M. W. 1961. The Pfizer handbook of microbial metabolites. McGraw-Hill Book Co., New York. 722 p.
7. Scott, P. M. and E. Somers. 1969. Biologically active compounds from field fungi. J. Agric. Food Chem. 17:430–436.
8. Shibata, S., S. Natori, and S. Udagawa. 1964. List of fungal products. Charles C. Thomas, Springfield, Illinois. 170 p.
9. Turner, W. B. 1971. Fungal metabolites. Academic Press, New York. 446 p.

Pigments

1. AUROFUSARIN. $C_{30}H_{18}O_{12}$ MW 570 Yellow prisms, m.p. >320 C d. $\lambda_{max}^{dioxane}$ 243, 267, 372.

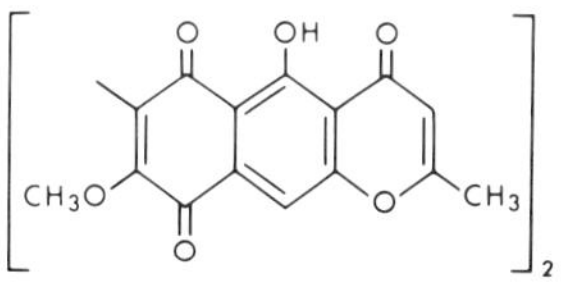

Source: Fusarium culmorum, F. graminearum (CMI 89367). *References:* Ashley, J. N., B. C. Hobbs, and H. Raistrick, *Biochem. J. 31,* 385 (1937); Baker, P. M., and J. C. Roberts, *J. Chem Soc.* C, 2234 (1966); Gray, J. S., G. C. J. Martin, and W. Ribgy, *J. Chem. Soc.* C, 2580 (1967); Shibata, S., E. Morishita, T. Takeda, and K. Sakata, *Tetrahedron Lett.* 4855 (1966).

2. BIKAVERIN, LYCOPERSIN, PASSIFLORIN, MYCOGONIN (7, 10-Dihydro-6, 11-dihydroxy-3, 8-dimethoxy-1-methyl-12H-benzo[b]xanthene-7, 10, 12-trione). $C_{20}H_{14}O_8$ MW 382 Red crystals, m.p. 322–324 C d. $\lambda_{max}^{CHCl_3}$ 254, 278, 518. Vacuolation factor, aging factor; active against *Leishmania brasiliensis,* 0.15 μg/ml.

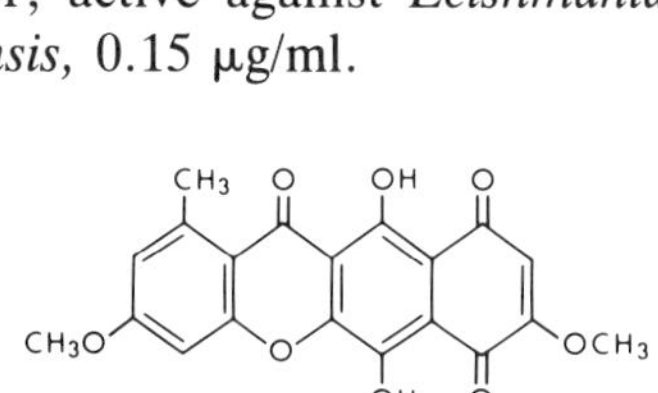

Source: Gibberella fujikuroi NRRL 2633 (Brian 917), *Fusarium oxysporum. References:* Balan, J., J. Fuska, I. Kuhr, and V. Kuhrova, *Folia Microbiol.* (Prague) *15,* 479 (1970); de Boer, J. J., D. Bright, G. Dallinga, and T. G. Hewitt, *J. Chem. Soc.* C, 2788 (1971); Brewer, D., G. P. Arsenault, J. L. C. Wright, and L. C. Vining, *J. Antibiot., 26,* 778 (1973); Kjaer, D., A. Kjaer, C. Pedersen, J. D. Bu'Lock, and J. R. Smith, *J. Chem. Soc.* C, 2788 (1971).

3. BOSTRYCOIDIN. $C_{15}H_{11}NO_5$ MW 285 Red plates, m.p. 243–244 C. λ_{max}^{EtOH} 251, 320, 475, 497, 525. LD_{50} mice, i.p., s.c., or per os, >250 mg/kg for 2–21 days. Inhibits acid-fast, gram-positive, and negative organisms.

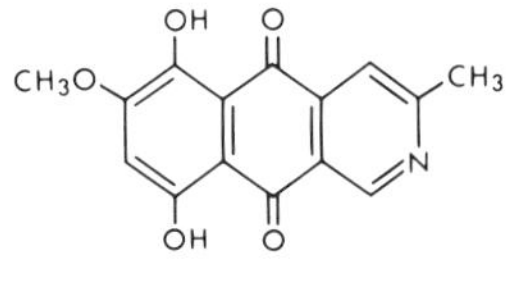

Source: Fusarium oxysporum (Fusarium bostrycoides), F. solani. References: Arsenault, G. P., *Tetrahedron Lett.,* 4033 (1965); Cajori, F. A., T. T. Otani, and M. A. Hamilton, *J. Biol. Chem., 208,* 107 (1954); Hamilton, M. A., M. S. Knorr, F. A. Cajori, *Antibiot. Chemother., 3,* 853 (1953).

4. CEPHALOCHROMIN. $C_{28}H_{22}O_{10}$ MW 518 Orange crystals, m.p. >300 C

d. λ_{max} 232, 270, 295, 330, 415. $[\alpha]_D$ + 523 ($CHCl_3$). Active against *Staphylococcus aureus* at a concentration of 1 ppm.

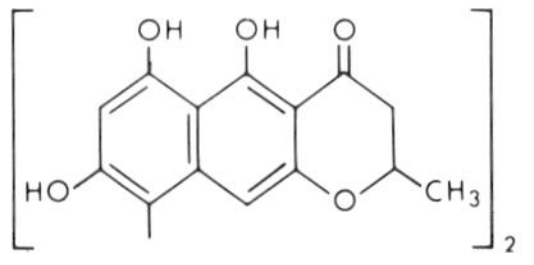

Source: Nectria flavo-viridis (Fusarium melanochlorum). References: Carey, S. T., and M. S. R. Nair, *Lloydia 38,* 448 (1975).

5. FUSAROSKYRIN (4, 4′, 5, 5′, 8, 8′-Hexahydroxy-2, 2′-dimethoxy-7, 7′-dimethyl-1, 1′-dianthraquinone). $C_{32}H_{22}O_{12}$ MW 598 Red crystals, m.p. >300 C. $\lambda_{max}^{CHCl_3}$ 256, 276, 305–310, 505. Cause of "purple speck" disease of soybean.

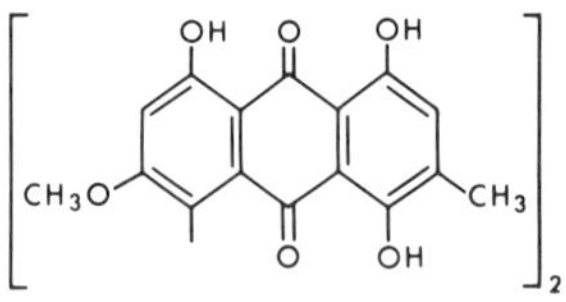

Source: Fusarium sp. *Reference:* Fujise, S., S. Hishida, M. Shibata, and S. Matsueda, *Chem. Ind.,* 1754 (1961).

6. FUSARUBIN, OXYJAVANICIN. $C_{15}H_{14}O_7$ MW 306 Red prisms, m.p. 218 C d. $\lambda_{max}^{CHCl_3}$ 303, 505. Inhibits *Staphylococcus.* Phytotoxic to tomato cuttings (40 μg/ml). LD_{50}, *Botrytis subtilis,* 2 μg/ml. Inhibits anaerobic decarboxylation of pyruvate, 30 μg/ml; oxidative decarboxylation of α-ketoglutarate, 77 μg/ml.

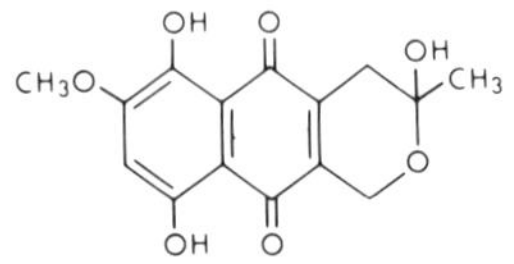

Source: Fusarium javanicum, F. solani. References: Arnstein, H. R. V., and A. H. Cook, *J. Chem. Soc.,* 1021 (1947); Kern, H., and S. Naef-Roth, *Phytopathol. Z. 60,* 316 (1967).

7. FUSCOFUSARIN $C_{30}H_{20}O_{11}$ MW 556 Brown powder, m.p. >300 C. λ_{max}^{EtOH} 225, 281, 346, 405.

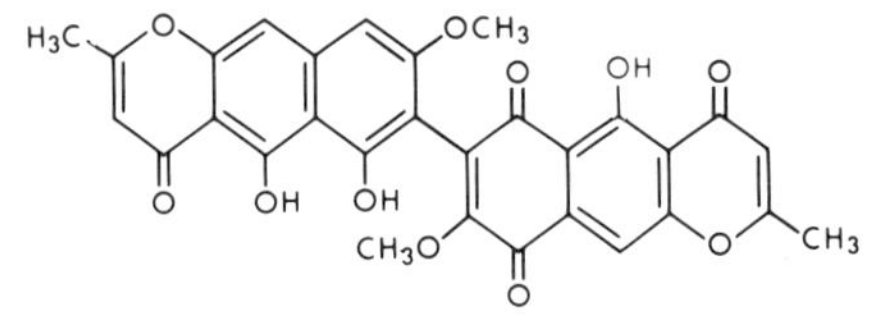

Source: Fusarium culmorum. Reference: Takeda, T., E. Morishita, and S. Shibata, *Chem Pharm. Bull. 16,* 2213 (1968).

8. ISOMARTICIN. $C_{18}H_{16}O_9$ MW 376 Brown prisms, m.p. 168–169 C. λ_{max}^{EtOH} 227, 306, 497. $[\alpha]_D^{25}$ + 26° ($CHCl_3$). Phytotoxic to tomato cuttings, 8 μg/g. LD_{50}, *Botrytis allii* Munn, >500 μg/ml. Inhibits *Bacillus subtilis,* 7 μg/ml. Inhibits anaerobic decarboxylation of pyruvate, 248 μg/ml; oxidative decarboxylation of α-ketoglutarate, 750 μg/ml.

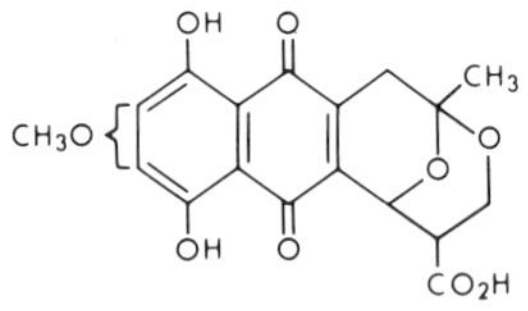

Source: Fusarium martii var. *pisi. References:* Kern, H., and S. Naef-Roth, *Phytopathol. Z., 53,* 45 (1965); Kern, H., and S. Naef-Roth, *Phytopathol. Z., 60,* 316 (1967).

9. JAVANICIN, SOLANIONE (5, 8-Dihydroxy-6-methoxy-2-methyl-3, 2′-oxopropyl-1, 4-naphthaquinone). $C_{15}H_{14}O_6$ MW 290 Red crystals, m.p. 208 C d. $\lambda_{max}^{CHCl_3}$ 307, 510. Antibiotic for gram-positive and acid-fast microorganisms. LD_{50} mice, i.p. >500 μg/kg. Phytotoxic to tomato cuttings, 60 μg/ml. Inhibits anaerobic decarboxylation of pyruvate, 29 μg/ml; oxidative decarboxylation of α-ketoglutarate, 87 μg/ml.

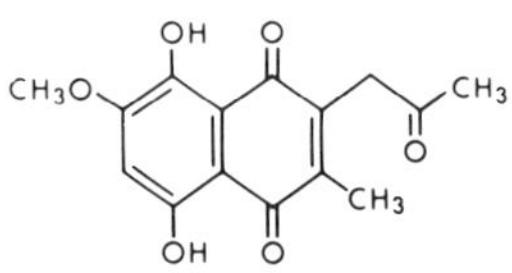

Source: Fusarium javanicum, F. solani. References: Arnstein, H. R. V., A. H. Cook, and M. S. Lacey, *Nature, 157,* 333 (1946); Arnstein, H. R. V., and A. H. Cook, *J. Chem. Soc.,* 1021 (1947); Kern, H., and S. Naef-Roth, *Phytopathol. Z., 60,* 316 (1967).

10. MARTICIN. $C_{18}H_{16}O_9$ MW 376 Dark red needles, m.p. 200–201 C. λ_{max}^{EtOH} 227, 305, 497. $[\alpha]_D^{25}$ + 132° ($CHCl_3$). Phytotoxic to tomato cuttings, 8 μg/g. LD_{50}, *Botrytis allii* Munn, >500 μg/ml. Inhibits *Bacillus subtilis,* 7 μg/ml. Inhibits anaerobic decarboxylation of pyruvate, 248 μg/ml; oxidative decarboxylation of α-ketoglutarate, 750 μg/ml.

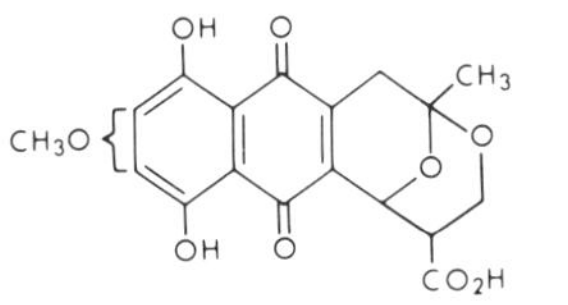

Source: Fusarium martii var. *pisi. References: Kern, H., and S. Naef-Roth, Phytopathol. Z., 53,* 45 (1965); Kern, H., and S. Naef-Roth, *Phytopathol. Z., 60,* 316 (1967).

11. NORBIKAVERIN. $C_{19}H_{12}O_8$ MW 368 m.p. >350 C d. $\lambda_{max}^{CHCl_3}$ 253, 273, 320, 515, 550.

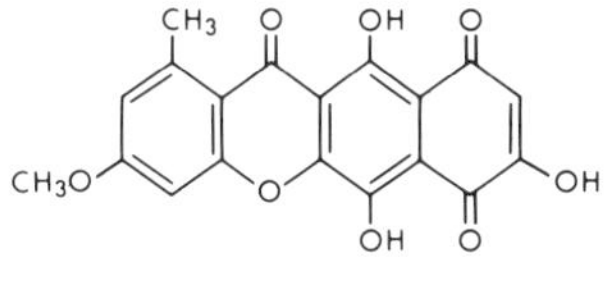

Source: Gibberella fujikuroi ACC 917. *Reference:* Kjaer, D., A. Kjaer, C. Pedersen, J. D. Bu'Lock, and J. R. Smith, *J. Chem. Soc.* C, 2788 (1971).

12. NORJAVANICIN (5, 8-Dihydroxy-6-methoxy-3, 2′-oxopropyl-1, 4-naphthaquinone). $C_{14}H_{12}O_6$ MW 276 Red needles, m.p. 200–204 C. Phytotoxic to tomato cuttings, 90 μg/ml. LD_{50}, *Botrytis allii* Munn, 10 μg/ml. Inhibits *Bacillus subtilis,* 1 μg/ml. Inhibits anaerobic decarboxylation of pyruvate, 3 μg/ml; oxidative decarboxylation of α-ketoglutarate, 31 μg/ml.

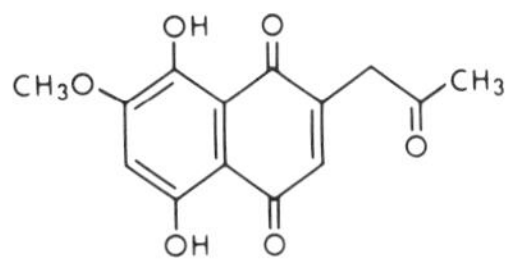

Source: Fusarium martii var. *pisi* (M808), *Fusarium* group *solani* M898. *References:* Kern, H., and S. Naef-Roth, *Phytopathol. Z., 60,* 316 (1967). Chilton, W. S., *J. Org. Chem., 33,* 4299 (1968).

13. NOVARUBIN. $C_{15}H_{14}O_7$ MW 306 Red crystals, m.p. 162 C. Phytotoxic to tomato cuttings, 35 μg/ml. LD_{50}, *Botrytis allii* Munn, 5 μg/ml. Inhibits *Bacillus subtilis,* 0.5 μg/ml. Inhibits anaerobic decarboxylation of pyruvate 4 μg/ml; oxidative decarboxylation of α-ketoglutarate, 31 μg/ml.

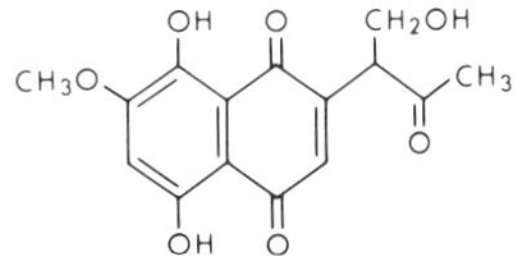

Source: Fusarium martii var. *pisi* M808, *Fusarium* group *solani* M898. *Reference:* Kern, H., and S. Naef-Roth, *Phytopathol. Z., 60,* 316 (1967).

14. 0-DEMETHYLANHYDROFUSARUBIN $C_{14}H_{10}O_6$ MW 274 Purple needles, m.p. 202–204 C. λ_{max}^{EtOH} 237, 285, 353, 492 sh, 546.

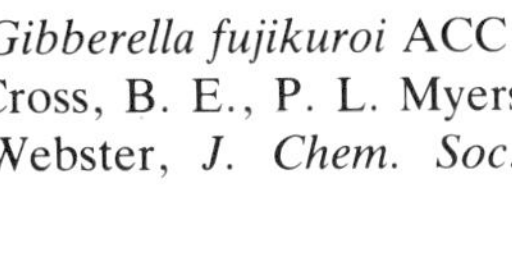

Source: Gibberella fujikuroi ACC 917. *Reference:* Cross, B. E., P. L. Myers, and G. R. B. Webster, *J. Chem. Soc.* C, 930 (1970).

15. RUBROFUSARIN (5, 6-Dihydroxy-8-methoxy-2-methyl-benzochromen-4-one). $C_{15}H_{12}O_5$ MW 272 Orange red needles, m.p. 210–211 C. λ_{max}^{EtOH} 225.5, 278, 327, 344, 310.

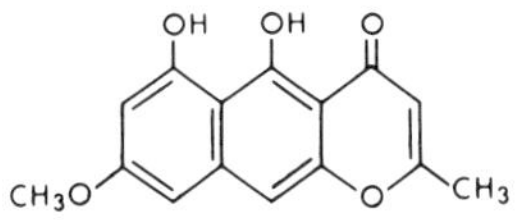

Source: Fusarium culmorum, Fusarium graminearum, Gibberella zeae. References: Ashley, J. N., B. C. Hobbs, H. Raistrick, *Biochem. J., 31,* 385 (1937); Mill, R. P., and F. F. Nord, *Arch. Biochem. Biophys., 4,* 419 (1944); Tanaka, H., T. Tamura, *Tetrahedron Lett.,* 151 (1961); Stout, G. H., D. L. Dreyer, L. H. Jensen, *Chem. Ind.,* 289 (1961); Stout, G. H., L. H. Jensen, *Acta. Crystallogr., 15,* 451 (1962); Tanaka, H., T. Tamura, *Agric. Biol. Chem., 26,* 767 (1962).

16. SOLANIOL (5, 8-Dihydroxy-3-(2-hydroxypropyl)-6-methoxy-2-methyl-1, 4-naphthaquinone). $C_{15}H_{16}O_6$ MW 292 Dark red needles, m.p. 190–194 C. d. $\lambda_{max}^{dioxane}$ 227, 304, 472, 500, 556. $[\alpha]_D^{25}$ + 122° (CH_3OH).

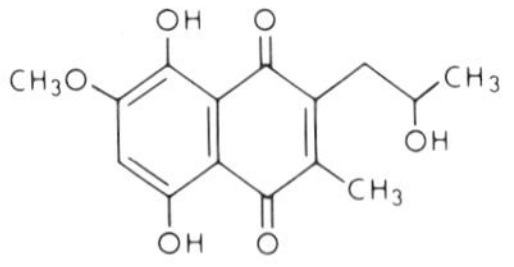

Source: Fusarium solani. Reference: Arsenault, G. P., *Tetrahedron, 24,* 4745 (1968).

Antibiotics

1. ANTIBIOTIC LL-Z1272 α. $C_{23}H_{31}O_3Cl$ MW 390 M.p. 72.5–73 C. λ_{max} 228, 293, 345. Anti-*Tetrahymena pyriformis* activity.

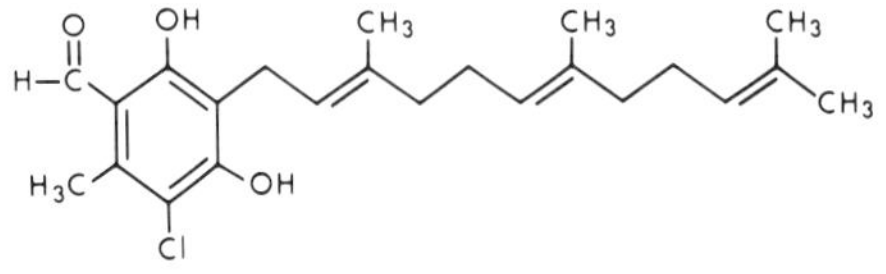

Source: Fusarium sp. *Reference:* Ellestad, G. A., R. H. Evans, Jr., and M. P. Kunstmann, *Tetrahedron, 25,* 1323 (1969).

2. ANTIBIOTIC LL-Z1272 β. $C_{23}H_{32}O_3$ MW 356 M.p. 97.5 C. λ_{max} 223, 233 sh., 297, 340.

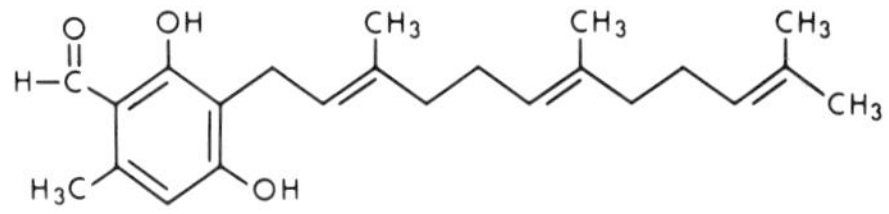

Source: Fusarium sp. *Reference:* Ellestad, G. A., R. H. Evans, Jr., and M. P. Kunstmann, *Tetrahedron, 25,* 1323 (1969).

3. ANTIBIOTIC LL-Z1272 γ ASCOCHLORIN. $C_{23}H_{29}O_4C1$ MW 404 M.p. 172–173. $[\alpha]_D^{25}$ −31 (CH_3OH). λ_{max} 230, 293, 347. Anti-*Tetrahymena pyriformis* activity.

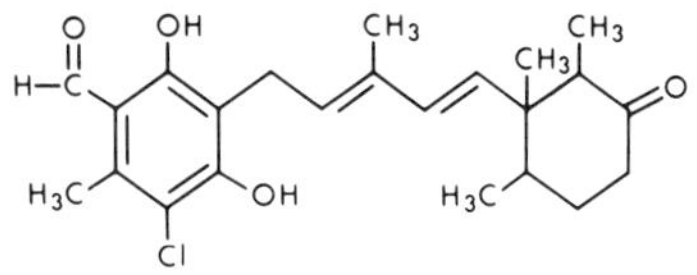

Source: Fusarium sp. *Reference* Ellestad, G. A., R. H. Evans, Jr., and M. P. Kunstmann, *Tetrahedron 25,* 1323 (1969).

4. ANTIBIOTIC LL-Z1272 δ. $C_{23}H_{31}O_4Cl$ MW 406 M.p. 129.5–130.5 C. $[\alpha]_D^{25}$ + 6° (CH_3OH). λ_{max} 231, 293, 346. Anti-*Tetrahymena pyriformis* activity.

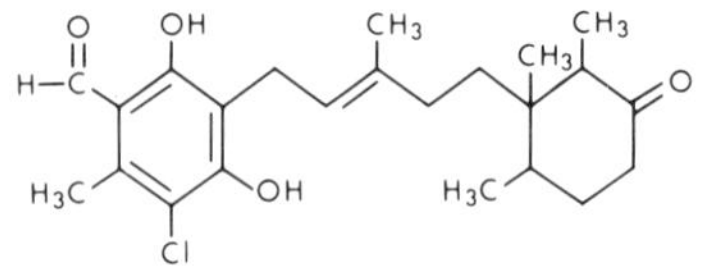

Source: Fusarium sp. *Reference:* Ellestad, G. A., R. H. Evans, Jr., and M. P. Kunstmann, *Tetrahedron, 25,* 1323 (1969).

5. ANTIBIOTIC LL-Z1272 ϵ. $C_{23}H_{32}O_4$ MW 372 M.p. 171.5–172.5 C. $[\alpha]_D^{25}$ + 6° (CH_3OH). λ_{max} 223, 233 sh, 295, 340 sh. Anti-*Tetrahymena pyriformis* activity.

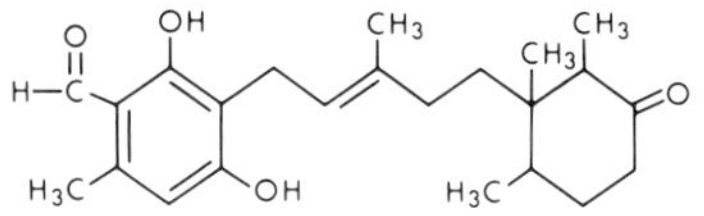

Source: Fusarium sp. *Reference:* Ellestad, G. A., R. H. Evans, Jr., and M. P. Kunstmann, *Tetrahedron, 25,* 1323 (1969).

6. ANTIBIOTIC LL-Z1272 ζ. $C_{25}H_{31}O_6Cl$ MW 462 M.p. 156.5–157 C. λ_{max} 239, 293, 347. $[\alpha]_D^{25}$ −15° (CH_3OH). Anti-*Tetrahymena pyriformis* activity.

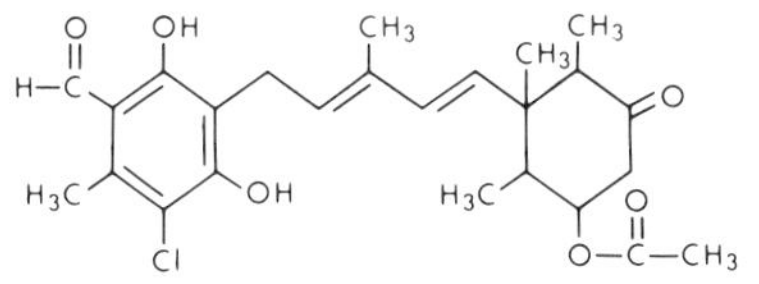

Source: Fusarium sp. *Reference:* Ellestad, G. A., R. H. Evans, Jr., and M. P. Kunstmann, *Tetrahedron, 25,* 1323 (1969).

Note: Clarification of Taxonomic Position of NRRL-3305 *Fusarium* sp. Producer of Orsellinic Sesquiterpenoid Antibiotics LL-Z1272 α, β, γ, δ, ε, and ζ.

The mould-producing antibiotic strain designated NRRL-3305 in the Agricultural Research Culture Collection was described as a *Fusarium* sp. by R. H. Evans et al. A reexamination of the strain NRRL-3305 by P. E. Nelson (*Private communications,* 1976) indicates a *Cylindocarpon* sp. rather than a *Fusarium* sp. *Reference:* Evans, R. H., Jr., M. P. Kunstmann, C. E. Holmund, and G. A. Ellestad (to American Cyanamid Company). 1970. Fusarium fermentation. U.S. Patent 3, 546, 073 (December 8).

7. ENNIATIN A, LATERITIIN I. $C_{36}H_{63}N_3O_9$ MW 681 M.p. 122 C. $[\alpha]_D^{18}$ −91.9° ($CHCl_3$). Active against gram-positive, gram-negative and acid-fast bacteria. Inhibits phytopathogenic fungi.

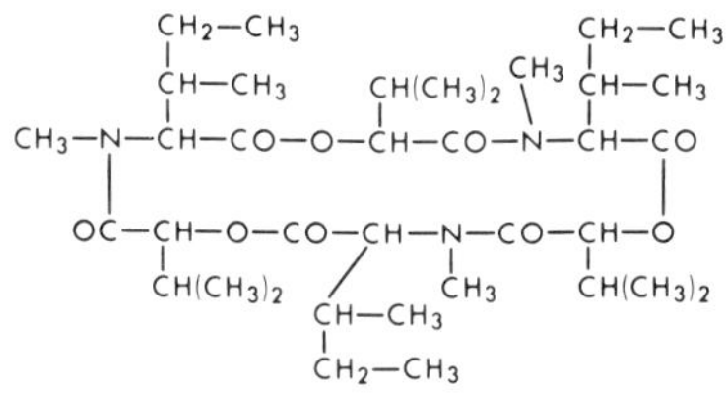

Source: Fusarium sp. ETH 1523/8. *References:* Gäumann, E., S. Roth, L. Ettlinger, Pl. A. Plattner, and U. Nager, *Experientia, 3,* 202 (1947); Plattner, Pl. A., U. Nager, and A. Boller, *Helv. Chim. Acta.. 31,* 594 (1948).

8. ENNIATIN B, LATERITIIN II. $C_{33}H_{57}N_3O_9$ MW 639 M.p. 174–176 C. $[\alpha]_D^{21}$ −108° ($CHCl_3$). Inhibits *Mycobacterium phlei,* 3 μg/ml; *Mycobacterium paratuberculosis,* 5 μg/ml. Activity against bacteria and phytopathogenic fungi weaker than Enniatin A.

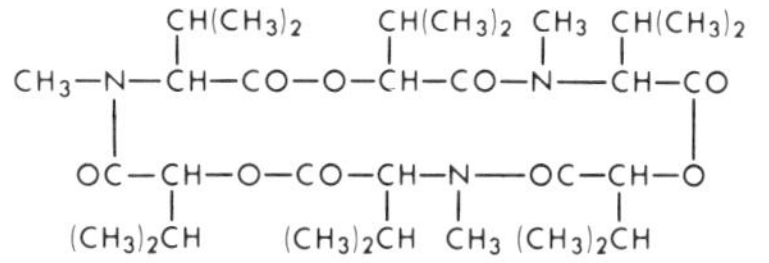

Source: Fusarium sp. ETH 4363 and ETH 1574. *References:* Plattner, Pl. A., and U. Nager, *Helv. Chim. Acta, 31,* 665 (1948); Plattner, Pl. A., U. Nager, and A. Boller, *Helv. Chim. Acta, 31,* 594 (1948); Plattner, Pl. A., and U. Nager, *Experientia, 3,* 325 (1947).

9. ENNIATIN C. $C_{24}H_{42}N_2O_6$ MW 454 M.p. 123 C. $[\alpha]_D^{22}$ −83° ($CHCl_3$).

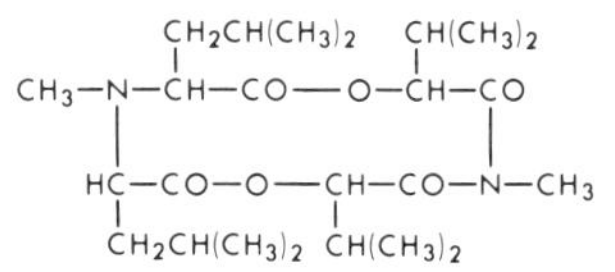

Source: Fusarium sp. *Reference:* Plattner, Pl. A., and U. Nager, *Helv. Chim. Acta, 31,* 2203 (1948).

10. EQUISETIN, *N*-METHYLTETRAMIC ACID. $C_{22}H_{31}NO_4$ MW 373 M.p. 65 C. λ_{max}^{EtOH} 235, 250, 292. Active against gram-positive, and acid-fast bacteria and *Neisseria perflava.* Lethal dose to mice, i.p., 63 mg/kg.

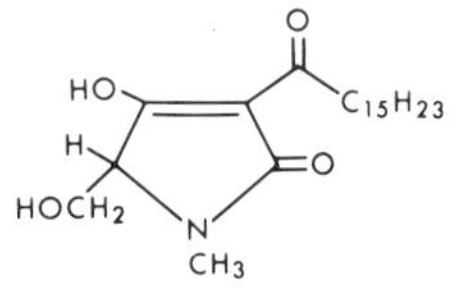

Source: Fusarium equiseti NRRL 5537. *Reference:* Burmeister, H. R., G. A. Bennett, R. F. Vesonder, and C. W. Hesseltine, *Antimicrob. Agents Chemother., 5,* 634 (1974).

11. LYCOMARASMIN. $C_9H_{15}N_3O_7$ MW 277 M.p. 227–229 C d. $[\alpha]_D^{20}$ −42° to −48° (H_2O, pH 7). Inhibits *Lactobacillus casei.* Tomato wilting toxin.

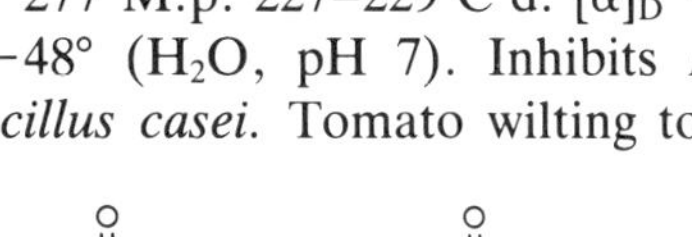

Source: Fusarium lycopersici. References: Plattner, Pl. A., N. Clauson-Kaas, *Experi-*

entia, 1, 195 (1945); Plattner, Pl. A., N. Clauson-Kaas, *Helv. Chim. Acta, 28,* 188 (1945); Hardegger von, E., P. Liechti, L. M. Jackson, A. Boller, and Pl. A. Plattner, *Helv. Chim. Acta, 46,* 60 (1963); Wooley, D. W., *J. Biol. Chem., 176,* 1291 (1948).

12. PHYTOTOXINS (Dehydrofusaric acid, Fusaric acid).
13. PIGMENTS (Bikaverin, Bostrycoidin, Cephalchromin, Fusarubin, Isomarticin, Javanicin, Marticin, Novarubin, Norjavanicin, Oxyjavanicin).
14. RAMIHYPHINS. MW 1192 Daltons White amphorous compound. $\lambda_{max}^{CH_3OH}$ 210. Structure Unknown. Active against pathogenic and saprophytic fungi. Stimulates branching of hyphae at subfungistatic concentrations.
Source: Fusarium sp. S-435. *Reference:* Barath, Z., H. Barathova, V. Betina, and P. Nemec, *Folia Microbiol* (Prague), *19,* 507 (1974).

Trichothecenes

1. ACETYL T-2 (3α, 4β, 15-Triacetoxy-8α-(3-methylbutryloxy)-12, 13-epoxytrichothec-9-ene). $C_{26}H_{36}O_{10}$ MW 508 Oil-like. Emetic, pigeon 18 mg/kg.

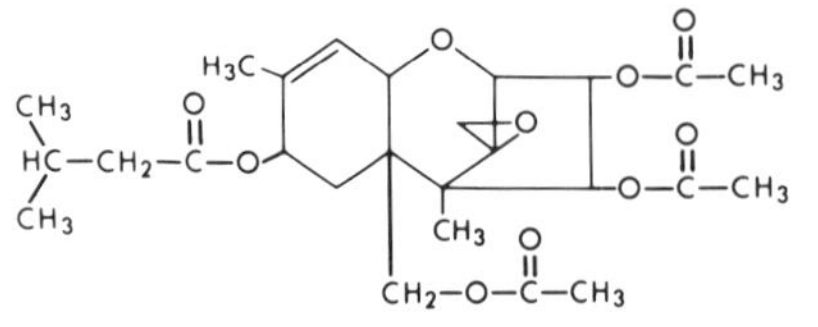

Source: Fusarium poae, NRRL 3287 = *F. tricinctum. Reference:* Kotsonis, F. N., R. A. Ellison, and E. B. Smalley, *Appl. Microbiol., 30,* 493 (1975).

2. CALONECTRIN (3α,15-Diacetoxy-12, 13-epoxytrichothec-9-ene). $C_{19}H_{26}O_6$ MW 350 Prisms m.p. 83–85 C. $[\alpha]_D^{27}$ + 14.6° ($CHCl_3$).

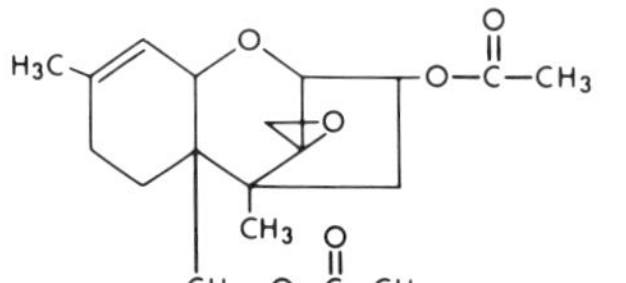

Source: Calonectria nivalis, Fusarium nivale, CMI 14764. *Reference:* Gardner, D., A. T. Glen, and W. B. Turner, *J. Chem. Soc.,* Perkins I, 2576 (1972).

3. DEACETYLCALONECTRIN (3α-Acetoxy-15-hydroxy-12, 13-epoxytrichothec-9-ene). $C_{17}H_{24}O_5$ MW 308 Prisms m.p. 184–186 C. $[\alpha]_D^{27}$ + 11.2 ($CHCl_3$).

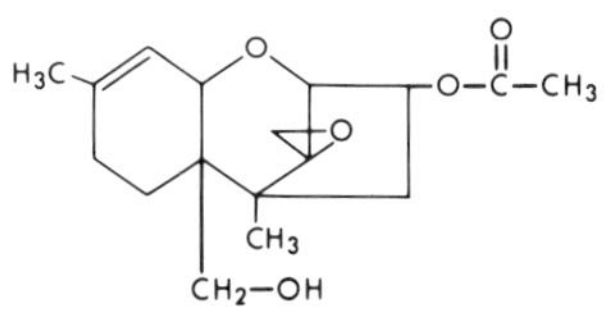

Source: Calonectria nivalis, Fusarium nivale, CMI 14764. *Reference:* Gardner, D., A. T. Glen, and W. B. Turner, *J. Chem. Soc.,* Perkin I, 2576 (1972).

4. DEOXYNIVALENOLMONOACETATE (3-Acetoxy-7, 15-dihydroxy-12, 13-epoxytrichothec-9-en-8-one). $C_{17}H_{22}O_7$ MW 338 M.p. 185.5–186 C. $[\alpha]_D^{25}$ + 40.5°. Emetic to laboratory animals. LD_{50} male mice, i.p. 76.7 mg/kg; female mice, i.p., 46.9 mg/kg; ducklings (10-day-old), s.c., 37 mg/kg. Emetic, ducklings (10-day-old), 10 mg/kg; dogs, s.c. 0.2 mg/kg. Inhibits *Tetrahymena pyriformis,* 29.0 μg/ml.

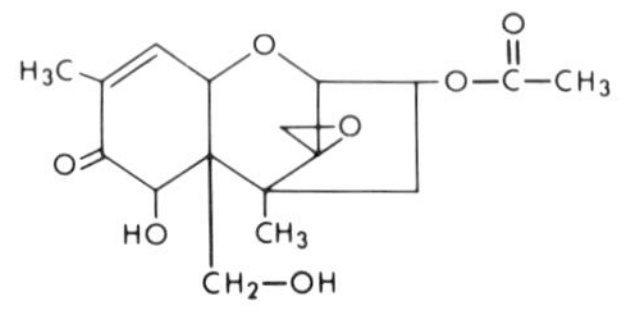

Source: Fusarium roseum. References: Blight, M. M., and J. F. Grove, *J. Chem. Soc.,* 1961 (1974); Yoshizawa, T., and N. Morooka, *Agric. Biol. Chem., 37,* 2933 (1973). Yoshizawa, T., and N. Morooka, *J. Food Hyg. Soc. Japan, 15,* 261 (1974).

5. DIACETOXYSCIRPENDIOL (4β, 15-Diacetoxy-3α, 7α-dihydroxy-12, 13-epoxytrichothec-9-ene). $C_{19}H_{26}O_8$ MW 382 M.p. 201–203 C.

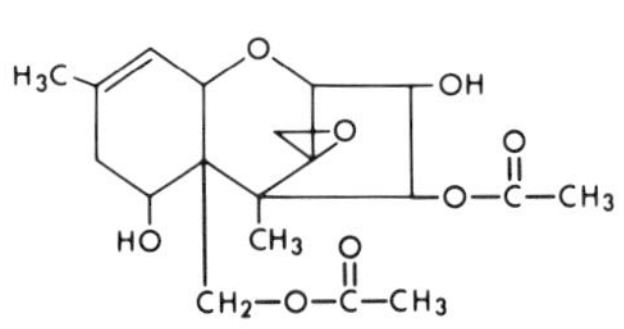

Source: Fusarium sp., Strain K-5036. *Reference:* Ishii, K., *Phytochemistry, 14,* 2469 (1975).

6. DIACETOXYSCIRPENTRIOL (4β, 15-Diacetoxy-3α, 7α, 8α-trihydroxy-12, 13-epoxytrichothec-9-ene). $C_{19}H_{26}O_9$ MW 398 M.p. 167–169 C.

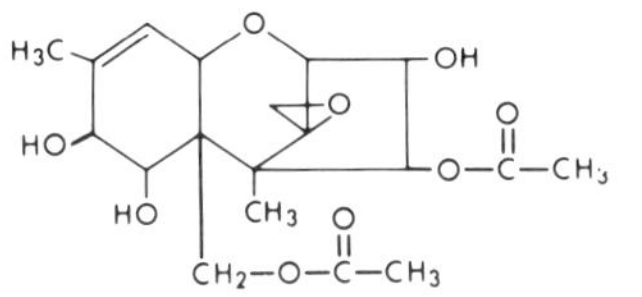

Source: Fusarium sp., Strain K-5036. *Reference:* Ishii, K., *Phytochemistry 14,* 2469 (1975).

7. DIACETOXYSCIRPENOL (4, 15-Diacetoxy-3-hydroxy-12, 13-epoxytrichothec-9-ene). $C_{19}H_{26}O_7$ MW 366 Prisms, m.p. 161–162 C $[\alpha]_D^{19}$ + 20° (acetone). Phytotoxin, petechial hemorrhagic; skin necrosis, emetic. Cress root elongation stimulated at 0.01–0.5 μg/ml; reduced at 10 μg/ml; inhibited at 100 μg/ml. Development of auxiliary shoots in *Vicia sativa L.* LD_{50} mice, i.p. 10 mg/kg; oral 7.3 mg/kg; chicken embryo 0.09 μg/egg. ID_{50} rabbit reticulocytes 0.03 mg/ml (whole cell); 5.0 μg/ml (cell free); rat liver, >50 μg/ml (cell free); tumor cells 0.03 μg/ml (whole cell). Emetic to ducklings, s.c., 0.2 mg/kg. Antiprotozoal activity 0.05 μg/ml. Inhibits *Penicillium digitatum,* 100 μg/ml.

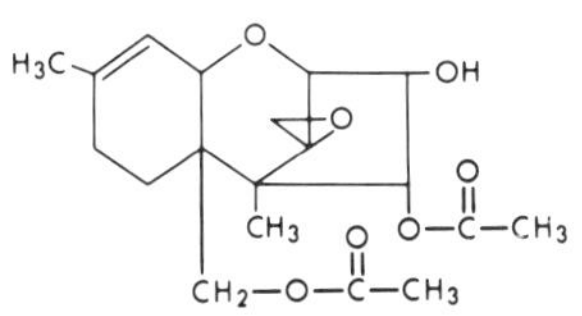

Source: Fusarium roseum, F. tricinctum, F. nivale, F. lateritum, F. scirpi CMI 15490, *F. intricans, F. equiseti* CMI 35100. *References:* Brian, P. W., A. W. Dawkins, J. F. Grove, H. G. Hemming, D. Lowe, and G. L. F. Norris, *J. Exp. Bot., 12,* 1 (1961); Bamburg, J. R., W. F. Marasas, N. V. Riggs, E. B. Smalley, and F. M. Strong, *Biotechnol. Bioeng., 10,* 445 (1968); Ueno, Y., *J. Food Hyg. Soc. Jpn., 14,* 403 (1973).

8. DIACETYLNIVALENOL (4β-15-Diacetoxy-3α, 7α-dihydroxy-12, 13-epoxytrichothec-9-en-8-one). $C_{19}H_{24}O_9$ MW 396 M.p. 135–136 C. $[\alpha]_D^{25}$ + 72° (acetone). Skin necrosis, emetic to laboratory animals. LD_{50} mice, i.p., 9.6 mg/kg; chicken embryo 0.9 mg/egg. ID_{50} rabbit reticulocytes 0.10 μg/ml (whole cell). Emetic, duckling, s.c., 0.4 mg.kg.

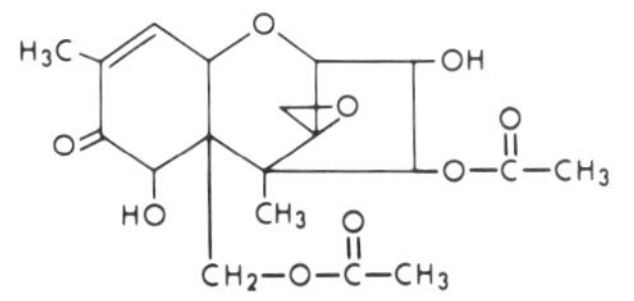

Source: Fusarium scirpi. References: Brian, P. W., A. W. Dawkins, J. F. Grove, H. G. Hemming, D. Lowe, and G. L. F. Norris, *J. Exp. Bot., 12,* 1 (1961); Ueno, Y., *J. Food Hyg. Soc. Jpn., 14,* 403 (1973).

9. FUSARENONE, NIVALENOL-4-ACETATE (4-Acetoxy-3α, 7α, 15-trihydroxy-12, 13-epoxytrichothec-9-en-8-one). $C_{17}H_{22}O_8$ MW 354 M.p. 91–92 C. $[\alpha]_D^{25}$ + 58 (CH_3OH). Damages hematopoietic tissue, proliferating cells of the intestinal epithelium testes. Emetic to laboratory animals. LD_{50} mice, i.p. 3.3 mg/kg; chicken embryo 2.6 mg/egg. ID_{50} rabbit reticulocytes 0.25 μg/ml (whole cell); 0.5 μg/ml (cell free); tumor cells 6.0 μg/ml (whole cell); rat liver 8.0 μg/ml (cell free). Emetic duckling, s.c., 0.4–0.5 mg/kg; cats, s.c., 1–2 mg/kg.

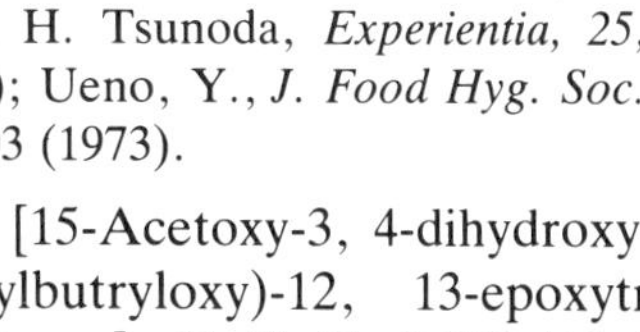

Source: Fusarium nivale. References: Ueno, Y., I. Ueno, T. Tatsuno, K. Ohokubo, H. Tsunoda, *Experientia, 25,* 1062 (1969); Ueno, Y., *J. Food Hyg. Soc. Jpn., 14,* 403 (1973).

10. HT-2 [15-Acetoxy-3, 4-dihydroxy-8-(3-methylbutryloxy)-12, 13-epoxytrichothec-9-ene]. $C_{21}H_{21}O_8$ MW 412 Gum. Skin necrosis, emetic. LD_{50} mice, i.p. 9

mg/kg; chicken embryo 0.5 μg/egg. ID_{50} rabbit reticulocytes, 0.03 μg/ml (whole cell). Emetic to ducklings, s.c., 0.1 mg/kg. Antiprotozoal activity 0.5 μg/ml. Skin necrotization 0.2 μg. Doses of 1.78 mg on skin of rat produced death; 0.17 mg caused severe edema and heavy scab formation; 0.71 and 0.35 mg produced same skin reaction as the lowest dose plus subcutaneous hemorrhaging.

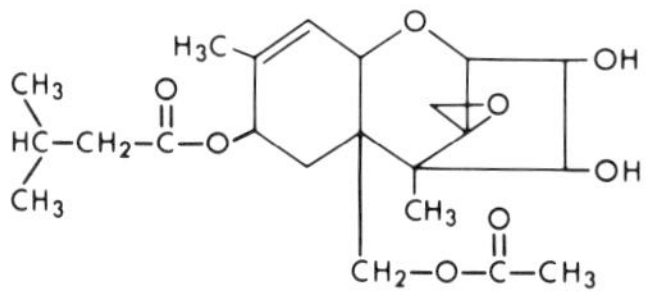

Source: Fusarium tricinctum. References: Bamburg, J. R., and F. M. Strong, *Phytochemistry, 8,* 2405 (1969); Ueno, Y., *J. Food Hyg. Soc. Jpn., 14* 403 (1973).

11. MONOACETOXYSCIRPENOL (15-Acetoxy-3α, 4β-dihydroxy-12, 13-epoxytrichothec-9-ene). $C_{17}H_{24}O_6$ MW 324 M.p. 172–173 C. Gastrointestinal hemorrhaging, bilateral inflammation of beak area of birds. LD_{50} rat, s.c., 0.752 mg/kg; female pig, i.p. 1 mg/kg. Turkey poults died within 7 days when fed 0.55 mg/day.

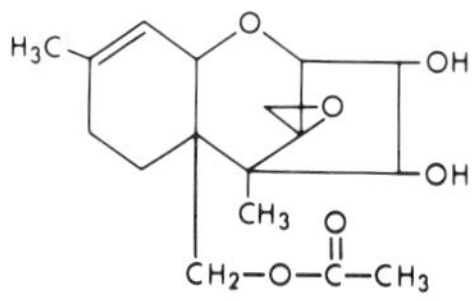

Source: Fusarium roseum. Reference: Pathre, S. V., C. J. Mirocha, C. M. Christensen, and J. Behrens, *J. Agric. Food Chem., 24,* 97 (1976).

12. NEOSOLANIOL (4β, 15-Diacetoxy-3α, 8α-dihydroxy-12, 13-epoxytrichothec-9-ene). $C_{19}H_{26}O_8$ MW 382 M.p. 171–172 C. Skin necrosis, emetic. LD_{50} mice, i.p., 14.5 mg/kg; chicken embryo 5 μg/egg. ID_{50} rabbit reticulocytes 0.25 μg/ml (whole cells); rat liver 20 μg/ml (cell free). Emetic to duckling, s.c., 0.2 mg/kg. Active against antiprotozoal 0.5 μg/ml. Skin necrotization to rabbit 1.0 μg.

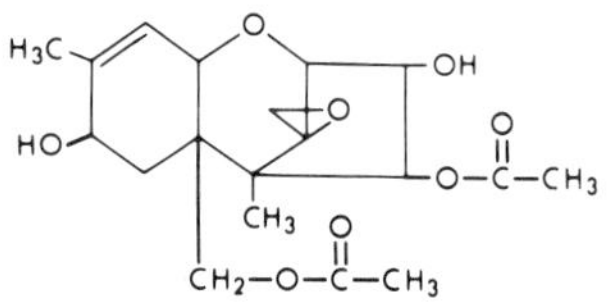

Source: Fusarium solani References: Ishii, K., K. Sakai, Y. Ueno, H. Tsunoda, and M. Enomoto, *Appl. Microbiol., 22,* 718 (1971); Ueno, Y., *J. Food. Hyg. Soc. Jpn., 14,* 403 (1973).

13. NIVALENOL (3α, 4β, 7α, 15-Tetrahydroxy-12, 13-epoxytrichothec-9-en-8-one). $C_{15}H_{20}O_7$ MW 312 M.p. 222–223 C. Skin irritant. Emetic to laboratory animals. Cell degeneration of bone marrow, lymph nodes, intestines, testes, thymus. LD_{50} mice, i.p. 4.1 mg/kg; chicken embryo 4.0 μg/egg. ID_{50} rabbit reticulocytes 3.0 μg/ml (whole cell); 0.5 μg/ml (cell free); tumor cell 6.0 μg/ml (whole cell); rat liver 8.0 μg/ml (cell free). Emetic, duckling s.c., 1 mg/kg; antiprotozoal activity 10 μg/ml; skin necrotization 10 μg.

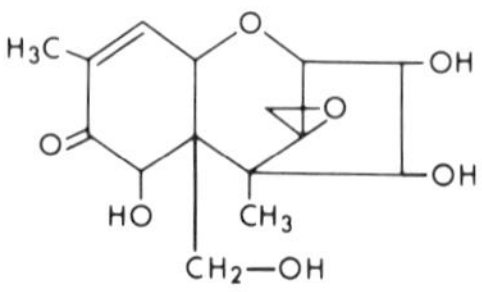

Source: Fusarium nivale. References: Tatsuno, T., Y. Fujimoto, and Y. Morita, *Tetrahedron Lett.,* 2823 (1969); Ueno, Y., *J. Food Hyg. Soc. Jpn., 14,* 403 (1973).

14. SCIRPENTRIOL (3, 4, 15-Trihydroxy-12-13-epoxytrichothec-9-ene). $C_{15}H_{22}O_5$ MW 282 M.p. 193 C. $[\alpha]_D^{23} \pm 3°$ (acetone). LD_{50} rat, i.p., 0.81 mg/kg. Inhibits cress root growth, 100 μg/ml.

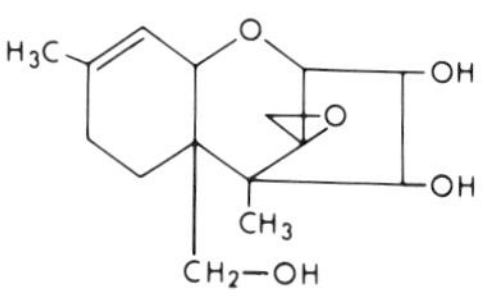

Source: Fusarium tricinctum. References: Brian, P. W., A. W. Dawkins, J. F. Grove, H. G. Hemming, D. Lowe, and G. L. F.

Norris, *J. Exp. Bot., 12,* 1 (1961); Pathre, S. V., C. J. Mirocha, C. M. Christensen, and J. Behrens, *J. Agric. Food Chem., 24,* 97 (1976).

15. T-2 [4β, 15-Diacetoxy-3α-hydroxy-8α-(3-methylbutryloxy)-12, 13-epoxytrichothec-9-ene]. $C_{24}H_{34}O_9$ MW 466 White needles, m.p. 151–152 C. $[\alpha]_D^{26}$ + 15° (EtOH). Hemorrhage, skin necrosis, shedding of intestinal mucosa in trout, oral lesions in chickens. LD_{50} mice, i.p., 3.04 mg/kg; chicken embryo 0.07 μg/egg; rat, oral, 6.1 mg/kg; trout, oral, 6.1 mg/kg. ID_{50} rabbit reticulocytes 0.03 μg/ml (whole cell); 0.15 μg/ml (cell free); rat liver, *ca.* 5 μg/ml (cell free). Emetic, pigeon 0.72 mg/kg; duckling, s.c., 0.1 mg/kg; cats, s.c., 0.1–0.2 mg/kg. Antiprotozoal activity 0.05 μg/ml. Skin necrotization 0.2 μg.

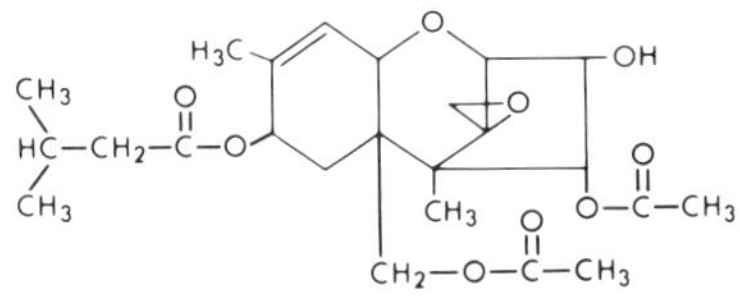

Source: Fusarium tricinctum. References: Bamburg, J. R., N. V. Riggs, and F. M. Strong, *Tetrahedron, 24,* 3329 (1968); Ellison, R. A., and F. N. Kotsonis, *Appl. Microbiol., 26,* 540 (1973); Ueno, Y., *J. Food. Hyg. Soc. Jpn., 14,* 403 (1973); Wyatt, R. D., B. A. Weeks, P. B. Hamilton, and H. R. Burmeister, *Appl. Microbiol., 24,* 251 (1972).

16. TRIACETOXYSCIRPENDIOL (4β, 8α, 15-Triacetoxy-3α,7α-dihydroxy-12, 13-epoxytrichothec-9-ene). $C_{21}H_{28}O_{10}$ MW 440 M.p. 185–186 C. LD_{50} rat, i.p., 1.2 mg/kg.

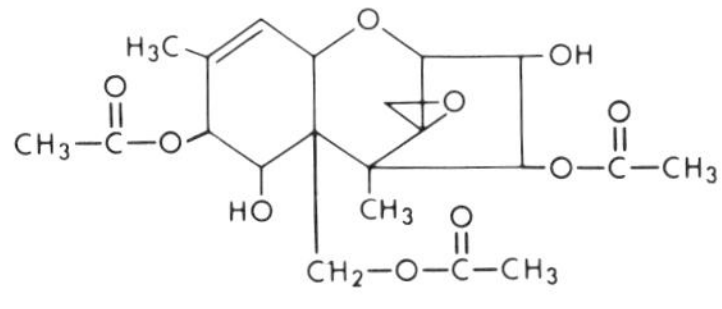

Source: Fusarium equiseti. Reference: Grove, J. F., *J. Chem. Soc.* C, 378 (1970).

17. VOMITOXIN, DEOXYNIVALENOL (3, 7, 15-Trihydroxy-12, 13-epoxytrichothec-9-en-8-one). $C_{15}H_{20}O$ MW 296 M.p. 151-153 C. $[\alpha]_D^{25}$ + 6.35° (EtOH). Emetic to swine, duckling, and refusal factor to swine. LD_{50} male mice, i.p., 70 mg/kg; female mice, i.p., 49.4 mg/kg; ducklings (10-day-old) 27 mg/kg. ID_{50} rabbit reticulocytes 2.0 μg/ml (whole cells). Emetic duckling (10-day-old) 10 mg/kg; dogs, s.c., 0.1 mg/kg; swine, intubation, 7 mg/60 lb. Feed refusal by swine, 40 mg/kg. Inhibits *Tetrahymena pyriformis* 4.6 μg/ml.

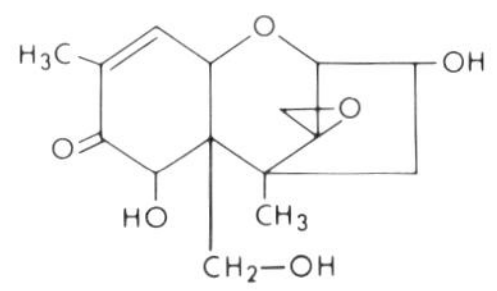

Source: Fusarium graminearum. References: Morooka, N., N. Uratsusi, T. Yoshizaiwa, and H. Yamamoto, *J. Food Hyg. Soc. Jpn., 13,* 368 (1972); Yoshizawa, T., and N. Morooka, *J. Food Hyg. Soc. Japan, 15,* 261 (1974); Vesonder, R. F., A. Ciegler, and A. H. Jensen, *Appl. Microbiol., 26,* 1008 (1973); Vesonder, R. F., A. Ciegler, A. H. Jensen, W. K. Rohwedder, and D. Weisleder, *Appl. Environ, Microbiol., 31,* 280 (1976).

Miscellaneous

1. 4-ACETAMIDO-2-BUTENOIC ACID. $C_6H_9NO_3$ MW 143 M.p. 140 C. Nontoxic to mice, i.p., 100 mg/kg. Noninhibitory to gram-positive or gram-negative bacteria, mycobacterium, yeast, and molds.

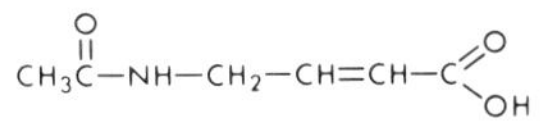

Source: Fusarium graminearum. Reference Vesonder, R. F., L. W. Tjarks, A. Ciegler, G. F. Spencer, and L. L. Wallen, *Phytochemistry, 16,* 1296 (1977).

2. 2-ACETYLQUINAZOLIN-4(3H)-ONE. $C_{10}H_8N_2O_2$ MW 188 Prisms, m.p. 205 C.

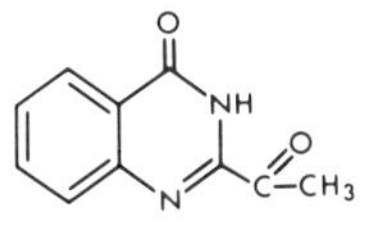

Source: Fusarium culmorum. Reference: Blight, M. M., and J. F. Grove, *J. Chem. Soc.,* 1691 (1974).

3. CULMORIN. $C_{15}H_{26}O_2$ MW 238 M.p. 179–180 C. $[\alpha]_D^{20}$ −14.45° ($CHCl_3$).

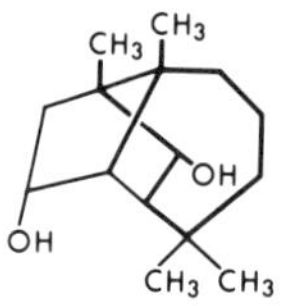

Source: Fusarium culmorum. References: Ashley, J. N., B. C. Hobbs, and H. Raistrick, *Biochem. J., 31,* 385 (1937); Barton, D. H. R., and N. W. Werstiuk, *Chem. Commun.,* 30 (1967).

4. CYCLONERODIOL [3-(2-Hydroxy-6-methylhept-5-en-2-yl)-1, 2-dimethycyclopentanol]. $C_{15}H_{28}O_2$ MW 240 Oil. $[\alpha]_D$ −20° ($CHCl_3$).

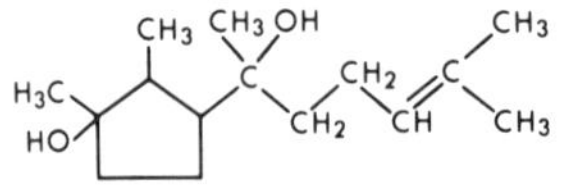

Source: Gibberella fujikuroi Reference: Cross, B. E., R. E. Markwell, and J. C. Steward, *Tetrahedron, 27,* 1663 (1971).

5. CYCLONEROTRIOL [6-(3-Hydroxy-2,3-dimethylcyclopentyl)-2-methylhept-2-ene-1, 6-diol]. $C_{15}H_{28}O_3$ MW 256 Prisms m.p. 113–114 C. $[\alpha]_D$ −27.8° (CH_3OH).

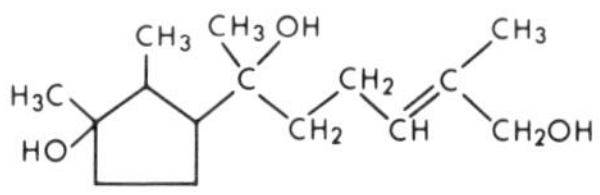

Source: Fusarium culmorum Reference: Hanson, J. R., P. B. Hitchcock, and R. Nyfeler, *J. Chem. Soc.,* 1586 (1975).

6. 7′-DEHYDROZEARALENONE. $C_{18}H_{20}O_5$ MW 316 Needles, m.p. 197–200 C. $[\alpha]_D^{24}$ -133.8 ($CHCl_3$). λ_{max} 229, 274, 313.

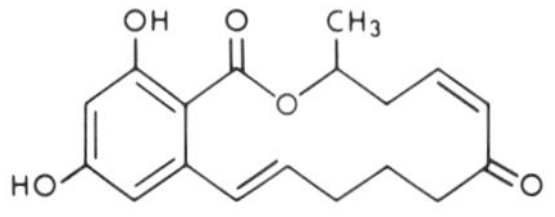

Source: Gibberella zeae. Reference: von Bolliger, G., and Ch. Tamm, *Helv. Chim. Acta, 55,* 3030 (1972).

7. N-DIHYDROJASMONOYLISOLEUCINE. $C_{18}H_{31}NO_4$ MW 325 M.p. 140–141.5 C. λ_{max} 3430, 1733, 1673.

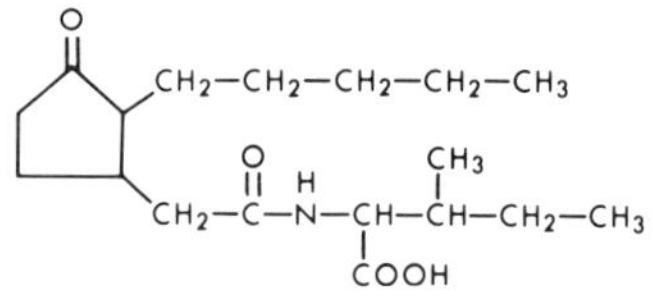

Source: Gibberella fujikuroi. Reference: Cross, B. E., and G. R. B. Webster, *J. Chem. Soc.,* 1839 (1970).

8. 1β, 7β-DIHYDROXYKAURENOLIDE $C_{20}H_{28}O_4$ MW 332 Gum. $[\alpha]_D^{20}$ −8° ($CHCl_3$).

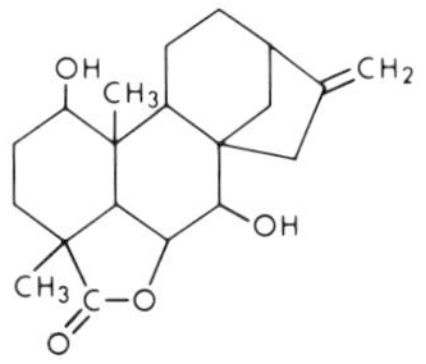

Source: Gibberella fujikuroi. Reference: Hedden, P., J. MacMillan, and M. J. Grinstead, *J. Chem. Soc.,* 2773 (1973).

9. 3β, 7β-DIHYDROXYKAURENOLIDE $C_{20}H_{28}O_4$ MW 332 M.p. 175–176 C. $[\alpha]_D^{24}$ −34.7° ($CHCl_3$).

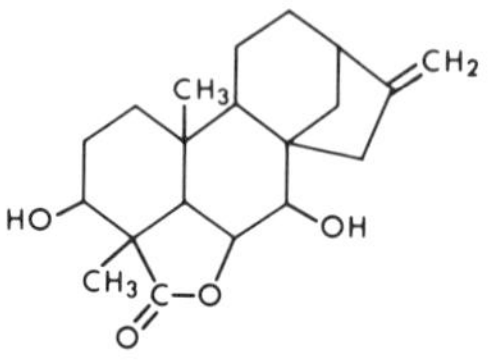

Source: Gibberella fujikuroi. References: Cross, B. E., J. R. Hanson, and R. H. B. Galt, *J. Chem. Soc.,* 3783 (1963); Bateson, J. H., and B. E. Cross, *Tetrahedron Lett.,* 3407 (1971).

10. 7β, 11α-DIHYDROXYKAURENOLIDE $C_{20}H_{28}O_4$ MW 332 M.p. 251–253 C. $[\alpha]_D^{20}$ −23° ($CHCl_3$).

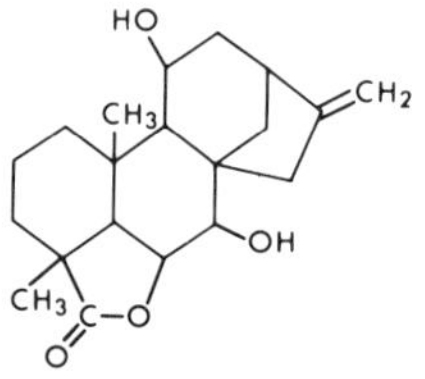

Source: Gibberella fujikuroi. Reference: Hedden, P., J. MacMillan, and M. J. Grinstead, *J. Chem. Soc.,* 2773 (1973).

11. 4β-7β-DIHYDROXY-18-NORKAURENOLIDE. $C_{19}H_{26}O_4$ MW 381 M.p. 211–213 C.

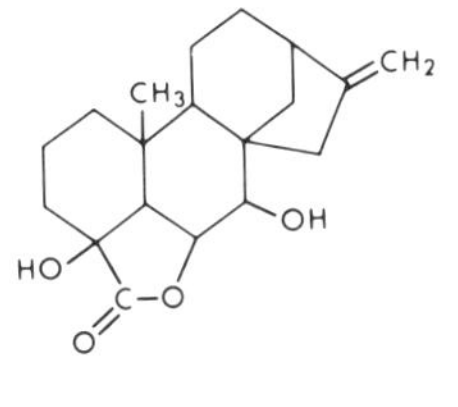

Source: Gibberella fujikuroi. Reference: Yamane, H., N. Murofushi, and N. Takahashi, *Agric. Biol. Chem., 38,* 207 (1974).

12. 7, 13-DIHYDROXYKAURENOLIDE $C_{20}H_{28}O_4$ MW 332 M.p. 259–262 C. $[\alpha]_D^{19}-1.7$ (pyridine).

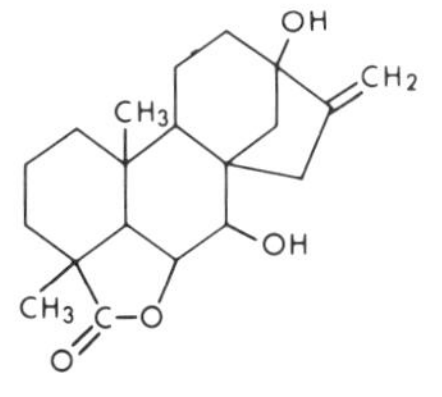

Source: Fusarium moniliforme. Reference: Serebryakov, E. P., A. V. Simolin, V. F. Kucherov, and B. V. Rosynov, Tetrahedron, 26, 5215 (1970).

13. 5-FORMYLZEARALENONE. $C_{19}H_{22}O_6$ MW346 Plates, m.p. 188–190 C. $[\alpha]_D^{24}-42$ ($CHCl_3$). λ_{max} 217, 254, 282, 317.

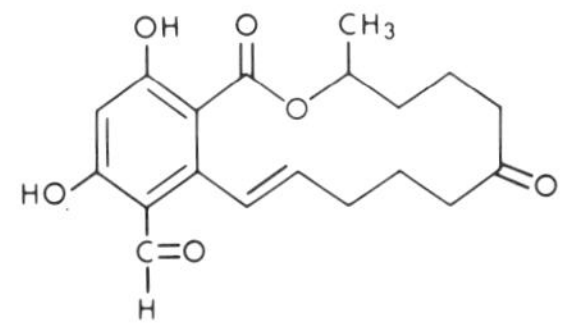

Source: Gibberella zeae. Reference: von Bolliger, G., and Ch. Tamm, *Helv. Chim. Acta, 55,* 3030 (1972).

14. FUSAMARIN [5-Butyl-3, 4-dihydro-6, 8-dihydroxy-3-(pent-3-enyl)isocoumarin]. $C_{18}H_{24}O_4$ MW 304 M.p. 159 C. $[\alpha]_D^{30}-11.39°$ (CH_3OH). λ_{max}^{EtOH} 220, 277, 310.

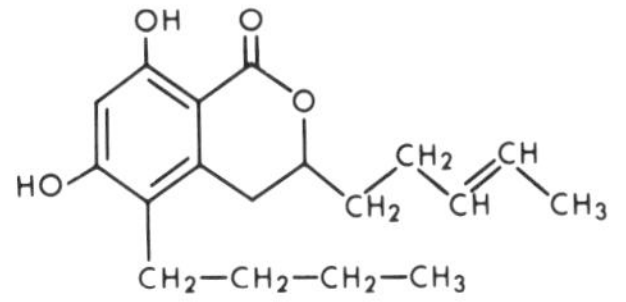

Source: Fusarium sp. *Reference:* Suzuki, Y., *Agric. Biol. Chem., 34,* 760 (1970).

15. 10-HYDROXYFUSARIC ACID, FUSARINOLIC ACID [5-(3-hydroxybutyl)pyridine-2-carboxylic acid; 5-(3-hydroxybutyl) picolinic acid]. $C_{10}H_{13}NO_3$ MW 195 Waxy plates m.p. 83–84 C. $[\alpha]_D^{27}$ + 12.5° (EtOH). λ_{max} 268.

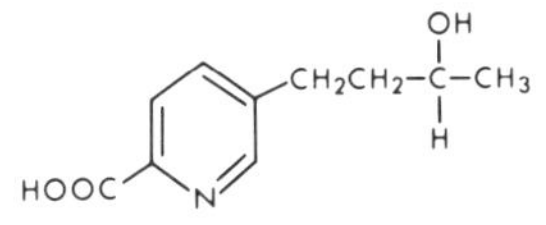

Source: Gibberella fujikuroi. References: Pitel, D. W., and L. C. Vining, *Can. J. Biochem., 48,* 623 (1970); Steiner, K., U. Graf, and E. Hardegger, *Helv. Chim. Acta, 54,* 845 (1971).

16. 4-HYDROXYMYOPORONE. $C_{15}H_{22}O_4$ MW 266 λ_{max} 251.

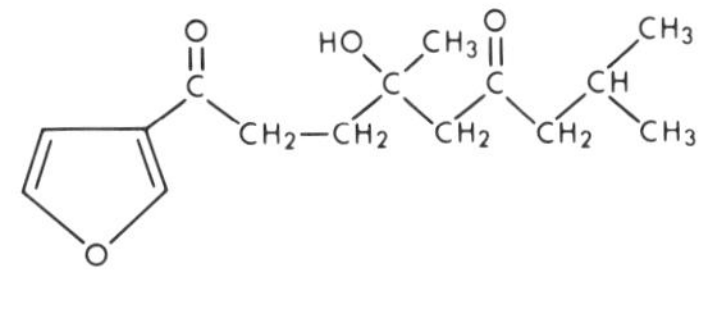

Source: Fusarium solani. Reference: Burka, L. T., L. Kuhnert, B. J. Wilson, and T. M. Harris, *Tetrahedron Lett.,* 4017 (1974).

17. 16α-HYDROXY-(-)-KAURAN-19-AL. $C_{20}H_{32}O_2$ MW 304 M.p. 169–173 C. $[\alpha]_D-68°$ (acetone).

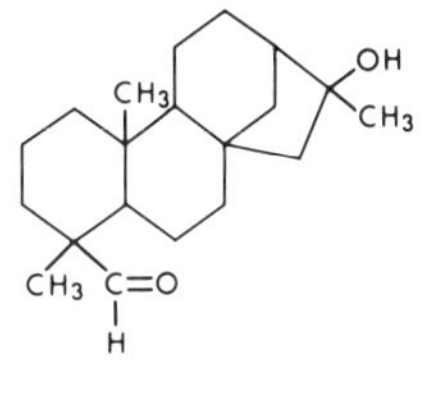

Source: Fusarium moniliforme. Reference: Serebryakov, E. P., A. V. Simolin, V. F. Kucherov, and B. V. Rosynov, *Tetrahedron, 26,* 5215 (1970).

18. 16 α-HYDROXY-(-)-KAURAN-19-OIC ACID. $C_{20}H_{32}O_3$ MW 320 M.p. 281–283 C. $[\alpha]_D -92°$

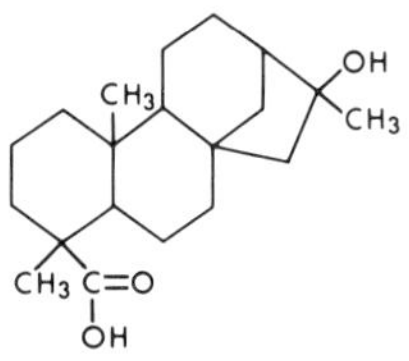

Source: Fusarium moniliforme. Reference: Serebryakov, E. P., A. V. Simolin, V. F. Kucherov, and B. V. Rosynov, *Tetrahedron, 26,* 5215 (1970).

19. 8′-epi-HYDROXYZEARALENONE. $C_{18}H_{22}O_6$ MW 334 M.p. 172–174 C. $[\alpha]_D^{24} -53.1$ (acetone). λ_{max} 237, 276, 315.

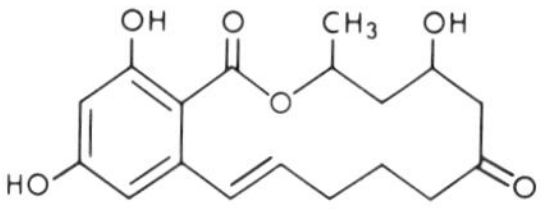

Source: Gibberella zeae. Reference: von Bolliger, G., and Ch. Tamm, *Helv. Chim. Acta, 55,* 3330 (1972).

20. 8′-HYDROXYZEARALENONE. $C_{18}H_{22}O_6$ MW 334 M.p. 210–212 C $[\alpha]_D^{24} -149.1$ (acetone). λ_{max} 236, 274, 314. Biological activity is unknown.

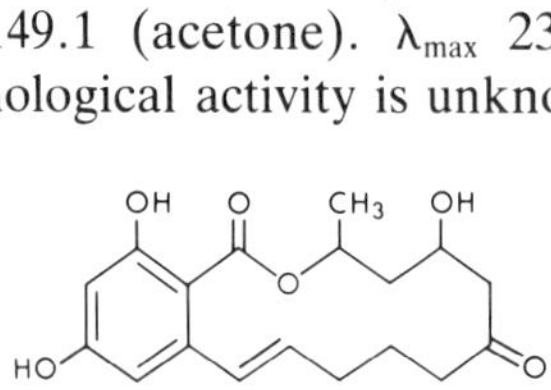

Source: Gibberella zeae. References: von Bolliger, G., and Ch. Tamm, *Helv. Chim. Acta, 55,* 3030 (1972).

21. 5-HYDROXYMETHYLFURAN-2-CARBOXYLIC ACID (Sumiris Acid). $C_6H_60_4$ MW 142 M.p. 164 C d.

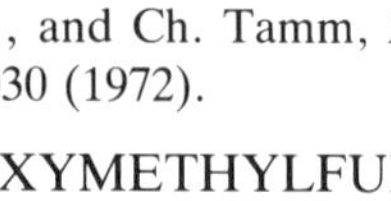

Source: Gibberella fujikuroi. Reference: Sumiki, Y., *J. Agric. Chem. Soc. Jpn., 7,* 819 (1931).

22. N-JASMONOYLISOLEUCINE. $C_{18}H_{29}NO_4$ MW 323

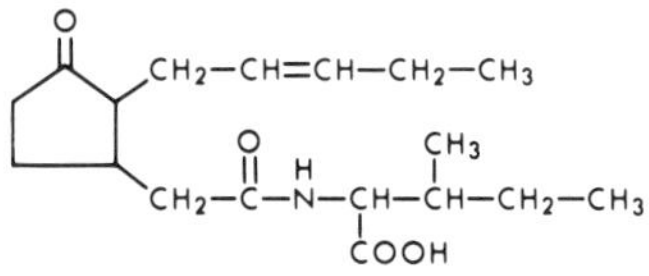

Source: Gibberella fujikuroi. Reference: Cross, B. E., and G. R. B. Webster, *J. Chem. Soc.,* 1839 (1970).

Mycotoxins

1. 2-ACETAMIDO-2, 5-DIHYDRO-5-OXO-FURAN. $C_6H_7NO_3$ MW 141 M.p. 115–116.5 C. LD_{50} mice, i.p., 43.16 mg/kg; per os 25 mg/kg. Gangrene of the tail of the cattle results when fed 3.8 mg/day for 90 days. Skin necrotization to rabbit when applied in olive oil; sensitivity increased 10 fold when applied in dimethylsulfoxide.

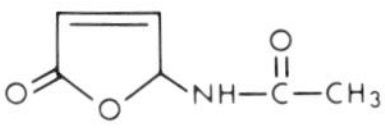

Source: Fusarium nivale. Reference: Yates, S. G., H. L. Tookey, J. J. Ellis, and H. J. Burkhardt, *Phytochemistry, 7,* 139 (1968).

2. 4-IPOMEANOL [1-(3-Fury1)-4-hydroxypentanone]. $C_9H_{12}O_3$ MW 168 Colorless oil with a pungent odor. $\lambda_{max}^{cyclohexane}$ 211, 243. Produces long oedma and pleural effusion in mice. Doses (1–6mg) i.p., to mice produce signs of illness immediately, followed quickly by dyspnoea which gradually increased in severity to time of death (from anoxia) in 5–8 hours. Comparable doses given orally produced state of latency in 2–3 hours followed by increasing respiratory distress leading to death within 24 hours.

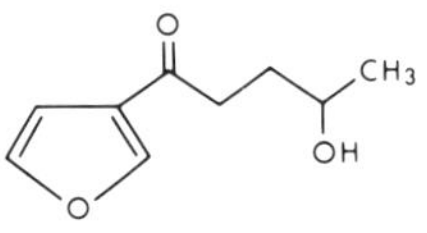

Source: Fusarium javanicum. Reference: Wilson, B. J., M. R. Boyd, T. M. Harris, and D. T. C. Yang, *Nature, 231* 52 (1971).

3. MONILIFORMIN (1-Hydroxycyclobut-1-ene-3, 4-dione). $C_4H_2O_3$ MW 98 M.p. 158 C.d. λ_{max}^{CH3OH} 229, 260. LD_{50} 1-day-old cockerels 4 mg/kg. Growth regulator and phytotoxin to plants.

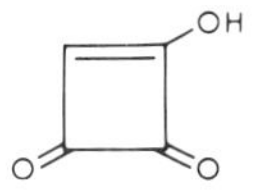

Source: Fusarium moniliforme. Reference: Cole, R. J., J. W. Kirskey, H. G. Cutler, B. L. Doupnik, and J. C. Peckham, *Science, 179,* 1324 (1973).

4. POAEFUSARIN (a) and SPOROFUSARIN (b), steroidal toxins of *Fusarium sporotrichiella,* were reported by L. E. Olifson as the etiologics of alimentary toxic aleukia (ATA). In 1973, C. J. Mirocha and S. Pathre showed a crude extract of poaefusarin (received from U. Bilal and L. Misurinko, USSR) to contain zearalenone and the trichothecenes T-2, neosolaniol and T-2 tetraol. *F. sporotrichiodes* received by these authors from the USSR produced T-2 and zearalenone.

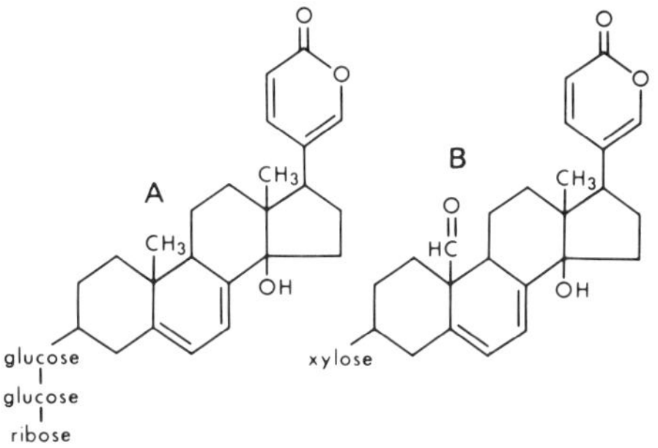

A. N. Leonov reports that Y. Yermakov and coworkers at his institute isolated T-2 toxin and an unidentified trichothecene from a *Fusarium* sp. morphologically similar to the strain Mirocha received from the U.S.S.R. Hence, the work of C. J. Mirocha and S. Pathre, confirmed by Y. Yermakov, as well as the isolation of trichothecenes from *F. poae* and *F. sporotrichiodes* by Ueno et al. in Japan and B. Yagen in Israel did not show the presence of steroidal toxins and causative principals of ATA appear to be trichothecenes.
References: Olifson, E., Ph.D. dissertation, I.T. at Industrial Nutrition, Moscow, 1965, pp. 1–36; Mirocha, C. J., and S. Pathre, *Appl. Microbiol, 26,* 719 (1973); Leonov, A. N., Current View of Chemical Nature of Factors Responsible for Alimentary Toxic Aleukia, Conference of Mycotoxins in Human and Animal Health, University of Maryland, University College, October 4–8, 1976; Ueno, Y., N. Sato, K. Ishii, K. Sakai, and M. Enomoto, *Jpn. J. Exp. Med., 42,* 461 (1972); Yagen, B., Toxins from a Strain Involved in Alimentary Toxic Aleukia, Conference on Mycotoxins in Human and Animal Health, University of Maryland, University College, October 4–8, 1976.

5. TRICHOTHECENES (Acetyl T-2, Deoxynivalenolmonoacetate, Diacetoxyscirpenol, Diacetylnivalenol, Fusarenone, HT-2, Monoacetoxyscirpenol, Neosolaniol, Nivalenol, T-2 Toxin, Triacetoxyscirpendiol, Vomitoxin).

6. ZEARALENOL $C_{18}H_{24}O_5$ MW 320 M.p. 131 C, 162–164 C. λ_{max}^{EtOH} 238.5, 275.5, 315. Synthetic zearalenol diastereoisomers have uterotrophic activity.

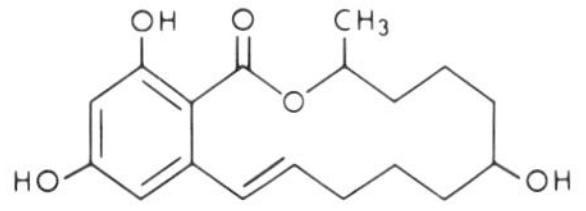

Source: Fusarium roseum (S-74-1C). *References:* Peters, A. C., *J. Chem. Med., 15,* 867 (1972); Stipanovic, R. D., and H. W. Schroeder, *Mycopathologia, 57,* 77 (1975).

7. ZEARALENONE (F-2). $C_{18}H_{22}O_5$ MW 318 M.p. 164–165 C. $[\alpha]_D -90°$ ($CHCl_3$). λ_{max}236, 274, 316. Potent estrogen, causing enlarged vulvae (1 mg/27 kg for 5 days) and mammary glands in swine; shrunken testes in young males.

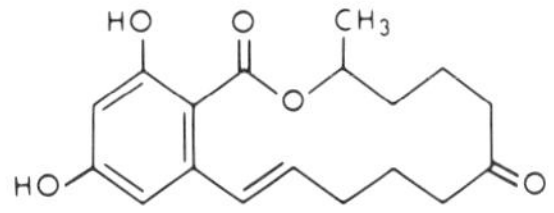

Source: Gibberella zeae. References: Stob, M., R. S. Baldwin, J. Tuite, F. N. Andrews, and K. G. Gillette, *Nature, 196,* 1318 (1962); Mirocha, C. J., C. M. Christensen, and G. H. Nelson, *Cancer Res., 28,* 2319 (1968).

Phytotoxins

1. ANTIBIOTICS (Enniatiin A, Enniatiin B, Lycomarasmin).
2. DEHYDROFUSARIC ACID (5-3′-Butenylpyridine-2-carboxylic acid; 5-3′-butenylpicolinic acid). $C_{10}H_{11}NO_2$ MW 177 M.p. 121 C. λ_{max} 230, 270. Plant growth inhibitor.

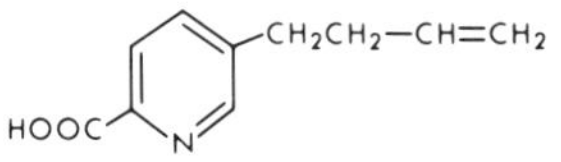

Source: *Gibberella fujikuori.* *References:* Stoll, Ch., *Phytopath. Z., 22,* 233 (1954); Stoll, Ch., and J. Renz, *Phytopath. Z., 29,* 380 (1957).

3. FUSARIC ACID (5-Butylpicolinic acid; 5-Butyl pyridine-2-carboxylic acid). $C_{10}H_{13}NO_2$ MW 179 Colorless crystals m.p. 108–109 C. Plant growth inhibitor. Weakly inhibits gram-positive and gram-negative bacteria. Hypotensive reagent to rabbits, cats, and dogs, i.p., 20 mg/kg. LD_{50} mice, i.v. 100 mg/kg; i.p. 80 mg/kg. Emetic to dogs, *per os* 10, 20, 40 mg/kg. Potent inhibitor of dopamine β-hydroxylase.

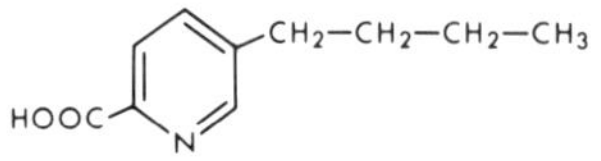

Source: *Fusarium* sp. *References:* Hidaka, H., *Nature, 231,* 54 (1971); Hidaka, H., T. Nagatsu, and K. Takeya, *J. Antibiot., 22,* 228 (1969); Yabuta, T., K. Kambe, and T. Hayashi, *J. Agric. Chem. Soc. Jpn., 10,* 1059 (1934).

4. MYCOTOXIN (Moniliformin, Diacetoxyscirpendiol, Diacetylnivalenol, Triacetoxyscirpendiol).
5. PICOLINIC ACID. $C_6H_5NO_2$ MW 123 M.p. 136–138 C. Phytotoxin, inhibits growth of maize and wheat roots; inhibits growth of pea, radish, and wheat seed.

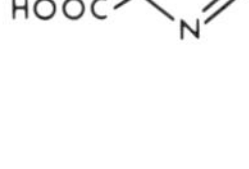

Source: *Fusarium laterium.* *Reference:* Berestetskii, O. A., S. P. Nadkernichnyi, V. G. Patyka, *Mikol. Fitopatol., 9,* 325 (1975).

6. PIGMENT (Fusarubin, Isomarticin, Javanicin, Norjavanicin, Novarubin).
7. TRICHOTHECENES (Diacetoxyscirpenol, Diacetylnivalenol, Triacetoxyscirpendiol).

Metabolites Derived from Phytoalexins

1. 3, 6a-DIHYDROXY-8, 9-METHYLENEDIOXYPETROCARPAN. $C_{16}H_{12}O_6$ MW 300 Needle-shaped crystals m.p. 178–181 C. $[\alpha]_D^{21}$ + 337°(EtOH). λ_{max} 281, 286.5, 309. Demethylation of pisatin.

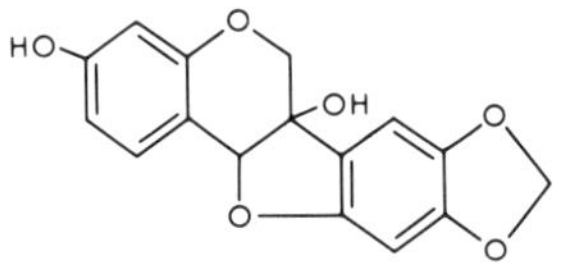

Source: *Fusarium solani.* *Reference:* Van Etten, H. D., S. G. Pueppkf, and T. C. Kelsey, *Phytochemistry, 14,* 1103 (1975).

2. 1a-HYDROXYPHASEOLLONE. $C_{20}H_{18}O_5$ MW 338 $[\alpha]_D^{24}$ −187°. λ_{max} 228.5, 280, 290. Metabolism of phaseollin, through oxidative transformation of phenolic group.

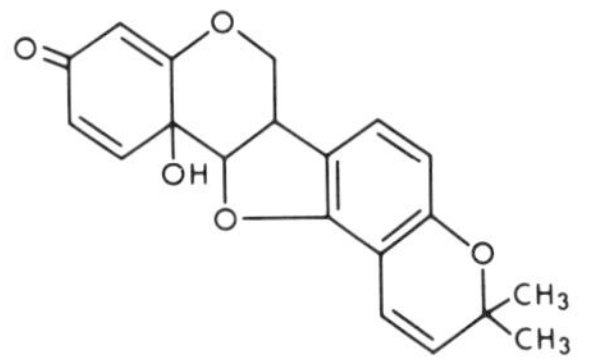

Source: *Fusarium solani.* *Reference:* Van Den Heuvel, J., H. D. Van Etten, J. W. Serum, D. L. Coffen, and T. H. Williams, *Phytochemistry, 13,* 1129 (1974).

33 Vascular Wilt Fusaria: Infection and Pathogenesis

W. E. MacHardy and C. H. Beckman

Introduction

The vascular wilt Fusaria attack many important agricultural and horticultural crops, including banana, cabbage, cotton, flax, muskemelon, onion, pea, tomato, watermelon, aster, carnation, chrysanthemum, gladiolus, and tulip. These Fusaria, unlike other *Fusarium* spp. have the unique ability to become established systemically in the water-conducting system of their host. Symptoms are quite variable, but include combinations of vein-clearing, leaf epinasty, wilting, chlorosis, necrosis, and abscission. Severely infected plants wilt and die, while plants affected to a lesser degree may become stunted and unproductive. The most prominent internal symptom is vascular browning.

In 1899 Smith (168) suggested that mechanical plugging of the vascular system was responsible for the wilting of cotton, cowpea, and watermelon plants infected with *Fusarium*. Later it was proposed that fungal toxins rather than plugging were the cause of symptoms that developed from vascular infections (20, 21, 88). Three-quarters of a century and hundreds of papers later there are still reports claiming that either vascular dysfunction and water stress or injury of cells due to fungal toxins is the major cause of symptoms. However, this does not mean we have gained little knowledge and understanding of vascular wilt pathogenesis over this time span. Although the causes of symptom induction and other aspects of Fusarium wilt still are not completely understood, enough is known to assemble a reasonably discernable picture of pathogenesis.

Numerous reviews (6, 45, 46, 47, 48, 49, 160, 173, 175, 190) and a monograph (185) have been written on Fusarium wilt within the past 25 years, but most either considered only one narrow aspect of pathogenesis, or covered many aspects but treated each as an independent event. In this chapter we present our interpretation of how the individual events that comprise vascular infection may interrelate and contribute to pathogenesis. These interpretations are presented in several flow charts in which evidence that seems to us substantial is indicated by solid arrows and that which is speculative is shown with dashed arrows. These flow charts and our discussion are based on a voluminous literature that is often contradictory. Some conclusions are difficult to assess, and must be weighed with caution because they were drawn from studies that deviated considerably from natural infections. Despite these drawbacks, a picture emerges that enables us to see both the achievements and the gaps and flaws and also suggests areas for future research. We expect that the picture will change and improve with time. Hopefully, this presentation will provide a suitable framework for evaluating the relevance and contribution of future work with respect to our overall conception of vascular wilt pathogenesis.

Definitions

Pathogenesis has been defined (62) as "the sequence of processes in *disease* development from the initial contact between a *pathogen* and its *host* to completion of the *syndrome.*" Resistance and susceptibility are relative terms used to indicate the degree of pathogenesis that has occurred in interactions between a host and a parasite. Resistance is associated with limited pathogenesis and symptom expression, whereas susceptibility is associated with extensive pathogenesis and symptom development. In resistant-type interactions, infection may be limited and the plant may show no obvious symptoms. In these cases infection has occurred, but pathogenesis, as defined above, has not occurred. With this type of resistance, the initial interactions between host and parasite may be similar to those associated with susceptible-type interactions, varying mainly in the rate, timing, and intensity of host responses. Thus, a fine line separates the limited infection of a resistant-type interaction from the extensive infection of a susceptible interaction. Considerable attention is thus given to the pathogenesis of highly resistant as well as highly susceptible host-parasite combinations. The term *host* in this chapter is applied to a susceptible plant that exhibits widespread symptoms when infected with a suitable forma specialis of *Fusarium oxysporum.* A *resistant host* is a cultivar or isoline of a susceptible host species that is genetically resistant to the specialized form of *F. oxysporum.* A *non-host* is a plant that does not exhibit symptoms when invaded by any of the specialized forms of *F. oxysporum.*

Ingress and Establishment

Penetration, Colonization and Distribution of *Fusarium* within the Roots of Susceptible Hosts.

The following is a generalized account of the penetration, ingress, and subsequent colonization and systematic distribution of *F. oxysporum* within the roots of susceptible host plants. It has been derived from studies of Fusarium wilt of cotton (104), flax (178), pea (92, 145), alfalfa (26), red clover (26), banana (158, 165, 171, 187, 188, 189), cabbage (4, 169), carnation (152), sweet potato (137), and onion (1).

The distinctive and essential feature of wilt pathogens is their colonization of vascular elements. The means by which they penetrate roots and enter vascular elements may differ, but are of two distinct types. Some pathogenic forms penetrate roots directly, whereas others must enter through wounds, or wounds greatly enhance vascular colonization.

The most common sites of direct penetration are located at or near the root tip of both taproots and lateral roots, although the hypocotyl and cotyledons can also be penetrated. The pathogen commonly penetrates root hairs or epidermal cells near the root cap, just behind the root tip, or within the zone of elongation. It then moves intercellularly and intracellulary through the root parenchyma tissue, enters the primary meristem, and invades the differentiating trachael vessels of the protostele. Within the zone of differentiation and more mature areas of the root, the endodermal cells and pericycle are penetrated intracellularly or intercellularly. Protoxylem vessels are entered through thin vessel walls between spiral thickenings. The large reticulate vessels are then invaded, and subsequent spread from vessel to vessel occurs through pits in the vessel wall. The number of infection sites, age of the plant, length of time the pathogen is in contact with the root surface, and inoculum density all influence symptom development.

Hepple (92) has shown that invasion in pea occurs through senescent cotyledons and progresses into the vascular stele through the cotyledonary traces; however, wounds that expose the vascular elements greatly enhance infection and disease incidence and severity.

This situation may well occur in other hosts, especially if the cotyledons remain in the soil. In still other hosts, e.g., banana, direct penetration seems to occur infrequently or not at all, (165, 187, 188) and wounds are essential to vascular infection.

Many microorganisms, including the cortical rot and decay Fusaria, invade the vascular system of plants in nature, yet few are able to successfully colonize this tissue and become distributed systemically (26, 125, 135, 154). It is interesting to compare root penetration and invasion by the vascular wilt Fusaria with that of the cortical rot and decay Fusaria. Initial attack by the cortical rot and decay Fusaria is characterized by (i) ability to penetrate more plant parts than the vascular Fusaria, (ii) more extensive intercellular and intracellular invasion of cortical tissue (usually accompanied by destruction), (iii) greater capacity to penetrate and destroy parenchyma, even parenchyma within the vascular cylinder, but (iv) less extensive invasion of xylem vessels, and (v) limited distribution in invaded vessels. Thus, in comparison to other *Fusarium* spp., the vascular wilt Fusaria have fewer infection courts and do not grow as extensively or cause as much destruction within host cortical tissue. However, the vascular wilt Fusaria can penetrate the endodermis and pericycle, colonize xylem vessels and become distributed upward more readily than the nonvascular Fusaria.

Only a few microorganisms are true vascular pathogens, and, indeed, one wonders why and how vascular pathogenesis came into being. It is much easier to appreciate cortical rot pathogenesis. Once the parasite has entered the host, it is in direct contact with a rather extensive, food-rich tissue of thin-walled parenchyma cells that require little enzyme specialization to degrade and consume. Thus, these parasites can readily use this tissue for growth and reproduction and can return to the soil in great numbers within a relatively short time. To vascular pathogens, in contrast, the cortex is a tissue to be traversed in order to reach the xylem vessels. Vascular Fusaria grow well and reproduce extensively within cortical tissues, but in nature these pathogens colonize this tissue sparsely. Rather, they grow and sporulate within xylem elements that, in comparison to the nutrient-rich cortical cells, are a near waste-land. The main substrates immediately accessible are cellulosic and pectinaceous substances exposed along the walls of vessel lumens and the very dilute amounts of carbohydrates, amino acids, and minerals comprising the vascular fluid. In addition, there is a low oxygen level and a fluctuating pH to content with in these vessels. Nevertheless, it is within this environment that the vascular wilt Fusaria are successful parasites, and only a very few fungi and bacteria have achieved this distinction.

Systemic Distribution of *Fusarium* within the Vascular System of Susceptible Hosts.

The continued development and spread of *F. oxysporum* within a susceptible host has been studied in greatest detail in Fusarium wilt of tomato (25, 61, 153, 164) and banana (12, 13, 14, 180). In tomato, hyphae of *F. oxysporum* f. sp. *lycopersici* are confined to protoxylem and metaxylem vessels in the lower nodes during the early stages of infection. Within the hypocotyl, lateral spread from vessel to vessel through pits soon results in extensive invasion of secondary as well as primary xylem elements. There follows an "explosive" stage of colonization in which an enormous number of spores is produced, followed by a rapid and extensive invasion of the primary and secondary xylem elements within the upper stem and petioles. The pathogen apparently "surges" forward within the xylem elements in a manner similar to that in banana roots, described below. It is not until the final stages of pathogenesis that the pathogen invades and proliferates in xylem parenchyma. Maraite (128) reported a similar "explosive" increase in colonization in muskmelon infected with *F. oxysporum* f. sp. *melonis*.

In banana, a monocot, the pattern of invasion and systemic distribution is strikingly similar to that in dicotyledonous hosts of *F. oxysporum.* Following entrance into the xylem elements of adventitious roots, *F. oxysporum* f. sp. *cubense* spreads into the rhizome stele and then invades the tracheary elements of the pseudostem. Extensive colonization of rhizome xylem occurs prior to invasion of the pseudostem and is analogous to the extensive colonization of the hypocotyl when susceptible tomato and cabbage plants are attacked. The upward distribution of *Fusarium* is accomplished principally through the production of microconidia, which are carried upward in the sap stream until they encounter an end wall or perforation plate that permits sap flow but screens out the microconidia. The microconidia then germinate, and the developing hyphae penetrate the porous barriers and sporulate above the obstruction. These "secondary" spores are swept upward again to the next barrier, and thus, by this repetitive process, *Fusarium* becomes distributed quickly along xylem vessels. Initial movement through the root xylem may be slow in young roots, requiring about 4 weeks to advance 75 cm, but in older, mature metaxylem vessels, the pathogen can advance in surges of 30 cm with every new generation of spores, e.g., every 2 to 3 days. Upon reaching the rhizome, the pathogen can become distributed within an 8-m-tall plant in less than 2 weeks.

Penetration, Colonization and Distribution of *Fusarium* within Resistant Hosts and Non-hosts

Studies comparing *Fusarium* penetration and invasion of resistant and susceptible hosts of cabbage (4, 169), flax (142, 178), pea (115), and tomato (164) indicate that roots of resistant plants may be penetrated as readily (or nearly so) as roots of susceptible plants. Differences in infection between resistant and susceptible hosts do not become obvious until the pathogen invades the cortex and stele. There is much less evidence of *Fusarium* within parenchymatous tissue of the cortex of resistant hosts, and continued invasion of the stele and xylem is severely restricted. In resistant cabbage and flax plants, the stele was rarely penetrated, and once within invaded xylem, *Fusarium* was restricted to a few vessels. The spread of *Fusarium* within the xylem of resistant banana and tomato hosts was also limited. In banana, infection was checked within the roots and rootlets or at the root base, and lesions were localized. Several studies have shown that in tomato cultivars resistant to *F. oxysporum* f. sp. *lycopersici,* invasion was restricted to vascular elements of the roots and lower stem (3, 29, 61, 68, 164, 170). Sometimes resistant roots were heavily infected, but symptoms did not appear as long as the invader remained confined to the primary xylem.

Several studies on a wide range of non-host plants (including several weeds, cereals, vegetables, and cotton) infected naturally or artifically with *F. oxysporum* have all indicated a pattern of penetration and invasion similar to that of resistant hosts: roots were colonized, sometimes extensively, but xylem invasion was not extensive and *Fusarium* spread was restricted to the roots and lower stems (5, 7, 84, 99, 114, 164, 185). In no instance did external symptoms develop when *Fusarium* was restricted to the roots and lower stem.

The Significance of Colonization Patterns in Susceptible and Resistant Hosts and Non-hosts

A comparison of the patterns of host colonization associated with resistant and susceptible interactions suggests that: (i) resistance is dependent upon limiting the invader to cortical tissues outside the root stele or confining the invader either to the roots or to a relatively few protoxylem and metaxylem vessels of the hypocotyl or lower stem; (ii) susceptibility is dependent upon entry of the pathogen into vascular elements of the root, and on extensive distribution and build-up in many vascular elements of the primary and secondary xylem; and (iii) moderate to severe external symptom appearance in susceptible hosts follows an "explosive" phase of systemic invasion. Symptom expression is a consequence of host-pathogen interactions that occur once the pathogen has invaded the upper stem and petioles. This is not to imply that systemic invasion will necessarily lead to symptom development, but we have not read of any instance of tolerance to Fusarium wilt fungi associated with intensive systemic host colonization following natural infections.

The tissues of the roots and of the hypocotyl are thus extremely important in the development of systemic infection. Apparently there are many cross-connections between vascular elements in this region, and therefore lateral and radical distribution of the pathogen can readily take place there. Thus, xylem of the upper root and hypocotyl serves as a distribution center (analogous to a central railroad switching yard), and an infection and build-up of a vascular pathogen in this region is particularly dangerous to the plant. It appears that once the outcome of this crucial battle is decided, the war for survival of a plant is either lost or won. Several additional lines,of evidence support this view, and they are discussed below.

Several grafting experiments involving various combinations of resistant and susceptible tomato rootstocks and scions, inoculated either within the scion or rootstock with *F. oxysporum* f. sp. *lycopersici,* indicate that (i) resistance lies within the root-hypocotyl system rather than within the stem and leaves, and (ii) susceptibility requires a build-up of the pathogen within the hypocotyl and lower stem. Symptoms did not develop in susceptible (S) scion / resistant (R) rootstock (i.e., $\frac{S}{R}$) combinations following root inoculation because of limited root invasion (89, 102, 162). The limited fungal development was by-passed when adventitious roots of the susceptible scion were inoculated; symptoms developed, presumably, because the fungus was then able to invade the lower stem and colonize it extensively. Symptoms developed in $\frac{S}{S}$ or $\frac{R}{S}$ combinations in which root inoculations led to extensive colonization within the susceptible rootstock (89, 103, 135). Keyworth (103) noted that in stem-inoculated $\frac{S}{S}$-plants, the stem was not invaded as extensively as in root-inoculated plants and there were fewer symptoms, again indicating the requirement for build-up of the pathogen in the rootstock. Scheffer (162) extended these studies by approach-grafting resistant and susceptible plants. When the stems of resistant scions were inoculated, the susceptible scion never showed symptoms (flaccidity) unless the fungus first moved downward and colonized the susceptible hypocotyl.

Virulence has been attributed to the capacity of a race to overcome host resistance mechanisms and build-up within the root-hypocotyl region. The colonization and build-up patterns of *F. oxysporum* f. sp. *lycopersici* race 1 and the more virulent race 2 were compared in the tomato cultivar Improved Pearson VF-11 (VF-11), which is resistant to race 1 but susceptible to race 2, and in the near-isoline Improved Pearson (IP), which is susceptible to both races (30, 61). Invasion and build-up within the hypocotyl and subsequent movement into the epicotyl of VF-11 was extensive with race 2 but limited with race 1. The extensive distribution and build-up of race 2 in the VF-11 cultivar was identical to that which occurred with both race 1 and race 2 in the susceptible tomato cultivar.

Consideration of all of the above evidence, taken from a wide range of experimental designs, has led to the development of one of the central themes of this chapter: limited distribution and build-up of *F. oxysporum* is a feature common to all resistant interactions and distinguishes them from susceptible interactions in which distribution and build-up are extensive.

Mechanisms of Host Resistance

Resistance to *F. oxysporum* could be achieved by mechanisms operative during the prevascular stage of pathogenesis, the vascular stage, or both (Fig. 33-1). Unfortunately, few studies have centered upon mechanisms by which plants resist initial penetration and ingress to vascular tissue. Mechanisms of resistance that precede and prevent or reduce vascular invasion are important in resistance of hops to *Verticillium,* another soil-borne vascular pathogen (174). These mechanisms include lignituber (papilla) formation, cell wall thickening, and rapid differentiation of the endodermis, and there is some indication that these mechanisms may also be important in limiting invasion by *Fusarium.* In resistant flax plants, the restricted distribution of *F. oxysporum* f. sp. *lini* within cortical tissue of roots was related to an accelerated cell division and cork wall formation in cells adjacent to invaded cells (178). The corky walls appeared to provide an effective barrier to further invasion of hyphae that may have been weakened by previous protoplastic reactions within invaded cells. Resistance in cabbage to *F. oxysporum* f. sp. *conglutinans,* however, was not associated with any morphological or visible wall structure differences between resistant and susceptible plants, including suberization or gum-like or other perceptible wall-occluding deposits within infected cortical tissue (4).

Invasion of the vascular system is limited in resistant infections, but the functioning of the host in accomplishing this is not clear. Resistance could be associated with short vessel length that can influence the rate of spore distribution, but no structural differences were found between xylem elements of resistant and susceptible cultivars of cotton (17), banana (13), and tomato (11, 121). Remaining possibilities include the production of physical barriers impenetrable to upward fungal spread and/or the accumulation of compounds that adversely affect fungal growth, reproduction, and later spread. These possibilities and related metabolic responses are discussed below.

Physical Barriers

Occlusion of vessel elements is apparently a common mechanism by which many plants limit vascular invasion (6). The occluding mechanism was operative in banana against various soil microorganisms and formae speciales of *F. oxysporum* isolated from other hosts. Sweet potato, broccoli, passion flower, squash, and sesame all had localized infection within 6 or 7 days after root inoculations with either a mixture of microorganisms or *F. oxysporum* f. sp. *lycopersici* (7). The resistance of cultivars to their respective *F. oxysporum* formae speciales has been correlated with rapid vascular occlusion by gels in cotton (22), by tyloses in tomato (11), and by gels, gums, tyloses, and vascular collapse in banana (12, 13, 14, 187). Gel formation is apparently a general phenomenon common to resistant-type responses (181).

Susceptibility was correlated with delayed occlusion in cotton (22) and tomato (11) and with gel breakdown and delayed tylosis in banana (13). In banana, the susceptibility of the cultivar Gros Michel to *F. oxysporum* f. sp. *cubense* at 28 C was correlated with delayed development of physical barriers and distribution and build-up of the pathogen. At 21 C or 30 C the localization of infections by gels and tyloses and the absence of symptoms was comparable to that in a resistant banana cultivar.

Langton (114) and Phillips et al. (153) both have shown that tomato cultivars susceptible to *F. oxysporum* f. sp. *lycopersici* had resistance mechanisms that were effective against this pathogen if these mechanisms were first made operative by pre-inoculation (3 days in advance) with the nonpathogen, *Cephalosporium.* The induced resistance was associated with occlusion of the primary xylem vessels by tyloses that occured immediately above the infection sites. The occlusions reduced the spread of the pathogen from primary

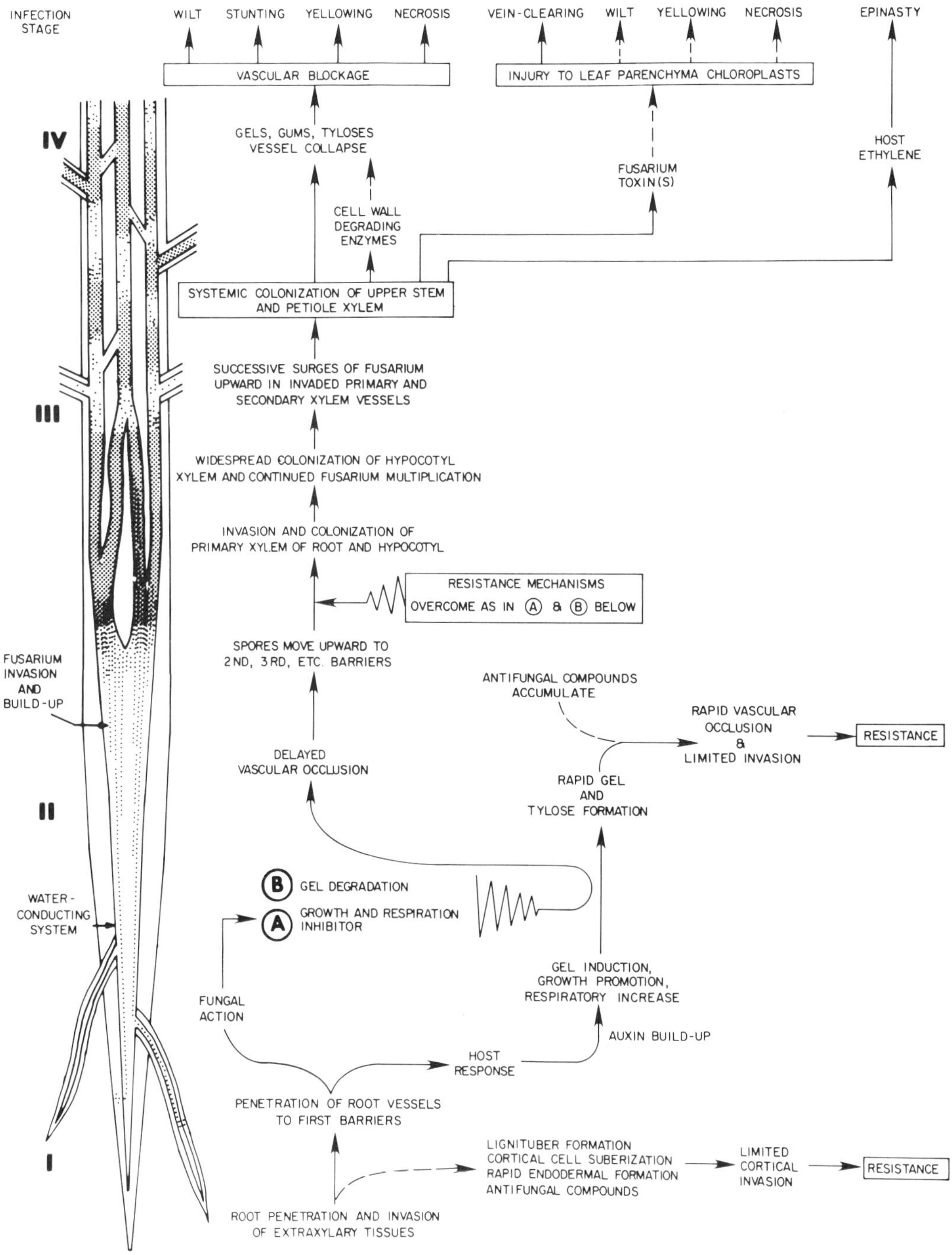

Fig. 33-1. Stages of infection, host resistance responses, and symptom expression in Fusarium vascular wilt pathogenesis. Resistance mechanisms are effective during STAGE I (pre-vascular invasion) and STAGE II (initial invasion of root and hypocotyl xylem vessels); susceptibility is dependent upon extensive colonization and build-up of *Fusarium* within the hypocotyl and extensive invasion of the upper stem and petioles via infected primary and secondary vessels (STAGE III); symptom induction is related to increased auxin and ethylene levels and possibly the activity of cell wall degrading enzymes and toxin action on leaf parenchyma (STAGE IV). Hydrolyzing enzymes and toxins may also assist with colonization activities by *Fusarium*. Dotted lines depict processes or sequences that have been proposed but have not been substantiated experimentally. The diagramatic representation of a plant's water-conducting system depicts the general pattern of colonization and multiplication along the plant axis by *Fusarium*.

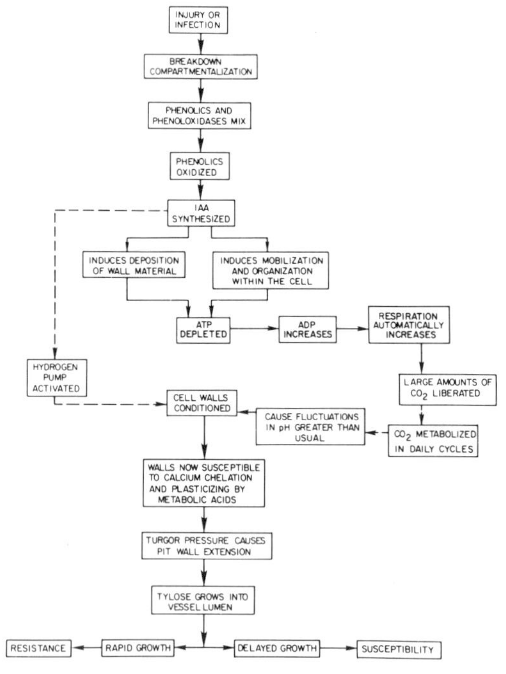

Fig. 33-2. Mechanism proposed for tylose induction and development in vascular wilt resistance. Dotted lines depict processes that have been proposed but have not been substantiated experimentally.

to secondary xylem, from the roots to the hypocotyl, and from the hypocotyl to the first internode (152). When plants were inoculated with *Fusarium* but not *Cephalosporium,* tyloses were sparse in primary vessels and hyphae were abundant in both primary and secondary vessels. An important point emerges from these studies: in susceptible or resistant plants, the potential for occlusion is present; what seems most critical in localization processes is not whether barriers are developed but how *rapidly* they are developed.

Metabolic Responses.

The events described above must occur within the vascular xylem during the first 3 to 5 days following infection if the host is to seal-off the invader effectively. The host must react quickly in a complex process of response that involves the mobilization of plant growth regulators and shifts in phenol metabolism, both requiring an increased energy flow that must be coupled to the increased work performed.

A suggested sequence of events by which all of the metabolic factors identified to date could interact to promote wound-healing or the sealing-off of infections is presented in Fig. 33-2. Host response is initiated by injury or infection. This stimulus causes a release of phenolics that are normally compartmented within specialized storage cells that appear randomly distributed in most tissues. The phenolics become enzymatically oxidized and, in the process, mediate the synthesis of indole acetic acid (IAA) and the inhibition of IAA oxidase. Indole acetic acid builds up, is translocated, and thus transmits a response signal to non-injured cells at a distance. These cells initiate the synthesis of protoplast and cell wall constituents. Increased respiration follows. Cell walls become plasticized (pliable) and swell, either by a process of CO_2 fixation that causes a fluctuation in hydrogen ions and a chelation of calcium or, alternatively, by a hydrogen pump system (28, 157). As a result of these events, end wall, perforation plate, and vascular pit membranes are softened and swell to produce vascular occluding gels. In addition, the vascular parenchyma cells grow to form tyloses that protrude through the softened pit membranes and enlarge within the vessel lumens until they are completely sealed. Finally, the interface of the infection and these occluding structures become infused with the oxidized phenolics. These phenolics polymerize with each other and with wall constituents to form a highly resistant and insulating barrier. The data upon which this model is based are presented below.

Respiration A rapid and intensive stimulation of oxidative processes was required for resistance mechanisms to be effective in cotton (108), cabbage (91), tomato (79), and radish (8). In radish, a dramatic respiratory burst 3 to 4 days after inoculation was a characteristic feature of resistant-type interactions. The timing of this burst of respiration in radish coincided with the timing of the occlusion response as determined in banana and tomato. The burst occurred when a radish cultivar resistant to *F. oxysporum* f. sp. *conglutinans* was infected with that organism or when either resistant or susceptible radish cultivars were inoculated with root flora, e.g., in any resistant-type reaction. The respiration burst failed to occur in a susceptible interaction, e.g., when a susceptible radish cultivar was inoculated with the host-specific, pathogenic *F. oxysporum* f. sp. *conglutinans*.

Plant growth regulators Plant growth regulators play a key role in our model of resistance responses to infection (Fig. 33-1, 33-2). Several studies have shown that synthetic plant growth regulators will induce resistance in otherwise susceptible tomato plants (34, 35, 39, 41, 50, 58). Indole acetic acid accumulates in plants infected with *Fusarium*. This build-up of IAA may result from the inhibition of IAA oxidase that degrades IAA (120, 131), and the process may be dependent on increases in phenolic substances. The concentration of phenols is known to regulate IAA through its inhibitory effects on IAA oxidase that degrades IAA, and the phenolic content does increase following infection by *Fusarium*. Mace and Solit (120) suggested that the phenol, dopamine, may function in resistance by helping to maintain high levels of IAA needed for localization responses such as tylose formation. Dopamine not only protects IAA from oxidation by inhibiting IAA oxidase, but may also be involved with the synthesis of IAA.

Increased levels of IAA in infected plants may also result from the activation of pathways that produce IAA (120, 132). In tomato, IAA build-up resulted from increased ability of leaf and stem extracts and stem slices to produce IAA from tryptophan. Indole acetic acid is produced when trytophan is oxidatively deaminated by the polyphenoloxidase (PPO) system (78), and this occurs in tomato (133), where PPO activity and tryptophan both increase following infection with *Fusarium*. Chlorogenic acid, and orthodihydroxy phenolic normally present in healthy tomato (but presumably compartmented within specialized parenchyma cells), also markedly enhanced IAA production in diseased plants. The increase corresponded in time with the large increase in PPO activity (130).

Ethylene is not believed to be directly involved in resistance reactions. There was no change in ethylene in a resistant tomato cultivar following infection by *F. oxysporum* f. sp. *lycopersici* (77), and adding ethylene enhanced disease development. Adding ethephon (an ethylene-releasing substance) lowered resistance and increased peroxidase activity 317% and PPO activity 146% (77).

Phenol metabolism Increase in the phenolic content is associated with infection of plants by vascular wilt Fusaria, and the timing of this increase seems critical to resistance. In tomato, several resistant host-*Fusarium* combinations responded with a burst in phenolic synthesis 3 days after inoculation compared to a gradual build-up in phenolics in susceptible host-*Fusarium* combinations (134). Three days was also the optimal time interval between pre-inoculation of a susceptible tomato cultivar with a non-pathogen and inoculation with *F. oxysporum* f. sp. *lycopersici* for maximal induced resistance. The correlation between 3 days for phenolic build-up and 3 days for resistance to develop suggests that the increased phenolic activity may contribute to the induced resistance. Further evidence linking phenolic activity to resistance was provided recently by Carrasco et al. (24), who demonstrated that pre-treatment with precursors of phenolic compounds 3 days before inoculation with *F. oxysporum* f. sp. *lycopersici* also induced resistance within a susceptible tomato cultivar, and the degree of induced resistance was positively correlated with the induced phenolic level. Thus, as with other processes, the influence of phenol metabolism on the course of pathogenesis is determined largely by the timing of events.

The increase in phenol levels following infection may arise from activation or shifts in phenol pathways, or from mobilization of stored phenols. Several studies indicate that phenol metabolism is affected in vascular wilt diseases. For example, when labeled shikimic acid was added to infected tomato plants, label increased in the non-hydrolysable material containing lignin of a resistant but not of a susceptible cultivar (66). Davis and Dimond (40) suggested that the shunting of phenolic metabolism to melanin formation may occur at the expense of other synthetic reactions; there was less lignin and fewer lignified cells in diseased than in healthy plants, although 'tanninase' and 'ligninase' were not detected in the vascular fluid of diseased plants. In diseased cotton, the pattern of phenolic accumulation was correlated with phenylalanine ammonia lyase (PAL) and B-glucosidase activity (172). The pathogen may also produce oxidases, and, thus, contribute to the oxidase pool. *Fusarium oxysporum* f. sp. *cubense* produced polyphenoloxidases and phenyloxidases in vitro (124), and *F. oxysporum* f. sp. *melonis* produced polyphenoloxidase when grown on autoclaved muskmelon tissue (128).

The mobilization of preformed phenols may also contribute to resistance responses. Specialized parenchyma cells scattered throughout the cortex and xylem of banana, cotton, and tomato are packed with phenolics (mainly dopamine) stored in vacuoles in an uncombined, reduced state (15, 16, 118, 123). These packaged phenols can be oxidized quickly once released and in contact with oxidative enzymes (15, 120, 124).

In tomato, phenolics (mainly o-dihydric phenolic glycosides) occur in conjugated forms rather than in a free stage (123). When the phenolics are released following cell injury, they must first undergo enzymatic hydrolysis by glycosidase. The free oxidizable phenolics are then thought to diffuse into vessel lumens where they polymerize and become incorporated into gel plugs to form gums (16, 181). The oxidative enzymes and phenolic substrates necessary for these reactions to occur are present in infected plants (40, 42, 184).

If the stored phenolics in vascular parenchyma do represent the major source of phenolics that undergo oxidation and polymerization to produce vascular browning, then browning as such is not correlated with resistance or susceptibility. Browning would only indicate injury to phenolic-storing cells. More exensive browning in a susceptible cultivar, as compared with a resistant one, reflects the wider distribution of the pathogen and, thus, a greater number of injured cells. There are no known differences between resistant and susceptible cultivars in the number of phenolic-storing cells. Again, the course of pathogenesis is determined largely by the timing of events. Rapid cell injury could result in quick release and activity of these packaged phenolics and could play several important roles necessary for effective containment of an invader. Delayed injury and phenolic release could permit build-up of the pathogen and lead ultimately to symptom development. Conversely, too rapid a release of phenolics could permit premature infusion of phenolics into gels and tylose walls, reduce the plasticity of vessel walls, and retard the sealing-off of infected vessels (16).

Calcium Vessel walls must be made plastic before they can swell to form gels. Using prepared membranes that simulated vessel walls, Beckman (9, 10, 16, 18) demonstrated that calcium bridges must first be disrupted before membranes would swell. Fluctuations in pH, which presumably exposed the calcium bridges to the action of carboxylic acids, and the subsequent removal of these bridges caused membranes to swell, thus making them vulnerable to disintegration by pectic enzymes. Calcium pectate gels are resistant to degradation by pectic enzymes produced by *F. oxysporum* f. sp. *lycopersici* (31). Walker (60) reported that calcium caused tomato plants to be slightly more resistant and that withholding calcium increased susceptibility to Fusarium wilt. Plant growth regulators also induced resistance of tomato to *Fusarium* attack if calcium was present (35, 60). The calcium and growth regulator treatments were thought to influence resistance through alteration of vessel wall pectic substances (59).

Antifungal Compounds

Host-produced chemicals that are inhibitory or toxic to *Fusarium* may also contribute to resistance, but considering all the evidence at hand, it seems to us unlikely that antifungal compounds are primary determinants of resistance to the vascular wilt Fusaria. Our conclusions are: (i) in natural infections or tap root inoculations, if localization is overcome and the invader builds up in primary and secondary tissues within the hypocotyl, susceptibility follows whether or not inhibitors are present in the upper stem; (ii) fungal inhibitors are not always present in healthy or diseased plants, or they may be present in low amounts, especially in the roots and lower stem; (iii) fungal inhibitors appear to be most effective in the upper stem, and may contribute to reduced build-up in resistant plants.

To date, the possible role of antifungal compounds in Fusarium wilt resistance has been examined extensively only in tomato. Much of the evidence that suggests the involvement of antifungal substances in resistance is indirect, and must be gleaned from studies that utilized cuttings or artificial inoculations along the stem (61, 82, 87, 102, 122). These studies have compared the growth and distribution of *F. oxysporum* f. sp. *lycopersici* within resistant and susceptible tomato cultivars following the uptake of microconidia either through the cut ends of stems or through cuts made along the stem. *Fusarium* did not develop or become distributed well within the stems of either susceptible or resistant cultivars following stem inoculations. There is some evidence that this poor colonization may be due to antifungal compounds. Bugbee (22) showed that tracheal fluid from occluded vessels of diseased cotton was toxic to *F. oxysporum* f. sp. *vasinfectum* in culture. Similarly, phytoalexin-like responses were found when tomato was inoculated with other vascular wilt Fusaria (37, 114). Stromberg and Corden (175) have suggested that susceptibility of tomato to *F. oxysporum* f. sp. *lycopersici* is due to detoxification of an antifungal compound that is present in sufficient quantity to limit invasion, but this is not consistent with other findings. Others have been unable to detect pre-formed inhibitory substances in extracts of a resistant tomato cultivar (170) or in roots and stems of tomato (81, 89, 114). Antifungal activity has been reported to be absent in healthy resistant and susceptible plants (54, 113), to be twice as active in resistant as in susceptible plants (87), or to be present in healthy resistant plants but absent in susceptible plants (82). A recent study (143) suggests that B-(1,3) glucanase, a lytic enzyme capable of degrading hyphal wall constituents, may contribute to resistance of a muskmelon cultivar to *F. oxysporum* f. sp. *melonis*.

Tomatine, a tomato alkaloid with antifungal activity, has been investigated intensively, but it does not appear to be a primary determinant of resistance to *Fusarium*. Irving (96) and Kern (100) found no correlation between the resistance status of cultivars and their tomatine content in whole stems of healthy plants. Kern (100) concluded that the tomatine content was not in high enough concentration to be important in resistance. McCance and Drysdale (136) reported that tomatine increased soon after inoculation to levels that were sufficient to adversely affect *F. oxysporum* f. sp. *lycopersici,* but they concluded that tomatine is not a primary determinant of resistance because the increase was similar in both resistant and susceptible cultivars. Smith (167) found no correlation between tomatine levels and resistance or susceptibility of tomato to *F. oxysporum* f. sp. *lycopersici* race 1 or race 2. Tomatine increased to the same fungitoxic level within a cultivar resistant to race 1 but susceptible to race 2 following inoculation with either race. There was also no correlation between the virulence of the races toward tomato and sensitivity to tomatine. Tomatine was equally inhibitory to the vegetative growth of race 1 and race 2 in vitro. Thus, it seems unlikely that the virulence of race 2 is determined either by its relative insensitivity to tomatine or its ability to limit tomatine production or detoxify tomatine within the resistant cultivar.

Despite inconsistencies in the literature linking antifungal compounds and resistance, numerous studies have demonstrated the presence of compounds inhibitory to *Fusarium* within vessels of diseased plants. How might such compounds contribute to a plant's defense system? One explanation may be that, in nature, resistance is accomplished by a two-component, sequential system that involves, first, the development of physical barriers that limit rapid upward distribution of the invader, and secondly, the accumulation of chemicals that have an adverse effect of fungal growth. These processes, in sequence, reduce the chance that an organism will colonize an invaded vessel or spread to adjacent vessels. Localization by physical barriers must occur; fungal inhibitors, if present, will reinforce the resistance process.

If one accepts this scheme, the most of the data that appears to discredit fungal inhibitor involvement in resistance is reconcilable. For example, it really is not necessary that an inhibitor be present in healthy plants or that there be more inhibitor in the resistant plant compared to a susceptible plant either before or following infection. All that is required for antifungal compounds to be effective in resistance is that they be present soon after infection so that they can accumulate to effective concentrations within infected, occluded vessels. Indeed, that such a two-component system may be operative in resistance to *Fusarium* and *Verticillium* has been suggested by results with cotton (22, 119), and tomato (136).

In summary, host resistance to invasion by *F. oxysporum* is an active process. It is characterized by enhanced metabolic activity that includes increased energy utilization, increased substrate pools, and increased enzymes and growth regulator activity needed for cell growth and the development of physical barriers that act in concert with antifungal compounds to contain the invader. As we shall see in the next section susceptibility to *F. oxysporum* occurs when these resistance responses are inhibited or delayed.

Pathogenesis in Susceptible *Fusarium*/Host Interactions

Cell Wall Degrading Enzymes

It has long been proposed that cell wall degrading enzymes acting upon exposed cellulosic and pectinaceous wall components are the means by which the vascular Fusaria obtain substrates required for vegetative growth and reproduction within invaded xylem vessels (80, 81, 155). Consequently, the literature abounds with in vitro and in vivo studies that have attempted to implicate these enzymes in events associated with (i) colonizing susceptible tissue, (ii) overcoming host resistance, and (iii) symptom induction. *Fusarium oxysporum* produces several enzymes that act upon the pectic and cellulose components of cell walls, and they are discussed below.

Pectin methylesterase The capacity to produce pectin methylesterase (PME) is general among the vascular Fusaria, having been detected both in vitro (23, 43, 64, 80, 81, 90,

149, 155, 179, 183, 191) and in vivo (23, 44, 54, 85, 109, 111, 112, 127, 130, 183). Pectin methylesterase was often not detected in vivo until later stages of infection (111, 130, 183). Whether the PME found in infected tissue was produced by the fungus or released from host walls is not clear.

Several studies in the 1950's indicated that although PME did not appear to play an important role in the initial establishment of *F. oxysporum* f. sp. *lycopersici* in a susceptible tomato cultivar, it may be an important factor in causing wilt, vascular discoloration, and vascular plugging (80, 81, 163, 191). Culture filtrates with high PME and low polygalacturonase activity produced typical wilt symptoms, decreased transpiration, and caused vascular plugging. Pierson et al. (155) demonstrated that the plugs stained with ruthenium red, indicating that they were composed, at least in part, of pectinaceous material. Pectin methylesterase free of polygalacturonase (PG) or depolymerase (DP) did not produce wilting, attack pectin, or macerate tomato tissue. Pectin methylesterase made PG more active, and it was suggested that the role of PME may be to de-esterify pectin prior to the action of DP or PG, because pectin-splitting enzymes act rapidly on de-esterified molecules (183). Paquin and Coulombe (149) reported that a virulent strain of *F. oxysporum* f. sp. *lycopersici* produced 2 to 3 times more PME than a less virulent strain, although both strains produced similar amounts of DP or PG when calculated on growth basis, and concluded that PME may have a major role in symptom development. Recent studies (54,111), however, report that only 5 to 10 percent of the PME increase in tomato following infection by *F. oxysporum* f. sp. *lycopersici* is of pathogen origin. Thus, the ability of the pathogen to produce PME may not be as important in pathogenesis as the ability of the pathogen to induce the host to produce PME or to release PME bound to cell walls.

Polygalacturonase Polygalacturonase is also produced by *F. oxysporum* in vitro (19, 23, 97, 149, 151, 166, 179, 191) and in vivo (23, 63, 85, 127, 130, 151, 183). Jones et al. (97) demonstrated in vitro that proteins isolated from cell walls completely inhibited PG production by *F. oxysporum* f. sp. *lycopersici* in 3 days and prevented cell wall degradation. Sugars also suppressed PG production in vitro (19, 63, 151). There are conflicting reports concerning PG activity within tomato. Patil and Dimond (151) reported an exo-PG was present in infected tomato, Deese and Stahman (43, 44) and Mussell and Green (141) reported endo-PG activity, and others (112, 130, 183) have been unable to detect any PG activity in infected tomato plants.

Pectin transeliminase Papavizas and Ayers (148) reported that 13 or 14 formae speciales of *F. oxysporum* and *F. solani* produced pectin transeliminase (PTE) adaptively in 1 week at 25 C. Ferraris et al. (63) found that pectin galactose transeliminase increased markedly in tomato following infection with *F. oxysporum* f. sp. *lycopersici,* and was equal in amount to that of PG.

Cellulase enzymes Cellulose is the most abundant substance comprising the inner cell walls exposed to vessel lumens. Since cellulose could serve as a carbohydrate for *F. oxysporum* following partial decomposition by cellulase enzymes, several studies have explored a possible role of cellulases in wilt pathogenesis. There is, however, little evidence to support the idea. Two enzymes, designated C_1 and Cx, are required to degrade cellulose. The C_1 enzyme acts upon native, insoluble cellulose to produce linear chains that are attacked by the Cx enzyme to produce cellobiose and glucose. Production of the Cx enzyme in vitro has been reported for several formae of *F. oxysporum* (64, 94, 179), but Matta and Dimond (130) were unable to detect cellulase activity at any time in wilting tomatoes, and others have found only low activity in tomato (85, 86). Hussain and Dimond (95) reported that *F. oxysporum* f. sp. *lycopersici* produced C_1 and Cz enzymes on living tomato tissue, and they demonstrated that a purified dialyzed cellulase solution from a culture of the tomato pathogen could induce wilt when applied to tomato cuttings. They proposed that cellulase may act on cell walls to release carbohydrates that would serve as pathogen substrates and that the larger fragments released into the vascular stream may contribute to the interferences in water movement associated with wilt. However, more recent evidence indicates that cellulase is unimportant in pathogenesis associated with wilt, although cellulases may be active in the final stages of disease when *F. oxysporum* ramifies the paravascular tissues that are dying and disintegrating.

Hemicellulases Few studies have explored the involvement of hemicellulases in Fusarium wilt pathogenesis, but these enzymes could be of greater significance than pectinase and cellulase enzymes in that the greater part of the amorphous matrix in which cellulose microfibrils are embedded is composed of hemicellulose-type materials, e.g., glucans, galactans, xylans, and mannans (192).

Role of cell wall degrading enzymes in pathogenesis Several lines of evidence indicate that cell wall degrading enzymes are not required for initial establishment of the pathogen in xylem elements. Mann (126) and McDonnell (138), using mutants unable to produce pectin-splitting enzymes, showed that pectolytic enzyme production was not necessary for successful establishment of *F. oxysporum* f. sp. *lycopersici* in tomato or for the development of wilt symptoms. Winstead and Walker (191) and Scheffer and Walker (163) demonstrated that filtrates of several Fusaria with high pectin methylesterase and low endopolygalacturonase activity caused comparable browning and plugging in several plant cultivars whether they were resistant or susceptible to the respective pathogens. They concluded that resistance cannot be explained on the basis of tolerance to pectic enzymes. Cooper et al. (31, 32) and Jones et al. (97), in perhaps the most thorough and comprehensive studies to date on cell wall degradation by *F. oxysporum* f. sp. *lycopersici,* also found no relation between cultivar resistance and the suceptibilities of cell walls to enzymatic degradation. Primary and secondary walls of resistant and susceptible tomato cultivars were degraded at similar rates by pectic and other cell wall degrading enzymes. The enzymes were highly inductive and extremely susceptible to catabolic repression, the first enzyme being induced by galacturonic acid. The authors suggested that the induction of enzyme synthesis may depend on the release of galacturonic acid by a constitutive enzyme; the process would then become autocatalytic.

There is also no direct evidence that cell wall degrading enzymes are necessary for overcoming host resistance (see B, Fig. 33-1), although enzyme production may be suppressed in a resistant host. Deese and Stahmann (43, 44) found that PG production was suppressed when *F. oxysporum* f. sp. *lycopersici* was grown on resistant tomato tissue and that both PME and endo-PG production were lower on infected resistant stem tissue or juice extracted from infected lower stems of a resistant cultivar as compared to a susceptible cultivar of tomato. Mussell and Green (140, 141) also concluded that resistance in tomato to *F. oxysporum* f. sp. *lycopersici* was associated with suppression of PG activity. The tomato wilt pathogen persisted in the stem of both resistant and susceptible cuttings following inoculation, but PG activity occurred mainly in the susceptible cultivar. Unfortunately, they did not quantify enzyme activity in relation to the amount of fungal growth and distribution in each cultivar, e.g., the greater PG activity may have reflected greater fungal growth and distribution in the susceptible cultivar. In a similar experiment, Langcake and Crysdale (112) did calculate enzyme activity per unit of fungal weight, and found that resistant and susceptible tissues were equally effective inducers of pectic enzymes and were equally degraded by these enzymes. Jones et al. (97) also reported that resistant and susceptible plants were equally effective inducers of PG production by *F. oxysporum* f. sp. *lycopersici.*

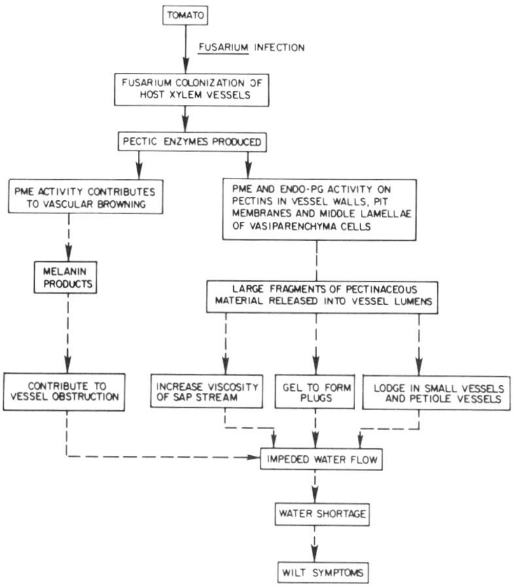

Fig. 33-3. Roles that have been suggested for cell wall degrading enzyme involvement in vascular wilt pathogenesis by *Fusarium*. Dotted lines depict processes that have been proposed but have not been substantiated experimentally.

In summary, numerous roles have been proposed by which wall-degrading enzymes contribute to symptom induction (Fig. 33-3), but these roles have not been substantiated. We propose another role: hydrolytic enzymes play a decisive role in permitting systemic distribution of the pathogen by weakening the vascular gels that normally contribute to localizing infections (see B, Fig. 33-1). Following systemic infection, these degraded gels may also contribute to vascular dysfunction and water stress.

Toxins

In our definition, based on that of Ludwig (117) and Wood (190), a toxin is a product of a microorganism or microorganism/host interaction that damages the host by acting directly on *living protoplasts* to influence disease development. According to this definition, cell wall degrading enzymes are not considered toxins. Growth regulators are not excluded, and could be considered vivotoxins if they met the criteria established by Dimond and Waggoner (52).

Two roles have been proposed for the involvement of toxins in Fusarium wilt pathogenesis: (i) toxins have a causal role in susceptibility, e.g., in overcoming host resistance, and (ii) toxins have a primary role in symptom induction. At present there is no direct evidence that *Fusarium* overcomes host rcsistance through the action of toxins on host protoplasts, and very weak evidence that toxins contribute to the Fusarium wilt syndrome.

A few studies have indicated that toxins may be involved in pathogenesis. Vein-clearing occurred in tomato 24 hr after inoculation, although *F. oxysporum* f. sp. *lycopersici* was not present in leaves showing vein-clearing even 36 hr after inoculation (65). Increased permeability has been reported in tomato inoculated with *F. oxysporum* f. sp. *lycopersici* (29), in cuttings treated with culture filtrates of *F. oxysporum* f. sp. *lycopersici* (176), and in cuttings that have been allowed to take up vascular fluids from wilted plants infected with *F. oxysporum* f. sp. *lycopersici* (83). Marte and Zazzerine (129) noted changes in infected leaf chloroplasts that were absent in plants lacking water, and suggested possible involvement of toxins. The changes appeared to accompany chlorosis. Several physiological changes occur in uninvaded leaves of diseased plants, e.g., respiratory increase, alterations in phenol metabolism, activation of polyphenoloxidases, and increases in growth regulator activity. Vascular browning also occurs far in advance of *Fusarium* invasion (36, 102). It has been suggested that these changes are due to the action of toxins. It is questionable, however, whether these changes are detrimental; they are, in fact, normal responses to infection, injury, or stress and represent part of a plant's resistance and repair capability.

Much of the evidence supporting the role of toxins in wilt pathogenesis is based on experiments in which cuttings were allowed to take up fungal filtrates or chemicals isolated from culture filtrates. Symptom development and alterations in selected physiological processes were then compared with inoculated, healthy, or water-stressed cuttings. This literature has been thoroughly reviewed for Fusarium wilt by Walker (186).

There are certain shortcomings inherent in filtrate experiments that must be considered when evaluating the contributions of these studies to our understanding of pathogenesis. One is that substrates available to the fungus in artificial media, stem tissue extracts, or even whole stem tissue probably differ greatly from those in infected vessels. These differences may result in the production of metabolites in vitro that are markedly different from those produced in vivo. Quantitative as well as qualitative differences may occur. Finally, many substances, if present in high enough concentration, will damage a plant and induce various symptoms. For example, Hodgson et al. (93) demonstrated that carbowaxes and various polysaccharides of both plant and microbial origin all induced wilting in tomato cuttings. Threfell (177) showed that wilting of tomato cuttings treated systemically with cell-free filtrates of *Verticillium cultures* was due to a heat-stable substance (not an enzyme) that blocked the ends of shoots. *Fusarium solani* f. sp. *eumartii* was also shown to produce a polysaccharide in culture that, when introduced into tomato cuttings, induced wilting by blocking the basal ends of the cuttings rather than by toxic action. *Fusarium* spp. isolated from *Hibiscus esculentus* produced substances in culture that caused wilt when added to *Hibiscus* cuttings as fungal filtrates, but in nature, infections by these isolates resulted in limited colonization and little or no expression of external symptoms (84). In spite of these shortcomings, most toxin studies have involved substances produced by forms of *F. oxysporum* in culture.

Gäumann (69) suggested that resistance or susceptibility is dependent upon whether or not the host can prevent the production of toxin by the pathogen. According to him, toxin production is controlled by the host and this, in turn, determines resistance or susceptibility. He has argued that sensitivity to fusaric acid is not an important primary determinant in resistance or susceptibility because fusaric acid causes injury in perhaps all higher plants; yet *F. oxysporum* f. sp. *lycopersici,* which produces fusaric acid, attacks only tomato. Clearly, other factors must be involved in determining host specificity and resistance or susceptibility.

In our view, there is a critical point in the localization of vascular infections at which a fungal toxin, such as fusaric acid, might determine whether a parasite is confined or becomes systemically distributed in vascular tissue of a suscept. This point is designated "A" in Fig. 33-1. Such a toxin could prevent the respiratory burst that is correlated with resistant-type reactions. It could also depress the rate of tylose development, thus preventing localization of infections. Fusaric acid and lycomarasmin, the two Fusarium metabolites that have received the most attention as possible wilt toxins, are discussed below.

Lycomarasmin Lycomarasmin was the first chemically defined wilt-inducing "toxin" (27). This compound and its effects were investigated thoroughly by Gäuman and his co-workers in Switzerland (67, 71, 72, 73, 74, 75, 76) and by Scheffer and Walker (163), Dimond and Waggoner (51), and Paquin and Waygood (150) in the United States and Canada. Lycomarasmin added to cuttings caused an irreversible wilting of leaves due to damage to cell membrane permeability. It has never been found in vivo, however, and there is overwhelming experimental evidence that lycomarasmin is not involved in the pathogenesis of natural infections (51, 68, 163).

Fusaric acid The evidence linking fusaric acid to pathogenesis is its presence in infected tissue and its correlation with host susceptibility and with the degree of fungal pathogenicity. Fusaric acid has been found in vivo in *Fusarium*-infected tomato (70, 101), pea (100), banana (147), cotton (98, 110), watermelon (144), and cowpea (156). Trace amounts were present in tomato, flax, cabbage, carnation, and watermelon after cross-inoculation with pathogens and nonpathogens of these hosts. Several studies relate fusaric acid production to pathogenicity. Davis (38) reported a possible correlation between pathogenicity of *F. oxysporum* f. sp. *niveum* and the amount of fusaric acid in diseased watermelon. On tissue of dead watermelon, a pathogenic strain of *F. oxysporum* f. sp. *niveum* produced 30 times more fusaric acid than two weak and 38 nonpathogenic strains. In susceptible cultivars of cowpeas infected with *F. oxysporum* f. sp. *udum*, the most susceptible cultivar yielded the largest amount of fusaric acid (156). *Fusarium oxysporum* f. sp. *lini* produced fusaric acid on susceptible flax tissue but not on two resistant flax tissues (179). A sweet potato isolate causing wilt produced fusaric acid, but an isolate not causing wilt did not produce fusaric acid in vitro.

Several lines of evidence argue against fusaric acid involvement in pathogenesis: (i) it was not always detected in vivo, either because it was not produced or it was detoxified, (ii) it was present in resistant as well as susceptible cultivars, and (iii) there was a lack of correlation between pathogenicity and the production of fusaric acid in vivo. Fusaric acid was not detected in *Fusarium*-infected cabbage or cabbage stem sections (90), carnation (38), or pea (100). Fisher (64) was unable to demonstrate fusaric acid production when isolates of *F. oxysporum* f. sp. *batatis* were grown on media with carbon sources similar to the pectin and cellulosic wall materials thought to be nutritionally important during pathogenesis. Also, although fusaric acid has been detected in several hosts, it is rapidly metabolized within host tissue, e.g., cabbage (90) and tomato (69, 100, 105, 161). All known transformations of fusaric acid result in detoxification (100). Davis (38) concluded that fusaric acid must play only a secondary role in promoting pathogenesis in flax and tomato because of a low, unchanging level of fusaric acid in these plants during disease progression. He also demonstrated the presence of fusaric acid in resistant cultivars of tomato, flax, cabbage, carnation, and watermelon shortly after inoculation. Kern (100) reported that a pathogenic strain of *F. oxysporum* f. sp. *lycopersici* produced very little fusaric acid, while a weakly pathogenic strain produced considerable fusaric acid when grown on disinfected tomato tissue. He also showed that although one pathogenic race of *F. oxysporum* f. sp. *pisi* produced considerable amounts of fusaric acid in pea, another pathogenic strain and a weakly pathogenic strain did not produce fusaric acid following inoculation. Kuo and Scheffer (107), using ultraviolet-induced mutants, found no correlation between pathogenicity of the mutants and their ability to produce fusaric acid. Following inoculation of tomato with high fusaric acid producing strains, fusaric acid was detected only in the roots and lower stem, and not in the petioles, upper stem, and leaves where its toxic effects were believed to occur. In summary, there is some evidence that fusaric may be involved in Fusarium wilt pathogenesis, but the evidence is inconsistant.

Disturbances in Internal Water Balance

Alterations in internal water relations during Fusarium wilt pathogenesis have been correlated with the appearance of foliar symptoms. There is increased resistance to water flow within diseased stems (106, 139, 163) and especially within petioles (25, 33, 55, 56, 57, 159, 163). This resistance to water flow is apparently not due to increased visocosity of vascular fluid following infection (55, 56, 116, 182). The remaining alternatives are (i) physical obstruction to mass flow along vascular elements and (ii) obstruction of water movement from vessels into surrounding tissues.

Several studies of water relations within infected stems have indicated disturbances in vascular function, but the results have not been definitive, probably because the most critical sites of vascular interference contributing to foliar symptoms are within the petiole (116, 139, 148, 159). Dunaway (55, 56, 57), in perhaps the best designed and most thorough study of wilting in *Fusarium*-infected tomato, investigated all aspects of water relations, including water uptake, transport, stomatal movements and transpirational water loss. He concluded that high resistance to water flow, especially within the petioles, is the *sole* cause of wilting. Infected leaves remained turgid, even though the resistance to water flow in the stem was more than 500 times that of healthy stems, provided the petiole and leaf resistances remained normal. Leaves wilted when petiole resistances reached infinity, but such leaves recovered when the petioles were removed and the leaves were placed in water. Roots did not have abnormally high resistance to water flow. Stomatal and transpirational measurements of infected plants revealed that (i) stomates closed, even in turgid leaves, (ii) transpirational water loss in all stages of disease development was always less than that of healthy or wounded plants, (iii) there was no alteration of leaf solute potential, (iv) leaves of diseased and healthy plants wilted at similar water potentials, and (v) wilting of diseased plants was reversible as was wilting induced in healthy plants. These data clearly indicate that wilting is not due to marked increased in cell permeability and subsequent loss of solute from the leaves, as suggested from results of filtrate studies. Transpirational and stomatal responses were identical to those of plants under water stress. Other researchers have also shown that infection results in stomatal closure (53) and that wilting is not due to alteration in leaf solute potential (2, 116).

Additional studies support Dunaway's conclusion that obstructions to water flow within petiole vascular bundles are responsible for wilt symptoms in Fusarium wilt of tomato. Dimond and Waggoner (53), simply by removing dysfunctional tissue at the base of petioles, demonstrated that osmotic properties of cells in symptomatic leaves were not impaired. They observed the recovery of full turgor when the leaves with re-cut petioles were placed in water. Scheffer and Walker (163) used dyes to demonstrate that leaf wilt was correlated with failure of the petiole bundles to conduct dye. Microscopic examination of diseased tissue revealed that plugging occurred and that the extent of plugging was correlated with disease severity. Some degree of browning was always present in the petioles of wilted leaves, and unilaterally wilted leaves had browning on the wilted side only.

Chambers and Corden (25, 33) attributed symptoms in Fusarium wilt of tomato to vessel collapse and the plugging of vessel pits within petioles that reduced lateral translocation of water. There appeared to be a correlation between the severity of these vascular symptoms and the expression of foliar symptoms. Epinasty, yellowing, and wilting were associated with mild, moderate, and severe vascular symptoms, respectively. They suggested that the brown resinous deposits that clogged pits in the walls of vessels had the effect of "waterproofing" the vessels, e.g., reducing or preventing water movement laterally through the pits into adjacent vessels. This restriction on lateral movement of water, in conjunction with occlusion of vessel lumens that restrict axial movement of water, would have a drastic effect on water transport, especially within petioles that are without cross-connections between vascular bundles.

The nature of vascular blockage in tomato wilt is not clear. Ludwig (116) claimed that wilting was due to a homogeneous hyaline material within the diseased vessels. Gothoskar et al. (79, 81) indicated that plugging was by pectic gels. Pierson et al. (155) provided support for pectic gels by demonstrating that plugs which developed in plants treated with pectic enzymes were similar to those which developed in infected plants. Chambers and Corden (25), however, were unable to detect either pectic plugs or resinous materials within vessel lumens of tomato, although vessel pits were plugged with a resinous material that caused vascular discoloration. They rarely found the plugs described by Pierson et al. and never found Ludwig's plugs. Other workers have concluded that although granular plugs occurred, they did not appear frequently enough to reduce the efficiency of water transport to a significant extent (25, 159).

Mycelial growth in vessels had long been considered as a possible contributor to increased resistance to water flow, but according to calculations by Waggoner and Dimond (182), mycelium alone can account for only a small portion of the large decrease in flow rate that develops in many wilt diseases. Nevertheless, in a later study Saaltink and Dimond (159) concluded that mycelial plugs seemed more important than granular plugs in the first stem internode. The plugs were usually appressed against a perforation plate. Twenty percent of the vessels were completely plugged and mycelium was present in 50–70 percent of the vessels. Others, however, have found little evidence of mycelial plugs within vessels (25, 83, 116, 163, 182).

Page (146) found that disturbances in the internal water status of *Fusarium*-infected banana plants were associated with vessel dysfunction. Reduced transpiration in laminae preceded external symptom expression, and the rate of water flow through healthy vessels. The water shortage that develops in infected banana plants may be a consequence of the tyloses, gums and vessel collapse that develops in colonized rhizomes (187, 188).

In summary, the wilt syndrome is mainly an expression of host responses to increased internal water stress caused by physical obstructions to water flow within and from xylem vessel elements. The water stress does not become critical until the later stages of pathogenesis, e.g., not until the pathogen has overcome the host's resistance mechanisms and colonized the rhizome (banana) or the upper stem and petioles (tomato). The water relations of Fusarium wilt pathogenesis are depicted in Fig. 33-1.

Summary and Conclusions

The bulk of studies on Fusarium wilt have been conducted on tomato, so the summary below is based largely on tomato wilt pathogenesis. Studies with other Fusarium wilts suggest that their pathogenesis is similar in many respects to that of tomato wilt.

Invasion of the roots of resistant hosts and non-hosts by *Fusarium* is sometimes restricted to a few cortical cells while at other times the vascular system is invaded but further fungal distribution is severely limited. We propose that the limited invasion of the vascular cylinder results from host responses such as lignituber formation, cell wall apposition, and inherently rapid endodermis formation as proposed and summarized by Talboys (174). We propose that limited distribution of the fungus within invaded vessels, whether entrance is gained by massive direct penetration or through wounds, is due to the temporary trapping of propagules of the pathogen on vessel end-walls, the rapid development of physical barriers that prevent further upward fungal movement, and by the accumulation of antifungal compounds that inhibit continued fungal development and lateral spread in the root vascular tissue. Resistance is an active process that requires increased utilization of energy, growth regulator activity, and possibly the metabolism of phenolic substrates.

Systemic invasion of susceptible hosts by *F. oxysporum* occurs when host resistance responses are delayed or otherwise altered, allowing the pathogen to spread upward to the hypocotyl. Within the hypocotyl there is extensive colonization of primary and secondary xylem because the xylem vessels there are centrally organized and interconnected. Here there is lateral distribution of the fungus that is followed by massive production of microconidia. These microconidia are, in turn, swept upward into the vascular elements of the upper stem, leaf traces, and petiole bundles.

With the exception of vein-clearing which sometimes appears in uninvaded tomato leaves soon after inoculation, most foliar symptoms appear *after* the pathogen has systemically invaded the xylem vessel elements of the host. Wilt develops as a response to severe internal water stress following vascular plugging, especially within the petioles. The evidence indicates that yellowing and necrosis of leaves and stunting of the plant as a whole also result from water stress. Gels, gums, tyloses, and vessel collapse have all been implicated in interference with water flow. Mycelial growth and cell wall-degrading enzymes of the pathogen may contribute to vascular plugging. Toxins (metabolites of the pathogen that permanently damage cell function) may be involved in the induction of symptoms such as chlorosis and necrosis as others have suggested, but we propose that these toxins may function chiefly by interfering with host resistance responses that normally restrict invasion.

Future Work

Work is needed to clarify how *F. oxysporum* disrupts the normal resistance responses in a susceptible host and why it does not disrupt these responses in a resistant host. Is the resistant host less sensitive to a "toxin" that disrupts the respiratory-dependent localization process? Is less toxin produced in a resistant host? Presumably, the various host-specific Fusaria all produce such inhibitory substances. How, then, do we have such specificity for the various host species?

We need to seek out as many different kinds of resistance mechanisms as we can and relate them to specific genes for resistance. This has not been done to date but must be done to make the best use of resistance within our armory of biological and chemical control weapons.

We might choose to explore the mechanisms in plants that provide for the general wound-healing-type responses and that are also involved in localization and resistance processes. How are these mechanisms turned on? What do parasites (of all sorts) do to turn on these responses and how are the responses turned off when the job is done? These questions have barely been touched upon!

If we are going to find answers to these questions, we must be prepared to make laborious, detailed, and integrated studies of those small sites where and when the crucial interactions that determine resistance or susceptibility take place. We must be prepared to undertake genetic and breeding studies designed specifically to answer scientific questions. We must use our very best knowledge, judgment, ingenuity, and techniques. We must incorporate and integrate histochemistry with biochemistry on a microscale and couple them with fine structure studies. Anything less will only add to the confusion of evidence that is already difficult to interpret. This prescription sounds like drudgery, and much of the gaining of knowledge is drudgery; but the quest can be exciting in terms of expanding knowledge if our efforts are judiciously placed.

Literature Cited

1. Abawi, G. S., and J. W. Lorbeer. 1971. Pathological histology of four onion cultivars infected by *Fusarium oxysporum* f. sp. *cepae*. Phytopathology 61: 1164–1169.
2. Ahmet, H. 1933. Untersuchen uber Tracheomycosen. Phytopathol. Z. 6:49–101.
3. Alon, H., J. Katan, and N. Kedar. 1974. Factors affecting penetrance of resistance to *Fusarium oxysporum* f. sp. *lycopersici* in tomatoes. Phytopathology 64:455–461.
4. Anderson, M. E., and J. C. Walker. 1935. Histological studies of Wisconsin Hollander and Wis-

consin Ballhead cabbage in relation to resistance to yellows. J. Agric. Res. 50:823–836.
5. Armstrong, G. M., and J. K. Armstrong. 1948. Nonsusceptible hosts as carriers of wilt Fusaria. Phytopathology 38:808–826.
6. Beckman, C. H. 1964. Host responses to vascular infection. Annu. Rev. Phytopathol. 2:231–252.
7. Beckman, C. H. 1966. Cell irritability and localization of vascular infections in plants. Phytopathology 56:821–824.
8. Beckman, C. H. 1967. Respiratory response of radish varieties resistant and susceptible to vascular infection by *Fusarium oxysporum* f. *conglutinans*. Phytopathology 57:669–702.
9. Beckman, C. H. 1969. Plasticizing of walls and gel induction in banana root vessels infected with *Fusarium oxysporum*. Phytopathology 59:1477–1483.
10. Beckman, C. H. 1971. The plasticizing of plant cell walls and tylose formation—a model. Physiol. Plant Pathol. 1:1–10.
11. Beckman, C. H., D. M. Elgersma, and W. E. MacHardy. 1972. The localization of Fusarial infections in the vascular tissue of single-dominant-gene resistant tomatoes. Phytopathology 62:1256–1260.
12. Beckman, C. H., and S. Halmos. 1962. Relation of vascular occluding reactions in banana roots to pathogenicity of root-invading fungi. Phytopathology 52:893–897.
13. Beckman, C. H., S. Halmos, and M. E. Mace. 1962. The interaction of host, pathogen, and soil temperature in relation to susceptibility to Fusarium wilt of bananas. Phytopathology 52:134–140.
14. Beckman, C. H., M. E. Mace, S. Halmos, and M. W. McGahan. 1961. Physical barriers associated with resistance in Fusarium wilt of bananas. Phytopathology 51:507–515.
15. Beckman, C. H., and W. C. Mueller. 1970. Distribution of phenols in specialized cells of banana roots. Phytopathology 60:79–82.
16. Beckman, C. H., W. C. Mueller, and M. E. Mace. 1974. The stabilization of artificial and natural cell wall membranes by phenolic infusion and its relation to wilt disease resistance. Phytopathology 64:1214–1220.
17. Beckman, C. H., G. E. VanderMolen, W. C. Mueller, and M. E. Mace. 1976. Vascular structure and distribution of vascular pathogens in cotton. Physiol. Plant Pathol. 9:87–94.
18. Beckman, C. H., and G. E. Zaroogian. 1967. Origin and composition of vascular gel in infected banana roots. Phytopathology 57:11–13.
19. Biehn, W. L., and A. E. Dimond. 1971. Effect of glactose on polygalacturonase production and pathogenesis by *Fusarium oxysporum* f. sp. *lycopersici*. Phytopathology 61:242–243.
20. Bisby, G. R. 1919. Studies on Fusarium diseases of potatoes and truck crops in Minnesota. Minn. Agric. Exp. Sta. Bull. 181. 21 p.
21. Brandes, E. W. 1919. Banana wilt. Phytopathology 9:339–389.
22. Bugbee, W. M. 1970. Vascular response of cotton to infection by *Fusarium oxysporum* f. sp. *vasinfectum*. Phytopathology 60:121–123.
23. Cappellini, R. A., and J. L. Peterson. 1976. Pectic enzymes associated with mimosa wilt. Bull. Torrey Bot. Club 103:227–229.
24. Carrasco, A., A. M. Boudet, and G. Marigo. 1978. Enhanced resistance of tomato plants to *Fusarium* by controlled stimulation of their natural phenolic production. Physiol. Plant Pathol. 12:225–232.
25. Chambers, H. L., and M. E. Corden. 1963. Semeiography of Fusarium wilt of tomato. Phytopathology 53:1006–1010.
26. Chi, C. C., W. R. Childers, and E. W. Hanson. 1964. Penetration and subsequent development of three *Fusarium* species in alfalfa and red clover. Phytopathology 54:434–437.
27. Clausson-Kass, N. P., P. A. Plattner, and E. Gaumann. 1944. Uber ein welkeerzeugendes Stoffwechselprodukt von *Fusarium lycopersici* Sacc. Ber. Schweiz Bot. Ges. 54:524–527. (Abstr., Rev. Appl. Mycol. 24:479)
28. Cleland, R. 1973. Auxin-induced hydrogen ion excretion from *Avena* coleoptiles. Proc. Nat. Acad. Sci. USA 70:3092–3093.
29. Collins, R. P., and R. P. Scheffer. 1958. Respiratory responses and systemic effects in *Fusarium*-infected tomato plants. Phytopathology 48:349–355.
30. Conway, W. S., and W. E. MacHardy. 1978. Distribution and growth of *Fusarium oxysporum* f. sp. *lycopersici* race 1 or race 2 within tomato plants resistant or susceptible to wilt. Phytopathology 68:938–942.
31. Cooper, R. M., B. Rankin, and R. K. S. Wood, 1978. Cell wall-degrading enzymes of vascular wilt fungi. II. Properties and modes of action of polysaccharidases of *Verticillium albo-atrum* and *Fusarium oxysporum* f. sp. *lycopersici*. Physiol. Plant Pathol. 13:101–134.
32. Cooper, R. M., and R. K. S. Wood. 1973. Induction of synthesis of extracellular cell-wall degrading enzymes in vascular wilt fungi. Nature (Lond.) 246:309–311.
33. Corden, M. E., and H. L. Chambers. 1966. Vascular dysfunction in Fusarium wilt of tomato. Amer. J. Bot. 53:284–287.
34. Corden, M. E., and A. E. Dimond. 1959. The effect of growth-regulating substances on disease resistance and plant growth. Phytopathology 49:68–72.
35. Corden, M. E., and L. V. Edgington. 1960. A calcium requirement for growth regulator-induced resistance to Fusarium wilt of tomato. Phytopathology 50: 625–626.
36. Davis, D. 1954. The use of intergeneric grafts to demonstrate toxins in the Fusarium wilt disease of tomato. Amer. J. Bot. 41:395–398.
37. Davis, D. 1968. Partial control of Fusarium wilt in tomato by formae of *Fusarium oxysporum*. Phytopathology 58:121–122.

38. Davis, D. 1969. Fusaric acid in selective pathogenicity of *Fusarium oxysporum*. Phytopathology 59:1391–1395.
39. Davis, D., and A. E. Dimond. 1953. Inducing disease resistance with plant growth-regulators. Phytopathology 43:137–140.
40. Davis, D., and A. E. Dimond. 1954. The source and role of phenols in Fusarium wilt symptoms. Phytopathology 44:485–486. (Abstr.).
41. Davis, D., and A. E. Dimond. 1956. Site of disease resistance induced by plant-growth regulators in tomato. Phytopathology 46:551–552.
42. Davis, D., P. E. Waggoner, and A. E. Dimond. 1953. Conjugated phenols in the Fusarium wilt syndrome. Nature (Lond.) 172:959–961.
43. Deese, D. C., and M. A. Stahmann. 1962. Pectic enzymes in *Fusarium*-infected susceptible and resistant tomato plants. Phytopathology 52:255–260.
44. Deese, D. C., and M. A. Stahmann. 1962. Wilt resistance in tomatoes. Pectic enzyme formation by *Fusarium oxysporum* f. *lycopersici* on susceptible and resistant tomato stems. J. Agric. Food Chem. 10:145–150.
45. Dimond, A. E. 1955. Pathogenesis in wilt diseases. Annu. Rev. Plant Physiol. 6:329–350.
46. Dimond, A. E. 1963. The physiology of wilt diseases, p. 91–103. *In* The physiology of fungi and fungus diseases, West Virginia Agric. Exp. Sta. Bull. 488T. 106 p.
47. Dimond, A. E. 1967. Physiology of wilt disease, p. 100–118. *In* G. J. Mirocha and I. Uritani (ed), The dynamic role of molecular constituents in plant-parasite interactions, Bruce Publishing Co., St. Paul, Minnesota. 372 p.
48. Dimond, A. E. 1970. Biophysics and biochemistry of the vascular wilt syndrome. Annu. Rev. Phytopathol. 8:301–322.
49. Dimond, A. E. 1972. The origin of symptoms of vascular wilt diseases, p. 289–310. *In* R. K. S. Wood, A. Balio, and A. Graniti (ed.), Phytotoxins in plant diseases. Academic Press, New York. 530 p.
50. Dimond, A. E., and M. E. Corden. 1957. Reduction and promotion of the development of Fusarium wilt of tomato by gibberellic acid. Phytopathology 47:519 (Abstr.).
51. Dimond, A. E., and P. E. Waggoner. 1953. The physiology of lycomarasmin production by *Fusarium oxysporum* f. *lycopersici*. Phytopathology 43:195–199.
52. Dimond, A. E., and P. E. Waggoner. 1953. On the nature and role of vivotoxins in plant disease. Phytopathology 43:229–235.
53. Dimond, A. E. and P. E. Waggoner. 1953. The water economy of Fusarium wilted tomato plants. Phytopathology 43:619–623.
54. Drysdale, R. B., and P. Langcake. 1973. Response of tomato to infection by *Fusarium oxysporum* f. *lycopersici*, p. 423–433. *In* R. S. W. Byrde and C. V. Cutting (ed.), Fungal pathogenicity and the plant's response, Academic Press, New York. 449 p.
55. Duniway, J. M. 1971. Resistance to water movement in tomato plants infected with *Fusarium*. Nature (Lond.) 230:252–253.
56. Duniway, J. M. 1971. Water relations of Fusarium wilt in tomato. Physiol. Plant Pathol. 1:537–546.
57. Duniway, J. M. 1973. Pathogen-induced changes in host water relations. Phytopathology 63:458–466.
58. Edgington, L. V. 1965. Nature of wilt disease resistance induced by growth regulators. Phytopathology 55:1056 (Abstr.).
59. Edgington, L. V., M. E. Corden, and A. E. Dimond. 1961. The role of pectic substances in chemically induced resistance to Fusarium wilt of tomato. Phytopathology 51:179–182.
60. Edgington, L. V., and J. C. Walker. 1958. Influence of calcium and boron nutrition on development of Fusarium wilt of tomato. Phytopathology 48:324–326.
61. Elgersma, D. M., W. E. MacHardy, and C. H. Beckman. 1972. Growth and distribution of *Fusarium oxysporum* f. sp. *lycopersici* in near-isogenic lines of tomato resistant or susceptible to wilt. Phytopathology 62:1232–1237.
62. Federation of British Plant Pathologists. 1973. A guide to the use of terms in plant pathology. Phytopathol. Pap. No. 17. Commonwealth Mycological Institute, Kew, Surrey, England. 55 p.
63. Ferraris, L., A. Garibaldi, and A. Matta. 1974. Polygalacturonase and polygalacturonate transeliminase production in vitro and in vivo by *Fusarium oxysporum* f. sp. *lycopersici*. Phytopathol. Z. 81:1–14.
64. Fisher, K. D. 1965. Hydrolytic enzyme and toxin production by sweet potato Fusaria. Phytopathology 55:396–398.
65. Foster, R. E. 1946. The first symptoms of tomato Fusarium wilt: clearing of the ultimate veinlets in the leaf. Phytopathology 36:691–694.
66. Fuchs, A., and F. W. DeVries. 1969. Metabolism of radioactively labeled quinic acid and shikimic acid in healthy and *Fusarium*-infected tomato plants. Neth. J. Plant Pathol. 75:186–192.
67. Gaumann, E. 1948. Uber den Mechanismum des infektiosen Welkens. Biol. Zentralbl. 67:22–26.
68. Gaumann, E. 1951. Some problems of pathological wilting in plants. Adv. Enzymol. 11:401–437.
69. Gaumann, E. 1957. Fusaric acid as a wilt toxin. Phytopathology 47:342–357.
70. Gaumann, E. 1958. The mechanisms of fusaric acid injury. Phytopathology 48: 670–686.
71. Gaumann, E., and O. Jaag. 1946. Uber das Problem der Welkekrankeiten bei Pflanzen. Experientia 2:215–220.
72. Gaumann, E., and O. Jaag. 1947. Die phsiologischen Grundlagen des parasitogenen Welkens. I, II, III. Ber. Schweiz Bot. Ges. 57:3–34; 132–148; 227–241.
73. Gaumann, E., and S. Naef-Roth. 1953. D'un cycle annuel de la sensibilité des tomates aux toxines. Compt. Rend. Acad. Sci., Paris 236:170–172.
74. Gaumann, E., and S. Naef-Roth. 1953. Uber den

jahreszeitlichen Gang der Welketoxin-Empfindlichkeit der Tomatenpflanzen. Phytopathol. Z. 20:449–458.

75. Gaumann, E., and S. Naef-Roth. 1955. Die Dosis-Effekt-Beziehungen bei Lycomarasmin und Komplexon III. Phytopathol. Z 23:141–146.
76. Gaumann, E., and S. Naef-Roth. 1959. Uber Lycomarasminsaure, ein Unwandlungsprodukt des Lycomarasmins. Phytopathol. Z. 34:426–431.
77. Gentile, I. A., and A. Matta. 1975. Production of and some effects of ethylene in relation to Fusarium wilt of tomato. Physiol. Plant Pathol. 5:27–35.
78. Gorden, S. A., and L. G. Paleg. 1961. Formation of auxin from tryptophan through action of polyphenols. Plant Physiol. 36:838–845.
79. Gothoskar, S. S., R. P. Scheffer, M. A. Stahmann, and J. C. Walker. 1955. Further studies on the nature of Fusarium resistance in tomato. Phytopathology 45:303–307.
80. Gothoskar, S. S., R. P. Scheffer, J. C. Walker, and M. A. Stahmann. 1953. The role of pectic enzymes in Fusarium wilt of tomato. Phytopathology 43:535–536.
81. Gothoskar, S. S., R. P. Scheffer, J. C. Walker, and M. A. Stahmann. 1955. The role of enzymes in the development of Fusarium wilt of tomatoes. Phytopathology 45:381–387.
82. Gottlieb, D. 1943. Expressed sap of tomato plants in relation to wilt resistance. Phytopathology 33:1111 (Abstr.).
83. Gottlieb, D. 1944. The mechanism of wilting caused by *Fusarium bulbigenum* var. *lycopersici*. Phytopathology 34:41–59.
84. Griffiths, D. A., and W. C. Lin. 1966. In vitro toxin production by nonwilt-inducing isolates of *Fusarium* from Malaya. Plant Dis. Rep. 50:261–263.
85. Grossman, F. 1968. Studies on the therapeutic effects of pectolytic enzyme inhibitors. Neth. J. Plant Pathol. 74, Suppl. I:91–103.
86. Grossman, F., and L. Rapp. 1973, Zellulase (C_x)-Aktivitat in gesunden und Fusarium-infizierten tomatenpflanzen. Phytopathol. Z. 76:90–93.
87. Hammerschlag, F., and M. E. Mace. 1975. Antifungal activity of extracts from Fusarium wilt-susceptible and -resistant tomato plants. Phytopathology 65:93–94.
88. Haskell, R. J. 1919. Fusarium wilt of potato in the Hudson River Valley, New York. Phytopathology 9:223–260.
89. Heinze, P. H., and C. F. Andrus. 1945. Apparent localization of Fusarium wilt resistance in the Pan America tomato. Amer. J. Bot. 32:62–66.
90. Heitefuss, R., M. A. Stahmann, and J. C. Walker. 1960. Production of Pectolytic enzymes and fusaric acid by *Fusarium oxysporum* f. *conglutinans* in relation to cabbage yellows. Phytopathology 50:367–370.
91. Heitefuss, R., M. A. Stahmann, and J. C. Walker. 1960. Oxidative enzymes in cabbage infected by *Fusarium oxysporum* f. *conglutinans*. Phytopathology 50: 370–375.
92. Hepple, S. 1963. Infection of pea plants by *Fusarium oxysporum* f. sp. *pisi* in naturally infested soil. Trans. Brit. Mycol. Soc. 46:585–594.
93. Hodgson, R., W. H. Peterson, and A. J. Riker. 1949. The toxicity of polysaccharides and other large molecules to tomato cuttings. Phytopathology 39:47–62.
94. Horst, R. K. 1965. Pathogenic and enzymatic variations in *Fusarium oxysporum* f. *callistephi*. Phytopathology 55: 848–851.
95. Husain, A., and A. E. Dimond. 1960. Role of cellulolytic enzymes in pathogenesis by *Fusarium oxysporum* f. *lycopersici*. Phytopathology 50:329–331.
96. Irving, G. W., Jr. 1947. The significance of tomatin in plant and animal science. J. Wash. Acad. Sci. 37:293–296.
97. Jones, T. M., A. J. Anderson, and P. Albersheim. 1972. Host-pathogen interactions. IV. Studies on the polysaccharide-degrading enzymes secreted by *Fusarium oxysporum* f. sp. *lycopersici*. Physiol. Plant Pathol. 2:153–166.
98. Kalyanasundaram, R., and C. S. Venkata Ram. 1956. Production and systematic translocation of fusaric acid in Fusarium-infected cotton plants. J. Indian Bot. Soc. 35:7–10.
99. Katan, J. 1971. Symptomless carriers of the tomato Fusarium wilt pathogen. Phytopathology 61:1213–1217.
100. Kern, H. 1972. Phytotoxins produced by Fusaria, p. 35–48. *In* R.K.S. Wood, A. Balio, and A. Graniti (ed.), Phytotoxins in plant diseases, Academic Press, New York. 530 p.
101. Kern, H., and D. Kluepfel. 1956. Die Bildung von Fusarinsaure durch *Fusarium lycopersici* in vivo. Experientia 12:181–182.
102. Keyworth, W. G. 1963. The reaction of monogenic resistant and susceptible varieties of tomato to inoculation with *Fusarium oxysporum* f. *lycopersici* into stems or through Bonny Best rootstocks. Ann. Appl. Biol. 52:257–270.
103. Keyworth, W. G. 1964. Hypersensitivity of monogenic resistant tomato scions to toxins produced in Bony Best rootstocks invaded by *Fusarium oxysporum* f. *lycopersici*. Ann. Appl. Biol. 54:99–105.
104. Khadr, A. S., and W. C. Snyder. 1967. Histology of early stages of penetration in Fusarium wilt of cotton. Phytopathology 57:99 (Abstr.).
105. Klupfel, D. 1957. Uber die Biosynthese und die Umwandlungen der Fusarinsaure in Tomatenpflanzen. Phytopathol. Z. 29:349–379.
106. Kuc, J. 1966. Resistance of plants to infectious agents. Annu. Rev. Microbiol. 20:337–370.
107. Kuo, M. S., and R. P. Scheffer. 1964. Evaluation of Fusaric acid as a factor in development of Fusarium wilt. Phytopathology 54:1041–1044.
108. Lakshmanan, M. 1959. Respiratory changes in *Fusarium*-infected cotton. Phytopathol. Z. 36:-406–418.
109. Lakshminarayanan, K. 1957. In vivo detection of

pectin methyl esterase in the Fusarium wilt of cotton. Naturwissenschaften 44:93.
110. Lakshminarayanan, K., and D. Subramanian. 1955. Is fusaric acid a vivotixin? Nature (Lond.) 176:697–698.
111. Langcake, P., P. M. Bratt, and R. D. Drysdale. 1973. Pectinmethylesterase in *Fusarium*-infected susceptible tomato plants. Physiol. Plant Pathol. 3:101–106.
112. Langcake, P., and R. B. Drysdale. 1975. The role of pectic enzyme production in the resistance of tomato to *Fusarium oxysporum* f. *lycopersici*. Physiol. Plant Pathol. 6:247–258.
113. Langcake, P., R. B. Drysdale, and H. Smith. 1972. Post-infectional production of an inhibitor of *Fusarium oxysporum* f. *lycopersici* by tomato plants. Physiol. Plant Pathol. 2:17–25.
114. Langton, F. A. 1969. Interactions of the tomato with two formae speciales of *Fusarium oxysporum*. Ann. Appl. Biol. 62:413–427.
115. Linford, M. B. 1931. Transpirational history as a key to the nature of wilting in the Fusarium wilt of pea. Phytopathology 21:791–796.
116. Ludwig, R. A. 1952. Studies on the physiology of hadromycotic wilting in the tomato plant. MacDonald. Agr. Coll. Tech. Bull. 20. 38 p.
117. Ludwig, R. A. 1960. Toxins, p. 315–357. *In* J. G. Horsfall and A. E. Dimond (ed.), Plant pathology, an advanced treatise, Vol. 2, The pathogen, Academic Press, New York. 715 p.
118. Mace, M. E. 1963. Histochemical localization of phenols in healthy and diseased banana roots. Physiol. Plantarum 16:915–925.
119. Mace, M. E. 1978. Contributions of tyloses and terpenoid aldehyde phytoalexins to Verticillium wilt resistance in cotton. Physiol. Plant Pathol. 12:1–11.
120. Mace, M. E., and E. Solit. 1966. Interactions of 3-indoleacetic acid and 3-hydroxytryamine in Fusarium wilt of banana. Phytopathology 56:245–247.
121. Mace, M. E., and J. A. Veech. 1970. Uptake and distribution of Fusarium spores in Fusarium wilt-susceptible or resistant tomatoes. Phytopathology 60:1302 (Abstr.).
122. Mace, M. E., and J. A. Veech. 1971. Fusarium wilt of susceptible and resistant tomato isolines: host colonization. Phytopathology 61:834–840.
123. Mace, M. E., J. A. Veech, and C. H. Beckman. 1972. Fusarium wilt of susceptible and resistant tomato isolines: histochemistry of vascular browning. Phytopathology 62:651–654.
124. Mace, M. E., and E. M. Wilson. 1964. Phenol oxidases and their relation to vascular browning in *Fusarium*-invaded banana roots. Phytopathology 54:840–842.
125. Malalasekera, R. A. P., F. R. Sanderson, and J. Calhoun. 1973. Fusarium diseases of cereals. IX. Penetration and invasion of wheat seedlings by *Fusarium culmorum* and *F. nivale*. Trans. Brit. Mycol. Soc. 60:453–462.
126. Mann, B. 1962. Role of pectic enzymes in the Fusarium wilt syndrome of tomato. Trans. Brit. Mycol. Soc. 45:169–178.
127. Maraite, H. 1969. Pathogénie du fletrissement du melon causé par *Fusarium oxysporum* f. sp. *melonis*. Ph.D. thesis, Univ. Louvain, Leuven, Belgium.
128. Maraite, H. 1973. Changes in polyphenoloxidases and peroxidases in muskmelon (*Cucumis melo* L) infected by *Fusarium oxysporum* f. sp. *melonis*. Physiol. Plant Pathol. 3:29–49.
129. Marte, M., and A. Zazzerini. 1973. Vacuolar precipitates in leaves of tomato plants affected by *Fusarium*. Phytopathol. Z. 77:252–257.
130. Matta, A., and A. E. Dimond. 1963. Symptoms of Fusarium wilt in relation to quantity of fungus and enzyme activity of tomato stems. Phytopathology 53:574–578.
131. Matta, A., and I. A. Gentile. 1964. Variation in auxin content induced in tomato by *Fusarium oxysporum* f. sp. *lycopersici*. (Trans. title) Riv. Pat. Veg. Pavia, Ser. 3, 4:208–237. (Abstr., Rev. Appl. Mycol. 44:233).
132. Matta, A., and I. A. Gentile. 1965. On the mechanism of accumulation of B-indoleacetic acid in tomato plants with *Fusarium oxysporum* f. sp. *lycopersici* infection (Trans. title) Phytopathol. Mediterr. 4:129–137. (Abstr., Rev. Appl. Mycol. 45:418).
133. Matta, A., and I. A. Gentile. 1968. The relation between polyphenoloxidase activity and ability to produce indoleacetic acid in *Fusarium*-infected tomato plants. Neth. J. Plant Pathol. 74, Suppl. 1:47–51.
134. Matta, A., I. A. Gentile, and I. Giai. 1969. Accumulation of phenols in tomato plants infected by different forms of *Fusarium oxysporum*. Phytopathology 59:512–513.
135. May, C. 1930. The effect of grafting on resistance and susceptibility of tomatoes to Fusarium wilt. Phytopathology 20:519–521.
136. McCance, D. J., and R. B. Drysdale. 1975. Production of tomatine and rishitin in tomato plants inoculated with *Fusarium oxysporum* f. sp *lycopersici*. Physiol. Plant Pathol. 7:221–230.
137. McClure, T. T. 1949. Mode of infection of the sweet-potato wilt *Fusarium*. Phytopathology 39:876–886.
138. McDonnell, K. 1958. Absence of pectolytic enzymes in a pathogenic strain of *Fusarium oxysporum* f. *lycopersici*. Nature (Lond.) 182:1025–1026.
139. Melhus, I. E., J. H. Muncie, and W. T. H. Ho. 1924. Measuring water flow interference in certain gall and vascular diseases. Phytopathology 14:580–584.
140. Mussell, H. W., and R. J. Green, Jr. 1968. Production of polygalacturonase by *Verticillium albo-atrum* and *Fusarium oxysporum* f. sp. *lycopersici* in vitro and in vascular tissue of susceptible and resistant host. Phytopathology 58:1061 (Abstr.).

141. Mussel, H. W., and R. J. Green, Jr. 1970. Host colonization and polygalacturonase production by two tracheomycotic fungi. Phytopathology 60:192–195.

142. Nair, P. N., and T. Kommedahl. 1957. The establishment and growth of *Fusarium lini* in flax tissues. Phytopathology 47:25 (Abstr.).

143. Netzer, D., and G. Kritzman. 1979. B-(1,3) Glucanase activity and quantity of fungus in relation to Fusarium wilt in resistant and susceptible near-isogenic lines of muskmelon. Physiol. Plant Pathol. 14:47–55.

144. Nishimurs, S. 1957. Pathochemical studies on watermelon wilt. 5. On the metabolic products of *Fusarium oxysporum* f. *niveum*. Ann. Phytopathol. Soc. Jap. 22:215–220.

145. Nyvall, R. F., and W. A. Haglund. 1976. The effect of plant age on severity of pea wilt caused by *Fusarium oxysporum* f. sp. *pisi* race 5. Phytopathology 66:1093–1096.

146. Page, O.T. 1959. Observations on the water economy of *Fusarium*-infected banana plants. Phytopathology 49:61–66.

147. Page, O. T. 1959. Fusaric acid in banana plants infected with *Fusarium oxysporum* f. *cubense*. Phytopathology 49:230.

148. Papavizas, G. C., and W. A. Ayers. 1966. Polygalacturonate trans-eliminase production by *Fusarium oxysporum* and *Fusarium solani*. Phytopathology 56: 1269–1273.

149. Paquin, R., and L. J. Coulombe. 1962. Pectic enzyme synthesis in relation to virulence in *Fusarium oxysporum* f. *lycopersici* (Sacc.) Snyder and Hansen. Can. J. Bot. 40:533–541.

150. Paquin, R., and E. R. Waygood. 1957. The effect of Fusarium toxins on the enzymic activity of tomato hypocotyl mitochondria. Can. J. Bot. 35:207–218.

151. Patil, S. S., and A. E. Dimond. 1968. Repression of polygalacturonase synthesis in *Fusarium oxysporum* f. sp. *lycopersici* by sugars and its effect on symptom reduction in infected tomato plants. Phytopathology 58:676–682.

152. Pennypacker, B. W., and P. E. Nelson. 1972. Histopathology of carnation infected with *Fusarium oxysporum* f. sp. *dianthi*. Phytopathology 62:1318–1326

153. Phillips, D. V., C. Leben, and C. C. Allison. 1967. A mechanism for the reduction of Fusarium wilt by a *Cephalosporium* species. Phytopathology 57:916–919.

154. Pierre, R. E., and R. E. Wilkinson. 1970. Histopathological relationship of *Fusarium* and *Thielaviopsis* with beans. Phytopathology 60:821–824.

155. Pierson, C. F., S. S. Gothoskar, J. C. Walker, and M. A. Stahmann. 1955. Histological studies on the role of pectic enzymes in the development of Fusarium wilt symptoms in tomato. Phytopathology 45:524–527.

156. Prasad, M., and S. K. Chaudhary. 1974. In vitro production of fusaric acid and its impact on growth and sporulation in *Fusarium oxysporum* f. *udum*. Phytopathol. Z. 80:279–282.

157. Rayle, D. L. 1973. Auxin-induced hydrogen-ion secretion in *Avena* coleoptiles and its implications. Planta (Berl.) 114:63–73.

158. Rishbeth, J. 1955. Fusarium wilt of bananas in Jamaica. I. Some observations on the epidemiology of the disease. Ann. Bot. 19:293–328.

159. Salttink, G. J., and A. E. Dimond. 1964. Nature of plugging material in xylem and its relation to rate of water flow in *Fusarium*-infected tomato stems. Phytopathology 54:1137–1140.

160. Sadisivan, T. S. 1961. Physiology of wilt disease. Annu. Rev. Plant Physiol. 12:449–468.

161. Sanwal, B. D. 1956. Investigations on the metabolism of *Fusarium lycopersici* Sacc. with the aid of radioactive carbon. Phytopathol. Z. 25:333–384.

162. Scheffer, R. P. 1957. Analysis of Fusarium resistance in tomato by grafting experiments. Phytopathology 47:328–331.

163. Scheffer, R. P., and J. C. Walker. 1953. The physiology of Fusarium wilt of tomato. Phytopathology 43:116–125.

164. Scheffer, R. P., and J. C. Walker. 1954. Distribution and nature of Fusarium resistance in the tomato plant. Phytopathology 44:94–101.

165. Sequeira, L., T. A. Steeves, M. W. Steeves, and J. M. Riedhart. 1958. Role of root injury in Panama disease infections. Nature (Lond.) 182:309–311.

166. Sherwood, R. T. 1966. Pectin lyase and polygalacturonase production by *Rhizoctonia solani* and other fungi. Phytopathology 56:279–286.

167. Smith, C. A. 1978. Tomatine: role in resistance of tomato to *Fusarium oxysporum* f. sp. *lycopersici* race 1 and race 2. M.S. Thesis. Univ. New Hampshire. Durham, New Hampshire. 66 p.

168. Smith, E. F. 1899. Wilt diseases of cotton, watermelon, and cowpea. U. S. Dept. Agric. Bur. Plant Ind. Bull. 17.

169. Smith, R., and J. C. Walker. 1930. A cytological study of cabbage plants in strains susceptible or resistant to yellows. J. Agric. Res. 41:17–35.

170. Snyder, W. C., K. F. Baker, and H. N. Hansen. 1946. Interpretation of resistance of Fusarium wilt in tomato. Science 103:707–708.

171. Stover, R. H. 1957. Ecology and pathogenicity studies with two widely distributed types of *Fusarium oxysporum* f. *cubense*. Phytopathology 47:535 (abstr.).

172. Subramanian, D., and C. B. Dolia. 1974. Physiology of wilt resistance. Indian. J. Genet. Plant Breed. 34:251–253.

173. Subramanian, D., and L. Saraswathi-Devi. 1959. Water is deficient, p. 313–348. *In* J. G. Horsfall and A. E. Dimond (ed.), Plant pathology, an advanced treatise, Vol. 1, The diseased plant, Academic Press, New York. 674 p.

174. Talboys, P. W. 1964. A concept of the host-para-

site relationship in Verticillium wilt diseases. Nature (Lond.) 202:361–364.

175. Talboys, P. W. 1968. Water deficits in vascular disease, p.255–311. *In* T. T. Kozlowski (ed.), Water deficits and plant growth, Vol. 1, Academic Press, New York. 333 p.

176. Thatcher, F. S. 1942. Further studies of osmotic and permeability relations in parasitism. Can J. Res. C 20:283–311.

177. Threlfall, R. J. 1959. Physiological studies on the Verticillium wilt disease of tomato. Ann. Appl. Biol. 47:57–77.

178. Tisdale, W. H. 1917. Flax wilt: a study of the nature and inheritance of wilt resistance. J. Agric. Res. 11:573–605.

179. Trione, E. J. 1960. Extracellular enzyme and toxin production by *Fusarium oxysporum* f. *lini.* Phytopathology 50:480–482.

180. Trujillo, E. E. 1963. Pathological-anatomical studies of Gros Michel banana affected by Fusarium wilt. Phytopathology 53:162–166.

181. Vander Molen, G. E., C. H. Beckman, and E. Rodehorst. 1977. Vascular gelation: a general response phenomenon following infection. Physiol. Plant Pathol. 11:95–100.

182. Waggoner, P. E., and A. E. Dimond. 1954. Reduction in water flow by mycelium in vessels. Amer. J. Bot. 41:637–640.

183. Waggoner, P. E., and A. E. Dimond. 1955. Production and role of extracellular pectic enzymes of *Fusarium oxysporum* f. *lycopersici.* Phytopathology 45:79–87.

184. Waggoner, P. E., and A. E. Dimond. 1956. Polyphenol oxidases and substrates in potato and tomato stems. Phytopathology 46:495–497.

185. Waite, B. H., and V. C. Dunlap. 1953. Preliminary host range studies with *Fusarium oxysporum* f. *cubense.* Plant Dis. Rep. 37:79–80.

186. Walker, J. C. 1971. Fusarium wilt of tomato. Phytopathological Monograph No. 6. Amer. Phytopathol. Soc., St. Paul, Minnesota. 56 p.

187. Wardlaw, C. W. 1930. The biology of banana wilt (Panama disease). I. Root inoculation experiments. Ann. Bot. 44:741–766.

188. Wardlaw, C. W. 1930. The biology of banana wilt (Panama disease). II. Preliminary observations on sucker infection. Ann. Bot. 44:917–956.

189. Wardlaw, C. W. 1931. The biology of banana wilt (Panama disease). III. An examination of sucker infection throught root-bases. Ann. Bot. 45:382–399.

190. Wood, R. K. S. 1967. Physiological plant pathology. Blackwell Scientific Publications, Lt., Oxford and Edinburgh. 570 p.

191. Winstead, N. N., and J. C. Walker. 1954. Production of vascular browning by metabolites from several pathogens. Phytopathology 44:153–158.

192. Zaroogian, G. E., and C. H. Beckman. 1968. A comparison of cell wall composition in banana plants resistant or susceptible to *Fusarium oxysporum* f. sp. *cubense.* Phytopathology 58:733–735.

34 Formae Speciales and Races of *Fusarium oxysporum* Causing Wilt Diseases

G.M. Armstrong and Joanne K. Armstrong

The classification of all species and varieties of the section Elegans (82) as *Fusarium oxysporum* has been considered in numerous publications (3, 15, 18, 35, 54, 64) since the idea was expressed by Hansford in 1926 (40), and a revision was proposed by Snyder and Hansen in 1940 (71). Our early work (13, 14) revealed that the initial concept of limited or highly selective pathogenicity of the formae speciales did not always apply since, for example, f. sp. *vasinfectum* could attack plants in more than one family. Later, this likewise was found to be true for formae speciales *batatas, tracheiphilum, apii,* and others (6).

A list of the formae speciales and races of *Fusarium oxysporum* with comments on the one by Gordon (35) has been published (3). A revised list is submitted here with the addition of new formae speciales and races. It differs in some respects from the one by Booth (18) who listed 72 formae speciales, retaining f. sp. *mathioli, raphani, pini, radici-lupini, and nicotianae.* We changed f. sp. *mathioli* and *raphani* to races of f. sp. *conglutinans,* and the reasons for the omission of the others from our list have been explained (3).

Formae speciales and races of *Fusarium oxysporum* causing wilt diseases.

The forma specialis, authority, reference, host plant(s), and synonomy where appropriate are given.

aechmeae (Gerl. & Saut.) Armst. & Armst. (3) — *Aechmea.* Syn.: *F. bulbigenum* Cke. & Mass. f. *aechmeae* Gerl. & Saut. (3); *F. oxysporum* f. sp. *aechmeae* (Saut & Gerl.) Gordon (35).

albedinis (Killian & Maire) Gordon (35)—date palm (*Phoenix*). Syn.: *Cylindrophora albedinis* (Killian & Maire) Malençon (35); *F. oxysporum* Schl. var. *albedinis* (Killian & Maire) Malençon (35).

anethi Gordon (35)—dill (*Anethum).*

apii (Nels. & Sherb.) Snyd. & Hans. (71)—celery (*Apium*) (2), garden pea (*Pisum*) (2), Mexican sunflower (*Tithonia*) (1). Syn.: *F. oxysporum* var. *orthoceras* (App. & Wr.) Bilai (35).

arctii Matuo, Matsuda, Ozaki & Kato. (52)—great burdock (*Arctium*).

asparagii Cohen (35)—asparagus (*Asparagus*).

basilicum (Dzidzariya) Armst. & Armst.—basil (*Ocimun* sp.). Syn.: *F. oxysporum* var. *basilicum* (29).

batatas (Wr.) Snyd. & Hans. (71), races 1 and 2 Armst. & Armst. (3)—sweet potato (*Ipomoea*), tobacco (*Nicotiana*). Syn.: *F. batatis* (Wr.) (18); *F. bulbigenum* var. *batatas* Wr. (18).

betae see *spinaciae* race 2

callistephi (Beach) Snyd. & Hans. (71) races 1 and 2 Hoffman (4), race 3 (f. sp. *rhois*) Armst. & Armst. (4), race 4 (Olsen, *Tagetes*) Armst. & Armst. (4)—China aster (*Callistephus*), staghorn sumac (*Rhus*), African marigold (*Tagetes*).

cannabis Noviello & Snyd. (35)—hemp (*Cannabis*).

carthami Klis. & Hous. (35), races 1,2 and 3 Klis. & Thomas (48,49), race 4 Klis. (47)—safflower (*Carthamus*).

cassiae Armst. & Armst. (3)—*Cassia tora* L.

cattleyae Foster (35)—orchid (*Cattleya*).

cepae (Hanz.) Snyd. & Hans. (71)—onion (*Allium*).

chrysanthemi Armst., Armst. & Littrell (11)—*Chrysanthemum*.

ciceris (Padw.) Matuo & Sato as *ciceri* (18)—gram (*Cicer*). Syn: *F. lateritium* (Nees) Snyd. & Hans. f. *ciceri* (Padw.) Erwin (3).

coffeae (Garcia) Wellman (35)—coffee (*Coffea*). Syn.: *F. bulbiginum* Cke. & Mass. var. *coffeae* Garcia (35).

conglutinans (Wr.) Snyd. & Hans. (71), races 1,2,3, and 4 Armst. & Armst. (3)—cabbage (*Brassica*), radish (*Raphanus*), stock (*Matthiola*), Syn.: *F. conglutinans* Wr. (18); *F. oxysporum* var. *orthoceras* (App. & Wr.) Bilai (35).

coriandrii Narula & Joshi (18)—coriander (*Coriandrum*).

cubense (E. F. Smith) Snyd. & Hans. (71), races 1 and 2 Stover & Waite, race 3 Waite (3)—banana (*Musa*). Syn.: *F. cubense* E. F. Smith (18); *F. oxysporum* Schl. var. *cubense* (E. F. Smith) Wr. (18); *F. oxysporum* Schl. f. 3. Reink. & Wr. (18); *F. oxysporum* E. F. Smith var. *inodoratum* Brandes (18); *F. oxysporum* Schl. f. 4 Wr. (18).

cucumerinum Owen (35), races 1,2, and 3 Armst., Armst. & Netzer (12)—cucumber (*Cucumis*).

cumini Prasad & Patel (61)—cumin (*Cuminum*). Syn.: *F. oxysporum* Schl. f. sp. *cumini* Patel, Prasad, Mathur & Mathur (18).

cyclaminis Gerlach (35)—*Cyclamen*.

dahliae Solov'era & Madumarov (72)—*Dahlia*.

delphinii Laskaris (35)—larkspur (*Delphinium*).

dianthi (Prill. & Del.) Snyd. & Hans. (71), three races Hood & Stewart (3)—carnation (*Dianthus*).

elaeagni (George., Mocan. & Orens.) Armst. & Armst. (3)—Russian olive (*Elaeagnus*). Syn.: *F. oxysporum* Schlecht. var. *orthoceras* (App. & Wr.) Bilai (3).

elaeidis Toovey (35)—oil palm (*Elaeis*).

eucalypti Arya & Jain (35)—*Eucalyptus*.

fabae Yu & Fang (35)—broad bean (*Vicia*).

fragariae Winks & Williams (3)—strawberry (*Fragaria*).

gerberae Gordon (35)—transvaal daisy (*Gerbera*).

gladioli (Massey) Snyd. & Hans. (71)—*Gladiolus*.

glycines Armst. & Armst. (3)—soybean (*Glycine*).

habae Raabe (3)—*Hebe*.

herbemontis (Tochetto) Gordon (35)—grape (*Vitis*).

laciniati Pandotra, Gupta & Sastry (60)—*Solanum laciniatum*.

lagenariae Matuo & Yamamoto (3)—calabash gourd (*Lagenaria*).

lathyri Bhide & Uppal (35)—lang (*Lathyrus*).

lentis (Vasudeva & Srinivasan) Gordon (35)—lentil (*Lens*).

lilii Imle (35)—lily (*Lilium*).

lini (Bolley) Snyd. & Hans. (71)—flax (*Linum*) races have not been clearly defined. Syn.: *F. oxysporum* var. *orthoceras* (App. & Wr.) Bilai (35).

luffae Kawai, Suzuki & Kawai (3)—vegetable sponge (*Luffa*).
lupini Snyd. & Hans. (71) races 1,2, and 3 Armst. & Armst. (3)—lupine (*Lupinus*) races A, B, and C Richter (3).
lycopersici (Sacc.) Snyd. & Hans. (71)—tomato (*Lycopersicon*). Syn.: *F. oxysporum* Schl. *lycopersici* Sacc. (18); *F. lycopersici* Bruschi (18); *F. bulbigenum* var. *lycopersici* (Bruschi) Wr. & Reink. (18).
Alexander and Tucker (3) first demonstrated physiologic specilization. Finley (3) first designated the isolates as races 1 and 2. Race 3 Tokeshi et al. (77). Noguez and Tokeshi changed this to race 2 (59).
mathioli Baker. See *conglutinans* race 3
medicaginis (Weimer) Snyd. & Hans. (71)—alfalfa (*Medicago*).
melongenae Matuo & Ishigami (35)—egg plant (*Solanum*).
melonis (Leach & Currence) Snyd. & Hans. (71), races 1,2, and 3 Risser & Mas (3), race 4 Risser et al. (69), races 5.6, and 7 Armst. & Armst. (8)—muskmelon (*Cucumis*).
narcissi Snyd. & Hans. (71)—*Narcissus*.
nelumbicola (Nis. & Wat.) Booth (18) as *nelumbicolum* (35)—lotus (*Nelumbo*). Syn.: *F. bulbigenum* Cke. & Mass. var. *nelumbicolum* Nis. & Wat. (35).
niveum (E. F. Smith) Snyd. & Hans. (71), races 1 and 2 Crall (22), race 3 Netzer & Dishon (57) —watermelon (*Citrullus*).
opuntiarum (Pettinari) Gordon (35)—cactus (*Opuntia*). Syn.: *F. oxysporum* Schl. var. *opuntiarum* Pettinari (35).
passiflorae Gordon apud Purss (18)—passion vine (*Passiflora*).
perniciosum (Hept.) Toole (3), races 1 and 2 Toole (3)—silk tree (*Albizzia*). Syn.: *F. perniciosum* Hept. (18); *F. oxysporum* Schl. forma *perniciosum* (Hept.) Snyd. & Campi (18); *F. vasinfectum* var. *perniciosum* Carrera (18).
phaseoli Kendr. & Snyd. (35) races 1 and 2 Ribeiro (65)—bean (*Phaseolus*).
phormi Wager (35)—New Zealand flax (*Phormium*).
pisi (van Hall) Snyd. & Hans., race 1 Snyd. (3), race 2 Snyd. (3), race 3 Schreuder (3), race 4 Bolton, Nuttall & Lyall (3), race 5 Haglund & Kraft (38), races 6, 7, 8, 9, 10, and 11 Armst. & Armst. (5). Kraft and Haglund recognize only races 1, 2, and 5 (51) and new race 6 (39)—garden pea (*Pisum*). Syn.: *F. vasinfectum* Atk. var. *pisi* van Hall (18); *F. orthoceras* App. & Wr. var. *pisi* Linford (18); *F. oxysporum* Schl. f. 8 Snyd. (18); *F. oxysporum* Schl. var. *orthoceras* (App. & Wr.) Bilai (18)
psidii Prasad, Mehta & Lal (35)—guava (*Psidium*).
pyracanthae new forma specialis Armst. & Armst.
McRitchie (53) in 1973 reported a wilt of the variegated cultivar of *Pyracantha* spp. (firethorn). We inoculated 23 species and cultivars of plants with the *Fusarium* and found it related to f. sp. *vasinfectum* (unpublished results). It is now proposed as a new forma specialis.
querci Gordon (35)—oak (*Quercus*).
raphani Kend. & Snyd. See *conglutinans* race 2
rauvolfiae Janard., Gnag. & Husain (18)—*Rauvolfia*.
rhois Snyd., Toole & Hept. see *callistephi* race 3
ricini (Wr.) Gordon (35)—castor bean (*Ricinus*). Syn.: *F. orthoceras* App. & Wr. var. *ricini* (35).
sedi Raabe (35)—*Sedum*.
sesami (Zaprom.) Castellani (18)—sesame (*Sesamum*). Syn.: *F. vasinfectum* Atk. var. *sesami* Zaprometoff (18).
sesbaniae Singh (35)—*Sesbania*.
spinaciae (Sherb.) Snyd. & Hans. (71), races 1 and 2 Armst. & Armst. (7)—spinach (*Spinacia*), beet (*Beta*). Syn.: *F. spinaciae* (Sherb.) (7); *F. conglutinans* var. *betae* Stew-

art (7); *F. conglutinans* (Wr.) var. *betae* Stewart (7); *F. oxysporum* f. sp. *betae* (Stewart) Snyd. & Hans. (7); *F. oxysporum* Schl. var. *orthoceras* (App. & Wr.) Bilai (7).
stachydis Gordon (35)—Japanese artichoke (*Stachys*).
tracheiphilum (E. F. Smith) Snyd. & Hans. (71), races 1 and 2 Armst. & Armst. (3), race 3 (Hare) Armst. & Armst. (3)—protepea (cowpea) (*Vigna*). Syn.: *F. tracheiphilum* E. F. Smith (18); *F. bulbigenum* Cke. & Mass var. *tracheiphilum* (E. F. Smith) Wr. (18); *F. oxysporum* f. sp. *conglutinans* (Wr.) Snyd. & Hans., race 3 Hare (3).
trifolii (Jacz.) Bilai (35)—clover (*Trifolium*). Syn.: *F. trifolii* Jaczewski (35).
tuberosi Snyd. & Hans. (71)—potato (*Solanum*), Syn.: *F. oxysporum* Schl. f. sp. *solani* (Raillo) Bilai (18) *F. oxysporum* Schl. var. *solani* Raillo (18)
tulipae Apt. (35)—tulip (*Tulipa*).
vanillae (Tucker) Gordon (35)—*Vanilla*. Syn.: *F. batatis* var. *vanillae* Tucker (35).
vasinfectum (Atk.) Snyd. & Hans. (71), races 1, 2, 3, and 4 Armst. & Armst. (3), race 5 Ibrahim (3), race 6 Armst. & Armst. (9)—cotton (*Gossypium*), alfalfa (*Medicago*), soybean (*Glycine*), tobacco (*Nicotiana*). Syn.: *F. vasinfectum* Atk. (18).
voandzeiae Armst., Armst. & Billington (10)—Bambarra groundnut (*Voandzeia*).
zingiberi Trujillo (3)—ginger (*Zingiber*).

Wilts caused by *Fusarium oxysporum* have been reported on *Eruca sativa* Mill. (21), *Plantago ovata* Forsk (70), *Solanum khasianum* Clarke (19), and *Pterocarpus angolensisoc,* an important timber tree in Zambia showing a serious decline (Mukwa wilt) (63). Further inoculations are needed to determine the classification of these Fusaria among the formae speciales. In 1972, reports indicated a possible vascular wilt of *Citrus* in Brazil (76). In 1974, successful inoculations were made of Rangpur lime seedlings (*Citrus limonia* Osbeck), and trees of Galego lime (*C. aurantifolia* Swingle) grafted on Rangpur lime showed decline (75). Apparently the same disease caused by *F. oxysporum* occurs in Florida (personal communication, Dr. L. W. Timmer, Jan. 1978). Vascular wilt of *Koa* caused by *Fusarium oxysporum koae,* f. sp. nov. (a manuscript which is to be sent to Phytopathology has been reviewed, courtesy Dr. D. E. Gardner, June 1979).

Classification and Nomenclature

The Tenth International Botanical Congress in 1964 decided that the nomenclature of formae speciales shall not be governed by the provision of this Code. In 1968, we (6) stated that many of the formae speciales are poorly defined, and, if they are eventually covered by the Code, a more detailed account of their pathogenicity, including the names of cultivars of the differential hosts and the symptomology of the disease, should be required to establish a new forma specialis. This may be a formidable task because the name given a *Fusarium* causing a wilt disease of a new host may be uncertain until (i) inoculations of the new host with virulent isolates of all formae speciales and races of *F. oxysporum* and (ii) inoculations of the differential hosts of each forma specialis with the supposedly new one are made. These ideals will seldom be reached, but there is a need to stabilize the status of formae speciales since they are of great importance to the plant pathologist and breeder. The deposition of a living pathogenic culture, either lyophilized or under liquid nitrogen, in a national collection is necessary so that it is available to future workers. Also equally necessary, where possible, is the deposition of seed of differential cultivars in a center(s) for the preservation of germ plasm. To attain a measure of standardization, perhaps procedures are now in order similar to those used by the investigators of the cereal rusts (not reviewed) and recently by the investigators of club root (28) in which a tabulation of methods of each investigator was made. The concepts and procedures that seem to be most satisfactory for the formae speciales of *F. oxysporum* might then be decided at a future conference of the Fusarium Research Workers.

Throughout our inoculation experiments, we have used uniform methods that aimed to give the full pathogenic potentiality of the fungus and also to give consistently reproducible results. We (8) stated that among the more important factors influencing the results of the inoculations were (i) variation in pathogens; (ii) nature and concentration of the inoculum; (iii) differential hosts of unknown genetic composition; (iv) media for growth of plants; (v) stage at which inoculated, as seedlings or more mature plants; (vi) inoculations with or without transplanting into natural or artificially infested media and with or without dipping roots in inoculum or cutting of roots; (vii) and variation in temperature. Especially noticeable in the numerous papers on the Fusarium wilts of the cucurbits were the contradictions in host ranges of f. sp. *cucumerinum, melonis,* and *niveum.* Also the validity of races was uncertain. Some of the complexities and parameters involved in the above factors are in reviews (25, 26, 27, 36, 78, 81).

Several systems for designating physiologic races of plant pathogens have been suggested. The system of Black et al. (16) for designating races of *Phytophthora infestans* (Mont.) de Bary has been accepted as probably ideal; in it a physiologic race is designated by the virulence genes it can overcome. Risser et al. (68) proposed this nomenclature for four races of f. sp. *melonis,* but, with the discovery of races 5 to 9 inclusive (8), the system became inadequate since the genes for resistance of the differential cultivars were unknown. At this time, we prefer to designate races by numbers in an arithmetic series. Gabe (33) proposed the system of Black et al. for races of f. sp. *lycopersici.* Habgood (37) proposed a system in which the differential hosts were arranged in a fixed order, and each was assigned a denary number, 1, 2, 4, 8, etc, these numbers corresponding with the binary series 2^0, 2^1. 2^2, 2^3, etc. Habgood's system (37) with some modifications has been accepted for *Plasmodiophora brassicae* Wor. (20). Gilmour (34) suggested an octal notation which is considered an improvement over the system of Habgood. No attempt is made to give the systems for identifying the various rusts.

Although the review by Johnson (43) on the genetics of host-parasite interactions deals chiefly with the cereal rusts, where the most specific information is available, much information can be gained from it by workers with a soil-borne fungus such as *F. oxysporum.* He is of the opinion that few classification schemes now in use can be considered to have any permanence, with the possible exception of that for wheat stem rust by Stakman et al. (73) and that for the determination of races of flax rust by Flor (31, 32). Nevertheless, there is a distinct need for a uniform scheme for the formae speciales and races considered here, regardless of its lack of permanence.

Hosts with wilts caused by formae speciales and races of *Fusarium oxysporum* in which genes for resistance have been investigated

No attempt is made to give all citations for each forma specialis but chiefly one or a few of the most recent ones where references to early work can be found.

Banana (*Musa* spp. and *M. balbisiana* Bolla). See Waite (80) for literature review and some problems.

Bean (*Phaseolus vulgaris* L.). Resistance was conditioned by a single dominant gene (66).

Cabbage (*Brassica oleracea* var. *capitata* L.). Cultivars with single dominant genes controlling high degrees of resistance and those of the multigenic type have been available for many years, chiefly developed at Wisconsin by Walker and associates (81).

Cotton (*Gossypium hirsutum* L.). Kappelman (45, 46) reviewed earlier studies, four of which indicated that resistance to *Fusarium* was inherited as a dominant characteristic. In studies of seven families from eight inbred lines in the greenhouse, additive gene effects most satisfactorily explained the inheritance of resistance to f. sp. *vasinfectum* (45). The use of an isolate of high virulence in greenhouse inoculations was necessary to obtain a high correlation with field tests where nematodes were present (46).

Cowpea (protepea) [*Vigna unguiculata* (L.) Walp.]. From crosses of resistant cultivar Iron with several other cultivars, all resistant to race 1, Hare (41) concluded that resistance to races 2 and 3 depended on two dominant genes.

Cucumber (*Cucumis sativus* L.). Resistance was controlled by a dominant gene, tentatively designated *Foc* (58). Resistance was controlled by three genes [see Netzer et al. (58) for citation].

Flax *Linum usitatissimum* L.). See Kommedahl et al. (50).

Garden pea (*Pisum sativum* L.). Dr. P. Matthews, John Innes Institute (personal communication November, 1973) found inheritance of resistance to be different with race. He noted a marked environmental sensitivity in one race, and that with three cultivars inoculated with another race, two systems, with resistance either dominant or recessive, appeared to be operating. Resistance attributable to a single dominant gene factor in the host was discussed by Haglund and Kraft (39).

Gram (*Cicer arietinum*). Resistance was recessive and controlled by a single gene (62).

Muskmelon (*Cucumis melo* L.). Mortensen (56) thought resistance was due to a principal dominant gene plus two complementary dominant genes. Messiaen et al. (55) concluded that resistance in Charentais was monogenic dominant. Risser (67) found that the gene Fom_1 conferred resistance in Doublon to race 1. A single dominant gene $Fom_{1,2}$ conferred resistance to races 1 and 2. Risser et al. (68) proposed a new nomenclature for races and resistant genes, i.e., races 0, 1, 2, and 1,2 and genes Fom_1 and Fom_2.

Safflower (*Carthamus tinctorius* L.). In the greenhouse, resistance to race 3 was due to a major and a minor gene. In the field with one cultivar only, the minor gene was ineffective as only a single gene was found (17).

Sweet potato [*Ipomoea batatas* (L.) Lam.]. Resistance was inherited in a multifactorial fashion (74).

Tomato (*Lycopersicon esculentum* L.). Crill et al. (23, 24) discussed the advantages of monogenic resistance [vertical resistance of Van der Plank (79)] over polygenic tolerance (horizontal resistance of Van der Plank) in the tomato-breeding program in Florida. Jones and Crill (44) found that the incidence of disease in some I-gene containing cultivars was considerably less than those of Missouri Accession 160 from which the I-gene was derived. "Consequently, an unidentified race 1 isolate mistakenly could be designated as race 2 unless known race 1 and 2 cultures are included in the tests to determine pathogenesis."

Tulip (*Tulipa gesneriana* L.). Resistance was mainly based on additive gene action (30).

Watermelon [*Citrullus lanatus* (Thunb.) Mansf.]. Henderson et al. (42) suggested that wilt resistance of Summit was controlled by a single dominant gene. They discussed the variable results obtained by others on the inheritance of resistance in watermelon.

Literature Cited

1. Armstrong, G. M., and J. K. Armstrong. 1966. Wilt of Mexican sunflower caused by the celery-wilt Fusarium. Plant Dis. Rep. 50:391–393.
2. Armstrong, G. M., and J. K. Armstrong. 1966. The celery-wilt Fusarium causes wilt of garden pea. Plant Dis. Rep. 51:888–892.
3. Armstrong, G. M., and J. K. Armstrong. 1968. Formae speciales and races of *Fusarium oxysporum* causing a tracheomycosis in the syndrome of disease. Phytopathology 58:1242–1246.
4. Armstrong, G. M., and J. K. Armstrong. 1971. Races of the aster-wilt *Fusarium*. Phytopathology 61:820–824.
5. Armstrong, G. M., and J. K. Armstrong. 1974. Races of *Fusarium oxysporum* f. sp. *pisi*, causal agents of wilt of pea. Phytopathology 64:849–857.
6. Armstrong, G. M., and J. K. Armstrong. 1975. Reflections on the wilt Fusaria. Annu. Rev. Phytopathol. 13:95–103.
7. Armstrong, G. M., and J. K. Armstrong. 1976. Common hosts for *Fusarium oxysporum* formae speciales *spinaciae* and *betae*. Phytopathology 66:542–545.
8. Armstrong, G. M., and J. K. Armstrong. 1978. Formae speciales and races of *Fusarium oxysporum* causing wilts of the Cucurbitaceae. Phytopathology 68:19–28.
9. Armstrong, G. M., and J. K. Armstrong. 1978. A new race (race 6) of the cotton-wilt *Fusarium* from Brazil. Plant Dis. Rep. 62:421–423.
10. Armstrong, G. M., J. K. Armstrong, and R. V. Billington. 1975. *Fusarium oxysporum* forma specialis *voandzeiae,* a new form species causing wilt of Bambarra groundnut. Mycologia 67:709–714.
11. Armstrong, G. M., J. K. Armstrong, and R. H. Littrell. 1970. Wilt of chrysanthemum caused by *Fusarium oxysporum* f. sp. *chrysanthemi,* forma specialis nov. Phytopathology 60:496–498.
12. Armstrong, G. M., J. K. Armstrong, and D. Netzer. 1978. Pathogenic races of the cucumber-wilt *Fusarium*. Plant Dis. Rep. 62:824–828.
13. Armstrong, G. M., B. S. Haekins, and C. C. Bennett. 1942. Cross inoculations with isolates of Fusaria from cotton, tobacco, and certain other plants subject to wilt. Phytopathology 32:685–698.
14. Armstrong, G. M., J. D. MacLachlan, and R. Weindling. 1940. Variation in pathogenicity and cultural characteristics of the cotton-wilt organism, *Fusarium vasinfectum*. Phytopathology 30:515–20.
15. Bilai, V. I. 1955. The Fusaria (Biology and Systematics) Ukr.SSR. Acad. Sci. Kiev. 319 p. (In Russian, R. A. M. 36:731–732. 1957.)
16. Black, G. W., C. Mastenbroeck, W. R. Mills, and L. C. Peterson. 1953. A proposal for an international nomenclature of races of *Phytophthora infestans* and of genes controlling immunity in *Solanum demissum* derivatives. Euphytica 2:173–179.
17. Bockelman, H. E. 1974. The inheritance of resistance to Fusarium wilt in cultivars of safflower (*Carthamus tinctorius* L.). Diss. Abstr. 35B(4):1727.
18. Booth, C. 1971. The genus *Fusarium*. Commonw. Mycol. Inst., Kew, Surrey, England. 237 pp.
19. Bordoloi, D. N., J. H. Hazarika, and D. Ganguly. 1972. Wilt of *Solanum khasianum* caused by *Fusarium oxysporum*. Indian Phytopathol. 24:788–791.
20. Buczaki, S. T., H. Toxopeus, P. Mattusch, T. D. Johnston, G. R. Dixon, and L. A. Hobolth. 1975. Study of physiological specialization in *Plasmodiophora brassicae:* Proposals for attempted rationalization through an international approach. Trans. Brit. Mycol. Soc. 65:295–303.
21. Chatterjee, C., and J. N. Rai. 1975. Fusarium wilt of *Eruca sativa*—observation on comparative pathogenicity of some strains of *Fusarium oxysporum*. Indian Phytopathol. 27:309–311.
22. Crall, J. M. 1964. Investigations of and control of Fusarium wilt of watermelon. Annu. Rep. Florida Agric. Exp. Stn. ending June 30, 1963. p. 371.
23. Crill, P., J. P. Jones, and D. S. Burgis. 1973. Failure of "horizontal resistance" to control Fusarium wilt of tomato. Plant Dis. Rep. 57:119–121.
24. Crill, P., J. P. Jones, D. S. Burgis, and S. S. Woltz. 1972. Controlling Fusarium wilt of tomato with resistant varieties. Plant Dis. Rep. 56:695–699.
25. Day, P. R. 1960. Variation in phytopathogenic fungi. Annu. Rev. Microbiol. 14:1–16.
26. Day, P. R. 1966. Recent developments in the genetics of the host-parasite system. Annu. Rev. Phytopathol. 4:245–268.
27. Day, P. R. 1973. Genetic variability of crops. Annu. Rev. Phytopathol. 11:293–312.
28. Dixon, G. R. 1976. Methods used in Western Europe and the U.S.A. for testing Brassica seedling resistance to club root (*Plasmodiophora brassicae*). Plant Pathol. 25:129–134.
29. Dzidzariya, O. M., 1968. Measures for the control of fusariosis of East Indian basil. Pishch. Prom. SSR. 1968:129–140. (Translated title, English summary, R. A. M. 48:882. 1969.)
30. Eijk, J. P. Van, F. Garretsen, and W. Eikelboom. 1978. Breeding for resistance to *Fusarium oxysporum* f. sp. *tulipae* in tulip (*Tulipa* L.) 2. Phenotypic and genotypic evaluation of cultivars. Euphytica 28:67–71.
31. Flor, H. H. 1946. Genetics of pathogenicity in *Melampsora lini*. J. Agric Res. 73:335–357.
32. Flor, H. H. 1947. Inheritance of reaction to rust in flax. J. Agric. Res. 74:241–262.
33. Gabe, H. L. 1975. Standardization of nomenclature for pathogenic races of *Fusarium oxysporum* f. sp. *lycopersici*. Trans. Brit. Mycol. Soc. 64:156–159.
34. Gilmour, J. 1973. Octal notation for designating physiologic races of plant pathogens. Nature 242:620.

35. Gordon, W. L. 1965. Pathogenic strains of *Fusarium oxysporum*. Can J. Bot. 43:1309–1318.
36. Graniti, W. L. 1976. Tissue and organ specificity in plant diseases. pp. 27–41. *In* R.K.S. Wood and A. Graniti (ed.), Specificity in plant diseases, Plenum Press, New York and London. 354 p.
37. Habgood, R. M. 1970. Designation of physiological races of plant pathogens. Nature 227:1268–1269.
38. Haglund, W. A., and J. M. Kraft. 1970. *Fusarium oxysporum* f. *pisi,* race 5. Phytopathology 60:1861–1862.
39. Haglund, W. A., and J. M. Kraft. 1979. *Fusarium oxysporum* f. sp. *pisi,* Race 6: Occurrence and Distribution. Phytopathology 69:818–820.
40. Hansford, C. G. 1926. The Fusaria of Jamaica. Kew Bull. 7:257–288.
41. Hare, W. W. 1957. Inheritance of resistance to Fusarium wilt in cowpeas. Phytopathology 47:312–313. (Abstr.).
42. Henderson, W. R., S. F. Jenkins, and J. O. Rawlings. 1970. The inheritance of Fusarium wilt resistance in watermelon, *Citrullus lunatus* (Thunb.) Mansf. J. Amer. Soc. Hort. Sci. 95:276–282.
43. Johnson, T. 1968. Host specialization as a taxonomic criterion. p. 543–556. *In* G. C. Ainsworth and A. S. Sussman (ed.), The fungi, Vol. III, Academic Press, New York. 738 p.
44. Jones, J. P., and P. Crill. 1974. Susceptibility of "resistant" tomato cultivars to Fusarium wilt. Phytopathology 64:1507–1510.
45. Kappelman, A. J., Jr. 1971. Inheritance of resistance to Fusarium wilt in cotton. Crop Sci. 11:672–674.
46. Kappelman, A. J., Jr. 1975. Correlation of Fusarium wilt of cotton in the field and greenhouse. Crop Sci. 15:270–272.
47. Klisiewicz, J. M. 1975. Race 4 of *Fusarium oxysporum* f. sp. *carthami.* Plant Dis. Rep. 59:712–714.
48. Klisiewicz, J. M., C. A. Thomas. 1970. Pathogenic races of *Fusarium oxysporum* f. sp. *carthami.* Phytopathology 60:83–84.
49. Klisiewicz, J. M., and C. A. Thomas. 1970. Race differentiation in *Fusarium oxysporum* f. sp. *carthami.* Phytopathology 60:1706.
50. Kommedahl, T., J. J. Christensen, and R. A. Frederiksen. 1970. A half century of research in Minnesota on flax wilt caused by *Fusarium oxysporum.* Minn. Agric. Exp. Stn. Tech. Bull. 273. 35 p.
51. Kraft, J. M., and W. A. Haglund. 1978. A reappraisal of the race classification of *Fusarium oxysporum* f. sp. *pisi.* Phytopathology 68:273–275.
52. Matuo, T., A. Matsuda, K. Ozaki, and K. Kato. 1975. *Fusarium oxysporum* f. sp. *arctii* causing wilt of great burdock. Ann. Phytopathol. Soc. Japan 41:77–80.
53. McRitchie, J. J. 1973. Pathogenicity and control of *Fusarium osyxporum* wilt of variegated *Pyracantha.* Plant Dis. Rep. 57:389–391.
54. Messiaen, C. M., and R. Cassini. 1968. Recherches sur les fusarioses IV. La systematique des *Fusarium.* Ann. Epiphyties 19:387–454.
55. Messiaen, C. M., G. Risser, and P. Pecaut. 1962. Étude des plantes resistantes au *Fusarium oxysporum* f. sp. *melonis* dans la varieté de melon cantaloup *Charentais.* Ann Amelior. Plantes 12:157–164.
56. Mortensen, J. A. 1959. The inheritance of Fusarium resistance in muskmelons. Diss. Abstr. 19:2209.
57. Netzer, D., and I. Dishon. 1973. Screening for resistance and physiological specialization of *Fusarium oxysporum* in watermelon and muskmelon. Abstr. No. 0941 in: Abstract of Papers, 2nd Int. Cong. Plant Pathol, 7–12 September, 1973, Minneapolis, Minnesota (unpaged).
58. Netzer, D., S. Niego, and E. Galun. 1977. A dominant gene conferring resistance to Fusarium wilt in cucumber. Phytopathology 67:525–527.
59. Noguez, M. A., and H. Tokeshi. 1974. Revisão da classificacão da race 3 de *Fusarium oxysporum* f. sp. *lycopersici.* Anais Esc. Sup. Agric. 'Luiz Queiroz' 31:419–430.
60. Pandotra, V. R., J. H. Gupta, and K. S. M. Sastry. 1971. A note on wilt disease of *Solanum laciniatum.* Indian J. Mycol. Plant Pathol. 1:86–87.
61. Patel, P. N., N. Prasad, R. L. Mathur, and B. H. Mathur. 1957. Fusarium wilt of cumin. Current Sci. 26:181–182.
62. Pathak, M. M., K. P. Singh, and S. A. Lal. 1975. Inheritance of resistance to wilt (*Fusarium oxysporum* f. *ciceri*) in gram. Indian J. Farm Sci. 3:10–11.
63. Piearce, G. D. 1979. A new vascular wilt disease and its relationship to widespread decline of *Pterocarpus angolensis* in Zambia. PANS 25:37–45.
64. Raillo, A. I. 1950. Pilze des Gattung *Fusarium.* (Fungi of the genus Fusarium). State Publ. Moscow. 415 pp. (R.A.M. 31:148. 1952.)
65. Ribeiro, R. de L. D. 1977. Race differentiation in *Fusarium oxysporum* f. sp. *phaseoli,* the causal agent of bean yellows. Proc. Amer. Phytopathol. Soc. 4:165. (Abstr.).
66. Ribeiro, R. de L. D., D. J. Hagedorn, and R. E. Rand. 1977. Inheritance of resistance in beans to a race of *Fusarium oxysporum* f. sp. *phaseoli* from Brazil. Proc. Amer. Phytopathol. Soc. 4:133. (Abstr.).
67. Risser, G. 1973. Etude de l'heredité de la résistance du melon (*Cucumis melo*) aux races 1 et 2 de *Fusarium oxysporum* f. *melonis.* Ann. Amelior. Plantes 23:259–263.
68. Risser, G., Z. Banihashemi, and D. W. Davis. 1976. A proposed nomenclature of *Fusarium oxysporum* f. sp. *melonis* races and resistance genes in *Cucumis melo.* Phytopathology 66:1105–1106.
69. Risser, G., P. Mas, and J. C. Rode. 1969. Mise en evidence et characteristation d'une quatrieme race de *Fusarium oxysporum* f. *melonis.* Ann Phytopathol. 1:217–222.

70. Russell, T. E. 1975. Plantago wilt. Phytopathology 65:359–360.
71. Snyder, W. C., and H. N. Hansen. 1940. The species concept in *Fusarium*. Amer. J. Bot. 27:64–67.
72. Solov'eva, A. I., and T. Madumarov. 1969. Nature of morphological indices of *Fusarium oxysporum* Schl. forms. Mikol. Fitopatol. 3:342–345. (Russian, Transl. title. Biol. Abstr. 51:127788. 1970)
73. Stakman, E. C., D. M. Stewart, and W. Q. Loegering. 1962. Identification of physiologic races of *Puccinia graminis* var. *tritici*. U.S. Dept. Agric. ARS. E617. 54 p.
74. Struble, F. B., L. S. Morrison, and H. B. Cordner. 1966. Inheritance of resistance to stem rot and to root knot nematode in sweet potato. Phytopathology 56:1217–1219.
75. Takatzu, A., and J. C. Dianese. 1974. Xylem infection of Rangpur lime by *Fusarium oxysporum* in Brazil. Ciencia Cultura 26:153–155.
76. Takatzu, A., N. Gimenes, and F. Galli. 1972. Nota previa sobre a possivel ocorrencia de murcha vascular em plantas citricas causada por *Fusarium* sp. Rev. Agric. 47:21–23.
77. Tokeshi, H., F. Galli, and C. Kurozwa. 1966. Nova raca de Fusarium do Tomateiro em São Paulo. Anais Esc. Sup. Agric. 'Luiz Queiroz' 23:217–227.
78. Toussoun, T. A., and P. E. Nelson. 1975. Variation and speciation in the Fusaria. Annu. Rev. Phytopathol. 13:71–82.
79. Van der Plank, J. E. 1968. Disease resistance in plants. Academic Press, New York and London. 206 pp.
80. Waite, B. H. 1977. Inoculation studies and natural infection of banana varieties with races 1 and 2 of *Fusarium oxysporum* f. *cubense*. Plant Dis. Rep. 61:15–19.
81. Walker, J. C. 1965. Use of environmental factors in screening for disease resistance. Annu. Rev. Phytopathol. 3:197–208.
82. Wollenweber, H. W., and O. A. Reinking. 1935. Die Fusarien, ihre Beschreibung, Schadwirkung and Bekämpfung. Paul Parey, Berlin. 355 p.

35 Anatomical Changes Involved in the Pathogenesis of Plants by *Fusarium*

Barbara White Pennypacker

Members of the genus *Fusarium* are world-wide in geographical distribution and equally wide ranging in the variety of plant species they attack. Few, if any economically important crops are safe from infection by this ubiquitous genus.

Although widespread in terms of geographical occurence and host distribution, the Fusaria are limited in the number of ways in which they cause disease. They may be grouped into two categories, the vascular wilt Fusaria and the cortical rot Fusaria, on the basis of their disease syndromes. The two groups are as distinct in the anatomical changes they cause in the host as they are in the external symptoms they incite. For this reason, the following review will consider the groups separately in explaining some of the histological changes that occur during the process of pathogenesis by the Fusaria.

The Vascular Wilts

Most of the Fusarium wilt diseases are caused by formae speciales of *Fusarium oxysporum*. Despite their host specificity, these fungi cause similar symptoms in most host species. Typical wilt symptoms include chlorosis, one-sided and total wilting, vascular discoloration, and plant death. A great deal of research has been directed toward understanding the actual mechanisms involved in Fusarium wilt, and consequently, the anatomical aspects of the disease are well documented.

The formae speciales of *F. oxysporum* causing wilt are soil-borne and infection of the host usually occurs through the root system. The pathogen has been found to penetrate the root cap and zone of elongation intercellularly in the root of banana, melon, china aster, radish, and cabbage (11, 39, 41, 45). Infection of the host in these areas gives the fungus access to undifferentiated vascular tissue and allows easy access to the newly differentiating xylem vessel elements. In some cases, as with sweet potato (29), the cut surface of the seed piece allows direct access to the host xylem tissue. Although a similar situation is present in banana, the fungus rarely penetrates the rhizome through its cut surface (48).

Once entry is gained into the vascular system, the pathogen remains confined to the xylem vessel elements and ramifies throughout the plant in these cells (11, 14, 32, 43, 45). Late in pathogenesis, the fungus penetrates the surrounding xylem parenchyma cells by moving through the half-bordered pit pairs between the xylem parenchyma cells and the xylem vessel elements (Fig. 35-1) (32, 33, 35, 39). Ullstrup (45) reported that *F. conglutinans* var. *callistephi* (=*F. oxysporum* f. sp. *callistephi*) was able to penetrate the vascular cambium and phloem of china aster, and Phipps (35) found that *F. oxysporum* f. sp.

perniciosum moved from the xylem elements to the vasicentric and ray parenchyma cells, eventually reaching the cambium and phloem in mimosa.

Fungal colonization of the host vascular system is often rapid and frequently facilitated by the formation of conidia within the xylem vessel elements (Fig. 35-2). Beckman et al. (8), Trujillo (44) and Elgersma et al. (21) reported extensive sporulation within the xylem vessel elements of banana and tomato. Similar sporulation has been observed in watermelon (31, 39), carnation (32), and cabbage (41).

Spore transport within the host has been studied extensively for *F. oxysporum* f. sp. *cubense* and *F. oxysporum* f. sp. *lycopersici*. Beckman et al. (8) and Trujillo (44) reported that the spores moved freely in the transpiration stream until they were stopped by the perforation plates of the xylem vessel elements. The conidia were unable to pass scalariform perforation plates and were always stopped by the vessel endings at the root-rhizome plexus (8). Once the spores were impeded by the perforation plates, they germinated, the germ tubes penetrated the perforation plate, and the hyphae formed subsequently produced conidiophores and conidia (7, 8). Trapping, penetration, and sporulation evidently continue as necessary, allowing the fungus to colonize the host rapidly. Blocking of conidia at the perforation plate was not found in carnation (32), muskmelon (39), or flax (43), probably due to the anatomical configurations of the perforation plates of these plants.

In addition to trapping by the perforation plates, conidia of *F. oxysporum* f. sp. *cubense* were often immobilized by gels formed in the xylem vessel elements (7). These gels were often transient in nature and in the absence of other host reactions, only impeded colonization for a short time (4).

Sporulation in plants infected by the Fusarium wilt organisms was intravascular except in mimosa, where Phipps (35) found that *F. oxysporum* f. sp. *perniciosum* sporulated freely in the xylem vessel elements and in the lenticels of infected mimosa.

The vascular wilt Fusaria have been reported to produce enzymes during the process of host colonization (19, 20, 22, 23). Histological evidence of cell maceration (Fig. 35-3) and vascular plugging (Fig. 35-4) has been found in carnation and tomato (32, 37) and is an indication of enzyme activity. Pectic enzymes appear to be the most prevalent enzymes produced by these pathogens. The mode of action of these enzymes has been extensively examined; however, a detailed discussion is beyond the scope of this review.

Up to this point, the Fusarium wilt diseases have been examined from the standpoint of fungal activity. The host plants react to infection by the wilt Fusaria in a variety of ways. Typical of host reaction to most formae speciales of *F. oxysporum* is the formation of gels, gums, and tyloses in the xylem vessel elements. There are two theories concerning the origin of vascular gels. Beckman (5) and Beckman & Zaroogian (9), working with banana, found that such gels always occur on the upper side of a perforation plate, indicating that a mechanism other than accumulation is involved. Evidence indicated that gels in banana are caused by the swelling of the perforation plates and end walls of the vessels (5, 6).

Researchers working with tomato reached a different conclusion concerning the origin of vascular gels (20, 23, 37, 46). They theorize that the enzymatic degradation of cell walls by pectic enzymes leads to the accumulation of large cell wall fragments in the xylem vessels. Pectic enzymes capable of causing such cell destruction have been identified in the infected tomato (20, 22) and pectic substances are present in the vascular gels in tomato, banana, and carnation (9, 32, 37). It would appear that the mechanism of gel formation is host dependent.

Regardless of how they originate, gels and gums play an important part in localizing the pathogen in some hosts by trapping conidia in the vessel elements (7). If the gels persist long enough to allow tyloses to form, as frequently occurs in banana, the pathogen is successfully contained (7). However, if the gels are short-lived or formation of tyloses is

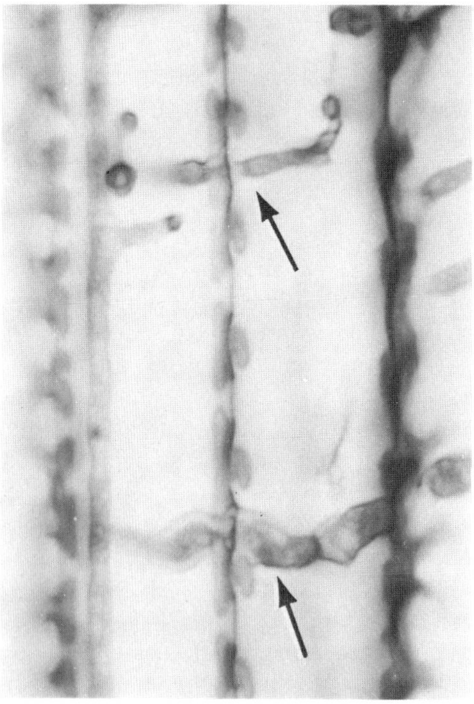
1

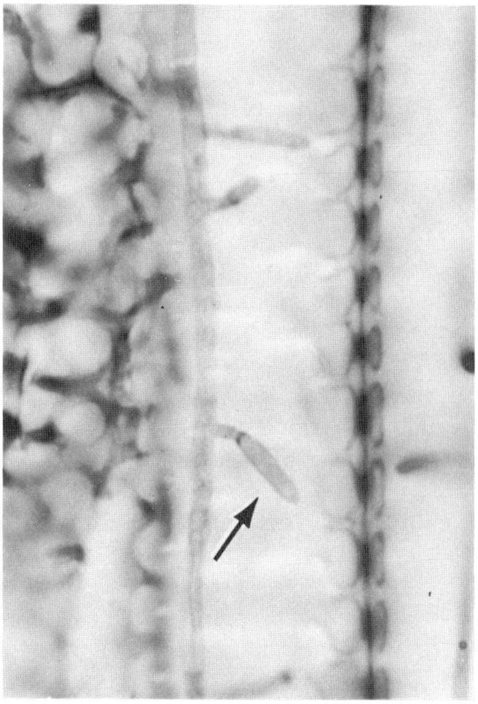
2

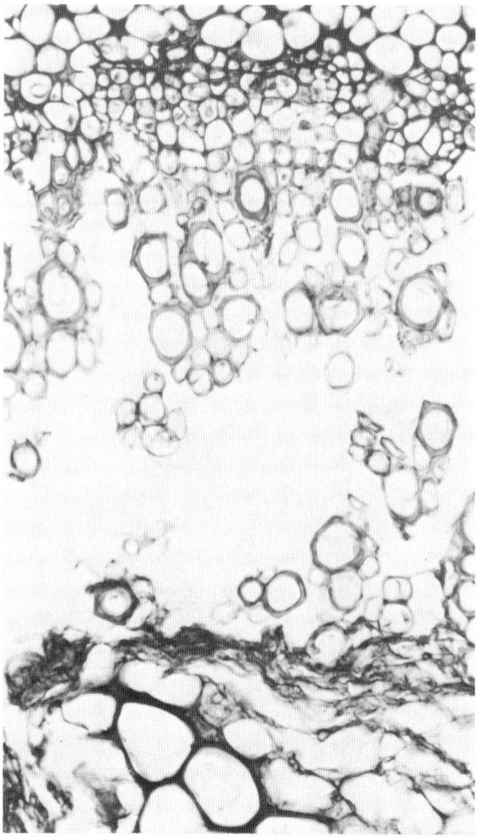
3

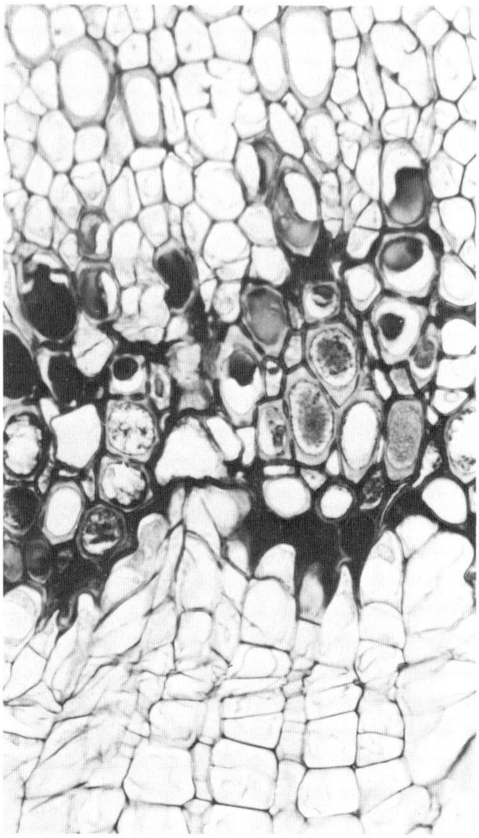
4

delayed or absent, conidia spread ahead of the vascular occlusion (4). Gums and gels have been reported in carnation, sweet potato, mimosa, watermelon, and cotton (12, 30, 31, 32, 35) and there are conflicting reports as to their occurrence in tomato (14, 37).

Tyloses are reported to play a role in localizing *F. oxysporum* f. sp. *cubense* in banana (7); however, although abundant, they fail to contain *F. oxysporum* f. sp. *batatas* in sweet potato (30). Tyloses formation appears to be host dependent and has been reported only in infected banana, melon and sweet potato.

The formation of hypertrophied, hyperplastic cells is a frequent host response to infection (Fig. 35-5). Wardlaw (47) reported such cells in the cortex of banana roots invaded by *F. cubense* (= *F. oxysporum* f. sp. *cubense*). The innermost hyperplastic cells were suberized and the combination of a wall of suberized, hyperplastic cells often limited penetration by the fungus. Chambers (14) and Pennypacker & Nelson (32) reported similar hyperplastic activity in the xylem parenchyma of infected vascular tissue in tomato and carnation respectively. As with tylose formation, the formation of hypertrophied, hyperplastic cells in response to invasion by the vascular wilt Fusaria is limited to a few host species.

In summary, the anatomical changes in the host which result from infection and colonization by the vascular wilt Fusaria consist primarily of the formation of gums and gels in almost all infected host species and the presence of tyloses and hypertrophied, hyperplastic cells in a few host plants. The type of xylem vessel perforation plate and the length of the xylem vessel elements were also involved in determining the rate at which the pathogen colonized the host.

The Cortical Rots

Many of the pathogenic species of the genus *Fusarium*, with the exception of those inciting vascular wilts, attack the cortex, storage tissue or seed of the host. These diseases, known by a variety of names including root rot, foot rot, stem rot, dry rot, scab, seedling blight, stand decline, and bulb rot, have as their initial symptom the development of a small lesion on the affected plant part. The initial lesions may enlarge and cause girdling, which then results in the stunting, chlorosis, and collapse of the plant.

Cortical rot diseases are caused by several species of *Fusarium*. The factor uniting these species is the mechanism by which they cause disease. These fungi are almost entirely confined to infecting and destroying the cortical regions of their hosts. Only during the late stages of pathogenesis do some of these pathogens invade the vascular tissues.

The cortical rot Fusaria are soil-, air-, or seed-borne and the mode of host penetration reflects this variation. The soil-borne cortical-rotting Fusaria generally infect the roots or hypocotyl of the host by penetrating the stomata, wounds, or directly through the root tip or meristematic region of the roots and through the hypocotyl (13, 15, 16, 26, 27, 40). Air-borne conidia of *Fusarium* usually gain entry to their host through wounds and sto-

Fig. 35-1. A portion of a longitudinal section through a stem of the carnation cultivar Improved White Sim infected with *Fusarium oxysporum* f. sp. *dianthi*. Note the hyphae in the pit pairs between the xylem vessel elements (arrows) (X1715).

Fig. 35-2. A portion of a longitudinal section through a stem of the carnation cultivar Improved White Sim infected with *Fusarium oxysporum* f. sp. *dianthi*. Note the conidium attached to the conidiophore (arrow) present in the xylem vessel element. Such intravascular sporulation is typical of the vascular wilt Fusaria. (X1505).

Fig. 35-3. A portion of a cross section through a stem of the carnation cultivar Improved White Sim infected with *F. oxysporum* f. sp. *dianthi*. Note that the maceration of the xylem parenchyma cells has left free floating xylem vessel elements (X360).

Fig. 35-4. A portion of a cross section through a stem of the carnation cultivar Improved White Sim infected with *F. oxysporum* f. sp. *dianthi*. Many of the xylem vessel elements are plugged with gums and gels (X380).

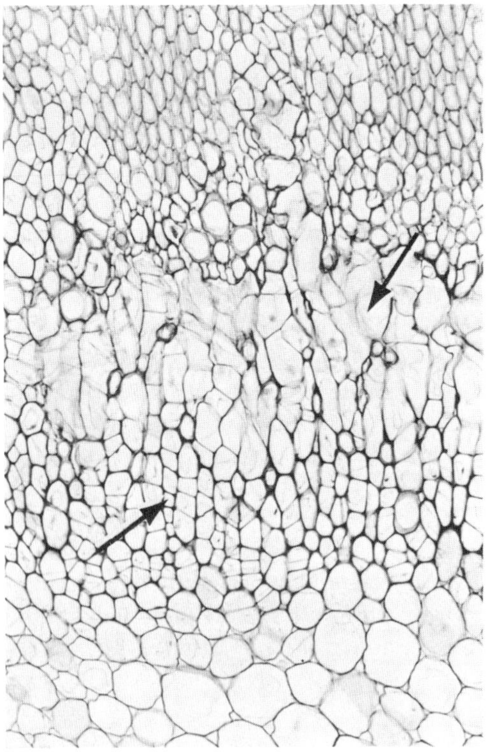
5

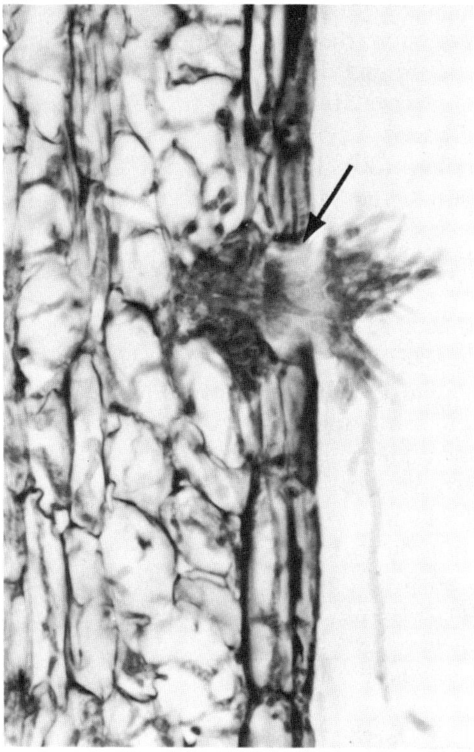
6

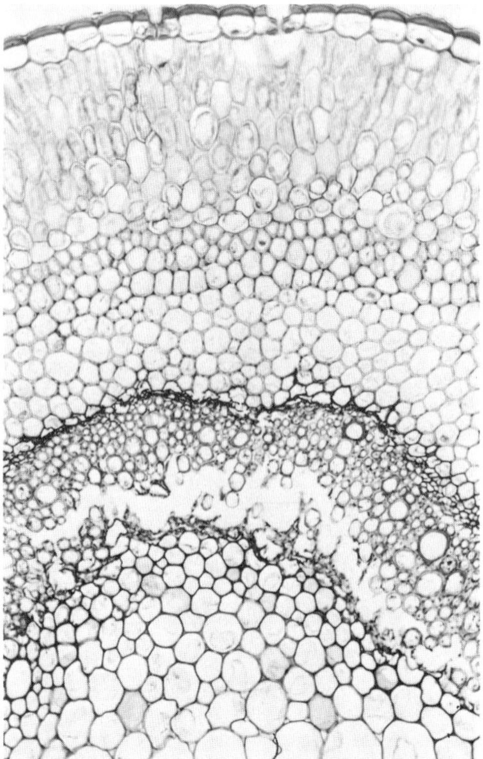
7

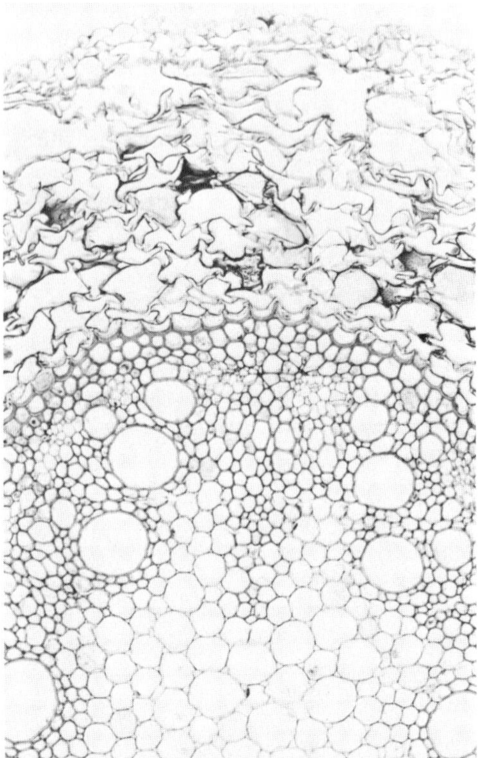
8

mata (18, 25, 34). Lawrence (28) and Bennett (10) reported that *F. moniliforme, F. culmorum*, and *F. avenaceum* were seed-borne and Lawrence found *F. moniliforme* in the tip cap or on the pericarp of sweet corn seed, which would allow early infection of the seedling.

Once in the host, these Fusaria are generally intercellular, at least initially. The hyphae of *F. solani* f. sp. *cucurbitae* and *F. solani* f. sp. *phaseoli* ramify through the middle lamellae of the cortical cells and in the schizogenous intercellular spaces (16, 26). In addition, Christou & Snyder (16) found that the hyphae of the bean root rot *Fusarium* branched dichotomously and actually enveloped the cortical cells in a digitate manner. Intercellular hyphal growth is frequently influenced by the conformation of the cortical cells. Bywater (13) reported that hyphal growth is radial when the cortical cells are isodiametric and longitudinal when the cells are elongated. Sparnicht & Roncadori (42) noted rapid growth of *F. oxysporum* and *F. roseum* in the mesophyll of cotton bracts and attributed this to the large intercellular spaces in the mesophyll, loosely packed parenchyma, and the absence of palisade parenchyma. Such hyphal growth patterns are reflected externally in the size and shape of the ensuing lesions (13).

The majority of the cortical Fusaria become intracellular in the cortical cells during the later stages of pathogenesis. Adams (1) and Pugh et al. (38) found that *Gibberella saubinetii* (= *F. roseum* 'Graminearum') colonized all parts of the wheat kernel both intercellularly and intracellularly and that it penetrated cells intracellularly via appressorial-like structures. Hadley (25) reported intracellular colonization of the phloem and fiber cells of carnation by means of fungal movement through the pit pairs. Phillips (34) noted constrictions in the hyphae where *F. roseum* f. sp. *cerealis* penetrated the cortical cells of carnation. *Fusarium sambucinum* (=*F. roseum* 'Sambucinum') was able to penetrate the suberized cortical cells of the potato tuber, although it was incapable of growing through wound cork cells (17). Some members of this category of Fusaria only move intracellularly when the host cells are dead or dying, as is the case with *F. solani* f. sp. *phaseoli* in bean (16).

Colonization of the various hosts attacked by these *Fusarium* species is frequently aided by the production of fungal enzymes which macerate cells and destroy the middle lamella. Such enzymes are of two types, the pectic enzyme produced by *F. solani* f. sp. *phaseoli* (3), and a cellulolytic enzyme implicated histochemically in yellow poplar canker caused by *F. solani* (2) and carnation stem rot caused by *F. roseum* 'Graminearum' (34). Histochemical evidence indicates that these enzymes are both involved in pathogenesis in yellow poplar canker and carnation stem rot.

Once the host cortex is colonized, frequent sporulation occurs. In contrast to the vascular wilt Fusaria, the cortical rot Fusaria rarely sporulate in the plant. Conidiophores usually penetrate the stomata of the stem and/or hypocotyl and sporodochia are produced externally (10, 18, 38) (Fig. 35-6). Chlamydospores frequently form in the dead cortical cells of infected lentils (27), beans (16), and red clover (40).

Fig. 35-5. A portion of a cross section through a stem of the carnation cultivar Improved White Sim infected with *F. oxysporum* f. sp. *dianthi*. Note the extensive hypertrophy and hyperplasia of the xylem parenchyma cells (arrows) (X180).

Fig. 35-6. A portion of a longitudinal section through a stem of the carnation cultivar Improved White Sim infected with *F. roseum* 'Graminearum'. Note the sporodochia protruding from a stomate (arrow) on the stem surface. External sporulation of this nature is common among the cortical rotting Fusaria. (X460).

Fig. 35-7. A portion of a cross section of a stem of carnation cultivar Improved White Sim infected with *F. oxysporum* f. sp. *dianthi*. Note the disintegration of the xylem parenchyma cells and that the pathogen has not penetrated beyond the vascular system of the host. (X160).

Fig. 35-8. A portion of a cross section through the mesocotyl of *Zea mays* infected with *F. moniliforme*. Note the collapse of the cortical cells and that the fungus has not affected the vascular system. (X150). (Histological specimen courtesy of Ellen B. Lawrence).

The host plant is not entirely passive following invasion by the cortical rot Fusaria. There is little vascular response as evidenced by the lack of gum, gels, and tyloses, probably because these pathogens rarely invade the vascular tissue except occasionally during the last stages of pathogenesis. In many cases, however, there is a definite host response in the form of hypertrophied, hyperplastic cortical cells. This type of reaction appears to be more prevalent in hosts attacked by the cortical rot Fusaria than in hosts invaded with vascular wilt Fusaria.

Graham (24) reported delineation of the lesion area by a periderm-like layer of cortical cells. These cells may play a part in restricting *F. oxysporum* var. *redolens* to the cortex of asparagus. Lignified cells combine with periderm cells in yellow poplar to contain infection by *F. solani* by blocking vertical and lateral hyphal penetration in the bark (2). The formation of a well-defined layer of cork cells by the potato tuber is essential in confining infection by *F. sambucinum*. Any factor which delays or inhibits such wound healing will allow the pathogen to penetrate the tuber beyond the point where it can be contained by periderm formation (17). Pierre & Wilkinson (36), studying beans resistant to *F. solani* f. sp. *phaseoli*, noted the development of hypertrophied and hyperplastic cortical cells internal to the lesion. Less resistant bean lines also formed a periderm-like layer, but it was located at the endodermis, which normally limits lateral spread of the pathogen.

In retrospect, the Fusaria causing cortical rots are similar in that they are mainly confined to the cortex of their host. Fungal growth in this tissue may be intercellular, intracellular, or a combination of both and is frequently facilitated by enzymatic degradation of the middle lamella and cell walls. Once the cortex is colonized, the fungus sporulates externally by means of sporodochia formed through the stomata and forms chlamydospores in the necrotic cells of the cortex. The pathogens may or may not be confined to the cortex by the endodermis and when they are not, vascular penetration occurs during the late stages of the disease. The host responds to infection in many cases by the production of hypertrophied, hyperplastic cells resembling a periderm. This response occasionally limits the spread of the pathogen. Lack of any host response leads to the complete maceration and collapse of the cortical cells and eventual death of the plant.

Summary

From the foregoing discussion, it can be seen that the vascular wilt Fusaria and the cortical rot Fusaria are quite dissimilar in the mechanisms by which they attack their hosts. The vascular wilt Fusaria, as the name implies, are confined to the host vascular system, although they may attack other tissues late in pathogenesis (Fig. 35-7). In contrast, the cortical rot Fusaria are generally confined to the cortical areas of their host, only penetrating the vascular system late in the disease process if at all (Fig. 35-8). Sporulation also shows a striking contrast, with conidia being restricted to the host xylem vessels in the wilt Fusaria and generally borne externally on sporodochia in the cortical rot Fusaria. The vascular wilt Fusaria usually penetrate directly through the meristematic regions of the host root, whereas the cortical rot Fusaria penetrate their hosts directly through stomates and wounds. Both groups of fungi appear to utilize enzymes, although only the cortical rot Fusaria appear to employ both pectic and cellulolytic enzymes. In addition to the dissimilarities in sporulation and colonization, the two groups of Fusaria invoke different anatomical responses in their hosts. Gums, gels, and occasionally tyloses generally follow invasion by the vascular wilt Fusaria, whereas hypertrophied and hyperplastic tissue formation is more usually the response to infection by the cortical rot Fusaria. Thus, it becomes evident that the members of the genus *Fusarium* employ at least two distinctly different methods in parasitizing the host plant.

Literature Cited

1. Adams, J. F. 1921. Observations on wheat scab in Pennsylvania and its pathological histology. Phytopathology 11:115–124.
2. Arnett, J. D., and W. Witcher. 1974. Histochemical studies of yellow poplar infected with *Fusarium solani*. Phytopathology 64:414–418.
3. Bateman, D. F. 1966. Hydrolytic and trans-eliminative degradation of pectic substances by extracellular enzymes of *Fusarium solani* f. *phaseoli*. Phytopathology 56:238–244.
4. Beckman, C. H. 1966. Cell irritability and localization of vascular infections in plants. Phytopathology 56:821–824.
5. Beckman, C. H. 1969. The mechanics of gel formation by swelling of simulated plant cell wall membranes and perforation plates of banana root vessels. Phytopathology 59:837–843.
6. Beckman, C. H. 1969. Plasticizing of walls and gel induction in banana root vessel infected with *Fusarium oxysporum*. Phytopathology 59:1477–1483.
7. Beckman, C. H., S. Halmos, and M. E. Mace. 1962. The interaction of host, pathogen, and soil temperature in relation to susceptibility to Fusarium wilt of bananas. Phytopathology 52:134–140.
8. Beckman, C. H., M. E. Mace, S. Halmos, and M. W. McGahan. 1961. Physical barriers associated with resistance in Fusarium wilt of bananas. Phytopathology 51:507–515.
9. Beckman, C. H., and G. E. Zaroogian. 1967. Origin and composition of vascular gel in banana roots. Phytopathology 57:11–13.
10. Bennett, F. T. 1928. On two species of *Fusarium*, *F. culmorum* (W. G. Sm.) Sacc. and *F. avenaceum* (Fries.) Sacc., as parasites of cereals. Ann. Appl. Biol. 15:213–244.
11. Brandes, E. W. 1919. Banana wilt. Phytopathology 9:339–390.
12. Bugbee, W. M. 1970. Vascular response of cotton to infection by *Fusarium oxysporum* f. sp. *vasinfectum*. Phytopathology 60:121–123.
13. Bywater, J. 1959. Infection of peas by *Fusarium solani* var. *martii* forma 2 and the spread of the pathogen. Trans. Brit. Mycol. Soc. 42:201–212.
14. Chambers, H. L., and M. E. Corden. 1963. Semeiography of Fusarium wilt of tomato. Phytopathology 53:1006–1010.
15. Chi, C. C., W. R. Childers, and E. W. Hanson. 1964. Penetration and subsequent development of three *Fusarium* species in alfalfa and red clover. Phytopathology 54:434–437.
16. Christou, R., and W. C. Snyder. 1962. Penetration and host-parasite relationships of *Fusarium solani* f. *phaseoli* in the bean plant. Phytopathology 52:219–225.
17. Cunningham, H. S. 1953. A histological study of the influence of sprout inhibitors on Fusarium infection of potato tubers. Phytopathology 43:95–98.
18. Dahl, A. S. 1934. Snowmold of turf grasses as caused by *Fusarium nivale*. Phytopathology 24:197–214.
19. Deese, D. C., and M. A. Stahmann. 1962. Pectic enzymes and cellulose formation by *Fusarium oxysporum* f. *cubense* on stem tissues from resistant and susceptible banana plants. Phytopathology 52:247–255.
20. Deese, D. C., and M. A. Stahmann. 1962. Pectic enzymes in *Fusarium*-infected susceptible and resistant tomato plants. Phytopathology 52:255–260.
21. Elgersma, D. M., W. E. MacHardy, and C. H. Beckman. 1972. Growth and distribution of *Fusarium oxysporum* f. sp. *lycopersici* in near-isogenic lines of tomato resistant or susceptible to wilt. Phytopathology 62:1232–1237.
22. Gothoskar, S., R. P. Scheffer, J. C. Walker, and M. A. Stahmann. 1953. The role of pectic enzymes in Fusarium wilt of tomato. Phytopathology 43:535–536.
23. Gothoskar, S. S., R. P. Scheffer, J. C. Walker, and M. A. Stahmann. 1955. The role of enzymes in the development of Fusarium wilt of tomato. Phytopathology 45:381–387.
24. Graham, K. M. 1955. Seedling blight, a fusarial disease of asparagus. Can. J. Bot. 33:374–400.
25. Hadley, B. A. 1973. A histopathological study of *Dianthus caryophyllus* stem stubs infected with *Fusarium roseum* 'Graminearum.' M.S. Thesis, Pa. State Univ., University Park. 42 p.
26. Hancock, J. G. 1968. Degradation of pectic substances during pathogenesis by *Fusarium solani* f. sp. *cucurbitae*. Phytopathology 58:62–69.
27. Kamel, M., M. N. Shatla, and M. Z. El Shanawani. 1973. Histopathological studies on hypocotyl of lentils infected by *Fusarium solani*. Z. Pflanzenkr Pflanzenshutz 80:547–550.
28. Lawrence, E. B. 1975. Histopathology of two inbred lines of sweet corn (*Zea mays*) naturally infected with *Fusarium moniliforme* and *Fusarium oxysporum*. M. S. Thesis, Pa. State Univ., University Park. 94 p.
29. McClure, T. T. 1949. Mode of infection of the sweet-potato wilt *Fusarium*. Phytopathology 39:878–886.
30. McClure, T. T. 1950. Anatomical aspects of the Fusarium wilt of sweet potatoes. Phytopathology 40:769–775.
31. Nishimura, S. 1971. Observations on wilting mechanism of *Fusarium*-infected watermelon plants. Rev. Plant Protection Res. 4:71–80.
32. Pennypacker, B. W., and P. E. Nelson. 1972. Histopathology of carnation infected with *Fusarium oxysporum* f. sp. *dianthi*. Phytopathology 62:1318–1326.
33. Peterson, J. L, and G. S. Pound. 1960. Studies on resistance in radish to *Fusarium oxysporum* f. *conglutinans*. Phytopathology 50:807–816.
34. Phillips, D. J. 1962. Histochemmical and morpho-

logical studies of carnation stem rot. Phytopathology 52:323–328.
35. Phipps, P. M., and R. J. Stipes. 1976. Histopathology of mimosa infected with *Fusarium oxysporum* f. sp. *perniciosum*. Phytopathology 66:839–843.
36. Pierre, R., and R. E. Wilkinson. 1970. Histopathological relationship of *Fusarium* and *Thielaviopsis* with beans. Phytopathology 60:821–824.
37. Pierson, C. F., S. S. Gothoskar, J. C. Walker, and M. A. Stahmann. 1955. Histological studies on the role of pectic enzymes in the development of Fusarium wilt symptoms in tomato. Phytopathology 45:524–527.
38. Pugh, G. W., H. Johann, and J. G. Dickson. 1933. Factors affecting infection of wheat heads by *Gibberella saubinetii*. J. Agric. Res. 46:771–797.
39. Reid, J. 1958. Studies on the fusaria which cause wilt in melons. 1. The occurrence and distribution of races of the muskmelon and watermelon Fusaria and a histological study of the colonization of muskmelon plants susceptible or resistant to Fusarium wilt. Can. J. Bot. 36:393–410.
40. Siddiqui, W. M., and P. M. Halisky. 1968. Histopathological studies of red clover roots infected by *Fusarium roseum*. Phytopathology 58:874–875.
41. Smith, R., and J. C. Walker. 1930. A cytological study of cabbage plants in strains susceptible or resistant to yellows. J. Agric. Res. 41:17–35.
42. Sparnicht, R. H., and R. W. Roncadori. 1972. Fusarium boll rot of cotton: pathogenicity and histopathology. Phytopathology 62:1381–1386.
43. Tisdale, W. H. 1917. Flaxwilt: a study of the nature and inheritance of wilt resistance. J. Agric Res. 11:573–605.
44. Trujillo, E. E. 1963. Pathological-anatomical studies of Gros Michel banana affected by Fusarium wilt. Phytopathology 53:162–166.
45. Ullstrup, A. J. 1937. Histological studies on wilt of china aster. Phytopathology 27:737–748.
46. Waggoner, P. E., and A. E. Dimond. 1955. Production and role of extracellular pectic enzymes of *Fusarium oxysporum* f. *lycopersici*. Phytopathology 45:79–87.
47. Wardlaw, C. W. 1930. The biology of banana wilt (Panama disease). I. Root inoculation experiments. Ann. Bot. 44:741–766.
48. Wardlaw, C. W. 1931. The biology of banana wilt (Panama disease). III. An examination of sucker infection through root-bases. Ann. Bot. 45:381–400.

V The Fungus *Fusarium:* Taxonomy

Prologue

T. A. Toussoun

This part is the last in this book. Naturally, it is not the least. The late Wat Dimock once said to me: "Settle on names for these fungi so the rest of us can get on with the work!" To some, taxonomy is merely the beginning. To others it is the end. Perhaps that is the reason the subject is where it is in this book.

Under "Classification" in Ainsworth and Bisby's *Dictionary of the Fungi* is the following: "The rank of species is basic (Art. 2) but as there is no objective definition of species for most fungi; what constitutes a species for one taxonomic mycologist may be a genus or a variety for another." And so it is for *Fusarium*. There are the "lumpers" who prefer broad limits in speciation and there are "splitters" who prefer more stringent limits. There are biologic species, and taxonomic species. Basically, I suppose, the use of specific characters in species delimitation depends on the gap we find in the variation between specimens, so it is these gaps that give us the possibility to use characters for delimitation and consequently form the basis for speciation. How different must this specimen be for it to merit a separate identity and of what taxonomic rank should it be? Is the difference in morphology or in behavior? Which is the more important of the two? How practical should one be? These are just some of the questions, and a full list would be endless. But enough of all this; I propose, instead, to be quite subjective at this stage and to offer the reader my biased key to those taxonomists who are the authors of this chapter, people whom I know as friends.

The first (actually he wrote the Introduction to this book) is Bill Snyder. He and his colleague, H. N. Hansen, created quite a stir 40 years ago with their revolutionary concepts in the taxonomy of *Fusarium*. The echoes are still rumbling around us and the changes they wrought are now part of the heritage of all who are interested in the genus. Read what he has to say under "Communication" and "Names." It is all there: the feelings of a plant pathologist, a pragmatist and a revolutionary in fungal taxonomy who states, "If the Rules are not modified in keeping with new facts and new philosophies, then in the name of progress we must move ahead without them." Snyder and Hansen have been accused of over-simplification, but one can use their system and still recognize as many differences as required, on the subspecific level.

There is no doubt that the giant in *Fusarium* taxonomy has been H. W. Wollenweber. His laboratory in Berlin-Dahlem is now under the direction of Wolfgang Gerlach, and his spirit lives on. I can assure you it is a most remarkable place, for it contains, neatly catalogued, every single drawing and every single note made by Wollenweber during his more than 50 years of detailed, painstaking work. As Wolf Gerlach told me, "Every one of Wollenweber's *Fusarium* species exists. To find them today all you need to do is go to the locations given by Wollenweber." As proof, he showed me two species isolated from the same water tap, and rediscovered 40 years after Wollenweber's initial observation. By no means do I wish to imply a status quo, for in these self-same files are new drawings and notes of old and new species studied by Wolf; and so for adherents of Wollenweber's classification, the most up-to-date treatment is Gerlach's contribution to this book. It represents 25 years of his own experience in addition to that of Wollenweber. These are meticulous, time-consuming studies of every isolate, on more than a dozen substrates under varied conditions. One may not necessarily agree with Wolf's speciation but every single specimen described does indeed exist.

One could level a few criticisms at the classifications I have just mentioned. Wollenweber's highly detailed (excessive according to some) treatment is in difficult German. Snyder & Hansen's proposals (too vague according to some) are scattered in several papers, in English. Messiaen & Cassini decided on a remedy for the French-speaking world and the article in this book is a condensed translation of their original paper. Both are plant pathologists and so they shared Snyder & Hansen's viewpoint. Indeed, Messiaen had earlier translated Snyder & Hansen's system into French, before attempting this work which contains their own philosophy. Cassini is a wheat and corn pathologist; Messiaen is interested in almost everything. If you question him closely about two species or varieties he might tell you, as he did me, that they differ in the same manner as Louis XIV and Louis XV chairs differ—style. Messiaen & Cassini's contribution is the best explanation of the Snyder & Hansen system there is. It is perhaps even an improvement in that it contains a key to the species. Their drawings (not reproduced in this book) are second only to Wollenweber's and Gerlach's. To those who work with *Fusarium* this is most important. Excellent photographs and/or drawings are an indication that the researcher knows whereof he speaks because he has worked long enough and hard enough to grow his fungi properly, the key prerequisite to successful taxonomy.

Colin Booth is a mycologist who understands the needs of those, such as plant pathologists, who must work with many organisms. Here is what he says in his book *The Genus Fusarium,* ". . . it should be appreciated that present-day nomenclature has largely developed as a working tool of plant pathologists and others as a usable system and is widely understood. In this publication, therefore, nomenclature follows the traditional system for *Fusarium* and it would be preferable to conserve all the names now in use. . . . " It is rare to find someone like Colin who sees both sides of the coin. His classification follows that of W. L. Gordon, which was in essence Wollenweber's system modified to accept Snyder & Hansen's views of *F. oxysporum* and *F. solani.* Colin has always been interested in the perfect states of fungi and that is why we asked him to contribute an article on that subject for this book. I think we did him somewhat of a disservice, for what may be his most significant contribution to *Fusarium* taxonomy is his emphasis on the morphology of the conidiogenous cell as being one of the dominant characters in *Fusarium* classification. In this book that subject is treated by Roger Goos, himself no outsider; having been with the United Fruit Co., Roger has a very good acquaintance with what is entailed in a study of Fusarium wilt of bananas.

I would like to conclude with a note on P. E. Nelson and T. A. Toussoun. They are not taxonomic mycologists and they do not have a system. They prefer to find a bridge that links the major classification systems now in use for *Fusarium.* Speaking for both, I have this plea: concentrate on the populations that make a species rather than on the individual. One hundred isolates of any given type should not be considered an inordinate number to study.

36 The Present Concept of *Fusarium* Classification

W. Gerlach

The foundation for "modern" taxonomic treatments of *Fusarium* was laid by Wollenweber and co-workers (2,53,73,95–106). Wollenweber devoted about 40 years of his professional life to the taxonomy of this genus, resulting in his "*Fusarium*-Monographie" (100,102) and the book "Die Fusarien" published with O. A. Reinking (104), which provides an excellent basis for further studies. Most of the numerous attempts to improve Wollenweber's system have recently been summarized and critically discussed by Booth (14) and Toussoun and Nelson (93). It is not necessary, therefore, to do it again in detail. It is already known from our earlier publications (25–35) and those of our co-workers (45,67,75–78) that we do not agree with the concept of Snyder and Hansen (81–83), accepted and later modified by others (61–64,84,85,92), because "on the whole it is neither justified from a taxonomic standpoint nor suitable for practical use and, in many cases, it does not follow the rules of nomenclature" (26). Our concept differs only in details when compared with the taxonomic treatments of Fusaria by Bilai (7,8), Booth (10–16), Bugnicourt (17), Doidge (21), Domsch, Gams and Anderson (22), Gordon (36–44), Jamalainen (49–52), Joffe and co-workers (54–57,69), Raillo (70,71), and Subramanian (87–91). Some of the species or varieties not mentioned, or considered as being synonymous with other Fusaria, by Booth (14), Gordon (37–43), Joffe (54) and others are considered as distinct taxa. The concept is based on Wollenweber's material perfectly preserved in our institute, on about 25 years of my own experience, and the studies of co-workers, involving a collection of some 10,000 herbarium specimens and isolates studied on different substrates and under varied conditions, and documented by drawings, photographs, and color slides. On a visit to our Institute, T. A. Toussoun called the collection and related materials a "*Fusarium* gold mine." In the following condensed review, the recognized sections, species, and varieties are listed according to Wollenweber and Reinking (104) for direct comparison, as well as a few newly described species and varieties. In addition to our comments ("distinct," "doubtful," etc.), available representative cultures (of our collection, Centraalbureau voor Schimmelcultures [CBS], Commonwealth Mycological Institute [IMI], or American Type Culture Collection [ATCC]) are listed especially in cases of those Fusaria which are "forgotten," seldom found, not generally known, or not mentioned or not accepted as distinct by Booth (14), and we also refer to descriptions given by other authors. This review should be understood as a complete catalogue of all Fusaria under discussion to indicate what in our opinion seems to be clear and what does not, as a base for further careful studies. In this text it is not possible to give full arguments why a species or variety is treated as distinct or not. This will be done in a "pictorial atlas" in preparation (32), a volume of about 400 pages with descriptions and full page illustrations (drawings and photographs) of nearly all Fusaria mentioned here, to be published at a later date.

The genus *Fusarium* Link 1809 ex Fr. 1821

Syn.: *Fusisporum* Link 1809 ex Fr. 1821
Selenosporium Corda 1837
Microcera Desm. 1848
Pionnotes Fr. 1849
Sporotrichella Karst. 1887
Lachnidium Giard 1891
? *Discocolla* Prill. et Delacr. 1894
? *Rachisia* Lindner 1913
Discofusarium Petch 1921
Pseudomicrocera Petch 1921
? *Fusidomus* Grove 1929

Type species: *Fusarium roseum* Link ex Gray 1821
(*F. sambucinum* Fuckel 1869)

The three specimens left by Link and illustrated by Wollenweber (96) in his "Fusaria autographice delineata" (no. 138,311,354) are three different fungi: one of the two original specimens of *Fusarium roseum* corresponds to *F. sambucinum* and the other to *F. graminearum;* the third specimen (*Fusisporium roseum* on a gramineous plant) may be *F. graminum* (100,102,104). In 1821 Gray designated the holotype specimen (cf. 96, no. 311) citing only stems of a malvaceous plant (?*Althaea rosea*) as a substrate in the validating text (22).

Section *Eupionnotes* Wollenw. 1913 (95) Stat. ascig.:*Nectria* Fr., *Plectosphaerella* Klebahn[1]

Fusarium aquaeductuum (Radlk. et Rabenh.) Lagerh. 1891 var. *aquaeductuum*
Stat. ascig.: *Nectria purtonii* (Grev.) Berk. 1860.
Distinct. Descriptions: Ref. 2,10,14,22,32,54,96,100,104.

Fusarium aquaeductuum (Radlk. et Rabenh.) Lagerh. var. *medium* Wollenw. 1931 (100).
Stat. ascig.: *Nectria episphaeria* (Tode ex Fr.) Fr. 1849.
Distinct. Descriptions: Ref. 10,14,32,33,96,100,104.

Fusarium cavispermum Corda 1837.
Distinct (CBS 172.31). Booth (14)—not mentioned.
Descriptions: Ref. 32,96(no.77,849,850),100,104.

Fusarium melanochlorum (Casp.) Sacc. 1886.
Stat. Ascig.: *Nectria flavo-viridis* (Fuckel) Wollenw. 1926 (99).
Distinct (62248 = CBS 202.65). Descriptions: Ref. 10,14,32,54,96,102,104.

Fusarium merismoides Corda 1838 var. *merismoides.*
Distinct. Descriptions: Ref. 14,17,22,30,32,37,51,54,91,96,100,104.

Fusarium merismoides Corda var. *acetilereum* Tubaki, C. Booth et Harada 1976 (94).
Not yet seen (IMI 181488, type culture). Description: Ref. 94.

Fusarium merismoides Corda var. *chlamydosporale* Wollenw. 1931 (100).
Distinct (CBS 179.31, type culture; 62256 = CBS 798.70). Booth (14) = *F. merismoides.* Descriptions: Ref. 32,96,(no.856), 100,104.

Fusarium merismoides Corda var. *crassum* Wollenw. 1931 (100).
Distinct (CBS 180.31, type culture; 62257, 62258). Booth (14) = *F. merismoides.*
Descriptions: Ref. 27,32,96,(no.857), 100,104.

Fusarium merismoides Corda var. *violaceum* Gerlach 1977 (30).
Distinct (62461 = CBS 634.76, type culture) or ? = *F. merismoides* var. *acetilereum.*
Descriptions: Ref. 30,32,33(p.730–731).

[1]*Plectosphaerella* Klebahn (non Kirschstein) = *Nectria* Fr. (66).

Fusarium dimerum Penzig 1882 var. *dimerum.*
Distinct. Descriptions: Ref. 14,22,32,37,54,91,96,100,104.
Fusarium dimerum Penzig var. *nectrioides* Wollenw. 1931 (100).
Doubtful (CBS 176.31, type culture). Booth (14) = *F. dimerum.*
Descriptions: Ref. 17,96(no.855),100,104.
Fusarium dimerum Penzig var. *pusillum*(Wollenw.) Wollenw. 1931(100).
Syn.: *F. pusillum* Wollenw. 1924 (96, no.550).
Doubtful (CBS 254.50). Booth (14) = *F. dimerum.*
Descriptions: Ref. 73,96(no.550,851,852),100,104.
Fusarium dimerum Penzig var. *violaceum* Wollenw. 1931 (100).
Distinct (63199 = CBS 632.76) Booth (14) = *F. dimerum.*
Descriptions: Ref. 96(no.854),100,104.
Fusarium flavum (Fr.) Wollenw. 1931(100).
Doubtful. Booth (14) = *F. dimerum.*
Descriptions: Ref. 32,96(no.546,547),100,104.
Fusarium tabacinum (Beyma) W. Gams apud Gams et Gerlagh 1968 (24).
Stat. ascig.: *Plectosphaerella cucumerina* (Lindf.) W. Gams apud Domsch et Gams 1972. [Syn.: *Micronectriella cucumeris* (Klebahn) C. Booth 1971 (14)[2]].
Distinct. Descriptions: Ref. 14,22,24,32,54.
Fusarium epistroma (Höhnel) C. Booth 1971 (14).
Stat. ascig.: *Nectria magnusiana* Rehm ex Sacc. 1878.
Distinct (IMI 85601). Descriptions: Ref. 10,14,32.

Section *Macroconia* Wollenw. 1926 (99) Stat. ascig.: *Nectria* Fr.

Fusarium buxicola Sacc. 1883.
Stat. ascig.: *Nectria desmazierii* Beccari et de Not. 1863.
Distinct. Descriptions: Ref. 10,14,32,54,96,102,104.
Fusarium expansum Schlecht. 1824.
Stat. ascig.: *Nectria stilbosporae* Tul. 1865.
Distinct. Booth (14)—not seen. Descriptions: Ref. 32,96(no.616,675,858),100,102,104.
Fusarium sphaeriae Fuckel 1869.
Stat. ascig.: *Nectria leptosphaeriae* Niessl 1886.
Distinct (CBS 717.74). Descriptions: Ref. 10,14,32,54,96,102,104.
Fusarium gigas Speg. 1884.
Distinct. Descriptions: Ref. 14,32,54,96(no.352),102,104.
Fusarium coccophilum (Desm.) Wollenw. et Reinking 1935 (104).
Stat. ascig.: *Nectria flammea* (Tul.) Dingley 1951 [Syn.: *Nectria coccophila* (Tul.) Wollenw. et Reinking 1935].
Distinct (62175 = CBS 793.70). Descriptions: Ref. 14,21,32,33,54,91,96,102,104.

Section *Spicarioides* Wollenw. et al. 1925 (106) Stat. ascig.: *Calonectria* de Not.

Fusarium decemcellulare Brick 1908.
Stat. ascig.: *Calonectria rigidiuscula* (Berk. et Br.) Sacc. 1878.
Distinct. Descriptions: Ref. 1 (no.21), 11,14,17,21,32,54,91,96,100,104.

Section *Submicrocera* Wollenw. 1926 (99) Stat. ascig.: *Calonectria* de Not.

Fusarium ciliatum Link 1825.
Stat. ascig.: *Calonectria decora* (Wallr.) Sacc. 1878.

[2]The type species of *Micronectriella* Höhnel is a bitunicate ascomycete and belongs to *Sphaerulina* Sacc. (6,65).

Distinct (CBS 132.35; 62172 = CBS 191.65 = IMI 112,499 = ATCC 16068).
Booth (14) ? = *F. aquaeductuum*. Descriptions: Ref. 21,32,96(no.54,437,438,872,1128), 102,104.
Fusarium cerasi Roll. et Terry 1892.
Excluded (102,p.123), = *Micropera drupacearum* Lév. (stat. ascig.: *Dermatea cerasi* de Not.). Booth (14) ? = *Septoria*.

Section *Pseudomicrocera* (Petch) Wollenw. 1926 (99) Stat. ascig.: *Calonectria* de Not.

Fusarium coccidicola P. Henn. 1903.
Syn.: *F. juruanum* P. Henn. 1904.
Stat. ascig.: *Calonectria diploa* (Berk. et Curt.) Wollenw. 1926
Distinct (62173). Descriptions: Ref. 14,32,33,54,91,96,102,104.
Fusarium orthoconium Wollenw. 1931 (100).
Doubtful. Booth (14)—not seen. Descriptions: Ref. 32,96(no.637),102,104.

Section *Arachnites,* Wollenw. 1917 (98) Stat. ascig.: *Nectria* Fr., *Monographella* Petrak

Fusarium kühnii (Fuckel) Sacc. 1886.
Doubtful. Booth (14) = *F. dimerum*. Descriptions: Ref. 32,96(no.68–70),100,104.
Fusarium nivale (Fr.) Ces. 1860 var. *nivale.*
Stat. ascig.: *Monographella nivalis* (Schaffnit) E. Müller 1977 (65) var. *nivalis.* [Syn.: *Calonectria nivalis* Schaffnit 1913; *Griphosphaeria nivalis* (Schaffnit) E. Müller et v. Arx 1955; *Micronectriella nivalis* (Schaffnit) C. Booth 1971 (14); *Calonectria graminicola* Berk. et Br.) Wollenw. 1913 (non *Nectria graminicola* Berk. et Br. 1859)].
Distinct. Descriptions:Ref. 1(no.309), 14,22,32,37,49,54,65,91,96,100,104.
Fusarium nivale (Fr.) Ces. var. *majus* Wollenw. 1931 (100).
Stat. ascig.: *Monographella nivalis* (Schaffnit) E. Müller var. *neglecta* (Krampe) W. Gams et E. Müller 1980. [Syn.: *Calonectria gramincola* (Berk. et Br.) Wollenw. var. *neglecta* Krampe 1926 (58)].
Distinct. Booth (14) = *F. nivale.* Descriptions: Ref. 32,58,96(no.808,882),100,104.
Fusarium nivale (Fr.) Ces. var. *oryzae* Zambettakis 1950 (107).
Doubtful. Booth (14)—unknown. Description: Ref. 107.
Fusarium larvarum Fuckel 1869 var. *larvarum.*
Stat. ascig.: *Nectria aurantiicola* Berk. et Br. 1873.
Distinct. Descriptions: Ref. 14,30,32,33,54,96,100,104.
Fusarium larvarum Fuckel var. *rubrum* Gerlach 1977 (30).
Distinct (62460 = CBS 638.76, type culture). Descriptions: Ref. 30,32,33(p.738–739).
Fusarium stoveri C. Booth 1971 (14).
Stat. ascig.: *Micronectriella stoveri* C. Booth 1964 (12).
Not seen, no culture received from Booth. Descriptions: Ref. 12,14.

Section *Sporotrichiella* Wollenw. apud Lewis 1913 (60) Monographic treatment: Seemüller 1968 (78)

Fusarium poae (Peck) Wollenw. apud Lewis 1913 (60).
Distinct. Descriptions: Ref. 1(no.308),14,22,32,37,50,54,60,78,91,96,102,104.
Fusarium chlamydosporum Wollenw. et Reinking 1925 (103) var. *chlamydosporum.*

Syn.: *F. fusarioides* (Frag. et Cif.) C. Booth 1971 (14).
Distinct. Descriptions: Ref. 14,21,22,30,32,54,73,78,91,96(no.883),102,103,104.
Fusarium chlamydosporum Wollenw. et Reinking var. *fuscum* Gerlach 1977 (30).
Distinct (62053 = CBS 635.76, type culture). Descriptions: Ref. 30,32.
Fusarium tricinctum (Corda) Sacc. 1886.
Syn.: *F. citriforme* Jamalainen 1943 (50).
Distinct. Descriptions: Ref. 14,22,32,50,54,78,91,96,102,104.
Fusarium sporotrichioides Sherb. 1915 (79) var. *sporotrichioides.*
Distinct. Descriptions: Ref. 14,22,32,37,50,54,78,79,91,96,102,104.
Fusarium sporotrichioides Sherb. var. *minus* Wollenw. 1931 (100).
Distinct (62426 = CBS 448.67). Booth (14) = *F. sporotrichioides.*
Descriptions: Ref. 32,71,78,96(no. 886),100,102,104.

Section *Roseum* Wollenw. 1913 (95) Stat. ascig.: *Gibberella* Sacc.

Fusarium graminum Corda 1837.
Distinct (62224, 62226, 62228). Booth (14) = *F. heterosporum,*
Raillo (71) = *F. avenaceum.*
Descriptions: Ref. 32,33,96(no.129,137,138,891),102,104.
Fusarium avenaceum (Fr.) Sacc. 1886 var. *avenaceum.*
Stat. ascig.: *Gibberella avenacea* R.J. Cook 1967 (20).
Distinct. Descriptions: Ref. 1(no.25),14,20,21,22,32,37,50,54,75,76,91,96,102,104.
Fusarium avenaceum (Fr.) Sacc. var. *pallens* Wollenw. 1924 (96).
Doubtful, not seen. Booth (14) = *F. avenaceum.*
Descriptions: Ref. 96(no.575),102,104.
Fusarium avenaceum (Fr.) Sacc. var. *volutum* Wollenw. et Reinking 1935 (104).
Perhaps distinct, not seen. Booth (14) = *F. avenaceum.*
Descriptions: Ref. 32,96(no.171,172,893),102,104.
Fusarium detonianum Sacc. 1886.
Distinct (CBS 174.31; 62194). Booth (14) = *F. avenaceum.*
Descriptions: Ref. 32,51,96(no.195,900),102,104.
Fusarium arthrosporioides Sherb. 1915 (79).
Distinct. Descriptions: 14,32,51,54,79,102,104.

Section *Arthrosporiella* Sherb. 1915 (79)

Fusarium semitectum Berk. et Rav. apud Berkeley 1875 var. *semitectum.*
Distinct. Descriptions: Ref. 14,22,32,33,37,54,73,91 (*F. incarnatum* [Roberge] Sacc.), 96,100,104.
Fusarium semitectum Berk. et Rav. apud Berkeley var. *majus* Wollenw. 1931 (100).
Distinct. Descriptions: Ref. 14,17,21,32,54,96,100,104.
Fusarium camptoceras Wollenw. et Reinking 1925 (103).
Distinct. Descriptions: Ref. 14,17,32,54,73,91,100,103,104.
Fusarium concolor Reinking 1934 (72).
Distinct (CBS 183.34, type culture). Descriptions: Ref. 14,32,37,54,72,96(no.1136), 102,104.
Fusarium diversisporum Sherb. 1915 (79).
Distinct (CBS 144.44; 62199 = CBS 795.70, as *F. anguioides).*
Booth (14) = *F. semitectum* var. *majus* (fide Raillo).
Descriptions: Ref. 32,33,73,79,96(no.118,553,911,912),100,104.
Fusarium anguioides Sherb. 1915 (79).
Doubtful (CBS 172.32), ? = *F. diversisporum* or *F. arthrosporioides.*

Booth (14) = *F. avenaceum.*
Descriptions: Ref. 17,32,51,73,79,96(no.902,903),102,104.

Section *Gibbosum* Wollenw. 1913 (95) Stat. ascig.: *Gibberella* Sacc.

Provisionally accepted in the sense of Gordon 1952 (37), but see Subramanian 1971 (91).
Fusarium equiseti (Corda) Sacc. 1886 sensu Gordon 1952 (37) var. *equiseti.*
Stat. ascig.: *Gibberella intricans* Wollenw. 1931 (100).
Distinct. Descriptions: Ref. 14,17,21,22,32,33,37,54,91 *(F. aurantiacum* Corda ex Fr.),96,100,104.
Fusarium acuminatum Ell. et Kellerm. 1895.
Syn.: *F. scirpi* Lamb. et Fautr. var. *acuminatum* (Ell. et Kellerm.) Wollenw. 1931 (100).
Stat. ascig.: *Gibberella acuminata* Wollenw. apud Wollenw. et Reinking 1935 (104). [Syn.: *Gibberella acuminata* C. Booth 1971 (14)].
Distinct. Descriptions: Ref. 14,21,22,32,33,37,51,54,91(*F. hippocastani* [Corda] Sacc.), 96,100,102,104.
Fusarium compactum (Wollenw.) Gordon 1952 (37).
Syn.: *F. scirpi* Lamb. et Fautr. var. *compactum* Wollenw. 1931 (100).
Distinct. Booth (14) = *F. equiseti.*
Descriptions: Ref. 21,32,54,96(no.923–925),100,104.
Fusarium longipes Wollenw. et Reinking 1925 (103).
Syn.: *F. scirpi* Lamb. et Fautr. var. *longipes* (Wollenw. et Reinking) Wollenw. 1931 (100).
Distinct (62061,62062; IMI 112211a). Booth (14) = *F. equiseti.*
Descriptions: Ref. 17,32,54,73,91,96(no.937),100,103,104.
Fusarium brachygibbosum Padwick 1945 (68).
Doubtful, not seen. Accepted by Subramanian (91); Booth (14) = *F. semitectum.*
Descriptions: Ref. 68,91.

Section *Fusarium* [Section Discolor Wollenw. 1913 (95)] Stat. ascig.: *Gibberella* Sacc.

Fusarium heterosporum Nees ex Fr. 1832 var. *heterosporum.*
Stat. ascig.: *Gibberella gordonii;* C. Booth 1971 (14).
Distinct. Descriptions: Ref. 14,32,54,96(no.292–304,602–604),100,104.
Fusarium heterosporum Nees ex Fr. var. *congoense* (Wollenw.) Wollenw. 1931 (100).
Syn.: *F. congoense* Wollenw. 1917 (98).
Doubtful, not seen; ? = *F. graminum* or *F. heterosporum.*
Booth (14) = *F. heterosporum* (fide Raillo).
Descriptions: Ref. 21,32,96(no.306,307,612,1140,1141),100,104.
Fusarium reticulatum Mont. 1843 var. *reticulatum.*
Stat. ascig.: ? *Gibberella cyanea* (Sollm.) Wollenw. 1931 (100).
Distinct (CBS 190.31; 63657 = CBS 473.76). Booth (14) = *F. heterosporum.*
Descriptions: Ref. 32,33,60,96(no.39,119–121,938,939,1138,1139),100,102,104.
Fusarium reticulatum Mont. var. *negundinis* (Sherb.) Wollenw. 1931 (100).
Syn.: *F. negundi* Sherb. apud. Hubert 1923 (47).
Probably distinct (CBS 183.31). Booth (14) = *F. heterosporum.*
Descriptions: Ref. 47,96(no.940),100,104.
Fusarium sambucinum Fuckel 1869 var. *sambucinum.*
Stat. ascig.: *Gibberella pulicaris* (Fr.) Sacc. 1877 var. *pulicaris.*
Distinct. Descriptions: Ref. 1(no.385),14,21,22,32,37,49,54,91(*F. maydis* Kalchbr.), 96,100,104.

Fusarium sambucinum Fuckel var. *coeruleum* Wollenw. 1917 (98).
Stat. ascig.: *Gibberella pulicaris* (Fr.) Sacc. var. *minor* Wollenw. 1931 (100).
Distinct. Descriptions: Ref. 14,32,37,54,96,100,104.
Fusarium suplhureum Schlecht. 1824
Syn.: *F. sambucinum* Fuckel f. 6 Wollenw. 1931 (100).
Stat. ascig.: *Gibberella cyanogena* (Desm.) Sacc. 1883.
Distinct. Descriptions: Ref. 2,14,21,32,33,37,91,95,96,100,104.
Fusarium trichothecioides Wollenw. apud. Jamieson et Wollenw. 1912 (53).
Doubtful, ? = *F. sulphureum.* Booth (14) — distinct.
Descriptions: Ref. 14,32,53,54,79,96(no.305),100,104.
Fusarium flocciferum Corda 1828.
Stat. ascig.: ? *Gibberella heterochroma* Wollenw. 1917 (98).
Distinct (CBS 821.68; 62217 = CBS 792.70).
Descriptions: Ref. 14,22,32,33,96,100,104.
Fusarium culmorum (W.G. Smith) Sacc. 1895.
Syn.: *F. culmorum* (W.G.Smith) Sacc. var. *cerealis* (Cooke) Wollenw. 1931 (100).
Distinct. Descriptions: Ref. 1(no.26),14,21,22,32,33,37,49,54,91*(F. cerealis* [Cooke] Sacc.),96,100,104.
Fusarium buharicum Jaczewski 1929 (48).
Distinct (CBS 178.35; 62165 = CBS 796.70 = IMI 141195).
Descriptions: Ref. 14,26,32,33,35,48,59,71,102.
Fusarium sublunatum Reinking 1934 (72) var. *sublunatum.*
Distinct (CBS 189.34, type culture), Booth (14) ? = *F. tumidum.*
Descriptions: Ref. 28,32,72,96(no.1150),102,104.
Fusarium sublunatum Reinking var *elongatum* (Reinking) Reinking apud Wollenw. et Reinking 1935 (104).
Syn.: *F. elongatum* Reinking 1934 (72).
Distinct (CBS 190.34, type culture). Booth (14) = *F. tumidum.*
Descriptions: Ref. 28,32,72,96(no.1151),102,104.
Fusarium graminearum Schwabe 1838.
Stat. ascig.: *Gibberella zeae* (Schw.) Petch 1936.
Distinct. Descriptions: Ref. 1(no.384),14,16,21,22,28,32,37,49,54,96,97,100,104.
Fusarium robustum Gerlach 1977 (28).
Distinct (63667 = CBS 637.76, type culture). Descriptions: Ref. 28,32
Fusarium lunulosporum Gerlach 1977 (29).
Distinct (62459 = CBS 636.76, type culture). Descriptions: Ref. 29,32.
Fusarium tumidum Sherb. 1928 (80) var. *tumidum.*
Distinct (63572 = CBS 486.76). Descriptions: Ref. 14,32,54,80,96(no.950–952),100,104.
Fusarium tumidum Sherb. var. *coeruleum* Bugnicourt 1939 (17).
Probably distinct (but CBS 128.40 = *F. eumartii*). Booth (14) = *F. tumidum.*
Descriptions: Ref. 17,32,102.
Fusarium tumidum Sherb. var. *humi* Reinking 1934 (72).
Perhaps distinct (but CBS 191.34 = *F. eumartii).* Booth (14) = *F. tumidum.*
Descriptions: Ref. 23,72,96(no.1152),102,104.
Fusarium macroceras Wollenw. et Reinking 1925 (103).
Distinct (CBS 146.25,type culture). Booth (14) = *F. trichothecioides.*
Descriptions: Ref. 32,73,96(no.949),100,103,104.
Fusarium bactridioides Wollenw. 1934 (101).
Distinct (CBS 177.35, type culture). Booth (14) = *F. trichothecioides.*
Descriptions: Ref. 32,96(no.1153),101,102,104.

Section *Lateritium* Wollenw. 1917 (98) Stat. ascig.: *Gibberella* Sacc.

Fusarium lateritium Nees ex Link 1824 var. *lateritium*
Syn.: *F. lateritium* Nees ex Link var. *minus* Wollenw. 1931 (100); *F. lateritium* Nees ex Link var. *mori* Desmazières 1837.
Stat ascig.: *Gibberella baccata* (Wallr.) Sacc. 1883 var. *baccata* [Syn.: *G. baccata* (Wallr.) Sacc. var. *moricola* (de Not.) Wollenw. 1931 (100)].
Distinct. Descriptions: Ref. 1(no.310),14,21,22,32,33,37,54,91,96,100,104.
Fusarium lateritium Nees ex Link var. *majus* (Wollenw.) Wollenw. 1931 (100).
Stat. ascig.: *Gibberella baccata* (Wallr.) Sacc. var. *major* Wollenw. 1931 (100).
Distinct (62246). Booth (14) = *F. lateritium.*
Descriptions: Ref. 32,96(no.26,286,593,594,962,963,1119,1120),100,104.
Fusarium lateritium Nees ex Link var. *longum* Wollenw. 1931 (100).
Distinct (63665 =CBS 633.76). Booth (14) = *F. stilboides.*
Descriptions: Ref. 21,32,96(no.964,965),100,104.
Fusarium lateritium Nees ex Link var. *buxi* C. Booth 1971 (14).
Stat. ascig.: *Gibberella buxi* (Fuckel) Winter apud Rabenh. 1887 [Syn.: *G. baccata* (Wallr.) Sacc. var. *major* Wollenw. 1931 (100) (fide Booth, 14, but see above!)].
Doubtful, not seen (IMI 133315). Description: Ref. 14.
Fusarium sarcochroum (Desm.) Sacc. 1882.
Stat ascig.: *Gibberella pseudopulicaris* Wollenw. 1931 (100).
Distinct (63714). Booth (14) = *F. sambucinum.*
Descriptions: Ref. 32,96(no.30,287–291,514,817),100,104,106.
Fusarium stilboides Wollenw. 1931 (100) var. *stilboides.*
Stat. ascig.: *Gibberella stilboides* Gordon ex C. Booth 1971 (14).
Distinct. Descriptions: Ref. 1(no.30),14,17,21,32,41,54,96(no.615,967,968),100,104.
Fusarium stilboides Wollenw. var. *minus* Wollenw. 1931 (100) emend. Bugnicourt 1939 (17).
? Distinct. Booth (14) = *F. stilboides.*
Descriptions: Ref. 17,73,96(no.966),100,102,103.
Fusarium inflexum R. Schneider apud Schneider et Dalchow 1975 (77).
Distinct (63203 = CBS 716.74, type culture). Descriptions: Ref. 32,77.
Fusarium xylarioides Steyaert 1948 (86).
Stat. ascig.: *Gibberella xylarioides* Heim et Saccas apud Heim 1950 (46).
Distinct (CBS 258.52, type culture). Descriptions: Ref. 1(no.24),9,14,32,46,54,86.
(Doubtful: the 'male' strain, described by Booth, 14).

Section *Liseola* Wollenw. et al. 1925 (106) Stat. ascig.: *Gibberella* Sacc., Monographic treatment: Nirenberg 1976 (67)

Fusarium verticillioides (Sacc.) Nirenberg 1976 (67).
Syn.: *Oospora verticillioides* Sacc. 1881; *F. moniliforme* Sheldon 1904; *F. moniliforme* Sheldon sensu Wollenw. et Reinking 1935 (104)pr.p.
Stat. ascig.: *Gibberella moniliformis* Wineland 1924.
Distinct (11782 = CBS 218.76 = IMI 202875). Descriptions: Ref. 32,67.
Fusarium fujikuroi Nirenberg 1976 (67).
Stat. ascig.: *Gibberella fujikuroi* (Sawada) Wollenw. 1931 (100) var. *fujikuroi.*
Distinct (12428 = CBS 221.76 = IMI 202879, type culture).
Descriptions: Ref. 32,67.
Fusarium proliferatum (Matsushima) Nirenberg 1976 (67) var. *proliferatum.*
Syn.: *Cephalosporium proliferatum* Matsushima 1971; *F. moniliforme* Sheldon sensu Wollenw. et Reinking 1935 (104) pr. p.

Distinct (11341 = CBS 217.76 = IMI 202873). Descriptions: Ref. 32,67.

Fusarium proliferatum (Matsushima) Nirenberg var. *minus* Nirenberg 1976 (67).
Syn.: *F. moniliforme* Sheldon var. *minus* Wollenw. 1931 (100) pr.p.; *F. moniliforme* Sheldon sensu Wollenw. et Reinking 1935 (104) pr.p.
Distinct (11730 = CBS 216.76 = IMI 202874, type culture).
Descriptions: Ref. 32,67.

Fusarium annulatum Bugnicourt 1952 (18).
Distinct (12294 = CBS 258.54 = IMI 202878, type culture). Booth (14) = *F. moniliforme*. Descriptions: Ref. 18,32,67.

Fusarium sacchari (Butler) W. Gams 1971 (23) var. *sacchari*.
Syn.: *Cephalosporium sacchari* Butler 1913; *F. moniliforme* Sheldon var. *subglutinans* Wollenw. et Reinking 1925 (103) pr.p.
Distinct (63340 = CBS 223.76 = IMI 202881). Descriptions: Ref. 23,32,67,91.

Fusarium sacchari (Butler) W. Gams var. *subglutinans* (Wollenw. et Reinking) Nirenberg 1976 (67).
Syn.: F. moniliforme Sheldon var. *subglutinans* Wollenw. et Reinking 1925 (103) pr.p.
Stat. ascig.: *Gibberella fujikuroi* (Swada) Wollenw. var. *subglutinans* Edwards 1933.
Distinct (10351 = CBS 215.76 = IMI 202872). Descriptions: Ref. 32,67.

Fusarium sacchari (Butler) W. Gams var. *elongatum* Nirenberg 1976 (67).
Syn.: *F. moniliforme* Sheldon var. *subglutinans* Wollenw. et Reinking 1925 (103) pr.p.
Distinct (12293 = CBS 220.76 = IMI 202877, type culture). Descriptions: Ref. 32,67.

Fusarium succisae (Schröter) Sacc. 1892.
Syn.: *F. moniliforme* Sheldon var. *anthophilum* (A. Braun) Wollenw. 1931 (100) pr.p.
Distinct (12287 = CBS 219.76 = IMI 202876). Booth (14) = *F. moniliforme*.
Descriptions: Ref. 32,67.

Fusarium anthophilum (A. Braun) Wollenw. 1917 (98).
Syn.: *Fusarium moniliforme* Sheldon var. *anthophilum* (A. Braun) Wollenw. 1931 (100) pr.p.
Distinct (63270 = CBS 222.76 = IMI 202880). Booth (14) = *F. moniliforme* var. *anthophilum* (fide Wollenw.). Descriptions: Ref. 32,67.

Fusarium moniliforme Sheldon var. *oryzae* Saccas 1951 (74).
Doubtful. Booth (14) = *F. moniliforme*. Description: Ref. 74.

Fusarium lactis Pirotta et Riboni 1879.
Doubtful (CBS 181.35), discussion by Nirenberg (67). Booth (14) = *F. moniliforme*.
Descriptions: Ref. 32,96(no.1155a,1155b,1156a,1156b),102,104.

Fusarium neoceras Wollenw. et Reinking 1925 (103).
Doubtful (CBS 147.25, type culture), discussion by Nirenberg (67).
Booth (14) = *F. moniliforme* var. *subglutinans;* Gams (23) = *F. sacchari*.
Descriptions: Ref. 32,73,96(no.977),100,103,104.

Section *Elegans* Wollenw. 1913 (95)

Fusarium oxysporum Schlecht. emend. Snyder et Hansen 1940 (81) var. *oxysporum* pr.p.
Syn.: See Booth (14), Domsch et al. (22), Subramanian (91).
Distinct. Numerous specialized forms or races (3,4,14,44,64,81,91).
Descriptions: Ref. 1(no.28,211–220),7,14,21,22,25,32,33,37,45,51,54,57,91,96,100,102, 104.

Fusarium oxysporum Schlecht. var. *meniscoideum* Bugnicourt 1939 (17).
Doubtful (CBS 221.49). Booth (14) = *F. oxysporum*. Descriptions: Ref. 17,32,102.

Fusarium redolens Wollenw. 1913 (95).
Syn.: *F. oxysporum* Schlecht. emend. Snyder et Hansen var. *redolens* (Wollenw.) Gordon 1952 (37); *F. solani* (Mart.) Sacc. var. *redolens* (Wollenw.) Bilai 1955 (7).

Distinct (62390 = CBS 248.61 = ATCC 16067). Two specialized forms (34,45,91).
Descriptions: Ref. 1(no.27),7,14,21,22,25,34,37,45,51,54,79,91,95,96(no.394,395,1022, 1193),100,104.

Fusarium udum Butler 1910.
Syn.: *F. unicinatum* Wollenw. 1917 (98); *F. lateritium* Nees ex Link var. *uncinatum* (Wollenw.) Wollenw. 1931 (100).
Distinct (62449, 62450, 62451). One specialized form (14).
Descriptions: Ref. 14,32,91,96(no.227,954),100,104.

Section *Martiella* Wollenw. 1913 (95) Stat. ascig.: *Nectria* Fr.

Fusarium javanicum Koorders 1907 var. *javanicum.*
Syn.: *F. javanicum* Koorders var. *radicicola* (Wollenw.) Wollenw. 1931 (100); *F. javanicum* Koorders var. *ensiforme* (Wollenw. et Reinking) Woolenw. 1931 (100); *F. solani* (Mart.) Sacc. emend. Snyder et Hansen 1941 (82) pr.p.
Stat. ascig.: *Nectria ipomoeae* Halsted 1892 [Syn.: *Hypomyces ipomoeae* (Halsted) Wollenw. 1913 (95)].
Distinct. Booth (14) = *F. solani* emend. Snyder et Hansen. Some specialized forms.
Descriptions: Ref. 17,21,22,32,54,55,56,73,96,102,104.

Fusarium coeruleum (Libert) Sacc. 1886.
Syn.: *F. solani* (Mart.) Sacc. var. *coeruleum* (Sacc.) C. Booth 1971 (14).
Distinct. Descriptions: Ref. 2,14,19,21,32,50,54,79,91,96,102,104.

Fusarium solani (Mart.) Sacc. 1881 var. *solani.*
Syn.: *F. solani* (Mart.) Appel et Wollenw. 1910 (2); *F. solani* (Mart.) Appel et Wollenw. var. *minus* Wollenw. 1917 (98); *F. solani* (Mart.) Appel et Wollenw. var. *martii* (Appel et Wollenw.) Wollenw. 1931 (100); *F. solani* (Mart.) Appel et Wollenw. var. *straitum* (Sherb.) Wollenw. 1931 (100); *F. solani* (Mart.) Appel et Wollenw. var. *aduncisporum* (Weimer et Harter) Wollenw. 1931 (100); *F. solani* (Mart.) Sacc. emend Snyder et Hansen 1941 (82) pr.p.
Stat. ascig.: *Nectria haematococca* Berk. et Br. var. *brevicona* (Wollenw.) Gerlach, comb. nov. [Syn.: *Hypomyces haematococcus* (Berk. et Br.) Wollenw. var. *breviconus* Wollenw., Fusar. delin. no.828, 1930].
Distinct. Booth (14) takes *F. solani* in the broader sense of Snyder et Hansen (82). Numerous specialized forms (14,64,82,91).
Descriptions: Ref. 1(no.29),2,14,17,19,21,22,32,33,37,50,54,56,73,79,91,96,102,104.

Fusarium eumartii Carpenter 1915 (19).
Syn.: *F. solani* (Mart.) Appel et Wollenw. var. *eumartii* (Carpenter) Wollenw. 1931 (100).
Stat. ascig.: *Nectria haematococca* Berk. et. Br. 1873 var. *haematococca* [Syn.: *Hypomyces haematococcus* (Berk. et Br.) Wollenw. 1926 (99)].
Distinct (62213, 62214, 62216, 62215 = CBS 487.76). Booth (14) = *F. solani* f. sp. *eumartii* (82). Descriptions: Ref. 17,19,32,33,96(no.422,830–832),99,102,104.

Fusarium illudens C. Booth 1971 (14).
Stat. ascig.: *Nectria illudens* Berkeley apud Hooker 1855.
Not seen, no culture received from Booth (IMI 15564).
Description: Ref. 14.

Section *Ventricosum* Wollenw. 1913 (95) Stat. ascig. *Nectria* Fr.

Fusarium ventricosum Appel et Wollenw. 1910 (2).
Syn.: *F. cuneiforme* Sherb. 1915 (79); ? *F. argillaceum* (Fr.) Sacc. 1886.

Stat. ascig.: *Nectria ventricosa* C. Booth 1971 (14) [Syn.: *Hypomyces solani* Reinke et Berth. 1879; *Hyponectria solani* (Reinke et Berth.) Petch 1937; *Nectriopsis solani* (Reinke et Berth.) C. Booth 1960 (11); non *Nectria solani* Reinke et Berth. 1879]. Distinct (CBS 205.31; IMI 112498; 62452). Descriptions: Ref. 2,11,14,22,32,54, 79,96,(no.429–431,833),100,104.

Fusarium caucasicum Letov 1929 (59).
? Distinct (CBS 179.35). Booth (14) ? = *F. coeruleum*.
Descriptions: Ref. 32,59,102.

Doubtful Species and Varieties of *Fusarium* Described Since 1940

Fusarium bicellulare Kirschstein 1941 (Hedwigia 80:136). Probably = *F. aquaeductuum* or *F. aquaeductuum* var. *medium*. Booth (14) — not known in culture.

Fusarium biseptatum Sawada 1959 (Spec. Publ. Coll. Agric. Nat. Taiwan Univ. 8:228). Published without Latin diagnosis. Booth (14) — not mentioned.

Fusarium buxicola Sacc. var. *chlamydosporum* Batikyan 1969 (Biol. Zh. Armen. 22:90). Published without Latin diagnosis. Booth (14) — not mentioned.

Fusarium castaneicola Yamamoto 1962 (Trans. Mycol. Soc. Japan 3:114). Stat ascig.: *Nectria castaneicola* Yamamoto et Oyasu 1957. Booth (14) — not mentioned.

Fusarium dominicanum Ciferri 1955 (Sydowia 9:115). Booth (14) — not known in culture.

Fusarium elegans Yamamoto et Maeda 1962 (Trans. Mycol. Soc. Japan 3:115). Stat. ascig.: *Nectria elegans* Yamamoto et Maeda 1957. Booth (14) — not mentioned.

Fusarium javanicum Koord. var. *chrysanthemi-leucanthemi* Batikyan 1969 (Biol. Zh. Armen. 22:86). Published without Latin diagnosis. Booth (14) — not mentioned.

Fusarium javanicum Koord. var. *sclerotii* Batikyan 1969 (Biol. Zh. Armen. 22:85). Published without Latin diagnosis. Booth (14) — not mentioned.

Fusarium kurdicum Petrak 1959 (Sydowia 13:96). Stat. ascig.: *Calonectria kurdica* Petrak 1959. Booth (14) — not mentioned.

Fusarium laricis Sawada 1950 (Rep. For. Exp. Stn. Tokyo 46:130). Booth (14) — not known in culture.

Fusarium lateritium Nees var. *microconidium* Batikyan et Abramyan 1969 (Biol. Zh. Armen. 22:60). Published without Latin diagnosis. Booth (14) — not mentioned.

Fusarium lunatum (Ell. et Ev.) v. Arx 1957 (Verh. Akad. Wet. Amst. 51:101). Syn.: *Gloeosporium lunatum* Ell. et Ev. 1891. Probably = *F. dimerum* or *F. dimerum* var. *violaceum;* suggestion accepted by v. Arx (5). Booth (14) = *Gloeosporium lunatum* Ell. et Ev.

Fusarium martiellae-discolorioides Batikyan 1969 (Biol. Zh. Armen. 22:87). Published without Latin diagnosis. Booth (14) — not mentioned.

Fusarium mindoanum Petrak 1950 (Sydowia 4:576). Booth (14) — not know in culture.

Fusarium otomycosis Ming Yo-nung et Yu Ta-fuh 1966 (Acta Microbiol. Sin. 12:178). Booth (14) = *F. solani*.

Fusarium phyllostachydicola Yamamoto 1962 (Trans. Mycol. Soc. Japan 3:118). Stat. ascig.: *Gibberella phyllostachydicola* Yamamoto 1957. Booth (14) — not mentioned.

Fusarium ramulicola Sawada 1959 (Spec. Publ. Coll. Agric. Nat. Taiwan Univ. 8:228). Published without Latin diagnosis. Booth (14) — not mentioned.

Fusarium retusum Wellman 1943 (Phytopathology 33:957). Booth (14) = ? *F. oxysporum* var. *redolens*.

Fusarium semitectum Berk. et Rav. var. *violaceum* Batikyan et Abramyan 1969 (Biol. Zh. Armen. 22:59). Published without Latin diagnosis. Booth (14) — not mentioned.

Fusarium splendens Matuo et Kobayashi 1960 (Trans. Mycol. Soc. Japan 2:13). Stat.

ascig.: *Hypocrea splendens* Phil. et Plowr. 1885. Published without Latin diagnosis. Booth (14) = ? *F. aquaeductuum.*

Fusarium ulmicola Dearness et House 1940 (Circ. N.Y.State Mus. 24:60). Published without Latin diagnosis. Booth (14) — not known in culture.

Fusarium wolgense Rodigin 1942 (Trudȳ bashkir. sel'.khoz. Inst. 3:102). Booth (14) — not mentioned.

Literature Cited

1. Anon. 1964–1977. C.M.I. Descriptions of pathogenic fungi and bacteria. Commonwealth Mycol. Inst., Kew, Surrey, England.
2. Appel, O., and H.W. Wollenweber. 1910. Grundlagen einer Monographie der Gattung *Fusarium* (Link). Arg. Kais. Biol. Anst. Land-Forstwirtsch. 8:1–217.
3. Armstrong, G.M., and J.K. Armstrong. 1968. Formae speciales and races of *Fusarium oxysporum* causing a tracheomycosis in the syndrome of disease. Phytopathology 58:1242–1246.
4. Armstrong, G.M., and J.K. Armstrong. 1975. Reflections on the wilt Fusaria. Annu. Rev. Phytopathol. 13:95–103.
5. Arx, J.A. von. 1959. CBS Baarn. Briefl. Mitteilung v. 26.1.
6. Arx, J.A. von, and E. Müller. 1975. A re-evaluation of the bitunicate ascomycetes with keys to families and genera. Studies Mycol. 9:1–159.
7. Bilai, V.I. 1955. The Fusaria (Biology and Systematics). Kiev: Akad. Nauk. Ukr. SSR. 320 p.
8. Bilai, V.I. 1970. Experimental morphogenesis in the fungi of the genus *Fusarium* and their taxonomy. Ann. Acad. Sci. Fenn. A., IV Biologica 168:7–18.
9. Blittersdorff, R. von, and J. Kranz. 1976. Vergleichende Untersuchungen an *Fusarium xylarioides* Steyaert (*Gibberella xylarioides* Heim et Saccas), dem Erreger der Tracheomykose des Kaffees. Z. Pflanzenkr. Pflanzensch. 83:529–544.
10. Booth, C. 1959. Studies of Pyrenomycetes. IV. *Nectria* (part 1). Commonwealth Mycol. Inst. Mycol. Paper 73:1–115.
11. Booth, C. 1960. Studies of Pyrenomycetes. V. Nomenclature of some Fusaria in relation to their nectrioid perithecial states. Commonwealth Mycol. Inst. Mycol. Paper 74:1–16.
12. Booth, C. 1964. Studies of Pyrenomycetes. VII. Commonwealth Mycol. Inst. Mycol. Paper 94:1–16.
13. Booth, C. 1966. Provisional key to Fusaria. Commonwealth Mycol. Inst., Kew, Surrey, England. 11 p.
14. Booth, C. 1971. The Genus *Fusarium*. Commonwealth Mycol. Inst., Kew, Surrey, England. 237 p.
15. Booth, C. 1975. The present status of Fusarium taxonomy. Annu. Rev. Phytopathol. 13:83–93.
16. Booth, C. 1977. *Fusarium*. Laboratory guide to the identification of the major species. Commonwealth Mycol. Inst., Kew, Surrey, England. 58 p.
17. Bugnicourt, F. 1939. Les *Fusarium* et *Cylindrocarpon* de l'Indochine. Encycl. Mycol. 11:1–206.
18. Bugnicourt, F. 1952. Une espèce fusarienne nouvelle, parasite du riz. Rev. Gén. Bot. 59:13–18.
19. Carpenter, C.W. 1915. Some potato tuber-rots caused by species of *Fusarium*. J. Agric. Res. 5:183–209.
20. Cook, R.J. 1967. *Gibberella avenacea* sp. n., perfect stage of *Fusarium roseum* f. sp. *cerealis* 'Avenaceum.' Phytopathology 57:732–736.
21. Doidge, E.M. 1938. Some South African Fusaria. Bothalia 3:331–483.
22. Domsch, K.H., W. Gams, and T.H. Anderson. 1980. Compendium of soil fungi. London: Academic Press (in press).
23. Gams, W. 1971. Cephalosporium-artige Schimmelpilze (Hyphomycetes). Gustav Fischer, Stuttgart. 262 p.
24. Gams, W., and M. Gerlagh. 1968. Beiträge zur Systematik und Biologie von *Plectosphaerella cucumeris* und der zugehörigen Konidienform. Persoonia 5:177–188.
25. Gerlach, W. 1961. *Fusarium redolens* Wr., seine Morphologie und systematische Stellung. Ein Beitrag zur Kenntnis der Elegans-Fusarien. Phytopathol. Z. 42:150–160.
26. Gerlach, W. 1970. Suggestions to an acceptable modern Fusarium system. Ann. Acad. Sci. Fenn. A, IV Biologica 168:37–49.
27. Gerlach, W. 1972. *Fusarien* aus Trinkwasserleitungen. Ann Agric. Fenn. 11:298–302.
28. Gerlach, W. 1977. *Fusarium robustum* spec. nov., der Erreger einer Stammfäule an *Araucaria angustifolia* (Bertol.) O. Kuntze in Argentinien? Phytopathol. Z. 88:29–37.
29. Gerlach, W. 1977. *Fusarium lunulosporum* spec. nov. von Grapefruit aus Südafrika, ein Fruchtfäuleerreger. Phytopathol. Z. 88:280–284.
30. Gerlach, W. 1977. Drei neue Varietäten von *Fusarium merismoides, F. larvarum* und *F. chlamydosporum*. Phytopathol. Z. 90:31–42.
31. Gerlach, W. 1977. *Fusarium* species inciting plant diseases in the tropics, p. 210–217. *In* J. Kranz, H. Schmutterer, and W. Koch (ed.), Diseases, pests and weeds in tropical crops, Paul Parey, Berlin.
32. Gerlach, W. 1982. The genus *Fusarium*—a picto-

rial atlas. Mitt. Biol. Bundesanst. Land-Forstwirtsch. Berlin-Dahlem (in press).
33. Gerlach, W., and D. Ershad. 1970. Beitrag zur Kenntnis der *Fusarium*- und *Cylindrocarpon*-Arten in Iran. Nova Hedwigia 20:725–794.
34. Gerlach, W., and H. Pag. 1961. *Fusarium redolens* Wr., seine phyto-pathologische Bedeutung und eine an *Dianthus*-Arten gefäßparasitäre Form (*F. redolens* Wr. f. *dianthi* Gerlach). Phytopathol. Z. 42:349–361.
35. Gerlach, W., and G. Scharif. 1970. Der Erreger einer Fußkrankheit an *Hibiscus cannabinus* in Iran —*Fusarium bucharicum* Jaczewski. Phytopathol. Z. 68:323–333.
36. Gordon, W.L. 1944. The occurrence of *Fusarium* species in Canada. I. Species of *Fusarium* isolated from farm samples of cereal seed in Manitoba. Can. J. Res., C. 22:282–286.
37. Gordon, W.L. 1952. The occurrence of *Fusarium* species in Canada. II. Prevalence and taxonomy of *Fusarium* species in cereal seed. Can. J. Bot. 30:209–251.
38. Gordon, W.L. 1954. The occurrence of *Fusarium* species in Canada. III. Taxonomy of *Fusarium* species in the seed of vegetable, forage, and miscellaneous crops. Can. J. Bot. 32:576–590.
39. Gordon, W.L. 1954. The occurrence of *Fusarium* species in Canada. IV. Taxonomy and prevalence of *Fusarium* species in the soil of cereal plots. Can. J. Bot. 32:622–629.
40. Gordon, W.L. 1956. The occurrence of *Fusarium* species in Canada. V. Taxonomy and geographic distribution of *Fusarium* species in soil. Can. J. Bot. 34:833–846.
41. Gordon, W. L. 1956. The taxonomy and habitats of the *Fusarium* species in Trinidad, B.W.I. Can. J. Bot. 34:847–864.
42. Gordon, W.L. 1959. The occurrence of *Fusarium* species in Canada. VI. Taxonomy and geographic distribution of *Fusarium* species in plants, insects, and fungi. Can. J. Bot. 37:257–290.
43. Gordon, W.L. 1960. The taxonomy and habitats of *Fusarium* species from tropical and temperate regions. Can. J. Bot. 38:643–658.
44. Gordon, W.L. 1965. Pathogenic strains of *Fusarium oxysporum*. Can. J. Bot. 43:1309–1318.
45. Hantschke, D. 1961. Untersuchungen über Welkekrankheiten der Edelnelke in Deutschland und ihre Erreger. Phytopathol. Z. 43:113–168.
46. Heim, R. 1950. La carbunculariose du caféier. Rev. Mycol. 15, Suppl. Colon. 2:89–98.
47. Hubert, E. 1923. The red stain in the wood of boxelder. J. Agric. Res. 26:447–457.
48. Jaczewski, A.A. 1929. Some diseases of cotton fibres. Microbiol. J. 9:159–167.
49. Jamalainen, E.A. 1943. Über die *Fusarien* Finnlands. I. Valt. Maatalouskoet. Julk. 122:1–26.
50. Jamalainen, E.A. 1943. Über die *Fusarien* Finnlands. II. Valt. Maatalouskoet. Julk. 123:1–25.
51. Jamalainen, E.A. 1944. Über die *Fusarien* Finnlands. III. Valt. Maatalouskoet. Julk. 124:1–24.
52. Jamalainen, E.A. 1970. Studies on Fusarium fungi in Finland. Ann. Acad. Sci. Fenn. A, IV Biologica 168:54–56.
53. Jamieson, C.O., and H.W. Wollenweber. 1912. An external dry rot of potato tubers caused by *Fusarium trichothecioides* Wollenw. J. Wash. Acad. Sci. 2:146–152.
54. Joffe, A.Z. 1974. A modern system of Fusarium taxonomy. Mycopathol. Mycol. Appl. 53:201–228.
55. Joffe, A.Z., and J. Palti. 1970. *Fusarium javanicum* Koorders in Israel. Mycopathol. Mycol. Appl. 42:305–314.
56. Joffe, A.Z., and J. Palti. 1972. *Fusarium* species of the Martiella section in Israel. Phytopathol. Z. 73:123–148.
57. Joffe, A.Z., J. Palti, and R. Arbel-Sherman. 1974. *Fusarium oxysporum* Schlecht. in Israel. Phytoparasitica 2:91–107.
58. Krampe, O. 1926. Fusarium als Erreger von Fuß- und Keimlings-krankheiten am Getreide. Angew. Bot. 8:217–261.
59. Letov, A.S. 1929. A new species of *Fusarium* on cotton plants. Mater. Mikol. Fitopatol. Ross. Leningrad 8:205–218.
60. Lewis, C.E. 1913. Comparative studies of certain disease-producing species of *Fusarium*. Maine Agric. Exp. Stn. Bull. 219:203–258.
61. Matuo, T. 1961. On the classification of Japanese Fusaria. Ann. Phytopathol. Soc. Japan 26:43–47.
62. Matuo, T. 1972. Taxonomic studies of phytopathogenic Fusaria in Japan. Rev. Plant Prot. Res. 5:34–45.
63. Messiaen, C.M. 1959. La systématique du genre *Fusarium* selon Snyder et Hansen. Rev. Pathol. Vég. Entomol. Agric. Fr. 38:253–266.
64. Messiaen, C.M., and R. Cassini. 1968. Recherches sur les fusarioses. IV. La systématique des *Fusarium*. Ann Épiphyt. 19:387–454.
65. Müller, E. 1977. Die systematische Stellung des "Schneeschimmels". Rev. Mycol. 41:129–134.
66. Müller, E., and J.A. von Arx. 1962. Die Gattungen der didymosporen Pyrenomyceten. Beitr. Kryptogamenfl. Schweiz 11 (2):1–922.
67. Nirenberg, H. 1976. Untersuchungen über die morphologische und biologische Differenzierung in der Fusarium-Sektion Liseola. Mitt. Biol Bundesanst. Land-Forstwirtsch. Berlin-Dahlem 169:1–117.
68. Padwick, G.W. 1945. Notes on Indian fungi. III. Commonwealth Mycol. Inst. Mycol. Paper 12:1–15.
69. Palti, J., and A.Z. Joffe. 1971. Causes of the Fusarium wilts of cucurbits in Israel and conditions favouring their development. Phytopathol. Z. 70:31–42.
70. Raillo, A. 1935. Diagnostic estimation of morphological and cultural characters in the genus *Fusarium*. Bull. Plant. Prot. II, Leningrad (Phytopathol.) 7:1–100.
71. Raillo, A. 1950. Griby roda *Fusarium*. State Publ. Moskva: Gos. izd-vo selk-hoz. lit-ry. 415 p.
72. Reinking, O.A. 1934. Interesting new Fusaria.

Zentralbl. Bakteriol. Parasitenkd. Infektionskr., II, 89:509–514.
73. Reinking, O.A., and H.W. Wollenweber. 1927. Tropical Fusaria. Phillippine J. Sci. 32:103–253.
74. Saccas, A. 1951. Etude morphologique, biologique et expérimentale d'un *Fusarium* ravageur des cultures de riz à la Station Centrale de Boukoko (A.E.F.). Rev. Pathol. Vég. Entomol. Agric. Fr. 30:65–96.
75. Schneider, R. 1958. Untersuchungen über Variabilität und Taxonomie von *Fusarium avenaceum* (Fr.) Sacc. Phytopathol. Z. 32:95–126.
76. Schneider, R. 1958. Untersuchungen über Variation und Patho-genität von *Fusarium avenaceum* (Fr.) Sacc. Phytopathol. Z. 32:129–148.
77. Schneider, R., and J. Dalchow. 1975. *Fusarium inflexum* spec. nov., als Erreger einer Welkekrankheit an *Vicia faba* L. in Deutschland. Phytopathol. Z. 82:70–82.
78. Seemüller, E. 1968. Untersuchunger über die morphologische und biologische Differenzierung in der Fusarium-Sektion Sporotrichiella. Mitt. Biol. Bundesanst. Land-Forstwirtsch. Berlin-Dahlem 127:1–93.
79. Sherbakoff, C.D. 1915. Fusaria of potatoes. N.Y. (Ithaca) Agric. Exp. Stn. Mem. 6:87–270.
80. Sherbakoff, C.D. 1928. An examination of Fusaria in the herbarium of the pathological collections, Bureau of Plant Industry, U.S. Department of Agriculture. Phytopathology 18:148 (Abstr.).
81. Snyder, W.C., and H.N. Hansen. 1940. The species concept in *Fusarium*. Amer. J. Bot. 27:64–67.
82. Snyder, W.C., and H.N. Hansen. 1941. The species concept in *Fusarium* with reference to section Martiella. Amer. J. Bot. 28:738–742.
83. Snyder, W.C., and H.N. Hansen. 1945. The species concept in *Fusarium* with reference to Discolor and other sections. Amer. J. Bot. 32:657–666.
84. Snyder, W.C., H.N. Hansen, and J.W. Oswald. 1957. Cultivars of the fungus, *Fusarium*. J. Madras Univ. B 27:185–192.
85. Snyder, W.C., and T.A. Toussoun. 1965. Current status of taxonomy in *Fusarium* species and their perfect stages. Phytopathology 55:833–837.
86. Steyaert, R.L. 1948. Contribution à l'étude des parasites des végétaux du Congo Belge. Bull. Soc. Roy. Bot. Belg., 80, Sér. 2, 30:11–49.
87. Subramanian, C.V. 1951. Is there a 'wild type' in the genus *Fusarium*? Proc. Nat. Inst. Sci. India 17:403–411.
88. Subramanian, C.V. 1952. Studies on south Indian Fusaria. II. Fusaria isolated from black cotton soils. Proc. Nat. Inst. Sci. India 18:557–584.
89. Subramanian, C.V. 1954. Studies on south Indian Fusaria. III. Fusaria isolated from some crop plants. J. Madras Univ. B 24:21–46.
90. Subramanian, C.V. 1955. The ecological and taxonomic problems in Fusaria. Proc. Indian Acad. Sci. B 41:102–109.
91. Subramanian, C.V. 1971. Hyphomycetes. An account of Indian species, except Cercosporae. New Delhi: Indian Council Agric. Res. 930 p.
92. Toussoun, T.A., and P.E. Nelson. 1968. A pictorial guide to the identification of *Fusarium* species according to the taxonomic system of Snyder and Hansen. Pa. State Univ. Press, Univ. Park, Pa. 51 p.
93. Toussoun, T.A., and P.E. Nelson. 1975. Variation and speciation in the Fusaria. Annu. Rev. Phytopathol. 13:71–82.
94. Tubaki, K., C. Booth, and T. Harada. 1976. A new variety of *Fusarium merismoides*. Trans. Brit. Mycol. Soc. 66:355–356.
95. Wollenweber, H.W. 1913. Studies on the Fusarium problem. Phytopathology 3:24–50.
96. Wollenweber, H.W. 1916–1935. Fusaria autographice delineata. Berlin: Selbstverlag. 1200 Tafeln.
97. Wollenweber, H.W. 1917. (1918). Conspectus analyticus Fusariorum. Ber. Dtsch. Bot. Ges. 35:732–742.
98. Wollenweber, H.W. 1917. Fusaria autographice delineata. Ann. Mycol. 15:1–56.
99. Wollenweber, H.W. 1926. Pyrenomyceten-Studien. II. Angew. Bot. 8:168–212.
100. Wollenweber, H.W. 1931. Fusarium-Monographie. Fungi parasitici et saprophytici. Z. Parasitenkd. 3:269–516.
101. Wollenweber, H.W. 1934. *Fusarium bactridioides* sp. nov., associated with *Cronatium*. Science N.S. 79:572.
102. Wollenweber, H.W. 1943. Fusarium-Monographie. II. Fungi parasitici et saprophytici. Zentralbl. Bakteriol. Parasitenkd. Infektionskr., II, 106:104–135, 171–202.
103. Wollenweber, H.W., and O.A. Reinking. 1925. Aliquot Fusaria tropicalia nova vel revisa. Phytopathology 15:155–169.
104. Wollenweber, H.W., and O.A. Reinking. 1935. Die Fusarien, ihre Beschreibung, Schadwirkung und Bekämpfung. Paul Parey, Berlin. 355 p.
105. Wollenweber, H.W., and O.A. Reinking. 1935. Die Verbreitung der Fusarien in der Natur. Friedländer und Sohn, Berlin. 80 p.
106. Wollenweber, H.W., C.D. Sherbakoff, O.A. Reinking, H. Johann, and A.A. Bailey. 1925. Fundamentals for taxonomic studies of *Fusarium*. J. Agric. Res. 30:833–843.
107. Zambettakis, C. 1950. Une fusariose du panicule du riz en Oubangui. Rev. Mycol. 15, Suppl. Colon. 2:106–111.

37 Taxonomy of *Fusarium*

C. M. Messiaen and R. Cassini

Taxonomic History

The Wollenweber System

The Wollenweber System culminated in 1935 with the publication of the monograph entitled "Die Fusarien" by Wollenweber and Reinking (61). The system is divided into 16 sections and 65 species which are themselves subdivided into 55 varieties and 22 forms. In order to identify a particular isolate the reader first looks up the key to the sections (analogous to subgenera) having latinized names. This key is based on characteristics which are relatively simple and which can be easily observed (with the exception of the characteristic of terminal vs. intercalary chlamydospores), although a few of these characteristics may be difficult to use. It is quite different when one comes to the separation of species presented within the sections, which are based on cultural characters subject to variation (presence or absence of sporodochia, pionnotes, sclerotia), and on tedious measurements of conidia, classed according to the number of septa present.

The System Proposed by Snyder and Hansen

As early as 1935 the Soviet researcher Raillo (46) questioned the Wollenweber and Reinking system insofar as species distinctions were concerned, while recognizing the value of the sections.

In 1940, Snyder and Hansen (54) presented similar objections and began their studies first with the section Elegans. They demonstrated that the species distinctions were without value, since a particular clone determined as belonging to a particular species could easily, through mutation, take on the characteristics of another. This variability studied by numerous single-spore cultures remained, nevertheless, within the section. Snyder and Hansen proposed therefore that all the forms belonging to the section Elegans be considered as a single species, *Fusarium oxysporum,* in the place of the 10 species, of which the most striking were *F. orthoceras, F. conglutinans, F. bulbigenum,* and *F. oxysporum.* In 1941 the same authors regrouped in a single species all the forms belonging to the section Martiella (*F. solani*) (55). In 1945 a similar study resulted in the complete reworking of the remaining sections (56).

The genus *Fusarium* thereby was reduced to 10 species (or rather nine, because after admitting *F. ciliatum* as a doubtful species, Snyder made no further reference to it in his later publications) (56). Within each species, the often large morphologic variations have no taxonomic value since by mutation or sexual recombination the diverse forms could be experimentally shown to derive from one another.

On the other hand, especially in *F. oxysporum* and *F. solani,* specialized forms having a particular virulence for this or that host plant can be distinguished such as *F. oxysporum* f. sp. *lycopersici* or *F. solani* f. sp. *phaseoli*. One can, in addition, sometimes distinguish within the specialized forms, races which attack certain host cultivars. Thus, *F. oxysporum* f. sp. *lycopersici* race 1 does not attack a cultivar of tomato having gene I while race 2 can. This nomenclature is analogous to the one utilized by specialists in the rusts. These specialized forms or races cannot be determined by observation of their morphology, they can only be determined through inoculation tests, that is to say physiologically. A collection of *F. oxysporum* f. sp. *lycopersici* race 1 can thus unite representatives of the old species *F. orthoceras, F. oxysporum, F. bulbigenum,* and *F. conglutinans.*

Objections presented to the Snyder and Hansen system The system of Snyder and Hansen was well received by most specialists for *F. oxysporum, F. solani,* and *F. moniliforme*. The grouping in one species, *F. roseum,* of the four Wollenweber sections Roseum, Arthrosporiella, Gibbosum and Discolor, has on the other hand been criticized. Gordon (23, 24, 25, 26, 27, 28, 29), who spent many years studying Fusaria isolated from soil, seed, plants, and insects in Canada and Trinidad, is the one whose work merits the greatest attention in this context. He adopted a mixed system following Snyder and Hansen for *F. oxysporum,* maintaining *F. coeruleum* next to *F. solani* in the section Martiella, and keeping instead of *F. roseum* the four sections of Wollenweber subdivided into about 15 species and forms. The publications of Gordon which dated from 1944 to 1965 have, unlike those of Snyder and Hansen, the advantage of presenting drawings and plates in color. However, as with Snyder and Hansen, Gordon did not present a key. Booth presents one but has returned to a system which is still closer to that of Wollenweber and Reinking. In a book published in 1971 (7) he follows Snyder and Hansen only for the section Elegans, which he equates to *F. oxysporum*. In all the other sections several species are described. To Snyder and Hansen's *F. roseum,* 17 species are presented, some of which contain varieties.

The cultivars of Snyder and Hansen and Oswald Refuting the objections of Gordon but apparently influenced by the criticism of their *F. roseum,* Snyder and Hansen and Oswald in 1957 (57) maintained their simplified classification but added the notion of the cultivar derived from the classification of cultivated plant varieties. This division of cultivars can be superimposed on the division of specialized forms without there being any confusion between the two points of view. Snyder, Hansen, and Oswald thus described the existence of the cultivar Redolens in *F. oxysporum,* of the cultivar Coeruleum in *F. solani,* Subglutinans in *F. moniliforme,* Dimerum and Gigas in *F. episphaeria* and, finally, that which without doubt satisfied those who opposed their system, the cultivars Culmoruum, Avenaceum, Graminearum, Sambucinum, Equiseti, and Acuminatum in *F. roseum.*

A Practical Approach

The mycological researcher who can afford to spend a good bit of time on *Fusarium* may perhaps hesitate between the two systems. That is what Gordon and Booth have done in presenting intermediary solutions. The researcher who is only occasionally occupied with *Fusarium* and who is more interested in the phytopathological point of view rather than the systematic one does not hesitate; it is Snyder and Hansen's system that he chooses. We will bring to their system only minimal modifications. We will mention here only those that are of a general nature.

If one admits the validity of the comparison that Snyder, Hansen, and Oswald (57) have made between the species of *Fusarium* and those of *Brassica, Triticum* or *Allium,* we feel that the notion of cultivar would correspond in the Fusaria to particular isolates or clones. The entities Graminearum or Culmorum correspond better to the botanical varieties *acephala, gemmifera, capitata, bullata,* and *botrytis,* of *Brassica oleracea* or to the varieties *agregatum* or *viviparum* in *Allium cepa.* We will replace therefore the term cultivar, which does not seem correct to us, with the simple designation of variety (variety in the botanical sense of the term). Snyder and Toussoun (58) rightly insist on the advantages there are in keeping the same specific epithet for a designation of the conidial state and the ascosporic state of the same fungus. We feel, however, that we should, as counseled by F. Moreau (Les Champignons, p. 944, 1954), respect the rules of Latin grammar. *Gibberella* and *Calonectria,* for example, are feminine while *Fusarium* is neuter. One should, therefore, write *Gibberella rosea* and *Fusarium roseum, Calonectria rigidiuscula,* and *Fusarium rigidisculum, Calonectria nivalis,* and *Fusarium nivale.*

Key to Conidial Species

Microconidia Present

A. Microconidia pyriform or somewhat spherical produced on bottle-shaped conidiophores. Colors comprised in zone 1 (39): *F. tricinctum.*
B. Elliptical microconidia.
  1. Macroconidia very large (> 50 μm, often with more than nine septa). Microconidia in chains on ramified conidiophores. Coloration comprised in zone 1 (39): *F. rigidiusculum.*
  2. Macroconidia of moderate size (rarely five, never seven septate).
    a. Colorations comprised in zone 1 (39). Microconidia in bunches on ramified conidiophores mixed in with macroconidia of irregular size and irregular curvature. Colonies powdery, secondary colonies frequently found at the bottom of slants: *F. roseum* var. *arthrosporioides.*
    b. Colorations comprised in zone 2 (39).
      i. Chlamydospores absent.
      Microconidia in chains: *F. moniliforme.*
      Microconidia in false heads: *F. moniliforme* var. *subglutinans.*
      ii. Chlamydospores present, microconidia in false heads on short conidiophores. Macroconidia < 4 μm wide: *F. oxysporum.*
    c. Colorations comprised in zone 3 (39). Microconidia produced in young cultures on long conidiophores in droplets of water often in mixture with macroconidia. The macroconidia dominate in the mature cultures. Width > 4 μm. Chlamydospores present: *F. solani.*

Microconidia Absent

A. Colorations comprised in zone 3 (39). See *F. solani* above (certain isolates produce very few microconidia, all the other characteristics remaining valid).
B. Colorations not comprised in zone 3 (39).
  1. Growth slow at 25 C (< 0.4 cm/day).
    a. Growth still slower at 18 C, sporodochia abundant or pionnotes continuous.
      i. Macroconidia very large (> 50 μm) with at least eight septations or more: *F. episphaeria.*
      ii. Macroconidia of medium size (3–5 septate) crescent-shaped, pointed at both ends: *F. episphaeria* var. *dimerum.*

---

[1]Zones comprise the following color ranges: zone 1—brown, beige, tawny, yellow, orange, red, pink, white; zone 2—violet, mauve, orange, pink, white; zone 3—blue-gray, violet-gray, gray-green, indigo, blue, green, yellow, beige, white.

iii. Two-celled crescent-shaped macroconidia: *F. episphaeria* var. *dimerum*.
iv. Macroconidia of medium size, foot-celled, narrowed to a beak at the top: *F. lateritium*.

b. Growth slow at 25 C (mycelium white with little sporulation), rapid at 18 C with the production of numerous conidia in pink pionnotal sheets, macroconidia obtuse at both ends: *F. nivale*.

2. Growth rapid (> 0.4 cm/day) at 25 and 18 C; macroconidia foot-celled, colorations comprised in zone 1 (39): *F. roseum*.
   a. Chlamydospores very abundant in culture, often packed by two or three, mycelium dirty white or buff, stroma sometimes carmine-red, macroconidia humpbacked, irrespective of their length or width: *F. roseum* var. *gibbosum*.
   b. Chlamydospores present but not abundant in culture.
      i. Macroconidia with a tip more sharply bent than the base, < 5 μm wide at their center, produced very abundantly, in microsporodochia often fused in a rose-orange spongy mass: *F. roseum* var. *sambucinum*.
      ii. Macroconidia > 5 μm wide, even when they only have 1 or 2 septa; extremties not very pointed. Stroma always carmine red: *F. roseum* var. *culmorum*.
   c. Chlamydospores absent.
      i. Macroconidia width 3 – 6 μm, stroma carmine: *F. roseum* var. *graminearum*.
      ii. Macroconidia very long (>50 μm) and thin (< 3 μm wide). Stroma often carmine red: *F. roseum* var. *avenaceum*.

Key to the Perfect States of *Fusarium*

The known perfect states of ***Fusarium*** are part of the Hypocreaceae in the Sphaeriales. The perithecia of *Hypomyces, Nectria, Gibberella,* and *Calonectria* which correspond to the *Fusarium* imperfect states have walls made up of several cellular layers, which are relatively soft when they are impregnated with water, sometimes without color but most often colored red, blue, or violet. They are either isolated and produced on the substrate, or grouped on fleshy stromas, rarely embedded in the tissue of the host (with the exception of *Calonectria nivalis)*.

Two-celled Ascospores

A. Perithecia red, sometimes without color, borne on a subiculum. Asci with eight ascospores finely striated at maturity, brown in color: *Hypomyces solani* (*F. solani*).
B. Perithecia dark red, borne on a fleshy stroma or borne singly on the substrate. Ascospores colorless: *Nectria episphaeria* (*F. episphaeria*).

Ascospores at First Two-Celled (in the asci), Multicellular at Maturity

A. Perithecia dark blue-violet, borne on the surface of the host. Asci eight-spored: *Gibberella moniliformis* (*F. moniliforme*).
B. Perithecia at first pink, then brown, borne within the tissues of the host in the case of the leaf sheath of Gramineae, borne externally if the substrate is compact, asci eight-spored: *Calonectria nivalis* (*F. nivale*).

Ascospores Multicellular When Young

A. Perithecia dark blue-violet, borne externally on the host, either isolated or on a fleshy stroma. Asci eight-spored: *Gibberella rosea* and *G. lateritia* (*F. roseum* and *F. lateritium*).

B. Perithecia yellow, rust colored, red or brown. Ascospores large, longitudinally striated, brown in color at maturity, asci often four-spored: *Calonectria rigidiuscula* (*F. rigidiusculum*).

Snyder and Toussoun (58) feel that it is not necessary to create other genera besides *Nectria, Hypomyces, Calonectria,* and *Gibberella* for the perfect states of *Fusarium*. One should, however, remember that the perithecia of *F. moniliforme* and *F. nivale* can create confusion from the fact that the ascospores segment late and do not become multicellular until the asci have disintegrated. This is what caused Saccardo to create the genus *Lisea* for the perfect state of *F. moniliforme* and more recently Muller and Von Arx (41) to create *Griphosphaeria nivalis,* the perfect state of *F. nivale,* arguing that the perithecia are included within the tissues of the host. Booth names it *Micronectriella nivalis* (7), but, on the other hand he does not maintain *Hypomyces* in *Fusarium,* the perfect state of *Fusarium solani* being *Nectria haematococca.*

Characteristics, Virulence, and Parasitic Specialization of the Various Species of *Fusarium*

Fusarium tricinctum (Cda.) Snyd. & Hans.

The most characteristic isolates of *F. tricinctum* produce a very large number of spherical or pear-shaped microconidia on bottle-shaped conidiophores. The colonies produce a very particular odor which is repungnant to some, agreeable to others. The cultures have a more or less pronounced carmine red pigment analogous to that of *F. roseum* var. *graminearum,* and are occasionally completely white. The macroconidia are rare and of irregular shape.

These cultures are often associated in nature with a mite, *Pediculopsis* (or *Siteroptes*) *graminum.* Snyder and Hansen (55) have proposed the designation *F. tricinctum* f. sp. *poae* for those clones capable of attacking the carnation. We will conserve this name in order to designate those isolates which can associate with the mites in attacking the inflorescences of carnations or of gramineae.

In Canada, Gordon (24) noted *F. tricinctum* as one of the major constituents of the microflora of cereal grain but gave no indication that this fungus was damaging and gave no indication of the presence of the mites. We have observed it, under the same conditions, in the north of France at rates no higher than 2–3%. We have also found it, also without mites, on dying branches of elder (*Sambucus nigra*), in company with *F. lateritium* and *F. roseum* var. *sambucinum.* Certain clones of *F. tricinctum* f. sp. *poae* are stable in culture. The majority evolve, however, gradually losing their carmine red pigment and their facility to sporulate. One occasionally finds *F. tricinctum* which produces far fewer microconidia and many more macroconidia than those that we have just described. The aspects of the colony and the form of the macroconidia suggest that they are transition forms between *F. tricinctum* and *F. roseum,* through the intermediary of *F. roseum* var. *arthrosporioides.* These forms correspond to *F. sporotrichioides* of Wollenweber and Reinking (61) and of Gordon (24).

Fusarium rigidiusculum (Brick) Snyd. & Hans.

This species is found only in tropical regions, on dying branches and fruits of trees and bushes. Gordon described it from Trinidad (28), Bugnicourt (10) in Indochina, and Roger (51) in the Ivory Coast.

Contrary to the other species of *Fusarium* which survive or even develop well (*F. nivale*) when the cultures are placed in the refrigerator between 4 and 8 C, *F. rigidiusculum* is very sensitive to cold and cannot survive more than a few weeks at such temperatures. Among cultivated plants *Hevea* and especially *Theobroma cacao* can be attacked. On the latter tree certain isolates cause dieback of branches and others provoke tumors called cushion galls (9).

According to Ford, Bourret, and Snyder (19), the isolates capable of causing cushion galls are completely distinct from the saprophytic clones that are homothallic, producing perithecia on various substrates. They are all heterothallic or incapable of producing the perfect state. They constitute the forma specialis *theobromae*. This distinction of saprophytic clones and the f. sp. *theobromae* seems to negate Hutchins' (34) interpretation of a virus which was not transmitted through the sexual state in order to explain the absence of virulence in clones which arose from perithecia.

Fusarium rigidiusculum is remarkable for the size of its macroconidia, habitually produced in sporodochia which are yellow on a substrate with an intense carmine red pigment. Variability is still possible, however, for we have received from Snyder and Toussoun cultures which do not have sporodochia and pigments.

The majority of the authors described the perithecia as having four-spored asci. Reichle and Snyder (49) studied seven clones coming from various regions and showed that two of these were homothallic and regularly produced four-spored asci, that the others were heterothallic, and that the crosses produced asci which had four to eight ascospores. This perfect state should be designated *Calonectria rigidiuscula* (Brick) Snyd. & Hans.

Fusarium moniliforme (Sheldon) Snyd. & Hans.

This species, like the preceding one, is particularly frequent in warm regions, but has a larger geographic distribution. Gordon finds it only rarely in Canada; the mycologists in the north of France are unaware of it. It is, on the other hand, frequent in the southern regions of France, where it plays a parasitic role which is not very strong, and where its perfect state is unknown. We have seen it as a kernel rot on corn ears (22). Contrary to what takes place with *F. roseum* var. *graminearum* or *culmorum*, the infected grains germinate without difficulty when the embryo has not been attacked. *Fusarium moniliforme* can also invade the foliar sheaths and nodal tissues of senescent corn plants without causing obvious damage, and it may exist in the corn plants in full vigor. In tropical and subtropical countries it has been noted to attack fruit (figs in particular). We have observed it in France on figs, following the invasion of the fruit by larva of the diptera, *Oscinosoma discretum*. On the Gramineae of warm and humid climates it has been found attacking the young leaves of sugar cane, corn, sorghum, and millet. It is on rice that *F. moniliforme* is the most clearly pathogenic (51). The isolates obtained from rice are those which are alone capable of synthesizing gibberellins. This gigantism of rice or 'bakanae' first described in Japan, then in Indochina and India, and more recently in Italy (16), has not yet been seen in France. It seems logical to regroup those clones which produce Gibberellin and cause 'bakanae' on rice as constituting a specialized form which we could call *F. moniliforme* f. sp. *oryzae gigantis*.

The other parasitic activities of *F. moniliforme* are not so specific. Clones which are isolated from figs or from asparagus are also capable of invading corn ears. On the rhizomes of asparagus one also finds *F. moniliforme* in company with *F. oxysporum* and *F. roseum* without being able to define a clearly parasitic role for *F. moniliforme*.

Certain clones of *F. moniliforme* which contrary to type form sporodochia and macroconidia in abundance produce their microconidia in false heads instead of in chains. They constitute the variety *subglutinans*, in which we should include the type *anthophilum*, whose macroconidia are strongly bent. *Anthophilum* described in France on Scabiosa (36) has been found in Trinidad on coconuts (28).

In our earlier French paper (39) we wrote: "Only the absence of chlamydospores distinguishes *F. moniliforme* var. *subglutinans* from *F. oxysporum.*" Booth (7) has shown this to be an error: the two species can be distinguished on the basis of their microconidiophore morphology, those of *F. oxysporum* are quite short and only rarely branched; those of *F. moniliforme* var. *subglutinans* are longer and frequently branched. These characters were figured in our original illustrations.

The perithecia of *F. moniliforme, Gibberella moniliformis* (Sheldon) Snyd. & Hans., are frequent in subtropical countries, but rare in Europe. We have never seen them in France, but they were recently described from Rumania (59).

Fusarium oxysporum (Schlecht.) Snyd. & Hans.

Worldwide in occurrence, this is the most frequent of all the Fusaria in soil — 51–66% of the total *Fusarium* isolated by Gordon (26, 27), 40 – 70% in the isolations made by Guillemat and Montegut (31, 32, 33). It is consequently a very active saprophyte. *Fusarium oxysporum* can also play a very prominent role in plant diseases, and its virulence can be as finely specialized as that of the Uredineae, for it is within this species that we find those forms which are capable of causing tracheomycoses. This parasitic specialization is totally independent of the variations in the morphologic aspect of the cultures. These morphologic variations of *F. oxysporum,* however, need to be described.

Morphologic variability in *F. oxysporum* There is a good deal of variation in spore morphology even within a specialized form or a race with respect to shape and size of macroconidia and proportion of microconidia to macroconidia. Certain isolates do not produce any macroconidia. In order to be sure that they do indeed belong to a *Fusarium* (i.e., produce macroconidia), they can be placed at lower temperatures around 18 C. However, one observes a greater intensity of the violet pigment, in those isolates that produce it, at 25 rather than at 18 C.

Independently of these variations due to conditions of the substrate and to the capability of producing the violet pigment, the isolates of *F. oxysporum* can be classified in various morphologic types from the macroscopic point of view. We will adopt the classification of Waite and Stover (60) which they proposed for *F. oxysporum* f. sp. *cubense* and which seems to be adaptable to the various isolates that we have been able to observe (isolated directly from soil or belonging to the formae speciales *melonis, niveum, lycopersici, gladioli, callistephi, apii, batatas*).

a. Sporodochial type: aerial mycelium, rather short but dense; numerous sporodochia rose-orange colored; violet sclerotia present or absent.

b. Sclerotial type: the sporodochia are replaced by violet to pinkish beige sclerotia.

c. Cottony type: sporodochia are absent; the extremely vigorous aerial mycelium fills the tube. Producing only microconidia at 25 C, these isolates may produce a few macroconidia at 18 C.

d. Ropy type: the aerial mycelium, less vigorous, is clumped in wicks. Sporodochia are absent.

e. Slimy pionnotal type: the aerial mycelium is only vigorous on the margin of the colonies, where its aspect is similar to that of the ropy type. In the center of the colony, the aerial mycelium is completely collapsed. On this smooth surface large numbers of macroconidia are produced in slimy pionnotes at least at 18 C.

In the laboratory, when one renews these cultures by single spore or by mass transfer, one observes, in an abrupt fashion in the first case and progressively in the second case, transformations of one type to the other.

The sporodochial types are sometimes stable, but more often they evolve in an irreversible manner toward the slimy pionnotal type, either directly or passing in stages through the sclerotial type and the ropy type. The cottony type, on the other hand, is much more stable and can remain unchanged for up to 10 years in a collection. In the specialized forms virulence need not be lost. The dynamics of this evolution have been recently studied by Follin and Laville (18) in *F. oxysporum* f. sp. *cubense.* These authors have systematically utilized in their renewals extremities of young hyphae, isolated microconidia (produced on young mycelium), macroconidia (produced on mycelium of middle age), chlamydospores (produced on old mycelium), or mycelial fragments taken from old cultures. Their results show that the original type remains stable when the transfers are made from young mycelium or microconidia, and that the evolution toward the slimy pionnotal type is rapid only in the case of transfers of chlamydospores or old mycelium, the macroconidia giving intermediate results. They also showed the ultimate state of this evolution, the type they call "shorn," whose growth is feeble and very slow and which can really be termed degenerate. They therefore believe, and quite rightly it seems to us, that this morphologic evolution which is irreversible in *F. oxysporum* should be termed a senescence.

In nature isolates obtained from soil or from plants where the attack remains subterranean (gladiolus) are generally of the sporodochial type. On the other hand, those that are made from the stems of plants attacked by vascular diseases are more often of the cottony, ropy, or slimy pionnotal types, which are therefore not anomalies due to the artificial growth medium.

In order to explain this diversity we may note: i) that the sporodochial types are those which in the presence of soil form the most chlamydospores; and ii) contrary to the other types, the sporodochial types have a slower linear growth on media containing saccharose as compared to media containing cellulose or methyl-cellulose, all the while producing the most sporodochia. The sporodochial types, therefore, seem best adapted to saprophytism.

From the practical point of view it does not make much difference whether one employs cultures of one type or of the other for the selection of resistant cultivars or for the study of the physiology of wilts: the essential point is that the isolates be virulent, and this quality is in general much more stable in *F. oxysporum* than the morphologic aspect (isolates which have been kept in culture for 10 years can still be extremely virulent). On the other hand, ecologists should note that in order to study the behavior of *F. oxysporum* in soil under natural conditions, they should preferably utilize non-senescent cultures (sporodochial or cottony).

Certain authors, such as Gordon (24), distinguish *F. redolens* from *F. oxysporum.* These are cultures whose macroconidia are wider and whose width can exceed 4 μm. Snyder, Hansen, and Oswald (57) have proposed a cultivar 'Redolens,' but this creation is, it seems to us, the least justified, since all the transitions between typical Redolens and ordinary *F. oxysporum* can be found.

There were at the time of our original paper about 66 specialized forms which had been described. Since then and according to Booth (7) there are now 72. The question is, does this represent a satisfactory picture of the specialization in *F. oxysporum?* Is the list not too long and are not certain authors a little bit too much in a hurry to describe a new specialized form when they have found a wilt on a new host? The principal contradictions seem to be between the extreme specialization of certain races *(F. oxysporum* f. sp. *lycopersici, pisi, melonis,* for example) and the wide host range of the f. sp. *vasinfectum* or *apii* as was shown by the work of Armstrong and Armstrong (1, 2, 3). We would like to make the following remarks.

i) In the Uredineae one observes side by side with species whose specialization is extremely advanced, such as *Puccinia recondita* or *Puccinia hordei,* others which apparently will attack plants which belong to a whole family *(Puccinia malvacearum).* One also observes in the Uredineae forms whose development (at two different vegetative stages, it is true) can occur on hosts as widely separated as pea and celery for *F. oxysporum* f. sp. *apii.* These anomalies, far from appearing absurd, may guide us in the future toward the study of a biochemical affinity between these distantly related host plants.

ii) The vegetative status of the plants utilized in tests may explain the differences in results and the contradiction between authors on the delimitation of specialized forms and races. *Fusarium oxysporum* f. sp. *niveum, melonis,* and *cucumerinum* appear to be quite distinct when one inoculates adult plants (38, 45), but much less distinct when one inoculates young plants or seedlings. It is the same case for races 1 and 2 of *F. oxypsorum* f. sp. *cubense.* It is of course a lot easier to make numerous tests on seedlings than on adult plants!

iii) The very marked, wide host range of f. sp. *vasinfectum* occurs in tropical regions, where perhaps the high temperatures overcome the defense reaction of plants to the vascular wilts, as we already know occurs with the viruses (tobacco mosaic on *N. glutinosa)* and the *Phytophthora* species (14).

Controversies on the genetic mechanisms in variation in *Fusarium oxysporum* The perfect form of *F. oxysporum* is unknown. Its relationship with *F. moniliforme,* which also has been noted serologically (37), could make us suppose that it originally was a *Gibberella,* but no one has found perithecia either in nature or in culture. Buxton (11, 12) and Buxton and Ward (13), having obtained from a culture of *F. solani* f. sp. *pisi* perithecia which they called *Nectria hematococca (= Hypomyces solani),* thought that they had been able to cross *Nectria* and *F. oxysporum. Fusarium oxysporum* and *F. solani* would therefore be related (37). Gordon (30), Snyder and Alexander (53), and Reichle, Snyder, and Matuo (50) have strongly criticized Buxton's results, which they considered to be erroneous. The morphologic variations and the parasitic variations in *F. oxysporum* must therefore be explained by a system other than the sexual one. Here also Buxton (11, 12) has presented interesting hypotheses which bring into action heterokaryosis (presence in the same thallus of nuclei of different genetic constitution, redistribution of these in branching hyphae and in conidia), parasexuality (exchange of chromosome segments between nuclei of a heterothallic thallus), and adaptive mutation (transformation of race 1 to race 2 of *F. oxysporum* f. sp. *pisi* under the influence of root exudates of resistant peas). Snyder (52) has severely criticized both these concepts and Buxton's results; he believes that heterokaryosis plays only a negligible role in the variation in *F. oxysporum* and that parasexuality plays an even lesser role. He observed in effect that the nuclei of the mycelial cells and of the conidia are always derived from a single initial nucleus. As far as the adapted mutations are concerned, if a simple contact of 3-weeks duration between conidia with the root exudates of the resistant host were sufficient to create new races of *F. oxysporum,* as was affirmed by Buxton, then the appearance of these in nature should be much more rapid and much more general than it really is.

We are therefore reduced to interpreting the biologic variation in *F. oxysporum* by mutations of the classic type, the survival of the mutants being determined by the varieties cultivated in rotations. Morphologic variation is probably due to classic mutations (pigments, size of conidia) and to a senescence that is irreversible in culture, as shown by Follin and Laville (18), but does not affect virulence, which is without doubt a much more fundamental hereditary characteristic than colony morphology in the specialized forms of *F. oxysporum.*

Fusarium solani (Mart.) (Appel & Wr.) Snyd. & Hans.

As in *F. oxysporum,* this species is worldwide in distribution and occurs under all climates. In the isolations made by Gordon (26, 27) it represents 5–10% of the *Fusarium* microflora of Canadian soils and 8–20% in those of Guillemat and Montegut (31, 32, 33). It is abundant on vigorous and senescent roots, especially of the Solanaceae. *Fusarium solani* also contains specialized forms. These are fewer in number and less specialized than *F. oxysporum.* They attack either at the least a whole species or at the most several genera of the same family. Races here are more determined by the types of symptoms than by parasitic specialization. The specialized forms of *F. solani* cause either root rots, basal stem rots, or rots of fruits or tubers.

This division in specialized forms can be criticized. The demarcation between the f. sp. *phaseoli* and *pisi* is rather vague according to Yang and Hagedorn (62), who found virulent isolates on both plants. With a smaller collection of isolates in Europe we have come to similar conclusions: vague demarcation between the f. sp. *pisi* and *phaseoli,* while the f. sp. *cucurbitae* race 1 and *radicicola* seem well separated. One could also suppose that the f. sp. *eumartii* and *radicicola* would be better designated as races 1 and 2 of the same specialized form, the host being the same and only the symptoms being different.

From the morphological point of view one can see in *F. solani,* as in *F. oxysporum,* very large variations in the shape and the size of the macroconidia, the production of microconidia (very few in certain isolates), and the abundance of sporodochia, though the clones that do not produce them in abundance are the exception in this case.

Having obtained perithecia from saprophytic clones *(Hypomyces solani)* (Rke. and Berth.) Snyd. & Hans. and of *Hypomyces solani* f. sp. *cucurbitae* (Snyd. & Hans.), Snyder and Hansen have shown that very large variations in the size of the conidia, pigmentation, and speed of growth can be observed among the cultures derived from single ascospores of a perithecium. These morphologic variations, which encompass the sections of Martiella and Ventricosum of Wollenweber, therefore have no systematic value and cannot be used as indicators of whether a clone belongs to this or that specialized form. One can observe, however, that the *F. solani* f. sp. *radicicola* has the 'coeruleum' look (conidia short with obtuse extremities, dark indigo pigments going to rose on certain media such as malt extract or potato dextrose agar, which confirms the value of media based on oatmeal for the examination of colors of colonies). On the other hand, the f. sp. *phaseoli* and even more the f. sp. *cucurbitae* race 1 present in general conidia that are rather long and pointed, the saprophytic forms being intermediate in these respects.

Snyder, Hansen, and Oswald (57) have recognized the existence of a cultivar Coeruleum. One could just as well take up the old designation Ensiforme for the clones that are very pale in color and whose spores are long. We do not think that these distinctions are useful.

The saprophytic forms are often homothallic; on the other hand, f. sp. *cucurbitae* races 1 and 2 and *phaseoli* are heterothallic, the perithecia being formed upon the confrontation of isolates having different compatibility types which often come from well-separated regions. One should not, however, be adamant that all pathogenic isolates of *F. solani* are heterothallic. One can observe in Guadeloupe (French West Indies) an *F. solani* that is homothallic, produces abundant red perithecia in culture, causes collar rot of eggplant, attacks potato stems, and rots potatoes as vigorously as *F. solani* f. sp. *radicicola.* Although homothallic, it behaves like a *F. solani* f. sp. *eumartii.*

Fusarium episphaeria **(Tode) Snyd. & Hans.**

Numerous saprophytic forms and parasites on insects have been regrouped in this species by Snyder and Hansen (56). The saprophytes found in all climates on various substrates and in the soil [1–5% of the isolations made by Gordon (26, 27)] correspond to the section Eupionnotes of Wollenweber & Reinking (61) and are composed of forms whose conidia are two-celled (which Snyder, Hansen, and Oswald (57) recognize as the cultivar Dimerum), and conidial forms with several cells, such as the old *F. merismoides* of Wollenweber. These various forms of *Fusarium* in fact form a homogeneous unit, characterized by the crescent shape of the macroconidia, a poorly marked foot-cell, and a pointed tip.

It is to this group that we can tie in the *F. biasolettianum* observed on vines in Italy (8) and in France (4, 47, 48) and forming reddish crusts at the points of sap exudation in the spring. The perithecia of this *Fusarium* were classified by Briosi and Farnetti (8) in *Hypomyces,* which shows the fragility of the distinction between *Hypomyces* and *Nectria* since we must allow now that the perfect state is *Nectria episphaeria* (Tode) Snyd. & Hans.

To this group of saprophytes, Snyder and Hansen add a certain number of Fusaria which are capable of parasitizing scale insects under humid conditions in tropical climates. They also form colonies of very slow growth, nearly completely sporodochial. These Fusaria generally have very long conidia (as long as but thinner than those of *F. rigidiusculum).* Having obtained mutants with short spores analogous to the type merismoides, Snyder and Hansen tie these large-spored forms to *F. episphaeria,* creating the cultivar Gigas (57) (and for us, therefore, the variety *gigas)* and the f. sp. *coccophilum* for these forms. The perfect state is *Nectria episphaeria* f. sp. *coccophila.*

Next to *F. episphaeria* f. sp. *coccophilum* we should recall the existence of the sections Microcera and Submicrocera of Wollenweber whose habitat is likewise that of scale insects or the stroma of Sphaeriales.

The macroconidia, likewise crescent-shaped, are notable by their extreme thinness, but could easily be integrated into *F. episphaeria,* given the great variability of this species. Unfortunately the perfect state here is *Calonectria.* Snyder and Hansen (56) considered *Fusarium ciliatum* (perfect state *Calonectria ciliata)* uniting the sections Microcera and Submicrocera, a doubtful species. In 1965 in their schematic diagram of the nine species of *Fusarium,* Snyder and Toussoun (58) omitted *F. ciliatum* without mentioning whether or not it was included in *F. episphaeria.*

Fusarium lateritium (Nees) Snyd. & Hans.

This *Fusarium* is found in temperate as well as tropical areas on varied substrates and even in soil. It is known especially as a secondary invader or as a more or less virulent parasite of tree twigs and vines. Wollenweber and Reinking (61) distinguish in this species the varieties *mori* (characterized by its host), *majus* (conidia slightly longer than the type, 35 μm instead of 30), and *longum* (42μm), *uncinatum* (bent spores), and *minus* (short conidia, few septations). Doidge (17) and Gordon (24) mention a larger variation in the length of the conidia (22–43 μm) without distinguishing varieties. *Fusarium lateritium* has been found in France on mulberry *(Morus alba)* (4). Recently Gambogi (20) has observed severe attacks on citrus species in Italy which had been damaged by frost. In the tropics numerous trees are attacked. We have seen it abundantly in the vicinity of Avignon on branches of elder (*Sambucus nigra),* on mulberry, and on dead or dying locust tree *(Robinia pseudoacacia),* as well as on madder *(Rubia peregrina).* In the region around Paris (following the work of Guyot and Montegut, *unpublished data)* we have made numerous collections of *F. lateritium* on Clematis, locust tree, and spindle tree, either from sporodochia or from perithecia of the *Gibberella* type. None of these isolates could be identified with the Wollenweber varieties (conidia measured from 19–33 μm in length).

Snyder and Hansen (56) underlined the strong affinities between *F. roseum* and *F. lateritium.* The perfect states *(G. rosea* and *G. lateritia)* are similar and the habitats of *F.*

lateritium overlap that of *F. roseum* var. *sambucinum*. Among the isolates that we have obtained from dying twigs around Avignon, some clearly show the characteristics of *F. lateritium* (slow-growing colonies, entirely sporodochial without aerial mycelium, macroconidia with beak-like tips), and others evidence those of *F. roseum* var. *sambucinum* (growth slightly more rapid, with numerous small aggregations of conidia on aerial mycelium, macroconidia bent into a hook at the tip). Still others show a combination of the characteristics: conidia with beak-like tips and growth of the *sambucinum* type, or conidia of the *sambucinum* type with sporodochial growth. The same situation obtains in the region around Paris for the forms isolated from *Clematis, Ulmus,* and *Euonymus,* obtained from sporodochia or from perithecia; they are difficult to relate in a precise manner to the descriptions given by Wollenweber and Reinking (61) for the diverse varieties of *G. baccata (=G. lateritia)* or *G. pulicaris (=G. rosea* corresponding to *F. roseum* var. *sambucinum).* From the drawings and the photographs of Gordon (24), one can equally note that his *F. sambucinum* f. 6 presents strong affinities with *F. lateritium.* A culture of *F. lateritium* from the laboratory of Snyder, which has come to us throught the intermediary of Bouhot (Versailles), likewise shows intermediary aspects.

We will conserve, therefore, the species *F. lateritium* only on a provisional basis. It would be very interesting to study the variability of this fungus in order to determine whether or not it could be transformed into a *F. roseum* variety *lateritium.*

If the pathogenesis of *F. lateritium* is rather weak on Citrus species, there are on the other hand about seven known formae speciales; of these, three attack *Cicer arietinum* and the tropical legumes *Crotalaria juncea* and *Cajanus indicus.* Since similar diseases on these hosts have also been ascribed to *F. oxysporum* or its synonym *F. othoceras,* a careful examination of these diseases is needed, in our opinion, so as to determine whether there really are two distinct vascular diseases on each host or whether Gordon made an error in placing these pathogens as formae speciales of *F. lateritium.*

***Fusarium nivale* (Fr.) Snyd. & Hans.**

This species, whose name brings forth a northerly and cold habitat, sometimes is found in tropical regions. In fact, Snyder and Hansen attach to *F. nivale* the old *F. larvarum,* which parasitizes scale insects along with *F. episphaeria* f. sp. *coccophilum;* Gordon has recently observed *F. larvarum* in Trinidad (28). This lumping needs to be looked at again, since Booth describes a *Nectria (N. aurantiicola)* as the perfect state of *F. larvarum,* which he considers to be synonymous with *N. episphaeria* (6). This would suggest a tie-in between *F. larvarum* and *F. episphaeria,* the former constituing a new variety of the latter with conidia having blunt tips. Much more typical are the parasites of the Gramineae in the temperate regions that constitute, according to Snyder and Hansen (56), the specialized form *F. nivale* f. sp. *graminicola.* It differs from the other Fusaria by its lower optimum temperature for growth (18 C) and its much weaker resistance to various factors which are easily overcome by the other species: high osmotic pressures (sugar or salt), methyl and ethyl alcohol.

The perfect state would be *Calonectria nivalis* Schaffnit, Snyder and Toussoun (58) having rejected the name *Griphosphaeria* proposed by Muller and Von Arx (41). The two characteristics that distinguish the perithecia of *F. nivale* are in fact not sufficient to create a new genus: if one takes into account the late septation of the ascospores, one should then, in order to be logical, take up equally the genus *Lisea* for the perfect state of *F. moniliforme.* The "enclosed within the tissues of the host" character of the perithecia varies in culture on sterilized plant substrates: the perithecia will form like those of *Gibberella* if the support is fairly stiff (straws without sheaths and a sufficient humidity). Finally, while the wall color is rather indefinite at maturity, the young perithecia are pink.

***Fusarium roseum* (Lk.) Snyd. & Hans.**

Fusarium roseum constitutes the largest regrouping that Snyder and Hansen (56) did in *Fusarium*, since it unites four sections of Wollenweber: Roseum, Arthrosporiella, Gibbosum, Discolor. One should emphasize that with the exception of Gibbosum these groupings by Wollenweber and Reinking (61) were rather arbitrary. For example, the cultures producing microconidia were dispersed in the sections Roseum, Arthrosporiella, and Discolor, obviously showing that this characteristic was not considered important at this stage of the determination, although it is the first question posed in the key to the sections. One can therefore agree with Snyder and Hansen when they consider without value the distinctions given by Wollenweber in the four sections, and it certainly isn't by following Gordon and Booth in their rather illogical system that we will make the situation better.

Taking note of the amplitude, perhaps excessive, of the simplification that they had proposed, Snyder, Hansen, and Oswald (57) distinguished six cultivars within *F. roseum:* Culmorum, Graminearum, Avenaceum, Sambucinum, Equiseti, and Acuminatum. The first three cultivars are considered as parasites constituting the f. sp. *cerealis* and the last three are considered saprophytes.

Unlike the cultivars of *F. oxysporum, F. moniliforme, F. solani,* and *F. episphaeria* which designate extreme types, the common clones bearing only the name of the species, it seems that the cultivars of *F. roseum* represent for Snyder, Hansen, and Oswald the principal morphologic tendencies within the species *F. roseum.* We will keep as botanical varieties four of these cultivars (Sambucinum, Graminearum, Avenaceum, and Culmorum). As numerous recent American mycologists have done, we will regroup Equiseti and Acuminatum in a single variety *gibbosum* corresponding to the old section of Wollenweber, which constitues a homogeneous entity. We will reunite in a final variety *arthrosporioides* the forms producing microconidia that were not included in the six cultivars of Snyder, Hansen, and Oswald (57).

Do the six varieties thus delimited constitute a fairly homogenous unit which could form a single species? Certainly not to the same extent as *F. oxysporum* and *F. solani,* but it is likewise impossible to give them specific rank since Snyder, Hansen, and Oswald have mentioned transitions by mutation from *sambucinum* to *culmorum, sambucinum* to *gibbosum, gibbosum* to *avenaceum,* and *graminearum* to *avenaceum.* The transition *graminearum* to *culmorum* has never been observed, however.

***Fusarium roseum* var. *arthrosporioides* (Sherb.) n. comb.** In some clones of the other varieties of *F. roseum,* unicellular conidia of a generally irregular form are occasionally produced, but there exists on the other hand a group of forms, characterized by the abundant production of ellipsoidal or slightly pear-shaped conidia, which by all other characters, in particular the coloration of the colonies, belong to *F. roseum.* Dispersed by Wollenweber in the sections Roseum, Arthrosporiella, and Discolor, these Fusaria are characterized by ramified conidiophores whose denticulate extremities successively produce unicellular or septate conidia in a manner similar to that of *Fusicladium pirinum.* The abundance of these conidiophores gives a powdery appearance to the colonies. The conidia are easily dislodged, and secondary colonies appear frequently at the botton of the tubes if shaken. On rich media sporodochia can be formed. Regardless of their length or width, the macroconidia are characterized by an irregular curvature. We propose uniting this assemblage of forms under *F. roseum* var. *arthrosporioides* (Sherb.) n. comb., the old *F. arthrosporioides* constituting a type which represents the mean in the variety.

Fusarium roseum var. *arthrosporioides* is found on various substrates and even in the soil, although fairly rarely according to Gordon (24, 25, 29) (in tropical regions short-spore isolates of the *semitectum* type are regularly found in soil and form chlamydospores). We have found it on cucurbits (melons, squash) and eggplant, where it causes a flower end rot. We have also seen it on *Arundo donax* as a weak parasite on broken or cut stems, and on corn, growing on sugary exudates at the orifice of galleries of the European corn borer. It would seem that *F. roseum* var. *arthrosporioides* is not a very important pathogen, but interest in it seems to center on its potential in bridging the gap between *F. tricinctum* and *F. roseum,* as is suggested by the drawings of Gordon (24). Here again a careful study on the variability of *F. tricinctum* and *F. roseum* var. *arthrosporioides* is necessary before considering *F. tricinctum* as a simple variety of *F. roseum.* One also finds intermediary forms between *F. roseum* var. *arthrosporioides* and *F. roseum* var. *avenaceum* which may be fairly virulent on cereals, in particular a clone causing death of *Agrostis alba* var. *stolonifera* which is cultivated in Italy on artificial substrates.

***Fusarium roseum* var. *sambucinum* (Fuckel) Snyd. & Hans.** We mentioned the habitat (twigs and stems) and the morphologic aspect of *F. roseum* var. *sambucinum* when we were comparing it with *F. lateritium,* to which it has strong affinities. Its perithecia, which belong to *Gibberella rosea* (Lk.) Snyd. & Hans., have been called *G. pulicaris* or *G. cyanogena.* According to Gordon, who has made a special study of *G. cyanogena* (it corresponds to his *F. sambucinum* f. 6), *F. roseum* var. *sambucinum* is heterothallic but *F. roseum* var. *graminearum* is homothallic. This should not stop us from considering the perfect states as belonging to *G. rosea,* similar conditions occurring in *Hypomyces solani* and *Calonectria rigidiuscula. Fusarium roseum* var. *sambucinum* is also found in the soil, where it can survive by means of chlamydospores. In a 1964 soil analysis of a plot in corn monoculture since 1958 in the region around Paris, we identified (in addition to 14% *F. oxysporum,* 7% *F. solani,* 34% *F. roseum* var. *gibbosum,* and 22% *F. roseum* var. *culmorum)* 10.6% of the isolates as belonging to the group *sambucinum-lateritium,* with all the intermediaries between these two types. In Canada, Gordon (29) reports that *F. roseum* var. *sambucinum* causes a powdery rot of potatoes.

***Fusarium roseum* var. *gibbosum* (Wr.) n. comb.** The old section Gibbosum of Wollenweber grouped together an assemblage of forms presenting very typical common characteristics: abundance of chlamydospores in culture and a typical curvature of the conidia regardless of their length or width. The other characteristics that distinguish the species within the section are subject to frequent mutations that can alter the color of the colonies (brown, beige, carmine, or dirty white) or the length of the conidia. Gordon (25), who kept the three species *F. equiseti, F. acuminatum,* and *F. compactum* in the Gibbosum section, described a remarkable mutation in 1954 which originated in *F. compactum.*

In 1957, Snyder, Hansen, and Oswald (57) recognized two cultivars, Equiseti (short spores) and Acuminatum (long spores), although a continuous variation is observable between the Equiseti and Acuminatum types. Actually the collaborators of Snyder seem to prefer grouping these two into one single cultivar, Gibbosum (42), and we will follow their example by considering that all the old members of the section Gibbosum of Wollenweber and Reinking (61) constitute *F. roseum* var. *gibbosum.*

This *Fusarium,* like *F. oxysporum* and *F. solani,* is a major constituent of the microflora of soils in all climates. In Canada, according to Gordon (26, 27), it represents 16–29% of the isolates of *Fusarium* from soil, whilst the groups Avenaceum, Graminearum, and Culmorum represent only 1–2%. Mycologists who isolate from soil brown-beige colonies with abundant mycelium strewn with chlamydospores (often in packets of two or three) should always think, before putting them in the *Mycelia sterilia,* that they could be a *Fusarium roseum* var. *gibbosum,* a fact which is not evident at first glance for a beginner. If the tubes are placed under alternate light at 18 C a few macroconidia will form which will confirm the diagnosis. It is also isolated from the collar and roots of dying plants, but apparently is not a strong pathogen. It seems to be most virulent on carnation (40). It is possible that *F. roseum* var. *gibbosum* is a weak pathogen of fruits and flowers in the tropics or in greenhouses of temperate countries: bananas in Central America, according to Badger (5); flowers of *Strelitizia reginae* in glasshouses in Italy, according to Garibaldi (21); and melons in France.

After a certain time in culture, the majority of *F. roseum* var. *gibbosum* mutate: the pigmentation of the mycelium weakens, changing from beige with brown zones to a dirty white, and the production of chlamydospores diminishes. Mutants often appear, producing sporodochia and a rose or carmine pigment. Perfect states belonging to the genus *Gibberella (Gibberella intricans, G. acuminata)* have been described in relation with *F. roseum* var. *gibbosum.* They seem to be fairly rare and should be brought into *Gibberella rosea* (Lk.) Snyd. & Hans.

***Fusarium roseum* var. *culmorum* (Schwabe) Snyd. & Hans.** This *Fusarium,* characterized by its short and fat conidia, is especially known as a damping-off agent and as a cause of necrosis of roots on cereals; it also attacks the heads and grains of cereals at maturity, and is especially known in this manner on corn around Paris. Thanks to its abundance of chlamydospores it survives well in soil. Gordon isolated it frequently from cultivated soils planted to cereals in Canada (1.65% of the total *Fusarium* population). In our observations, these proportions can range from 7% (Versailles) or 15% (Clermont-Ferrand) in cultivated soils alternating corn and winter cereals or monoculture of corn, to as high as 22% in soils of Grignon. The populations of *F. roseum* var. *culmorum* seem to be less stable in soil than the variety *gibbosum,* however, reaching a maximum immediately after cereal harvest and a minimum at the end of winter (in the case of corn). *Fusarium roseum* var. *culmorum* is not strictly dependent on cereals, since it is capable of attacking a large number of plants, including asparagus, *Allium,* and carnation. We have also found it on rotten carrots and on *Arundo donax.* Gordon (29) has observed it on many substrates.

Although usually stable in culture, *F. roseum* var. *culmorum* can, according to Snyder and Hansen (56) and Oswald (44), produce mutants with smaller conidia that could cause confusion with *F. roseum* var. *sambucinum;* these conidia also do not have the usually characteristic carmine red coloration. We have observed a mutation of this type. The perfect state has never been observed.

***Fusarium roseum* var. *graminearum* (Schwabe) Snyd. & Hans.** Unlike *F. roseum* var. *culmorum,* this fungus seems to be characteristic of oceanic climates where winter is mild and summer is hot and humid. It is abundant on cereals in France, particularly in the southwest regions, where one routinely finds its perithecia on stands and leaves of corn as well as on the blooms of wheat. If one makes isolations from these perithecia by monoascospore culturing, some of them (about 60%) are capable of forming perithecia on artificial media with large variations in both abundance and appearance; these, therefore, can be called homothallic. On the other hand, we do not know whether those single ascospore cultures which did not produce perithecia are heterothallic or are neuters. The capability of forming perithecia generally does not last long in culture, no more than 1–2 years following continuous culture on agar media. The aerial mycelium (brilliantly colored in red and brown with golden reflections on recently isolated clones) becomes whitish and production of conidia drops to almost nothing. One can also observe, after repeated single spore culture, an evolution in the opposite direction, toward an abundantly sporulating type characterized by conidia that present a constriction at the level of the septations. This type produces numerous chlamydospores in the mycelium and in the conidia, and is also sometimes capable of producing perithecia. Certain clones isolated directly from soil in the southwest of France can produce conidia abundantly, but are incapable of producing perithecia. Their conidia are a little bit larger and more curved than those that come from perithecia, but they are nevertheless identifiable as *F. roseum* var. *graminearum* rather than variety *culmorum* because of the general shape of the conidia and the appearance of the colonies.

Starting with *F. roseum* var. *graminearum*, Snyder, Hansen, and Oswald (57) have obtained mutants with long spores analogous to *F. roseum* var. *avenaceum*, and mutants that produce chlamydospores but are nevertheless not similar to *F. roseum* var. *culmorum*.

In summary, the great variability of *F. roseum* var. *graminearum* in culture distinguishes it from the variety *culmorum*, which is much more stable. One finds *F. roseum* var. *graminearum* on *Allium* sp., bean, and in particular in the mixed cultures of corn and bean traditional to the southwest of France. In the eastern USA, *F. graminearum* causes severe losses to greenhouse carnations; ascospores produced at the base of dead carnations cause stub die back (43).

***Fusarium roseum* var. *avenaceum* (Sacc.) Snyd. & Hans.** This variety brings together those cultures with very long and very thin conidia and which do not produce chlamydospores. Do these cultures form a homogenous group? This is doubtful because, according to Snyder and Hansen (56) and Oswald (44), cultures of the type Avenaceum can be derived on the one hand from *F. roseum* var. *graminearum* (mutants with long spores), and the other from *F. roseum* var. *gibbosum* (loss of chlamydospores). This might explain the rather large variations, both macroscopic and microscopic, in *F. roseum* var. *avenaceum:* certain cultures produce a carmine red pigment, others produce none; some produce sporodochia, others produce none; some produce spores which are curved, others are abruptly bent.

As with the variety *graminearum, F. roseum* var. *avenaceum* is not very stable in culture and one can observe loss of the carmine red pigment, variations in the quantity of sporodochia produced, and even a senescence which is analogous and leads to the slimy pionnotal type in *F. oxysporum* often accompanied by a production of conidia which are doubly bent. As with the two preceding varieties, *F. roseum* var. *avenaceum* can invade cereals in various stages of their development as well as carnation and *Allium* sp. It is likewise isolated from roots and collars of dying leguminous plants such as pea and bean. Lansade (35) finds it on potatoes and Gordon (29) on various substrates. A perfect state *Gibberella* has been recently described on wheat by Cook (15). It has been designated as *Gibberella avenacea* by Cook; according to us this would be better called *G. rosea* var. *avenacea* especially if one keeps, as Cook does, *F. roseum* var. *avenaceum* for the conidial form.

Specialized forms in *Fusarium roseum* We have tried to ascertain the existence of formae speciales by inoculating a number of hosts with a collection of *F. roseum* strains originating from various substrates; the results are summarized in Table 37-1. We think that only the situation found with the strains of *F. roseum* var. *culmorum* isolated from and virulent towards *Allium* sp. could lead to the definition of a specialized strain. The mechanism of this specialization is very rudimentary (resistance in vivo and in vitro to allicine, the garlic phytonicide). The delimitation of a f. sp. *cerealis* would be different on maize and wheat seedlings. On potato tubers the very aggressive *F. roseum* var. *sambucinum* strains we have observed have been isolated from elder or elm twigs, and they were also very aggressive on pea seedlings. Therefore a *F. roseum* var. *sambucinum* does not need to have been isolated from pea or from potato in order to behave as virulent on these hosts: the conditions for the delimitization of a specialized form are not realized. But it is a general rule, in temperate countries (Canada, France), that a collection of *Fusarium* strains isolated from rotten potatoes will be composed of a number of strains of *F. roseum* var. *sambucinum,* some *arthrosporiella-avenaceum* intermediates, some var. *culmorum,* var. *graminearum,* and var. *avenaceum* and also, of course, *F. solani* f. sp. *radicicola.* Situations which are different but of a similar nature can be defined for pea or carnation, but they are too complex to be expressed by the concept of "forma specialis."

Table 37-1 The reaction of different host plants to inoculation with several varieties of *Fusarium roseum*

Host Plant	Varieties of *F. roseum*						
	sambucinum	culmorum	graminearum	avenaceum	X[a]	arthrosporioides	gibbosum
Young garlic plants	[b]	Severe[c]					
Maize seedlings		Severe	Severe	Mild			
Wheat seedlings	Moderately Severe	Severe	Severe	Moderately Severe	Severe		
Pea seedlings	Severe	Severe	Severe	Severe	Severe		
Potato tubers	Severe	Moderate	Moderate	Severe	Severe		
Carnation cuttings	Moderate		Severe	Severe			Mild

[a]X = a strain intermediate between *arthrosporioides* and *avenaceum* isolated from *Agrostis* sp.
[b]No entry means the host plant showed no reaction.
[c]Only with clones isolated from *Allium* sp.

Conclusion All those who have worked on the taxonomy of *Fusarium* since 1945 have taken a stance half way between the ideas of Wollenweber and Reinking (61) and of Snyder & Hansen (54, 55, 56). In 1945 we accepted the latter and have had in general no difficulty in classifying the Fusaria we isolated from France and North Africa as well as those we received from other countries. Only those *F. roseum* isolates which produce microconidia made us hesitate at first. Clearly therefore, we chose to make only minor modifications (botanical varieties instead of cultivars, rules of latin grammar, etc.) to the Snyder & Hansen system, the sole exception being a new subdivision within *F. roseum* which would group the microconidia-producing isolates.

Contrary to what Gordon (24) and Booth (7) have done, we carefully avoided abandoning the identity of *F. roseum* and giving specific rank to the entities *avenaceum, culmorum, gibbosum,* etc. There is no doubt in our minds that it is preferable to identify isolates combining characteristics of two of these varieties as "atypical *F. roseum,*" rather than hesitating between the sections Roseum, Discolor, Arthrosporiella, or Gibbosum without ever arriving at a clear-cut answer.

The only problems that remain in the system we have proposed are at the boundaries between *F. lateritium* and *F. roseum* var. *sambucinum,* which make us wish for critical studies which would allow *F. lateritium* to be designated as a variety of *F. roseum.* A similar but less urgent study is needed for *F. tricinctum.* These are the borderline cases which make systematics the difficult art that it is and whose solution is always such a pleasure to attain. It is to the credit of Snyder & Hansen and their collaborators that they have continually expressed the idea that systematics should be a useful tool to all biologists interested in organisms, rather than a subject created for the delectation of a few specialists.

Insofar as pathogenic specialization is concerned, one should guard against treating all the *Fusarium* species in the same manner. Among those species most often found in temperate regions, the specialization of *F. oxysporum* is comparable to that of the Uredineae and completely justifies the creation of specialized forms and of races. That of *F. solani,* less pronounced, does not go beyond the specialized form (used in its the widest sense) adapted to a genus or family. Finally, the specificity of *F. roseum* appears to be no greater than that of *Rhizoctonia solani* and in our view does not justify specialized forms.

Literature Cited

1. Armstrong, G. M., and J. K. Armstrong. 1959. *Physalis alkekengi* a new host for the U.S. cotton, cassia, and sweet potato wilt Fusaria. Plant Dis. Rep. 43:509–510.
2. Armstrong, G. M., and J. K. Armstrong. 1960. American, Egyptian and Indian cotton wilt Fusaria, their pathogenicity and relationship with other wilt Fusaria. U.S. Dep. Agric. Tech. Bull. 1219.
3. Armstrong, G. M., and J. K. Armstrong. 1961. Some clovers susceptible to the Cassia, cotton and other wilt Fusaria. Phytopathology 51:642 (Abstr.).
4. Arnaud, G., and M. Arnaud. 1931. Traité de pathologie vegetale. Paul Lechevalier, Paris.
5. Badger, A. M. 1965. Influence of relative humidity on fungi causing crown rot of boxed bananas. Phytopathology 55:668–692.
6. Booth, C. 1959. Studies on Pyrenomycetes. IV. *Nectria* (part 1). Mycol. Paper 73. Commonw. Mycol. Inst., Kew, Surrey, England.
7. Booth, C. 1971. The genus *Fusarium*. Commonw. Mycol. Inst., Kew, Surrey, England. 237 p.
8. Briosi, G., and R. Farneti. 1901. Intorno ad un nuovo tipi di licheni. Atti. Ist. Bot. Pavia.
9. Brunt, A. A., and A. L. Wharton. 1962. Etiology of a gall disease of Cocoa in Ghana caused by *Calonectria rigidiuscula*. Ann. Appl. Biol. 50:283–289.
10. Bugnicourt, F. 1939. Les *Fusarium* et *Cylindrocarpon* de l'Indochine. Paul Lechevalier, Paris. 206 p.
11. Buxton, E. W. 1958. A change of pathogenic race in *Fusarium oxysporum* f. *pisi* induced by root exudate from a resistant host. Nature, London 181:1222–1224.
12. Buxton, E. W. 1958. Mechanisms of variation in *Fusarium oxysporum* in relation to host-parasite interactions, p. 183–191. *In* C. S. Holton et al. (ed.), Plant pathology, problems and progress 1908–1958, Univ. Wisc. Press, Madison. 588 pp.
13. Buxton, E. W., and V. Ward. 1962. Genetic relationship between pathogenic strains of *Fusarium oxysporum, Fusarium solani* and an isolate of *Nectria haematococca*. Trans. Brit. Mycol. Soc. 45:261–273.
14. Chamberlain, D. W., and J. W. Gerdeman. 1966. Heat-induced sensibility of soybeans to *Phytophthora megasperma* var. *sojae, Phytophthora cactorum* and *Helminthosporium sativum*. Phytopathology 56:70–73.
15. Cook, R. J. 1967. *Gibberella avenacea* sp. n., perfect stage of *Fusarium roseum* f. sp. *cerealis* 'Avenaceum.' Phytopathology 57:732–736.
16. Corbetta, G., and F. Togliani. 1955. Infezioni fungine e disinfezione della semente. Il Riso. 4:6–8.
17. Doidge, E. M. 1938. Some South African Fusaria. Bothalia, Pretoria 3:331–483.
18. Follin, J. C., and E. Laville. 1966. Variations chez le *Fusarium oxysporum*. Comportement des lignées issuées des differents organes de multiplication. Fruits 21:261–268.
19. Ford, E. J., J. A. Bourret, and W. C. Snyder. 1967. Biologic specialization in *Calonectria* (*Fusarium*) *rigidiuscula* in relation to green point gall of cocoa. Phytopathology 57:710–712.
20. Gambogi, P. 1966. Decadimento degli agrumi causato da *Fusarium lateritium*. Communication au 1er Congres de l'Union Phytopathol. Mediterr. Bari.
21. Garibaldi, A. 1964. Attachi di *Fusarium roseum* su fiori di *Strelitzia* reginae. Riv. Ortoflorofrutticolt. Ital. 89:531–534.
22. Gaudineau, M., and C. M. Messiaen. 1954. Quelques maladies cryptogamiques sur épis, tiges et feuilles de Mais. Ann. Epiphyties 3:273–299.
23. Gordon, W. L. 1944. The occurrence of *Fusarium* species in Canada. I. Species of *Fusarium* isolated from farm samples of cereal seeds in Manitoba. Can. J. Res., C, 22:282–286.
24. Gordon, W. L. 1952. The occurrence of *Fusarium* species in Canada. II. Prevalence and taxonomy of *Fusarium* species in cereal seed. Can. J. Bot. 30:209–251.
25. Gordon, W. L. 1954. The occurrence of *Fusarium* species in Canada. III. Taxonomy of *Fusarium* species in the seed of vegetable, forage and miscellaneous crops. Can. J. Bot. 32:576–590.
26. Gordon, W. L. 1954. The occurrence of *Fusarium* species in Canada. IV. Taxonomy and prevalence of *Fusarium* species in the soil of cereal plots. Can J. Bot. 32:622–629.
27. Gordon, W. L. 1956. The occurrence of *Fusarium* species in Canada. V. Taxonomy and geographic distribution of *Fusarium* species in soil. Can. J. Bot. 34:833–846.
28. Gordon, W. L. 1956. The taxonomy and habitats of *Fusarium* species in Trinidad B.W.I. Can. J. Bot. 34:847–864.
29. Gordon, W. L. 1959. The occurrence of *Fusarium* species in Canada. VI. Taxonomy and distribution of *Fusarium* species on plants, insects and fungi. Can. J. Bot. 37:257–290.
30. Gordon, W. L. 1960. Is *Nectria haematococca* the perfect stage of *Fusarium oxysporum*? Nature 183:903.
31. Guillemat, J., and J. Montegut. 1956. Contribution à l'étude de la microflore fongique des sols cultivés. Ann. Epiphyties 7:471–540.
32. Guillemat, J., and J. Montegut. 1957. Deuxième contribution à l'étude de la microflore fongique des sols cultivés. Ann. Epiphyties 8:185–208.
33. Guillemat, J., and J. Montegut. 1958. Troisième contribution à l'étude de la microflore fongique des sols cultivés. Ann Epiphyties 9:27–54.
34. Hutchins, L. M. 1965. Loss of gall-inducing capacity on Cacao when *Calonectria rigidiuscula* passes from the conidial to the perfect stage. Plant Dis. Rep. 49:565–567.

35. Lansade, M. 1950. Recherches sur la Fusariose on pourriture seche de la pomme de terre. Ann. Epiphyties 1:157–207.
36. Lemesle, R. 1935. Mycocedie florale produite par le *Fusarium moniliforme* var. *anthophilum* sur le *Scabiosa succisa*. Rev. Gen. Bot. 47:337–362.
37. Madoshing, C. 1964. A serological comparison of three *Fusarium* species. Can. J. Bot. 42:1143–1146.
38. Mas, P., and G. Risser. 1966. Caracterisation, symptomes et virulence de diverses races de *Fusarium oxysporum* Schl. f. sp. *melonis* Sn. et. Hans. I. Congr. Union Phytopathol. Mediterr., Bari:503–508.
39. Messiaen, C. M., and R. Cassini. 1968. Recherches sur les Fusarioses. IV. La Systématique des *Fusarium*. Ann. Epiphyties 19:387–454.
40. Moreau, M. 1957. Le dépérissement des Oeillets. Paul Lechevalier, Paris. 309 p.
41. Müller, E., and J. A. Von Arx. 1955. Einige Beitrage zur Systematik und Synonymie der Pilze. Phytopathol. Z. 24:353–372.
42. Nash, S. M., and J. V. Alexander. 1965. Comparative survival of *Fusarium solani* f. *cucurbitae* and *F. solani* f. *phaseoli* in soil. Phytopathology 35:963–966.
43. Nelson, P. E., B. L. White, and T. A. Toussoun. 1971. Occurrence of perithecia of *Gibberella* sp. on carnation. Phytopathology 61:743–744.
44. Oswald, J. W. 1942. Taxonomy and pathogenicity of fungi associated with root rot of cereals in California with special reference to the Fusaria and their variants. Ph.D. Thesis. Univ. Calif., Berkeley.
45. Owen, J. H. 1956. Cucumber wilt, caused by *Fusarium oxysporum* f. *cucumerium* n. f. Phytopathology 46:153–157.
46. Raillo, A. I. 1935. Diagnostic estimation of morphological and cultural characters in the genus *Fusarium*. Bull. Plant Prot. II, Leningrad (Phytopathol.) 7:1–100.
47. Ravaz, L. 1914. Sève noire et sève rouge. Progrès agricole et viticole.
48. Ravaz, L. 1928. Pleurs de sang. Progrès agricole et viticole.
49. Reichle, R. E., and W. C. Snyder. 1964. Heterothallism and ascospore number in *Calonectria rigidiuscula*. Phytopathology 54:1297–1299.
50. Reichle, R. E., W. C. Snyder, and C. Matuo. 1964. *Hypomyces* stage of *Fusarium solani* f. *pisi*. Nature, London 203:664–665.
51. Roger, L. 1953. Phytopathologie des pays chauds, Vol. 2. Paul Lechevalier, Paris. p. 1462–1544, 2142–2196.
52. Snyder, W. C. 1961. Parasexuality in the fungi. Recent Adv. Bot. Univ. Toronto Press. p. 371–374.
53. Snyder, W. C., and J. V. Alexander. 1961. Perfect stages of *Fusarium oxysporum* and *Fusarium solani* f. *pisi* still unknown. Nature, London 189:596.
54. Snyder, W. C., and H. N. Hansen. 1940. The species concept in *Fusarium*. Amer. J. Bot. 27:64–67.
55. Snyder, W. C., and H. N. Hansen. 1941. The species concept in *Fusarium* with reference to section Martiella. Amer. J. Bot. 28:738–742.
56. Snyder, W. C., and H. N. Hansen. 1945. The species concept in *Fusarium* with reference to Discolor and other sections. Amer. J. Bot. 32:657–666.
57. Snyder, W. C., H. N. Hansen, and J. W. Oswald. 1957. Cultivars of the fungus, *Fusarium*. J. Madras Univ. B 27:185–192.
58. Snyder, W. C., and T. A. Toussoun. 1965. Current status of taxonomy in *Fusarium* species and their perfect stages. Phytopathology 55:833–837.
59. Tirmnicu, M. 1965. Formele periteciale speciilor de *Gibberella* gasite pe porumb in Romania. An. Sect. Protect. Plant. 3:47–52.
60. Waite, B. H., and R. H. Stover. 1960. VI. Variability and the cultivar concept in *Fusarium oxysporum* f. *cubense*. Can J. Bot. 38:985–994.
61. Wollenweber, H. W., and O. A. Reinking. 1935. Die Fusarien, ihre Beschreibung, Schadwirkung aund Bekampfung. Paul Parey, Berlin. 335 p.
62. Yang, S. M., and D. J. Hagedorn. 1965. Pathogenicity studies with *Fusarium solani* from beans and peas. Phytopathology 55:1085.

38 Perfect States (Teleomorphs) of *Fusarium* Species

C. Booth

Introduction

The presence of the perfect or perithecial state (teleomorph)* in *Fusarium* appears, superficially at least, to be of more value to the taxonomist than to the fungus. In fact, some of the most successful *Fusarium* species (anamorphs) such as *F. oxysporum* and *F. culmorum* appear to have lost their sexual ability and to have adopted other methods of facilitating genetic adaptions. Many species which still retain their sexual capability apparently find little use for it. In the first place there is evidence of a definite mutational trend from homothallic to heterothallic strains and for these heterothallic strains to develop impaired mating systems. Thus in many species of *Gibberella* and in heterothallic strains of *Nectria haematococca (Hypomyces solani)* perithecial production is rare. In fact, available evidence suggests that perithecia are seldom produced in local populations in nature by these strains because opposite mating strains are often widely separated geographically (11, 22, 26). Burnett (4), in a sweeping summarisation of the evidence for *Nectria haematococca* f. sp. *cucurbitae,* stated that in nature no two morphological and physiologically compatible types of species exist together.

In Pyrenomycete fungi in general, the perithecial state, apart from providing a means of genetic interchange, provides a resting stage in which the ascospores can, and often do, remain viable for years. This has been found to be untrue in the genera of the Hypocreales which produce *Fusarium* conidial states. In these genera the chlamydospore or stromatic tissue has a much greater survival value than the ascospore. In fact, it is seldom possible to germinate ascospores if the perithecium has been dry for 4–6 months.

If one attempts to assess the value of the ascospore as a method of dispersal, the issue is somewhat confused. All macroconidia of *Fusarium* species are slime spores and one would expect them to be distributed by soil particles or by water. Records of isolations from spore traps seldom state whether the culture was from a microconidium, macroconidium, or ascospore. Nelson et al. (19) isolated abundant ascospores of *Gibberella zeae* from the air in greenhouses in which carnations were cultivated. In fact, there is little doubt that ascospores belonging to perithecial states of *Fusarium* species form a percentage of the air spora wherever perithecia are formed abundantly on crop residues.

To the taxonomist endeavouring to establish a species concept in the form genus *Fusarium,* where almost always identifications have to be made from cultures of the unstable conidial state, the discovery or collection of a perithecial state provides a definite starting point in the study of this variation. Each perithecium contains in its ascospores the range

*The terms teleomorph and anamorph were proposed by Hennebert and Weresub (12) for the perfect and imperfect states respectively. Perithecial state as used in this text has a more precise meaning.

of variation in morphology, pigmentation, and growth rate that the species is capable of attaining without further mutation.

Perithecial States

The known perithecial states of *Fusarium* species all belong to the Hypocreales. This order, as redefined by Rogerson (24), is one of five orders placed in the subclass Ascomycotina, the Euascomycetes or Pyrenomycetes. The order is characterised by the definite perithecial wall which is soft, fleshy or membranous, and usually brightly coloured. They have a definite ostiole, the canal of which is lined by periphyses. True apically free paraphyses are absent, but apical paraphyses which grow down from the roof of the centrum have been observed in a number of species in the early stages of development: asci are unitunicate, typically clavate and typically nonamyloid; ascospores diverse but usually 1–3 septate. Species often have two conidial states, and these, as in some species of *Fusarium,* may be referred to as microconidia and macroconidia. Where no marked difference in size exists but where two forms are produced, it is better to refer to them as primary and secondary conidia. Almost always the dominant conidial form is enteroblastic (phialidic), but the secondary form may be holoblastic. A third form, the chlamydospore common in some species, is a thick-walled thallic spore.

Initially only four genera of the Hypocreales were thought to have *Fusarium*-type conidial states. These are *Nectria, Gibberella, Calonectria,* and *Micronectriella,* but the latter genus has now been transferred to *Monographella* in the Amphisphaeriales. The first three are all superficial, with or without a stroma (depending largely on the substrate), and are somewhat arbitrarily divided into three genera. *Calonectria* is separated from *Nectria* because it has two or more transverse septa in the ascospores, and from *Gibberella* because it lacks the purple pigmentation of this genus.

Micronectriella was used by Booth (2) to accommodate the perithecial forms of a number of the more primitive *Fusarium* species. These species, *F. tabacinum, F. nivale,* and *F. stoveri,* are comparatively slow-growing species without any clear separation of the conidia into microconidia and macroconidia. The perithecia are immersed when they develop in nature, although they may form superficially in culture on straw on the agar surface. Müller (17) removed *Micronectriella nivalis* to the Amphisphaeriales as *Monographella nivalis;* no mention was made of the other two apparently related species and these still require critical study and possible redisposition.

The removal of *Micronectriella* leaves the following as recognized perithecial states:

Calonectria De Not., Comm. Critt. 2: 477, 1867.
Gibberella Sacc., Michelia 1: 43,317, 1877.
Nectria Fries, Sum. Veg. Scan. 387, 1849.

Perithecial Names with Their *Fusarium* States

Calonectria diploa (Berk. & Curt.) Wr. = *F. juruanum* P. Henn. Syn.: *Nectria diploa* Berk. & Curt. Both perithecia and conidial pustules develop on scale insects.

Calonectria rigidiuscula (Berk. & Br.) Sacc. = *F. decemcellulare* Brick Syn.: *Nectria rigidiuscula* Berk. & Br. Both homothallic and heterothallic strains were recognised by Reichle and Snyder (23). They found the homothallic strains invariably formed 4-spored asci, whereas the two heterothallic strains have 2–8-spored asci. Both occur in nature, but the 4-spored strain is most commonly found. Ford et al. (8) produced evidence that the heterothallic strains are more pathogenic.

Gibberella acuminata Wr. = *F. acuminatum* Ell. & Ev. Syn.: *Gibberella saubinetii* (Dur. & Mont.) Sacc. f. *dahliae* Sacc. Wollenweber apparently took up the name from the specimen collected by Bizzozero 1880 and issued in Saccardo's Mycothea Veneta 1467 for the perithecial state of *Fusarium acuminatum.* The genetic connection was confirmed by Gordon (11), who found it to be bisexual but self-sterile.

Gibberella avenacea Cook and *F. avenaceum* (Fr.) Sacc. This species was described by R.J. Cook (5), but following a query (in litt.) by Miss J. Dingley, further examination of the material and our improved knowledge of *F. avenaceum* suggest that it is a form of *Gibberella zeae* rather than the perithecial state of *F. avenaceum.*

Gibberella baccata (Wallr.) Sacc. = *F. lateritium* Nees Syn.: *Sphaeria baccata* Wallroth.; *Gibberella lateritium* (Nees) Snyd. & Hans., nom. illegit. Gordon (11) described this species as bisexual but self-sterile. We have found homothallic strains that have produced perithecia on sterilised straw after 3–4 weeks, but no perithecia developed on agar cultures over the same period. Perithecia appear to be widely distributed geographically on woody hosts.

Gibberella buxi (Fuckel) Wint. = *F. lateritium* var. *buxi* Booth. This is a form of *Gibberella baccata* occurring on *Buxus sempervirens* with wider ascospores. Cultural characteristics suggest that it is heterothallic.

Gibberella cyanogena (Desm.) Sacc. = *F. sulphureum* Schlecht. Syn.: *Sphaeria cyanogena* Desm., 1848; *Gibbera saubinetii* Dur. & Mont., 1856; *Gibberella saubinetii* (Dur. & Mont.) Sacc., 1879. Perithecia widely distributed on both woody and dead herbaceous hosts. Gordon (10) described it as bisexual, self-sterile, and interfertile. El-Ani (7) found male mutant strains were formed similar to those in *Nectria haematococca.* He also found that all strains had four chromosomes.

Gibberella fujikuroi (Sawada) Wr. = *F. moniliforme* Sheld. Syn.: *Lisea fujikuroi* Sawada, 1919; *Gibberella moniliformis* (Sheld.) Wineland, 1924. This species has been intensively investigated, and strains have been found to be bisexual and self-sterile. Hsieh, Smith, and Snyder (13) found that crossing attempts with 60 isolates of *Fusarium moniliforme* produced 19 compatable isolates that formed three different mating groups, these were not interfertile and perithecia were not formed when two cultures belonging to different mating groups were mated, irrespective of their sex or compatability characteristics. These mating strains may give additional support to Nirenberg (20), who regards Sawada's *Gibberella fujikuroi* to be distinct from *Gibberella moniliformis* Wineland (28). The latter she accepted as the perithecial state of *F. moniliforme* Sheld. but took up as an earlier name *Oospora verticillioides* Sacc. which she transferred as *Fusarium verticillioides* (Sacc.) Nirenberg.

This *Oospora* name dates from 1882 and thus antedates *Fusarium moniliforme.* Norton and Chen (21) first drew attention to it as a possible earlier name for *F. moniliforme,* but due to its inadequate description, which did not mention the *Fusarium* state, and also because of the lack of type material, I consider it would only lead to confusion to drop the name *F. moniliforme* from the *Fusarium* literature, especially as it is over 50 years since the possible earlier name was pointed out but not taken up. [See Booth, (2) p. 27 and (3) pp. 91–92.] For the conidial state of *Gibberella fujikuroi* Dr. Nirenberg took up the name *Fusarium fujikuroi* Nirenberg sp. nov.

Gibberella fujikuroi (Saw.) Wollenw. v. **subglutinans** Edwards. The conidial state of this species is generally referred to as *Fusarium moniliforme* var. *subglutinans* Wollenw. &

Reinking. Nirenberg transferred it to *Fusarium sacchari* (Butl.) Gams var. *subglutinans* (Wollenw. & Reink.) Nirenberg.

Gibberella gordonii Booth = *F. heterosporum* Nees ex Fr. This species is bisexual and self-sterile and the type specimen was produced on straw in culture by W.L. Gordon (11) and is deposited in Herb. IMI as No. 94015.

Gibberella intricans Wollenw. = *F. equiseti* (Corda) Sacc. Perithecia of this species were obtained by Wollenweber from bananas from the West Indies and Central America and were described by him in 1930. No further record appears to exist of its occurrence and the genetic connection has not been confirmed.

Gibberella pulicaris (Fries) Sacc. = *F. sambucinum* Fuckel. Syn.: *Sphaeria pulicaris* Fries. Although this is the type species of the genus, little work appears to have been published on mating experiments or genetic aspects. The species was found to be heterothallic, and single-ascospore cultures produced perithecia in approximately 50% of the cultures when mated.

Gibberella stilboides Gordon ex Booth = *F. stilboides* Wollenw. Perithecia were first produced by Gordon from the mating of a strain (WLG 3688) from coffee in Tanzania with a strain (WLG 4849) from coffee in Malawi. Although related to *F. lateritium,* this species has a very distinct pigmentation, and no success was achieved by cross mating of opposite strains of *F. lateritium* with those of *F. stilboides.*

Gibberella xylarioides Neim & Saccas = *F. xylarioides* Steyaert. This species causes tracheomycosis of coffee *(Coffea liberica* and *C. canephora)* and records of its occurrence shows a spread across Africa from the Ivory Coast in 1948 to Ethiopia in recent years. The most frequently occurring strain is the female, which produces perithecial initials readily in culture on agar and which has variable curved conidia. In mating experiments at the CMI the only success was achieved in using a strain with long 'lateritium'-like spores, which suggests there is a sex-linked morphological difference between the male and female strains.

Gibberella zeae (Schw.) Petch = *F. graminearum* Schwabe. Syn.: *Sphaeria zeae* Schweinitz. The perithecia of this species are the most frequently collected *Gibberella* species. It is bisexual and self-fertile, and perithecia will develop from single-ascospore or occasionally from single-conidium cultures. In artificial culture there is a cultural factor present which prohibits perithecial production. Especially in single-ascospore isolates, if no perithecia develop after 3–4 weeks and the aerial mycelium is then scraped off and the surface of the agar well washed with sterile water, perithecia will often develop in the surface of the agar. If the aerial mycelium is removed from half the plate only and this half washed, then perithecia will develop only on the washed agar.

Hypocrea

Species of this genus that have been studied in culture produced a *Trichoderma* conidial state or none at all. In 1960 Matuo and Kobayashi (16) reported that *Hypocrea splendens* produced a *Fusarium* conidial state which they named *Fusarium splendens.* In their photographs the perithecial stroma was atypical of *Hypocrea* species; as suggested by Doi (6), Matuo & Kobayashi have most likely collected a *Nectria* hyperparasite. Doi was unable to obtain any conidial state in his isolates from *Hypocrea splendens.*

Hypomyces (Fries) L. & R. Tulasne, Ann. Sci. Nat. Bot. IV. 13: 11, 1860. Syn: *Hypocrea* subgenus *Hypomyces* Fries, Syst. Orbis. Veget. 105, 1825. The genus *Hypomyces* erected by the Tulasnes was based on the Fries subgenus and contained nectrioid perithecia formed on a byssoid mycelium parasitic on other fungi. Maire (15), who redefined the genus, stressed the importance of the generally warted, fusiform or spindle-shaped ascospores with apiculate ends. Species with these characteristics form a uniform group with a *Cladobotryum* or *Verticillium* conidial state and they may produce aleuriospores (chlamydospores) or variable forms such as *Trichothecium* but not *Fusarium.* The genus was taken up for perithecial states of *Fusarium solani* because Wollenweber (29) placed in *Hypomyces* all nectrioid species with chlamydospores. This is not a valid separation because the presence or absence of chlamydospores in ascospore cultures of *Nectria* is not a generic distinction. *Nectria* is the correct genus for these *Fusarium*-producing forms.

Micronectriella cucumeris (Kleb.) Booth = *F. tabacinum* (Beyma) Gams. Syn.: *Plectosphaerella cucumeris* Klebahn 1930. Homothallic and self-fertile perithecia often develop abundantly in fresh isolates. As *Micronectriella* is now untenable (see below), the name *Plectosphaerella cucumeris* Klebahn should be used.

Monographella nivalis (Schaffnit) Müller, Revue de Mycologie 41: 129–134, 1977. Syn.: *Calonectria nivalis* Schaffnit, 1913; *Griphosphaeria nivalis* (Schaffnit) Müller & von Arx, 1955; *Micronectriella nivalis* (Schaffnit) Booth, 1971; *Calonectria graminicola* Wollenw. (non Berk. et Br.), 1913. Anamorph: *Gerlachia nivalis* (Ces. ex Sacc.) W. Gams & E. Müller, Neth. J. Plant Path. 86: 45–53. 1980; *Fusarium nivale* Ces. ex Sacc., Syll, Fungi 10: 726, 1892.

The controversy which has arisen in the past regarding the correct name for this species has been centered on the septation of the ascospores (they begin as 1-septate but become 3-septate when mature) and also on the immersed perithecia as found in nature. When Ihssen (14) first found the perithecia he took up the name *Nectria graminicola* Berk. & Br. Wollenweber, observing that the ascospores became 3-septate, transferred the species to *Calonectria* as *C. graminicola* (Berk. & Br.) Wollenw. However, after Weese had shown that Ihssen's fungus was not the same species as Berkeley and Broome had described as *Nectria graminicola,* Schaffnit (25) re-described it as *Calonectria nivalis.* Subsequent controversy first related to the position of the perithecium in the host tissue, and this led to the transfer into *Griphosphaeria* as *G. nivalis* (Schaffn.) Müller and von Arx (18), which was later transferred by Booth (2) into *Micronectriella.* Further controversy has primarily been directed towards which order this fungus belongs to. Although Booth wished to keep it in the Hypocreales, Müller (17) found that Hohnels genus *Micronectriella,* although erected for *Calonectria*-like species with immersed perithecia, was in fact based on *Micronectriella pterocarpi,* which has a bitunicate ascus and is related to *Sphaerulina* in the Dothideales. In his examination of 'nivalis' perithecia he found they had a structure more closely related to the Amphisphaeriales than to the Hypocreales and, although not necessarily more than a species character, the ascus has an amyloid apical ring which is atypical in the Hypocreales.

Because of this evidence Müller (17) transferred it to the genus *Monographella* as *M. nivalis.* This transfer was further supported by the developmental studies of Subramanian and Bhat (27). Further conidiogenesis studies by Gams and Müller (9) suggest that the conidia are formed from annellate conidiogenous cells and they therefore transferred *Fusarium nivale* to a new genus *Gerlachia* as *G. nivalis* (Ces. ex Sacc.) W. Gams and E. Müller.

Micronectriella stoverii Booth = *Fusarium stoveri* Booth. This species was frequently isolated by R.H. Stover when he was investigating 'Sigatoka' spot of bananas in Honduras. This name now appears untenable, but until fresh isolates have been critically studied, it appears pointless to make an automatic transfer to the genus *Monographella.*

Nectria Fries, Sum. Veg. Scand. 387, 1849. All the macro or secondary conidial forms of *Nectria* are enteroblastic, phialidic; these conidial forms have been assigned to numerous form genera (1). Three sections of the genus have species which produce *Fusarium* conidial states. The Haematococca group, based on the structure of the perithecial wall, is the most important to Fusarium workers. The others are the Episphaeria group, which are parasites on specific Sphaeriaceous fungi, and the Aurantiicola group parasitic on scale insects.

Nectria aurantiicola Berk. & Br., 1873 = *F. larvarum* Fuckel. Syn.: *Sphaerostilbe aurantiicola* (Berk. & Br.) Petch, 1920; *Corollomyces aurantiicola* (Berk. & Br.) Hohnel, 1912. This species, together with *Nectria flammea,* is frequently found as a parasite of scale insects, especially on *Citrus.*

Nectria desmazieresii Beccari & de Not, 1863 = *F. buxicola* Sacc. A slow-growing species associated with premature leaf fall of *Buxus sempervirens* and not collected from any other host.

Nectria episphaeria (Tode ex Fr.) Fr., 1849 = *F. aquaeductuum* Lagerh. aggr. Perithecia common as parasites of *Diatrype stigma,* where they tend to develop near or over the ostiole. Although the conidia are referred to *Fusarium aquaeductuum* Lagerh., it is obvious that a complex of closely related species, grading into the typical *F. aquaeductuum* found in sewage farms or polluted water, are involved here.

Nectria flavo-viridis (Fuckel) Wollenweber, 1926 = *F. melanochlorum* Casp. Syn.: *Sphaerostilbe flavo-viridis* Fuckel, 1871. This species produces a vivid green pigment in cultures on potato sucrose agar. It is slow growing and found on the bark of dead wood associated with Sphaeriaceous fungi.

Nectria haematococca Berk. & Br., 1873 = *F. solani* (Mart.) Sacc. Syn.: *Hypomyces solani* Reinke & Berth., 1879. This is the correct name for perithecia of *Fusarium solani.* Many distinct strains occur belonging to different mating groups which are not interfertile. *Hypomyces solani,* which was often used in American literature for this species, is not acceptable (see under *Hypomyces*).

Nectria illudens Berkeley, 1855 = *F. illudens* Booth, 1971. This is one of the 'Haematococca' group. The principal collections of this rare species are from New Zealand.

Nectria leptosphaeriae Niessl in Krieger, 1886 = *F. sphaeriae* Fuckel. This is another species of the Episphaeria group found only on the ascocarps of *Leptosphaeria* spp. No records are available of any genetic work carried out on this species. No perithecia developed in single ascospore isolates.

Literature Cited

1. Booth, C. 1959. Studies of pyrenomycetes IV. Mycol. Paper 73. Commonw. Mycol. Inst., Kew, Surrey, England. 115 p.
2. Booth, C. 1971. The genus *Fusarium.* Commonw. Mycol. Inst., Kew, Surrey, England. 237 p.
3. Booth, C. 1975. The present status of *Fusarium* taxonomy. Annu. Rev. Phytopathol. 13:83–93.
4. Burnett, J. H. 1975. Mycogenetics. J. Wiley, London and New York. 375 p.
5. Cook, R. J. 1967. *Gibberella avenacea* sp. n., perfect stage of *Fusarium roseum* f. sp. *cerealis* 'Avenaceum.' Phytopathology 57:732–736.
6. Doi, Y. 1968. On two similar species of Hypocrea and their specific segregations by culture tests of associative effects. Mem. Nat. Sci. Mus. Tokyo 1:41–47.
7. El-Ani, A. S. 1956. Cytogenetics of sex in *Gibberella cyanogena* (Desm.) Sacc. Science 123:850.
8. Ford, E. J., J. A. Bourret, and W. C. Snyder. 1967. Biologic specialization in *Calonectria (Fusarium) rigidiuscula* in relation to green point gall of cocoa. Phytopathology 57:710–712.
9. Gams, W., and E. Müller. 1980. Conidiogenesis of *Fusarium nivale* and *Rhynchosporium oryzae* and its taxonomic implications. Netherlands J. Plant Pathol. 86:45–53.

10. Gordon, W. L. 1954. Geographical distribution of mating types in *Gibberella cyanogena* (Desm.) Sacc. Nature 173:505.
11. Gordon, W. L. 1961. Sex and mating types in relation to the production of perithecia by certain species of *Fusarium. Proc. Can. Phytopathol. Soc. 28:11.*
12. Hennebert, G. L., and L. K. Weresub. 1977. Terms for states and forms of fungi. *In* B. Kendrick (ed.), The whole fungus: The sexual-asexual synthesis, Nat. Mus. Can. and Kananaskis Foundation, Canada. 793 p.
13. Hsieh, W. H., S. N. Smith, and W. C. Snyder. 1976. Mating groups in *Fusarium moniliforme.* Phytopathology 67:1041–1043.
14. Ihssen, G. 1910. *Fusarium nivale* Sorauer, de Erreger der Schneeschimmelkrankheit und sein Zusammenhand mit *Nectria graminicola* Berk. & Br. Cbl. f. Bakteriol. 27:48–66.
15. Maire, R. 1911. Remarques sur quelques Hypocreacees. Ann. Mycol. Berl., 9:323–324.
16. Matuo, T., and T. Kobayashi. 1960. A new *Fusarium,* the conidial state of *Hypocrea splendens* Phil. & Plowr. Trans. Mycol. Soc. Japan 2:13–15.
17. Müller, E. 1977. Die Systematische Stellung des "Schneeschimmels." Rev. Mycol. 41:129–134.
18. Müller, E., and J. S. von Arx. 1955. Einige Beitrage zur Systematik und synonymie der Pilze. Phytopathol. Z. 24:353–372.
19. Nelson, P. E., B. L. White, and T. A. Toussoun. 1971. Occurrence of perithecia of *Gibberella* sp. on carnation. Phytopathology 61:743–744.
20. Nirenberg, H. 1976. Utersuchungen über die morphologische und biologische Differenzierung in der Fusarium-Sektion Liseola. Mitteilungen Biol. Bundesanstalt Land- und Forstwirtschaft Berlin-Dahlem. Heft 169:1–117.
21. Norton, J. B. S., and C. C. Chen. 1920. Another corn seed parasite. Science 52:250–251.
22. Reichle, R. E., W. C. Snyder, and T. Matuo. 1964. *Hypomyces* stage of *Fusarium solani* f. *pisi.* Nature 203:664–665.
23. Reichle, R. E., and W. C. Snyder. 1964. Heterothallism and ascospore numbers in *Calonectria rigidiuscula.* Phytopathology 54:1297–1299.
24. Rogerson, C. T. 1970. The Hypocrealean fungi (Ascomycetes, Hypocreales). Mycologia 62:865–910.
25. Schaffnit, E. 1913. Zur systematik von *Fusarium nivale* bzw. seiner hoheren fruchtform. Mycol. Centralblatt 2:253–258.
26. Schipper, B., and W. C. Snyder. 1967. Mating type and sex of *Fusarium solani* f. *cucurbitae* race 1 in the Netherlands. Phytopathology 57:328 (Abstr.).
27. Subramanian, C. V., and D. J. Bhat. 1978. Developmental morphology of Ascomycetes III. *Monographella nivalis.* Rev. Mycol. 42:293–307.
28. Wineland, G. O. 1924. An ascigerous stage and synonymy for *Fusarium moniliforme.* J. Agric. Res. 28:909–922.
29. Wollenweber, H. W. 1913. Studies on the *Fusarium* problem. Phytopathology 3:24–50.

39 Conidiogenous Cells in the Fusaria

Roger D. Goos

Introduction

Classification of the Fungi Imperfecti has for many years been based on the concepts introduced by Saccardo (15) in Volume IV of his *Sylloge Fungorum,* in which major emphasis was placed on the color and arrangement of the conidiophores, and the color, shape, and septation of the conidia. These concepts were employed by all early students of the Fusaria and form the basis of the classifications proposed by Appel and Wollenweber (1), Sherbakoff (16), Wollenweber and Reinking (29), and, to a considerable extent, those of Snyder and Hansen (17, 18, 19) and Gordon (7). It has only been in the last 25 years, following the appearance of the now classic paper by Hughes (8), that developmental morphology has become accepted as a major criterion for the delimitation of species within this group of fungi. Recent monographs of the Hyphomycetes (2, 10, 21) have used conidium ontogeny as a major criterion for delimiting species, with conidium and conidiophore morphology serving as secondary characteristics. Pirozynski (14) summarized the current attitude well when he stated ". . . in the Fungi Imperfecti it is the conidiogenous cell, and particularly the area which participates in conidium production and release, that overshadows the conidium structure as a taxonomic criterion."

Use of conidium ontogeny as a criterion for the classification of the Fungi Imperfecti is not a recent development, although its widespread application is. Costantin (4) proposed a system for the Hyphomycetes based in part on conidium ontogeny which was never accepted, possibly because of its limited scope. A major contribution to this line of thought was made by Vuillemin (26, 27, 28) who proposed a classification based upon the mode of conidium development, which he recognized as being of two basic types: thallospores, which are formed directly from pre-existing elements of the thallus, and conidiospores (conidia vera), which are produced as newly formed elements. It is of interest that some 50 years later, Vuillemins's concepts were adapted by the Kananaskis Conference on Fungi Imperfecti (9) as distinguishing the two basic modes of conidium development. Vuillemin's views were reviewed and emended by Mason (12, 13), who contributed additional concepts and definitions. It was Hughes (8), however, who organized these concepts and put them into a workable system. Hughes stated ". . . there are only a limited number of methods whereby conidia can develop from other cells and . . . morphologically related imperfect states will only be brought together when the precise methods of conidium origin take first place in the delimitation of the major groupings." On the basis of conidium and conidiophore ontogeny, Hughes divided the Hyphomycetes into eight sections, which grouped together morphologically similar fungi primarily on the basis of conidium ontogeny. Hughes basic concepts have found general acceptance among mycologists, although modifications to his system have been proposed by Tubaki (24, 25), Subramanian (20), and others (9).

Table 39-1. Types of conidiogenous cells found in *Fusarium* species. [Based on Booth's system (3)]

Type of conidiogenous cell	Section or species where found
Simple phialide	Section Liseola Section Spicarioides Section Sporotrichiella Section Lateritium Section Elegans Section Martiella Section Arachnites Section Episphaeria Section Coccophilum Section Gibbosum Section Discolor
Proliferating phialide (Percurrent growth)	*F. tabacinum* (Sect. Arachnites) *F. ventricosum* (Sect. Martiella) *F. merismoides* (Sect. Episphaeria)
Polyphialide (sympodial growth)	*F. moniliforme* var. *subglutinans* (Sect. Liseola) *F. dimerum* (Sect. Arachnites) *F. merismoides* (Sect. Episphaeria)
Polyblastic cell	Section Arthrosphoriella *F. fusarioides* (microconidia) *F. sporotrichioides* (microconidia) *F. camptoceras* (macroconidia) *F. avenaceum* (macroconidia) *F. semitectum* (macroconidia) *F. semitectum* var. *majus* (macroconidia)

Conidium morphology has been the major character used in the identification of the Fusaria, and little attention has been given to the conidiogenous cell. As pointed out by Booth (3), this has probably been because it has been assumed that the conidiogenous cell in all Fusaria was a phialide, and therefore consideration of the conidiogenous cell would contribute little to species determination. Booth pointed out, however, that within this genus, conidia may arise blastogenously or from phialides, and that two modifications of the phialide may occur. In Booth's view, the structure of the microconidiophore provides the easiest means for distinguishing certain species of the Fusaria.

Present Concepts of the Conidiogenous Cell in the Fusaria

The four types of conidiogenous cells recognized by Booth as occurring in the Fusaria are listed in Table 39-1; sections and species in which they occur are also listed. As can be seen in Fig. 39-1, in most species the conidiogenous cell is a phialide, or some modification of a phialide (a). The *proliferating phialide* (f,g) is distinguished by percurrent growth through the conidiogenous area, resulting in elongation of the conidiophore, and the production of a new conidiogenous area at a higher level. Such growth may occur repeatedly. A *polyphialide* is formed when more than one conidiogenous area develops on the same cell (successively or synchronously), resulting in multiple conidiogenous sites on a single cell (b,c). A *polyblastic cell* gives rise to a series of single conidia at each of several successively formed conidiogenous sites (d,e). These conidia are holoblastic, formed by the blowing-out of the wall of the conidiogenous cell. Booth (3) noted that blastoconidia develop as dry spores and they may therefore be amenable to wind dispersal. It should also be noted that polyblastic cells and phialides occur in the same species, and that

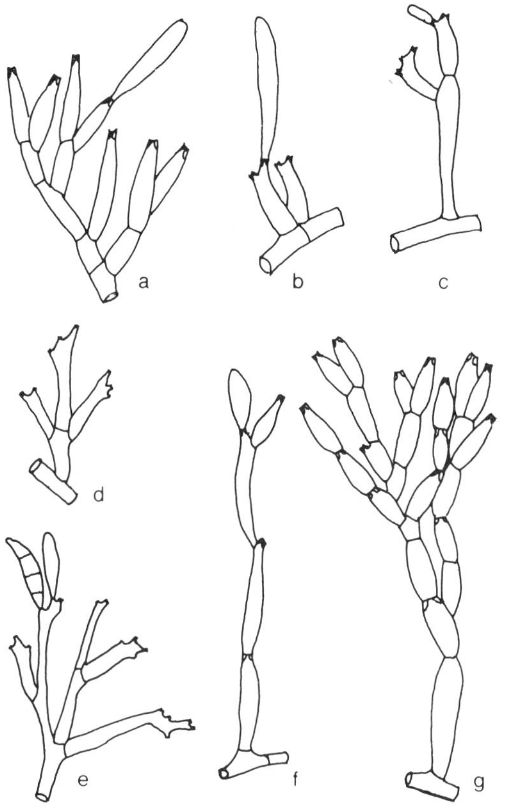

Fig. 39-1. Conidiogenous cells of *Fusarium*. (a) Simple phialides of *F. solani*. Polyphalides of *F. merismoides* (b) and *F. ~~moniliforme var.~~ subglutinans* (c). (d,e) Polyblastic cells of *F. semitectum*. Proliferating phialides of *F. tabacinum* (f) and *F. arthrosporioides* (g). All figures redrawn from Booth (3).

conidium production in those species where polyblastic cells occur is not restricted to polyblastic cells.

Studies of the Phialide in the Fusaria

Conidium formation in the Fusaria has been studied by several authors. Dickson and Johann (5) described the process in *Giberella saubinetii [Fusarium sulphureum* Booth, (3)] as follows: "The conidia were pushed off the conidiophore before septation was completed, and new conidia formed in their place." This is a simplistic description of conidium production from phialides.

As noted by Hughes (8), in many phialides the first conidium is truly endogenous and develops within the unbroken outer wall of that part of the phialide distal to the neck. After rupture of the outer wall, the first conidium is liberated and a collarette becomes evident. The position of the break determines the length of the collarette, which in most Fusaria is inconspicuous. In *Fusarium semitectum,* Hughes (8) observed that the first-formed conidium secedes by a break in the wall at the point where the initial is separated from the phialide by constriction, resulting in the differentiation of a minute cylindrical collarette. A barely visible frill was sometimes seen just above the base of the first-formed conidium, which was thought to represent that part of the wall which was connected to the collarette.

Observations by Goos and Summers (6) would seem to support Hughes' interpretation. Using fluorescent antibody-staining techniques, with antibodies produced in response to conidial suspensions, these authors observed that conidial and phialide walls stained intensely, while germ tubes and hyphae responded weakly to the stain, indicating antigenic dissimilarity between the walls of the conidia and hyphae, but antigenic similarity between the walls of the conidia and phialides. They further observed that newly formed conidia frequently possessed brightly staining "caps," which they interpreted as being the broken outer wall, as described by Hughes.

Subramanian (22) reported a very similar sequence of events in *F. solani* and *F. decemcellulare*. As a succession of conidia is produced from the open end of the phialide, the wall of each conidium is apparently formed *de novo*. In these two species, it appears that a given phialide may produce macroconidia or microconidia, but not both (22).

The ultrastructural study of conidiogenesis in the Fusaria appears to be limited to the single report by Marchant (11) on *F. culmorum*. In this study it was found that the macroconidia were not produced as conventional phialoconidia, but that the first as well as subsequent conidia were produced endogenously. Marchant interpreted the process as being thallic, rather than the blastic process associated with phialoconidium production. Marchant's observations show interesting parallels with those of Subramanian on *F. decemcellulare,* and raise questions concerning the observations described earlier on *F. oxysporum* (6) and *F. semitectum* (8). The need for further ultrastructural studies on conidiogenesis within the Fusaria seems obvious.

Conclusions

Use of conidiogenous cells as a taxonomic criterion within the Fungi Imperfecti is being increasingly accepted, and no contemporary taxonomist can afford to ignore the application of this characteristic in delimiting genera and species. That a similar trend should occur in delimiting species within the Fusaria seems logical and desirable. Booth's acceptance of this characteristic as a taxonomic criterion in delimiting species within the Fusaria, his most conspicuous contribution to *Fusarium* taxonomy according to Toussoun and Nelson (23), is useful in introducing new concepts into classification of these fungi. Its usefulness as a taxonomic character in the Fusaria will be determined only as Booth's system is put to the test, but its introduction should be welcomed as affording an additional criterion for species identification in a difficult group of economically important fungi.

Literature Cited

1. Appel, O., and H. W. Wollenweber. 1910. Grundlagen Einer Monographie der Gattung *Fusarium* (Link). Arb. Kaiserl. Biol. Anstalt Land-U. Forstw. 8:1–207.
2. Barron, G. L. 1968. The genera of Hyphomycetes from soil. Williams Wilkens Co., Baltimore. 364 p.
3. Booth, C. 1971. The genus *Fusarium*. Commonwealth Mycol. Inst. Kew, Surrey, England. 237 p.
4. Costantin, J. 1888. Les Mucedinées simples. Librairie Paul Klincksieck, Paris. 210 p.
5. Dickson, J. G., and Helen Johann. 1920. Production of conidia in *Gibberella saubinetii*. J. Agric. Res. 19:235–237.
6. Goos, R. D., and D. F. Summers. 1964. Use of fluorescent antibody techniques in studies on the morphogenesis of fungi. Mycologia 56:701–707.
7. Gordon, W. L. 1952. The occurrence of *Fusarium* species in Canada. II. Prevalence and taxonomy of *Fusarium* species in cereal seed. Can. J. Bot. 30:209–251.
8. Hughes, S. J. 1953. Conidiophores, conidia, and classification. Can. J. Bot. 31:577–659.
9. Kendrick, W. B. (ed.) 1971. Taxonomy of the fungi imperfecti. Univ. Toronto Press, Toronto. 309 p.
10. Kendrick, W. B., and J. W. Carmichael. 1973. Hyphomycetes, p. 323–509. *In* G. C. Ainsworth, F. K. Sparrow and A. S. Sussman (ed.), The Fungi: an advanced treatise, Vol. IV A, Academic Press, New York.
11. Marchant, R. 1975. An ultrastructural study of "phialospore" formation in *Fusarium culmorum* grown in continuous culture. Can. J. Bot. 53:1978–1987.
12. Mason, E. W. 1933. Annotated account of fungi received at the Imperial Mycological Institute, List II (fasc. 2). Commonwealth Mycol. Inst. Mycol. Pap. 3. 67 p.
13. Mason, E. W. 1937. Annotated account of fungi received at the Imperial Mycological Institute, List

2 (fasc. 3). Commonwealth Mycol. Inst. Pap. 4. 31 p.
14. Pirozynski, K. A. 1971. Characters of conidiophores as taxonomic criteria, p. 37–49. *In* B. C. Kendrick (ed.), Taxonomy of the fungi imperfecti, Univ. Toronto Press; Toronto.
15. Saccardo, P. A. 1886. Sylloge fungorum, Vol. IV. Pavia. p. 1–807.
16. Sherbakoff, C. D. 1915. Fusaria of potatoes. N. Y. Cornell Agric. Exp. Stn. Mem. 6:89–270.
17. Snyder, W. C., and H. N. Hansen. 1940. The species concept in *Fusarium*. Amer. J. Bot. 27:64–67.
18. Snyder, W. C., and H. N. Hansen. 1941. The species concept in *Fusarium* with reference to section Martiella.' Amer. J. Bot. 28:738–742.
19. Snyder, W. C., and H. N. Hansen. 1945. The species concept in *Fusarium* with reference to Discolor and other sections. Amer. J. Bot. 32:657–666.
20. Subramanian, C. V. 1962. The classification of the Hyphomycetes. Bull. Bot. Surv. India. 4:249–259.
21. Subramanian, C. V. 1971. Hyphomycetes. Indian Council Agric. Res. New Delhi. 930 p.
22. Subramanian, C. V. 1971. The Phialide, p. 92–119. *In* W. B. Kendrick (ed.), Taxonomy of the fungi imperfecti, Univ. Toronto Press, Toronto.
23. Toussoun, T. A., and P. E. Nelson. 1975. Variation and speciation in the Fusaria. Annu. Rev. Phytopathol. 13:71–82.
24. Tubaki, K. 1958. Studies on Japanese Hyphomycetes. V. Leaf and stem group with a discussion of the classification of Hyphomycetes and their perfect stages. J. Hattori Bot. Lab. 20: 142–244.
25. Tubaki, K. 1963. Taxonomic studies of Hyphomycetes. Ann Rept. Inst. Fermentation, Osaka. 1:25–54.
26. Vuillemin, P. 1910a. Materiaux pour une classification rationnelle des fungi imperfecti. Compt. Rend. Acad. Sci., Paris. 150:882–884.
27. Vuillemin, P. 1910b. Les Conidiospores. Bull. Soc. Sci. Nancy 11:129–172.
28. Vuillemin, P. 1911. Les Aleuriospores. Bull. Soc. Sci. Nancy 12: 151–175.
29. Wollenweber, H. W., and O. A. Reinking. 1935. Die Fusarien, ihre Beschreibung, Schadwirkung und Bekampfung. Paul Parey, Berlin. 355 p.

Fm infections in humans - 210

281

Fusarium: Diseases, Biology, and Taxonomy

Fm. p 60